AF413082

Tropospheric Aerosol – Formation, Transformation, Fate and Impacts

University of Leeds, United Kingdom
22–24 July 2013

FARADAY DISCUSSIONS

Volume 165, 2013

RSCPublishing

The Faraday Division of the Royal Society of Chemistry, previously the Faraday Society, founded in 1903 to promote the study of sciences lying between Chemistry, Physics and Biology.

EDITORIAL STAFF

Editor
Philip Earis

Deputy editor
Heather Montgomery

Development editor
Rowan Frame

Senior publishing editor
Susan Weatherby

Publishing editors
Jessica Brand, Helen Potter

Publishing assistants
Victoria Bache, Sian Gordon, Ruba Miah

Publisher
Niamh O'Connor

Faraday Discussions (Print ISSN 1359-6640, Electronic ISSN 1364-5498) is published 8 times a year by the Royal Society of Chemistry, Thomas Graham House, Science Park, Milton Road, Cambridge, UK CB4 0WF. Volume 163 ISBN-13: 978-1-84973-692-3

2013 annual subscription price: print+electronic £765, US $1428; electronic only £727, US $1356. Customers in Canada will be subject to a surcharge to cover GST. Customers in the EU subscribing to the electronic version only will be charged VAT. All orders, with cheques made payable to the Royal Society of Chemistry, should be sent to RSC Order Department, Royal Society of Chemistry, Thomas Graham House, Science Park, Milton Road, Cambridge, CB4 0WF, UK. Tel +44 (0) 1223 432398; E-mail orders@rsc.org

If you take an institutional subscription to any RSC journal you are entitled to free, site-wide web access to that journal. You can arrange access *via* Internet Protocol (IP) address at www.rsc.org/ip. Customers should make payments by cheque in sterling payable on a UK clearing bank or in US dollars payable on a US clearing bank.

PRINTED IN THE UK

Faraday Discussions documents a long-established series of *Faraday Discussion* meetings which provide a unique international forum for the exchange of views and newly acquired results in developing areas of physical chemistry, biophysical chemistry and chemical physics.

SCIENTIFIC COMMITTEE, Volume 165

Chair
Professor Gordon McFiggans (University of Manchester, UK)

Professor Alastair Lewis (The University of York, UK)
Professor Jonathan Reid (University of Bristol, UK)
Professor Astrid Kiendler-Scharr (Forschungszentrum Juelich, Germany)
Professor Hartmut Herrmann (University of Leipzig, Germany)

FARADAY STANDING COMMITTEE ON CONFERENCES

Chair
A Mount (Edinburgh, UK)

W A Brown (UCL, UK)
I Hamley (Reading, UK)
J Hirst (Nottingham, UK)
G Hutchings (Cardiff, UK)
C Percival (Manchester, UK)

THE
AEROSOL
SOCIETY
the science of airborne particles

Tropospheric Aerosol – Formation, Transformation, Fate and Impacts

Faraday Discussions

www.rsc.org/faraday_d

A General Discussion on Tropospheric Aerosol – Formation, Transformation, Fate and Impacts was held in Leeds, UK on 22nd, 23rd and 24th July 2013.

RSC Publishing is a not-for-profit publisher and a division of the Royal Society of Chemistry. Any surplus made is used to support charitable activities aimed at advancing the chemical sciences. Full details are available from www.rsc.org

CONTENTS

ISSN 1359-6640; ISBN 978-1-84973-692-3

Cover
Image supplied courtesy of Gerard Capes, University of Manchester.

INTRODUCTORY LECTURE

PAPERS AND DISCUSSIONS

CONCLUDING REMARKS

ADDITIONAL INFORMATION

Faraday Discussions

RSC Publishing

PAPER

Introductory lecture: Atmospheric organic aerosols: insights from the combination of measurements and chemical transport models

Spyros N. Pandis,[*a] Neil M. Donahue,[b] Benjamin N. Murphy,[b] Ilona Riipinen,[c] Christos Fountoukis,[d] Eleni Karnezi,[b] David Patoulias[a] and Ksakousti Skyllakou[a]

Received 24th August 2013, Accepted 28th August 2013

DOI: 10.1039/c3fd00108c

The formation, atmospheric evolution, properties, and removal of organic particulate matter remain some of the least understood aspects of atmospheric chemistry despite the importance of organic aerosol (OA) for both human health and climate change. Here, we summarize our recent efforts to deal with the chemical complexity of the tens of thousands of organic compounds in the atmosphere using the volatility–oxygen content framework (often called the 2D-Volatility Basis Set, 2D-VBS). Our current ability to measure the ambient OA concentration as a function of its volatility and oxygen to carbon (O : C) ratio is evaluated. The combination of a thermodenuder, isothermal dilution and Aerosol Mass Spectrometry (AMS) together with a mathematical aerosol dynamics model is a promising approach. The development of computational modules based on the 2D-VBS that can be used in chemical transport models (CTMs) is described. Approaches of different complexity are tested against ambient observations, showing the challenge of simulating the complex chemical evolution of atmospheric OA. The results of the simplest approach describing the net change due to functionalization and fragmentation are quite encouraging, reproducing both the observed OA levels and O : C in a variety of conditions. The same CTM coupled with source-apportionment algorithms can be used to gain insights into the travel distances and age of atmospheric OA. We estimate that the average age of OA near the ground in continental locations is 1–2 days and most of it was emitted (either as precursor vapors or particles) hundreds of kilometers away. Condensation of organic vapors on fresh particles is critical for the growth of these new particles to larger sizes and eventually to cloud condensation nuclei (CCN) sizes. The semivolatile organics currently simulated by CTMs are too volatile to condense on these tiny particles with high curvature. We show that chemical aging reactions converting these semivolatile

[a]Department of Chemical Engineering, University of Patras, Patra, Greece. E-mail: spyros@chemeng.upatras.gr; Tel: +30-2610-969510

[b]Carnegie Mellon University, Pittsburgh, Pennsylvania, USA

[c]Stockholm University, Stockholm, Sweden

[d]FORTH, Patra, Greece

compounds to extremely low volatility compounds can explain the observed growth rates of new particles in rural environments.

1   Introduction

In the past 20 years ambient air pollution, most notably particulate matter (PM), has come to be recognized as a risk factor contributing to declines in respiratory and cardiovascular health[1,2] and increased risk of acute morbidity and mortality.[3–5] At the same time, atmospheric aerosol particles influence the Earth's radiation balance directly by scattering and absorbing solar radiation, and indirectly by acting as CCN. The role of these particles in the energy balance of our planet is one of the major uncertainties in the global change problem.[6]

Organic compounds are important components of atmospheric fine particles ($PM_{2.5}$, particles < 2.5 μm in diameter). Throughout the troposphere they constitute roughly half of the $PM_{2.5}$ mass, residing mostly on accumulation mode particles.[7] However, while the inorganic fraction of $PM_{2.5}$ is relatively well understood, significant uncertainty limits our understanding of almost all aspects of OA, including its sources, atmospheric transformation, and fate.

OA has been traditionally viewed as a relatively inert, non-volatile mixture of compounds from numerous primary sources (primary organic aerosol, POA), coated by secondary compounds derived from gas-phase oxidation of volatile precursors (secondary organic aerosol, SOA).[8] The chemical complexity of the OA (there are thousands or even tens of thousands of complex large organic compounds in typical ambient aerosol),[9] its unknown chemical composition (less than 20% of the OA mass has been quantified even by the best studies),[10] the unknown physical and chemical properties of the majority of the known OA components, and the difficulty of describing mathematically such a complex system in atmospheric CTMs have seriously limited scientific progress in both the air quality and climate change areas. CTMs in both regional and global scales are often not able to reproduce the observed OA levels, their chemical characteristics (degree of oxidation), their diurnal variation, *etc.*[11,12] As a result, we are currently unable to accurately evaluate the effects of different strategies of reduction of OA concentrations in polluted areas and to quantify the effect of OA on the energy balance of the planet.

Recent work[13] has changed the above picture of OA. Most of the emitted organic particulate matter from combustion sources like transportation, biomass burning, *etc.* evaporates after emission. The resulting semi-volatile organic vapors can then react in the gas phase with the hydroxyl radical, OH, and other atmospheric oxidants forming low volatility oxidation products that can recondense to the particulate phase in timescales of several hours or even days. This evaporation/reaction/condensation process causes significant changes to the chemical nature of primary OA (it becomes highly oxygenated), its size distribution, its distribution in space and its physical and chemical properties. The resulting products will probably have quite different health effects than the original compounds. This theory can explain why the aerosol in large urban centers is dominated by oxygenated compounds (Oxygenated Organic Aerosol, OOA) and not hydrocarbon-like OA (HOA).[14]

Similar challenges exist in our efforts to understand secondary OA. Initially the SOA precursors were assumed to form practically non-volatile SOA with a constant yield.[15] This description is actually still used in most climate models.[12] The next step was to assume the production of two surrogate products that form a pseudo-ideal solution (Odum *et al.*, 1996).[16] This approach is used currently by the majority of the available regional CTMs. However, the resulting models fail to reproduce observed OA concentrations in the US, Europe, Asia, *etc.*, especially in periods (warm sunny summer days) when SOA is expected to be the dominant OA component. Potential gaps in understanding include homogeneous reactions of the SOA vapors,[17] heterogeneous reactions (oxidation of the organics in the particulate phase,[18] oligomerization,[19] the role of NO_x and UV levels in the formation of SOA, *etc.* These processes are not well understood and are not included explicitly in most CTMs.

Our existing mental model of OA (primary and secondary) is insufficient to describe the thousands of semivolatile organic compounds, moving between the gas and particulate phases, continuously reacting, generating even more organic products, and interacting with each other and the inorganic compounds, including water, in the particulate phase. This results in problems with source apportionment (quantification of the OA sources) and also predicting the response of OA during potential changes in emissions of anthropogenic pollutants. Significant uncertainties in the role of the organics in the formation and growth of new particles to cloud condensation nuclei remain. Efforts to approach the problem at the individual compound level have had limited success. Each organic compound has its own complex behavior that is difficult to extrapolate and the study of each one of them requires a few person-years.

It is clear that we need a robust framework on which to place past and future measurements in both the lab and the field so that we can then use these results in CTMs. Our efforts to develop and apply this framework are summarized in the following sections of this paper.

2   The 2-D volatility–oxygen content framework

Describing the chemical complexity of ambient OA in chemical transport models requires the simplification of the system to its most important dimensions. One of the promising ways to map this chemical space with tens of thousands of dimensions to a manageable space is using as dimensions the volatility and oxygen content of the OA (expressed as the O : C ratio)[20] (Fig. 1). The volatility determines the partitioning of the OA compounds between the gas and particulate phases and also the ability of a molecule to participate in the formation of new particles. The oxygen content of organic particles is directly connected to their ability to absorb water and act as cloud condensation nuclei (CCN). Our hypothesis is that we can capture the most essential elements of the OA system using these two dimensions. The corresponding challenges are to find ways to easily experimentally characterize OA in these two dimensions, describe the corresponding processes in this coordinate system, and then translate all of this into computational modules that can be used in the existing CTMs. Obviously, OA will have a distribution of volatilities and distribution of O : C ratios so we will need to work with these distributions and not just the average values.

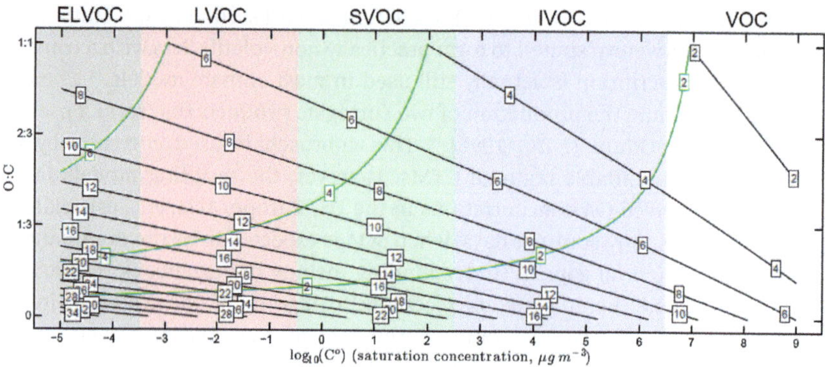

Fig. 1 The 2D Volatility Basis Set space with the volatility (expressed as the logarithm of the saturation concentration) as the x-axis and the O : C ratio as the y-axis (based on Fig. 4 of Donahue et al.[21]). The black isopleths are the number of carbon atoms and the green isopleths the number of oxygen atoms.

The organic compounds can be separated in five groups by volatility (Fig. 1):

(i) *Extremely Low Volatility Organic Compounds* (ELVOCs with $C^* < 3.2 \times 10^{-4}$ µg m^{-3}, volatility bins 10^{-4} µg m^{-3}, 10^{-5} µg m^{-3}, *etc.*). These compounds, if created by gas-to-particle conversion processes, can be very important for the formation and growth of new particles created *in situ* in the atmosphere by nucleation.

(ii) *Low Volatility Organic Compounds* (LVOCs, with 3.2×10^{-4} µg m$^{-3} < C^* < 3.2 \times 10^{-1}$ µg m^{-3}, volatility bins 10^{-3}, 10^{-2} and 10^{-1} µg m^{-3}). These compounds are in the particulate phase in typical atmospheric concentration levels.

(iii) *Semi Volatile Organic Compounds* (SVOCs, with 3.2×10^{-1} µg m$^{-3} < C^* < 3.2 \times 10^{2}$ µg m^{-3}, volatility bins 1 , 10 and 100 µg m^{-3}). The corresponding compounds exist in both the gas and particulate phases under typical ambient conditions.

(iv) *Intermediate Volatility Organic Compounds* (IVOCs, with 3.2×10^{2} µg m$^{-3} < C^* < 3.2 \times 10^{6}$ µg m^{-3}, volatility bins 10^{3}, 10^{4}, 10^{5} and 10^{6} µg m^{-3}). These exist in the gas phase in the atmosphere but can be easily converted to condensible compounds producing secondary organic aerosol.

(v) *Volatile Organic Compounds* (VOCs, with $C^* > 3.2 \times 10^{6}$ µg m^{-3}). Most of the emissions of gas-phase organics fall in this traditional category.

This grouping facilitates discussion of the vast array of atmospheric organic compounds.

The O : C of atmospheric organic compounds varies from 0 (for hydrocarbons) to 2 (for carbon dioxide, oxalic acid, *etc.*). However, there are few organic compounds with O : C exceeding unity. Donahue *et al.*[21] used group contribution methods to determine the mean chemical properties of the compounds in this 2-D VBS showed in Fig. 1. They argued that the compounds in the space shown have 0–8 oxygens.

Donahue *et al.*[22] and Kroll *et al.*[23] have suggested that the average oxidation state of carbon, OSc, may be used as the principal y-axis of the 2D-VBS. In practice, the O : C and OSc are highly correlated allowing their use almost interchangeably. In this work we will use the O : C as the preferred axis.

3   Measurement of OA in the 2D-VBS

Measurement of the OA in this framework requires the measurement of the OA concentration distribution as a function of volatility and O : C content. Measuring the OA volatility distribution or the OA O : C content distributions on their own is challenging. Some recent efforts are described in the following sections.

3.1   Measurement of the volatility distribution

The most commonly used technique for the measurement of the OA volatility is the use of a thermodenuder (TD) together with a technique for the measurement of the OA concentration. Thermodenuders heat the aerosol sample to a fixed temperature, allow the particles to evaporate, remove the organic vapors from the sample stream and cool the sample back to ambient temperature.[24–26] Their result is the mass fraction remaining (MFR) of the OA as a function of temperature (Fig. 2). While the MFR provides some qualitative information about the volatility of the OA (the further the MFR curve is to the right the lower the volatility of the OA), it generally depends not only on the volatility of the OA and the characteristics of the TD but also on the mass concentration of the OA, its size distribution, the enthalpies of evaporation of the OA components and any mass transfer delays in the evaporation of the particles.[27]

Estimation of the volatility distribution of the OA requires combination of the thermodenuder MFR with a model describing the OA evaporation. Fitting of the model predictions to the MFR measurements allows estimation of the OA volatility distribution and the other unknown parameters (enthalpies, resistances in mass transfer other than gas-phase diffusion). Given the nature of the MFR curve (a monotonically decreasing curve going from one to zero, almost always with a roughly sigmoidal shape) and the multiple unknown parameters there can be multiple "good" solutions to the problem. For example, Fig. 2 shows the predicted thermograms (MFR *versus* temperature) for two organic aerosols with very

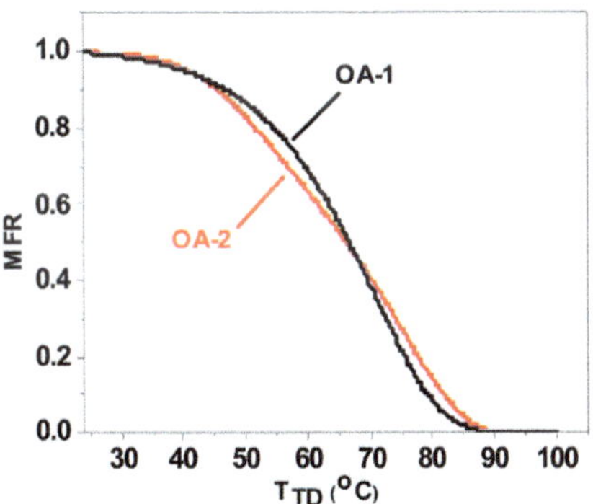

Fig. 2   Predicted thermograms (MFR *versus* T_{TD}) for OA with different properties. A single component aerosol OA-1 (effective saturation concentration $C^* = 1\ \mu g\ m^{-3}$, enthalpy of vaporization $\Delta H_{vap} = 80\ kJ\ mol^{-1}$ and effective mass accommodation coefficient $a_m = 1$) gives practically the same thermogram as one with a multi component aerosol OA-2 (consisting of 50% $C^* = 0.1\ \mu g\ m^{-3}$, 25% 1 $\mu g\ m^{-3}$, 25% 10 $\mu g\ m^{-3}$, $\Delta H_{vap} = 150\ kJ\ mol^{-1}$ and $a_m = 0.05$).

different properties. Any reasonable experimental error in the measurements will make them practically indistinguishable.

Karnezi *et al.*[28] proposed an algorithm for the estimation of the volatility distribution from the TD measurements and the corresponding uncertainty. They argued that in a lot of cases the uncertainty of the estimated volatility distributions is quite high. Fig. 3a depicts a thoretically predicted thermogram and pseudo-measurements constructed by the addition of random experimental error to these predictions. The results of the parameter estimation using just these measurements are shown in Fig. 3b, together with the correct volatility distribution. The estimated volatility distribution is quite uncertain and the best guess has the wrong shape.

Karnezi *et al.*[28] investigated a number of ways to improve the accuracy of the volatility distribution. These included additional measurements, measurements at multiple residence times, *etc.* The best approach, based on their analysis, was the performance of isothermal dilution measurements (Fig. 3c) in a small smog chamber together with the thermodenuder measurements. Isothermal dilution has the advantage that it produces results that do not depend on the enthalpy of evaporation, while the use of a small chamber allows access to much longer

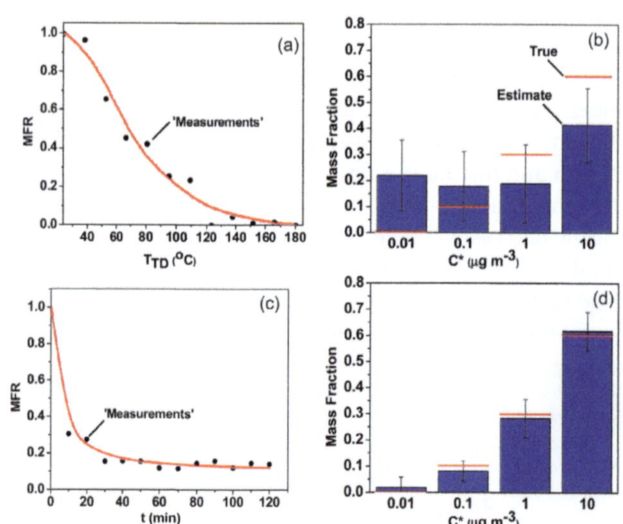

Fig. 3 (a) The red line is the thermogram corresponding to the true properties of the aerosol (consisting of 10% $C^* = 0.1$ μg m^{-3}, 30% 1 μg m^{-3}, 60% 10 μg m^{-3}, $\Delta H_{vap} = 50$ kJ mol^{-1} and $a_m = 1$) and the black dots correspond to the 'measured' MFR *versus* T_{TD} for the aerosol. (b) Estimated and true volatility distribution, using TD measurements, for an OA (consisting of 10% $C^* = 0.1$ μg m^{-3}, 30% 1 μg m^{-3}, 60% 10 μg m^{-3}, $\Delta H_{vap} = 50$ kJ mol^{-1} and $a_m = 1$). The error bars represent the uncertainty of the estimated value. The vaporization enthalpy was estimated with a value equal to 91 kJ mol^{-1} and the accommodation coefficient with a value equal to 0.44. (c) Isothermal dilution measurements (MFR as function of time). The red line corresponds to the aerosol consisting of 10% $C^* = 0.1$ μg m^{-3}, 30% 1 μg m^{-3}, 60% 10 μg m^{-3}, $\Delta H_{vap} = 50$ kJ mol^{-1} and $a_m = 1$ and the black dots correspond to the 'measured' MFR *versus* time. (d) Estimated and true volatility distribution, combining TD and isothermal dilution measurements, for an OA (consisting of 10% $C^* = 0.1$ μg m^{-3}, 30% 1 μg m^{-3}, 60% 10 μg m^{-3}, $\Delta H_{vap} = 50$ kJ mol^{-1} and $a_m = 1$). The error bars represent the uncertainty of the estimated value. The vaporization enthalpy was estimated to be equal to 50 kJ mol^{-1} and the accommodation coefficient with a value equal to 0.88.

timescales than a thermodenuder. Combination of the information of the two methods produced in most cases an accurate volatility distribution with modest uncertainty (Fig. 3d).

One of the major potential problems in thermodenuder-based estimates of the OA volatility distribution is the potential for chemical changes to the organic aerosol compounds during their heating to high temperatures. Comparisons of the AMS spectra of the remaining OA at the same MFR with that of isothermal dilution experiments can provide a valuable test of the validity of the thermodenuder measurements. Based on our existing measurements, problems may start to arise at temperatures over 150 °C in thermodenuders.

3.2 Measurement of the O : C

The OA O : C can be measured with a High Resolution AMS (HR-AMS) using the detailed mapping of the AMS spectrum to fragments of organic molecules. However, the AMS provides on its own only the average O : C of the OA and not a distribution. There may be compounds with much higher and much lower O : C compared to the average in the sampled aerosol. Some information about the O : C distribution, but not the actual distribution, can be obtained by combination of the AMS with a thermodenuder.[29] This can allow the estimation of the average O : C of each volatility bin in the VBS framework.

Despite the high-resolution of the mass spectra and the corresponding information there are uncertainties in the measured O : C due to the uncertainty introduced by the need to separate the water from the dehydration of the organic molecules from the water vapor from the ambient air, *etc.*

3.3 Combination of physical and chemical characteristics

The combination of the rich dataset that the AMS spectra provide with a factor analysis technique (Positive Matrix Factorization or PMF)[30] can provide additional insights into the sources and processing of ambient OA that can be mapped onto the 2D-VBS. PMF can separate the fresh primary organic aerosol from hydrocarbon combustion sources like transportation (Hydrocarbon-like OA or HOA), the relatively oxygenated OA from biomass combustion (BBOA), and the oxygenated OA (OOA) (Fig. 4).

PMF often results in two OOA types, one of moderate oxygen content that correlates with semivolatile inorganic PM components and has been named semivolatile-OOA (SV-OOA) and one of higher oxygen content that correlates with low volatile components like sulfates (LV-OOA). The significance of these OOA types is still under debate. They could represent different aerosol types (*e.g.*, from different precursors or chemical pathways) or they could be the two extremes of the oxidation of OA observed in the corresponding study.

PMF can be performed in measurements of a thermodenuder-AMS combination, providing in this way thermograms of the corresponding aerosol types. These thermograms can then be used (as discussed in Section 3.1) together with a thermodenuder model to estimate the volatility distribution of each aerosol type and place them in the 2D-VBS space (Fig. 4). While this still does not provide a full 2D-VBS distribution due to our inability to measure the O : C distribution, it does provide a rich characterization of the ambient OA in the corresponding space. These measurements can be used both to provide insights about the sources and

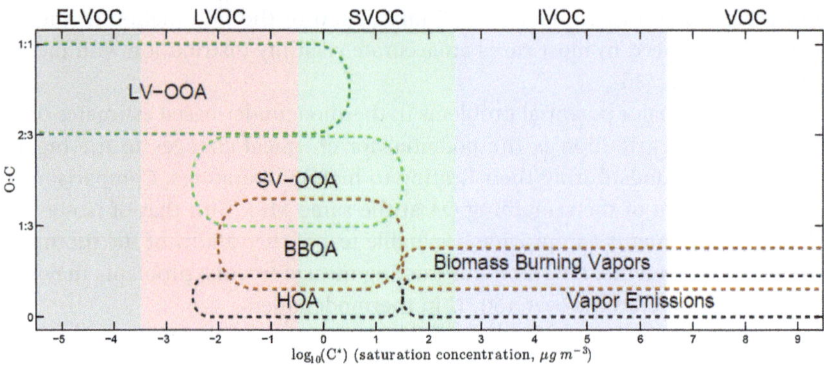

Fig. 4 Volatility and O : C ratio for atmospheric organic compounds. The volatility ranges discussed in the text (ELVOC, LVOC, SVOC, IVOC and VOC) are indicated by colour bands. Also shown are the approximate locations of factors (HOA, BBOA, SV-OOA, and LV-OOA) estimated from the PMF analysis of ambient OA mass spectra together with the locations of primary emissions of biomass burning vapors and other organic vapour emissions.

processing of ambient OA and also to test our understanding during the evaluation of chemical transport models.

4 Simulation of OA in the 2D-VBS

Application of the 2D-VBS in CTMs requires discretization of the volatility–oxygen content space into bins. Murphy $et\ al.$[31] used 12 logarithmically spaced bins along the axis of effective saturation concentration ($C^* = 10^{-5}$ to 10^6 $\mu g\ m^{-3}$ at 300 K). They included, therefore, ELVOCs, LVOCs, SVOCs and IVOCs in the system while they simulated the VOCs explicitly using the SAPRC gas-phase chemistry mechanism. The 0.0 to 1.2 O : C range was simulated using 13 bins spaced by 0.1. The 2D-VBS was therefore discretized using 156 bins.

Emissions of primary organic aerosol (LVOCs, SVOCs and IVOCs) were assumed to have an O : C of 0.1 and to have a volatility distribution based on the available measurements.[32] The first generation products of anthropogenic and biogenic VOCs were assumed to have volatility distributions consistent with the available smog chamber studies and a volatility dependent O : C ratio based on the literature.[33–36,29] More details can be found in Murphy $et\ al.$[37,31]

4.1 Combining functionalization and fragmentation

One of the major features of the 2D-VBS is its ability to describe the multiple generations of reactions of the first generation products of the VOCs as well as the SVOCs and IVOCs. Murphy $et\ al.$[31] focused on the gas-phase oxidation of these compounds by the hydroxyl radical, OH. After the unique first-generation step, we hypothesize that aging by reaction with OH can be written as follows:

$$\text{ROG} + \text{OH} \rightarrow \sum_{i=1}^{N_O} \sum_{j=1}^{N_{C^*}} \alpha_{i,j} P_{i,j}$$

where ROG is a specific reactive organic gas and $\alpha_{i,j}$ is the stoichiometric yield of the product, $P_{i,j}$ with i oxygen atoms added relative to the organic gas parent and experiences a change in volatility by j volatility bins relative to the parent. N_{C*} is the total number of volatility intervals. The number of oxygen atoms added per generation of oxidation, i, will vary depending on the particular chemical pathway followed. Murphy et al.[31] following the analysis of Donahue et al.[21] adopted the functionalization scheme shown in Fig. 5. The transformation must be mapped from the kernel, which describes the number of added oxygens, to the y-axis of the 2D-VBS, which is O : C. Each point in the 2D-VBS has a different average carbon number[21] and thus will inform the translation from number of added oxygen atoms to change in O : C.[37] In this scheme all OA components (anthropogenic and biogenic) were allowed to age chemically.

Fragmentation takes place at the same time as functionalization. Murphy et al.,[31] following the analysis of Donahue et al.,[21] assumed that the O : C of the larger fragments will be similar to the O : C of the reactant compounds. As a result these fragments will move directly to the right in the 2D-VBS space, increasing in volatility but not in O : C. The smallest fragments will have a much larger O : C than the reactant. Murphy et al.[31] assigned mass in all of the volatility bins to the right of the midpoint (from the reactant volatility bin to the highest modeled bin) O : C values that move diagonally towards the highest modeled O : C. The fragmentation branching ratio was parameterized following Donahue et al.[22] as: $\beta_{\mathrm{frag}} = (\mathrm{O} : \mathrm{C})^{(1/6)}$. Based on this dependence on O : C, compounds with O : C exceeding 0.4 will most likely fragment.

This scheme was tested in a Lagrangian CTM against measurements collected during the EUCAARI campaings using AMS measurements from Finokalia (Greece), Cabauw (Netherlands) and Mace Head (Ireland) during the summer of 2008 and winter of 2009. The resulting model tended to underpredict the OA during the summer (average fractional bias $= -0.13$ and average fractional error $= 0.17$) in all sites while during the winter it overpredicted OA concentration in Cabauw and Mace Head (average fractional bias $= 0.59$ and average fractional error $= 0.61$) and underpredicted in Finokalia (average fractional bias $= -0.34$ and average fractional error $= 0.34$). While this performance for OA mass concentration is encouraging the scheme seriously underpredicted the O : C in all locations in both seasons (average fractional bias $= -0.56$ and average fractional error $= 0.56$). The scheme drives the OA components to much lower volatility rapidly, after one or may be two oxidation steps (Fig. 5) and the corresponding

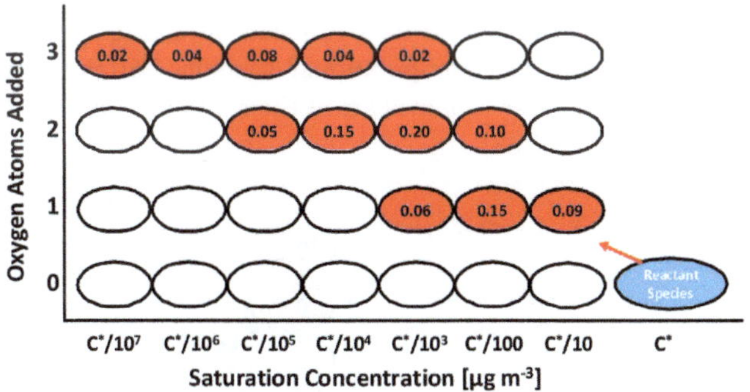

Fig. 5 Functionalization scheme in the 2D-VBS proposed by Murphy et al.[31] The numbers in the red circles are the yields of the corresponding products.

compounds move almost completely to the particulate phase. While in the particulate phase they are protected to a large extent from attacks by OH[38] and additional oxidation and therefore their O : C does not increase further. Fragmentation reduces the OA concentration, balancing to some extent the significant increases caused by this relatively aggressive functionalization scheme, but it does not increase the corresponding O : C.

This detailed functionalization/fragmentation scheme of Murphy *et al.*[31] had encouraging performance but additional improvements are needed so that it can capture the evolution of OA.

4.2   A simple scheme

Murphy *et al.*,[31,37] recognizing the challenges of capturing the complexities of fragmentation and functionalization, proposed a simple aging scheme for the 2D-VBS to capture the net change of the aging OA volatility–O : C distribution. This scheme outlined in Fig. 6 assumes that the net change caused by the various processes is a reduction of the average volatility by a factor of 10 and the addition of one oxygen atom for half the products and two for the other half. This scheme is an extension of the 1D-VBS having the same volatility reduction rate during the gas-phase reactions with OH (a rate constant of 1×10^{-11} cm^3 molec^{-1} s^{-1} for anthropogenic SOA and 4×10^{-11} cm^3 molec^{-1} s^{-1} for the products of anthropogenic IVOCs and SVOCs). Murphy *et al.*[31,37] assumed that for biogenic SOA the aging resulted in an increase in O : C but no net change in volatility.

The evaluation of this simple scheme against the EUCAARI measurements was surprisingly good. The model reproduced the summer OA concentration measurements in all sites well (average fractional bias = 0.33 and average fractional error = 0.25) and had a tendency towards overprediction in the winter (average fractional bias = 0.47 and average fractional error = 0.52). It did quite well in reproducing the O : C observations in all sites during both seasons (average fractional bias = 0.01 and average fraction error = 0.18). This suggests that this simple scheme appears to capture the net impact of chemical aging on the OA concentration and O : C ratio, at least for the specific conditions of the EUCAARI dataset. It clearly represents an oversimplification of what is actually taking place, but it can still provide useful guidance for the development of future 2D-VBS aging schemes.

4.3   The road forward

The above examples illustrate the efforts to translate our limited understanding of the OA evolution to practical computational schemes for the simulation of atmospheric OA. Even if these schemes are at their infancy, they already provide

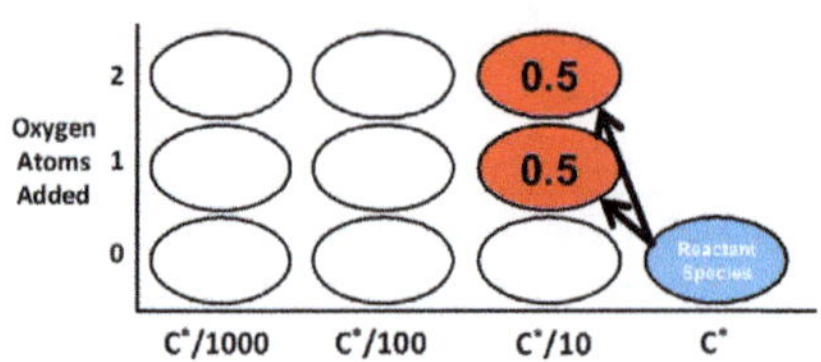

Fig. 6   Simple aging parameterization proposed by Murphy *et al.*[37]

encouraging results. Future steps should involve more detailed evaluation and constraining of these schemes with ambient observations not only of the OA concentration and the O : C but also the OA volatility distribution, laboratory studies of the chemical aging of anthropogenic and biogenic SVOCs, IVOCs, and subsequent analysis of the results using this modeling framework, and finally efforts to project detailed reaction schemes onto the 2D-VBS.

5 The age of OA

The OA in the atmosphere appears to reach a highly oxidized state (LV-OOA in Fig. 4). However, the time required for the transition from fresh POA, VOCs, or IVOCs to LV-OOA is uncertain. It is for sure less than the lifetime of particles in the atmosphere (estimated to be of the order of week). Wagstrom and Pandis[39] proposed an algorithm for the calculation of the age distribution of the different pollutants at a given place including OA and its components. The algorithm is based on the Particle Source Apportionment Technology (PSAT), a technique to follow material emitted from different sources inside a CTM.[40] The age of pollutants is calculated by defining as a "source" the emissions during specified time intervals (instead of the traditional use of source types or source areas). The simulation period is then discretized (say in periods of six hours) and PSAT calculates the contribution of the different temporal sources to the concentration of a pollutant in a given place and time. From these, the age distribution of the pollutant is calculated.

The average age of the OA at the ground level during a summertime period in the Eastern US is depicted in Fig. 7.

The age is based on the time of emission of either the corresponding particles or the precursor vapors. So the age includes the period required for the formation of secondary particulate matter. The average age varies from approximately a day in areas with significant emissions of VOCs to approximately three days in more

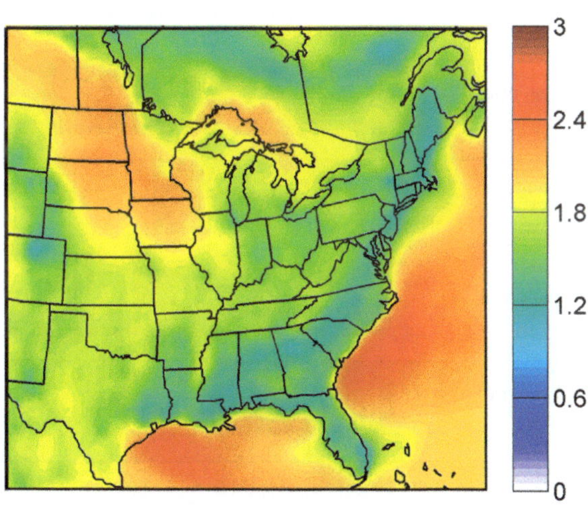

Fig. 7 Estimated average age (days) of OA at the ground level during July. The estimates were made using PSAT and PMCAMx following the approach of Wagstrom and Pandis.[39] The OA was simulated using the VBS approach.

remote areas (*e.g.*, over the ocean). It is significantly less than the average lifetime of particles as one would expect. These results strongly suggest that the conversion of material to OOA requires 1–2 days during the summertime.

5.1   Source–receptor relationships

The fact that a lot of the OA around us (even in major urban areas) has been in the atmosphere for a significant period of time suggests that a significant fraction of this PM has been emitted elsewhere. As an example, Skyllakou *et al.*[41] examined the contribution of different source areas to the OA in Paris, a megacity, during the summer. They classified the OA into three categories based on its origin: local sources referred to the contributions from the urban agglomeration (an area of around 50 km around the city center), mid-range those coming from distances between 50 and 500 km, and long range those coming from more than 500 km from Paris. Using PSAT they estimated the contributions shown in Fig. 8. While more than half of the fresh POA was local, both the anthropogenic and biogenic SOA came mostly from midrange sources, with significant contributions from long-range transport. For total OA, the local contribution was estimated to be quite small (around 10%), with half of the remaining OA coming from mid-range and the other half from long-range sources. These estimates were consistent with the highly oxygenated nature of OA measured during that period in the MEGA-POLI project[42] and estimates based on the small differences of OA concentrations in sites inside and around Paris.[43]

6   OA and ultrafine particles

Studies during the last decade have revealed that sulfuric acid plays a dominant role in the formation of new particles in the atmosphere by nucleation.[44] However, the growth of the fresh ultrafine particles to larger sizes in the boundary layer

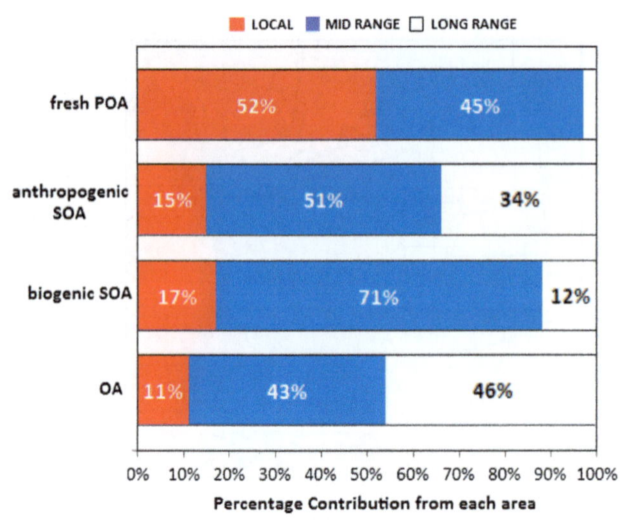

Fig. 8 Percent contributions from each transport category (local, mid-range and long range) for different types of organic aerosol in Paris during the summer.

cannot be explained in most cases by condensation of sulfuric acid monomers alone.[45] Low volatility organics appear to be involved and to actually dominate the growth in periods of high biogenic activity. Exceptions to this are areas with high sulfur dioxide concentrations, like parts of the northeastern US.[46] However, the gas-phase organics simulated by current CTMs are in the SVOC volatility range. The curvature of the fresh particles (they are a few nanometers in diameter) together with the corresponding Kelvin effect makes condensation of these vapors on these fresh particles negligible and cannot explain the corresponding growth.

There have been a number of efforts to explain how organic vapors can help these nanometer size particles grow. Riipinen et al.[45] estimated that a significant fraction of the SOA produced (around 50%) should behave as if it were nonvolatile. Something like that is, however, not compatible with the first-generation yields of SOA from biogenic precursors observed in the lab. The yields are quite low for low SOA concentrations. One potential explanation is that these ELVOCs are produced as second-generation products from the oxidation for SVOCs in the gas phase by OH.[38] To test this hypothesis we coupled the VBS approach with the DMAN particle dynamics model of Jung et al.[47] and evaluated it under typical nucleation conditions in Europe. We found, after a number of tests that a reaction converting the gas-phase fraction of the $C^* = 1$ μg m^{-3} to ELVOCs with $C^* = 10^{-3}$ μg m^{-3} with a reasonable rate constant of 1×10^{-11} cm^3 molec^{-1} s^{-1} (currently used in the VBS for secondary OA components) can explain the observations. The predicted growth of new particles for a typical nucleation event is shown in Fig. 9. The new particles grow to approximately 50 nm with an average growth rate of 5 nm h^{-1}.

The composition of the growing particles is shown in Fig. 10. In the beginning they consist mostly of ammonium sulfate and bisulfate as sulfuric acid and ammonia were the nucleating vapors in the simulation. However, the system soon runs out of sulfuric acid (most of the available sulfur dioxide has reacted) and organics are responsible for the growth of the particles to larger sizes. By the end of the day these new particles consist mostly of organic particulate matter. The organic material that is required for the corresponding growth is quite small, just 30 ng m^{-3}. Of course at the same time there is considerable condensation of

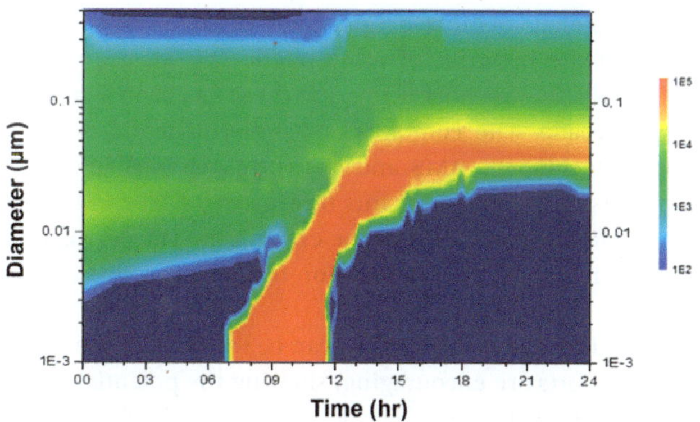

Fig. 9 Predicted evolution of the aerosol number distribution (particles cm^{-3}) for a typical nucleation event in Northern Europe.

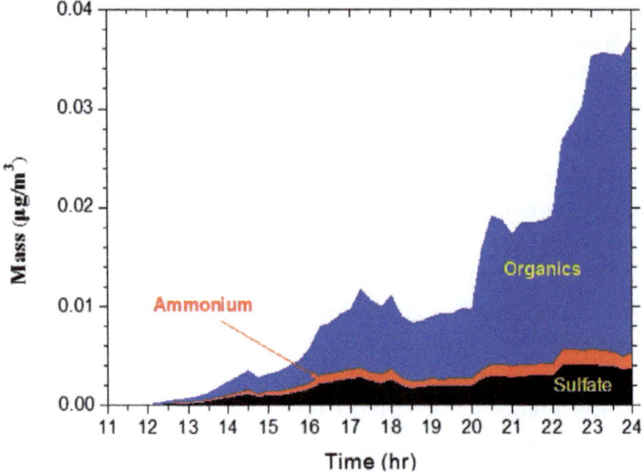

Fig. 10 Predicted chemical composition of the new particles formed during a nucleation event.

organic PM to the larger particles in the accumulation mode. The volatility distribution of the material condensing on the growing fresh particles is interesting. Less than half of this material is the ELVOCs produced from the assumed aging reaction. The other material is SVOCs that mix with these ELVOCs, accelerating in this way the overall growth rate.

7   Conclusions

Most of the OA that we are breathing, even in urban areas, has started its atmospheric life (as organic vapors or particles) a day or two ago in an area probably a few hundred kilometers away. Coupling of the available CTMs with source apportionment algorithms like PSAT used in this work suggest that the average age of the OA that we are breathing is 1–2 days and that its average transport distance is 300–1000 km depending on the conditions. This allows the OA to chemically age, become quite oxygenated and to have different properties from its original form.

The two dimensional volatility–oxygen content space can be a useful tool for the description of the behavior of atmospheric OA. Measurement of the OA distribution in this space is challenging, but there has been significant progress during the last decade. The combination of thermodenuder with isothermal volatility measurements, the AMS and the corresponding aerosol evaporation models is a promising approach for this measurement. The PMF analysis of the corresponding AMS measurements of the evaporating OA can also provide valuable information about the characterization of the OA in this two-dimensional space.

The evolution of the OA in the 2D-VBS can be included in the available CTMs. The existing first efforts are encouraging, showing the potential of these models to reproduce the observed OA concentrations and O:C levels. However, the best performing approach is still the simplest one, simulating only the net effect of functionalization and fragmentation processes. Additional studies of these processes in the laboratory and mapping of the corresponding results in the 2D-VBS space is required to improve the corresponding models.

The low volatility and semivolatile components of ambient OA (LVOCs and SVOCs) are the most important for its mass concentration. However, the extremely low volatility components (ELVOCs) can be critical for the growth of new particles formed by nucleation and therefore for the CCN concentration. Ambient observations strongly indicate that the growth of new particles by organics is indeed taking place and can dominate the growth in environments with modest or low sulfur dioxide concentrations. Chemical aging reactions converting the SVOCs to ELVOCs through reactions with OH can explain the observed growth and are consistent with the available observations.

Acknowledgements

This research was supported by the US Department of Energy and the FP7 IDEAS project ATMOPACS.

References

1 C. A. Pope, *J. Aerosol Med.*, 2000, **13**, 335–354.
2 S. F. van Eeden, A. Yeung, K. Quinlam and J. C. Hogg, *Proc. Am. Thorac. Soc.*, 2005, **2**, 61–67.
3 D. W. Dockery, *Environ Health Perspect*, 2001, **109**(Suppl. 4), 483–486.
4 R. D. Brook, *et al.*, *Circulation*, 2004, **109**, 2655–2671.
5 K. A. Miller, *et al.*, *N. Engl. J. Med.*, 2007, **356**, 447–458.
6 IPCC, Intergovernmental Panel on Climate Change, *Climate Change 2007: The Scientific Basis*, Cambridge University Press, Cambridge, 2007.
7 J. H. Seinfeld and S. N. Pandis, *Atmospheric Chemistry and Physics: From Air Pollution to Climate Change*, J. Wiley, New York, 2nd edn, 2006.
8 N. M. Donahue, A. L. Robinson and S. N. Pandis, *Atmos. Environ.*, 2009, **43**, 97–109.
9 A. H. Goldstein and I. E. Galbally, *Environ. Sci. Technol.*, 2007, **41**, 1514–1521.
10 M. Hallquist, *et al.*, *Atmos. Chem. Phys.*, 2009, **9**, 5155–5236.
11 P. H. McMurry, M. F. Shepherd, and J. S. Vickery, *Particulate Matter Science for Policy Makers*, Cambridge University Press, Cambridge, 2004.
12 M. Kanakidou, *et al.*, *Atmos. Chem. Phys.*, 2005, **5**, 1053–1123.
13 A. L. Robinson, *et al.*, *Science*, 2007, **315**, 1259–1262.
14 Q. Zhang, *et al.*, *Geophys. Res. Lett.*, 2007, **34**, L13801.
15 S. N. Pandis, *et al.*, *Atmos. Environ.*, 1992, **26**, 2266–2282.
16 J. R. Odum, *et al.*, *Environ. Sci. Technol.*, 1996, **30**, 2580–2595.
17 M. Claeys, *et al.*, *Science*, 2004, **303**, 1173–1176.
18 Y. Rudich, *et al.*, *Annu. Rev. Phys. Chem.*, 2007, **58**, 321–352.
19 M. Kalberer, *et al.*, *Science*, 2004, **303**, 1659–1662.
20 J. L. Jimenez, *et al.*, *Science*, 2009, **326**, 1525–1529.
21 N. M. Donahue, *et al.*, *Atmos. Chem. Phys.*, 2011, **11**, 3303–3318.
22 N. M. Donahue, *et al.*, *Atmos. Chem. Phys.*, 2012, **12**, 615–634.
23 J. H. Kroll, *et al.*, *Nat. Chem.*, 2011, **3**, 133–139.
24 H. Burtscher, *et al.*, *J. Aerosol Sci.*, 2001, **32**, 427–442.
25 B. Wehner, *et al.*, *J. Aerosol Sci.*, 2002, **33**, 1087–1093.
26 W. J. An, R. K. Pathak, B. H. Lee and S. N. Pandis, *J. Aerosol Sci.*, 2007, **38**, 305–314.
27 I. Riipinen, *et al.*, *Atmos. Environ.*, 2010, **44**, 597–607.
28 E. Karnezi, I. Riipinen and S. N. Pandis, *Atmos. Meas. Tech.*, 2013, submitted for publication.
29 E. Kostenidou, *et al.*, *Environ. Sci. Technol.*, 2009, **43**, 4884–4889.
30 V. A. Lanz, *et al.*, *Atmos. Chem. Phys.*, 2007, **7**, 1503–1522.
31 B. N. Murphy, *et al.*, *Atmos. Chem. Phys.*, 2012, **12**, 10797–10816.
32 A. P. Tsimpidi, *et al.*, *Atmos. Chem. Phys.*, 2010, **10**, 525–546.
33 J. A. Huffman, *et al.*, *Environ. Sci. Technol.*, 2009, **43**, 5351–5357.
34 J. E. Shilling, *et al.*, *Atmos. Chem. Phys.*, 2009, **9**, 771–782.
35 N. L. Ng, *et al.*, *Atmos. Chem. Phys.*, 2010, **10**, 4625–4641.

36 Q. Chen, *et al.*, *Environ. Sci. Technol.*, 2011, **45**, 4763–4770.
37 B. N. Murphy, *et al.*, *Atmos. Chem. Phys.*, 2011, **11**, 7859–7873.
38 N. M. Donahue, *et al.*, *Environ. Chem.*, 2013, **10**, 151–157.
39 K. M. Wagstrom and S. N. Pandis, *J. Geophys. Res.*, 2009, **114**, D14303.
40 K. M. Wagstrom, *et al.*, *Atmos. Environ.*, 2008, **42**, 5650–5659.
41 K. Skyllakou, *et al.*, *Atmos. Chem. Phys.*, 2013, submitted for publication.
42 F. Freutel, *et al.*, *Atmos. Chem. Phys.*, 2013, **13**, 933–959.
43 M. Beekmann, *et al.*, *Atmos. Chem. Phys.*, 2013, submitted for publication.
44 M. Sipilä, *et al.*, *Science*, 2010, **327**, 1243–1246.
45 I. Riipinen, *et al.*, *Atmos. Chem. Phys.*, 2011, **11**, 3865–3878.
46 J. G. Jung, P. J. Adams and S. N. Pandis, *Aerosol Sci. Technol.*, 2008, **42**, 495–504.
47 J. G. Jung, P. J. Adams and S. N. Pandis, *Atmos. Environ.*, 2006, **40**, 2248–2259.

Faraday Discussions

RSCPublishing

Quantitative and time-resolved nanoparticle composition measurements during new particle formation†

Bryan R. Bzdek,[a] Andrew J. Horan,[a] M. Ross Pennington,[a] Joseph W. DePalma,[a] Jun Zhao,[b] Coty N. Jen,[b] David R. Hanson,[c] James N. Smith,[de] Peter H. McMurry[b] and Murray V. Johnston[*a]

Received 12th March 2013, Accepted 12th April 2013

DOI: 10.1039/c3fd00039g

The chemical composition of 20 nm diameter particles was measured with the Nano Aerosol Mass Spectrometer (NAMS) in a rural/coastal environment during days when new particle formation (NPF) occurred and days when NPF did not occur. NAMS provides a quantitative measure of nanoparticle elemental composition with high time resolution. These measurements show that nanoparticle chemical composition is dynamic on both types of days and that changes in nanoparticle chemical composition do not necessarily correlate with changes in aerosol mass or number concentration. On NPF days, NAMS can distinguish between elements associated with particle formation and early mass growth from those associated with later mass growth. In the early stage of NPF, the particle phase sulphur mole fraction (S) increases simultaneously with the increase in gas phase sulphuric acid. This composition change occurs *before* the mode diameter has grown into the NAMS-measured size range and is quantitatively described by sulphuric acid condensation. The nitrogen mole fraction (N) also increases during this time period. The N/S mole ratio is approximately 2, indicating that particulate sulphate is fully neutralized. As the mode diameter passes into and through the NAMS-measured size range, N increases at a faster rate than S (N/S mole ratio

---

aDepartment of Chemistry and Biochemistry, University of Delaware, Newark, Delaware 19716, U.S.A. E-mail: bbzdek@udel.edu; ajhoran@udel.edu; mrpenn@udel.edu; jdepalma@udel.edu; mvj@udel.edu; Fax: +302 831-6335; Tel: +302 831-8014

bDepartment of Mechanical Engineering, University of Minnesota, 111 Church Street SE, Minneapolis, Minnesota 55455, U.S.A. E-mail: zhaoj@umn.edu; jenxx006@umn.edu; mcmurry@me.umn.edu; Tel: +612 624-2817

cChemistry Department, Augsburg College, 2211 Riverside Avenue, Minneapolis, Minnesota 55454, U.S.A. E-mail: hansondr@augsburg.edu; Fax: +612 330-1649; Tel: +612 330-1620

dAtmospheric Chemistry Division, National Center for Atmospheric Research, 1850 Table Mesa Drive, Boulder, Colorado 80305, U.S.A. E-mail: jimsmith@ucar.edu; Fax: +303-497-1400; Tel: +303 497-1468

eApplied Physics Dept., University of Eastern Finland, P.O. Box 1627, 70211, Kuopio, Finland. Fax: +358 162585; Tel: +358 2 9445 1111

† Electronic supplementary information (ESI) available: Nanoparticle event summary for 2012 and air mass histories. See DOI: 10.1039/c3fd00039g

increases above 2), indicating that a separate, nitrogen-based growth process exists, possibly involving aminium salts, inorganic nitrate and/or organonitrates. Carbonaceous matter is the most abundant component (~50% by mass) of the growing nanoparticles, but it is the inorganic species that are preferentially enhanced during NPF relative to other times of day. Concurrent measurements of cloud condensation nucleation activity during NPF events suggest that these newly formed particles are hygroscopic. Nanoparticle composition on non-NPF days also shifts toward a more inorganic composition during the daytime, but the chemical species are different from NPF days and the particles are less hygroscopic. Incorporation of S into growing nanoparticles is adequately explained by existing models, but currently no models exist to satisfactorily explain incorporation of nitrogen-containing species or carbonaceous matter.

1   Introduction

Atmospheric new particle formation (NPF) is the process of creating aerosol by gas-to-particle conversion.[1,2] NPF is a two-step process involving nucleation of new nanoparticles followed by their growth to larger sizes.[3] Once the particles have grown sufficiently, they may substantially impact cloud condensation nuclei concentrations[4-6] and thus influence precipitation patterns and global climate.[7-10]

Nucleation is thought to be driven primarily by sulphuric acid,[11-13] as the nucleation rate typically varies in proportion to $[H_2SO_4]^p$ ($1 \leq p \leq 2$).[14,15] However, other species such as bases[16] (ammonia[17-19] or amines[20-22]) and organic molecules[23-26] may also be important. With respect to growth, a variety of chemical species are implicated as playing important roles. Sulphuric acid is clearly important, but by itself cannot quantitatively explain measured nanoparticle growth rates.[27-30] Therefore, other growth channels must exist. Possible species include nitrogen-containing compounds and carbonaceous matter.[30-36]

It is important to understand the chemical processes underlying nanoparticle growth[37] because a nucleated nanoparticle can only become climatically relevant if its growth rate is higher than its loss rate due to scavenging.[29,38] Understanding growth mechanisms is challenging because of the limitations of analytical measurement technology in this size range.[39,40] In particular, most methods used to determine nanoparticle composition lack the time resolution needed to quantify the dynamics of NPF. While time resolved measurements are routine for much larger accumulation mode particles,[41] the sources and processes governing composition may be much different. High time resolution measurements of nanoparticle chemical composition in urban locations have been used to distinguish primary from secondary nanoparticles in a manner not possible with accumulation mode particle measurements.[42,43]

In this work, the chemical composition of nanoparticles during NPF is measured with the Nano Aerosol Mass Spectrometer (NAMS), which provides quantitative elemental composition measurements of size-selected particles in the 10–30 nm range with high time resolution.[27,42-50] NAMS permits time- and composition-resolved ambient measurements of nanoparticles inaccessible by most instrumentation. It is shown that nanoparticle composition during NPF is dynamic, and composition changes associated with early stages of growth are distinguished from those associated with later stages of growth. Additionally, time-resolved nanoparticle composition on NPF days is compared to that on non-NPF days to determine more generally how the two types of days differ. The results provide insight into the chemical processes governing atmospheric NPF and its possible role in climate.

2   Experimental

The field campaign was conducted at the Hugh R. Sharp Campus of the University of Delaware in Lewes, Delaware, U.S.A. (38°47′02″N, 75°09′39″W) from 23 July to 31 August 2012. The field site is located 800 m south of the Delaware Bay, which is at the outlet of the Delaware River to the Atlantic Ocean and 3 km west of the Atlantic Ocean. A large salt marsh sits adjacent to the site (<50 m) on the west. This field site was used previously for a campaign to measure gas phase sulphur emissions[51,52] as well as to study nanoparticle chemical composition during NPF in the autumn of 2007.[44]

Nanoparticles were sampled through an inlet approximately 6 m above the ground. Nanoparticle chemical composition measurements were accomplished with NAMS, a single particle mass spectrometer that gives quantitative elemental composition measurements in the 10–30 nm size range.[47,48,50] Nanoparticles entering NAMS were charged with a unipolar charger[53] and then focused through two digital ion guides and size-selectively captured in an ion trap. Trapped particles were irradiated with a high-energy pulsed laser beam to reach the "complete ionization limit", thereby creating a laser-induced plasma and quantitatively disintegrating the nanoparticles into multiply charged positive atomic ions that were mass analysed by time-of-flight. The integrated area underneath the ion signals gave the elemental composition. Relative to previous versions of NAMS, replacement of an aerodynamic lens with a second digital ion guide substantially improved focusing of 20 nm diameter nanoparticles into the ion trap, thereby improving the time resolution of composition measurements. Deconvolution of overlapping signal intensities was accomplished by the method of Zordan *et al.* (2010).[54] The particle size range to be analysed was selected by the frequency applied to the ring electrode of the ion trap. During this campaign, NAMS was set to analyse the composition of 20 ± 3 nm mass normalized diameter particles, which corresponds approximately to 18 ± 3 nm mobility diameter particles.[55]

Nanoparticle chemical composition was averaged over one or multiple 5-min bins such that a minimum of 20 particles were included in the average. Averaging the measured compositions of 20 particles simultaneously maximizes time resolution while minimizing uncertainty from variations in the dynamics of the laser plume.[56] For the time periods discussed here, particle concentrations were such that at least 20 particles were analysed in almost every 5-min bin during the daytime. Based on measurements with standard aerosols, elemental compositions measured by NAMS are generally within 10% of expected values after averaging a sufficient number of particles (*e.g.* 20).[54,56]

Mass fractions for apportioned molecular species were determined from the elemental mole fractions by a method described previously.[44] These results, together with diameter growth rates inferred from scanning mobility particle sizer

(SMPS) measurements, were used to determine the sulphuric acid vapour concentration ($[H_2SO_4]_{NAMS}$) required to explain particulate sulphur content:[27]

$$[H_2SO_4]_{NAMS} = \frac{2GR_{MEAS}}{\nu_1 \bar{c}_1 \Gamma_m} \times \frac{\rho_{particle}}{\rho_{sulphate}} \tag{1}$$

where GR_{MEAS} is the measured growth rate (nm h^{-1}), ν_1 is the volume occupied by a hydrated H_2SO_4 molecule (cm^3), $\bar{c}_1$ is the mean thermal speed of the condensing monomer (nm h^{-1}), $\rho_{particle}$ is the density of ambient particles during the time period of interest (estimated to be 1.5 g cm^{-3}), $\rho_{sulphate}$ is the density of condensed phase sulphuric acid (1.8 g cm^{-3}) and $\Gamma_m = \dfrac{1}{MF_{sulphate}}$, where $MF_{sulphate}$ is the average sulphate mass fraction for the time period of interest. Eqn (1) does not include the Fuchs–Sutugin correction for mass flux, which only affects the calculated sulphuric acid concentration by <5%, much smaller than the uncertainties of GR and Γ_m measurements (approximately 10% to 30%). Eqn (1) also assumes that the mass accommodation coefficient is close to 1.[29,57] Note that the parameter Γ_m obtained from NAMS measurements is related to the more familiar Γ_v obtained from gas phase sulphuric acid measurements[29] by:

$$\Gamma_v = \Gamma_m \times \frac{\rho_{sulphate}}{\rho_{particle}} \tag{2}$$

Gas phase sulphuric acid concentrations were measured independently using the Cluster-Chemical Ionization Mass Spectrometer (Cluster CIMS), which has been described previously.[58,59] Agreement between $[H_2SO_4]_{NAMS}$ and $[H_2SO_4]_{CIMS}$ would imply that eqn (1) and (2) are valid models for the uptake of S onto freshly nucleated particles.

Ambient gas phase ammonia and amine concentrations were measured using the Ambient pressure Proton transfer Mass Spectrometer (AmPMS).[60] Size-resolved cloud condensation nuclei (CCN) activation was measured at 5 different mobility diameters from 50 to 175 nm using a condensation particle counter (CPC; model 3760, TSI, Inc., St. Paul, Minnesota, U.S.A.) and a cloud condensation nucleus counter (CCNc; Droplet Measurement Technologies, Boulder, Colorado, U.S.A.) downstream of a differential mobility analyzer (DMA; model 3081, TSI, Inc., St. Paul, Minnesota, U.S.A.) following the setup described by Levin *et al.*[61] Calibrations with ammonium sulfate particles were performed weekly during the campaign. After inverting the CPC and CCNc data following the procedures described in Levin *et al.*, we constructed CCN activation curves and found the critical supersaturation (S_c) at each sampled diameter.

Particle size distributions were measured using an SMPS (electrostatic classifier model 3080, CPC model 3788, TSI, Inc., St. Paul, Minnesota, U.S.A.). Growth rates from 10–25 nm diameter were determined by plotting the mode particle diameter against local time, fitting the result linearly and taking the derivative. During the year 2012 (with the exception of July–September 2012), nanoparticle number concentrations were monitored with 15-minute time resolution on a daily basis using an ultrafine particle monitor (model 3031, TSI, Inc., St. Paul, Minnesota, U.S.A.) at an adjacent Delaware Department of Natural Resources and Environmental Control monitoring site. This instrument provided nanoparticle number concentrations for six size bins: 20–30 nm, 30–50 nm, 50–70 nm, 70–100 nm, 100–200 nm and >200 nm.

3   Results and discussion

3.1   Nanoparticle events in Lewes

The Lewes field site is frequently impacted by nanoparticle events where the concentration of sub-50 nm diameter particles increases quickly over a short period of time. During a previous one-month-long campaign to study NPF at this site in October–November 2007,[44] regional NPF was observed on nearly half of the measurement days. During 2012, nanoparticle number concentrations were monitored with an ultrafine particle monitor. Nanoparticle events were most frequent in the late winter/early springtime months (February–April; average of 17 nanoparticle events/month) and were least frequent in November–December (average of 2 nanoparticle events/month) (see Fig. S1 of the Electronic Supplementary Information, ESI†). These nanoparticle events could be divided into two groups. One group consisted of an increase in nanoparticle number concentration for a fixed duration with little to no growth to larger sizes. These events were classified as local events. The second group was described by an increase in nanoparticle number concentration in the 20–30 nm size bin followed by growth to larger sizes (30–50 nm and 50–70 nm size bins). These events were classified as regional events when the initial burst of nanoparticles was followed by growth to larger sizes. These two event classes had very different air mass histories, as determined by HYSPLIT.[62] Air masses during local events typically transported to the site from the south, having passed over areas with substantial development and a coal-fired power plant. On the other hand, air masses during regional events typically transported to the site from the northwest, having passed over largely rural areas, the Delaware Bay and salt marshes. As examples of these two classes of events, air mass back trajectories for 11–14 August 2012 are shown in Fig. S2.† Both regional event days (12 and 13 August) have air masses coming from the northwest, whereas both non-event days (11 and 14 August) have air masses coming from the south.

3.2   Nanoparticle chemical composition on regional new particle formation days

Nanoparticle elemental composition data acquired by NAMS for two regional NPF days during the field campaign are presented in Fig. 1. Fig. 1a shows the SMPS-measured aerosol size distribution for these two events. The horizontal dotted lines indicate the size range of particles analysed by NAMS during this study. The NPF event on 12 August 2012 is a strong event (*i.e.* high nanoparticle number concentration) with a low condensational sink and fast growth to larger sizes. The event on 13 August 2012 is a weak NPF event with a much larger condensational sink but a similar growth rate to larger sizes. For both events, air masses came from the northwest, although later in the day on 13 August there was some recirculation of the airmass around the site (see ESI†).

Fig. 1b shows NAMS-measured S and Si elemental mole fractions plotted together with Cluster CIMS-measured gas phase sulphuric acid concentration. Fig. 1c shows NAMS-measured N mole fraction as well as "excess" nitrogen (Excess N), which we define here as twice the S mole fraction subtracted from the N mole fraction (Excess Nitrogen $= N - 2S$). If Excess N < 0, then not enough nitrogen is available to neutralize sulphuric acid, even if all of the N exists as

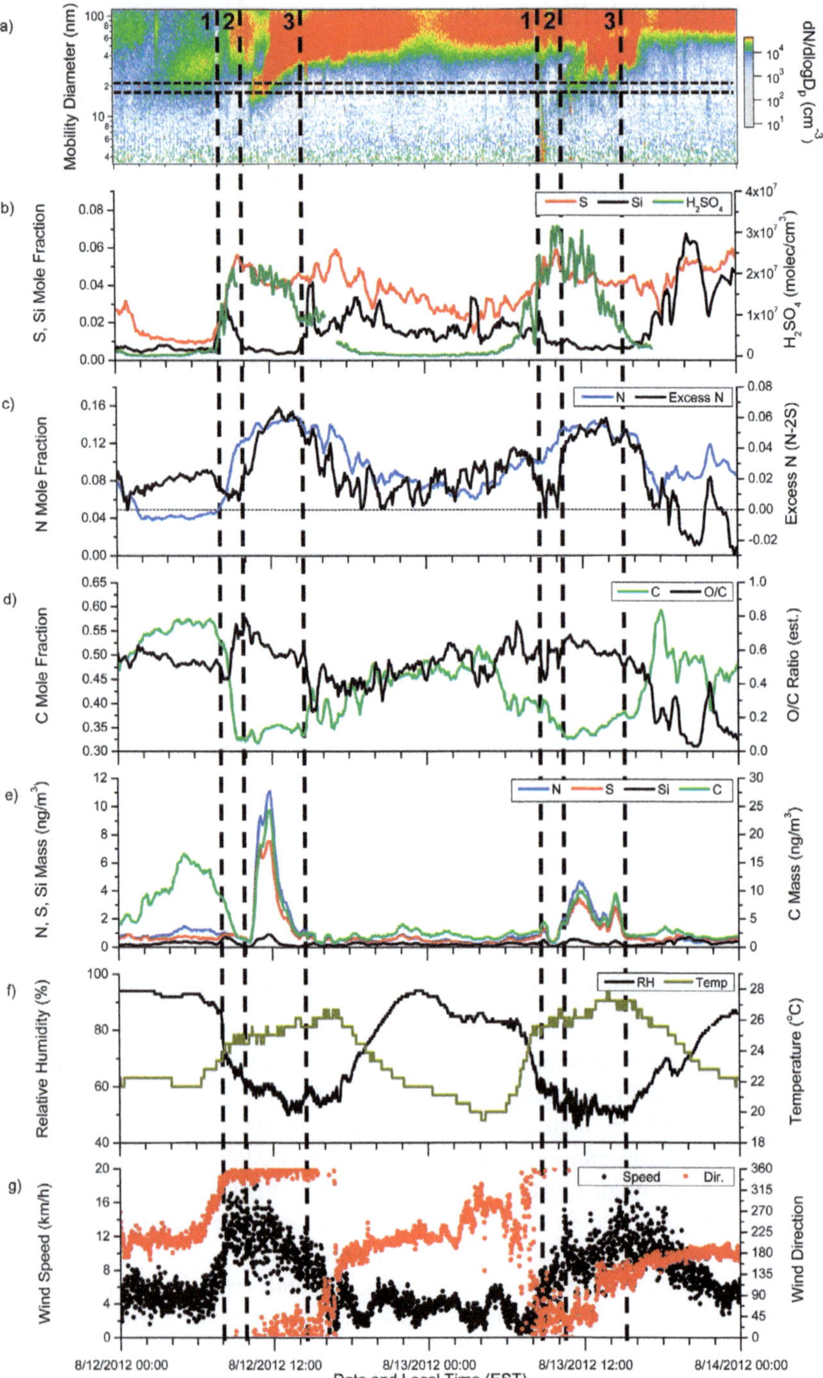

Fig. 1 a) Aerosol size distributions, b) S and Si mole fraction as well as gas phase sulphuric acid concentration, c) N and Excess N mole fractions, d) C mole fraction and estimated O/C molar ratio, e) N, S, Si and C elemental mass, f) relative humidity and temperature and g) wind speed and wind direction for 12 and 13 August 2012. The horizontal dotted lines in a) indicate the NAMS-measured size range. The horizontal dotted line in c) indicates Excess N = 0. Vertical dotted lines indicate 1) when gas phase

cation forming species (ammonia, amines). Therefore, the particle necessarily is acidic. Implicit in this statement is the assignment of all elemental S as sulphate, the validity of which has been discussed in detail elsewhere.[27] If Excess N = 0 (dotted line in Fig. 1c) then exactly enough N exists to completely neutralize sulphuric acid. If Excess N > 0, then more nitrogen is present in the particle than required to neutralize sulphuric acid, meaning that a nitrogen-containing species other than ammonium or aminium sulphate must exist in the particle. Based on the uncertainty in the NAMS elemental composition measurement and the range of elemental mass fractions observed for S and N, we assign an absolute uncertainty of 0.01 to the Excess N mole fraction.

Fig. 1d shows the C mole fraction and its estimated O/C molar ratio. The O/C ratio was estimated using a molecular apportionment algorithm described previously.[44] Briefly, oxygen is first apportioned to the inorganic components of the aerosol (as SiO_2, SO_4^{2-}, NH_4^+ and NO_3^-) and the remaining oxygen mole fraction is divided by the carbon mole fraction (which is not apportioned to any other molecular species). The absolute uncertainty in the estimated O/C molar ratio ranges from 0.06–0.14 depending on the day and time. This range does not include systematic error arising from assumptions in the molecular apportionment algorithm. Fig. 1e shows elemental mass for N, S, Si and C. Elemental mass was determined by multiplying elemental mass fractions by the SMPS-measured aerosol mass in the 17–21 nm size range. Fig. 1f shows relative humidity (RH) and ambient temperature (°C), and Fig. 1g shows wind speed (km h^{-1}) and wind direction. For each day, the three vertical dotted lines indicate 1) the time when gas phase sulphuric acid increases, 2) the time when the mode diameter of the particle size distribution associated with the event begins to move into the NAMS-measured size range and 3) the time when the mode diameter grows past the NAMS-measured size range.

The event on 12 August is a strong regional event with fast growth (growth rate from 10–25 nm = 8.9 ± 0.8 nm h^{-1}). Fig. 2 shows an expanded view of the key stages of the event. At the onset of NPF (vertical line 1), the increase in S mole fraction coincides with the rapid increase in gas phase sulphuric acid concentration (Fig. 1b; Fig. 2a). However, this increase does *not* correspond to an increase in aerosol mass in the NAMS-measured size range (Fig. 1e; Fig. 2d). While condensation of sulphuric acid over this (short) time period is sufficient to change the composition of pre-existing particles, it is not enough to significantly increase the aerosol mass concentration in the relevant size range. Nonetheless, the change in composition is important since it gives insight into the early steps of nanoparticle formation and growth, which must primarily involve sulphuric acid.

Later in the day, the gas phase sulphuric acid concentration decreases quickly, but the S mole fraction does not. This lack of correlation between the two is characteristic of a regional event. Initially, the sulphuric acid and S mole fraction are correlated because a local process (condensation of gas phase molecules) drives the composition change of pre-existing particles at the site. Near the end

---

sulphuric acid increases, 2) when the mode diameter of the event passes into the NAMS-measured size range and 3) when the mode diameter moves out of the NAMS-measured size range. Nanoparticle composition data were subjected to 6-point smoothing; the sulphuric acid concentration was subjected to 10-point smoothing.

the day, the particles being analysed have been transported to the site during the NPF event and their composition reflects growth processes along the way. Therefore in a regional event, the decrease in S mole fraction necessarily lags the decrease in gas phase sulphuric acid because the transported particles were previously exposed to a high gas phase sulphuric acid concentration.

A significant increase in the Si mole fraction is observed at the start of the event, coincident with the sulphuric acid and S mole fraction increases. However, the Si mole fraction quickly decreases as the mode diameter moves into the NAMS-measured size range. This dependence is consistent with a local, photochemical source of Si. This source is unknown but could involve siloxane emissions from plastic tubing and containers used at the site. A photochemical source of particle phase Si has been observed previously in a study of nanoparticles in Pasadena, California, U.S.A.[46] Later in the day, particles that have not been exposed to the local source of Si are transported to the site, and their composition reflects growth processes in the regional event.

Fig. 1c and 2b show N mole fraction and Excess N. The N mole fraction increases simultaneously with the increase in gas phase sulphuric acid and S mole fraction. The strong correlation between N and S suggests that cation forming species (ammonia, amines) are being incorporated into the particle to neutralize sulphate. During this time period, the N/S molar ratio approaches 2, indicating sulphate is neutralized. However, simply assigning N to sulphate neutralization is an oversimplification. Excess N, which is an indicator of species that cannot be associated with sulphate neutralization, does not increase until much later when the mode diameter moves into the NAMS-measured size range. The increase in Excess N correlates with the increase in aerosol mass (Fig. 1e; Fig. 2d), suggesting that it is associated with later (as opposed to earlier) stages of particle growth. Because NAMS only provides elemental composition, the molecular identity of these species was not directly measured. However, potential molecular species include ammonium or aminium nitrate, aminium-organic acid salts and organonitrates. The increase in Excess N occurs simultaneously with an increase in ambient temperature and a decrease in RH (Fig. 1f). If the identity of Excess N is ammonium nitrate, this dependence contradicts expected nitrate partitioning to aerosol, which is usually more abundant during periods of low temperature and high RH.[27,63–67] Therefore, if ammonium nitrate is the identity of Excess N, that would mean a substantial gas phase nitric acid concentration (not measured in this campaign) must exist during the event. Gas phase ammonia concentrations, measured with AmPMS, were less than 200 pptv during this event. Based on the Aerosol Inorganics Model (AIM),[68,69] particulate phase ammonium nitrate would require a gas phase nitric acid concentration on the order of 10–40 ppbv, which seems unlikely. Therefore, equilibrium partitioning of ammonium nitrate to the particle phase during this time period is also unlikely.

Fig. 1d and 2c show C mole fraction and its estimated O/C molar ratio. During the NPF event, the C mole fraction decreases concurrently with the increase in S mole fraction and gas phase sulphuric acid. The C mole fraction does not increase again until after the aerosol mass has passed through the NAMS-measured size range. Although the C mole fraction does not increase during NPF, its aerosol mass does (Fig. 1e), indicating that carbonaceous matter contributes substantially to particle growth. The O/C ratio initially increases about the same time as the sulphuric acid and S mole fraction increase, suggesting highly oxidized organic

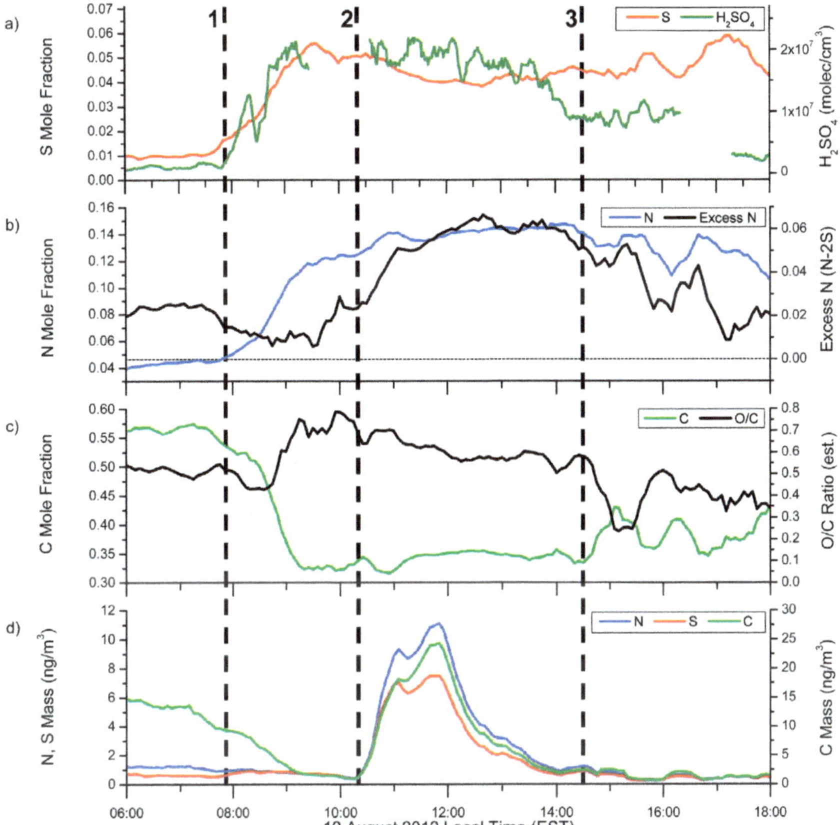

Fig. 2 Expanded view of the event on 12 August 2012 showing a) S mole fraction and gas phase sulphuric acid concentration, b) N and Excess N mole fractions, c) C mole fraction and estimated O/C molar ratio and d) N, S and C elemental mass. The horizontal dotted line in b) indicates Excess N = 0. Vertical dotted lines indicate 1) when gas phase sulphuric acid increases, 2) when the mode diameter of the event passes into the NAMS-measured size range and 3) when the mode diameter moves out of the NAMS-measured size range. Nanoparticle composition data were subjected to 6-point smoothing; the sulphuric acid concentration was subjected to 10-point smoothing.

molecules may be important in the early stage of nanoparticle growth.[70] As the aerosol mass passes into and through the NAMS-measured size range, the O/C ratio slowly decreases. Once the aerosol mass has passed through the NAMS-measured size range, the O/C ratio slowly increases again, suggesting that the carbonaceous matter is gradually aging.[71,72] We calculated an average mass-weighted O/C ratio for the event by converting elemental mass (in ng m^{-3}) over the period of the event to elemental mole fraction and then apportioning to molecular components. This procedure gave an estimated O/C ratio of 0.63 over the event.

A weak regional event occurred on 13 August, which is shown in Fig. 1 and expanded in Fig. 3. On this day, growth was similar to the previous day (growth rate from 10–25 nm on 13 August = 7.1 ± 1.0 nm h^{-1}) but there was a substantial condensational sink. Many of the same trends are observed for this event as for the event on 12 August. When the gas phase sulphuric acid concentration

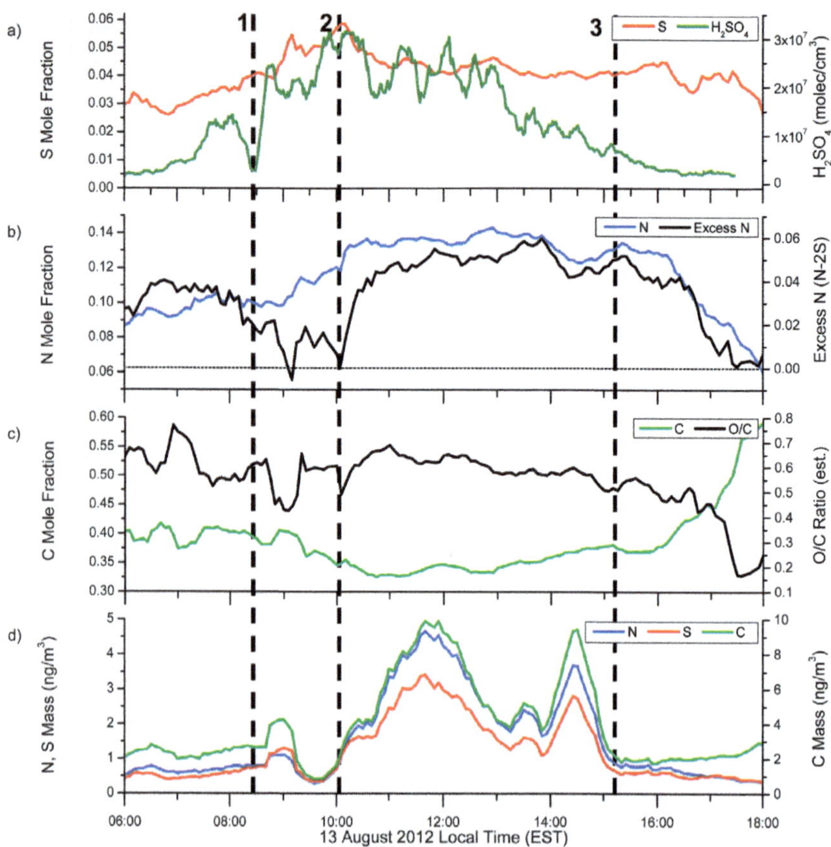

Fig. 3 Expanded view of the event on 13 August 2012 showing a) S mole fraction and gas phase sulphuric acid concentration, b) N and Excess N mole fractions, c) C mole fraction and estimated O/C molar ratio and d) N, S and C elemental mass. The horizontal dotted line in b) indicates Excess N = 0. Vertical dotted lines indicate 1) when gas phase sulphuric acid increases, 2) when the mode diameter of the event passes into the NAMS-measured size range and 3) when the mode diameter moves out of the NAMS-measured size range. Nanoparticle composition data were subjected to 6-point smoothing; the sulphuric acid concentration was subjected to 10-point smoothing.

increases, S mole fraction also increases. However, the change in particle phase S mole fraction is not as dramatic as for the event on 12 August because a substantially higher S mole fraction in the background aerosol existed prior to the event on 13 August. Nonetheless, the same general trend is observed in that the S mole fraction increases simultaneously with the increase in gas phase sulphuric acid, whereas there is a substantial delay between the drop in gas phase sulphuric acid and S mole fraction at the end of the event. A local photochemical source of Si is indicated at the beginning of the event by the short-lived increase in Si mole fraction. A much higher increase in Si mole fraction in the evening of 13 August after the event has ended suggests the presence of a separate, non-photochemical source.

As on 12 August, the N mole fraction on 13 August increases simultaneously with the S mole fraction, suggesting that this nitrogen-containing species is associated with sulphate neutralization, and the N/S ratio of approximately 2 indicates full neutralization. Again as on 12 August, Excess N on 13 August increases later in time, simultaneous with the increase in aerosol mass. Furthermore, the increase in Excess N is coincident with an increase in temperature and a decrease in RH, opposite of what would be expected for ammonium nitrate partitioning. The C mole fraction also decreases relative to N and S, and the mass-weighted O/C ratio is 0.61, although the time-dependent trends of O/C ratio are not as pronounced as for 12 August.

Fig. 4 presents mass-weighted average molecular mass fractions for both 12 and 13 August during the daytime. O/C ratios are given by the inset in the carbonaceous matter. The other two days presented in the figure will be discussed later. The average composition and O/C ratio are essentially the same during both events. An effective gas phase sulphuric acid concentration can be calculated from the average sulphate mass fraction and growth rate over each event as described in eqn (1).[27,29] The results of this calculation are shown in Table 1. For both events, the sulphuric acid concentration calculated from the sulphate mass fraction matches within experimental error the measured sulphuric acid concentration, showing that uptake of sulphuric acid is approximately collision limited. In other words, the sulphate mass fraction during NPF can be quantitatively explained by condensation of gas phase sulphuric acid molecules. This agreement is significant and has been observed only on two other occasions.[30,45] The observed closure between gas and particle phase measurements validates existing models to describe S uptake.[29] Additionally, the Γ_v values (NAMS-measured Γ_m adjusted for particle density, as discussed in eqn (2) in the experimental section) reported in Table 1 are consistent with those measured in Atlanta, Georgia, U.S.A., and Boulder, Colorado, U.S.A., as well as values for some NPF events in Mexico City, Mexico, and Hyytiälä, Finland,[29,45] indicating that the contribution of S to particle growth is similar to that observed in those locations.

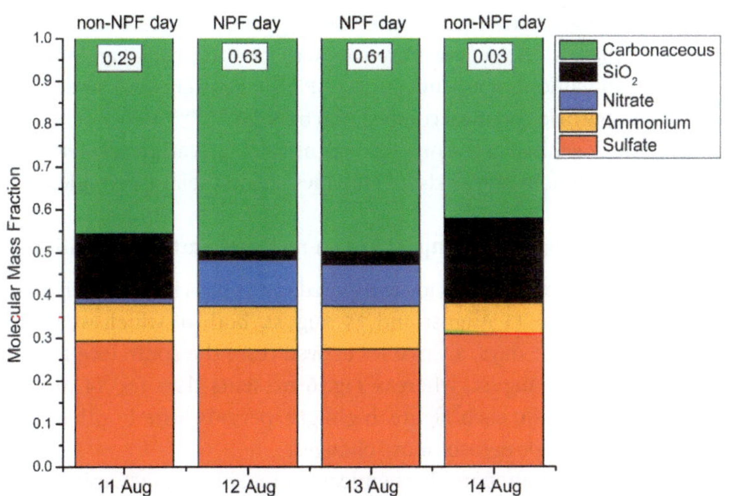

Fig. 4 Average apportioned mass fractions during the daytime for all four days. Insets in the carbonaceous matter indicate estimated O/C molar ratios. Uncertainties for the O/C ratios range from 0.06 to 0.14 over these days and do not include systematic error arising from assumptions in the apportionment algorithm.

Table 1 Particle growth rate from 10–25 nm diameter, NAMS-measured sulphate mass fraction, ratio of total particle volume growth to growth from sulphuric acid (Γ_v), calculated [H_2SO_4] from the sulphate mass fraction, and experimentally measured [H_2SO_4] for the events on 12 and 13 August 2012.

	12 August	13 August
Growth Rate (nm h^{-1})^a	8.9 ± 0.8	7.1 ± 1.0
Sulphate Mass Fraction^a	0.27 ± 0.08	0.27 ± 0.08
$\Gamma_v{}^b$	4.4 ± 1.3	4.4 ± 1.3
[H_2SO_4]$_{NAMS}$ (molec cm^{-3})^c	$2.6 \pm 0.8 \times 10^7$	$2.1 \pm 0.6 \times 10^7$
[H_2SO_4]$_{CIMS}$ (molec cm^{-3})^a	$1.9 \pm 1.0 \times 10^7$	$2.2 \pm 1.1 \times 10^7$

^{a} Experimental measurement. ^{b} Assumes $\rho_{particle} = 1.5$ g cm^{-3}. ^{c} Calculated from eqn (1).

Nanoparticles in Lewes are fully neutralized, consistent with our previous study in this location,[44] but contrasting with our study in a remote boreal forest in the early spring.[45] As a result, in Lewes the apportioned nitrate (proxy for Excess N) is substantial, whereas in the boreal forest no apportioned nitrate was present. With respect to carbonaceous matter, the O/C ratios in Lewes are higher than those measured in the boreal forest, indicating that the Lewes carbonaceous matter is more highly oxidized, though the O/C ratios measured at the end of the boreal forest campaign (mid spring) were close to those reported here.[45] Relative to the previous campaign to study NPF in Lewes,[44] O/C ratios are lower in this study, but the difference could be a result of the time of year (fall *vs.* summer) and the extent of anthropogenic contributions in the local environment (more vehicular emissions in the summer).

More generally, the Lewes data show four different chemical components important to particle growth during NPF: sulphate, cation forming nitrogen (ammonia or amine), carbonaceous matter and nitrate (or, more precisely, Excess N). The relative contributions of these pathways to mass growth are quantitatively determined, as shown in Fig. 4. In contrast, only three of these components were important to mass growth in the boreal forest: sulphate, cation forming nitrogen and carbonaceous matter. The general observation that nanoparticles are preferentially enhanced in inorganic components during NPF is consistent with several previous studies of nanoparticle chemical composition by NAMS.[27,44,45] However, it should be emphasized that over half of the mass growth arises from carbonaceous matter and Excess N, for which no models exist to suitably explain their contributions.

3.3   Nanoparticle chemical composition on non-new particle formation days

In this section, the two NPF days are compared to the days immediately preceding and succeeding them: 11 August and 14 August, both of which were non-NPF days. For both of these days, air masses arrived from the south (see ESI†). Fig. 5 presents data for 11 August, whereas Fig. 6 presents data for 14 August. The vertical dotted lines in each figure highlight periods where either chemical composition changes or aerosol mass changes.

On both days, the nanoparticle chemical composition is dynamic. The dotted lines in Fig. 5 and 6 highlight specific time points to illustrate that these changes in nanoparticle chemical composition are not necessarily correlated with changes in aerosol mass or number concentration. For example, the first vertical dotted line on Fig. 6 shows a time period when chemical composition is static while aerosol

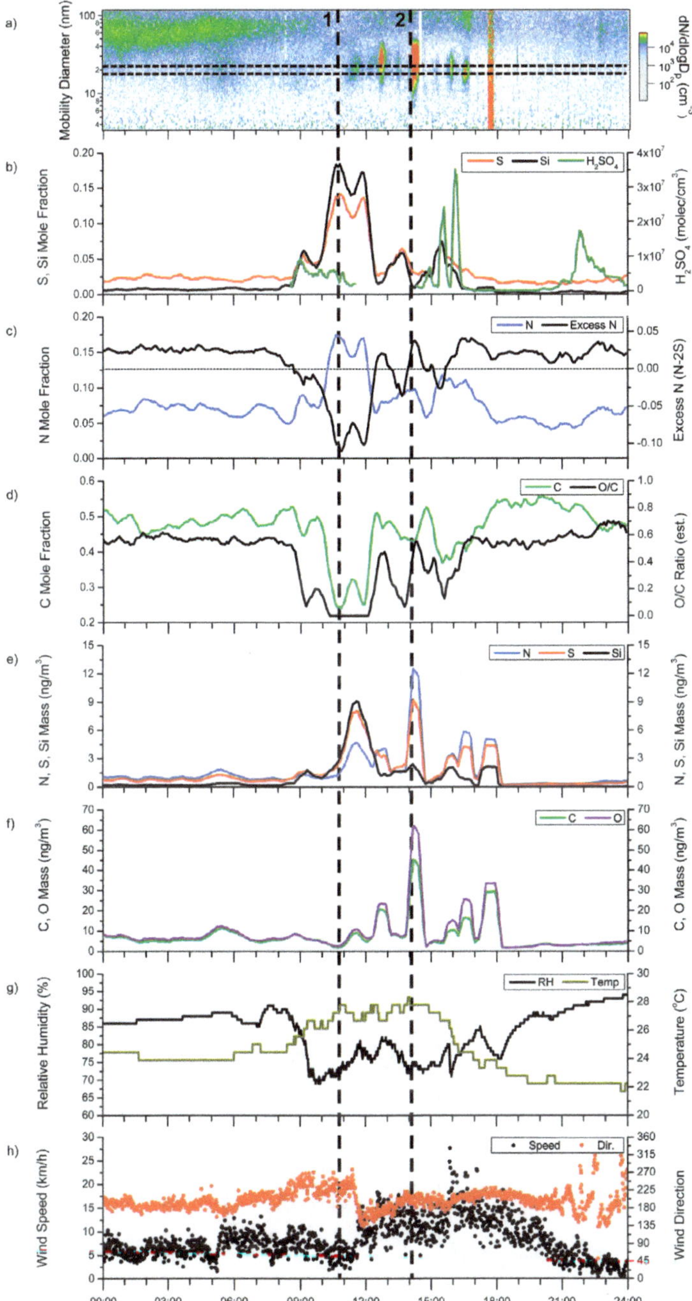

Fig. 5 a) Aerosol size distributions, b) S and Si mole fraction as well as gas phase sulphuric acid concentration, c) N and Excess N mole fractions, d) C mole fraction and estimated O/C molar ratio, e) N, S and Si elemental mass, f) C and O elemental mass, g) relative humidity and temperature and h) wind speed and wind direction for 11 August 2012. The horizontal dotted lines in a) indicate the NAMS-measured size range. The horizontal dotted line in c) indicates Excess N = 0. Vertical dotted lines indicate periods of contrasting nanoparticle composition. Nanoparticle composition data were subjected to 6-point smoothing; the sulphuric acid concentration was subjected to 10-point smoothing.

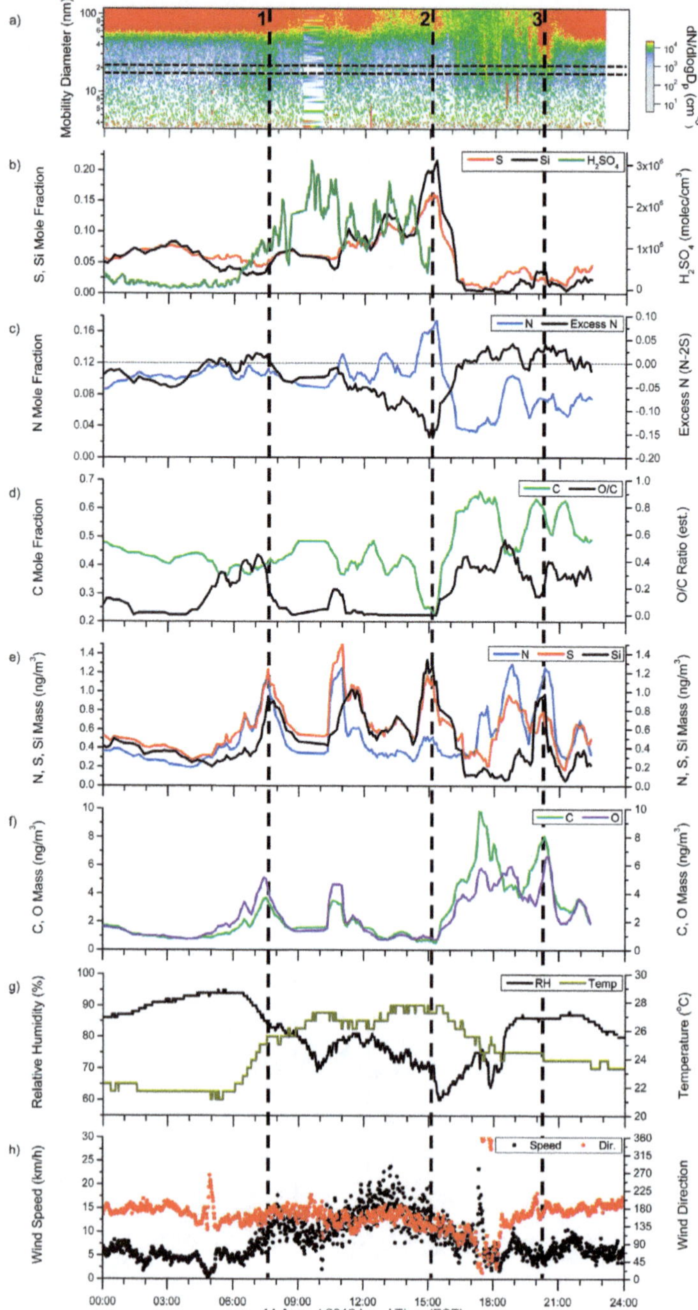

Fig. 6 a) Aerosol size distributions, b) S and Si mole fraction as well as gas phase sulphuric acid concentration, c) N and Excess N mole fractions, d) C mole fraction and estimated O/C molar ratio, e) N, S and Si elemental mass, f) C and O elemental mass, g) relative humidity and temperature and h) wind speed and wind direction for 14 August 2012. The horizontal dotted lines in a) indicate the NAMS-measured size range. The horizontal dotted line in c) indicates Excess N = 0. Vertical dotted lines indicate periods of contrasting nanoparticle composition. Nanoparticle composition data were subjected to 6-point smoothing; the sulphuric acid concentration was subjected to 10-point smoothing.

mass changes substantially. On the other hand, the first dotted line on Fig. 5 shows particle composition changing before and during a change in aerosol mass.

Despite the dynamic nature of the aerosol on these days, some general trends are clear. Inorganic components are preferentially enhanced relative to carbonaceous matter during the daytime, which is evidenced by the substantial increases in S, Si and N mole fractions. The Si mole fraction is particularly noteworthy because it is so high relative to NPF days. Si also has been reported previously in urban and suburban nanoparticle studies.[46,49] On these days, Si and S are highly correlated ($r = 0.98$ and 0.89 for 11 August and 14 August, respectively), and the daytime enhancement of each suggests a photochemical source. By comparison, Si is a minor component of the aerosol on NPF days and it is not correlated with S (except at the very beginning of the event, after which the locally produced Si was presumably overwhelmed by other, regionally produced species). Changes in S mole fraction do not respond quickly to changes in gas phase sulphuric acid because the concentration of sulphuric acid is generally an order of magnitude lower than during event days, making the condensation rate slow.

Also interesting about the non-NPF days are the trends in N mole fraction and, more importantly, Excess N. Although the N mole fraction increases during the daytime of both non-NPF days, Excess N is <0 during the daytime, whereas it is generally >0 during the nighttime. This diurnal dependence is consistent with expected ammonium nitrate partitioning to the particle phase at night as well as organonitrate formation.[73]

Fig. 4 presents apportioned mass fractions for the daytimes of 11 and 14 August in comparison to the NPF days of 12 and 13 August. Inorganics represent a similar fraction of the daytime aerosol mass on all four days. However, the identities of the species making up the inorganic fraction are different. On non-NPF days, Si (taken as SiO_2) is a substantial component of the nanoparticle mass, whereas on NPF days SiO_2 is replaced by ammonium nitrate (a proxy in the apportionment algorithm for Excess N). The aerosol is neutralized on NPF days, whereas it is acidic on non-NPF days. Notably, the average sulphate mass fraction is similar across all four days, despite substantial differences in gas phase sulphuric acid concentrations. The carbonaceous mass fraction is significantly more oxidized on NPF days relative to non-NPF days, most likely because local combustion sources dominate the aerosol on non-NPF days.

On NPF days, the critical supersaturation (S_c) of 50 nm diameter particles measured with the CCNc during the daytime have a hygroscopicity similar to that of pure ammonium sulphate ($S_c = 0.4\%$), whereas those measured in the daytime on non-NPF days were less hygroscopic ($S_c = 0.5$–1.0%). The greater hygroscopicity on NPF days is consistent with more highly oxidized carbonaceous matter as measured by NAMS. The lower hygroscopicity on non-NPF days is consistent with less oxidized carbonaceous matter, although the chemical form and effect of Si on particle hygroscopicity is unknown.

4   Conclusions

NAMS was used to obtain highly time-resolved nanoparticle chemical composition during days with and without NPF. Inorganic components are shown to be preferentially enhanced on both NPF and non-NPF days. Nanoparticle chemical composition is dynamic on both types of days, and these changes in chemical composition need not be correlated with changes in aerosol mass or number concentration. The composition measurements on NPF days substantially improve our understanding of the chemical processes underlying NPF. Because of its high time resolution and quantitative capabilities, NAMS is capable of distinguishing species likely to be associated with particle formation and early growth from those likely to be associated with later stages of growth.

At the onset of NPF, the particle phase S mole fraction increases simultaneously with the gas-phase sulphuric acid concentration, and both increase well before the mode diameter of the aerosol passes into the size range of the chemical composition measurement. Later in the event, the S mole fraction no longer correlates with the change (this time a decrease) in sulphuric acid concentration. These time dependent changes are likely to be general characteristics of NPF as they have been observed in a remote boreal forest as well. The measured S mole fraction during NPF is quantitatively explained by condensation of gas-phase sulphuric acid, indicating that existing models to describe S uptake into growing nanoparticles are appropriate.

Nitrogen is shown to be a key chemical component for NPF. Although some nitrogen is associated with neutralization of sulphuric acid at the onset of NPF, a substantial portion of the particle phase N mole fraction is associated with a (or several) separate chemical species. This so-called Excess N increases when the mode diameter of the aerosol passes through the size range of chemical composition measurement, highlighting its role in particle growth but not particle formation. Carbonaceous matter contributes ~50% of the particle mass growth.

In this location, nanoparticles on non-NPF days are more acidic and have carbonaceous matter that is less oxidized than on NPF days. Inorganic components are enhanced during the daytime on both types of days, but the identities of these components are different. Non-NPF days are influenced mainly by local emissions, whereas NPF days are influenced by regional processes.

Future work should focus on elucidating the complex roles of nitrogen and carbon in NPF. Excess nitrogen on NPF days does not appear to follow the expected partitioning trend of ammonium nitrate. Additionally, carbonaceous matter was incompletely characterized, as O/C molar ratios could only be indirectly estimated. Uncertainties associated with nitrogen and carbon highlight the need for direct molecular composition measurements. Models do exist that reasonably describe the incorporation of S into growing nanoparticles (*e.g.* eqn (1) and 2). To date, no models currently do the same for N and C uptake. This manuscript strengthens our confidence in models previously assumed to be correct and exemplifies the need to accurately model uptake of the other species.

Acknowledgements

This work was supported by the National Science Foundation under grant no. AGS1205304. B.R.B. acknowledges a STAR Graduate Fellowship (FP-91731501) awarded by the U.S. Environmental Protection Agency (EPA). The views expressed in this publication are solely those of the authors, and the U.S. EPA does not endorse any products or commercial services mentioned in this publication. Meteorological and ultrafine particle monitor measurements were performed by the Delaware Department of Natural Resources and Environmental Control. The authors thank Julia Eichhorn for assistance in the acquisition and analysis of NAMS data. The National

Center for Atmospheric Research is sponsored by the National Science Foundation. J.N.S. acknowledges funding from the Finnish Academy Grant No. 251007 and U.S. Department of Energy grant no. DE-SC0006861. The authors thank George W. Luther III for providing access to the Lewes site and facilitating the measurement campaign.

References

1 M. Kulmala, H. Vehkamaki, T. Petaja, M. Dal Maso, A. Lauri, V. M. Kerminen, W. Birmili and P. H. McMurry, *J. Aerosol Sci.*, 2004, **35**, 143–176.
2 R. Zhang, A. Khalizov, L. Wang, M. Hu and W. Xu, *Chem. Rev.*, 2012, **112**, 1957–2011.
3 M. Kulmala, J. Kontkanen, H. Junninen, K. Lehtipalo, H. E. Manninen, T. Nieminen, T. Petäjä, M. Sipilä, S. Schobesberger, P. Rantala, A. Franchin, T. Jokinen, E. Järvinen, M. Äijälä, J. Kangasluoma, J. Hakala, P. P. Aalto, P. Paasonen, J. Mikkilä, J. Vanhanen, J. Aalto, H. Hakola, U. Makkonen, T. Ruuskanen, R. L. Mauldin, J. Duplissy, H. Vehkamäki, J. Bäck, A. Kortelainen, I. Riipinen, T. Kurtén, M. V. Johnston, J. N. Smith, M. Ehn, T. F. Mentel, K. E. J. Lehtinen, A. Laaksonen, V.-M. Kerminen and D. R. Worsnop, *Science*, 2013, **339**, 943–946.
4 V. M. Kerminen, M. Paramonov, T. Anttila, I. Riipinen, C. Fountoukis, H. Korhonen, E. Asmi, L. Laakso, H. Lihavainen, E. Swietlicki, B. Svenningsson, A. Asmi, S. N. Pandis, M. Kulmala and T. Petaja, *Atmos. Chem. Phys.*, 2012, **12**, 12037–12059.
5 C. Kuang, P. H. McMurry and A. V. McCormick, *Geophys. Res. Lett.*, 2009, **36**, L09822, DOI: 10.1029/2009GL037584.
6 J. Merikanto, D. V. Spracklen, G. W. Mann, S. J. Pickering and K. S. Carslaw, *Atmos. Chem. Phys.*, 2009, **9**, 8601–8616.
7 R. J. Charlson, S. E. Schwartz, J. M. Hales, R. D. Cess, J. A. Coakley, J. E. Hansen and D. J. Hofmann, *Science*, 1992, **255**, 423–430.
8 S. S. Lee and G. Feingold, *Geophys. Res. Lett.*, 2010, **37**, L23806, DOI: 10.1029/2010gl045596.
9 U. Lohmann and J. Feichter, *Atmos. Chem. Phys.*, 2005, **5**, 715–737.
10 D. Rosenfeld, U. Lohmann, G. B. Raga, C. D. O'Dowd, M. Kulmala, S. Fuzzi, A. Reissell and M. O. Andreae, *Science*, 2008, **321**, 1309–1313.
11 M. Sipila, T. Berndt, T. Petaja, D. Brus, J. Vanhanen, F. Stratmann, J. Patokoski, R. L. Mauldin, A. P. Hyvarinen, H. Lihavainen and M. Kulmala, *Science*, 2010, **327**, 1243–1246.
12 R. J. Weber, J. J. Marti, P. H. McMurry, F. L. Eisele, D. J. Tanner and A. Jefferson, *J. Geophys. Res.-Atmos.*, 1997, **102**, 4375–4385.
13 L. H. Young, D. R. Benson, F. R. Kameel, J. R. Pierce, H. Junninen, M. Kulmala and S. H. Lee, *Atmos. Chem. Phys.*, 2008, **8**, 4997–5016.
14 C. Kuang, P. H. McMurry, A. V. McCormick and F. L. Eisele, *J. Geophys. Res.-Atmos.*, 2008, **113**, D10209, DOI: 10.1029/2007JD009253.
15 T. Nieminen, H. E. Manninen, S. L. Sihto, T. Yli-Juuti, R. L. Mauldin, T. Petaja, I. Riipinen, V. M. Kerminen and M. Kulmala, *Environ. Sci. Technol.*, 2009, **43**, 4715–4721.
16 M. Chen, M. Titcombe, J. Jiang, C. Jen, C. Kuang, M. L. Fischer, F. L. Eisele, J. I. Siepmann, D. R. Hanson, J. Zhao and P. H. McMurry, *Proc. Natl. Acad. Sci. U. S. A.*, 2012, **109**, 18713–18718.
17 S. M. Ball, D. R. Hanson, F. L. Eisele and P. H. McMurry, *J. Geophys. Res.-Atmos.*, 1999, **104**, 23709–23718.
18 D. R. Benson, M. E. Erupe and S. H. Lee, *Geophys. Res. Lett.*, 2009, **36**, L15818, DOI: 10.1029/2009GL038728.
19 J. Kirkby, J. Curtius, J. Almeida, E. Dunne, J. Duplissy, S. Ehrhart, A. Franchin, S. Gagne, L. Ickes, A. Kurten, A. Kupc, A. Metzger, F. Riccobono, L. Rondo, S. Schobesberger, G. Tsagkogeorgas, D. Wimmer, A. Amorim, F. Bianchi, M. Breitenlechner, A. David, J. Dommen, A. Downard, M. Ehn, R. C. Flagan, S. Haider, A. Hansel, D. Hauser, W. Jud, H. Junninen, F. Kreissl, A. Kvashin, A. Laaksonen, K. Lehtipalo, J. Lima, E. R. Lovejoy, V. Makhmutov, S. Mathot, J. Mikkila, P. Minginette, S. Mogo, T. Nieminen, A. Onnela, P. Pereira, T. Petaja, R. Schnitzhofer, J. H. Seinfeld, M. Sipila, Y. Stozhkov, F. Stratmann, A. Tome, J. Vanhanen, Y. Viisanen, A. Vrtala, P. E. Wagner, H. Walther, E. Weingartner, H. Wex, P. M. Winkler, K. S. Carslaw, D. R. Worsnop, U. Baltensperger and M. Kulmala, *Nature*, 2011, **476**, 429–433.
20 T. Berndt, F. Stratmann, M. Sipila, J. Vanhanen, T. Petaja, J. Mikkila, A. Gruner, G. Spindler, R. L. Mauldin, J. Curtius, M. Kulmala and J. Heintzenberg, *Atmos. Chem. Phys.*, 2010, **10**, 7101–7116.

21 H. Yu, R. McGraw and S. H. Lee, *Geophys. Res. Lett.*, 2012, **39**, L02807, DOI: 10.1029/2011gl050099.

22 J. H. Zollner, W. A. Glasoe, B. Panta, K. K. Carlson, P. H. McMurry and D. R. Hanson, *Atmos. Chem. Phys.*, 2012, **12**, 4399–4411.

23 A. Metzger, B. Verheggen, J. Dommen, J. Duplissy, A. S. H. Prevot, E. Weingartner, I. Riipinen, M. Kulmala, D. V. Spracklen, K. S. Carslaw and U. Baltensperger, *Proc. Natl. Acad. Sci. U. S. A.*, 2010, **107**, 6646–6651.

24 R. Y. Zhang, I. Suh, J. Zhao, D. Zhang, E. C. Fortner, X. X. Tie, L. T. Molina and M. J. Molina, *Science*, 2004, **304**, 1487–1490.

25 R. Y. Zhang, L. Wang, A. F. Khalizov, J. Zhao, J. Zheng, R. L. McGraw and L. T. Molina, *Proc. Natl. Acad. Sci. U. S. A.*, 2009, **106**, 17650–17654.

26 G.-L. Hou, W. Lin, S. H. M. Deng, J. Zhang, W.-J. Zheng, F. Paesani and X.-B. Wang, *J. Phys. Chem. Lett.*, 2013, **4**, 779–785.

27 B. R. Bzdek, C. A. Zordan, M. R. Pennington, G. W. Luther and M. V. Johnston, *Environ. Sci. Technol.*, 2012, **46**, 4365–4373.

28 C. Kuang, M. Chen, J. Zhao, J. Smith, P. H. McMurry and J. Wang, *Atmos. Chem. Phys.*, 2012, **12**, 3573–3589.

29 C. Kuang, I. Riipinen, S. L. Sihto, M. Kulmala, A. V. McCormick and P. H. McMurry, *Atmos. Chem. Phys.*, 2010, **10**, 8469–8480.

30 J. N. Smith, M. J. Dunn, T. M. VanReken, K. Iida, M. R. Stolzenburg, P. H. McMurry and L. G. Huey, *Geophys. Res. Lett.*, 2008, **35**, L04808, DOI: 10.1029/2007GL032523.

31 K. C. Barsanti, P. H. McMurry and J. N. Smith, *Atmos. Chem. Phys.*, 2009, **9**, 2949–2957.

32 M. Dall'Osto, D. Ceburnis, C. Monahan, D. R. Worsnop, J. Bialek, M. Kulmala, T. Kurten, M. Ehn, J. Wenger, J. Sodeau, R. Healy and C. O'Dowd, *J. Geophys. Res.-Atmos.*, 2012, **117**, DOI: 10.1029/2012jd017522.

33 J. N. Smith, K. C. Barsanti, H. R. Friedli, M. Ehn, M. Kulmala, D. R. Collins, J. H. Scheckman, B. J. Williams and P. H. McMurry, *Proc. Natl. Acad. Sci. U. S. A.*, 2010, **107**, 6634–6639.

34 L. Wang, A. F. Khalizov, J. Zheng, W. Xu, Y. Ma, V. Lal and R. Zhang, *Nat. Geosci.*, 2010, **3**, 238–242.

35 L. Wang, V. Lal, A. F. Khalizov and R. Zhang, *Environ. Sci. Technol.*, 2010, **44**, 2461–2465.

36 N. M. Donahue, E. R. Trump, J. R. Pierce and I. Riipinen, *Geophys. Res. Lett.*, 2011, **38**, DOI: 10.1029/2011gl048115.

37 I. Riipinen, T. Yli-Juuti, J. R. Pierce, T. Petaja, D. R. Worsnop, M. Kulmala and N. M. Donahue, *Nat. Geosci.*, 2012, **5**, 453–458.

38 I. Riipinen, S. L. Sihto, M. Kulmala, F. Arnold, M. Dal Maso, W. Birmili, K. Saarnio, K. Teinila, V. M. Kerminen, A. Laaksonen and K. E. J. Lehtinen, *Atmos. Chem. Phys.*, 2007, **7**, 1899–1914.

39 B. R. Bzdek and M. V. Johnston, *Anal. Chem.*, 2010, **82**, 7871–7878.

40 B. R. Bzdek, M. R. Pennington and M. V. Johnston, *J. Aerosol Sci.*, 2012, **52**, 109–120.

41 A. S. Wexler and M. V. Johnston, *J. Air Waste Manage. Assoc.*, 2008, **58**, 303–319.

42 J. P. Klems, M. R. Pennington, C. A. Zordan, L. McFadden and M. V. Johnston, *Environ. Sci. Technol.*, 2011, **45**, 5637–5643.

43 J. P. Klems, C. A. Zordan, M. R. Pennington and M. V. Johnston, *Anal. Chem.*, 2012, **84**, 2253–2259.

44 B. R. Bzdek, C. A. Zordan, G. W. Luther and M. V. Johnston, *Aerosol Sci. Technol.*, 2011, **45**, 1041–1048.

45 M. R. Pennington, B. R. Bzdek, J. W. DePalma, J. N. Smith, A.-M. Kortelainen, L. Hildebrandt Ruiz, T. Petaja, M. Kulmala, D. R. Worsnop and M. V. Johnston, *Atmos. Chem. Phys. Discuss.*, 2013, **13**, 14115–14140.

46 M. R. Pennington, J. P. Klems, B. R. Bzdek and M. V. Johnston, *J. Geophys. Res.-Atmos.*, 2012, **117**, D00V10, DOI: 10.1029/2011JD017061.

47 S. Y. Wang and M. V. Johnston, *Int. J. Mass Spectrom.*, 2006, **258**, 50–57.

48 S. Y. Wang, C. A. Zordan and M. V. Johnston, *Anal. Chem.*, 2006, **78**, 1750–1754.

49 C. A. Zordan, S. Wang and M. V. Johnston, *Environ. Sci. Technol.*, 2008, **42**, 6631–6636.

50 M. R. Pennington and M. V. Johnston, *Int. J. Mass Spectrom.*, 2012, **311**, 64–71.

51 G. W. Luther and H. A. Stecher, *J. Geophys. Res.-Atmos.*, 1997, **102**, 16215–16217.

52 H. A. Stecher, G. W. Luther, D. L. MacTaggart, S. O. Farwell, D. R. Crosley, W. D. Dorko, P. D. Goldan, N. Beltz, U. Krischke, W. T. Luke, D. C. Thornton, R. W. Talbot, B. L. Lefer, E. M. Scheuer, R. L. Benner, J. G. Wu, E. S. Saltzman, M. S. Gallagher and R. J. Ferek, *J. Geophys. Res.-Atmos.*, 1997, **102**, 16219–16236.

53 P. H. McMurry, A. Ghimire, H. K. Ahn, H. Sakurai, K. Moore, M. Stolzenburg and J. N. Smith, *Environ. Sci. Technol.*, 2009, **43**, 4653–4658.

54 C. A. Zordan, M. R. Pennington and M. V. Johnston, *Anal. Chem.*, 2010, **82**, 8034–8038.

55 M. V. Johnston, S. Wang and M. S. Reinard, *Appl. Spectrosc.*, 2006, **60**, 264A–272A.

56 J. P. Klems and M. V. Johnston, *Anal. Bioanal. Chem.*, 2013, DOI: 10.1007/s00216-013-6800-x.

57 A. Jefferson, F. L. Eisele, P. J. Ziemann, R. J. Weber, J. J. Marti and P. H. McMurry, *J. Geophys. Res.-Atmos.*, 1997, **102**, 19021–19028.

58 J. Zhao, F. L. Eisele, M. Titcombe, C. Kuang and P. H. McMurry, *J. Geophys. Res.-Atmos.*, 2010, **115**, D08205, DOI: 10.1029/2009JD012606.

59 J. Zhao, J. N. Smith, F. L. Eisele, M. Chen, C. Kuang and P. H. McMurry, *Atmos. Chem. Phys.*, 2011, **11**, 10823–10836.

60 D. R. Hanson, P. H. McMurry, J. Jiang, D. Tanner and L. G. Huey, *Environ. Sci. Technol.*, 2011, **45**, 8881–8888.

61 E. J. T. Levin, A. J. Prenni, M. D. Petters, S. M. Kreidenweis, R. C. Sullivan, S. A. Atwood, J. Ortega, P. J. DeMott and J. N. Smith, *J. Geophys. Res.-Atmos.*, 2012, **117**, DOI: 10.1029/2011jd016854.

62 R. R. Draxler and G. D. Rolph, *HYSPLIT (HYbrid Single-Particle Lagrangian Integrated Trajectory) Model access via NOAA ARL READY*, NOAA Air Resources Laboratory, Silver Spring, MD, 2013, website http://ready.arl.noaa.gov/HYSPLIT.php.

63 F. Drewnick, J. J. Schwab, J. T. Jayne, M. Canagaratna, D. R. Worsnop and K. L. Demerjian, *Aerosol Sci. Technol.*, 2004, **38**, 92–103.

64 J. L. Jimenez, J. T. Jayne, Q. Shi, C. E. Kolb, D. R. Worsnop, I. Yourshaw, J. H. Seinfeld, R. C. Flagan, X. F. Zhang, K. A. Smith, J. W. Morris and P. Davidovits, *J. Geophys. Res.-Atmos.*, 2003, **108**, 8425, DOI: 10.1029/2001JD001213.

65 S. H. Lee, D. M. Murphy, D. S. Thomson and A. M. Middlebrook, *J. Geophys. Res.-Atmos.*, 2003, **108**, 8417, DOI: 10.1029/2001JD001455.

66 M. P. Tolocka, D. A. Lake, M. V. Johnston and A. S. Wexler, *Atmos. Environ.*, 2004, **38**, 3215–3223.

67 K. S. Woo, D. R. Chen, D. Y. H. Pui and P. H. McMurry, *Aerosol Sci. Technol.*, 2001, **34**, 75–87.

68 S. L. Clegg, P. Brimblecombe and A. S. Wexler, *J. Phys. Chem. A*, 1998, **102**, 2137–2154.

69 A. S. Wexler and S. L. Clegg, *J. Geophys. Res.-Atmos.*, 2002, **107**, 4207, DOI: 10.1029/2001JD000451.

70 M. Ehn, E. Kleist, H. Junninen, T. Petaja, G. Lonn, S. Schobesberger, M. Dal Maso, A. Trimborn, M. Kulmala, D. R. Worsnop, A. Wahner, J. Wildt and T. F. Mentel, *Atmos. Chem. Phys.*, 2012, **12**, 5113–5127.

71 N. M. Donahue, K. M. Henry, T. F. Mentel, A. Kiendler-Scharr, C. Spindler, B. Bohn, T. Brauers, H. P. Dorn, H. Fuchs, R. Tillmann, A. Wahner, H. Saathoff, K. H. Naumann, O. Mohler, T. Leisner, L. Muller, M. C. Reinnig, T. Hoffmann, K. Salo, M. Hallquist, M. Frosch, M. Bilde, T. Tritscher, P. Barmet, A. P. Praplan, P. F. DeCarlo, J. Dommen, A. S. H. Prevot and U. Baltensperger, *Proc. Natl. Acad. Sci. U. S. A.*, 2012, **109**, 13503–13508.

72 W. A. Hall, M. R. Pennington and M. V. Johnston, *Environ. Sci. Technol.*, 2013, **47**, 2230–2237.

73 J. L. Fry, D. C. Draper, K. J. Zarzana, P. Campuzano-Jost, D. A. Day, J. L. Jimenez, S. S. Brown, R. C. Cohen, L. Kaser, A. Hansel, L. Cappellin, T. Karl, A. H. Roux, A. Turnipseed, C. Cantrell, B. L. Lefer and N. Grossberg, *Atmos. Chem. Phys. Discuss.*, 2013, **13**, 1979–2034.

Faraday Discussions

RSC Publishing

PAPER

Regional and global impacts of Criegee intermediates on atmospheric sulphuric acid concentrations and first steps of aerosol formation

Carl J. Percival,*[a] Oliver Welz,[b] Arkke J. Eskola,[b] John D. Savee,[b] David L. Osborn,[b] David O. Topping,[a] Douglas Lowe,[a] Steven R. Utembe,[a] Asan Bacak,[a] Gordon M[c]Figgans,[a] Michael C. Cooke,[c] Ping Xiao,[c] Alexander T. Archibald[†,c] Michael E. Jenkin,[cd] Richard G. Derwent,[e] Ilona Riipinen,[f] Daniel W. K. Mok,[g] Edmond P. F. Lee,[gh] John M. Dyke,[h] Craig A. Taatjes[b] and Dudley E. Shallcross*[c]

Received 26th March 2013, Accepted 1st May 2013

DOI: 10.1039/c3fd00048f

Carbonyl oxides ("Criegee intermediates"), formed in the ozonolysis of alkenes, are key species in tropospheric oxidation of organic molecules and their decomposition provides a non-photolytic source of OH in the atmosphere (Johnson and Marston, *Chem. Soc. Rev.*, 2008, **37**, 699, Harrison *et al.*, *Sci. Total Environ.*, 2006, **360**, 5, Gäb *et al.*, *Nature*, 1985, **316**, 535, ref. 1–3). Recently it was shown that small Criegee intermediates, C.I.'s, react far more rapidly with SO_2 than typically represented in tropospheric models, (Welz, *Science*, 2012, **335**, 204, ref. 4) which suggested that carbonyl oxides could have a substantial influence on the atmospheric oxidation of SO_2. Oxidation of SO_2 is the main atmospheric source of sulphuric acid (H_2SO_4), which is a critical contributor to aerosol formation, although questions remain about the fundamental nucleation mechanism (Sipilä *et al.*, *Science*, 2010, **327**, 1243, Metzger *et al.*, *Proc. Natl. Acad. Sci. U. S. A.*, 2010 **107**, 6646, Kirkby *et al.*, *Nature*, 2011, **476**, 429,

---

[a]*The Centre for Atmospheric Science, The School of Earth, Atmospheric and Environmental Science, The University of Manchester, Simon Building, Brunswick Street, Manchester, M13 9PL, UK. E-mail: c.percival@ manchester.ac.uk*

[b]*Combustion Research Facility, Sandia National Laboratories, 7011 East Ave., MS 9055, Livermore, California 94551, USA*

[c]*Biogeochemistry Research Centre, School of Chemistry, The University of Bristol, Cantock's Close BS8 1TS, UK. E-mail: d.e.shallcross@bristol.ac.uk*

[d]*Atmospheric Chemistry Services, Okehampton, Devon, EX20 1FB, UK*

[e]*rdscientific, Newbury, Berkshire, UK*

[f]*Department of Applied Environmental Science, University of Stockholm, Stockholm, Sweden*

[g]*Department of Applied Biology and Chemical Technology, Hong Kong Polytechnic University, Hung Hom, Hong Kong*

[h]*School of Chemistry, University of Southampton, Highfield, Southampton SO17 1BJ, UK*

ref. 5–7). Non-absorbing atmospheric aerosols, by scattering incoming solar radiation and acting as cloud condensation nuclei, have a cooling effect on climate (Intergovernmental Panel on Climate Change (IPCC), *Climate Change 2007: The Physical Science Basis*, Cambridge University Press, 2007, ref. 8). Here we explore the effect of the Criegees on atmospheric chemistry, and demonstrate that ozonolysis of alkenes *via* the reaction of Criegee intermediates potentially has a large impact on atmospheric sulphuric acid concentrations and consequently the first steps in aerosol production. Reactions of Criegee intermediates with SO_2 will compete with and in places dominate over the reaction of OH with SO_2 (the only other known gas-phase source of H_2SO_4) in many areas of the Earth's surface. In the case that the products of Criegee intermediate reactions predominantly result in H_2SO_4 formation, modelled particle nucleation rates can be substantially increased by the improved experimentally obtained estimates of the rate coefficients of Criegee intermediate reactions. Using both regional and global scale modelling, we show that this enhancement is likely to be highly variable spatially with local hot-spots in *e.g.* urban outflows. This conclusion is however contingent on a number of remaining uncertainties in Criegee intermediate chemistry.

Introduction

Ozonolysis of alkenes proceeds *via* a 1,3-cycloaddition of ozone across the olefinic bond to produce a primary ozonide, the decomposition of which forms a carbonyl moiety and a Criegee intermediate(C.I.).[1]

$$O_3 + \text{alkene} \rightarrow \text{C.I.} + \text{products} \tag{1}$$

The fate of the C.I. determines the end products of the ozonolysis reaction and can have substantial effects on the atmosphere.[4] For example, these Criegee intermediates are central to the tropospheric budgets of secondary organic aerosols (SOA), ozone, NO_x, NO_y and HO_x.[9] Recent field measurements[2] have shown that alkene ozonolysis dominates night-time OH production and is responsible for up to 46% of urban daytime summer OH levels. Furthermore, formation of SOA can be initiated by C.I. reactions that form acids and hydroperoxides.[10]

The pioneering work of Cox and Penkett[11] showed that the gas-phase oxidation of SO_2 is enhanced during ozone–alkene reactions, leading to an increase in SO_3 formation and production of H_2SO_4 aerosol in their laboratory systems. Based on rate coefficients derived from indirect studies, Calvert and Stockwell[12] speculated that the interaction of C.I. with SO_2 could compete with OH oxidation of SO_2, but only in polluted urban environments. Quantitative effects of Criegee chemistry have remained uncertain[13] because of a lack of direct measurements of C.I. reactions. However, recent measurements of the reaction of the simplest gas-phase Criegee intermediate, CH_2OO, with SO_2 showed it was up to 10 000 times faster than models assumed.[4] Such a large rate coefficient suggests that reaction with Criegee intermediates would compete with OH and in some places dominate urban oxidation of SO_2 and could have a significant impact on SO_2 oxidation elsewhere near the surface. In previous work that inference was derived from a steady-state model.[4] However, demonstrating the tropospheric impact of Criegee intermediates on sulphate aerosol formation requires detailed modelling that considers the kinetic limiting factors on H_2SO_4 formation and on nucleation, and moreover, the models depend on other Criegee species also reacting rapidly with SO_2.

Until recent determinations of CH_2OO kinetics,[4] no direct measurements existed of any reaction of Criegee intermediates. Taatjes *et al.*,[14] have also reported kinetics of the reaction of the next Criegee Intermediate, CH_3CHOO, and have shown that this also reacts faster with SO_2 than previously assumed. Taatjes *et al.*,[14] also directly probed reaction products for the first time and were able to show that SO_3 is a primary reaction product for the reaction of C.I. with SO_2. In this work they were able to resolve the two distinct conformers, *syn-* and *anti-*, and that the *anti*-conformer is far more reactive towards water and SO_2 than is the *syn*-CH_3CHOO conformer. Both conformers react rapidly with SO_2 and NO_2, strengthening the proposal that Criegee reactions play a significant role in tropospheric sulfate and nitrate chemistry. Boy *et al.*,[15] report a rate coefficient of 10^{-13} cm^3 molecule^{-1} s^{-1} for the reaction of C.I. with SO_2. However, a rate constant of 10^{-11} cm^3 molecule^{-1} s^{-1} is supported by both experimental and theoretical work. The theoretical results of Kurtén *et al.*,[16] show that the SO_2 reaction with acetone oxide, which has no alpha-hydrogen, has a barrierless entrance channel and a low-lying transition state for SO_3 + carbonyl formation, similar to that for CH_2OO + SO_2. Jiang *et al.*,[17] theoretically characterized CH_2OO + SO_2 and the Criegee + SO_2 reactions for the C.I.'s from limonene ozonolysis. In a similar vein to Kurtén *et al.*,[16] they do not report barriers on the entrance channel for either reaction. In light of these findings, it would suggest that the rate coefficient with C.I. with SO_2 would be larger than the 10^{-13} cm^3 molecule^{-1} s^{-1} implied in the manuscript by Boy *et al.*[15] Indeed, in more recent work Vreecken *et al.*,[18] have also reported a theoretical rate coefficient of the order of 10^{-11} which has been further supported by recent experimental work.[19] In this work we will investigate the potential impact of increased reactivity on H_2SO_4 production in the troposphere using three types of atmospheric model.

First, the STOCHEM-CRI is a global model that alerts us to major regions where C.I. Chemistry may be important and provides some global estimates of the contribution to H_2SO_4 production. Second, we are aware that the global model will underestimate the level of alkene in urban areas and in urban outflow and in addition will not have the spatial resolution to determine the impact on plume scales. Hence we have used the WRF-CHEM model set up for Europe to investigate the impact of higher spatial resolution, where greater structure and impact does emerge. Finally, analysis of the global model suggests that C.I. Chemistry may be important in the Amazon region. Although WRF-CHEM modelling would be beneficial here we do not have high resolution emission or meteorological fields to drive this model. Hence we have used a Photochemical Trajectory model in this instance, where we can obtain trajectories and integrate an MCM Chemistry scheme (more detailed than CRI) to look at this region in more detail. The speed of integration allows us to look at a wide range of sensitivity studies in terms of emissions and water vapour levels.

Methodology

1 The global chemistry transport model CRI-STOCHEM

The following papers describe model results and their comparison with a variety of field data: Archibald *et al.*,[20,21] Cooke *et al.*,[22] Henne *et al.*,[23] Bacak *et al.*,[24] Utembe *et al.*[25,26] The description of the model is broken down into two parts; the meteorological module and the chemical module, which will be presented in subsections 1.2 and 1.3, respectively. The model was originally described in Collins *et al.*,[27] with updates described by Derwent *et al.*,[28] with a new extended chemical scheme and an organic aerosol module described in Watson *et al.*,[29] Jenkin *et al.*,[30] and Utembe *et al.*[31]

1.1 General aspects. STOCHEM is a global 3-dimensional chemistry transport model in which 50 000 constant mass air parcels are advected using a Lagrangian approach allowing the chemistry and transport processes to be uncoupled. STOCHEM is an "offline" model with the transport and radiation codes driven by archived meteorological data, generated by the UKMO Unified Model (UM) at a climate resolution of 1.25° longitude $\times$ 0.83° latitude $\times$ 12 vertical levels 20.[32] Archived meteorological data contains pressure, temperatures, humidities, winds, tropopause heights, cloud, precipitation, boundary layer and surface parameters.[33] The chemical mechanism used in STOCHEM is described in section 1.3.

1.2 Meteorological parameterisations

1.2.1 Vertical coordinate. The vertical coordinate used in STOCHEM is a hybrid pressure coordinate, η, originally described in Simmons and Burridge[34] and used in the Meteorological Office Unified model. The hybrid pressure coordinate is calculated using eqn (I)[27]

$$\eta = (P/P_{\mathrm{s}}) + A(1/P_0 - 1/P_{\mathrm{s}}) \tag{I}$$

where P is the pressure of a given air parcel, P_{s} is the surface pressure, P_0 is the reference pressure (1000 hPa) and A is a coefficient with the dimensions of pressure.

1.2.2 Advection scheme. The 50 000 constant mass Lagrangian air parcels are advected using wind fields generated by the UK Meteorological office unified model (UM).[35] The archived meteorological data used is for the year 1998 at a resolution of 1.25^0 longitude $\times$ 0.83^0 latitude $\times$ 12 vertical levels between the surface and about 100 mb. The archived meteorology contains 6 hourly wind fields containing horizontal winds (v_x and v_y), and vertical winds (v_η). The advection time step is set to three hours so the meteorological wind fields (6 hourly) are linearly interpolated with respect to time. The 3-dimensional spatial interpolation uses a cubic polynomial in the vertical direction and a bi-linear method is used for the horizontal components.[36] A fourth order Runge–Kutta method is used to solve the ordinary differential equation to calculate the new position of each Lagrangian cell at the end of each time step. The temperature, pressure and humidity of the cells are calculated by 3-dimensional interpolation in an analogous manner to the wind fields after the advection step.

1.3 Chemical module

1.3.1 Chemical mechanism. The chemical mechanism used is the common representative intermediates mechanism version 2 and reduction 5 (CRI v2-R5). Its generation and validation are described in detail in the papers Watson *et al.*,[29] Jenkin *et al.*,[30] and Utembe *et al.*[31] The CRI scheme was devised from a reduction of the Master Chemical Mechanism (MCM) and is a near-explicit chemical mechanism originally described by Jenkin *et al.*[37] The construction is broken down into three parts. The initiation reactions of a VOC are photolysis, reaction with OH, reaction with NO_3 and reaction with ozone. Not all VOCs undergo all reactions, for instance no reaction between ozone and an alkane has been observed. The products of the initiation reactions are predominantly peroxy

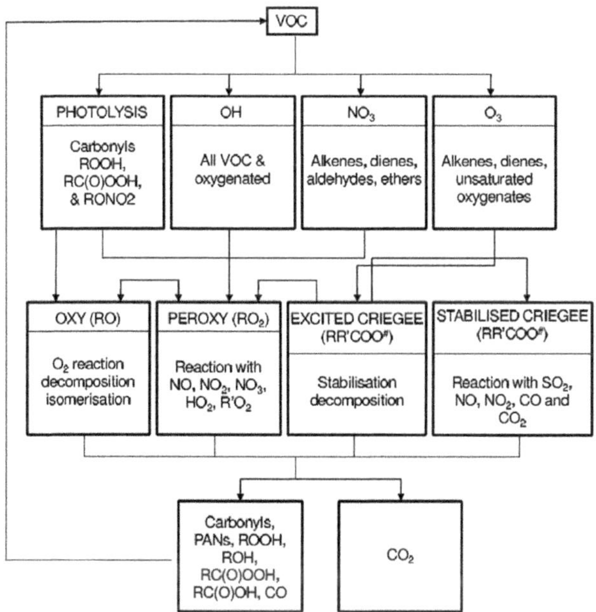

Fig. 1 Chemistry used in the construction of the MCM. The oxidation is broken down into initiation, propagation (reaction of intermediates) and termination (formation of products) reactions.

radicals which are oxidised to give a range of species as shown in Fig. 1. The products formed by this scheme are oxidised themselves until they are lost from the system or form CO_2. The MCM v3.1 is the third version of the master chemical mechanism which has undergone improvements because of developments in the understanding of mechanistic processes. The MCM v3.1 describes the degradation of 130 emitted VOCs. Due to its near-explicit nature the MCM v3.1 consists of 5900 species competing in 13 500 photochemical reactions. Such a large mechanism is impractical for use in a global or regional scale model due to computational limitations. Therefore, a series of reduced mechanisms were produced using the MCM v3.1 as a reference benchmark.[29–31] A key assumption of the reduction methodology is that the potential for ozone formation of a given VOC is related to the total number of reactive bonds (*i.e.* C–H and C–C bonds).[38] This allowed a series of generic intermediates to be defined which are "common representatives" for a large set of species. The resultant reduced mechanism used here (CRI v2-R5) represents the degradation of methane and 22 emitted VOCs using 196 species competing in 555 photochemical reactions and gives excellent agreement with the MCM v3.1 over a full range of NO_x levels.[29,30] The secondary organic aerosol (SOA) scheme that was then developed from this mechanism is described in detail in Utembe *et al.*[31]

1.3.2 Calculation of C.I.. During the integration the concentration of SCI (stabilized Criegee Intermediates) are calculated using the steady state assumption where the total production rate (the sum of the terms k_x[alkene$_x$][O_3]-$\Gamma_{SCI(x)}$, where k_x is the rate coefficient for the reaction of alkene$_x$ with ozone and $\Gamma_{SCI(x)}$ is the fraction of stabilized C.I. formed) is divided by the total loss rate ($k_5 + k_4[NO_2] + k_3[SO_2] + k_2[H_2O]$), where k_5 is the unimolecular loss rate for an SCI (assigned a value of 200 s^{-1}), and k_4 (6×10^{-12} cm^3 molecule^{-1} s^{-1}), k_3 (4×10^{-11} cm^3 molecule^{-1} s^{-1}) and k_2 (1×10^{-16} cm^3 molecule^{-1} s^{-1}) are rate coefficients for reaction of the C.I. with NO_2, SO_2 and H_2O.

1.3.3 Photolysis. The photolysis rate of a given species is calculated using the following integral.

$$J_A = \int_0^\infty F(\lambda)\sigma_A(\lambda)\phi_A(\lambda)d\lambda \tag{II}$$

where J_A is the photolysis rate of compound A, $F(\lambda)$ is the spherically integrated actinic flux at a given wavelength, $\sigma_A(\lambda)$ is the absorption cross section of compound A at a given wavelength λ and $\Phi_A(\lambda)$ is the quantum yield for dissociation of A at a given wavelength. The integral is restricted to tropospherically relevant wavelengths and for computational efficiency the integral is replaced by a sum over all wavelengths. The cross section and quantum yields are taken from the recommendations of either the Jet Propulsion Laboratory (JPL) kinetic evaluation reports[39,40] or the International Union of Pure and Applied Chemistry (IUPAC) data evaluations.[41] Spherically integrated actinic fluxes are calculated over 106 wavelength intervals within the range 200–660 nm for a given air parcel using a variant of the two stream model developed by Hough.[42] This representation uses measured values for the solar flux with attenuation of the photons being calculated for a range of factors. These attenuation factors include absorption in and above the stratosphere by O_2 and O_3, absorption by O_2 and O_3 in the troposphere, absorption and scattering (reflection and refraction) by aerosol and clouds, scatter by other gas phase molecules in the troposphere and the surface albedo. The method is a two stream representation as upwards and downwards fluxes are calculated to evaluate the incident light at a given point. The photolysis rate for each reaction is calculated explicitly for each air parcel at a time resolution of one hour. The one hourly photolysis rates are linearly interpolated with respect to time to achieve five minute resolution values which are used in the chemical integration.

1.3.3 Emissions. There are three types of emissions featured in STOCHEM; surface emissions, stratospheric sources and 3-dimensional emissions. Emissions from the surface from biomass burning, vegetation, oceans, soil and 'other' surface emissions are distributed using monthly two dimensional source maps at a resolution of 5° longitude by 5° latitude.[43] The anthropogenic surface distributions for CH_4, CO, NO_x and VOCs have been developed for the year 2000 by the International Institute for Applied Systems Analysis (IIASA) and are described in detail by Cofala *et al.*[44] Emissions totals for CO, NO_x and NMVOCs are taken from the Precursor of ozone and their Effects in the Troposphere (POET) inventory[45] for the year 1998. Emission totals for CH_4 have been taken from the inverse model study of Mikaloff-Fletcher *et al.*,[46] except for the ocean emissions which are from Houweling *et al.*[47] The anthropogenic and biomass burning emissions of the aromatic species *ortho*-xylene, benzene and toluene were taken from Henze *et al.*[48] Biomass burning emissions of ethyne, formaldehyde and acetic acid are produced using scaling factors from Andreae and Merlet[49] per mole CO emitted. The POA anthropogenic global total was taken from the AEROCOM dataset using the totals from fossil fuels and biofuel burning.[50]

The POA biomass burning global total was taken from the GFED v2.1 dataset.[51] Isoprene is unique among the biogenic hydrocarbons because of its direct relationship to photosynthetic activity of plants.[52] This is not the case with other biogenically emitted species such as terpenes whose emissions continue at night.

Therefore the surface emissions of isoprene within STOCHEM are emitted at a rate proportional to the cosine of the solar zenith angle during the day, with no emissions at night. The rate is adjusted so an appropriate amount is emitted per month and thus per year adding up to the global total of 501 Tg/yr. The Lagrangian cells within the model are kept below 100 hPa whereas in reality there would be exchange between the troposphere and stratosphere. In order to represent this air exchange a simulated net downward flux of ozone and HNO_3 into the top level of the model is calculated using 3-hourly vertical wind fields and monthly ozone fields from Li and Shine.[53] These emissions are distributed on a resolution of 5° longitude by 5° latitude at a vertical coordinate of $\eta = 0.1$. Based on the work of Murphy and Fahey[54] the HNO_3 flux is calculated as one thousandth of the ozone flux by mass of N. Lightning and aircraft emissions are not surface emissions and are distributed in a 3-dimensional manner within the model. The distributions for lightning emissions are parameterized based on the work of Price and Rind[55] with the emissions being distributed evenly between the convective cloud top height and the surface. The lightning emissions are input on a resolution of 5° longitude by 5° latitude at a vertical resolution of $\eta = 0.1$. The emissions are scaled so that the global total NO_x emission from lightning is 5 Tg(N)/yr. The NO_x emissions from civil and military aircraft are taken from NASA inventories for 1992.[56] The implementation of the emissions from aircraft is the same as for lightning with an annual total of 0.85 Tg(N)/yr. All emissions are converted into units of molecules per second per grid square and are implemented as an additional term in the production flux for a given chemical species. The emission flux is implemented during the chemistry timestep.

1.3.4 Dry and wet deposition. When Lagrangian cells are located in the boundary layer the species within these air parcels can be lost by wet and dry deposition. The rate of dry deposition is dependent on whether the air parcel is over land or ocean with appropriate species dependent deposition velocities. This representation can be extended to include differences between types of land surface, such as savannah and forest. Dry deposition velocities for each species are calculated using eqn (III) and IV

$$F_A = c_A \nu_d / H \qquad \text{(III)}$$

where F_A is the dry deposition flux of species A, c_A is the concentration of species A, ν_d is the deposition velocity of species A and H is the height above the ground, and

$$\nu_d = (\nu_1 \times \nu_a)/(\nu_1 + \nu_a) \qquad \text{(IV)}$$

where ν_d and ν_1 represent the species dependent deposition velocity at 50 m and at 1 m, respectively and ν_a is the relative aerodynamic deposition velocity between 1 m and 50 m. In this implementation the assumption is made that species deposition velocities at 50 m are representative of a given species deposition velocity throughout the mixed boundary layer. Soluble species can be removed from the atmosphere by precipitation referred to commonly as wet deposition. Species dependent scavenging coefficients for convective and dynamic precipitation are taken from Penner *et al.*[56] These coefficients are combined with precipitation rates and scavenging profiles to calculate loss rates of each species from an air parcel. Precipitation rates at a given point in the atmosphere are calculated using the dynamic precipitation, convective precipitation and convective cloud top heights read in from the meteorology.

1.3.5 Aerosol module. An organic aerosol module suitable for use with the CRI v2 and all reduced variants has been developed by Utembe *et al.*[31] POA is directly emitted into the atmosphere and is removed by wet and dry deposition. The method, briefly described here for SOA formation has been used previously with the MCM[57-61] giving good agreement with both field and chamber measurements. For a more detailed description of the methodology see Johnson *et al.*[58] The formation of SOA is represented by a species dependent dynamic equilibrium between the gas and aerosol phases, eqn (V).

$$C_a/C_g = K_p C_{om} \tag{V}$$

where C_a is the concentration of a given species in the aerosol phase, C_g is the concentration of a given species in the gas phase, K_p is the partitioning coefficient ($\mu g\ m^{-3}$) and C_{om} is the total mass concentration of condensed organic material. K_p is evaluated using the absorptive partitioning theory of Pankow,[62] eqn (VI).

$$K_P = \frac{7.501 \times 10^{-9} RT}{MW_{om} \zeta P_L^\circ} \tag{VI}$$

where R is the gas constant ($8.314\ \mathrm{J\ K^{-1}\ mol^{-1}}$), T is the temperature (in K), MW_{om} is the mean molecular weight of the absorbing particulate organic material ($\mathrm{g\ mol^{-1}}$), ζ is the activity coefficient of the species in the condensed phase and P° is its liquid vapour pressure (in Torr). ζ is set to unity for all species as it is expected that aerosol particles comprise mixtures of similar types of molecules. The vapour pressures for many of the species in the MCM have not been measured and therefore an estimation method was employed. An extended semi-empirical form of the Clausius–Clapeyron equation was applied along with esti-mated values of species boiling temperature (Tb) and entropy change of vapourisation at that temperature ($\Delta S_{vap}(T_b)$). T_b was estimated for a given species using the fragmentation method of Stein and Brown.[63] $\Delta S_{vap}(T_b)$ values were estimated using the methodology of Baum *et al.*[64] Utembe *et al.*[32] used a photo-chemical trajectory model to produce a reduced representation of secondary organic aerosol (SOA) formation which is compatible with all six versions of the CRI v2, including reduction 5 used in these studies. The TORCH-2003 campaign in the south-east UK in late July and early August of 2003 resulted in the production of a comprehensive measurement dataset of VOCs, NO_x, Ozone and Organic Aerosol.[59,60] A photochemical trajectory model was initially run with the MCM v3.1 coupled with the parameterized gas to aerosol absorptive partitioning method (eqn (5)) for about 2000 closed shell species for evaluation against the measurements made during TORCH-2003.[31] The MCM v3.1 organic aerosol representation was thus reduced to 365 low volatility species which represented >95% by mass of the simulated SOA. The MCM v3.1 aerosol module was used as a reference benchmark for the production of an optimized SOA module suitable for use with the CRI v2 and its derivatives.[31] The resultant aerosol module includes an emission of primary organic aerosol (POA) and 14 SOA species representing the gas phase absorption of low volatility products. There are three species repre-senting aromatic hydrocarbons, ten terpene-derived species and one isoprene-derived species. The large set of terpene derived species is needed as terpene derivatives were found to be important SOA precursors over the full range of pollution conditions. In contrast, SOA derived from aromatic hydrocarbons and isoprene are found to only be significant at high and low pollution levels, respectively, allowing a more limited number of surrogates to be used.[31]

2   WRF-CHEM model

The model used is version 3.2 of the Weather Research and Forecasting model with Chemistry (WRF-Chem), detailed in Grell *et al.*,[65] and Fast *et al.*,[66] with adaptions described in Lowe *et al.*[67] WRF-Chem is a regional 3-D model, which has been run over a Western European domain at a resolution of 15 km, using archived European Centre for Medium-Range Weather Forecasts (ECMWF) met data obtained from the ECMWF Data Server, MOZART global model outputs for our boundary conditions,[68] anthropogenic emissions are based on a combination of the TNO[69] and NAEI emission databases, more detail of domain setup in Lowe *et al.*[67] Biogenic emissions are calculated 'online' using a modified Gunther scheme in WRF-Chem. In the default Gunther scheme, emissions of mono-terpenes are allocated to isoprene and other VOCs. Since CRI-STOCHEM explicitly simulates monoterpenes, the Gunther scheme has been modified to allocate monoterpenes emissions to monoterpenes.

Gas-phase chemistry is modeled using the common representative interme-diates mechanism (CRIv2-R5),[25,27,29] as described for CRI-STOCHEM. The aerosol phase is modeled using the 8-bin sectional MOSAIC scheme.[70] Gas-phase reac-tions, gas-aerosol interactions, and dynamical processes are operator-split over a 4-minute timestep, though each has an adaptive step solver within that outer-step (see Grell *et al.*,[65] for more details).

Predictions of nucleation rates are made using SO_2, OH, O_3, NO_2, C_2H_4 concentrations, aerosol surface area, and met fields output from WRF-Chem. Base steady-state H_2SO_4 concentrations are calculated using SO_2 + OH produc-tion, and loss rates to existing aerosol. We do not consider transport, or dry or wet deposition, of gaseous H_2SO_4 in these calculations, and so the steady-state H_2SO_4 concentrations are higher than those calculated on-line within the model (which does include these processes), but on average only by a factor of 2, at most by a factor of 6, and the spatial distribution patterns are the same. Steady state Criegee intermediate concentrations were calculated using $10 \times C_2H_4$ (see later) and O_3 field data, unimolecular decay, with loss *via* reactions with SO_2, NO_2 and H_2O. This concentration was then used to calculate the steady state concentration of H_2SO_4 produced using rate coefficients for SO_2 + C.I. as outlined above (again with loss to existing aerosol). Nucleation rates for H_2SO_4 were calculated using the base and base + C.I. steady-state H_2SO_4 concentrations, which are compared in order to get the increase in nucleation.

3   PTM (Photochemical Trajectory Model) model

A version of the PTM[71,72] model was set up running the MCMv3.1 chemical mechanism in a simple box model with no dry deposition, exchange with the free troposphere and no wet deposition. The mixing ratios of 31 long-lived species with lifetimes longer than 1 h were set initially to those for the Amazon basin using the output from a global chemistry transport model STOCHEM running the CRI chemical mechanism. The STOCHEM output was taken from the lowest model level for summer conditions. All other chemically reactive species were allowed to follow their own diurnal cycles or to accumulate during the model integrations. The photolysis rates in the box model were set for July 1st conditions in the Amazon basin. The box model was then integrated with the mixing ratios of the 31 long-lived species reset automatically to their initial values after every FACSIMILE time step. The outputs from the constrained box model were the fluxes required to maintain the mixing ratios of the 31 species at their initial values. These fluxes were averaged over the first 5 days of each box model integration.

In the base case model for the Amazon basin, the time-integrated flux required to maintain the SO_2 mixing ratio at 1 ppb was found to be 0.013 ppb per day, largely through the well-known route involving the oxidation of SO_2 by hydroxyl OH radicals and we have included the novel chemistry of the Criegee Intermediate chemistry was added as follows:

$$\text{C.I.} + H_2O \rightarrow \text{products} \tag{2}$$

$$\text{C.I.} + SO_2 \rightarrow SO_3 + \text{products} \tag{3}$$

$$\text{C.I.} + NO_2 \rightarrow \text{products} \tag{4}$$

$$\text{C.I.} \rightarrow \text{products} \tag{5}$$

the time integrated flux through SO_2 increased somewhat over the base case. The magnitude of the increase in time-integrated flux depended on the rate coefficients chosen for $k_1 - k_4$.

In all experiments, we have chosen 4.0×10^{-11} cm^3 molecule^{-1} s^{-1} for k_3 based on Welz et al.,[4] and Taatjes et al.,[14] measurements of CH_3CHOO, by using their values for CH_2OO and CH_3CHOO. The value of k_4 was set to 1.0×10^{-15} cm^3 molecule^{-1} s^{-1}, based on its assignments in MCMv3.1. The values used for k_2 ranged from 1.0×10^{-17} cm^3 molecule^{-1} s^{-1} to 3.0×10^{-14} cm^3 molecule^{-1} s^{-1}.

4   Nucleation calculations

Nucleation mechanisms are not well understood,[73] and whilst it is known that H_2SO_4 is an important determinant, a contribution from other species is required to explain observed production rates.[74-76] The identity of the possible species remains unclear, though candidates include NH_3[77,78] and volatile organic compounds (VOCs) such as organic acids,[79-82] 1,3,5-trimethylbenzene and amines.[83] Kirkby et al.,[7] reported that, while atmospherically relevant mixing ratios of ammonia increase the nucleation rate of sulphuric acid particles by more than 100–1000 fold, the concentrations of ammonia and sulphuric acid are insufficient to account for observed boundary-layer nucleation. Although it has been difficult to replicate H_2SO_4 dependent ambient particle production rates under laboratory conditions, Sipila et al.,[5] showed that nucleation rate scales with the gaseous concentration of H_2SO_4, by a factor of 1.0 to 2.1, and thus changes in the sulphuric acid concentrations are expected to be directly reflected in the nucleation rate. For this study we used the nucleation rate equation stated in that paper:

$$J = k[H_2SO_4]^n \tag{VII}$$

where J is the nucleation rate (cm^{-3} s^{-1}), a kinetic constant and $[H_2SO_4]$ the concentration of H_2SO_4 (molecules cm^{-3}) and n varies from $n = 1$ to 2.1 with an average of 1.5, a value that is in good agreement with the majority of atmospherically observed nucleation rates presented in Kirkby *et al.*[7] Using this equation we can perform difference comparisons on estimated nucleation rates using the concentration of H_2SO_4 calculated in STOCHEM with and without the effect of the Criegee intermediate for $n = 2$ (in agreement with ambient observations reported by Kuang *et al.*[84]) or 1.5 (the average of atmospherically observed nucleation rates reported by Kirkby *et al.*[7]). Nucleation mechanisms involving H_2SO_4 can be broadly classified as binary neutral, binary ion induced, ternary neutral, ternary ion induced, but each may be expected to be strongly dependent on H_2SO_4 concentration.[5] No single mechanism will dominate the production rate of stable nucleation mode particles at all locations. We therefore stress that the absolute magnitude of the atmospheric implications will depend on whether the nucleation mechanism exhibits a comparable H_2SO_4 dependence, but that the impacts of C.I. oxidation of SO_2 will be at least qualitatively represented by our calculations. Furthermore, to quantify the impact of the changing nucleation rates on atmospheric aerosol loadings and CCN concentrations, their survival probabilities are needed. These probabilities depend on the rate at which they grow to larger sizes as compared with the rate at which they are lost by coagulation. Our results are thus indicative on the potential increase in the nucleation rate but further work is needed to quantify the effect of this increase on aerosol and CCN loadings in the models used here.

Results and discussion

There are many potential loss reaction partners for C.I., but, besides unimolecular decomposition of C.I., three stand out in terms of the combination of rate coefficient[2] and concentration, *i.e.*, H_2O, SO_2 and NO_2

$$C.I. + H_2O \rightarrow products \tag{2}$$

$$C.I. + SO_2 \rightarrow SO_3 + products \tag{3}$$

$$C.I. + NO_2 \rightarrow products \tag{4}$$

$$C.I. \rightarrow products \tag{5}$$

Therefore, the steady state concentration of C.I. will be

$$[C.I.]_{\text{steady state}} = \frac{k_1[O_3][\text{alkene}]}{k_2[H_2O] + k_3[SO_2] + k_4[NO_2] + k_5} \tag{VIII}$$

First, the rate coefficients k_3 and k_4 have been determined recently,[4] *i.e.* $k_3 = (3.9 \pm 0.7) \times 10^{-11}$ cm^3 molecule^{-1} s^{-1} and $k_4 = (6 \pm 2) \times 10^{-12}$ cm^3 molecule^{-1} s^{-1} for CH_2OO and $k_3 = (2.7 \pm 0.5) \times 10^{-11}$ cm^3 molecule^{-1} s^{-1} for CH_3CHOO.[14] Second, work by Welz *et al.*,[4] and Leather *et al.*,[85] have shown that k_2 is $< 4 \times 10^{-15}$ cm^3 molecule^{-1} s^{-1} for CH_2OO and it could be as small 1.5×10^{-17} cm^3 molecule^{-1} s^{-1}. For this estimation we assume a value of $k_2 = 1 \times 10^{-16}$ cm^3 molecule^{-1} s^{-1} for the reaction of all types of C.I.'s with H_2O, which represents a significant increase over previous estimates of around 1×10^{-17} cm^3 molecule^{-1}

s^{-1}.[86] Third, the rate coefficient k_1 will vary with the alkene in question, for ethene $k_1 = 1.6 \times 10^{-18}$ cm^3 molecule^{-1} s^{-1},[86] for propene $k_1 = 1.0 \times 10^{-17}$ cm^3 molecule^{-1} s^{-1},[86] and on alkyl substitution k_1 will increase significantly, *e.g.* for 2, 3, dimethyl 2-butene[87] $k_1 = 1.2 \times 10^{-15}$ cm^3 molecule^{-1} s^{-1} and a group rate coefficient should reflect the mix of alkenes (see later). However, for this estimate we assume a value of $k_1 = 5 \times 10^{-17}$ cm^3 molecule^{-1} s^{-1}. Having arrived at values for the various rate coefficients, it is worth inspecting the denominator; given that [H$_2$O] will vary from ~2–6 $\times$ 10^{17} molecule cm^{-3} under the temperature and pressure chosen, the loss rate due to reaction (2) is 20–60 s^{-1} and will dominate the bimolecular reactive loss. In order to produce a loss rate of 20 s^{-1} SO$_2$ levels of at least 20 ppb and NO$_2$ levels of around 130 ppb are required, not impossible, but only found in heavily polluted urban environments. Unimolecular loss (k_5) is assumed to range between 100–200 s^{-1}, therefore, we assume a total loss rate of about 200 s^{-1}, or a k_2 value of 10^{-16} cm^3 molecule^{-1} s^{-1}, the overall loss of C.I. will in fact be governed mainly by this assumed k_5. The level of ozone and alkene is highly variable across the surface of the Earth. However, in polluted environments and away from the surface where deposition and titration by NO can occur, a typical O$_3$ level is 30 ppb (noting that it could be much higher than this under photochemical smog conditions). In polluted urban areas the level of alkenes (that are measured) can be very high; recent measurements in China and India exceeding 100 ppb and elsewhere typically being 30 ppb. Noting that this is a measurement of a small selection of alkenes (typically C$_2$–C$_6$) and that a much larger pool of carbon must exist,[88] 30 ppbv can be regarded as an underestimate of

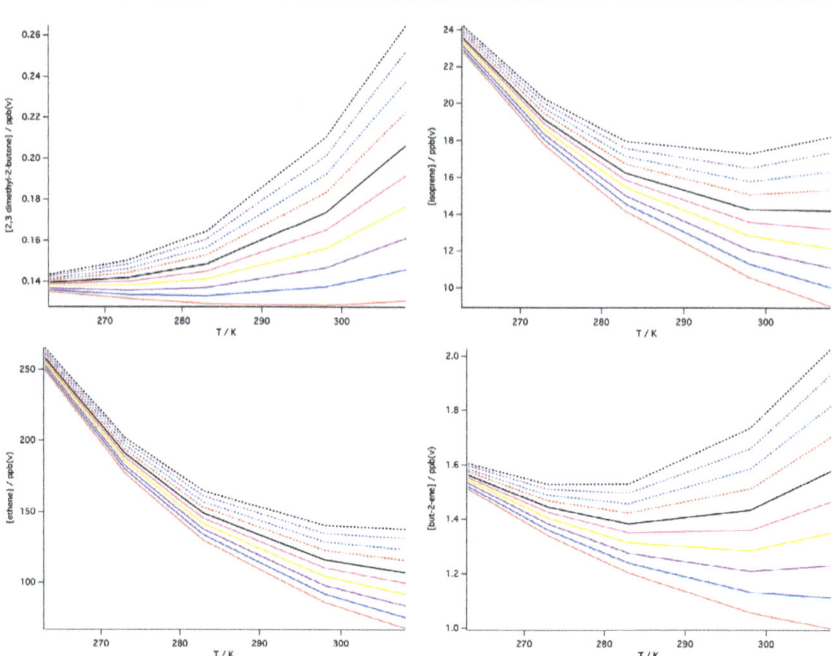

Fig. 2 A plot of alkene concentration against temperature, given an ozone concentration of 30 ppb, that yields sufficient C.I. so that the loss of SO$_2$ *via* reaction with OH and C.I. are equivalent. The various lines show the variation as a function of relative humidity (RH). —10% RH, —20% RH, —30% RH, —40% RH, —50% RH, —60% RH, --- 70% RH--80% RH, ---90% RH, ---100% RH.

Table 1 A summary of measured alkene concentrations for various urban campaigns. Here k_x = rate coefficient for the ozonolysis of alkene $\times 10^{17}$ cm^3 molecule^{-1} s^{-1}, τ is is the atmospheric lifetime with respect to ozonolysis, assuming an ozone concentration of xppb and [ethene]$_{\text{equivalent}} = [x] \times k_x/k_{\text{ethene}}{}^a$

Alkene	k_x	Urban Mexico City[32]			Indus. Mexico City[32]			Chinese Min[33,34,35]			Chinese Max[33,34,35]			Kathmandu[33,34,35]		
		[Alkene]	τ	[ethene] equivalent	[Alkene]	τ	[ethene] equivalent	[Alkene]	τ	[ethene] equivalent	[Alkene]	τ	[ethene] equivalent	[Alkene]	τ	[ethene] equivalent
ethene	0.17	23.33	0.37	23.33	32.08	0.27	32.08	2.1	4.13	2.10	34.8	0.25	34.80	48.4	0.179	48.40
propene	1.04	5.93	0.24	36.28	10.96	0.13	67.05	0.2	7.00	1.22	8.2	0.17	50.16	12.8	0.109	78.31
iso-butene	0.92	3.04	0.52	26.45	5.28	0.30	28.57	0.1	15.86	0.54	4	0.40	21.65			
2-methyl-1-butene	1.33	1.37	0.80	10.72	3.82	0.29	29.89									
2-methyl-2-butene	40.80	0.89	0.04	213.60	1.69	0.02	405.60									
trans-2-butene	13.10	1.05	0.11	80.91	2.48	0.04	191.11	0.01	11.11	0.77	3.4	0.03	262.00			
trans-2-pentene	20.90	0.74	0.09	90.98	1.36	0.05	167.20	0.02	3.48	2.46	5.3	0.01	651.59			
cis-2-butene	19.00	0.83	0.09	92.76	1.31	0.06	146.41	0.02	3.83	2.34	2.7	0.03	301.76			
1,3,butadiene	0.66	0.55	4.04	2.14	0.63	3.53	2.45	0.02	111.12	0.08	2.5	0.90	9.71			
isoprene	1.40	0.33	3.15	2.72	0.32	3.25	2.63	0.04	25.99	0.33	1.7	0.61	14.00	0.3		0.249 2.47
cis-2-pentene	20.90	0.37	0.19	45.49	0.7	0.10	86.05	0.02	3.48	2.46	9.4	0.01	1155.65			
1-pentene	0.88	0.48	3.45	2.49	1.18	1.41	6.11									
1-hexene	1.01	0.16	9.01	0.95	0.48	3.00	2.85									
1-butene	0.92							0.07	123.78	0.37	2.4		12.99	2.5		4.119 13.52
1,2,propadiene	0.02															
2-methyl 2-butene	40.80															
2-methyl-1-pentene	1.31															
2-methyl-2-pentene	45.00															
trans-2-hexene	15.70															
cis-2-hexene	14.40															
1-methyl	83.20															
cycpentene																
alpha pinene	8.20															

Table 1 (*Contd.*)

Alkene	k_x	Urban Mexico City[32]			Indus. Mexico City[32]			Chinese Min[33,34,35]			Chinese Max[33,34,35]			Kathmandu[33,34,35]		
		[Alkene]	τ	[ethene] equivalent	[Alkene]	τ	[ethene] equivalent	[Alkene]	τ	[ethene] equivalent	[Alkene]	τ	[ethene] equivalent	[Alkene]	τ	[ethene] equivalent
n-alkenes																
Total		39.07	0.01	618.81	62.29	0.01	1168.01	2.60	0.70	12.57	74.4	0.003	2514.30	64	0.053	142.71

Alkene	Taipei[33,34,35]			Seoul[36]			Karachi[37]			Delhi[38]	Dallas[39]			Birmingham[2]			LA[40]
	[Alkene]	τ	[ethene] equivalent	[Alkene]	τ	[ethene] equivalent	[Alkene]	τ	[ethene] equivalent	[Alkene]	[Alkene]	τ	[ethene] equivalent	[Alkene]	τ	[ethene] equivalent	[Alkene]
ethene	14.1	0.62	14.1	5.20	1.67	5.20	19	0.46	19.00		3.31	2.62	3.31	1.20	7.64		
propene	4.60	0.30	28.14	1.10	1.27	6.73	5.5	0.25	33.65		1.82	0.77	11.13	0.16	9.25	0.99	
iso-butene	2.70	0.59	14.61				1.2	1.32	6.49					0.075	15.44	0.06	
2-methyl-1-butene																	
2-methyl-2-butene																	
trans-2-butene							0.3	0.37	23.12		0.47	0.24	36.22	0.04	2.03	4.52	
trans-2-pentene														0.046	1.06	8.65	
cis-2-butene							0.2	0.38	22.35		0.24	0.32	26.82	0.051	2.3	3.98	
1,3,butadiene							0.8		3.11					0.06	39.18	0.23	
isoprene							0.8		6.59		0.65	1.60	5.35	0.067	16.41	0.56	
cis-2-pentene											0.14	0.50	17.21	0.067	2.44	3.76	
1-pentene											0.18	9.21	0.93				
1-hexene																	
1-butene	0.90	1.76	4.87				1.1	1.44	5.95		0.32	4.96	1.73	0.07	23.96	0.38	
1,2,propadiene							0.2	475.70	0.02								

Table 1 (*Contd.*)

Alkene	Taipei[33,34,35]			Seoul[36]			Karachi[37]			Delhi[38]	Dallas[39]			Birmingham[2]			LA[40]
	[Alkene]	τ	[ethene] equivalent	[Alkene]	τ	[ethene] equivalent	[Alkene]	τ	[ethene] equivalent	[Alkene]	[Alkene]	τ	[ethene] equivalent	[Alkene]	τ	[ethene] equivalent	[Alkene]
2-methyl																	
2butene																	
2-methyl-1-																	
pentene																	
2-methyl-2-pentene																	
trans-2-hexene																	
cis-2-hexene																	
1-methyl																	
cycpentene																	
alpha pinene																	
n-alkenes																	29.00
Total	22.3	0.14	61.7	6.3	0.72	11.93	29.1	0.08	120.28	1300	7.13	0.08	102.71	1.79	0.37	24.87	29.00

Alkene	Seoul[36]			HK[41,42]			Porto Alegre Brazil[43]			Borneo[98]			
	[Alkene]	τ	[ethene] equivalent	[Alkene]	τ	[ethene] equivalent	[Alkene]	τ	[ethene] equivalent	[alkene]	τ	[ethene] equivalent	k
ethene	7.6	1.14	7.60				58.2	0.149	58.2				
propene	3.3	0.42	20.19				28.3	0.306	173.13	0.047	31.5	0.29	
iso-butene							16.5	0.525	89.29				
2-methyl -1-butene							8.9	0.974	69.63				
2-methyl -2-butene							17	0.510	4080.00				
trans-2-butene	1.8	0.06	138.71				6.3	1.375	485.47				
Trans-2-pentene	0.6	0.12	73.76				12.8	0.677	1573.65				

Table 1 (*Contd.*)

Alkene	Seoul[36]			HK[41,42]			Porto Alegre Brazil[43]			Borneo[98]			
	[Alkene]	τ	[ethene] equivalent	[Alkene]	τ	[ethene] equivalent	[Alkene]	τ	[ethene] equivalent	[alkene]	τ	[ethene] equivalent	k
cis-2-butene	0.9	0.09	100.59				4.6	1.884	514.12				
1,3,butadiene							4.6	1.884	17.86				
isoprene	0.3	3.47	2.47				0.7	12.378	5.76	1.1	1.00	9.17	
cis-2-pentene	0.9	0.077	110.65				5.5	1.575	676.18				
1-pentene	0.5	3.32	2.59				4.6	1.884	23.82				
1-hexene							2.3	3.767	13.66				
1-butene	3.2	0.50	17.32				7.8	1.111	42.21				
1,2,propadiene													
2-methyl 2-butene	1	0.04	240.00										
2-methyl-1-pentene							2.2	3.938	16.95				
2-methyl-2-pentene							4	2.166	1058.83				
trans-2-hexene							3.7	2.342	341.71				
trans-2-hexene							1.6	5.415	135.53				
1-methyl cycpentene							4.3	2.015	2104.47				
alpha pinene				10	0.87	482.36	5	1.733	241.18	0.024	8.02	1.14	
n-alkenes													
mono terpene										0.24	12.83	0.71	8.00
camphene										0.066	518.42	0.02	0.05
terpinene										0.202	0.02	525.18	865.00
limonene										0.071	43.37	0.21	0.5
methacrolein										0.106	133.26	0.07	0.11

Table 1 (*Contd.*)

Alkene	Seoul[36]			HK[41,42]			Porto Alegre Brazil[43]			Borneo[98]			
	[Alkene]	τ	[ethene] equivalent	[Alkene]	τ	[ethene] equivalent	[Alkene]	τ	[ethene] equivalent	[alkene]	τ	[ethene] equivalent	k
methyl vinyl ketone										0.45	8.13	1.13	0.42
sesquiterpens										0.17	0.16	58.69	58.00
Total	20.1	0.01	713.87	10	0.87	482.36	198.9	0.04	11721.64	2.45	0.02	596.69	

Alkene	Boston[93]			
	ppb	k	lifetime	[ethene] equivalent
ethene	3.430	0.17	2.64	3.43
propene	1.030	1.00	1.49	6.06
2 methyl propene	0.660	5.00	0.47	19.41
2 methyl 13 butadiene	0.324	1.00	4.75	1.91
2 methyl 2 butene	0.246	20.00	0.31	28.94
244 trimethyl 1 pentene	0.145	5.00	2.12	4.26
1-butene	0.173	1.00	8.93	1.01
trans2pentene	0.168	20.00	0.46	19.76
1,3,butadiene	0.160	1.00	9.62	0.94
2-methyl 1-butene	0.124	5.00	2.48	3.65
a pinene	0.050	8.00	3.85	2.35
trans-2-butene	0.115	20.00	0.67	13.53
1-pentene	0.086	1.00	17.90	0.51
c9 alkenes	0.047	20.00	1.65	5.49
cyclohexene	0.070	1.00	22.00	0.41
cis-2-pentene	0.078	20.00	0.99	9.18
cis-2-butene	0.095	20.00	0.81	11.18

Table 1 (*Contd.*)

Alkene	Boston[93]			
	ppb	k	lifetime	[ethene]equivalent
trans-2-hexene	0.058	20.00	1.32	6.86
4 methyl 1 pentene	0.068	1.00	22.64	0.40
1 octene	0.041	1.00	37.33	0.24
c10 alkenes	0.032	20.00	2.41	3.76
c6 alkenes	0.052	20.00	1.49	6.08
1 hexene	0.050	1.00	30.79	0.29
b pinene	0.028	1.00	54.99	0.16
3 methyl 1 pentene	0.035	1.00	43.99	0.21
1 nonene	0.022	1.00	69.29	0.13
cyclopentene	0.040	1.00	38.49	0.24
trans-3-hexene	0.027	20.00	2.89	3.14
2 methyl 1 pentene	0.027	5.00	11.55	0.78
c5 alkenes	0.030	20.00	2.57	3.53
c8 alkenes	0.019	20.00	4.11	2.21
c9 alkenes	0.016	20.00	4.95	1.83
1 1dimethylcyclohexene	0.016	3.00	31.58	0.29
cis-2-hexene	0.020	20.00	3.85	2.35
trans-3-heptene	0.007	20.00	10.78	0.84
2 methyl 2 pentene	0.017	20.00	4.62	1.96
cis-3-hexene	0.010	20.00	7.70	1.18
trans-3-heptene	0.006	20.00	12.32	0.74
c7 alkenes	0.007	20.00	10.78	0.84
c8 alkenes	0.004	20.00	20.53	0.44
c9 alkenes	0.003	20.00	23.10	0.39
c7 alkenes	0.003	20.00	26.95	0.34

Table 1 (*Contd.*)

Alkene	Boston[93]			
	ppb	k	lifetime	[ethene]equivalent
4 methylhexene	0.001	1.00	1077.80	0.01
c8 alkenes	0.001	20.00	61.59	0.15
Total	7.64		0.05	171.41

Alkene	Seoul[36]			HK[41,42]			Porto Alegre Brazil[43]		
	[Alkene]	τ	[ethene] equivalent	[Alkene]	τ	[ethene] equivalent	[Alkene]	τ	[ethene] equivalent
ethene	7.6	1.14	7.60				58.2	0.149	58.2
propene	3.3	0.42	20.19				28.3	0.306	173.13
iso-butene							16.5	0.525	89.29
2 methyl-1-butene							8.9	0.974	69.63
2 methyl-2-butene							17	0.510	4080.00
trans-2-butene	1.8	0.06	138.71				6.3	1.375	485.47
trans-2-pentene	0.6	0.12	73.76				12.8	0.677	1573.65
cis-2-butene	0.9	0.09	100.59				4.6	1.884	514.12
1,3,butadiene							4.6	1.884	17.86
isoprene	0.3	3.47	2.47				0.7	12.378	5.76
cis-2-pentene	0.9	0.077	110.65				5.5	1.575	676.18
1-pentene	0.5	3.32	2.59				4.6	1.884	23.82
1-hexene							2.3	3.767	13.66
1-butene	3.2	0.50	17.32				7.8	1.111	42.21
1,2,propadiene									
2 methyl 2 butene	1	0.04	240.00						

Table 1 (*Contd.*)

Alkene	Seoul[36]			HK[41,42]			Porto Alegre Brazil[43]		
	[Alkene]	τ	[ethene] equivalent	[Alkene]	τ	[ethene] equivalent	[Alkene]	τ	[ethene] equivalent
2 methyl-1-pentene							2.2	3.938	16.95
2 methyl-2-pentene							4	2.166	1058.83
trans-2-hexene							3.7	2.342	341.71
trans-2-hexene							1.6	5.415	135.53
1-methyl cycpentene							4.3	2.015	2104.47
alpha pinene				10	0.87	482.36	5	1.733	241.18
n-alkenes									
Total	20.1	0.01	713.87	10	0.87	482.36	198.9	0.04	11721.64

[a] A summary of measured alkene concentrations for various campaigns. Here k_x = rate coefficient for the ozonolysis of alkene $\times 10^{17}$ cm^{3}molecule^{-1}s^{-1}, τ is the atmospheric lifetime in days with respect to ozonolysis, and [ethene]$_{equivalent}$ = $[x] \times k_x/k_{ethene}$.

the 'alkene' level in these areas. Therefore, combining all these elements together, one arrives at an estimate for $[\text{C.I.}]_{\text{steady state}}$ of 3.75×10^5 molecule cm^{-3} in a polluted environment. In order to obtain a 'low' estimate we can assume that $k_2 = 1 \times 10^{-15}$ cm^3 molecule^{-1} s^{-1}, $[H_2O]$ is 6×10^{17} molecule cm^{-3}, alkene and ozone levels are 20 ppb, which leads to a lower estimate for $[\text{C.I.}]_{\text{steady state}}$ of 4.2×10^4 molecule cm^{-3}.

Potential impact on SO$_2$ oxidation

Oxidation of SO$_2$ is initiated by OH radicals

$$OH + SO_2 \rightarrow HOSO_2 \tag{6}$$

HOSO$_2$ is thought to be taken up rapidly by aerosol. As demonstrated by Taatjes *et al.*,[14] oxidation of SO$_2$ by C.I.'s will lead to SO$_3$ that will also be removed heterogeneously rapidly. Recent theoretical studies[18] have suggested that the products of the reaction of SO$_2$ with C.I. at high pressure could be a secondary ozonide. First, if the rate coefficients for small C.I.'s reported recently at 4 Torr[4,14] are not at the high pressure limit, then the rate coefficients will be larger at 760 Torr. Second, although the stabilisation of the potential adduct may be possible at 760 Torr, rather than decomposing to form a carbonyl and SO$_3$, as seen at 4 Torr, the likely fate of the adduct is formation of H$_2$SO$_4$ *via* reaction with water, in an analogous way to SO$_3$. In order to determine whether the oxidation of SO$_2$ through reaction with C.I. competes with reaction (6) we must inspect the ratio of rates of initiation

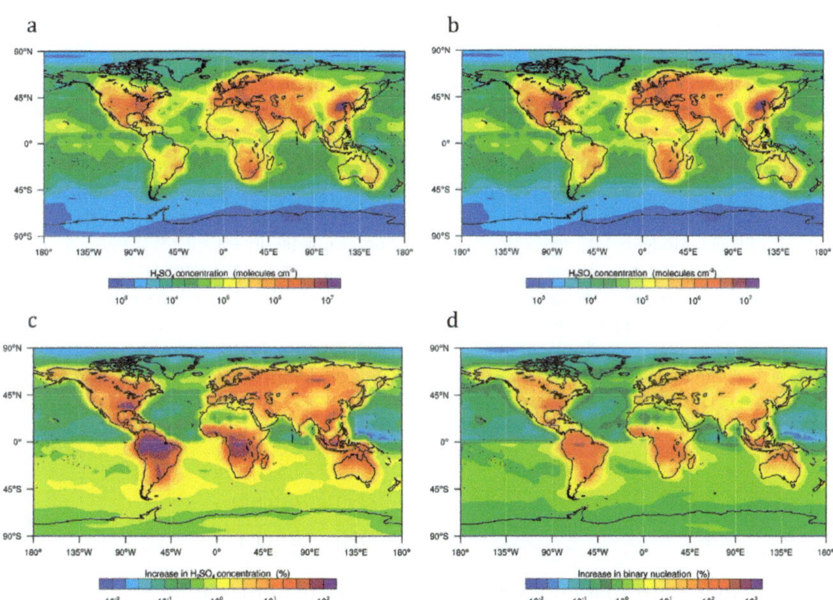

Fig. 3 Global plot of the impact of C.I. on H$_2$SO$_4$ production, panel a = [H$_2$SO$_4$/ molecule cm^{-3}] from OH oxidation of SO$_2$, panel b = [H$_2$SO$_4$/ molecule cm^{-3}] from OH and C.I. oxidation of SO$_2$, panel c increase in [H$_2$SO$_4$] from oxidation of SO$_2$ by C.I. and panel d increase in nucleation rate from oxidation of SO$_2$ by C.I., assuming an exponent of 1.5.

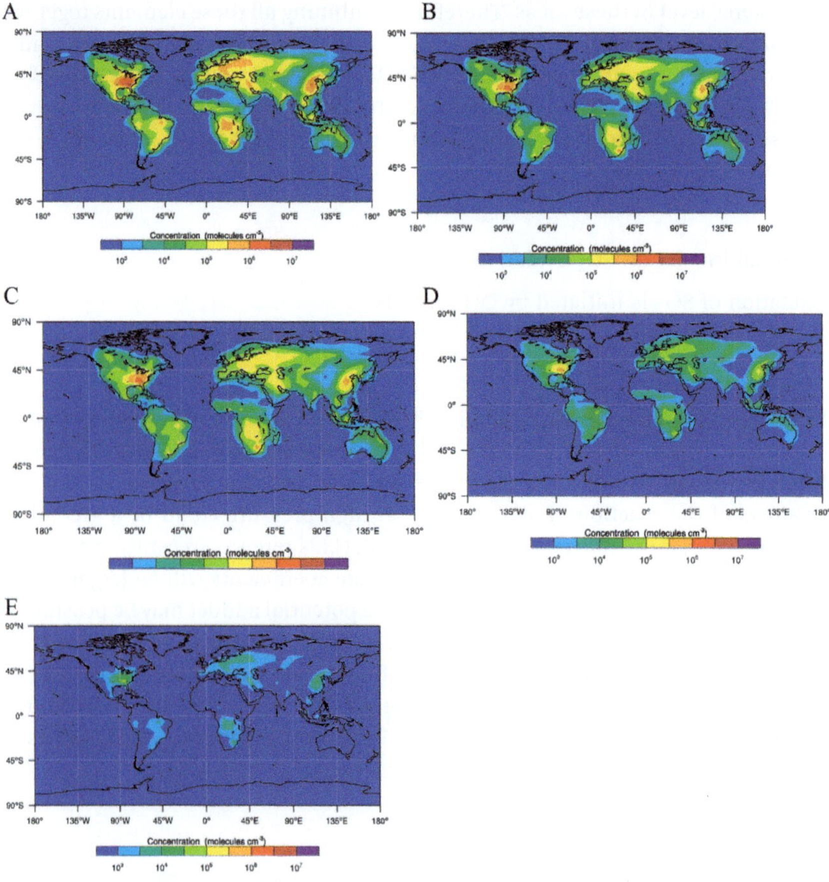

Fig. 4 Global plot of the impact of C.I. reaction rate coefficient on H_2SO_4 production. Panel a Model H_2SO_4 derived from the reaction of Criegee Intermediates (derived from the model) with SO_2, using the following parameters. $k(CI + SO_2) = 4 \times 10^{-11}$ cm^3 molecule^{-1} s^{-1}, $k(CI + H_2O) = 1 \times 10^{-16}$ cm^3 molecule^{-1} s^{-1} and thermal loss rate for C.I. = 200 s^{-1}. Panel b Model H_2SO_4 derived from the reaction of Criegee Intermediates (derived from the model) with SO_2, using the following parameters. $k(CI + SO_2) = 4 \times 10^{-11}$ cm^3 molecule^{-1} s^{-1}, $k(CI + H_2O) = 1 \times 10^{-15}$ cm^3 molecule^{-1} s^{-1} and thermal loss rate for C.I. = 200 s^{-1}. Panel c Model H_2SO_4 derived from the reaction of Criegee Intermediates (derived from the model) with SO_2, using the following parameters. $k(CI + SO_2) = 4 \times 10^{-11}$ cm^3 molecule^{-1} s^{-1}, $k(CI + H_2O) = 1 \times 10^{-17}$ cm^3 molecule^{-1} s^{-1} and thermal loss rate for C.I. = 200 s^{-1}. Panel d Model H_2SO_4 derived from the reaction of Criegee Intermediates (derived from the model) with SO_2, using the following parameters. $k(CI + SO_2) = 4 \times 10^{-12}$ cm^3 molecule^{-1} s^{-1}, $k(CI + H_2O) = 1 \times 10^{-16}$ cm^3 molecule^{-1} s^{-1} and thermal loss rate for C.I. = 200 s^{-1}. Panel e Model H_2SO_4 derived from the reaction of Criegee Intermediates (derived from the model) with SO_2, using the following parameters. $k(CI + SO_2) = 4 \times 10^{-13}$ cm^3 molecule^{-1} s^{-1}, $k(CI + H_2O) = 1 \times 10^{-16}$ cm^3 molecule^{-1} s^{-1} and thermal loss rate for C.I. = 200 s^{-1}.

$$\text{Ratio}(SO_2) = \frac{k_3[SO_2][\text{C.I.}]}{k_6[SO_2][\text{OH}]} = \frac{k_3[\text{C.I.}]}{k_6[\text{OH}]} \tag{IX}$$

Therefore, for oxidation *via* reaction (3) to be equal to that for reaction (6)

$$[\text{C.I.}] = \frac{k_6[\text{OH}]}{k_3} \tag{X}$$

   This journal is © The Royal Society of Chemistry 2013

The rate coefficient k_6 is both temperature and pressure dependent, in this calculation we assume standard atmospheric pressure and a temperature of 290 K (average surface temperature) hence k_6 is[39,40,41] $\sim 9 \times 10^{-13}$ cm^3 molecule^{-1} s^{-1} and we use new kinetic data for $k_3 = (3.3 \pm 0.7) \times 10^{-11}$ cm^3 molecule^{-1} s^{-1} (the average of the rate coefficient for the reaction of CH_2OO and CH_3CHOO with SO_2 to represent k_3 for all C.I.). Assuming a daytime [OH] level of 1×10^6 molecule cm^{-3}, with nighttime levels ranging[89] from 1×10^5 molecule cm^{-3} to 1×10^4 molecule cm^{-3} leads to concentrations of C.I. required for parity of 2.4×10^4 molecule cm^{-3} for daytime and 2.4×10^3 molecule cm^{-3} to 2.4×10^2 molecule cm^{-3} for nighttime, *i.e.* the oxidation of SO_2 by C.I. will not only compete, but in some cases dominate that *via* oxidation by OH. Fig. 2 shows a plot of the concentration of alkene required to generate sufficient levels of the C.I. so that the rate of oxidation of SO_2 *via* reaction with C.I. is equivalent to that *via* reaction 3 for a range of species.

Table 1 summarises speciated alkene measurements for various urban campaigns with the median level of the limited number of species measured being 29 ppb. Where the concentration of particular alkene is measured, it is possible to estimate the atmospheric lifetime with respect to ozonolysis. The reactivity of ozonolysis of alkenes varies by over 3 orders of magnitude[90] and thus the specific VOC will have a significant impact on their atmospheric lifetime. The average atmospheric lifetime of alkenes is 0.22 days, with a median of 0.07 days. It is clear that the ozonolysis of alkenes is important; indeed it is comparable with deposition, which is on average 2 days.[91] Estimates of the ethene equivalent level of alkene (*i.e.*, the amount of ethene that would exhibit the same reactivity as the actual spectrum of alkenes) shows that in most cases 100s ppb of ethene equivalent alkene are present and thus it is likely that C.I. chemistry will have a significant impact on H_2SO_4 production. Indeed, studies in the USA of urban air[92]

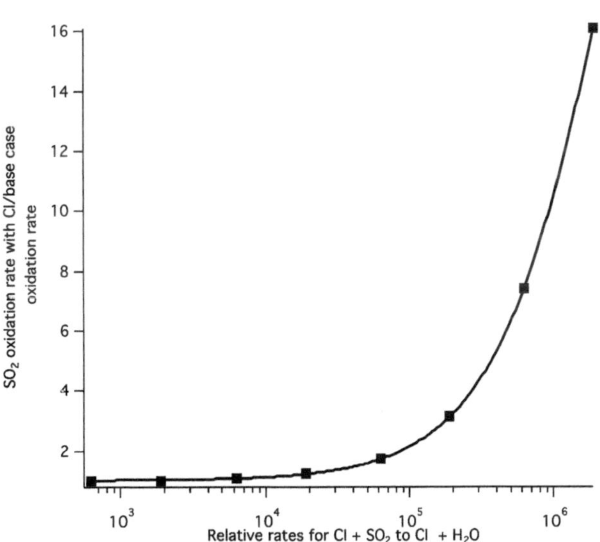

Fig. 5 A plot of the intergrated flux of SO_2 *via* the C.I. + SO_2 reaction as a function of reltive rate of C.I.+SO_2 to C.I. + H_2O.

and analysis of exhaust emissions[93] suggest that the ethene contribution to Criegee intermediate production is about 2%, and of the measured alkenes (still an underestimate of the total) highly substituted but-2-enes and pent-2-enes contribute a significant fraction. Hence, the use of $10 \times$ ethene for the WRF-CHEM model integrations becomes apparent.

In order to quantitatively assess the impact of the C.I. and reaction (4) on atmospheric H_2SO_4 formation three model approaches were adopted; a global model (CRI-STOCHEM) and a regional model (WRF-Chem) were integrated and the total C.I. level derived, from which the impact on SO_2 oxidation was estimated.

Fig. 3 shows comparisons of H_2SO_4 concentration with and without reaction (4). The more significant enhancement of SO_2 oxidation in urban areas, discussed earlier, would not be resolved at the scale of the global model ($5° \times 5°$). Our results suggest increases in predicted H_2SO_4 concentration range from 0.01 to $\sim$200%, depending on the location. It is noteworthy that over equatorial and tropical areas, reaction (4) is playing an especially significant role. It is noteworthy that significant enhancements in SO_2 oxidation are predicted in the Amazon by the global model; here low gas-phase SO_2 levels and high BVOC are observed,[94]

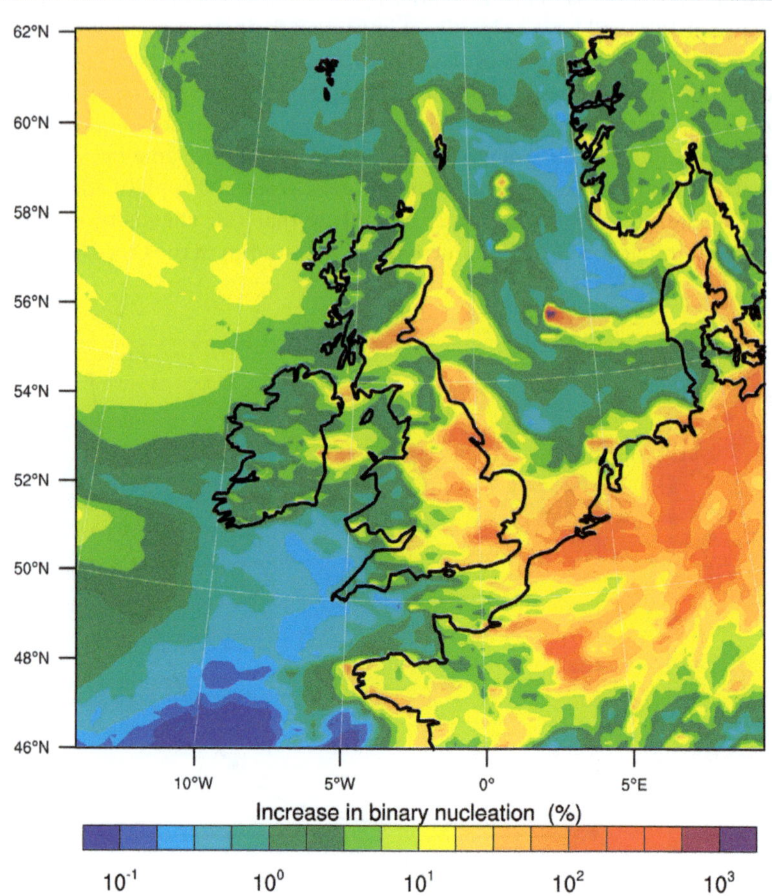

Fig. 6 WRF-CHEM plots of the increase nucleation rate from oxidation of SO_2 by C.I., assuming an exponent of 1.5.

consistent with the mechanism postulated in this paper. As Martin *et al.* recommend,[94] measurements of H_2SO_4 are required to quantify nucleation in this region.

STOCHEM-CRI has been integrated under a number of scenarios as shown in Fig. 4; using a base case where $k(SCI + SO_2) = 4 \times 10^{-11}$ cm^3 molecule^{-1} s^{-1}, $k(SCI + H_2O) = 1 \times 10^{-16}$ cm^3 molecule^{-1} s^{-1} and $k(SCI \rightarrow$ decomposition$) = 200$ s^{-1}. A series of integrations has looked at the sensitivity to $k(SCI + SO_2)$ which has been reduced from the base case value of 4×10^{-11} cm^3 molecule^{-1} s^{-1} to 4×10^{-12} cm^3 molecule^{-1} s^{-1} and then 4×10^{-13} cm^3 molecule^{-1} s^{-1}, whilst keeping all other parameters constant. Furthermore, integrations have been performed where $k(SCI + H_2O)$ has been varied up and down in magnitude from the base case, *i.e.* 1×10^{-15} cm^3 molecule^{-1} s^{-1}; 1×10^{-16} cm^3 molecule^{-1} s^{-1}; 1×10^{-17} cm^3 molecule^{-1} s^{-1}. The enhancement in model H_2SO_4 caused by C.I. chemistry has been calculated for each integration. It emerges that the most sensitive parameter is the rate coefficient for $k(SCI + SO_2)$, a dramatic decrease in the production of H_2SO_4 is observed when this rate coefficient is reduced by a factor of 10 and then 100. Indeed, at $k(SCI + SO_2) = 4 \times 10^{-13}$ cm^3 molecule^{-1} s^{-1}, the impact on H_2SO_4 is very weak. The impact of $k(SCI + H_2O)$ on H_2SO_4 is an interesting one, there is a decrease between $k(SCI + H_2O) = 1 \times 10^{-17}$ cm^3 molecule^{-1} s^{-1} and $k(SCI + H_2O) = 1 \times 10^{-16}$ cm^3 molecule^{-1} s^{-1} but not a significant one. However, increasing $k(SCI + H_2O)$ to 1×10^{-15} cm^3 molecule^{-1} s^{-1} does reduce H_2SO_4 levels, particularly in the tropics. Here, $k(SCI + H_2O)$ begins to become comparable with thermal decomposition in importance.

The impact of the C.I. + H_2O rate coefficient was further investigated using the PTM model. Fig. 5 presents the time-integrated fluxes through SO_2 for each assignment of $k(CI + H_2O)$. When $k(CI + H_2O)$ is set at 1.0×10^{-17} cm^3 molecule^{-1} s^{-1} as in the MCMv3.1, the time-integrated flux through SO_2 is 16.1 times that in the base case. However, as $k(CI + H_2O)$ increases from 1.0×10^{-17} cm^3 molecule^{-1} s^{-1}, the time-integrated flux decreases accordingly. Ultimately, when $k(CI + H_2O)$ is much greater than 5×10^{-14} cm^3 molecule^{-1} s^{-1} then the time-integrated flux is effectively the same as in the base case and no additional oxidation routes for SO_2 can be detected.

Initially, the MCMv3.1 used the relative rate coefficients for C.I. + SO_2 and C.I. + H_2O of 4375 based on the value of 7.0×10^{-14} cm^3 molecule^{-1} s^{-1} for the former and 1.6×10^{-17} cm^3 molecule^{-1} s^{-1} for the latter. If these relative rates can be used in conjunction with the new rate coefficient for C.I. + SO_2, then this would imply C.I. + H_2O of about 6.9×10^{-15} cm^3 molecule^{-1} s^{-1}. With this combination of rate coefficients, then the influence of C.I.'s on the oxidation rate of SO_2 would be small but discernible, generating an increase of about 8% at most. However, it is unlikely that the rate coefficient k_1 is as high as this because Welz *et al.*,[4] determined an upper limit of 4.0×10^{-15} cm^3 molecule^{-1} s^{-1}. If an assumed a value for $k(CI + H_2O)$ of 1.0×10^{-16} cm^3 molecule^{-1} s^{-1} results in a relative rate coefficient for C.I. + SO_2 and C.I. + H_2O of 3.0×10^5. With these rate coefficients, the influence of C.I.'s on the oxidation rate of SO_2 is large and up to a factor of 4, at least. If an assumed a value for $k(CI + H_2O)$ is as low as 1.5×10^{-17} cm^3 molecule^{-1} s^{-1}, giving a relative rate coefficient for C.I. + SO_2 and C.I. + H_2O of 2.0×10^6, the influence of the SO_2 oxidation rate may be as high as factor of 16.

Through field measurements noted previously, it is clear in urban areas and in urban outflow the level of 'equivalent ethene' alkene measured (which will be an underestimate) leads to significant C.I. steady state levels. Hence, in these areas, SO_2 oxidation by C.I. will compete and even dominate over oxidation by OH. Further global model simulations where model alkene levels are doubled leads to an approximate doubling of the level of C.I.

Using the parameters presented in Sipilä *et al.*,[5] Fig. 3d displays the increase in nucleation rates J as a result of reaction (4). The increase in boundary layer nucleation rate ranges from 0.02 to ~820% for $n = 2$, and from 0.01% to ~450% for $n = 1.5$. It is acknowledged that, in the atmosphere the nucleation rate will be buffered and increased nucleation does not result in a similar magnitude increase in observable particles or cloud condensation nuclei. Furthermore, assuming that the dominant nucleation mechanism exhibits an H_2SO_4 dependence comparable to that observed elsewhere,[7] outflow from urban areas (WRF-Chem) is shown to be an important region where nucleation enhancement takes place.[66] Indeed, even in northern Europe, where SO_2 levels are low, this mechanism is playing a significant role in the nucleation rate, with increases of up to 75%, as shown in Fig. 6. Here, the combination of elevated alkene levels (both anthropogenic and biogenic) and increasing levels of ozone downwind through photochemical production, combine to provide conditions where this enhancement prospers. A recent combined laboratory and field study by Mauldin *et al.*[95] demonstrates that up to 50% of the observed H_2SO_4 in northern European boreal environments arises from reaction of SO_2 with an unknown organic oxidant, postulated to be a C.I. Those field studies further support the conclusions of this paper.

In a recent paper Pierce *et al.*,[96] investigated the effect of C.I. chemistry on global sulphuric acid concentrations and resulting aerosol production with the global chemical transport model GEOS-Chem. The predicted increases in the sulphuric acid concentrations were relatively similar to the ones predicted with this study, with some differences in the regional distribution of the enhancements. The aerosol dynamic simulations in GEOS-Chem suggested that C.I. chemistry would not have a significant impact on aerosol production globally, mostly due to the dampening due the growth *vs.* coagulation processes discussed above. It is notable, however, that analogous to our results using the global model, the spatial resolution of GEOS-Chem would not resolve the chemistry occurring in urban areas where high alkene plumes meet elevated O_3, as shown in our WRF-CHEM results. Our results with the regional model indicate that while the enhancement of aerosol loadings might be modest on the global scale, it might be more important on a regional-scale hotspots like urban outflows. To quantify this importance, however, further regional-scale simulations with full aerosol microphysics and *e.g.* interactions with the SOA are needed.

Conclusions

We have established that C.I. reactions are a dominant removal mechanism for SO_2 in the urban environment and the regional and global model analysis shows that they are major contributors to sulphate aerosol nucleation, particularly on regional scale. While some pioneering studies on the global importance of Criegee chemistry on aerosol loadings have been conducted, our results indicate that more work on the regional-scale simplications on aerosol production from Criegee-related chemistry is needed. These conclusions depend on assumptions regarding remaining uncertain parameters in C.I. chemistry, notably the rate coefficients for reactions of Criegee intermediates with water and the products (and their tropospheric fate) of C.I. reactions with SO_2 at atmospheric pressure. Quantifying these parameters directly would enable a more secure description of the link between ozonolysis and sulfate aerosol in the troposphere.

Acknowledgements

The participation of OW, AJE, JDS, DLO, and CAT are funded by the Division of Chemical Sciences, Geosciences, and Biosciences, the Office of Basic Energy Sciences, the U. S. Department of Energy. DES, JMD, GM^{c}F and CJP thank NERC for funding. The Advanced Light Source is supported by the Director, Office of Science, Office of Basic Energy Sciences, of the U. S. Department of Energy under Contract DE-AC02-05CH11231 at Lawrence Berkeley National Laboratory. Sandia is a multi-program laboratory operated by Sandia Corporation, a Lockheed Martin Company, for the National Nuclear Security Administration under contract DE-AC04-94-AL85000. DKWM, JMD, CJP and EPFL acknowledge support from the Research Grant Council (RGC) of the Hong Kong Special Administrative Region (HKSAR, Grant No Polyu 5019/11P).

References

1 D. Johnson and G. Marston, *Chem. Soc. Rev.*, 2008, **37**, 699–716.
2 R. M. Harrison, *et al.*, *Sci. Total Environ.*, 2006, **360**, 5–25.
3 S. Gäb, E. Hellpointner, W. V. Turner and F. Korte, *Nature*, 1985, **316**, 535–536.
4 O. Welz, *Science*, 2012, **335**, 204–207.
5 M. Sipilä, *et al.*, *Science*, 2010, **327**, 1243–1246.
6 A. Metzger, *et al.*, *Proc. Natl. Acad. Sci. U. S. A.*, 2010, **107**, 6646–6651.
7 J. Kirkby, *et al.*, *Nature*, 2011, **476**, 429–433.
8 Intergovernmental Panel on Climate Change (IPCC). *Climate Change 2007: The Physical Science Basis*; Cambridge University Press, 2007.
9 J. H. Kroll, *et al.*, *J. Phys. Chem. A*, 2001, **105**, 1554–1560.
10 K. J. Heaton, R. L. Sleighter, P. G. Hatcher, W. A. Hall 4th and M. V. Johnston, *Environ. Sci. Technol.*, 2009, **43**, 7797–7802.
11 R. A. Cox and S. A. Penkett, *Nature*, 1971, **230**, 321–322.
12 J. G. Calvert and W. R. Stockwell, *Environ. Sci. Technol.*, 1983, **17**, 428–443.
13 S. Hatakeyama and H. Akimoto, *Res. Chem. Intermed.*, 1994, **20**, 503–524.
14 C. A. Taatjes, *et al.*, *Science*, 2013, **340**, 177–180.
15 M. Boy, *et al.*, *Atmos. Chem. Phys. Discuss.*, 2012, **12**, 27693–27736.
16 T. Kurtén, *et al.*, *J. Phys. Chem. A*, 2011, **115**, 8669–8681.
17 L. Jiang, *et al.*, *J. Phys. Chem. A*, 2010, **114**, 12452–12461.
18 L. Vreecken, *et al.*, *Phys. Chem. Chem. Phys.*, 2012, **14**, 14682–14695.
19 P. T. M. Carlsson, C. Keunecke, B. C. Kruger, M.-C. Maaß and T. Zeuch, *Phys. Chem. Chem. Phys.*, 2012, **14**, 15637–15640.
20 A. T. Archibald, *et al.*, *Atmos. Chem. Phys.*, 2010, **10**, 1–23.
21 A. T. Archibald, *et al.*, *Atmos. Sci. Letts.*, 2010, **11**, 33–38.
22 M. C. Cooke, *et al.*, *International Journal of Oil, Gas and Coal Technology*, 2010, **3**, 88–103.
23 S. Henne, *et al.*, *Environ. Sci. Technol.*, 2012, **46**, 1650–1658.
24 A. Bacak, *et al.*, *Atmos. Environ.*, 2011, **45**, 6414–6422.
25 S. R. Utembe, *et al.*, *Atmos. Environ.*, 2010, **44**, 1609–1622.
26 S. R. Utembe, *et al.*, *Atmos. Environ.*, 2011, **45**, 1604–1614.
27 W. Collins, *et al.*, *J. Atmos. Chem.*, 1997, **26**, 223–274.
28 R. G. Derwent, *et al.*, *Clim. Change*, 2008, **88**, 385–401.
29 L. A. Watson, *et al.*, *Atmos. Environ.*, 2008, **42**, 7196–7204.
30 M. E. Jenkin, *et al.*, *Atmos. Environ.*, 2008, **42**, 7185–7195.
31 S. R. Utembe, *et al.*, *Atmos. Environ.*, 2009, **43**, 1982–1990.
32 T. C. Johns, *et al.*, *Clim. Dyn.*, 1997, **13**, 103–134.
33 D. S. Stevenson, *et al.*, *Geophys. Res. Lett.*, 1998, **25**, 3819–3822.
34 A. J. Simmons and D. M. Burridge, *Mon. Weather Rev.*, 1981, **109**, 758–766.
35 M. J. Cullen, *Meteorol. Mag.*, 1993, **122**, 81–94.

36 W. J. Collins, *et al.*, *Atmos. Environ.*, 1997, **34**, 255–267.

37 M. E. Jenkin and K. C. Clemitshaw, *Atmos. Environ.*, 2000, **34**, 2499–2527.

38 M. E. Jenkin, *et al.*, *Atmos. Environ.*, 2002, **36**, 4725–4734.

39 W. B. DeMore *et al.*, *Chemical Kinetics and Photochemical Data for Use in stratospheric Modeling, Evaluation Number 10*. Jet Propulsion Laboratory, California Institute of Technology, Pasadena, CA, 1992.

40 S. P. Sander *et al. Chemical Kinetics and Photochemical Data for Use in Atmospheric Studies: Evaluation Number 15*. Jet Propulsion Laboratory, California Institute of Technology, Pasadena, CA, 2006.

41 R. Atkinson, *et al.*, *Atmos. Chem. Phys.*, 2006, **6**, 3625–4055.

42 A. Hough, The calculation of photolysis rates for use in global tropospheric modelling studies. AERE Report R-13259, 1988.

43 J. Olivier*et al.Description of EDGAR Version 2.0*: A set of global emission inventories of greenhouse gases and ozone-depleting substances for all anthropogenic and most natural sources on a per country basis and on 1 degree × 1 degree grid. Netherlands Environmental Assessment Agency, 1996.

44 J. Cofala *et al.*, *Scenarios of global anthropogenic emissions of air pollutants and methane up to 2030*. Technical report, International Institute for Applied systems Analysis, Laxenburg, Austria, 2005.

45 C. Granier, *et al.*, POET, a database of surface emissions of ozone precursors, http://www.aero.jussieu.fr/projet/ACCENT/POET.php, 2005.

46 Mikaloff-Fletcher, *et al.*, *Global Biogeochem. Cycles*, 2004, **18**, GB4004.

47 S. Houweling, *et al.*, *J. Geophys. Res.*, 2000, **105**, 8981–9002.

48 D. Henze, *et al.*, *Atmos. Chem. Phys.*, 2008, **8**, 2405–2420.

49 M. Andreae and P. Merlet, *Global Biogeochem. Cycles*, 2001, **15**, 955–966.

50 F. Dentener, *et al.*, *Atmos. Chem. Phys.*, 2006, **6**, 4321–4344.

51 T. J. Randerson, *et al.*, *Global fire emissions database, version 2 (gfedv2.1)*. data set. available on-line [http://daac.ornl.gov/].Technical report, Oak Ridge National Laboratory Distributed Active Archive Center, Oak Ridge, TN, 2007.

52 J. H. Seinfeld and S. N. Pandis, *Atmospheric Chemistry and Physics*. Wiley, NY, NY, 2006.

53 D. Li and K. P. Shine, *A 4-dimensional ozone climatology for UGAMP models*, Internal Rep. 35, Technical report, University of Reading, UK, 1995.

54 D. M. Murphy and D. W. Fahey, *J. Geophys. Res.*, 1994, **99**, 5325–5332.

55 C. Price and D. Rind, *J. Geophys. Res.*, 1992, **97**, 9919–9933.

56 J. E. Penner *et al.*, *IPCC Special Report on Aviation and the Global Atmosphere*. Technical report, Intergovernmental Panel on Climate Change, 1999.

57 M. E. Jenkin, *Atmos. Chem. Phys.*, 2004, **4**, 1741–1757.

58 D. Johnson, *et al.*, *Environ. Chem.*, 2004, **1**, 150–165.

59 D. Johnson, S. R. Utembe and M. E. Jenkin, *Atmos. Chem. Phys.*, 2006, **6**, 419–431.

60 D. Johnson, *et al.*, *Atmos. Chem. Phys.*, 2006, **6**, 403–418.

61 S. R. Utembe, *et al.*, *Faraday Discuss.*, 2005, **130**, 311–326.

62 J. F. Pankow, *Atmos. Environ.*, 1994, **28**, 185–188.

63 S. E. Stein and R. L. Brown, *J. Chem. Inf. Model.*, 1994, **34**, 581–587.

64 E. J. Baum, *Chemical Property Estimation: Theory and Application*. CRC Press, Boca Raton, FL, 1998.

65 G. A. Grell, *Atmos. Environ.*, 2005, **39**, 6957–6975.

66 J. D. Fast, *et al.*, *J. Geophys. Res.*, 2006, **111**, D21305.

67 D. Lowe, *et al.*, WRF-Chem developments for organic and N2O5 chemical modelling, in preparation.

68 L. K. Emmons, *et al.*, *Geoscientific Model Development*, 2010, **3**, 43–67.

69 J. H. Kuenen *et al.*, *High resolution European emission inventory for the years 2003–2007*, TNO report TNO-060-UT-2011-00588, Utrecht, Netherlands, 2011.

70 R. A. Zaveri, *et al.*, *J. Geophys. Res.*, 2008, **113**, D13204.

71 R. G. Derwent, M. E. Jenkin, S. M. Saunders and M. J. Pilling, *Atmos. Environ.*, 1988, **32**, 2429–2441.

72 S. R. Utembe, M. E. Jenkin, R. G. Derwent, A. C. Lewis, J. R. Hopkins and J. F. Hamilton, *Faraday Discuss.*, 2005, **130**, 311–326.

73 M. E. Erupe, A. A. Viggiano and S.-H. Lee, *Atmos. Chem. Phys.*, 2011, **11**, 4767–4775.

74 P. Paasonen, *et al.*, *Atmos. Chem. Phys.*, 2010, **10**, 11223–11242.

75 K. C. Barsanti, P. H. McMurry and J. N. Smith, *Atmos. Chem. Phys.*, 2009, **9**, 2949–2957.

76 J. N. Smith, *et al.*, *Proc. Natl. Acad. Sci. U. S. A.*, 2010, **107**, 6634–6639.

77 S. M. Ball, D. R. Hanson, F. L. Eisele and P. H. McMurry, *J. Geophys. Res.*, 1999, **104**, 23709–23718.

78 D. Benson, A. Markovich and S.-H. Lee, *Atmos. Chem. Phys.*, 2011, **11**, 4755–4766.

79 B. Bonn and G. K. Moortgat, *Geophys. Res. Lett.*, 2003, **30**, DOI: 10.1029/2003GL017000.

80 I. Kavouras, G. N. Mihalopoulos and E. G. Stephanou, *Nature*, 1998, **395**, 683–686.
81 C. D. O'Dowd, P. Aalto, K. Hämeri, M. Kulmala and T. Hoffmann, *Nature*, 2002, **416**, 497–498.
82 R. Zhang, *et al.*, *Science*, 2004, **304**, 1487–1490.
83 K. C. Barsanti, P. H. McMurry and J. N. Smith, *Atmos. Chem. Phys.*, 2009, **9**, 2949–2957.
84 C. Kuang, *et al.*, *J. Geophys. Res.*, 2008, **113**, D10209.
85 K. E. Leather, *et al.*, *Atmos. Chem. Phys.*, 2012, **12**, 469–479.
86 J. G. Calvert *et al.*, *The Mechanism of Atmospheric Oxidation of the Alkenes*, Oxford University Press, New York, 2000.
87 C. R. Greene and R. Atkinson, *Int. J. Chem. Kinet.*, 1992, **24**, 803–8011.
88 A. C. Lewis, *et al.*, *Nature*, 2000, **405**, 778–781.
89 K. M. Emmerson and N. Carslaw, *Atmos. Environ.*, 2009, **43**, 3220–3226.
90 K. E. Leather, M. R. McGillen and C. J. Percival, *Phys. Chem. Chem. Phys.*, 2010, **12**, 2935–2943.
91 J. B. A. Muller, *et al.*, *Atmos. Meas. Tech.*, 2010, **3**, 163–176.
92 W. P. L. Carter, *Atmos. Environ.*, 1994, **24A**, 481–518.
93 AQIRP, 1995, Effects of gasoline T50, T90 and sulfur on exhaust emissions of current and future technology vehicles. Auto/Oil Air Quality Improvement Research Program, Technical Bulletin No. 18.
94 S. T. Martin, *et al.*, *Rev. Geophys.*, 2010, **48**, RG2002.
95 R. L. Mauldin III, *et al.*, *Nature*, 2012, **488**, 193–196.
96 J. R. Pierce, *et al.*, *Atmos. Chem. Phys. Discuss.*, 2012, **12**, 33127–33163.
97 C. E. Jones, J. R. Hopkins and A. C. Lewis, *Atmos. Chem. Phys.*, 2011, **11**, 6971–6984.

PAPER

Comparing simulated and experimental molecular cluster distributions†

Tinja Olenius,*[a] Siegfried Schobesberger,[a] Oona Kupiainen-Määttä,[a] Alessandro Franchin,[a] Heikki Junninen,[a] Ismael K. Ortega,[a] Theo Kurtén,[b] Ville Loukonen,[a] Douglas R. Worsnop,[ac] Markku Kulmala[a] and Hanna Vehkamäki[a]

Received 8th March 2013, Accepted 15th April 2013

DOI: 10.1039/c3fd00031a

Formation of secondary atmospheric aerosol particles starts with gas phase molecules forming small molecular clusters. High-resolution mass spectrometry enables the detection and chemical characterization of electrically charged clusters from the molecular scale upward, whereas the experimental detection of electrically neutral clusters, especially as a chemical composition measurement, down to 1 nm in diameter and beyond still remains challenging. In this work we simulated a set of both electrically neutral and charged small molecular clusters, consisting of sulfuric acid and ammonia molecules, with a dynamic collision and evaporation model. Collision frequencies between the clusters were calculated according to classical kinetics, and evaporation rates were derived from first principles quantum chemical calculations with no fitting parameters. We found a good agreement between the modeled steady-state concentrations of negative cluster ions and experimental results measured with the state-of-the-art Atmospheric Pressure interface Time-Of-Flight mass spectrometer (APi-TOF) in the CLOUD chamber experiments at CERN. The model can be used to interpret experimental results and give information on neutral clusters that cannot be directly measured.

1 Introduction

Atmospheric new particle formation has been observed in numerous sites around the world[1] and is estimated to be a significant source of aerosol particles and cloud condensation nuclei.[2–4] Atmospheric aerosol particles influence the global climate and have adverse effects on visibility and human health.[5,6] Currently, the effect of aerosols forms the dominant uncertainty in predicting radiative forcing and global temperature change.[7] Sulfuric acid is known to be involved in

[a]*Department of Physics, University of Helsinki, FIN-00014 Helsinki, Finland. E-mail: tinja.olenius@helsinki.fi*
[b]*Department of Chemistry, University of Helsinki, FIN-00014 Helsinki, Finland*
[c]*Aerodyne Research Inc., Billerica, Massachusetts 01821, USA*

† Electronic Supplementary Information (ESI) available: See DOI: 10.1039/c3fd00031a

atmospheric particle formation (see for example ref. 8–10), but the identity and role of other compounds remains uncertain. Atmospheric bases, such as ammonia and amines, as well as other organic compounds and ions have been proposed to enable or enhance sulfuric acid driven particle formation.[11-14] Nevertheless, the understanding of the initial formation mechanisms remains incomplete.

New particle formation starts with gas phase molecules clustering into small molecular clusters. The main limitations in the experimental examination of the initial clusters are their low concentrations and the insufficient size resolution or selective sensitivity of instruments. Understanding the formation and growth of the clusters requires information on both size distribution and chemical composition, which are challenging to determine experimentally for clusters consisting of only few molecules. Implementation of high-resolution mass spectrometry, together with electrical mobility spectrometry, enables the concentration measurements and chemical characterization of charged small particles from the molecular scale upward.[15,16] The experimental observation of electrically neutral clusters is possible down to approximately 1 nm in mobility diameter.[17] Nevertheless, all available methods for determining the chemical composition of the clusters require charging, which is likely to have an effect on the measurement result for example *via* changing the composition of the clusters.[18]

The challenges involved in the experimental studies make theoretical methods giving information on both charged and neutral clusters necessary for exploring atmospheric cluster formation. In this work we used the kinetic collision and evaporation model ACDC[19] to study the steady-state concentrations and kinetics of a set of small molecular clusters under conditions relevant to the CLOUD experiment at CERN.[20] We show a good agreement between the modeled steady-state concentrations of negatively charged clusters and experimental concentrations measured with a mass spectrometer in the CLOUD experiment. After showing that the model is able to reproduce the experimental results for the concentrations of charged clusters, we demonstrate how the model can also give information on electrically neutral clusters and thus improve our knowledge of atmospheric particle formation mechanisms.

2   Methods

2.1   Theoretical methods and model system

The model system is a set of molecular clusters consisting of sulfuric acid (H_2SO_4) and ammonia (NH_3) molecules. Electrically neutral, and negatively and positively charged clusters are included. In principle the clusters can contain 0–5 sulfuric acid and 0–5 ammonia molecules, forming a so-called 5 × 5 simulation box. However, as some of these clusters can be predicted to be highly unstable because of an unfavorable composition, they were omitted from the simulation to reduce the computational burden. These unstable clusters include negatively charged clusters containing more or as many ammonia molecules as acid molecules, positively charged clusters containing more acid than ammonia molecules, and neutral and positive clusters that have a very high ammonia : acid ratio. All the clusters included in the simulation are listed in Table 1.

The steady-state concentrations of the cluster set were solved with the kinetic model ACDC (Atmospheric Cluster Dynamics Code).[19] Since the current version of

Table 1 Molecular clusters included in the model system. The first column and the second column indicate the number of sulfuric acid molecules and all the possible numbers of ammonia molecules in the cluster, respectively

a) Electrically neutral clusters

	Number of NH_3 molecules
0 acids	1–4
H_2SO_4	0–4
$(H_2SO_4)_2$	0–4
$(H_2SO_4)_3$	0–4
$(H_2SO_4)_4$	0–5
$(H_2SO_4)_5$	0–5

b) Negatively charged clusters

	Number of NH_3 molecules
HSO_4^-	0–1
$(H_2SO_4)\cdot HSO_4^-$	0–1
$(H_2SO_4)_2\cdot HSO_4^-$	0–2
$(H_2SO_4)_3\cdot HSO_4^-$	0–3
$(H_2SO_4)_4\cdot HSO_4^-$	0–4

c) Positively charged clusters

	Number of NH_3 molecules + H^+
0 acids	1–3
H_2SO_4	1–3
$(H_2SO_4)_2$	1–4
$(H_2SO_4)_3$	2–5
$(H_2SO_4)_4$	3–5
$(H_2SO_4)_5$	5

the model is described in detail in ref. 21, the basic features are discussed only briefly here. The code writes out the birth–death equations, in other words, the time derivatives of the cluster concentrations, for a given set of clusters assuming all possible collision and evaporation processes. The time evolution of the cluster concentrations is simulated using the Matlab ode15s solver for sets of differential equations until the time-independent steady-state is achieved. The clusters are formed in collisions of smaller clusters and evaporations of larger clusters, and they are destroyed by colliding with other clusters and by fragmenting into smaller clusters. Hence the time derivative can be written as

$$\frac{dC_i}{dt} = \frac{1}{2}\sum_{j<i}\beta_{j,(i-j)}C_jC_{i-j} + \sum_j\gamma_{(i+j)\to i,j}C_{i+j} - \sum_j\beta_{i,j}C_iC_j - \frac{1}{2}\sum_{j<i}\gamma_{i\to j,(i-j)}, \quad (1)$$

where C_i is the concentration of cluster i, $\beta_{i,j}$ is the collision coefficient of clusters i and j, and $\gamma_{k\to i,j}$ is the evaporation coefficient of cluster k evaporating into clusters i and j. As we included both electrically neutral and charged clusters in the simulation, the clusters can also be created and destroyed by ionization and recombination which occur *via* collisions with generic ionizing species. These species, assumed to have the mass and radius of O_2^- (mass 32.00 u, radius 2.23 Å) for negative polarity and H_3O^+ (mass 19.02 u, radius 1.96 Å) for positive polarity, are added into the system with a constant production rate. The negative generic ion can collide with neutral clusters containing sulfuric acid and ionize them negatively by converting one sulfuric acid molecule into a bisulfate ion (HSO_4^-), or it can collide with positively charged clusters and neutralize them by removing the extra proton (H^+). Similarly, the positive generic ion can ionize base-containing neutral clusters positively, and neutralize negatively charged clusters by donating a proton. The negative and positive generic ions can also be destroyed by recombining with each other. Charged clusters and/or generic ions of the same polarity cannot collide with each other.

For collisions between electrically neutral clusters, the collision coefficients are calculated according to the kinetic gas theory:

$$\beta_{i,j} = \left(\frac{3}{4\pi}\right)^{1/6} \left[6k_\mathrm{B}T\left(\frac{1}{m_i}+\frac{1}{m_j}\right)\right]^{1/2} \left(V_i^{1/3}+V_j^{1/3}\right)^2, \tag{2}$$

where k_B is the Boltzmann constant, T is the temperature and m_i and V_i are the mass and volume of cluster i. The volume is calculated from the atomic masses and the liquid densities of the compounds in the cluster. For collisions between neutral and charged clusters, the collision rate is calculated from the polarizability and the dipole moment of the neutral cluster obtained from the quantum chemical calculations (Table S2 in the supplementary information†).[22,23] The recombination coefficient of positive and negative ions, both charged molecular clusters and generic small ions, was assumed to be 1.6×10^{-6} cm^{-3} s^{-1}.[24]

Cluster evaporation rates are calculated as in ref. 25:

$$\gamma_{(i+j)\rightarrow i,j} = \beta_{i,j}\frac{P_\mathrm{ref}}{k_\mathrm{B}T}\exp\left(\frac{\Delta G_{i+j}-\Delta G_i-\Delta G_j}{k_\mathrm{B}T}\right), \tag{3}$$

where ΔG_i is the Gibbs free energy of formation of cluster i at a reference pressure P_ref. The formation free energies were computed with a quantum chemical multi-step method B3LYP/CBSB7//RI-CC2/aug-cc-pV(T + d)Z. The free energies of most of the neutral and positive clusters are published in our previous work,[23,25] and the free energies and other thermochemical parameters of the negative clusters, as well as the remaining neutral and positive clusters, are given in Table S1 in the supplementary information.† As the free energies depend on the temperature, in this work they were calculated separately for each simulated temperature directly from the vibrational frequencies and rotational constants.

In this study we used two additional sink terms reported in the experimental studies: a wall loss term corresponding to the sticking of the clusters onto the walls of the experiment chamber and a dilution term. The dilution sink has a constant value of 1.06×10^{-4} s^{-1}, and the wall loss factor, determined by experimental means (see for example ref. 26), has been parameterized to depend on the size of the cluster as

$$S_i = 0.774 \text{ m}^{-1}\mathrm{s}^{-1/2} \times \sqrt{\frac{k_\mathrm{B}TC_\mathrm{C}(d_i)}{3\pi\eta d_i}}, \tag{4}$$

where η is the viscosity of air, and d_i and $C_\mathrm{C}(d_i)$ are the mobility diameter and the slip correction factor for cluster i, respectively. The mobility diameter is defined as

$$d_i = (2r_{i,m}+0.3 \text{ nm}) \times \sqrt{1+\frac{28.8 \text{ u}}{m_i}}, \tag{5}$$

where $r_{i,m}$ and m_i are the mass radius and mass (in atomic mass units u) of the cluster.[16] Eqn (4) is derived for electrically neutral clusters, and we assumed that the size-dependent wall loss coefficient for charged clusters and generic ions is enhanced by a factor of five. This empirical factor is based on the fact that the measured total ion concentration in the CLOUD experiments at a known ion production rate can be explained by an ion loss term enhancement of this magnitude. The physical reason for the enhanced loss rate of ions compared to neutral clusters is the non-uniform ion deposition in the CLOUD chamber by pions and also by muons from an ionizing beam (see Sect. 2.2; the muons penetrate the beam stopper and contribute to experiments where the beam is disabled).

When cluster collisions result in clusters outside of the simulation box, they are allowed to grow out of the system if their composition is favorable and they can be assumed to grow further instead of re-evaporating immediately. The boundary conditions are based on the relative stability of the clusters in terms of the acid : base ratio (see for example ref. 25). Neutral clusters are allowed to leave the system if they contain at least six acid and five ammonia molecules; the requirement for negatively charged clusters is at least six acids (including the negative ion) and one base, and for positively charged clusters at least five acids and six bases (including the positive ion). If these conditions are not satisfied, the cluster is brought back to the system by evaporating monomers out of it until it reaches the nearest boundary of the simulation box.

2.2 Measurements and conditions in the CLOUD experiment

Concentrations of negatively charged sulfuric acid–ammonia clusters were measured in the CLOUD experiment at CERN (for more details, see ref. 20) using an Atmospheric Pressure interface Time-Of-Flight mass spectrometer (APi-TOF).[15] The instrument obtains the ion signal as counts per unit time as a function of mass-to-charge ratio. The signal was converted to concentrations by applying a mass dependent conversion function. For deriving this function, its mass-dependency was first derived by dedicated laboratory experiments: ammonium bisulfate particles were size-selected by a high-resolution differential mobility analyzer and sampled by both the APi-TOF and an electrometer that served as the reference, covering most of the size-range the APi-TOF can detect. For the data from the CLOUD experiment, we compared the ion signal measured by the APi-TOF with the results of an ion mobility spectrometer (Neutral cluster and Air Ion Spectrometer, NAIS, Airel Ltd.)[27] that sampled in parallel and measured ion concentrations as a function of mobility-equivalent size. Losses in each instrument's sampling line were taken into account, and the corrected ion concentrations from the NAIS were used as reference for determining the final setup-specific conversion function, and for assessing uncertainties.

The APi-TOF data was measured under steady-state conditions with constant sulfuric acid and ammonia concentrations at temperatures of 248, 278 and 292 K. The sulfuric acid concentration was measured with a Chemical Ionization Mass Spectrometer (CIMS)[28,29] and varied between approximately 10^6 and 10^9 cm^{-3} in individual experiments. The ammonia concentration was measured with a Long-Path Absorption Photometer (LO-PAP)[30] and a Proton Transfer Reaction Mass Spectrometer (PTR-MS).[31] Ammonia was not added to the experiment chamber intentionally in all the runs, but a trace amount of ammonia (below the experimental detection limit of 35 ppt) is known to have been present in the chamber anyway, because clusters containing ammonia molecules were detected by the API-TOF. In the experiments with intentionally added ammonia, the ammonia mixing ratio was in the range of approximately 100 to 1000 ppt, corresponding to 2.6×10^9 cm^{-3} to 2.6×10^{10} cm^{-3}.

We ran the simulations using the same range of sulfuric acid and ammonia concentrations that were reported in the experiment. In the case of ammonia, we fixed the concentration of single ammonia molecules. For comparison with the experiments with no added ammonia, the simulated ammonia concentration was set to either 5 ppt or to the experimental detection limit of 35 ppt. In the case of sulfuric acid, the total concentration of all electrically neutral clusters consisting of one acid molecule and any number of ammonia molecules was fixed instead of the concentration of single acid molecules. This was considered to correspond better to the measured acid concentration since according to theoretical calculations, the CIMS can also detect acid molecules clustered with base molecules as acid monomers.[18] However, under the conditions studied in this work the sum of all neutral clusters containing one acid was practically equal to the concentration of pure acid monomers.

Ions were produced in the experiments either by only natural ionization due to galactic cosmic rays (GCR), or additionally by a pion beam which allowed higher total ion production rates. In the simulations, we used the rates 3 ion pairs s^{-1} cm^{-3}, corresponding to natural GCR ionization, and 40 ion pairs s^{-1} cm^{-3}, corresponding to the ionization at the average pion beam intensity of 60 kHz.

3   Results and discussion

3.1   Comparison of modeled and measured concentrations of negatively charged clusters

The data presented and discussed in the Results and discussion section are for 278 K and ionization rate due to GCR (3 ion pairs s^{-1} cm^{-3}). Results for 248 K, 292 K and beam-augmented ionization are qualitatively similar and are presented in the supplementary information.† Fig. 1 shows the modeled and experimental distributions of negatively charged clusters containing 1–5 sulfuric acid

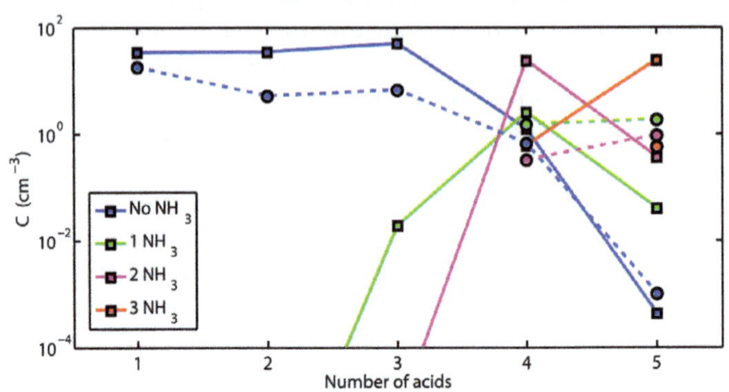

Fig. 1 Modeled (solid lines) and experimental (dashed lines) concentrations of negatively charged clusters containing sulfuric acid and varying number of ammonia molecules as a function of number of acid molecules (including the ion) at [H$_2$SO$_4$] = 5 × 10^7 cm^{-3} and [NH$_3$] = 100 ppt. Note that the smallest observed cluster containing three ammonia molecules is the pentamer.

molecules (including the bisulfate ion), denoted from now on as mono-, di-, tri-, tetra- and pentamers, respectively, under representative experimental conditions at $[H_2SO_4] = 5 \times 10^7$ cm^{-3} and $[NH_3] = 100$ ppt. Concentrations of clusters with different numbers of ammonia molecules are shown as separate lines. The shapes of the modeled and experimental distributions of pure acid clusters are qualitatively similar with the exception that according to the measurements, the monomer clearly has a higher concentration compared to the dimer than what is predicted by the model (note that this might not be clearly visible in the logarithmic scale). The model also predicts higher concentrations than observed for the ammonia-containing tetra- and pentamers. These discrepancies can be due to cluster fragmentation processes inside the instrument; at least the ratio of the dimer and monomer concentrations has been observed to be easily affected by the instrument settings. Moreover, it is possible that trimers can break into monomers and dimers in the atmospheric pressure interface, and hence the measured concentrations of the smallest clusters should be examined with caution. The fragmentation issues are discussed in more detail in Section 3.3.

    3.1.1 Absolute concentrations. Fig. 2 shows the cluster concentrations as a function of sulfuric acid concentration. As stated in the Methods section, we assumed that the measured acid concentration $[H_2SO_4]$ can be taken to be the total concentration of all neutral clusters containing one acid and any number of base molecules. However, according to the simulation results for neutral clusters the most abundant one-acid cluster under the studied conditions is in fact the pure acid monomer, and the contribution of acid monomers clustered with ammonia molecules is less than one percent. Therefore the acid axis in Figs. 2 and 3 practically corresponds to the true acid monomer concentration. Concentrations of pure i-mers containing only sulfuric acid (including the ion) and total concentrations of i-mers, containing none or any number of ammonia molecules in addition to the acid, are shown as separate lines and markers and denoted with subscripts "pure" and "tot", respectively. Overall the concentrations show rather good agreement in the slopes. The experimental mono- and dimer concentrations are dependent on the ammonia concentration, whereas the modeled concentrations are independent of it (the lines corresponding to different ammonia concentrations fall on top of each other). The modeled and measured absolute concentrations are of the same order of magnitude except for the mono- and dimers. The modeled concentrations of tri-, tetra- and pentamers are slightly higher than the measured ones for the experiments with intentionally added ammonia. For the experiments with no added ammonia the experimental data points are located approximately between the model curves for the ammonia mixing ratios of 5 and 35 ppt. According to both the simulation and experimental results, there are no ammonia molecules attached to mono-, di- and trimers. For tetra- and pentamers, the total concentration C_{tot}, including both pure and ammonia-containing clusters, is higher than the concentration of pure acid i-mers C_{pure} when ammonia is added to the chamber. The ratio $C_{\text{pure}}/C_{\text{tot}}$ is larger for experimental results, which could be explained by evaporation of ammonia molecules inside the instrument: a fraction of clusters that contain ammonia could have been converted to pure acid clusters before their detection. Given that the mono- and dimers can be formed in the fragmentation of some larger clusters, the fact that their measured concentrations are lower at higher ammonia concentration is likely due to clusters being stabilized by ammonia.

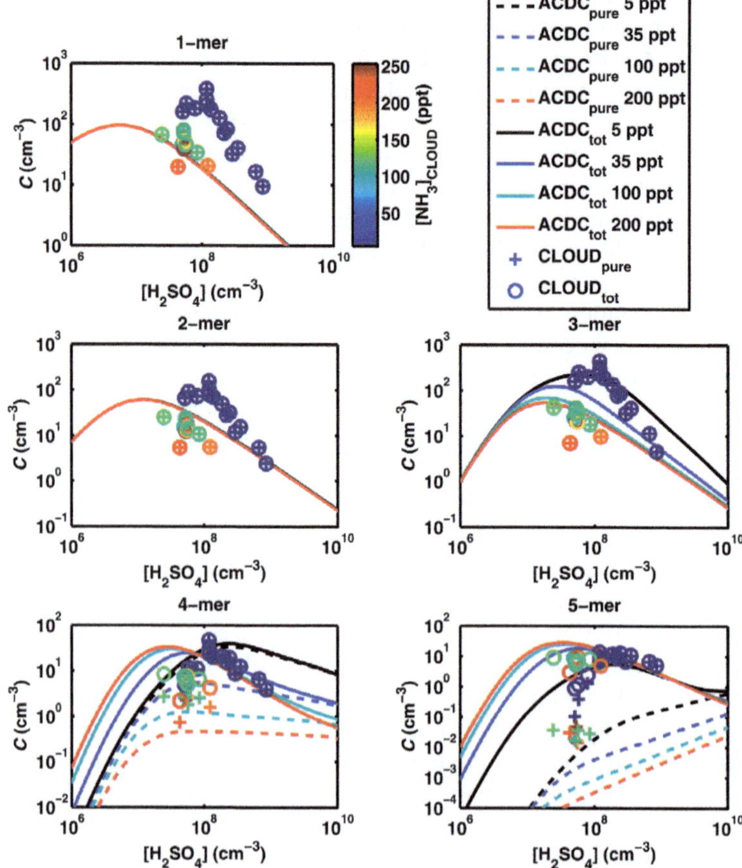

Fig. 2 Modeled and experimental concentrations of negatively charged sulfuric acid mono-, di-, tri-, tetra- and pentamers as a function of sulfuric acid concentration. Concentrations of i-mers containing only sulfuric acid and total concentrations of pure i-mers plus i-mers containing ammonia molecules are shown as separate lines and markers. Note the different scales in the y-axes.

Although it seems that the model is also able to cover the experiments at background-level ammonia concentrations, certain caution should be used in the interpretation of these results. ACDC is an acid–base model and, as stated in Section 2.1, allows clusters to leave the simulation box if they contain a certain amount of acid and base molecules. On the other hand, it is possible that during the background-level experiments the ammonia concentration in the chamber may have been so low that it did not significantly enhance cluster formation. If the growth in reality occurred mostly *via* different mechanisms, presumably binary acid–water clustering, the simulated conditions are not directly comparable to the experiments. This should be kept in mind especially when comparing with the experiments where the detected clusters generally do not contain ammonia.

3.1.2 Relative concentrations of clusters with i acids and i + 1 acids. In addition to the concentrations, the model also enables the monitoring of fluxes between the clusters. This feature can be used for example to find out the main sources and sinks of any cluster, or to examine the cluster growth mechanisms.

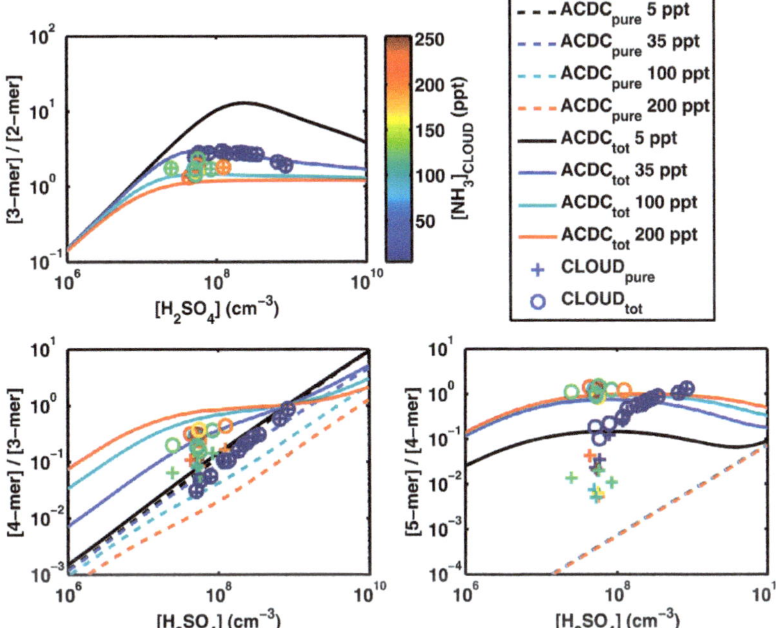

Fig. 3 Ratios of the concentrations of negatively charged clusters as a function of sulfuric acid concentration. Ratios of pure sulfuric acid *i*-mer concentrations and ratios of total *i*-mer concentrations including clusters both with and without ammonia are shown as separate lines and markers. Note the different scales in the *y*-axes.

We studied the formation pathways of the negative clusters by tracing the net fluxes, in other words the difference between collision and evaporation (or ionization and recombination) fluxes, into and out of the clusters. According to the simulations, negative clusters are formed by the subsequent addition of acid molecules, or also ammonia molecules in the case of tetra- and pentamers, onto a bisulfate ion core. The growth starts from the single bisulfate ion and proceeds to the pure tetramer by collisions with single acid molecules. The tetramer successively gains two ammonia molecules before growing into a pentamer by the addition of an acid. The pentamer gains one more ammonia and then grows out of the simulation box. Based on these growth pathways the concentrations of different negative clusters should be related to each other. The ratios of the concentrations C_{i+1}/C_i, where $i = 2$–4 is the number of acids, are presented in Fig. 3. The ratio of the dimer and monomer concentrations is not shown because of the known problems with the experimental concentrations, discussed above in Section 3.1. The behavior of the modeled and experimental concentration ratios with respect to the variation in the acid and ammonia concentrations shows a qualitatively good agreement, except for the ratio of the penta- and tetramer concentrations at low ammonia mixing ratios, which seems to be due to the mismatch in the pentamer concentrations below $[\mathrm{H_2SO_4}] = 10^8$ cm^{-3} (Fig. 2).

   3.1.3   Distributions of ammonia-containing clusters. Fig. 4 shows the fractions of clusters containing different numbers of ammonia molecules for negatively charged tetra- and pentamers at $[\mathrm{H_2SO_4}] = 5 \times 10^7$ cm^{-3} and $[\mathrm{NH_3}] = 100$ ppt. According to the modeling results, under these conditions the major

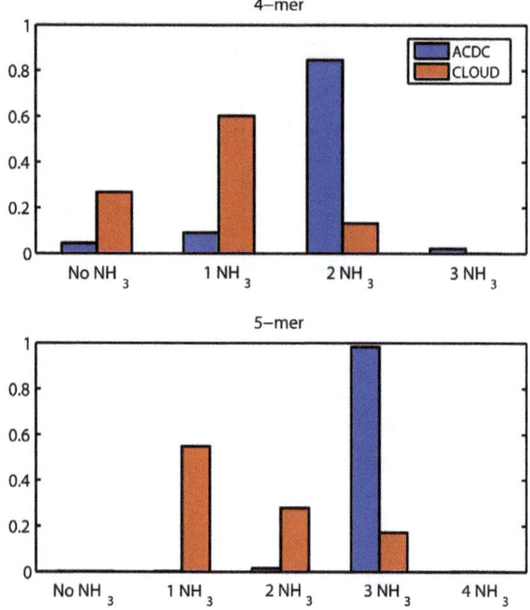

Fig. 4   Modeled and experimental fractions of negatively charged tetra- and pentamers containing different numbers of ammonia molecules at $[H_2SO_4] = 5 \times 10^7$ cm^{-3} and $[NH_3] = 100$ ppt.

fraction (85%) of tetramers contains two ammonias, with minor contributions of pure acid tetramers and tetramers clustered with one or three ammonias, whereas most of the experimentally observed tetramers contain one or no ammonias. Similarly, the model predicts that practically all the pentamers (98%) contain three ammonias, but the major fraction of observed pentamers contains only one ammonia, with smaller fractions containing two or three ammonias. The discrepancy could be due to uncertainties in the quantum chemical formation free energies or the possible evaporation of ammonia molecules before detection. According to the quantum chemical results, the evaporation rate of ammonia from the tetramers with one and two ammonias is higher than that of acid. The ammonia also evaporates faster from the pentamers containing three and two ammonias, but for the pentamer with only one ammonia, the evaporation rate of acid is higher. This could explain why pure pentamers were not observed. It is also possible that in reality the clusters contain water molecules (see Section 3.3), which can decrease the number of ammonias in the clusters. Finally, recent experimental findings[32] suggest that there may be a kinetic barrier for addition of ammonia into acid–ammonia clusters, which also might explain the discrepancies in the modeled and measured ammonia contents.

3.2   Modeled concentrations of electrically neutral clusters

As the ACDC model is capable of reproducing the experimental results for the concentrations of charged clusters qualitatively and even quantitatively under some conditions, it can be used to give reliable estimates of the concentrations of electrically neutral clusters. Modeling results for the distribution of neutral clusters at $[H_2SO_4] = 5 \times 10^7$ cm^{-3}, $[NH_3] = 100$ ppt and 278 K (conditions

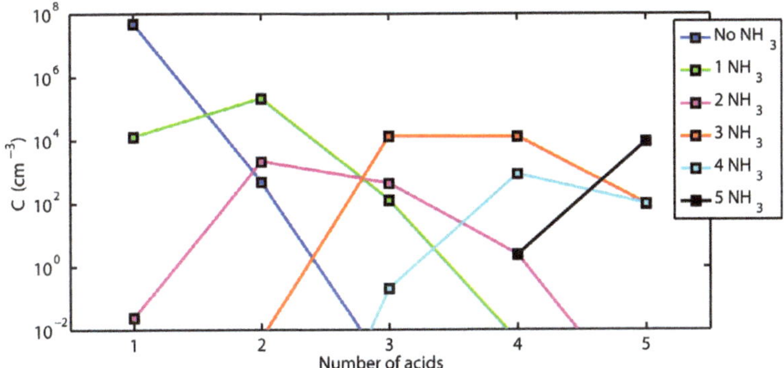

Fig. 5 Modeled concentrations of electrically neutral clusters containing sulfuric acid and a varying number of ammonia molecules as a function of the number of acid molecules at $[H_2SO_4] = 5 \times 10^7$ cm^{-3} and $[NH_3] = 100$ ppt.

corresponding to Fig. 1) are presented in Fig. 5. Under these conditions, the number of ammonia molecules in the most abundant neutral clusters is either equal to or one less than the number of acid molecules. This is a general feature in the acid–ammonia system, and follows from the stabilizing effect of ammonia in neutral sulfuric acid clusters.

The relative stability of the clusters can be predicted from the evaporation rates derived from the formation free energies,[25] but the relative abundances cannot be estimated solely based on them, as kinetic effects also play a role. Although the most stable (and abundant) neutral clusters contain approximately the same number of acid and ammonia molecules, the exact relative concentrations of the clusters also depend on the concentrations of acid and ammonia *via* the collision rates of monomers with the clusters. In addition, the dynamics of the neutral clusters are affected by the presence of ionizing species and charged clusters. Finally, external loss terms, such as the wall loss term in this study, affect the concentrations of all the clusters and ions. Assessment of the concentrations and kinetics of neutral clusters requires that the various dynamic effects are taken into account. To better understand these effects, we studied how the neutral clusters are formed under the conditions corresponding to Fig. 1 ($[H_2SO_4] = 5 \times 10^7$ cm^{-3}, $[NH_3] = 100$ ppt and $T = 278$ K). The growth occurs by the successive addition of acid and ammonia molecules, which is expected from the distribution of the neutral clusters (Fig. 5). However, we found that under other conditions a notable fraction of the neutral clusters can be formed by the recombination of charged clusters, either with other charged clusters of the opposite polarity, or with the generic ionizing species. These formation routes become significant for example at low acid concentration or higher temperature. This is likely due to the fact that under the aforementioned conditions the evaporation processes of neutral clusters are relatively significant. Ions are more tightly bound due to electrostatic interaction, and consequently the ionic formation routes of neutral clusters are more prominent when the growth along neutral pathways is limited by a too high evaporation rate of some cluster compared to its collision rate with any species that can grow it further. The dependence of the formation routes of the neutral clusters on the ambient conditions reflects how interactions with ions can affect the concentrations of neutral clusters.

3.3 Sources of uncertainty

From the experimental point of view, the greatest source of uncertainty, also qualitative, is probably due to fragmentation processes possibly occurring inside the APi-TOF. These include the breaking of the clusters in collisions, and the evaporation of molecules from the clusters. The first-mentioned process is due to the accelerations of the ions in the electric fields created by the various ion guiding and focusing elements. The ions are therefore subjected to more energetic collisions with gas molecules than in the ambient, and one or more molecules may be lost from cluster ions. It can be noted that the magnitude of this type of fragmentations depends on the instrumental settings, and that their effect is probably most significant for the smallest clusters, such as the mono- and dimers. Relative concentrations of the larger clusters are less affected. The evaporation-type fragmentation can occur if some molecules in the cluster are relatively loosely bound. These molecules can be lost if the evaporation is fast enough to take place during the time between the cluster entering the instrument and its detection. Water is the only compound that is known for sure to evaporate so rapidly that it will not be detected in the clusters, although there is water vapor in the chamber (and thus water molecules attached to a significant fraction of the clusters). Overall the de-clustering inside the instrument is still poorly understood. However, its effect must be limited since comparisons with ion mobility spectrometers show good agreement, as shown previously by Ehn *et al.* (ref. 16) and as was also seen when comparing the APi-TOF with the NAIS ion data for this study.

Another source of uncertainty is the data conversion from ion count rates to ion concentrations. Variations in the comparison between the APi-TOF and the NAIS during the measurement campaign were taken into account by calculating a mean and a standard deviation. The mean was used for the conversions, and the uncertainties of the conversion function were assessed from the standard deviation. A representative example of the error bars calculated from the standard deviation by applying the propagation of uncertainty is presented in Fig. S1 in the supplementary information.† Although the error bars are relatively large, they are not enough to fully cover the differences between experimental and modeling results.

Some of the corrections for sampling losses, which were included in the conversion, were based on simplified theoretical assumptions,[33] which could also lead to some systematic error. In any case, the resulting uncertainties and errors are mostly in the absolute numbers, whereas the shape of the cluster distribution would hardly be affected.

From the model perspective, the most significant error source is likely the uncertainties in the Gibbs free energies that can be estimated to be around ± 1 kcal mol^{-1}. Therefore test simulations were performed with either stabilizing or de-stabilizing all the clusters by 1 kcal mol^{-1}. This changes the absolute concentrations up to around an order of magnitude, as can be expected, while the qualitative behavior remains the same. Possible uncertainties in the collision frequencies are more difficult to assess. On the other hand, the same methods for calculating the collision coefficients between neutral molecules and ionic clusters have been applied in the modeling study by Kupiainen *et al.* (ref. 23), which reported good agreement with experimental data. The relative abundance of different clusters can also be somewhat affected by the boundary conditions determining which clusters are allowed to grow out from the system. However, the currently used conditions are reasonable based on the existing experimental and modeled distributions of acid–ammonia clusters.

The modeled clusters do not contain water molecules. Kinetic modeling of the full multicomponent acid–base–water system is impossible, as the collision frequency of water with the clusters (as well as the water evaporation rate) is approximately ten orders of magnitude higher than that of acid or base molecules. This results in a numerically extremely stiff set of equations, which cannot be solved practically for our system. Water can, nevertheless, be implemented in the model implicitly by calculating the effective collision and evaporation rates of the hydrated clusters.[34] However, the computational effort of the quantum chemical calculations required to obtain the formation free energies of the hydrates is significant as multiple additional molecules are added to the clusters, both due to configurational sampling and to the time required by each energy or force evaluation. In light of the other considerable sources of uncertainty in the comparison of the modeled and experimental data (see above), estimating the effect of water was considered to be beyond the scope of this study.

4   Conclusions

We modeled the kinetics of a set of neutral and charged molecular clusters consisting of sulfuric acid and ammonia molecules and studied the steady-state concentrations and growth pathways of the clusters. Evaporation rates of the clusters were derived from quantum chemical formation free energies with no free parameters. We compared the concentrations of negatively charged clusters to experimental results measured with the APi-TOF (Atmospheric Pressure interface Time-Of-Flight mass spectrometer) in the CLOUD experiment[20] and found good agreement. After demonstrating the validity of the model in the case of charged clusters, we showed how the model can be used to obtain information on neutral clusters that cannot be directly measured.

The behavior of the concentrations of negative clusters with respect to variations in the acid and ammonia concentrations is well predicted by the model, although there are some discrepancies between the measured and modeled absolute cluster concentrations. However, taking into account the experimental uncertainties that are mainly due to the still poorly understood cluster fragmentation inside the mass spectrometer, the agreement is impressive. The model also correctly predicts that the smallest negative ammonia-containing clusters have four acid molecules. The discrepancies in the number of ammonia molecules in the clusters might be explained by the evaporation of ammonia inside the instrument. As the instrumental effects are not well known, the modeling results can help in the interpretation of the measurements. The dynamic model and the high-resolution experiments are complementary methods to study the properties of atmospheric molecular clusters, and provide necessary tools to understand the growth mechanisms of these clusters.

Acknowledgements

This work was supported by FP7-MOCAPAF project no. 257360 (ERC Starting Grant), FP7-ATMNUCLE project no. 227463 (ERC Advanced Grant), FP7 Marie Curie Initial Training Network CLOUD-ITN no. 215072, Vilho, Yrjö and Kalle

Väisälä Foundation, Maj and Tor Nessling Foundation project no. 2011200 and Academy of Finland Center of Excellence project no. 1118615 and LASTU project no. 135054. The authors thank CSC- IT Center for Science in Espoo, Finland for the computing time, tofTools development team for providing the TOF-MS analysis tools and the CLOUD consortium for providing the experimental data.

References

1 M. Kulmala, H. Vehkamäki, T. Petäjä, M. Dal Maso, A. Lauri, V.-M. Kerminen, W. Birmili and P. H. McMurry, *J. Aerosol Sci.*, 2004, **35**, 143–176.

2 D. V. Spracklen, K. S. Carslaw, M. Kulmala, V.-M. Kerminen, G. W. Mann and S.-L. Sihto, *Atmos. Chem. Phys.*, 2006, **6**, 5631–5648.

3 D. V. Spracklen, K. S. Carslaw, M. Kulmala, V.-M. Kerminen, S.-L. Sihto, I. Riipinen, J. Merikanto, G. W. Mann, M. P. Chipperfield, A. Wiedensohler, W. Birmili and H. Lihavainen, *Geophys. Res. Lett.*, 2008, 35.

4 J. Merikanto, D. V. Spracklen, G. W. Mann, S. J. Pickering and K. S. Carslaw, *Atmos. Chem. Phys.*, 2009, **9**, 8601–8616.

5 J. C. Cabada, A. Khlystov, A. E. Wittig, C. Pilinis and S. N. Pandis, *J. Geophys. Res.*, 2004, **109**, D16S03 1–13.

6 A. Nel, *Science*, 2005, **308**, 804–806.

7 IPCC, *The Intergovernmental Panel on Climate Change, Climate Change 2007: The Physical Science Basis*, Cambridge University Press, New York, 2007.

8 R. J. Weber, J. J. Marti, P. H. McMurry, F. L. Eisele, D. J. Tanner and A. Jefferson, *J. Geophys. Res.*, 1997, **102**, 4375–4385.

9 S.-L. Sihto, M. Kulmala, V.-M. Kerminen, M. Dal Maso, T. Petäjä, I. Riipinen, H. Korhonen, F. Arnold, R. Janson, M. Boy, A. Laaksonen and K. E. J. Lehtinen, *Atmos. Chem. Phys.*, 2006, **6**, 4079–4091.

10 C. Kuang, P. H. McMurry, A. V. McCormick and F. L. Eisele, *J. Geophys. Res.*, 2008, **113**, D10209.

11 T. Kurtén, V. Loukonen, H. Vehkamäki and M. Kulmala, *Atmos. Chem. Phys.*, 2008, **8**, 4095–4103.

12 V. Loukonen, T. Kurtén, I. K. Ortega, H. Vehkamäki, A. A. H. Padua, K. Sellegri and M. Kulmala, *Atmos. Chem. Phys.*, 2010, **10**, 4961–4974.

13 A. B. Nadykto and F. Yu, *Chem. Phys. Lett.*, 2007, **435**, 14–18.

14 F. Yu and R. P. Turco, *Geophys. Res. Lett.*, 2000, **27**, 883–886.

15 H. Junninen, M. Ehn, T. Petäjä, L. Luosujärvi, T. Kotiaho, R. Kostiainen, U. Rohner, M. Gonin, K. Fuhrer, M. Kulmala and D. R. Worsnop, *Atmos. Meas. Tech.*, 2010, **3**, 1039–1053.

16 M. Ehn, H. Junninen, S. Schobesberger, H. E. Manninen, A. Franchin, M. Sipilä, T. Petäjä, V.-M. Kerminen, H. Tammet, A. Mirme, S. Mirme, U. Hõrrak, M. Kulmala and D. R. Worsnop, *Aerosol Sci. Technol.*, 2011, **45**, 522–532.

17 M. Kulmala, J. Kontkanen, H. Junninen, K. Lehtipalo, H. E. Manninen, T. Nieminen, T. Petäjä, M. Sipilä, S. Schobesberger, P. Rantala, A. Franchin, T. Jokinen, E. Järvinen, M. Äijälä, J. Kangasluoma, J. Hakala, P. P. Aalto, P. Paasonen, J. Mikkilä, J. Vanhanen, J. Aalto, H. Hakola, U. Makkonen, T. Ruuskanen, R. L. Mauldin III, J. Duplissy, H. Vehkamäki, J. Bäck, A. Kortelainen, I. Riipinen, T. Kurtén, M. V. Johnston, J. N. Smith, M. Ehn, T. F. Mentel, K. E. J. Lehtinen, A. Laaksonen, V.-M. Kerminen and D. R. Worsnop, *Science*, 2013, **339**, 943–946.

18 T. Kurtén, T. Petäjä, J. Smith, I. K. Ortega, M. Sipilä, H. Junninen, M. Ehn, H. Vehkamäki, L. Mauldin, D. R. Worsnop and M. Kulmala, *Atmos. Chem. Phys.*, 2011, **11**, 3007–3019.

19 M. J. McGrath, T. Olenius, I. K. Ortega, V. Loukonen, P. Paasonen, T. Kurtén, M. Kulmala and H. Vehkamäki, *Atmos. Chem. Phys.*, 2012, **12**, 2345–2355.

20 J. Kirkby, J. Curtius, J. Almeida, E. Dunne, J. Duplissy, S. Ehrhart, A. Franchin, S. Gagné, L. Ickes, A. Kurten, A. Kupc, A. Metzger, F. Riccobono, L. Rondo, S. Schobesberger, G. Tsagkogeorgas, D. Wimmer, A. Amorim, F. Bianchi, M. Breitenlechner, A. David, J. Dommen, A. Downard, M. Ehn, R. C. Flagan, S. Haider, A. Hansel, D. Hauser, W. Jud, H. Junninen, F. Kreissl, A. Kvashin, A. Laaksonen, K. Lehtipalo, J. Lima, E. R. Lovejoy, V. Makhmutov, S. Mathot, J. Mikkilä, P. Minginette, S. Mogo, T. Nieminen, A. Onnela, P. Pereira, T. Petäjä, R. Schnitzhofer, J. H. Seinfeld, M. Sipilä, Y. Stozhkov, F. Stratmann, A. Tome, J. Vanhanen, Y. Viisanen, A. Vrtala, P. E. Wagner, H. Walther, E. Weingartner, H. Wex, P. M. Winkler, K. S. Carslaw, D. R. Worsnop, U. Baltensperger and M. Kulmala, *Nature*, 2011, **476**, 429–433.

21 T. Olenius, O. Kupiainen, I. K. Ortega, T. Kurtén and H. Vehkamäki, Free energy barrier in the growth of sulfuric acid–ammonia and sulfuric acid–dimethylamine clusters, *J. Chem. Phys.*, 2013, in press.

22 T. Su and M. T. Bowers, *J. Chem. Phys.*, 1973, **58**, 3027–3037.

23 O. Kupiainen, I. K. Ortega, T. Kurtén and H. Vehkamäki, *Atmos. Chem. Phys.*, 2012, **12**, 3591–3599.

24 H. Israël, *Atmospheric Electricity, vol. I.*, Israel Program for Sci. Transl. & NSF, Jerusalem, 1970.

25 I. K. Ortega, O. Kupiainen, T. Kurtén, T. Olenius, O. Wilkman, M. J. McGrath, V. Loukonen and H. Vehkamäki, *Atmos. Chem. Phys.*, 2012, **12**, 225–235.

26 J. Voigtländer, J. Duplissy, L. Rondo, A. Kürten and F. Stratmann, *Atmos. Chem. Phys.*, 2012, **12**, 2205–2214.

27 S. Mirme, A. Mirme, A. Minikin, A. Petzold, U. Hõrrak, V.-M. Kerminen and M. Kulmala, *Atmos. Chem. Phys.*, 2010, **10**, 437–451.

28 H. Berresheim, T. Elste, C. Plass-Dülmer, F. L. Eisele and D. J. Tanner, *Int. J. Mass Spectrom.*, 2000, **202**, 91–109.

29 T. Petäjä, R. L. Mauldin III, E. Kosciuch, J. McGrath, T. Nieminen, P. Paasonen, M. Boy, A. Adamov, T. Kotiaho and M. Kulmala, *Atmos. Chem. Phys.*, 2009, **9**, 7435–7448.

30 F. Bianchi, J. Dommen, S. Mathot and U. Baltensperger, *Atmos. Meas. Tech.*, 2012, **5**, 1719–1725.

31 M. Norman, A. Hansel and A. Wisthaler, *Int. J. Mass Spectrom.*, 2007, **265**, 382–387.

32 B. R. Bzdek, J. W. DePalma, D. P. Ridge, J. Laskin and M. V. Johnston, *J. Am. Chem. Soc.*, 2013, **135**, 3276–3285.

33 P. Gormley and M. Kennedy, *Proc. R. Irish Acad.*, 1949, **52A**, 163–169.

34 P. Paasonen, T. Olenius, O. Kupiainen, T. Kurtén, T. Petäjä, W. Birmili, A. Hamed, M. Hu, L. G. Huey, C. Plass-Duelmer, J. N. Smith, A. Wiedensohler, V. Loukonen, M. J. McGrath, I. K. Ortega, A. Laaksonen, H. Vehkamäki and M. Kulmala, *Atmos. Chem. Phys.*, 2012, **12**, 9113–9133.

Faraday Discussions

RSC Publishing

PAPER

How do organic vapors contribute to new-particle formation?

Neil M. Donahue,[*a] Ismael K. Ortega,[b] Wayne Chuang,[a] Ilona Riipinen,[abcd] Francesco Riccobono,[f] Siegfried Schobesberger,[b] Josef Dommen,[f] Urs Baltensperger,[f] Markku Kulmala,[b] Douglas R. Worsnop[be] and Hanna Vehkamaki[b]

Received 21st March 2013, Accepted 8th May 2013

DOI: 10.1039/c3fd00046j

Highly oxidised organic vapors can effectively stabilize sulphuric acid in heteronuclear clusters and drive new-particle formation. We present quantum chemical calculations of cluster stability, showing that multifunctional species can stabilize sulphuric acid and also present additional polar functional groups for subsequent cluster growth. We also model the multi-generation oxidation of vapors associated with secondary organic aerosol formation using a two-dimensional volatility basis set. The steady-state saturation ratios and absolute concentrations of extremely low volatility products are sufficient to drive new-particle formation with sulphuric acid at atmospherically relevant rates.

1 Background

New-particle formation is the dominant source of particle number in the atmosphere,[1-4] and a sufficient number of nucleated particles survive and grow large enough to make new-particle formation a globally important source of cloud condensation nuclei (CCN).[5,6] CCN levels and their changes over time in turn constitute a major source of uncertainty in climate forcing.

There is strong evidence that new-particle formation in the atmosphere is associated with sulphuric acid vapor,[4,7-9] suggesting that sulphuric acid (H_2SO_4, or SA) is almost always present in the smallest molecular clusters. An exception is costal regions with abundant iodine.[10] Observed new-particle formation rates cover the range $0.1 < J_{1.5} < 1000$ cm^{-3} s^{-1} over a sulphuric acid vapor

[a]Carnegie Mellon University Center for Atmospheric Particle Studies, 5000 Forbes Ave., Pittsburgh, PA, USA. E-mail: nmd@andrew.cmu.edu; Tel: +01 412-268-4415

[b]Division of Atmospheric Science, Department of Physics, University of Helsinki, FI-00014 Helsinki, Finland

[c]Department of Applied Environmental Science, University of Stockholm, Stockholm, Sweden

[d]Bert Bolin Centre for Climate Research, University of Stockholm, Stockholm, Sweden

[e]Aerodyne Research, Inc., Billerica, Massachusetts, USA

[f]Paul Scherrer Institute, Laboratory of Atmospheric Chemistry, CH-5232 Villigen, Switzerland

concentration $10^6 <$ [SA] $<10^8$ molec cm^{-3}, where $J_{1.5}$ is the appearance rate of particles at 1.5 nm mobility diameter.[11,12] The two are highly correlated, with $J_{1.5}$ vs. [H$_2$SO$_4$] showing a log–log slope between 1 and 2, but also showing variability of more than 1 order of magnitude at any given [SA]. Over this same range, the sulphuric acid binary collision rate (J_{SA}) ranges from $10^2 < J_{SA} < 10^6$ cm^{-3} s^{-1}. There are also strong indications that organic compounds may be involved in the nucleation[13,14] and growth[15] of the smallest clusters. Here we shall explore the role of organics, focusing on the potential for gas-phase oxidation to generate organic compounds that can enhance new-particle formation and growth.

A significant challenge regarding oxidised organic compounds is the huge number of different compounds present in the atmosphere.[16,17] Thus, if organics are involved in nucleation and/or growth of the smallest molecular clusters, it is important to identify the critical properties of those organics, and probably to understand how the diversity of compounds and properties influences the role of organics.

The diversity of organic properties is also important to the formation and attributes of secondary organic aerosol (SOA). One approach to confronting the diversity of ambient organic compounds is to group them based on their vapor pressure (volatility), using a formalism known as the Volatility Basis Set (VBS).[18] The VBS emerged from a paper in the Atmospheric Chemistry Faraday Discussion (130) that identified the continuous effect of oxidation chemistry on the organic volatility distribution as a potentially critical part of the SOA atmospheric life-cycle.[19] That work also identified multi-generation gas-phase oxidation of organics as a potentially important source of extremely low volatility compounds (ELVOC),[19] which has subsequently been verified in experimental studies designed to isolate later-generation aging effects.[20,21] Those ELVOC will play a prominent role in this work.

In the VBS, organics are grouped in decadally spaced "volatility bins" for both SOA production[22,23] and Primary Organic Aerosol (POA) emission,[24,25] and the chemical evolution can be modeled with simplified mechanisms.[26] However, the single dimension of volatility is ill suited to describing progressive oxidation, and so a second dimension describing oxidation has been discussed, using either the mean oxidation state of carbon $\overline{OS}_C = 2\,O:C\,-\,H:C$,[17,27] or the oxygen to carbon ratio (O:C).[28-30] The 2D-VBS ansatz is that pure-component volatility (C^o) and carbon oxidation state ($\overline{OS}_C$) alone can reasonably describe the properties of the highly complex mixture of organic molecules comprising organic aerosol, though there may be hundreds of thousands of different organic compounds involved. A corollary, however, is that the diversity of properties of the organics (for example the volatility distribution) is itself fundamentally important to organic-aerosol behaviour.

While the VBS was formulated to represent the bulk equilibrium thermodynamics of OA partitioning, it also can be used to represent the dynamics of organic condensation, including the effect of curvature on net condensation for very small particles.[31] It is not obvious that saturation vapor pressure alone should describe the interactions of oxidised organic compounds and sulphuric acid in the smallest clusters containing only a few molecules, where specific hydrogen bonding interactions are likely to dominate. However, here we shall consider the possibility that volatility and oxidation state together are also a reasonable indicator of the degree to which organics can contribute to new-particle formation.

2   Theory

The essential features of condensation to a bulk phase were presented in the units common to organic-aerosol dynamics by Donahue *et al.*[31] The net flux of a compound labeled i to unit surface area of a particle is:

$$F_i = v_\mathrm{D}\alpha_i[C_i^\mathrm{v} - a_iKC_i^\mathrm{o}] = v_\mathrm{D}C_i^\mathrm{o}\alpha_i[S_i - a_iK(D_\mathrm{p})] \tag{1}$$

where v_D is a characteristic condensation velocity, α_i is the mass accommodation coefficient of the molecule on the surface (typically assumed to be 1), C_i^v is the vapor-phase mass concentration, and C_i^o is the saturation concentration. S_i is the saturation ratio, $S_i = C_i^\mathrm{v}/C_i^\mathrm{o}$, and a_i is the activity of the species at the condensed-phase surface.

The velocity (for diameter change) in eqn (1) for a 200 g mole^{-1} compound with a density of 1.4 g cc^{-1} is: $v_\mathrm{D} = 0.226$ (nm hr^{-1})/(ng m^{-3}). We use units of nm hr^{-1} because growth rates during new-particle formation events are expressed in those units, typically ranging from 1–10 nm hr^{-1} depending on particle size and ambient conditions.[4,32] In addition, the condensation sink for vapors to particle surfaces ranges from 0.1–100 h^{-1} under a wide range of conditions in the atmosphere.[11,33,34] Eqn (1) is fundamentally about evaporation, and an important scaling can be found by assuming that a surface monolayer changes diameter by about 0.8 nm; this means that a species with a concentration of $C^\mathrm{o} \sim 4$ ng m^{-3} will cause condensation at a rate of approximately 1 monolayer per hour. If this species is in equilibrium, it also indicates a characteristic molecular evaporation lifetime of 1 h from a flat surface.

The term $K(D_\mathrm{p})$ describes the Kelvin effect, the increase in the equilibrium saturation concentration, for particles with diameter D_p. This scales with a "Kelvin Diameter", D_K. Assuming a surface tension $\sigma = 0.03$ N m^{-1}:

$$K = \exp(D_\mathrm{K}/D_\mathrm{p}); \ D_\mathrm{K} = (4\sigma M)/(RT\rho) \simeq 7 \text{ nm} \tag{2}$$

Thus, at the bulk limit, 3 nm diameter spheres (containing roughly 64 molecules with a molar weight of 200 g mole^{-1} and a density of 1.5 g cc^{-1}) will increase volatility by 1 order of magnitude and 1.5 nm spheres (with roughly 8 molecules) would increase volatility by 2 orders of magnitude, were bulk behavior a sensible descriptor for such a tiny object.

The Kelvin term is central to classical nucleation theory,[35–37] as any sphere in a single-component system with $S_i > K(D_\mathrm{crit})$ will grow without limit, and nucleation is governed by the equilibrium distribution of particles with $D_\mathrm{p} = D_\mathrm{crit}$ at that given S_i. Furthermore, for new-particle formation in the atmosphere, supersaturation is governed in almost all cases by a steady state between chemical production of condensible vapors and condensational loss of those vapors to larger background aerosol particles, which dominate the overall surface area.

While the surface tension $\sigma = 0.03$ N m^{-1} is typical of many organic compounds, independent of polarity,[38] its role here is more for illustration than calculation. We do not know the structure of condensing organic compounds, and we shall argue their interactions with sulphuric acid molecules are of paramount importance; thus D_K effectively subsumes uncertain properties into a single rough scaling factor. It is more important that accurate quantum chemical calculations are becoming feasible for molecular clusters similar in size to that expected for $K(D_\mathrm{p}) \simeq 100$.[39] It is thus now possible to probe interactions at the Kelvin scale computationally and to eliminate unphysical extrapolations of bulk properties.

Multi-component nucleation is more complicated, but at the classical limit largely reduces to calculation of the activity coefficient γ_i for small clusters with rapidly changing composition. However, the logical starting point is for a more or less ideal mixture, where $\gamma_i = 1$. This is an especially sensible when considering highly oxidised, low-volatility organic vapors that one might find in the atmosphere. There are thousands of different molecules[16,17] associated with organic aerosols, and so we must resort to broad average properties to render the problem tractable (and, conversely, if average properties are not valid and instead only a few select molecular types participate in new-particle formation, then the formation rate would be inconsequential).

For bulk organic aerosol conditions, organics with O:C > 0.33 and low volatility ($C^\mathrm{o} < 300$ ng m^{-3}) will on average have activity coefficients very near 1 in the organic mixture.[28] Assuming a very rich mixture of molecules, we can usefully define an *effective* saturation ratio

$$S_i^{\mathrm{eff}} = \frac{1}{C_i^\mathrm{o}} \sum_j^{C_j^\mathrm{o} \le C_i^\mathrm{o}} C_j^{\mathrm{vap}} \tag{3}$$

To the extent that these compounds do form an ideal mixture, an effective supersaturation over some critical threshold, broadly between 10 and 1000 (a few molecules), would ensure "quasi-homogeneous" nucleation of purely organic particles.

More scaling is in order for these units. A mass concentration of 3 ng m^{-3} is equivalent to a mass fraction of 3 pptm, and if these molecules have a molar weight of order 300 g mole^{-1}, that corresponds to a mole fraction of 3×10^{-14}, or a number density of roughly 10^7 molecules cm^{-3} at STP. In this discussion we will assume a collisional rate constant for relatively heavy molecules of order $k_\mathrm{col} \simeq 10^{-10}$ cm^3 molec^{-1} s^{-1} = 3.6×10^{-7} cm^3 molec^{-1} hr^{-1}. Thus molecules and small clusters will collide with condensible molecules that have a concentration of 10^7 molecules cm^{-3} with a frequency of 10^{-3} s^{-1}, or 3.6 h^{-1}.

If we consider that all compounds likely to be involved in nucleation are of sufficiently low volatility that they stick to background aerosols with nearly unit probability, then the steady-state concentration of these "nucleator" vapors will be given by the simple steady-state relation:

$$C_i^{\mathrm{ss}} = \frac{P_i}{\mathrm{CS}}; \quad S_i^{\mathrm{ss}} = \frac{P_i}{\mathrm{CS}\, C_i^\mathrm{o}} \tag{4}$$

If we consider an arbitrary threshold of 0.3 ng m^{-3} (10^6 molec cm^{-3}) for a saturation concentration and $S = 10$ for a critical saturation ratio for organics to participate in new-particle formation, along with a condensation sink (CS) of 10 h^{-1}, condensible organic vapors would need to have a production term $P_i = 30$ ng m^{-3} hr^{-1}. This is a small fraction of the secondary organic aerosol formation rate.

Organic vapors will certainly nucleate quasi-homogeneously when S^eff rises to some critical value. However, under atmospheric conditions that homogeneous organic nucleation rate may be negligible. Because of the strong indications that sulphuric acid is involved in atmospheric new-particle formation, it is likely that the activity coefficient for mixtures of sulphuric acid and the relevant organics is significantly less than 1, allowing nucleation at much lower values of S^{eff}. It is also almost certain that water molecules are present in the small clusters. However, ambient observations reveal no strong dependence of observed new-particle formation attributable to relative humidity,[40] nor does bulk SOA show a strong humidity dependence in the laboratory[41] or and the field (below 80% RH).[42] Consequently, because the treatment here is semi-empirical, we shall overlook the role of water vapor until there is empirical evidence of an important role governing the new-particle formation rate.

3   New-particle formation and sulphuric acid

Stipulating that sulphuric acid vapors are involved in the formation of nucleating clusters, the variability in observed new-particle formation rates at a given [SA] is thus likely driven by varying levels of stabilizing agents for sulphuric acid clusters. Those stabilizing agents may be bases (ammonia or amines)[3,43] or oxidised organic vapors,[14,44] among other things. Here we are focused on the potential role of hydrogen bonding oxidised organic vapors. If two sulphuric acid molecules are typically involved in very small clusters, we can define an operational new-particle formation rate:

$$J_{\text{npf}} = \tfrac{1}{2} f_{\text{npf}}\, k_{\text{col}}\, [\text{SA}]^2; \quad 0.00015 \leq f_{\text{npf}} \leq 0.015 \tag{5}$$

where f_{npf} describes the fraction of the sulphuric acid dimer formation rate that ends up with new-particle formation (the apparent survival probability of sulphuric acid dimers).[3] Consequently, in this discussion we shall explore two related hypotheses: organic vapors are (or can be) intimately involved in each step of new-particle formation and growth; or, inorganic sulphate salts form seeds for heterogeneous nucleation of organic vapors at very small sizes ($D_{\text{p}} \ll 3$ nm).

New-particle formation rates are typically specified for mobility diameters of 1.5 nm[4,14] or 1.7 nm,[2] based on a combination of the minimum size cutoff of state-of-the-art particle counters and inference about the "critical cluster" size for nucleation. The mobility diameter is larger than the physical diameter by approximately 0.3 nm due to the finite size of air molecules through which particles move during mobility measurement;[45] in this discussion we use physical size. Observed growth rates over this size range are of order 1–2 nm hr^{-1}.[4] Very small clusters thus must remain suspended in air for a long time – tens of minutes or more – before growing *via* condensation. Thus, the relevant timescales for evaporation and coagulational loss are of that order. Furthermore, clusters with short evaporation lifetimes (minutes or less) have no hope of growing.

4   Coagulational loss

Coagulation losses above the operational nucleation threshold but below the minimum size of microphysics modules in chemical transport models are typically treated *via* the parameterization of Kerminen and Kulmala.[46] However, coagulation loss below this threshold is thus part of the new-particle formation process leading up to the empirically observed rate at (mobility) diameters of 1.5 or 1.7 nm. Here we shall argue that coagulation is the dominant contributor to f_{npf} and thus that there is very little room for significant evaporation of small clusters.

For this discussion we shall consider an example case with fairly vigorous new-particle formation (for a boreal-forest environment), with $[\mathrm{H_2SO_4}] = 10^7$ molec cm^{-3}, CS = 10 h^{-1}, and different levels of condensible organic vapors. New particle formation rates under these conditions span the range $1 < J_{1.5} < 30$ cm^{-3} s^{-1} cm^{-3} s^{-1},[4,14] while $J_{\mathrm{SA}} = 10^4$ cm^{-3} s^{-1}, so $10^{-4} \lesssim f_{\mathrm{npf}} \lesssim 3 \times 10^{-3}$, the arrival frequency of sulphuric acid vapors is 10^{-3} s$^{-1} = 3.6$ h^{-1}, and the growth rate from sulphuric acid condensation alone is approximately 0.3 nm hr^{-1}. If additional vapors are condensing more rapidly than sulphuric acid (with a ratio $\Gamma \simeq 3$),[47] the timescale for small cluster growth (doubling of size) *via* condensational growth is of order 1/3 h, so the "loss" term for growth is of order 3 h^{-1}.

On its face, this would suggest that of order 0.1% of the sulphuric acid collisions result in new-particle formation. However, ambient measurements also indicate that the smallest clusters grow slowly, at roughly 1 nm hr^{-1} or less.[4] Consequently, one must account for the loss of these smallest clusters to larger particles (*i.e.* coagulation). As shown in Fig. 1, the loss rate of small clusters is appreciable. If the first nanometer of growth (to particles with a physical diameter of roughly 1.5 nm) takes as much as 2 h, and loss to the condensational sink averages 4 h^{-1}, then the survival probability of these clusters against loss to larger (accumulation mode) particles is $\exp(-8) = 3.3 \times 10^{-4}$. Thus, if growth of the smallest clusters is governed by sulphuric acid vapor it appears likely that most of the nucleating fraction f_{npf} may be determined by coagulation of the smallest (most diffusive) clusters. This is entirely consistent with the findings from Chen *et al.*[3] showing that sulphuric acid forms a semi-stable dimer and a stable heterodimer, and there is evidently no other evaporation loss.

It follows from the preceding argument that evaporation must be negligible for all or almost all of the new-particle formation process, compared with the condensation timescales. Otherwise the empirical nucleating fraction would be even lower than observed. Pure sulphuric acid nucleation is far to slow to account for observations,[2] almost certainly because evaporation of the smallest SA clusters is too rapid. Thus for oxidised organics to play a substantial role in new-particle

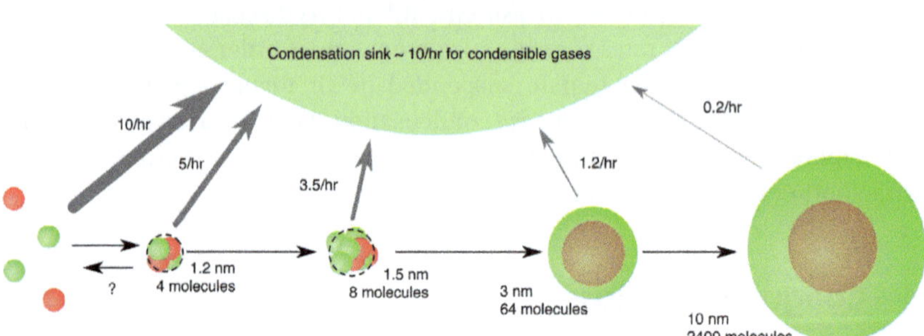

Fig. 1 Nucleation process for organics (green) and sulphuric acid (red). Diameters are physical diameters (mobility diameters are ~0.3 nm larger). Typical growth rates, especially for $D_p \lesssim 3$ nm, are of order 1–2 nm hr^{-1}. Coagulational loss is the dominant process below this size; thus, even with no evaporation of small clusters, only a small fraction reach the operational threshold for particle formation (1.2 nm $\lesssim D_p \lesssim 1.5$ nm).

formation, they must form very stable clusters with sulphuric acid, with evaporation lifetimes of many minutes under standard conditions.

5   Organic aging and new-particle formation

Here we shall consider oxidation of *cis*-pinonic acid and other first-generation pinene oxidation products by OH in the gas phase to produce second-generation gas-phase products with multiple polar functional groups capable of forming stable heterodimers with sulphuric acid but also capable of continued growth to larger, equally stable clusters. Previous studies have shown that "aging" of α-pinene secondary organic aerosol (SOA) can significantly enhance bulk SOA levels[21,48,49] and that gas-phase oxidation of specific first-generation tracers such as *cis*-pinonic acid by OH radicals can lead to rapid formation of highly oxidised species such as 3-methyl-1,2,3-butanetricarboxylic acid (MBTCA).[20,50]

In Fig. 2 we place these two species in a two-dimensional volatility oxidation space (2D-VBS) used to describe complex oxidation and phase partitioning of organic systems.[21,27,28,51] The 2D-VBS is shifted toward very low volatility species, with contours showing a range of carbon and oxygen numbers. Depending on the method used to estimate its vapor pressure, MBTCA falls in the range $0.1 \leq C^o \leq 3$ ng m^{-3}.

While MBTCA is a relatively minor product of *cis*-pinonic acid oxidation,[20,50] there is strong evidence that aging of SOA species in general produces highly oxidised species,[45,51] and the tendency of the "oxidised organic aerosol" (OOA) to decarboxylate after vaporization at 600 °C indicates many of the oxygenated functional groups are organic acids or even diacids.[52] It is not certain that OOA

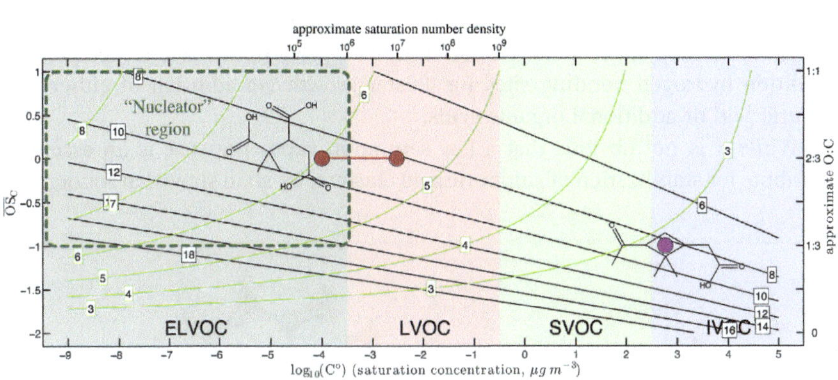

Fig. 2 Two dimensional volatility-oxidation space for organics. The *x*-axis is saturation concentration (in µg m^{-3}) with a secondary axis showing the approximate saturation number density. The *y*-axis is the average oxidation state of carbon atoms in a molecule ($\overline{OS}_C = 2\,O:C\ -H:C$), with a secondary axis showing approximate O : C for species with alternating =O and –OH functional groups. Contours show average carbon number (n_C, black) and average oxygen number (n_O, green), while colored regions indicate broad classes. Intermediate volatility organics (IVOC) are entirely in the gas phase under ambient conditions; Semivolatile organics (SVOC) are significantly in both phases, Low volatility organics (LVOC) have small gas-phase fractions, and Extremely low volatility organics (ELVOC) are only in the gas-phase because of steady-state disequilibration from gas-phase formation. One such process is gas-phase oxidation of *cis*-pinonic acid by OH radicals to produce the triacid MBTCA, as shown. This oxidised, ELVOC region may be a "nucleator region" where organics can assist sulphuric acid nucleation; alternatively, organics with $n_O \gtrsim 5$–6 may be needed, but these definitions are similar.

formation occurs wholly *via* gas-phase oxidation,[53] but observational[54,55] and modeling[30] studies suggest that gas-phase oxidation can be a major contributor.

This is potentially a recipe for generating supersaturation of organics that can participate in nucleation; low yields of very low volatility species can still lead to high supersaturations. While the fractional yields of ELVOC products may be low, oxidation may still produce substantial supersaturation. The "bins" in the 2D-VBS are separated by decades in volatility; if the mass yield into a less volatile bin is more than 10% of the yield into its more volatile neighbor, the resulting supersaturation in that bin will be higher. Thus product yields must decrease by more than one order of magnitude per decade for the resulting supersaturations to decrease. However, below some limit, products may be supersaturated, even homogeneously nucleating, but irrelevant. As we shall argue below, a practical lower limit to the organic vapor number density is of order 10^6 cm^{-3}, so species with $S_{\mathrm{eff}} \simeq 100$ would have $C^\circ \simeq 3 \times 10^{-6}$ µg m^{-3} – a very small value, but less than halfway through the "nucleator region" shown in Fig. 2.

5.1   Quantum chemistry modeling

Quantum chemistry calculations using the method combination B3LYP/CBSB7// RI-CC2/Aug-cc-pV(T + d)Z[39] show that organic compounds with a significant number of polar functional groups can form stable clusters with sulphuric acid, while less oxidised organics are substantially less effective in this capacity. For example, MBTCA and sulphuric acid can form a stable, 302 amu cluster, as shown in Fig. 3, with heterodimer evaporation rates four orders of magnitude lower than pure sulphuric acid dimer clusters (and 5 orders of magnitude lower than pure MBTCA dimer clusters). The key elements to forming stable clusters that can also grow are first and foremost the ability to form strong hydrogen bonds, as shown in Fig. 3, most notably *via* carboxylic acid moieties, but also multiple such moieties so that stable heterodimers are not a "dead end" but rather present addition hydrogen bonding sites for cluster growth *via* addition of either sulphuric acid or additional organic acids.

While it is not obvious that a low saturation vapor pressure is an essential attribute for stabilization of sulphuric acid clusters, we shall show that for organic

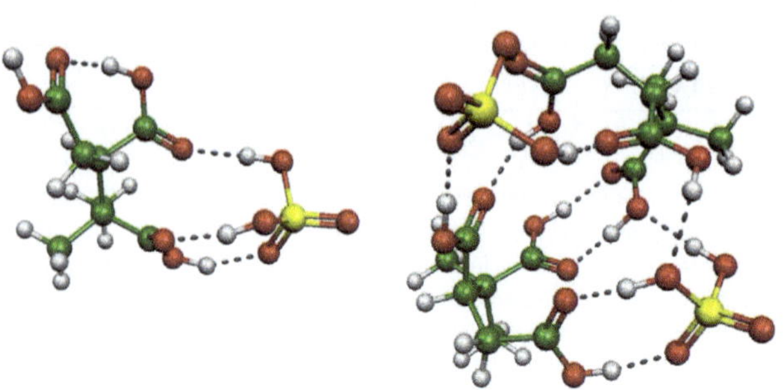

Fig. 3   Stable clusters of MBTCA and sulphuric acid. Both 1 : 1 and 2 : 2 MBTCA sulphuric acid heteromer clusters are substantially more stable than pure clusters of either constituent. Strong hydrogen bonding networks stabilize the clusters, with evaporation rates up to 4 orders of magnitude lower than sulphuric acid dimer evaporation rates, and the large number of hydrogen bonding moieties on the molecules enables continued growth *via* condensation of similar species.

vapors, stabilizers that facilitate both clustering and subsequent growth for the most part do have a low saturation vapor pressure as well. Though ammonia is quite effective, this requires a dissociation reaction to make ammonium bisulphate clusters, which only occurs in clusters with a few sulphuric acid and ammonia molecules.[2,43] Stable clusters involving organic acids and sulphuric acid do not require a chemical reaction; they rely on hydrogen bonds only. Consequently, if three acid moieties are required to form a stable cluster that can also grow, the resulting molecules can hardly avoid having a very low saturation vapor pressure.

It also appears that low volatility organic compounds produced *via* atmospheric oxidation reactions do have the appropriate attributes for cluster formation and growth. There is strong evidence that oxidised organic aerosol (OOA) comprises a high degree of carboxylic acid functionality, so being in the upper left region of the 2D-VBS may be a necessary and sufficient condition for a vapor to participate in new-particle formation. Thus, while we do not expect a perfect correlation between low volatility and the ability to stabilize sulphuric acid clusters, the region shown as "nucleator region" in Fig. 2 is roughly the region where molecules should have more than 5–6 oxygen atoms, and many should thus form strong hydrogen bonds with multiple bonding partners.

The boundaries of this region are also quite uncertain. The low O:C boundary is motivated by the evident role of polar moieties, but because gas-phase production of less oxidised ELVOC species is unlikely to be significant, that boundary is functionally irrelevant. The boundary at the ELVOC-LVOC demarkation is also vague, and in practice will be treated as adjustable in parameterizations; however, as we shall show below calculations suggest that reasonable conditions can lead to effective supersaturations near 100 at the boundary shown here.

5.2   2D-VBS modeling

A final question to address is whether reasonable gas-phase oxidation chemistry can produce a sufficient quantity of gas-phase vapors in steady state with appropriate properties to participate in atmospheric new-particle formation. Consequently, we implemented an oxidation scheme for *cis*-pinonic acid vapors in the 2D-VBS model described in Donahue *et al.*[21] Here, we assume that the first-generation oxidation products of *cis*-pinonic acid all retain 10 carbon atoms (they all fall of the "functionalization" branch of the oxidation chemistry) but later generation products behave as described in previous publications, with a probability of product fragmentation rising with increasing O:C.[21,51] We consider oxidation of 3–20 ppb of *cis*-pinonic acid for $10^5 < [OH] < 10^7$ cm^{-3}. The condensation sink at the Hyytiälä boreal-forest research station ranges between 2 h^{-1} (10% cumulative probability) and 11 h^{-1} (90%), with a median value of 6.5 h^{-1}.[11,56] Here we assume a condensation sink of 10 h^{-1}, which is in the range of the Hyytiälä data and also close to the condensation sink of the CLOUD experiment at CERN.[2]

In Fig. 4 we show the resulting modeled effective organic saturation ratios given in eqn (3) for different OH levels and *cis*-pinonic acid = 20 ppbv. For this simulation, organics in the "nucleator region" have a supersaturation above 100 for all [OH], with supersaturation increasing with increasing OH as expected. This means the total gas-phase number density of species in this region is $>10^8$ cm^{-3}. Thus, under conditions similar to those encountered in remote environments,

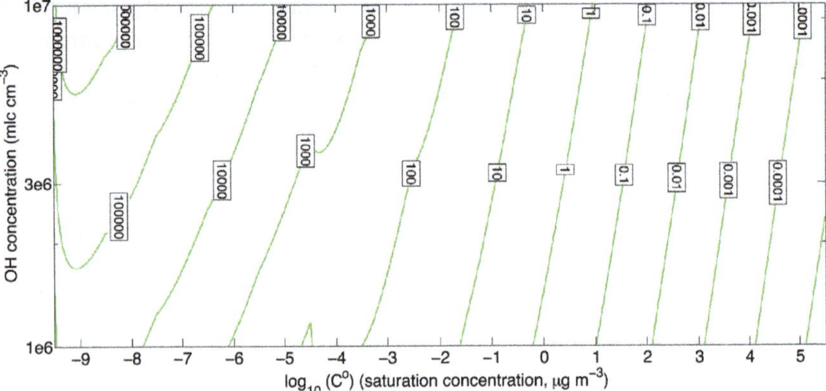

Fig. 4 Modeled effective saturation ratios for products of *cis*-pinonic acid oxidation as a function of C^o and [OH] under typical conditions. The ELVOC ("nucleator") limit is indicated with a dashed vertical line; products in this region have cumulative saturation ratios over 100 for all conditions (the total number density is $>10^8$ molec cm^{-3}).

gas-phase oxidation of first-generation SOA vapors can produce a sufficient steady-state burden of oxidised condensible products to contribute to new-particle formation.

It is likely important that the oxidised organic vapor concentrations exceed the sulphuric acid vapor concentrations (for the example case under consideration here, where [SA] $= 10^7$ cm^{-3}). Quantum calculations show that small clusters with organic : SA > 1 are much less stable than those with 1 : 1 stoichiometry. However, as there is no chemical reaction taking place but rather formation of an H-bond network, there is no reason to expect a strict stoichiometric equivalence. Thus it seems likely that nucleation involving sulphuric acid and organics under conditions with excess organic vapors will occur with clusters containing more organic species than sulphuric acid, within some range defined by the stability of the small clusters.

The model results shown in Fig. 4 show $S_{eff} > 100$ for condensible organic vapors (oxidised ELVOC). This also means that, regardless of the composition of the nucleating core, once the cluster reaches approximately 1.5 nm diameter, the core will be an effective heterogeneous nucleus for continued condensation of organic vapors. With increasing particle size, progressively more volatile organic vapors will be able to condense, as described in Donahue *et al.*[31] As the particles grow, progressively smaller values of S_{eff} will be required to support continued growth, and even more volatile constituents will rapidly reach an equilibrium activity as Raoult law mixing offsets the Kelvin term. Thus, at a relatively small particle size, dependent on conditions, subsequent particle growth is likely to decouple from the sulphuric acid condensation rate; if condensible organic vapor concentrations are larger than sulphuric acid vapor concentrations, the particle composition will thus become progressively more dominated by oxidised organic compounds.

6   Conclusions

Quantum chemical calculations show that highly oxidised polycarboxylic acids can make strong hydrogen bonds with sulphuric acid and form stable molecular clusters, leading to new-particle formation in the atmosphere. In addition, a

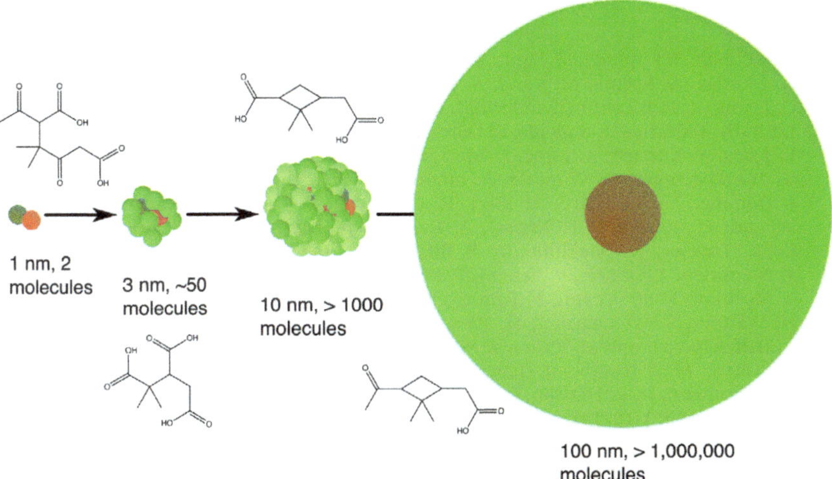

Fig. 5 Overall process of new-particle formation and growth involving association of oxidised organics and sulphuric acid. Initial stable cluster formation requires rare but highly oxidised, extremely low volatility organics. As clusters grow, progressively more common but also more volatile species can contribute to growth, until at some size near 10 nm the particle organic composition begins to resemble bulk organic-aerosol composition. In this illustration, a highly oxidised C_{10} keto-diacid and/or MBTCA (C_8 triacid) may stabilize the smallest clusters (dark green spheres) while more volatile species like pinic acid (C_9 diacid) may contribute to growth below 10 nm. Volatile species such as *cis*-pinonic acid (C_{10} keto acid) partition only to larger particles, if at all.

2D-VBS model of SOA "aging" – simplified as gas-phase oxidation of *cis*-pinonic acid by OH radicals – generates a sufficient supersaturation of condensible oxidised organic vapors to support nucleation and growth in this mixed organic–sulphuric acid system.

Schematically, the wide range of properties exhibited by oxidised organic vapors is a central element governing the behavior of growing particles, as shown in Fig. 5. Rare but strongly binding vapors can stabilize clusters with sulphuric acid down to the 1 : 1 heterodimer, while progressively more volatile but also more common organics can help drive subsequent growth. While large vapor concentrations of these organics are not required, roughly pptv levels are. Thus, it is unlikely that certain special compounds drive the process but rather that, collectively, a large body of oxidised ELVOC compounds are formed *via* gas-phase oxidation chemistry at a sufficient rate to drive the process.

Acknowledgements

This work was supported by funding from the U.S. National Science Foundation (grants AGS1136479 and CHE1012293), the Academy of Finland (CoE Project No. 1118615, LASTU Project No. 135054), ERC Project Nos. 257360- MOCAPAF and 27463-ATMNUCLE, the Swiss National Science Foundation (project nos. 200020_135307 and 206620_130527), and the Swedish Research Council, Vetenskapsrådet (grant 2011-5120). The authors are indebted to the CLOUD consortium for countless stimulating discussions and to the CSC Centre for Scientific Computing in Espoo, Finland for computer time.

References

1 M. Kulmala, I. Riipinen, M. Sipila, H. E. Manninen, T. Petaja, H. Junninen, M. Dal Maso, G. Mordas, A. Mirme, M. Vana, A. Hirsikko, L. Laakso, R. M. Harrison, I. Hanson, C. Leung, K. E. J. Lehtinen and W.-M. Kerminen, *Science*, 2007, **318**, 89–92.

2 J. Kirkby, J. Curtius, J. Almeida, E. Dunne, J. Duplissy, S. Ehrhart, A. Franchin, S. Gagne, L. Ickes, A. Kuerten, A. Kupc, A. Metzger, F. Riccobono, L. Rondo, S. Schobesberger, G. Tsagkogeorgas, D. Wimmer, A. Amorim, F. Bianchi, M. Breitenlechner, A. David, J. Dommen, A. Downard, M. Ehn, R. C. Flagan, S. Haider, A. Hansel, D. Hauser, W. Jud, H. Junninen, F. Kreissl, A. Kvashin, A. Laaksonen, K. Lehtipalo, J. Lima, E. R. Lovejoy, V. Makhmutov, S. Mathot, J. Mikkila, P. Minginette, S. Mogo, T. Nieminen, A. Onnela, P. Pereira, T. Petaja, R. Schnitzhofer, J. H. Seinfeld, M. Sipila, Y. Stozhkov, F. Stratmann, A. Tome, J. Vanhanen, Y. Viisanen, A. Vrtala, P. E. Wagner, H. Walther, E. Weingartner, H. Wex, P. M. Winkler, K. S. Carslaw, D. R. Worsnop, U. Baltensperger and M. Kulmala, *Nature*, 2011, **476**, 429–U77.

3 M. Chen, M. Titcombe, J. Jiang, C. Jen, C. Kuang, M. L. Fischer, F. L. Eisele, J. I. Siepmann, D. R. Hanson, J. Zhao and P. H. McMurry, *Proc. Natl. Acad. Sci. U. S. A.*, 2012, **109**, 18713–18718.

4 M. Kulmala, J. Kontkanen, H. Junninen, K. Lehtipalo, H. E. Manninen, T. Nieminen, T. Petäjä, M. Sipilä, S. Schobesberger, P. Rantala, A. Franchin, T. Jokinen, E. Järvinen, M. Äijälä, J. Kangasluoma, J. Hakala, P. P. Aalto, P. Paasonen, J. Mikkilä, J. Vanhanen, J. Aalto, H. Hakola, U. Makkonen, T. Ruuskanen, R. L. Mauldin, J. Duplissy, H. Vehkamäki, J. Bäck, A. Kortelainen, I. Riipinen, T. Kurtén, M. V. Johnston, J. N. Smith, M. Ehn, T. F. Mentel, K. E. J. Lehtinen, A. Laaksonen, V.-M. Kerminen and D. R. Worsnop, *Science*, 2013, **339**, 943–946.

5 J. R. Pierce and P. J. Adams, *Atmos. Chem. Phys.*, 2007, **7**, 1367–1379.

6 D. V. Spracklen, K. S. Carslaw, M. Kulmala, V.-M. Kerminen, S.-L. Sihto, I. Riipinen, J. Merikanto, G. W. Mann, M. P. Chipperfield, A. Wiedensohler, W. Birmili and H. Lihavainen, *Geophys. Res. Lett.*, 2008, **35**, L06808.

7 S. L. Sihto, M. Kulmala, V. M. Kerminen, M. Dal Maso, T. Petaja, I. Riipinen, H. Korhonen, F. Arnold, R. Janson, M. Boy, A. Laaksonen and K. E. J. Lehtinen, *Atmos. Chem. Phys.*, 2006, **6**, 4079–4091.

8 C. Kuang, P. McMurry, A. McCormick and F. Eisele, *J. Geophys. Res.*, 2008, **113**, D10209.

9 M. Sipilä, T. Berndt, T. Petäjä, D. Brus, J. Vanhanen, F. Stratmann, J. Patokoski, R. L. Mauldin, A.-P. Hyvärinen, H. Lihavainen and M. Kulmala, *Science*, 2010, **327**, 1243–1246.

10 G. McFiggans, C. S. E. Bale, S. M. Ball, J. M. Beames, W. J. Bloss, L. J. Carpenter, J. Dorsey, R. Dunk, M. J. Flynn, K. L. Furneaux, M. W. Gallagher, D. E. Heard, A. M. Hollingsworth, K. Hornsby, T. Ingham, C. E. Jones, R. L. Jones, L. J. Kramer, J. M. Langridge, C. Leblanc, J.-P. LeCrane, J. D. Lee, R. J. Leigh, I. Longley, A. S. Mahajan, P. S. Monks, H. Oetjen, A. J. Orr-Ewing, J. M. C. Plane, P. Potin, A. J. L. Shillings, F. Thomas, R. von Glasow, R. Wada, L. K. Whalley and J. D. Whitehead, *Atmos. Chem. Phys.*, 2010, **10**, 2975–2999.

11 M. Dal Maso, M. Kulmala, I. Riipinen, R. Wagner, T. Hussein, P. Aalto and K. Lehtinen, *Boreal Environment Research*, 2005, **10**, 323–336.

12 I. Riipinen, S.-L. Sihto, M. Kulmala, F. Arnold, M. Dal Maso, W. Birmili, K. Saarnio, K. Teinila, V. M. Kerminen, A. Laaksonen and K. E. J. Lehtinen, *Atmos. Chem. Phys.*, 2007, **7**, 1899–1914.

13 B. Bonn, M. Kulmala, I. Riipinen, S. L. Sihto and T. M. Ruuskanen, *J. Geophys. Res.*, 2008, **113**, D12209.

14 A. Metzger, B. Verheggen, J. Dommen, J. Duplissy, A. S. H. Prevot, E. Weingartner, I. Riipinen, M. Kulmala, D. V. Spracklen, K. S. Carslaw and U. Baltensperger, *Proc. Natl. Acad. Sci. U. S. A.*, 2010, **107**, 6646–6651.

15 I. Riipinen, T. Yli-Juuti, J. R. Pierce, T. Petäjä, D. R. Worsnop, M. Kulmala and N. M. Donahue, *Nat. Geosci.*, 2012, **5**, 453–458.

16 A. H. Goldstein and I. E. Galbally, *Environ. Sci. Technol.*, 2007, **41**, 1515–1521.

17 J. H. Kroll, N. M. Donahue, J. L. Jimenez, S. Kessler, M. R. Canagaratna, K. Wilson, K. E. Alteri, L. R. Mazzoleni, A. S. Wozniak, H. Bluhm, E. R. Mysak, J. D. Smith, C. E. Kolb and D. R. Worsnop, *Nat. Chem.*, 2011, **3**, 133–139.

18 N. M. Donahue, A. L. Robinson, C. O. Stanier and S. N. Pandis, *Environ. Sci. Technol.*, 2006, **40**, 2635–2643.

19 N. M. Donahue, K. E. Huff Hartz, B. Chuong, A. A. Presto, C. O. Stanier, T. Rosenørn, A. L. Robinson and S. N. Pandis, *Faraday Discuss.*, 2005, **130**, 295–309.

20 L. Müller, M.-C. Reinnig, K. H. Naumann, H. Saathoff, T. F. Mentel, N. M. Donahue and T. Hoffmann, *Atmos. Chem. Phys.*, 2012, **12**, 1483–1496.

21 N. M. Donahue, K. M. Henry, T. F. Mentel, A. K. Scharr, C. Spindler, B. Bohn, T. Brauers, H. P. Dorn, H. Fuchs, R. Tillmann, A. Wahner, H. Saathoff, K. H. Naumann, O. Möhler, T. Leisner, L. Müller, M.-C. Reinnig, T. Hoffmann, K. Salow, M. Hallquist, M. Frosch, M. Bilde, T. Tritscher, P. Barmet, A. P. Praplan, P. F. DeCarlo, J. Dommen, A. S. H. Prévôt and U. Baltensperger, *Proc. Natl. Acad. Sci. U. S. A.*, 2012, **109**, 13503–13508.
22 A. A. Presto and N. M. Donahue, *Environ. Sci. Technol.*, 2006, **40**, 3536–3543.
23 T. E. Lane, N. M. Donahue and S. N. Pandis, *Atmos. Environ.*, 2008, **42**, 7439–7451.
24 M. K. Shrivastava, T. E. Lane, N. M. Donahue, S. N. Pandis and A. L. Robinson, *J. Geophys. Res.*, 2008, **113**, D18301.
25 A. P. Grieshop, M. A. Miracolo, N. M. Donahue and A. L. Robinson, *Environ. Sci. Technol.*, 2009, **43**, 4750–4756.
26 A. L. Robinson, N. M. Donahue, M. K. Shrivastava, A. M. Sage, E. A. Weitkamp, A. P. Grieshop, T. E. Lane, J. R. Pierce and S. N. Pandis, *Science*, 2007, **315**, 1259–1263.
27 N. M. Donahue, J. H. Kroll, A. L. Robinson and S. N. Pandis, *Atmos. Chem. Phys.*, 2012, **12**, 615–634.
28 N. M. Donahue, S. A. Epstein, S. N. Pandis and A. L. Robinson, *Atmos. Chem. Phys.*, 2011, **11**, 3303–3318.
29 B. N. Murphy, N. M. Donahue and S. N. Pandis, *Atmos. Chem. Phys.*, 2011, **11**, 7859–7873.
30 B. N. Murphy, N. M. Donahue, C. Fountoukis, M. Dall'Osto, C. O'Dowd, A. Kiendler-Scharr and S. N. Pandis, *Atmos. Chem. Phys.*, 2012, **12**, 10797–10816.
31 N. M. Donahue, E. R. Trump, I. Riipinen and J. R. Pierce, *Geophys. Res. Lett.*, 2011, **38**, L16801.
32 I. Riipinen, J. R. Pierce, T. Yli-Juuti, T. Nieminen, S. Häkkinen, M. Ehn, H. Junninen, K. Lehtipalo, T. Petäjä, J. Slowik, R. Chang, N. C. Shantz, J. P. D. Abbatt, W. R. Leaitch, V.-M. Kerminen, D. R. Worsnop, S. N. Pandis, N. M. Donahue and M. Kulmala, *Atmos. Chem. Phys.*, 2011, **11**, 3865–3878.
33 M. Kulmala, A. Toivonen and J. Mäkelä, *Tellus, Ser. B*, 1988, **50**, 449–462.
34 M. Kulmala, L. Laakso, K. E. J. Lehtinen, I. Riipinen, M. Dal Maso, T. Anttila, V.-M. Kerminen, U. Hõrrak, M. Vana and H. Tammet, *Atmos. Chem. Phys.*, 2004, **4**, 2553–2560.
35 H. Reiss, *J. Chem. Phys.*, 1950, **18**, 840–849.
36 H. Vehkamaki and I. Riipinen, *Chem. Soc. Rev.*, 2012, **41**, 5160–5173.
37 H. Vehkamaki, M. J. McGrath, T. Kurten, J. Julin, K. E. J. Lehtinen and M. Kulmala, *J. Chem. Phys.*, 2012, **136**, 094107.
38 G. Korosi and E. Kovats, *J. Chem. Eng. Data*, 1981, **26**, 323–332.
39 I. K. Ortega, O. Kupiainen, T. Kurtén, T. Olenius, O. Wilkman, M. J. McGrath, V. Loukonen and H. Vehkamäki, *Atmos. Chem. Phys.*, 2012, **12**, 225–235.
40 A. Hamed, H. Korhonen, S.-L. Sihto, J. Joutsensaari, H. Järvinen, T. Petäjä, F. Arnold, T. Nieminen, M. Kulmala, J. N. Smith, K. E. J. Lehtinen and A. Laaksonen, *J. Geophys. Res.*, 2011, **116**, D03202.
41 N. Prisle, G. J. Engelhart, M. Bilde and N. M. Donahue, *Geophys. Res. Lett.*, 2010, **37**, L01802.
42 C. J. Hennigan, M. H. Bergin, J. E. Dibb and R. J. Weber, *Geophys. Res. Lett.*, 2008, **35**, L18801.
43 T. Kurtén, V. Loukonen, H. Vehkamäki and M. Kulmala, *Atmos. Chem. Phys.*, 2008, **8**, 4095–4103.
44 R. Zhang, L. Wang, A. F. Khalizov, J. Zhao, J. Zheng, R. L. McGraw and L. T. Molina, *Proc. Natl. Acad. Sci. U. S. A.*, 2009, **106**, 17650–17654.
45 M. Ehn, E. Kleist, H. Junninen, T. Petäjä, G. Lönn, S. Schobesberger, M. Dal Maso, A. Trimborn, M. Kulmala, D. R. Worsnop, A. Wahner, J. Wildt and T. F. Mentel, *Atmos. Chem. Phys.*, 2012, **12**, 5113–5127.
46 V.-M. Kerminen and M. Kulmala, *J. Aerosol Sci.*, 2002, **33**, 609–622.
47 C. Kuang, M. Chen, J. Zhao, J. Smith, P. H. McMurry and J. Wang, *Atmos. Chem. Phys.*, 2012, **12**, 3573–3589.
48 K. M. Henry and N. M. Donahue, *J. Phys. Chem. A*, 2012, **116**, 5932–5940.
49 K. Salo, M. Hallquist, Å. M. Jonsson, H. Saathoff, K.-H. Naumann, C. Spindler, R. Tillmann, H. Fuchs, B. Bohn, F. Rubach, T. F. Mentel, L. Müller, M. Renning, T. Hoffmann and N. M. Donahue, *Atmos. Chem. Phys.*, 2011, **11**, 11055–11067.
50 R. Szmigielski, J. D. Surratt, Y. Gomez-Gonzalez, P. Van der Veken, I. Kourtchev, R. Vermeylen, F. Blockhuys, M. Jaoui, T. E. Kleindienst, M. Lewandowski, J. H. Offenberg, E. O. Edney, J. H. Seinfeld, W. Maenhaut and M. Claeys, *Geophys. Res. Lett.*, 2007, **34**, L24811.
51 J. L. Jimenez, M. R. Canagaratna, N. M. Donahue, A. S. H. Prévôt, Q. Zhang, J. H. Kroll, P. F. DeCarlo, J. Allan, H. Coe, N. L. Ng, A. C. Aiken, K. D. Docherty, I. M. Ulbrich, A. P. Grieshop, A. L. Robinson, J. Duplissy, J. D. Smith, K. R. Wilson, V. A. Lanz,

C. Hueglin, Y. L. Sun, A. Laaksonen, T. Raatikainen, J. Rautiainen, P. Vaattovaara, M. Ehn, M. Kulmala, J. M. Tomlinson, D. R. Collins, M. J. Cubison, E. J. Dunlea, J. A. Huffman, T. B. Onasch, M. R. Alfarra, P. I. Williams, K. Bower, Y. Kondo, J. Schneider, F. Drewnick, S. Borrmann, S. Weimer, K. Demerjian, D. Salcedo, L. Cottrell, R. Griffin, A. Takami, T. Miyoshi, S. Hatakeyama, A. Shimono, J. Y. Sun, Y. M. Zhang, K. Dzepina, J. R. Kimmel, D. Sueper, J. T. Jayne, S. C. Herndon, A. M. Trimborn, L. R. Williams, E. C. Wood, C. E. Kolb, U. Baltensperger and D. R. Worsnop, *Science*, 2009, **326**, 1525–1529.

52 A. C. Aiken, P. F. Decarlo, J. H. Kroll, D. R. Worsnop, J. A. Huffman, K. S. Docherty, I. M. Ulbrich, C. Mohr, J. R. Kimmel, D. Sueper, Y. Sun, Q. Zhang, A. Trimborn, M. Northway, P. J. Ziemann, M. R. Canagaratna, T. B. Onasch, M. R. Alfarra, A. S. H. Prevot, J. Dommen, J. Duplissy, A. Metzger, U. Baltensperger and J. L. Jimenez, *Environ. Sci. Technol.*, 2008, **42**, 4478–4485.

53 C. D. Cappa and J. L. Jimenez, *Atmos. Chem. Phys.*, 2010, **10**, 5409–5424.

54 L. Hildebrandt, G. J. Englehardt, C. Mohr, E. Kostenidou, V. A. Lanz, A. Bougiatioti, P. F. DeCarlo, A. S. H. Prévôt, U. Baltensperger, N. Mihalopoulos, N. M. Donahue and S. N. Pandis, *Atmos. Chem. Phys.*, 2010, **10**, 4167–4186.

55 L. Hildebrandt, E. Kostenidou, N. Mihalopoulos, D. R. Worsnop, N. M. Donahue and S. N. Pandis, *Geophys. Res. Lett.*, 2010, **37**, L23801.

56 D. M. Westervelt, J. R. Pierce, I. Riipinen, W. Trivitayanurak, A. Hamed, M. Kulmala, A. Laaksonen, S. Decesari and P. J. Adams, *Atmos. Chem. Phys. Discuss.*, 2013, **13**, 8333–8386.

Faraday Discussions

RSC Publishing

PAPER

Modeling the influence of alkane molecular structure on secondary organic aerosol formation

Bernard Aumont,[*a] Marie Camredon,[a] Camille Mouchel-Vallon,[a] Stéphanie La,[a] Farida Ouzebidour,[a] Richard Valorso,[a] Julia Lee-Taylor[b] and Sasha Madronich[b]

Received 6th March 2013, Accepted 21st May 2013

DOI: 10.1039/c3fd00029j

Secondary Organic Aerosols (SOA) production and ageing is a multigenerational oxidation process involving the formation of successive organic compounds with higher oxidation degree and lower vapor pressure. Intermediate Volatility Organic Compounds (IVOC) emitted to the atmosphere are expected to be a substantial source of SOA. These emitted IVOC constitute a complex mixture including linear, branched and cyclic alkanes. The explicit gas-phase oxidation mechanisms are here generated for various linear and branched C_{10}–C_{22} alkanes using the GECKO-A (Generator for Explicit Chemistry and Kinetics of Organics in the Atmosphere) and SOA formation is investigated for various homologous series. Simulation results show that both the size and the branching of the carbon skeleton are dominant factors driving the SOA yield. However, branching appears to be of secondary importance for the particle oxidation state and composition. The effect of alkane molecular structure on SOA yields appears to be consistent with recent laboratory observations. The simulated SOA composition shows, however, an unexpected major contribution from multifunctional organic nitrates. Most SOA contributors simulated for the oxidation of the various homologous series are far too reduced to be categorized as highly oxygenated organic aerosols (OOA). On a carbon basis, the OOA yields never exceeded 10% regardless of carbon chain length, molecular structure or ageing time. This version of the model appears clearly unable to explain a large production of OOA from alkane precursors.

1   Introduction

Long carbon chain hydrocarbons ($C_{>10}$) are emitted in the atmosphere from biomass and as unburnt byproducts of biomass and fossil fuel combustion. These emissions form a complex mixture, including a large fraction of linear, cyclic and

---

[a]LISA, UMR CNRS 7583, Université Paris Est Créteil et Université Paris Diderot, France. E-mail: bernard. aumont@lisa.u-pec.fr

[b]National Center for Atmospheric Research, Boulder, CO, USA

branched alkanes.[1–4] These organic compounds of intermediate volatility (IVOC) are emitted largely in the condensed phase and then volatized by dilution[5] in the atmosphere. Gas phase oxidation of IVOC leads to the production of organic species of low enough volatility to condense and contribute to secondary organic aerosols (SOA) production.[6] Laboratory studies have shown that IVOC alkanes are potential SOA contributors.[7–12] Modeling studies have also identified IVOC as substantial SOA precursors at the continental scale and in the plumes of megacities.[13–17]

SOA production in the Mexico City plume was explored using a detailed chemical scheme: the Generator of Explicit Chemistry and Kinetics of Organics in the Atmosphere – GECKO-A.[17] In these simulations, n-alkanes were used as surrogate species of IVOC emissions. The model was found to be able to explain typical SOA levels but the prediction of the atomic O/C ratio fell short.[17,18] SOA formation from the gas phase oxidation of intermediate volatility n-alkanes was recently examined using the GECKO-A tool.[19] For the n-alkane series, most SOA contributors were found to be reduced enough to be categorized as hydrocarbon-like organic aerosols (HOA), although of secondary origin.[19] Simulated results with GECKO-A suggest that gas phase n-alkane oxidation cannot explain the large fraction of highly oxygenated organic aerosols (OOA) observed in the atmosphere. Branched alkanes are a major fraction of the emitted hydrocarbons in engine exhausts.[4] These branched alkanes are more prone to fragment in the early stage of the oxidation than their corresponding linear analogues.[7,9,11] This enhanced fragmentation is expected to alter both the SOA yields and the mean SOA oxidation state.

The aim of this study is to examine the influence of the parent alkane molecular structure on SOA yield and oxidation states. The study focuses on the effect of the degree of branching of the carbon backbone for various C_{10}–C_{22} alkane series. The mechanism self-generator GECKO-A[20] is used to generate highly detailed oxidation schemes, and box model simulations are performed to explore the formation and ageing of the particle phase subjected to continuous gas phase oxidation of the organic matter. Section 2 briefly presents the GECKO-A modeling and results are discussed in Section 3.

2   Model description

2.1   The GECKO-A modeling tool

The number of species involved in the multigenerational gas phase oxidation of hydrocarbons grows exponentially with the size of the carbon skeleton.[20,21] For a C_8 hydrocarbon, the explicit description of the gas phase oxidation processes up to the final production of CO_2 involves more than one million species, far exceeding the size of chemical mechanisms that can be managed manually. The chemical mechanism generator GECKO-A is a computer program designed to overcome this difficulty. This tool is used here to develop an oxidation scheme for alkane series with various degrees of branching.

Reaction pathways and rate constants are assigned during the chemical mechanism generation on the basis of experimental data and Structure Activity Relationships (SAR). The protocol implemented in GECKO-A to self generate the mechanisms is described by Aumont *et al.*,[20] with chemistry updates described by Aumont *et al.*[19] The relative contribution of fragmentation *vs.* functionalization oxidation routes is a key ratio in the context of SOA formation.[22] Evolution along the fragmentation or functionalization routes is mostly dictated by fate of the alkoxy radicals produced as intermediates during the gas phase oxidation.[7] In the previous version of GECKO-A,[19] the alkoxy radical chemistry was based on estimations provided by the SAR developed by Atkinson.[23] Vereecken and Peeters[24] have recently reported a SAR based on quantum chemical calculations to estimate the barrier heights for alkoxy radical decomposition for a large set of organic moieties. In particular, the SAR includes rates for leaving groups, such as the nitrate moieties, which are prevalent in the oxidation mechanisms of long-chain alkanes. The recent SAR provided by Vereecken and Peeters[24] for alkoxy radical decomposition has therefore been implemented in GECKO-A.

Phase partitioning is described in the model assuming that the condensed phase behaves as an ideal well mixed homogeneous liquid phase. Gas/particle phase equilibrium is described by Raoult's law:[25]

$$P_i = x_i P_i^{\mathrm{vap}} \tag{1}$$

where P_i^{vap} is the vapor pressure of the species i, x_i its mole fraction in the condensed phase and P_i its equilibrium partial pressure. At phase equilibrium, the ratio ξ_i of the species in the condensed phase is given by:[26]

$$\xi_i^{\mathrm{aer}} = \frac{N_{i,\mathrm{aer}}}{N_{i,\mathrm{aer}} + N_{i,\mathrm{gas}}} = \left(1 + \frac{M_{\mathrm{OA}} P_i^{\mathrm{vap}}}{C_{\mathrm{OA}} RT} \times 10^6\right)^{-1} = \left(1 + \frac{C_i^*}{C_{\mathrm{OA}}}\right)^{-1} \tag{2}$$

where $N_{i,j}$ is the number concentration in phase j (molecule of species i per cm^3 of air), T is the T/K, R the ideal gas constant (atm m^3 K^{-1} mol^{-1}), C_{OA} the aerosol mass concentration (μg m^{-3} of air), M_{OA} the mean organic molar mass in the aerosol (g mol^{-1}) and C_i^* is an effective saturation mass concentration (μg m^{-3} of air). The Nannoolal *et al.*[27] group contribution method was used to estimate the vapor pressure of each non radical species included in the mechanism, as described by Valorso *et al.*[28] This method was used in conjunction with the Nannoolal *et al.*[29] method to estimate boiling points. Contributions for functional groups not provided by Nannoolal *et al.*[27,29] were taken from Compernolle *et al.*[30]

The explicit description of *n*-decane oxidation is expected to lead to a mechanism including about 10^8 species,[20] one order of magnitude above the size of chemical schemes that can be both generated and solved. Simplifications are therefore required to reduce the schemes down to a manageable size. Usually, simplifications in GECKO-A are performed based on the lumping of structural isomers.[17,28,31] The description of oxidative trajectories along fragmentation or functionalization routes is a key aspect of the modeling study performed here and requires a detailed representation of alkoxy radical chemistry. The chemical fate of the alkoxy radical (O$_2$ reaction, H-shift isomerisation, or C–C bond breaking) depends on the chemical structure in the vicinity of the alkoxy moiety.[23,24] No lumping is therefore performed to keep the molecular structures of the successive generations of secondary organic species. Mechanism reduction has been performed as follows. Species produced with a maximum yield below 5×10^{-6} are not treated further in the mechanism and the chemical removal of these species is considered as a final sink. This approximation does not significantly alter the mass budget, the cumulative loss of carbon atoms at the end of the simulation being less than 1% of the initial organic carbon. Furthermore, gas phase oxidation becomes negligible for non-volatile species, where most of the species mass occurs in the condensed phase. According to eqn (2) and for C_{OA} representative of polluted conditions (10 µg m^{-3}), a species is almost exclusively in the condensed phase at thermodynamic equilibrium when P^{vap} is below 10^{-13} atm. The gas phase chemistry is therefore omitted for such low volatility species. Finally, to reduce the mechanisms to a manageable size, we assume high NO$_x$ conditions, *i.e.* the peroxy + peroxy reactions are ignored. This approximation is appropriate under most polluted conditions. Nevertheless, this high NO$_x$ condition remains hardly representative of the atmospheric oxidation on timescales exceeding one day. Restricting the study to high NO$_x$ conditions is therefore a severe simplification.[19] Simulations performed here are clearly exploratory and intended to examine some, though not all, aspects of the alkane molecular structure on SOA formation. The numbers of species ultimately considered in the mechanisms are listed in Table 1 for various alkane series.

Time integration of the set of ordinary differential equations associated with gas phase oxidation is solved using the "2-step" solver.[32,33] Phase equilibrium is enforced at each time step, as described by Camredon *et al.*[34] Reactions in the condensed phase are not considered in this model configuration. Ageing of the particles is driven by gas phase chemistry that progressively shifts the various gas/particles equilibria as oxidation progresses.

Table 1 Number of species generated by GECKO-A for the gas phase oxidation mechanisms and number of non radical species for which gas/particle equilibrium is considered

Parent alkane	# species	# phase equilibria
C_{10} *alkanes*		
n-decane	1.7×10^5	5.1×10^4
2-methyl nonane	1.6×10^5	5.0×10^4
3,3 diethyl hexane	1.0×10^5	3.3×10^4
4,5-dimethyl octane	2.4×10^5	7.3×10^4
3,4,5-trimethyl heptane	2.9×10^5	9.1×10^4
C_{14} *alkanes*		
n-tetradecane	6.2×10^5	2.1×10^5
2-methyl tridecane	7.9×10^5	2.6×10^5
3,3 diethyl decane	5.0×10^5	1.7×10^5
6,7-dimethyl dodecane	8.6×10^5	2.9×10^5
5,6,7-trimethyl undecane	8.0×10^5	2.7×10^5
C_{18} *alkanes*		
n-octadecane	8.4×10^5	3.2×10^5
2-methyl heptadecane	1.4×10^6	5.3×10^5
3,3 diethyl tetradecane	1.0×10^6	3.6×10^5
8,9-dimethyl hexadecane	1.2×10^6	4.5×10^5
7,8,9-trimethyl pentadecane	1.1×10^6	4.1×10^5
C_{22} *alkanes*		
n-docosane	6.5×10^5	2.5×10^5
2-methyl uncosane	1.3×10^6	5.1×10^5
3,3 diethyl octadecane	1.5×10^6	5.5×10^5
10,11-dimethyl eicosane	1.2×10^6	4.5×10^5
9,10,11-trimethyl nonadecane	1.3×10^6	4.6×10^5

2.2 Simulation conditions

Simulations were conducted under conditions similar to those described by Aumont *et al.*[19] The simulations are run for constant environmental conditions. The temperature is set to 298 K. Photolysis frequencies are calculated using the TUV model[35] for mid latitude conditions and for a zenith angle of 45°. A constant OH source of 2×10^7 radicals cm^{-3} s^{-1} is included to initiate oxidation in the model. The NO_x mixing ratio is set to 1 ppb. Note that the simulated SOA formation is almost insensitive to the prescribed NO_x value, the fate of the organic peroxy radical being exclusively the reaction with NO as described above (*i.e.* high NO_x conditions). In this study, C_{OA} is set to a constant value of 10 µg m^{-3}, representative of polluted tropospheric conditions. The composition is assumed to be non-volatile organic matter with a mean molar mass M_{OA} of 250 g mol^{-1}, within the expected values for atmospheric organic aerosol.[25,36,37] Oxidation of the parent hydrocarbon may obviously lead to the formation of SOA and therefore alters the prescribed aerosol composition and mass concentration. The hydrocarbon initial mixing ratio is set to an arbitrary value of 10 ppt carbon (*ca.* $C_0 = 6.5 \times 10^{-3}$ µg m^{-3}), a value low enough to not modify the prescribed aerosol properties. For these conditions ($C_0 \ll C_{OA}$), SOA yields are independent of C_0.

3 Results

3.1 SOA yields

The time profiles of the simulated SOA yields are given in Fig. 1 for various C_{10}, C_{14}, C_{18} and C_{22} alkane series. Five homologous series are considered: *n*-alkane (Fig. 1a), 2 methyl series (Fig. 1b), 3,3-diethyl series (Fig. 1c), 2 vicinal methyl groups in the middle of the backbone (Fig. 1d) and 3 vicinal methyl groups in the middle of the backbone (Fig. 1e). Yields (denoted Y_C hereafter) are here defined as the ratio between the carbon atoms in the condensed phase to the initial carbon load included in the parent backbone. The time scales are expressed as the number of lifetimes N_τ of the parent hydrocarbon defined as:[31]

$$N_\tau = \ln\frac{C_0}{C_t} = \frac{t}{\tau} \tag{3}$$

where C_0 is the initial parent compound concentration, C_t its concentration at the simulated time t and τ its e-folding lifetime. Rate constants for OH radical reactions for the alkanes studied here range from *ca.* 1×10^{-11} cm^3 s^{-1} (C_{10} alkane) to *ca.* 3×10^{-11} (C_{22} alkane).[38,39] Using a typical atmospheric OH concentration of 10^6 radical cm^{-3}, the corresponding atmospheric lifetimes range from a few hours (C_{22} alkanes) to one day (C_{10} alkanes).

SOA yields typically grow with decay of the parent alkane and reach a maximum for $N_\tau \sim 5$. This time scale is however somewhat longer for the C_{10} alkanes due to an enhanced contribution of the later generation oxidation products to SOA formation.[19] As expected, Fig. 1 shows that (i) Y_C increases with the size of the carbon backbone in each homologous series and (ii) for a given carbon chain length, Y_C decreases with the branching degree of the parent hydrocarbon. These trends are consistent with experimental observations.[7,9,11] Furthermore, Fig. 1 shows that backbone branching does not substantially change SOA time profiles. This behavior suggests that, for a given carbon chain length, major SOA contributors arise from species produced in the same generations (see Section 3.3).

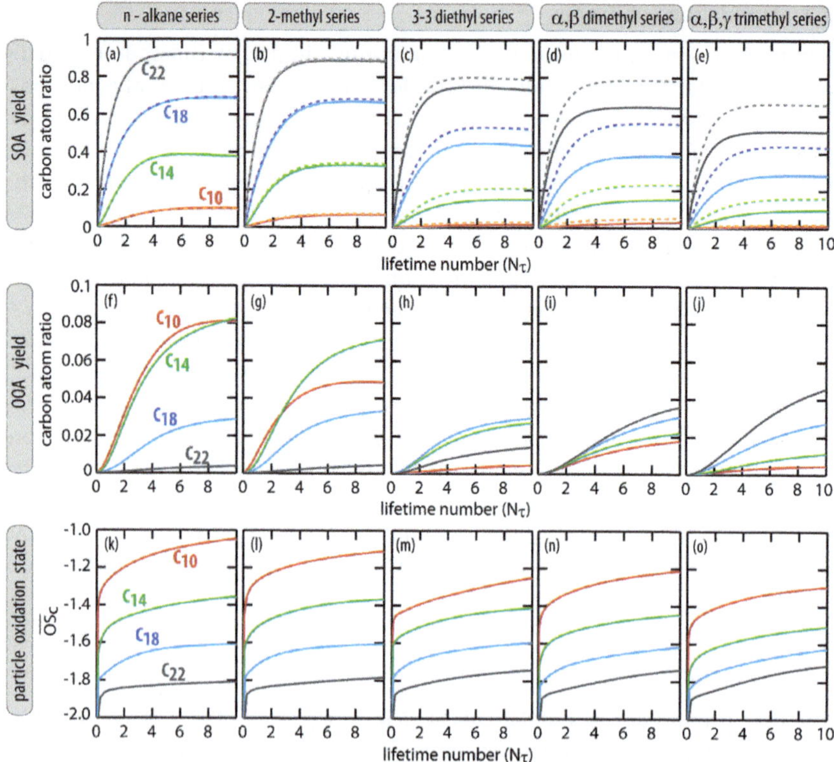

Fig. 1 SOA yields Y_C (top row), OOA yields $Y_{C,OOA}$ (middle row) and particle oxidation states OS_C (bottom row) for the n-alkanes (1st column), 2-methyl alkanes (2nd column), 3,3 dimethyl alkanes (3rd column), α,β dimethyl alkanes (4th column), α,β,γ trimethyl alkanes (5th column). Red, green, blue and grey lines are for C_{10}, C_{14}, C_{18}, C_{22} parent compounds, respectively. Dotted lines in panels a–e denote results obtained using the Atkinson (2007) SAR for alkoxy radical decomposition. The time scale is defined as multiples of lifetimes of the initial hydrocarbon (see text).

Fig. 1a–e shows that simulations performed with the two SAR implemented in GECKO-A for alkoxy radical decomposition[23,24] provide similar SOA yields for straight and lightly branched alkanes. However, discrepancies between the results obtained with the different SAR increase with the number of branches on the carbon backbone. SOA production is found to be substantially lower in the simulations including the SAR by Vereecken and Peeters. For example, the maximum SOA yield obtained for 6,7,8-trimethyl pentadecane (Fig. 1e) decreases from 0.43 using the Atkinson SAR[23] to 0.28 using the Vereecken and Peeters SAR.[24] As stated above, the SAR developed by Vereecken and Peeters[24] includes group contributions for a comprehensive set of moieties generated during the oxidation of alkanes. In the subsequent sections, we therefore use as a benchmark the results obtained with the mechanisms generated with the Vereecken and Peeters SAR for alkoxy radical decomposition.

Fig. 2 shows the SOA yields simulated at $N_\tau = 5$ (*i.e.* around the maximum yield) for various carbon chain lengths as a function of the number of branching methyl groups. Results given in Fig. 2 are for structures with methyl branches centered in the middle of the backbone (*e.g.* 7-methyl tridecane for the C_{14} alkane

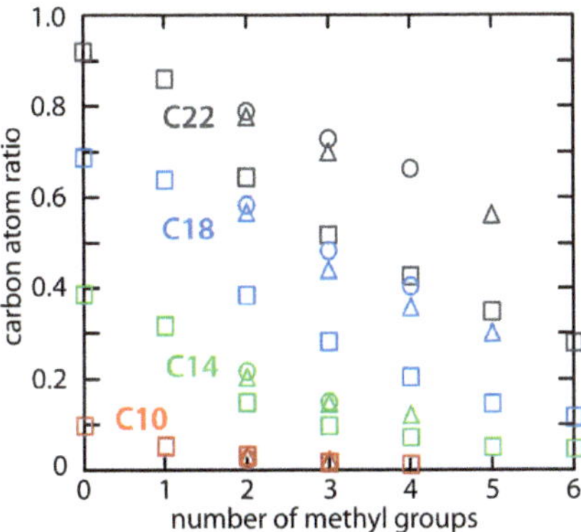

Fig. 2 Maximum SOA yields Y_C as a function of the number of branching methyl groups. Red, green, blue, and grey symbols are for C_{10}, C_{14}, C_{18} and C_{22} alkanes, respectively. Methyl branches are located in the middle of the carbon backbone. For species with more than 2 branches, squares denote for structure having no methylene carbon (–CH_2–) between each branch, circles denote structures having one methylene groups between each branch and triangles denote structures with 2 methylene carbons between each branch.

with one branch). Various structures are considered with either 0, 1 or 2 methylene (*i.e.* –CH_2–) groups between each branch (*e.g.* 5,6,7-, 4,6,8- and 3,6,9- trimethyl undecane for the C_{14} alkanes with 3 branches). The SOA yields decrease monotonically with the number of branching methyl groups. For example, Y_C for C_{10} alkanes decreases from 0.10 for *n*-decane to 0.03 and 0.01 for the dimethyl octane and tetramethyl hexane, respectively. Similarly, 6 branching methyl groups decrease Y_C by a factor of 4 (C_{22} alkanes) to 10 (C_{14} alkanes). The position of branching methyl groups is also found to have a substantial effect, with vicinal methyl branching leading to the largest effect. For example, Y_C changes from 0.58 to 0.38 for 7,10- and 8,9- dimethyl hexadecane, respectively.

Comparison of our modeling results with experimental observations is not straightforward owing to different simulation conditions. The branching effects simulated here agree nevertheless reasonably well with some experimental values. For example, the SOA yields for dodecane and 2-methyl undecane have been recently determined under high NO_x conditions in several experimental studies.[7,9,11] For our modeling conditions, the 2-methyl branching reduces the SOA yield by 22% in comparison to the linear structure for C_{12} alkanes. This value is consistent with the reduction in the 15–50% range observed experimentally. Similarly, Lim and Ziemann[7] report a 68% reduction of the SOA yield for 2,3 dimethyl decane in comparison to dodecane. A similar reduction (55%) was observed in the simulation results performed for the same couple of species (not shown here). Furthermore, Tkacik *et al.*[9] have reported that the SOA yields for C_{14} alkanes drop by almost a factor of 5 (from 0.38 to 0.08) after a single methyl branching in the middle of the carbon chain. The model was unable to explain this large effect: the simulated decrease for Y_C for the linear and same branched C_{14} structure is 18% only (see Fig. 2).

3.2   OOA yields

Highly oxygenated organic aerosols (OOA) are a major component identified *in situ* by aerosol mass spectrometer (AMS) measurements of submicron aerosols.[40–42] OOA is usually considered as SOA and is characterized by an atomic O/C ratio larger than 0.25[22] or equivalently by a mean carbon oxidation state (OS_C) greater than about -1.25.[43,19] We define here OOA yields $Y_{C,OOA}$ as the carbon-weighted concentration ratio of the species in the condensed phase with OS_C greater than -1.25 to the initial concentration of the parent compound. $Y_{C,OOA}$ can equivalently be defined as:

$$Y_{C,OOA} = f_{OOA} \times Y_C \tag{4}$$

where f_{OOA} is the fraction of the condensed carbon included in species with OS_C greater than -1.25. Fig. 1 shows the time profile of $Y_{C,OOA}$ and OS_C for the particle phase for the various homologous series. The dominant factor controlling OS_C appears to be the size of the parent backbone, OS_C decreasing clearly with the number of carbon atoms. Branching of the backbone is of secondary importance. This result is consistent with the experimental observations by Tkacik *et al.*[9] in their study of SOA formation from various linear, cyclic and branched alkanes under high NO_x conditions. In these experiments, the dominant factor driving O/C ratio in the particle was found to be the carbon number of the parent alkane as well. Furthermore, the particle O/C ratios reported by Tkacik *et al.*[9] for various C_{10} to C_{19} alkanes are typically in the 0.1–0.3 range, corresponding roughly to an OS_C in the -1.7 to -1.1 range.[43,19] Simulated values reported in Fig. 1 are within a similar range.

Fig. 3 shows f_{OOA} and Y_C as a function of the carbon chain length for an oxidation progress corresponding to $N_\tau = 5$. Within each homologous series, f_{OOA} increases when the size of the carbon backbone decreases. This outcome is consistent with the expectations that the oxygen content of the SOA contributors must increase when the size of the parent backbone decreases to bring their vapor pressures to a low enough value to condense.

For the conditions tested here, $Y_{C,OOA}$ remains below 0.1 whatever the chain length or molecular structure of the parent alkane (see Fig. 1). For short carbon chains ($C_{<16}$), branching of the carbon skeleton decreases both $Y_{C,OOA}$ and OS_C. This effect grows when the carbon backbone decreases. For a C_{10} molecule, $Y_{C,OOA}$

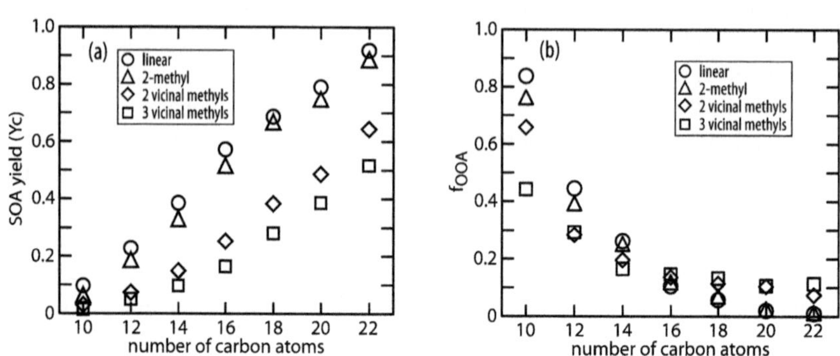

Fig. 3 SOA yields Y_C (left panel) and fraction f_{OOA} of the condensed carbon in highly oxygenated organic aerosols (right panel) as a function of the carbon chain length of the parent alkane. Symbol shapes denote distinct homologous series: linear structure (circles), 2-methyl branching (triangles), 2 vicinal methyl branching groups (diamonds) and 3 vicinal methyl branching groups (diamonds). Y_C and f_{OOA} are given for $N_\tau = 5$.

decreases by almost a factor of 2 for one branching methyl group and reaches one order of magnitude for 2 vicinal methyl groups (see Fig. 1). For short carbon chains, fragmentation almost exclusively leads to volatile products which ultimately end up as CO_2. Branching therefore decreases both Y_C and f_{OOA} (see Fig. 2 and 3). For long carbon chains ($C_{>16}$), the opposite trend is found with both $Y_{C,OOA}$ and OS_C increasing with the branching degree of the backbone. For a C_{22} backbone, $Y_{C,OOA}$ is increased by one order of magnitude from a linear structure to a structure with 3 vicinal methyl groups (see Fig. 1). Fragmentation leads to species with a long enough carbon skeleton to ultimately produce highly oxygenated low volatility species. Although branching decreases the overall SOA yields, it increases f_{OOA} significantly, ultimately leading to a substantial increase of the OOA yield.

3.3   Number of generations

Reaction sequences involved during gas phase oxidation lead to species with similar chemical structures but of increasing volatility when the number of carbon atoms in the parent alkane decreases. Within a given homologous series, the number of generations required for the production of low volatility species must therefore grow when the carbon skeleton of the parent alkane decreases in length. Fig. 4 show the composition of the particles at $N_\tau = 5$ discriminated on a generational basis for various homologous series. For the condition simulated here ($C_{OA} = 10 \ \mu g \ m^{-3}$), most SOA contributors are 1st and 2nd generation species for C_{22} parent alkanes but 3rd and 4th generation for a C_{10} parent alkane. As expected, Fig. 4 shows that the condensed species are on average produced after a growing number of oxidation steps when the size of carbon backbone decreases. Note that branching does not substantially change the relative contribution of the species produced by each generation to the SOA composition, similar distributions being observed for the various molecular structures with the same number of carbon atoms in the parent alkane (see Fig. 4).

Fig. 4 also shows the fraction of SOA contributors having the same carbon skeleton as the parent compounds, *i.e.* the fraction produced by oxidation along the functionalization routes only. The remaining fraction (with fewer carbon atoms) includes all the species produced along routes involving at least one fragmentation step. Fig. 4 shows that at least 80% of the SOA composition is made of species produced by oxidative trajectories involving functionalization steps only. As expected, the contribution of the fragmentation products to SOA increases with the degree of branching of the backbone. However, that contribution remains weak, typically of the order of 10% for parent alkanes with 3 vicinal methyl groups. Most fragmentation products follow trajectories evolving toward CO_2 without leading to the production of low volatility (*i.e.* highly oxygenated) products in significant yields.

3.4   SOA composition

As discussed in the previous section, SOA composition is dominated by species retaining the carbon skeleton of the parent hydrocarbons. Furthermore, the number of generations required to produce the low volatility species is found to be rather insensitive to the number of branching groups of the parent alkane. These outcomes suggest that most SOA contributors are produced by analogous

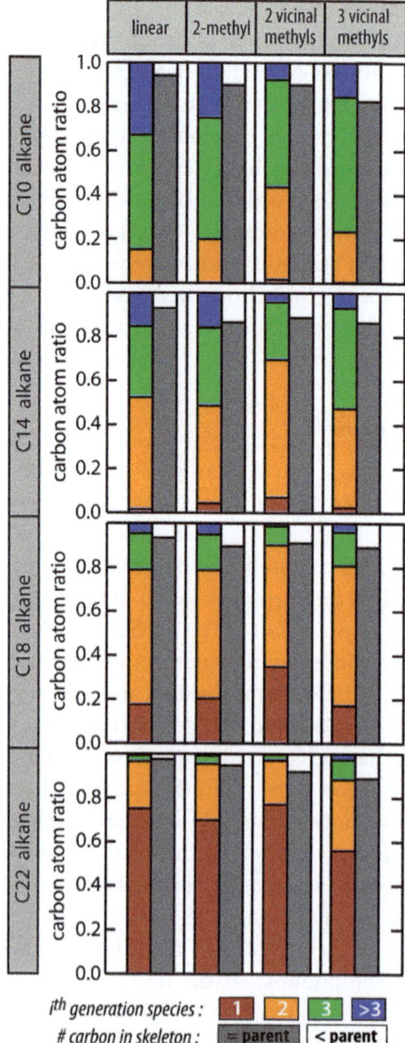

Fig. 4 Contribution of the species from a given generation to the particle composition (color histograms) and fraction of the species in the particle phase having the same carbon skeleton as the parent compound (grey histograms). Four homologous series are considered: linear (1st column), 2-methyl series (2nd column), 2 and 3 methyl vicinal methyl groups in the middle of the backbone (3rd and 4th column) and for C_{10} (first row) to C_{22} alkanes (4th row). Results are given for $N_\tau = 5$. Budget is performed on a carbon atom basis.

oxidation sequences along the functionalization routes. For the conditions simulated here, oxidation is driven by OH reaction. Fig. 5 shows the main reactions involved in the functionalization trajectories between 2 successive generations. Briefly, OH reaction on a paraffinic carbon atom leads to the abstraction of a hydrogen atom, leading to an alkyl radical and then to a peroxy radical (RO_2) after O_2 additions. For the high NO_x conditions simulated here, peroxy radicals react with NO to produce an organic nitrate moiety or an alkoxy radical (RO). Three reactions pathways are considered for alkoxy radicals: (1) C–C bond

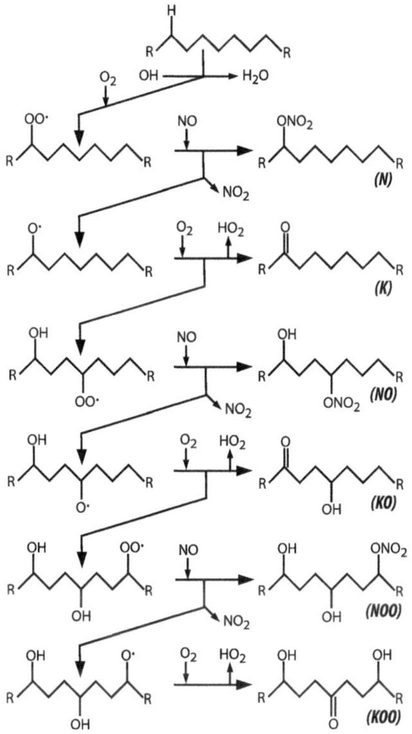

Fig. 5 The main reactions involved in the functionalization trajectories between 2 successive generations.

breaking (2) O_2 reaction (a minor pathway for long carbon chains) leading to a ketone moiety and (3) hydrogen shift isomerization, leading to an 1–4 hydroxy alkyl radical, which then adds O_2 to produce a new peroxy radical that reacts with NO as described above. Note that channel (1) is along the fragmentation route and is omitted in Fig. 5 for clarity. Finally, it is assumed the alkyl radical produced on a carbon already bearing an oxygenated functional group does not add O_2 but decomposes to form a ketone moiety. This concerns especially –C(OH)()– structures produced either by OH reaction or alkoxy H-shift isomerisation on an alcohol group and –C(ONO_2)()– structures produced by OH reaction on a carbon bearing the nitrate group. Oxidation along the functionalization trajectories leads to multifunctional species bearing various combination of nitrate moiety (denoted N hereafter), ketone moiety (denote K hereafter) and hydroxyl moiety (denoted O hereafter). For example, the first generation products of a long carbon chain alkane will be organic nitrate (N), ketone (K, although with a negligible yield for the parent alkane examined here), hydroxynitrate (NO), hydroxyketone (KO), dihydroxynitrate (NOO) and dihydroxyketone (KOO) (see Fig. 5). These first generation products are next oxidized to produce the set of second generation species. For the functional routes, this second generation products arise either from the oxidation of the already existing moieties, *i.e.* the oxidation of nitrate (N) or hydroxy (O) moiety to ketone (K) moiety or from the addition of a new set of functional groups on the carbon backbone.

Fig. 6 presents the composition of the SOA at $N_\tau = 5$ for various homologous series of alkanes. Position isomers (*i.e.* species with the same functional groups but with distinct position on the backbone) are lumped into same functional family for the purpose of presentation. Fifteen families of compounds are considered, accounting for 60% to 95% of the SOA mass budget, depending on the parent compounds considered. As already discussed above (see Section 3.3), SOA produced from the oxidation of C_{22} alkanes are dominated by first generation products, in particular N, KO, NO species with a C_{22} backbone. The number of generations

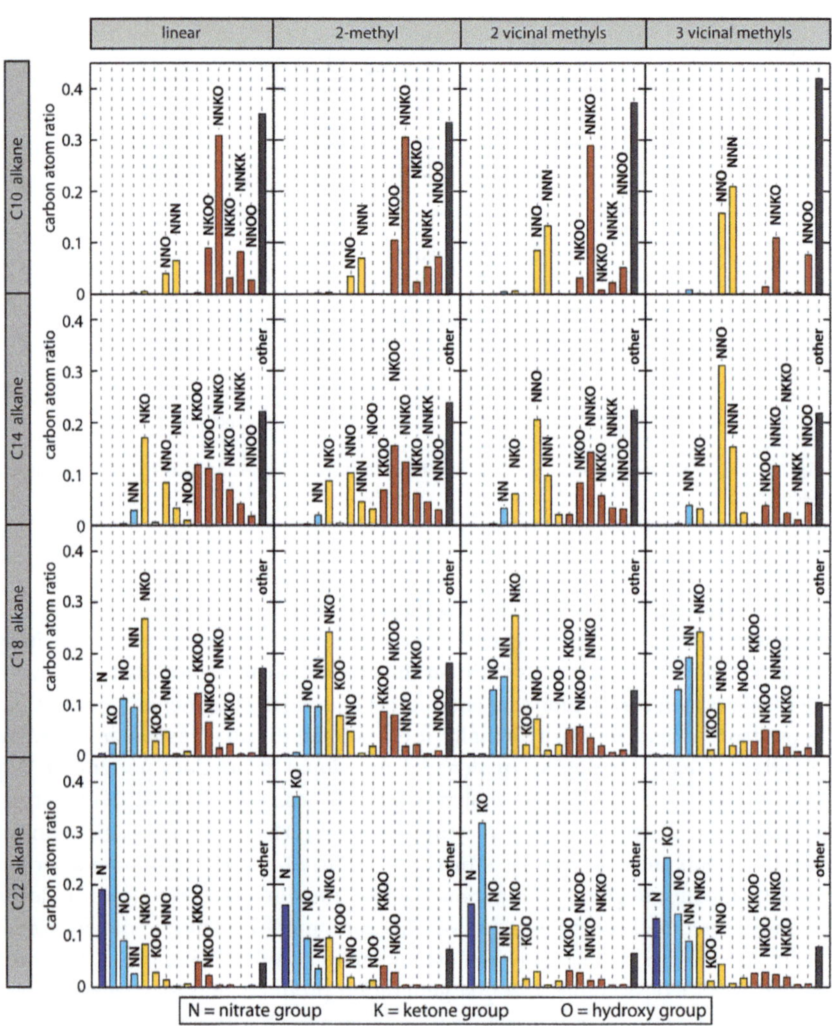

Fig. 6 Major contributors to the SOA composition. Position isomers are lumped into the same histogram. Colors discriminate contributors by number of functional groups borne by the species: dark blue = monofunctional species, light blue = difunctional species, orange = trifunctional species and red = species with more than 3 functional groups. N, K, O denotes species bearing the nitrate, ketone and hydroxy group, respectively. Four homologous series are considered: linear (1st column), 2-methyl series (2nd column), 2 and 3 methyl vicinal methyl groups in the middle of the backbone (3rd and 4th column) and for C_{10} (first row) to C_{22} alkanes (4th row). Results are given for $N_\tau = 5$. Budget is performed on a carbon atom basis.

required to produce low volatility species increases when the carbon skeleton of the parent alkane decreases. These successive oxidation steps increase the diversity of the species produced and ultimately the complexity of the SOA composition. For the oxidation of $C_{\leq 14}$ alkanes, the simulated SOA contributors are typically species bearing at least 3 functional groups, with diverse combinations of N, K, O moieties added on the carbon skeleton along the various functionalization trajectories. Fig. 6 shows a progressive evolution of the SOA composition with the carbon chain length of the parent compound. However, the chemical identities of the SOA contributors show similar patterns for parent alkanes having the same total number of carbons but various branching configuration. This outcome is consistent with the statement above that oxidative sequences leading to the formation of SOA contributors are nearly identical whether the molecular structure of the parent alkane is branched or not. Note that this finding also agrees with experimental observations. In particular, Lim and Ziemann[7] have shown that the SOA reaction products of linear and branched C_{12} alkane are similar.

3.5   Organic nitrates in the condensed phase

Fig. 6 shows that most simulated SOA contributors bear at least one nitrate moiety. For the C_{10} alkanes, this functional group is almost ubiquitous in the constituents of the particle phase. For example, 99% of the condensed molecules bear a least one nitrate moiety at $N_\tau = 5$ in the n-decane oxidation case. The occurrence of the nitrate moiety decreases with the parent alkane chain length. At $N_\tau = 5$, the fraction of condensed phase species bearing at least one nitrate group is 84%, 75% and 46% for the C_{14}, C_{18} and C_{22} n-alkane oxidation, respectively.

For our modeling conditions, organic nitrates ($RONO_2$) result exclusively from the following competing reactions:

$$RO_2 + NO \rightarrow RONO_2 \tag{R1}$$

$$RO_2 + NO \rightarrow RO + NO_2 \tag{R2}$$

In the current version of GECKO-A, the branching ratio for these 2 channels is estimated using the SAR developed for the SAPRC99 mechanism[44] and depends on the carbon chain length, the presence of functional groups on the carbon backbone and the nature of the RO_2 radical (*i.e.* primary, secondary or tertiary). The nitrate yield typically increases from a few percent for C_3 peroxy radicals to $\sim$25% for C_{10} peroxy radicals and reach a maximum value of $\sim$30% for $C_{>15}$ peroxy radicals. The presence of a hydroxy group on the carbon skeleton of the peroxy radical was found to significantly decrease the nitrate branching ratio,[45] behavior not included in the original SAR. In GECKO-A, we therefore reduced the nitrate yield by a factor 2 for hydroxy peroxy radicals. With this parameterization, the major first generation products generated by GECKO-A for the C_{22} n-alkane are produced with the following yields: KO = 0.58, N = 0.28, NO = 0.10, and KOO = 0.2. These values agree well with the yields recommended by Ziemann[46] based on experimental observations for long carbon chain alkanes: KO = 0.55, N = 0.30, NO = 0.15.

The SOA contributors given in Fig. 6 for the C_{22} species broadly result from the above yield distribution when first and second generation species are combined and explain the occurrence of the nitrate moiety in the 40–50% range. Additional factors explain the increase of the nitrate moiety prevalence when the size the parent alkane decreases. As stated in Section 3.3, more generations are required to produce SOA contributors when the carbon chain length decreases. The nitrate moiety tends to be preserved during these successive oxidation steps. Indeed, the nitrate group reduces significantly the reactivity of the alpha carbon atom toward H abstraction by OH radical (or RO radical *via* 1–5 H-shift isomerisation). For example, the SAR developed by Kwok and Atkinson[38] shows that OH reaction on a $-CH(ONO_2)-$ group is decreased by an order of magnitude compared to the equivalent $-CH_2-$ group. Oxidation of the nitrate moiety to a carbonyl moiety is therefore a slow process. Furthermore, channel (R1) follows a functionalization route while channel (R2) either leads to a functionalization or a fragmentation route, depending on the fate of the alkoxy radical (RO) produced. Although the functionalization route dominates the evolution of first generation alkoxy radicals, the probability of fragmentation increases greatly for subsequent generations as the number of functional groups borne by the carbon backbone increases. For the highly oxygenated organic compounds produced after a few generations, channel (R1) (*i.e.* nitrate formation) mostly favors SOA production by retaining the carbon skeleton (in addition to lowering the vapor pressure of the species) while channel (R2) mostly favors fragmentation of the carbon skeleton and ultimately full oxidation into CO_2.

Organic nitrates have been identified in the condensed phase during laboratory experiments devoted to SOA formation from oxidation of alkanes and alkenes[46–48] and during field observations.[49] However, the simulated fractions of nitrogen containing species in the particle phase appears surprisingly high and seems to be overestimated. The high NO_x conditions used here to generate the oxidation scheme explain only partially the large simulated organic nitrate yields. As stated in Section 2.1, oxidation of anthropogenic hydrocarbons under the high NO_x assumption remains hardly representative of atmospheric conditions beyond 1 day (corresponding to $N_\tau \sim 1$ for C_{10} alkanes). Nevertheless, the fraction of nitrogen containing species in the condensed phase at $N_\tau = 0.5$ is 0.94 and 0.68 for *n*-decane and *n*-tetradecane, respectively. Multifunctional organic nitrates therefore appear as major SOA constituents even in the early stage of alkane oxidation. Even under more realistic NO_x conditions, a large fraction of nitrogen containing species is simulated in the condensed phase. This has been for example recently shown in a GECKO-A modeling study examining the SOA budget in the Mexico City plume, where multifunctional organic nitrates were identified as major SOA contributors.[17] The causes of an organic nitrate overestimation in the simulated SOA are unclear. Hydrolysis in aerosol of organic nitrate has been proposed[49–51] and recently observed to occur in aerosol particles in a laboratory study.[52] Our understanding of the gas-phase chemistry of multifunctional nitrates is questionable as well. It was for example recently observed that carbonyl nitrates undergo fast photolysis and that the Kwok and Atkinson SAR[38] fell short to describe the OH reactivity of these multifunctional nitrates.[53] Additional experimental studies are clearly required to improve our understanding of the behavior of multifunctional organic nitrates in the gas and condensed phases.

4 Conclusion

The GECKO-A modeling tool was used to generate explicit gas phase oxidation mechanisms for various homologous series of C_{10}–C_{22} alkanes. Box model simulations were performed under high NO_x conditions to assess the sensitivity of SOA yield and composition to the molecular structure of the parent alkane. As expected, simulation results show that within each homologous series, the SOA yield increases with the carbon chain length of the parent alkane while the mean oxidation state and the SOA fraction that can be categorized as OOA both decrease. For a given chain length, the SOA yield decreases with the degree of branching of the parent alkane. For short backbones ($C_{<16}$), the branching inhibits the production of OOA due to an enhancement of the chemical transformation leading up to the full oxidation of the carbon to CO_2. This trend is reversed for long backbones ($C_{>16}$) for which the branching favors the formation of long enough fragmented species to produce low volatility and highly oxygenated species after subsequent oxidation along the functionalization routes. The particle composition is dominated by species having the same carbon skeleton as the parent compound, *i.e.* by species formed by the functionalization mechanisms. The SOA contributors are thus position and skeleton isomers for particles produced from alkanes with the same number of carbon atoms but various branching degree. In other words, our simulation results suggest that the branching substantially decreases SOA yields but does not greatly changes SOA composition.

In the version of the model used in this study, the condensed phase is assumed to be chemically inert, and gas/particles equilibrium enforced at each time step. SOA ageing is thus only induced by gas phase oxidation. Simulated results for the various alkane homologous series have shown that most SOA contributors are far too reduced to be categorized in the OOA component. On a carbon basis, the OOA yields never exceeded 10% regardless of carbon chain length, molecular structure, or ageing time. This version of the model appears clearly unable to explain a large production of OOA from alkane precursors. Including a detailed representation of the complex mixture of alkane isomers emitted in the atmosphere will thus not likely decrease substantially the mismatch between the simulated and observed particle oxidation states. Processes and/or other organic compound types not included in the model are therefore likely responsible for the large oxygen content usually observed for particles collected *in situ*. Processing of organic matter in the condensed phase is a likely process. For example, carboxylic acid is a major moiety observed in the condensed phase in both laboratory studies and field measurements.[54,55,51] However, this group does not contribute substantially to the carbon budget simulated with detailed schemes.[17,56] The gas phase oxidation of aliphatic compounds leads to a substantial production of 1–4 hydroxy ketones (see Section 3.5). These molecular structures are known to be converted to dihydrofurans in the condensed phase and may then volatilize.[7,46,57] These intermediates have been suggested as a source of carboxylic acid after the subsequent gas phase ozonolysis.[55,51] Oxidation in the condensed aqueous phase may also contribute to increase the oxygen content, especially through the formation of carboxylic moieties in the aqueous phase or the processing (oligomerization) of low molecular weight oxygenated intermediate like glyoxal.[58,59] For example, the inclusion of parameterizations to represent the aqueous phase processing of glyoxal in a box model was recently found to substantially increase the O/C ratio simulated with GECKO-A for the Mexico City plume and help to explain the mismatch between field observations and modeling outputs.[18] The integration of an explicit representation of multiphase mechanisms in the GECKO-A tool is the subject of ongoing studies.

Acknowledgements

BA acknowledges support from the Primequal program of the French Ministry of Ecology, Sustainable Development and Energy and the Sustainable Development Research Network (DIM-R2DS) of the Ile-de-France region and the French ANR within the project ONCEM. JLT and SM were supported in part by grant DE-FG02-ER65323 from the US Department of Energy Office of Science.

References

1 M. P. Fraser, G. R. Cass, B. R. T. Simoneit and R. A. Rasmussen, *Environ. Sci. Technol.*, 1997, **31**, 2356–2367.

2 J. J. Schauer, M. J. Kleeman, G. R. Cass and B. R. T. Simoneit, *Environ. Sci. Technol.*, 1999, **33**, 1578–1587.

3 J. J. Schauer, M. J. Kleeman, G. R. Cass and B. R. T. Simoneit, *Environ. Sci. Technol.*, 2002, **36**, 1169–1180.

4 D. R. Gentner, G. Isaacman, D. R. Worton, A. W. H. Chan, T. R. Dallmann, L. Davis, S. Liu, D. a. Day, L. M. Russell, K. R. Wilson, R. Weber, A. Guha, R. a. Harley and A. H. Goldstein, *Proc. Natl. Acad. Sci. U. S. A.*, 2012, **109**, 18318–23.

5 A. L. Robinson, N. M. Donahue, M. K. Shrivastava, E. A. Weitkamp, A. M. Sage, A. P. Grieshop, T. E. Lane, J. R. Pierce and S. N. Pandis, *Science*, 2007, **315**, 1259–1262.

6 N. M. Donahue, A. L. Robinson and S. N. Pandis, *Atmos. Environ.*, 2009, **43**, 94–106.

7 Y. B. Lim and P. J. Ziemann, *Environ. Sci. Technol.*, 2009, **43**, 2328–2334.

8 A. A. Presto, M. A. Miracolo, N. M. Donahue and A. L. Robinson, *Environ. Sci. Technol.*, 2010, **44**, 2029–2034.

9 D. S. Tkacik, A. a. Presto, N. M. Donahue and A. L. Robinson, *Environ. Sci. Technol.*, 2012, **46**, 8773–81.

10 A. T. Lambe, T. B. Onasch, D. R. Croasdale, J. P. Wright, A. T. Martin, J. P. Franklin, P. Massoli, J. H. Kroll, M. R. Canagaratna, W. H. Brune, D. R. Worsnop and P. Davidovits, *Environ. Sci. Technol.*, 2012, **46**, 5430–7.

11 C. D. Cappa, X. Zhang, C. L. Loza, J. S. Craven, L. D. Yee and J. H. Seinfeld, *Atmos. Chem. Phys.*, 2013, **13**, 1591–1606.

12 L. D. Yee, J. S. Craven, C. L. Loza, K. A. Schilling, N. L. Ng, M. R. Canagaratna, P. J. Ziemann, R. C. Flagan and J. H. Seinfeld, *J. Phys. Chem. A*, 2012, **116**, 6211–6230.

13 K. Dzepina, C. D. Cappa, R. M. Volkamer, S. Madronich, P. F. DeCarlo, R. A. Zaveri and J. L. Jimenez, *Environ. Sci. Technol.*, 2011, **45**, 3496–3503.

14 A. P. Tsimpidi, V. A. Karydis, M. Zavala, W. Lei, L. Molina, I. M. Ulbrich, J. L. Jimenez and S. N. Pandis, *Atmos. Chem. Phys.*, 2010, **10**, 525–546.

15 A. Hodzic, J. L. Jimenez, S. Madronich, M. R. Canagaratna, P. F. DeCarlo, L. Kleinman, and J. Fast, *Atmospheric Chemistry and Physics*, **10**, pp. 5491–5514.

16 S. H. Jathar, S. C. Farina, A. L. Robinson and P. J. Adams, *Atmos. Chem. Phys.*, 2011, **11**, 7727–7746.

17 J. Lee-Taylor, S. Madronich, B. Aumont, A. Baker, M. Camredon, A. Hodzic, G. S. Tyndall, E. Apel and R. A. Zaveri, *Atmos. Chem. Phys.*, 2011, **11**, 13219–13241.

18 E. M. Waxman, K. Dzepina, B. Ervens, J. Lee-Taylor, B. Aumont, J.-L. Jimenez, S. Madronich and R. Volkamer, *Geophys. Res. Lett.*, 2013, **40**, 978–982.

19 B. Aumont, R. Valorso, C. Mouchel-Vallon, M. Camredon, J. Lee-Taylor and S. Madronich, *Atmos. Chem. Phys.*, 2012, **12**, 7577–7589.

20 B. Aumont, S. Szopa and S. Madronich, *Atmos. Chem. Phys.*, 2005, **5**, 2497–2517.

21 J. H. Kroll, N. Donahue, J. L. Jimenez, S. H. Kessler, M. R. Canagaratna, E. R. Wilson, K. E. Altieri, M. L. R., A. S. Wozniak, H. Bluhm, E. R. Mysak, J. D. Smith, C. E. Kolb and D. R. Worsnop, *Nat. Chem.*, 2011, **3**, 133–139.

22 J. L. Jimenez, M. R. Canagaratna, N. M. Donahue, A. S. H. Prevot, Q. Zhang, J. H. Kroll, P. F. DeCarlo, J. D. Allan, H. Coe, N. L. Ng, A. C. Aiken, K. S. Docherty, I. M. Ulbrich, A. P. Grieshop, A. L. Robinson, J. Duplissy, J. D. Smith, K. R. Wilson, V. A. Lanz, C. Hueglin, Y. L. Sun, J. Tian, A. Laaksonen, T. Raatikainen, J. Rautiainen, P. Vaattovaara, M. Ehn, M. Kulmala, J. M. Tomlinson, D. R. Collins, M. J. Cubison, E. J. Dunlea, J. A. Huffman, T. B. Onasch, M. R. Alfarra, P. I. Williams, K. Bower, Y. Kondo, J. Schneider, F. Drewnick, S. Borrmann, S. Weimer, K. Demerjian, D. Salcedo, L. Cottrell, R. Griffin, A. Takami, T. Miyoshi, S. Hatakeyama, A. Shimono, J. Y. Sun, Y. M. Zhang, K. Dzepina, J. R. Kimmel, D. Sueper, J. T. Jayne, S. C. Herndon,

A. M. Trimborn, L. R. Williams, E. C. Wood, A. M. Middlebrook, C. E. Kolb, U. Baltensperger and D. R. Worsnop, *Science*, 2009, **326**, 1525–1529.

23 R. Atkinson, *Atmos. Environ.*, 2007, **41**, 8468–8485.

24 L. Vereecken and J. Peeters, *Phys. Chem. Chem. Phys.*, 2009, **11**, 9062–9074.

25 J. F. Pankow, *Atmos. Environ.*, 1994, **28**, 189–193.

26 N. M. Donahue, A. L. Robinson, C. O. Stanier and S. N. Pandis, *Environ. Sci. Technol.*, 2006, **40**, 2635–2643.

27 Y. Nannoolal, J. Rarey and D. Ramjugernath, *Fluid Phase Equilib.*, 2008, **269**, 117–133.

28 R. Valorso, B. Aumont, M. Camredon, T. Raventos-Duran, C. Mouchel-Vallon, N. L. Ng, J. H. Seinfeld, J. Lee-Taylor and S. Madronich, *Atmos. Chem. Phys.*, 2011, **11**, 6895–6910.

29 Y. Nannoolal, J. Rarey, D. Ramjugernath and W. Cordes, *Fluid Phase Equilib.*, 2004, **226**, 45–63.

30 S. Compernolle, K. Ceulemans and J. F. Muller, *Atmos. Chem. Phys.*, 2010, **10**, 6271–6282.

31 C. Mouchel-Vallon, P. Bräuer, M. Camredon, R. Valorso, S. Madronich, H. Herrmann and B. Aumont, *Atmos. Chem. Phys.*, 2013, **13**, 1023–1037.

32 J. G. Verwer and M. Vanloon, *J. Comput. Phys.*, 1994, **113**, 347–352.

33 J. G. Verwer, J. G. Blom, M. VanLoon and E. J. Spee, *Atmos. Environ.*, 1996, **30**, 49–58.

34 M. Camredon, B. Aumont, J. Lee-Taylor and S. Madronich, *Atmos. Chem. Phys.*, 2007, **7**, 5599–5610.

35 S. Madronich and S. Flocke, in *Solar Ultraviolet Radiation – Modeling, Measurements and Effects*, ed. C. Zerefos, Springer-Verlag, Berlin, 1997, pp. 23–48.

36 A. Hodzic, J. L. Jimenez, S. Madronich, M. R. Canagaratna, P. F. DeCarlo, L. Kleinman and J. Fast, *Atmos. Chem. Phys.*, 2010, **10**, 5491–5514.

37 N. M. Donahue, S. A. Epstein, S. N. Pandis and A. L. Robinson, *Atmos. Chem. Phys.*, 2011, **11**, 3303–3318.

38 E. S. C. Kwok and R. Atkinson, *Atmos. Environ.*, 1995, **29**, 1685–1695.

39 J. G. Calvert, R. G. Derwent, J. J. Orlando, G. Tyndall, and T. J. Wallington, *The mechanisms of atmospheric oxidation of the alkanes*, Oxford University Press, 2008.

40 A. C. Aiken, P. F. Decarlo, J. H. Kroll, D. R. Worsnop, J. A. Huffman, K. S. Docherty, I. M. Ulbrich, C. Mohr, J. R. Kimmel, D. Sueper, Y. Sun, Q. Zhang, A. Trimborn, M. Northway, P. J. Ziemann, M. R. Canagaratna, T. B. Onasch, M. R. Alfarra, A. S. H. Prevot, J. Dommen, J. Duplissy, A. Metzger, U. Baltensperger and J. L. Jimenez, *Environ. Sci. Technol.*, 2008, **42**, 4478–85.

41 Q. Zhang, D. R. Worsnop, M. R. Canagaratna and J. L. Jimenez, *Atmos. Chem. Phys.*, 2005, **5**, 3289–3311.

42 Q. Zhang, J. L. Jimenez, M. R. Canagaratna, J. D. Allan, H. Coe, I. Ulbrich, M. R. Alfarra, A. Takami, A. M. Middlebrook, Y. L. Sun, K. Dzepina, E. Dunlea, K. Docherty, P. F. DeCarlo, D. Salcedo, T. Onasch, J. T. Jayne, T. Miyoshi, A. Shimono, S. Hatakeyama, N. Takegawa, Y. Kondo, J. Schneider, F. Drewnick, S. Borrmann, S. Weimer, K. Demerjian, P. Williams, K. Bower, R. Bahreini, L. Cottrell, R. J. Griffin, J. Rautiainen, J. Y. Sun, Y. M. Zhang and D. R. Worsnop, *Geophys. Res. Lett.*, 2007, **34**, L13801.

43 N. M. Donahue, J. H. Kroll, S. N. Pandis and A. L. Robinson, *Atmos. Chem. Phys.*, 2012, **12**, 615–634.

44 W. P. L. Carter, *Documentation of the SAPRC-99 chemical mechanism for the VOC reactivity assessment*, Final Report to the California Air Resources Board under Contracts 92-329 and 95-308, 2000.

45 C. E. Jordan, P. J. Ziemann, R. J. Griffin, Y. B. Lim, R. Atkinson and J. Arey, *Atmos. Environ.*, 2008, **42**, 8015–8026.

46 P. J. Ziemann, *Int. Rev. Phys. Chem.*, 2011, **30**, 161–195.

47 Y. B. Lim and P. J. Ziemann, *Aerosol Sci. Technol.*, 2009, **43**, 604–619.

48 A. Matsunaga and P. J. Ziemann, *J. Phys. Chem. A*, 2009, **113**, 599–606.

49 D. A. Day, S. Liu, L. M. Russell and P. J. Ziemann, *Atmos. Environ.*, 2010, **44**, 1970–1979.

50 K. Sato, *Atmos. Environ.*, 2008, **42**, 6851–6861.

51 L. M. Russell, R. Bahadur and P. J. Ziemann, *Proc. Natl. Acad. Sci. U. S. A.*, 2011, **108**, 3516–3521.

52 S. Liu, J. E. Shilling, C. Song, N. Hiranuma, R. A. Zaveri and L. M. Russell, *Aerosol Sci. Technol.*, 2012, **46**, 1359–1369.

53 R. Suarez-Bertoa, B. Picquet-Varrault, W. Tamas, E. Pangui and J. F. Doussin, *Environ. Sci. Technol.*, 2012, **46**, 12502–12509.

54 S. Decesari, M. C. Facchini, S. Fuzzi and E. Tagliavini, *J. Geophys. Res.*, 2000, **105**, 1481–1489.

55 S. Liu, D. a. Day, J. E. Shields and L. M. Russell, *Atmos. Chem. Phys.*, 2011, **11**, 8321–8341.

56 B. Aumont, S. Madronich, I. Bey and G. S. Tyndall, *J. Atmos. Chem.*, 2000, **35**, 59–75.

57 R. Atkinson, J. Arey and S. M. Aschmann, *Atmos. Environ.*, 2008, **42**, 5859–5871.

58 M. Hallquist, J. C. Wenger, U. Baltensperger, Y. Rudich, D. Simpson, M. Claeys, J. Dommen, N. M. Donahue, C. George, A. H. Goldstein, J. F. Hamilton, H. Herrmann, T. Hoffmann, Y. Iinuma, M. Jang, M. E. Jenkin, J. L. Jimenez, A. Kiendler-Scharr, W. Maenhaut, G. McFiggans, T. F. Mentel, A. Monod, A. S. H. Prevot, J. H. Seinfeld, J. D. Surratt, R. Szmigielski and J. Wildt, *Atmos. Chem. Phys.*, 2009, **9**, 5155–5236.

59 B. Ervens, A. G. Carlton, B. J. Turpin, K. E. Altieri, S. M. Kreidenweis and G. Feingold, *Geophys. Res. Lett.*, 2008, **35**, L02816, DOI: 10.1029/2007gl031828.

Faraday Discussions

RSC Publishing

PAPER

Organic aerosol formation photo-enhanced by the formation of secondary photosensitizers in aerosols

Kifle Z. Aregahegn, Barbara Nozière* and Christian George

Received 20th March 2013, Accepted 26th April 2013

DOI: 10.1039/c3fd00044c

Secondary organic aerosols (SOA), which are produced by the transformations of volatile organic compounds in the atmosphere, play a central role in air quality, public health, visibility and climate, but their formation and aging remain poorly characterized. This study evidences a new mechanism for SOA formation based on photosensitized particulate-phase chemistry. Experiments were performed with a horizontal aerosol flow reactor where the diameter growth of the particles was determined as a function of various parameters. In the absence of gas-phase oxidant, experiments in which ammonium sulfate seeds containing glyoxal were exposed to gas-phase limonene and UV light exhibited a photo-induced SOA growth. Further experiments showed that this growth was due to traces of imidazole-2-carboxaldehyde (IC) in the seeds, a condensation product of glyoxal acting as an efficient photosensitizer. Over a 19 min irradiation time, 50 nm seed particles containing this compound were observed to grow between 3.5 and 30 $\pm$ 3% in the presence of either limonene, isoprene, α-pinene, β-pinene, or toluene in concentrations between 1.8 and 352 ppmv. The other condensation products of glyoxal, imidazole (IM) and 2,2-bi1H-imidazole (BI), also acted as photosensitizer but with much less efficiency under the same conditions. In the atmosphere, glyoxal and potentially other gas precursors would thus produce efficient photosensitizers in aerosol and autophotocatalyze SOA growth.

1   Introduction

Organic compounds are ubiquitous in ambient aerosols, accounting for up to 50% of the fine particulate mass.[1–4] The organic fraction plays an important role in determining their formation, growth, and removal of ambient aerosols.[5] Secondary Organic Aerosols (SOA), which are produced by the transformations of Volatile Organic Compounds (VOCs) in the atmosphere, have been the subject of sustained attention due to their importance in air quality, visibility, public health and climate.[6] Therefore, a good understanding of the formation and evolution of

Université de Lyon, université Lyon 1, CNRS, UMR5256, IRCELYON, Institut de recherches sur la catalyse et l'environnement de Lyon, Villeurbanne, F-69626, France

SOA in the atmosphere is necessary. However, gaining such an understanding poses a number of fundamental and practical challenges. For instance, it is difficult to directly observe SOA in the atmosphere as well as to differentiate between secondary and primary organic compounds that are present within SOA.[7] Investigations of SOA formation and growth processes are thus mainly based on laboratory experiments. Physical absorptive models based on volatility, such as the gas-to-particle partitioning model[8,9] and the Volatility Basis Set (VBS) model[10] are successful in describing SOA growth in a smog chamber but still largely underestimate SOA mass in the atmosphere.[11-14] Particulate-phase chemical reactions have recently been proposed as contributors to aerosol growth, thereby reducing such discrepancies.[15-18] It has been suggested that ionic reactions in the condensed phase, leading to the formation of C–C or C–O–C[15-17] bonds, can be catalyzed by strong acids,[18] amino acids,[19] or ammonium ions.[20] In particular, larger-than-expected yields of glyoxal SOA have recently been attributed to such reactions.[16,17] Indeed glyoxal is now expected to be an important source for SOA in the atmosphere due to its ubiquity and the fact that it is an oxidation product of a wide variety of biogenic and anthropogenic VOCs.[16,21-23]

Smog chamber experiments have shown, however, that a light-induced process is responsible for glyoxal SOA growth beyond that due to ionic reactions occurring in the dark.[17,24] The process(es) responsible for this light-induced SOA growth remain largely unknown, although several studies[25-27] have demonstrated that condensed phase glyoxal chemistry leads to the formation of light absorbing products. Recent work from our group has shown the existence of a photo-chemical growth pathway for aerosols containing a photosensitizer such as HULIS or 4-benzoyl benzoic acid (4-BBA). We have proposed that photosensitized reactions occurring on the particle surface give rise to the observed growth. Quite significantly, this growth does not require the presence of gas-phase oxidants. We believe that such a mechanism could also play an important role in the light-induced growth of glyoxal based SOA.[21,28]

In this work, we investigated photo-induced SOA growth associated with condensed phase glyoxal chemistry. Importantly, this allows us to identify atmospherically relevant photosensitizers. Aerosol flow tube experiments were performed in order to determine parameters affecting the observed SOA growth (precursor structure and concentration, seed composition, carrier gas, and UV-light intensity) and thereby elucidate a potential mechanism.

2 Experimental

SOA formation was investigated using a horizontal aerosol flow tube made of Pyrex (8 cm id × 152 cm length), thermostated and kept at 293 ± 1 K using a circulating water bath. This flow tube was surrounded by 7-UV lamps (Philips CLEO) with a continuous emission spectrum over 300–420 nm and a total irradiance of 3.7×10^{16} photon cm^2 s^{-1}. A complete description of this aerosol flow tube setup is given elsewhere.[28] First a monodispersed aerosol ("seeds") was generated from aqueous solutions using an atomizer (TSI 3076) and a Differential Mobility Analyzer (DMA) (TSI Model 3081, impactor size 0.0588 cm, aerosol flow = 0.3–0.5 lpm, sheath flow = 3–5 lpm). The seed aerosol particles were then dried with a silica gel diffusion drier and size selected to either 40 nm or 50 nm diameters using DMA. This monodispersed aerosol typically contained 5 000 particles cm^{-3} and corresponded to a total mass of a few μg m^{-3}. The seed particles were then directed into the flow tube reactor where they were exposed to a constant VOC flow and UV light. The VOC concentration was generated with a permeation tube in a temperature controlled oven (Dynacal, Valco instruments Co.inc. using VICI metronics dynacalibrator, model 150). The carrier gas used in all the experiments was synthetic air from a cylinder (purity 99.9990%) (total flow = 0.4–0.5 lpm), except when stated otherwise (section 3.3). The temperature and Relative Humidity (RH) inside the reactor were monitored at the inlet and at the outlet using a SP UFT75 sensor. The particle size distribution and concentration were monitored at the exit of the flow tube using a Scanning Mobility Particle Sizer (SMPS) (TSI 3080), consisting of a DMA, (TSI 3081) and a Condensation Particle Counter (CPC, TSI 3776). In the size measurements the bin width was 0.8 nm. For the particles studied in this work (typical size ∼50 nm), this represented 1.6% uncertainty on each size measurement and a little less than 3% uncertainty on the determination of each diameter growth factor.

Several series of experiments were performed to characterize the photosensitized SOA growth where the composition of the seeds, type and concentration of VOC, relative humidity, irradiation time, and residence time of the aerosol in the reactor were varied. The seeds were made of ammonium sulfate (Aldrich ≥ 99%) and/or organic acid (Succinic acid, Aldrich ≥ 99%) to which was added glyoxal (0.1 M, Aldrich, 40% w/w in water) or a photosensitizer such as 1H-Imidazole-2-carboxaldehyde ("IC", Aldrich 98%), 1H-Imidazole ("IM", Aldrich ≥ 99%) and 4-Benzoylbenzoic acid ("4BBA", Aldrich 99%) in a 1 : 10 : 1 weight ratio. The seed solutions were prepared with 18 MΩ distilled water (Millipore). The VOCs investigated for this study were limonene, α-pinene, β-pinene, isoprene, toluene, cyclohexene, ethanol, acetaldehyde, acetone and n-butanol.

To verify that the SOA produced in this work resulted specifically from the presence of photosentisizers, "blank" experiments were routinely performed in which the seed particles did not contain any photosensitizer, and all other conditions were the same. Furthermore, control experiments were carried out to 1) investigates the potential production of organic particles from the irradiation of the VOCs in the absence of seed particles, 2) check for the potential generation of gas-phase oxidant during the experiments, and 3) determine the photo-stability of the seed particles. For these, a NO_x detector (Thermo 42C chemiluminescent analyser, detection limit of 0.4ppbv) and an O_3 monitor (Thermo 49C, detection limit is 1ppbv) were connected to the outlet of the reactor.

3   Results and discussion

The control experiments showed that no particles were produced when VOCs were irradiated in the absence of seed particles in the reactor. They also showed that neither NO_x species nor O_3 were produced inside the reactor during the irradiation experiments, thus confirming the absence of gas-phase oxidants and thereby excluding gas-phase oxidation of the VOC in the experiments.

3.1   Autophotocatalytic SOA growth from glyoxal

In a first series of experiments, seed particles were produced from atomizing aqueous solutions containing a mixture of ammonium sulfate (0.95 mM),

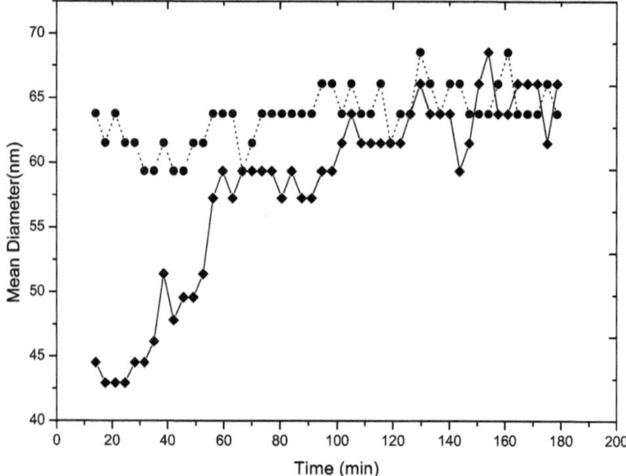

Fig. 1 Diameter growth of particles containing ammonium sulfate (0.95 mm), succinic acid (10.6 mm), Glyoxal (Gly) (0.1 M) at pH = 9 and exposed to UV-A light and 1.8 ppmv of limonene for 19 min. in the flow reactor. Experiments with freshly mixed seeds (solid line) are compared with those where the solution was left to react for 24 h after the addition of glyoxal (dotted line).

succinic acid (10.6 mM) and glyoxal (0.1 M). The dried and size selected particles were exposed to 1.8 ppmv of limonene and near-UV light in the reactor, and the particle diameter obtained at the end of the reactor was monitored in real time (Fig. 1). In this experiment, succinic acid was used as a proxy for the organic fraction of atmospheric particles, thus allowing for more "realistic" conditions. In some experiments, the seeds were produced from freshly mixed solutions while in others the solutions were left to react for 24 h after the addition of glyoxal with continuous stirring, all other conditions being identical. In most glyoxal experiments, the pH of the seed solution was adjusted to 9 by adding small amount of 2M NaOH solution. In some cases, the pH was left at 5.6 to study the effect of this parameter on SOA growth.

With the freshly mixed solutions, a significant growth of the particle diameter was observed after an induction time of 20–45 min following the addition of glyoxal (time 0 in the experiments), and reached a constant value after about 100 min. Beyond this time and even after 24 h, the particle mean diameter remained constant (Fig. 1). The seeds produced with the aged mixtures exhibited

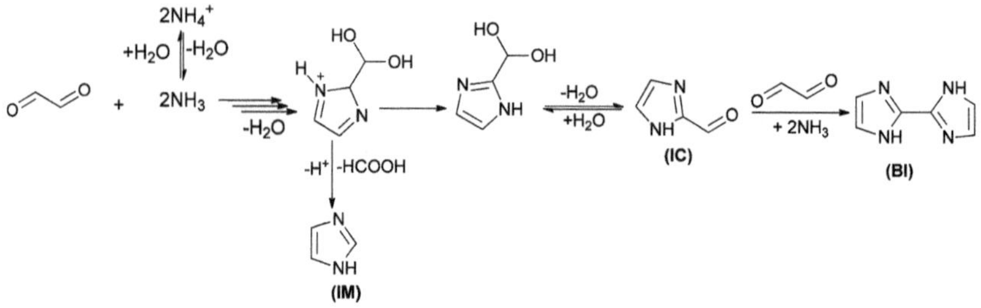

Fig. 2 Formation mechanisms for the different imidazoles produced in the Glyoxal/AS/water system proposed by Yu et al.[30] and Kampf et al.[25]

Table 1 Experimental conditions and SOA Diameter Growth Factors (DGF, %) obtained for 1.8 ppmv of limonene with different photosensitizers (Gl = glyoxal, SuA = succinic acid and AS = Ammonium sulfate)

Composition	Photosensitizer concentration (mM)	pH of solution	Dry particle diameter Dp_o (nm)	Particle diameter after irradiation Dp (nm)	Diameter growth factor **DGF (%)**
IC	1.3	7	49.6	Particle evaporates in the flow tube	—
AS	0	5.4	49.6	49.6	**0**
AS–SuA	0	3.5	49.6	49.6	**0**
GL–AS–SuA	100	5.6	49.6	57.3	**13.4**
	100	9	49.6	66.1	**24.9**
IC–AS–SuA	1.3	5.6	49.6	68.5	**27.6**
	1.3	9	49.6	68.5	**27.6**
IM–AS–SuA	1.8	5.6	47.8	49.6	**3.6**
BI–AS–SuA	0.93	5.6	49.6	51.4	**3.5**
4BBA–AS–SuA	0.4	5.6	49.6	55.2	**10.1**

some photochemical properties directly and led to particle growth in the flow tube, identical to those obtained with the freshly mixed seeds after 100 min, within the uncertainties (Fig. 1). The total diameter growth observed with the freshly mixed seeds (lowest to highest average mean diameter) was about $24.9 \pm 3\%$. The 20–45 min induction time observed before particle growth indicated that glyoxal was not directly responsible for the SOA growth, but rather some of its reaction products. SOA growth was also found to be larger with the seeds at pH = 9 ($24.9 \pm 3\%$) than at pH = 5.6 ($13.4 \pm 3\%$), indicating that this parameter directly affected the rate of formation of the relevant products.

The condensation of glyoxal with ammonia in condensed-phase ammonium is known to produce light-absorbing compounds,[20,24–26,29,30] potentially acting as photosensitizers. These products have been identified as imidazole (IM),[31] imidazole-2-carboxaldehyde (IC),[24,27] and 2,2-bi1H-imidazole (BI)[25] and their proposed formation mechanism in aerosols shown Fig.2.[25,27]

In atmospheric related studies, the yields of this imidazole and its derivatives are generally low and unlikely to contribute significantly to SOA mass.[20,24–27] However, these compounds have strong molar absorptivities and can thus significantly affect the optical properties of solutions or aerosols even at small concentrations.[24–27,29,30] These compounds could thus be the photosensitizers responsible for SOA growth in the experiments in this work. This is reinforced by the timescale and effect of pH on the observed SOA growth, which are in very good agreement with the rate constant of the reaction for glyoxal in NH_4^+ salts[20] producing the imidazoles and with the timescales for particle-phase imidazole formation in other flow tube experiments.[29,30]

To confirm the role of imidazoles as photosensitizers and determine their efficiency in SOA formation a second series of experiments was performed in which either (IM), (IC) or (BI) was added to the ammonium sulfate seeds containing succinic acid. These seeds were flown into the reactor to be exposed to 1.8 ppmv of limonene and UV-A light at a relative humidity (RH) of 15%. Table 1 summarizes the seed composition and particle growth obtained in these experiments.

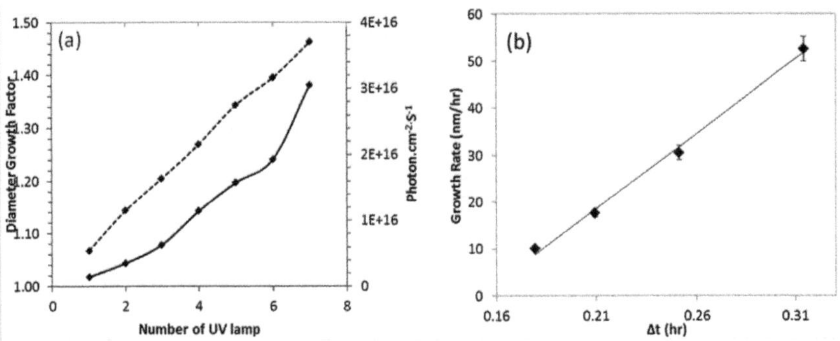

Fig. 3 Effect of light intensity and irradiation time on SOA growth in the reactor: (a) growth factor as function of number of lamps (solid line) and lamp intensities (dotted line), (b) growth rate as function of time of exposure (concentration × irradiation time). In both cases, the seeds contained IC (1.3 mM)-SuA (10.5 mM)-AS (0.95 mM) and constant limonene concentration.

In Table 1, the Diameter Growth Factor $(DGF(\%) = ((Dp - Dp_o)/Dp) \times 100$ where Dp is the particle diameter after irradiation and Dp_o is particle diameter in the dark) is calculated for different seed particles and photosensitizers.

The aqueous solutions containing only a photosensitizer (IC) evaporated over the timescale of our experiments, thus did not produce any stable seeds to perform SOA experiments. The seeds containing only ammonium sulfate or ammonium sulfate/succinic acid but no glyoxal or photosensitizer and exposed to limonene and UV-A light ("control experiments") led to no particle growth. This confirmed that the SOA growth observed in the other experiments was not due to the gas-phase oxidation of the VOCs. By contrast, all the experiments where the seeds contained a photosensitizer displayed some particle growth, evidencing the

Table 2 Diameter growth factors of SOA particle composed of IC–SuA–AS in Milli-Q water. The bin width on each particle size measurement is 0.8nm

Type of VOC	Vapour pressure[a] @20 °C (Bar)	Concentration (ppm)	pH of the solution	Residence time Rt (min)	Dry particle diameter Dp_o (nm)	Particle diameter after irradiation Dp (nm)	Diameter growth factor DGF (%)
Limonene	0.0017	1.8	5.6	19	49.6	68.5	**27.6**
Isoprene	0.622	4	5.6	19	49.6	51.4	**3.13**
		200	5.6	19	49.6	63.8	**22.26**
α-Pinene	0.0051	63	5.6	19	49.6	61.5	**19.35**
β-Pinene	0.00267	63	5.6	19	49.6	59.4	**16.5**
Toluene	0.028	352	5.6	19	49.6	51.4	**3.5**
Cyclohexene	0.2133	6	5.6	19	49.6	49.6	**0**
		>500	5.6	19	49.6	49.6	**0**
Ethanol	0.0551	688	5.6	19	49.6	49.6	**0**
n-Butanol	0.0063	470	.6	19	49.6	49.6	**0**
Acetaldehyde	0.321	2.4	5.6	19	49.6	49.6	**0**
Acetone	0.066	4	5.6	19	49.6	49.6	**0**

[a] CAMEO chemical datasheet.

critical role of this photosensitizer. The most efficient particle growth was observed with IC and was even slightly larger (27.6%) than in the glyoxal experiments (24.9%). With the two other imidazoles studied (IM) and (BI), the SOA growth observed was small, <5% or below the detection limit. IC was thus identified as the glyoxal product most likely to act as a photosensitizer in the glyoxal experiments. This efficiency could be due to a combination of the aromatic group of the imidazole and the carbonyl group, the latter being known to have strong photosensitizing properties.[32]

Thus, in the atmosphere, the partitioning and reaction of glyoxal in ammonium sulfate particles would produce an efficient photosensitizer, IC, which, in turn would trigger SOA growth by photo-induced reactions in or at the surface of the particles. SOA growth from gas-phase glyoxal in the atmosphere would thus be autophotocatalyzed.

3.2   Investigating the mechanism

The effect of various experimental conditions on the particle growth was characterized by further experiments. The effect of light intensity on SOA growth was studied in the reactor by using different numbers of UV lamps. The results showed that particle growth increased linearly with light intensity (Fig. 3a).

Particle growth was also found to depend on VOC concentration and the residence time (Rt) of the particles in the flow tube. For instance, an increase in isoprene concentration from 4 to 200 ppmv led to an increase in particle size from 3.13 to 22.3% (Table 2). These parameters affected the SOA growth from limonene, α-pinene, and β-pinene in the same way.

The growth rate of SOA, GR, was also determined as the ratio $\Delta D_m/\Delta t$, where ΔD_m is the difference of the mean particle diameter in the dark and under irradiation, and Δt their residence time in the reactor. Different residence times between 5 and 20min were thus tested. The results are presented in Fig. 3b and

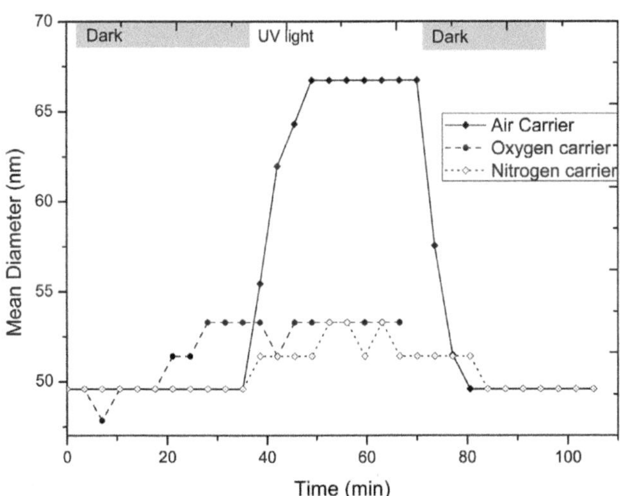

Fig. 4   Particle growth obtained from seeds containing imidazole-2-carboxaldehyde, succinic acid and ammonium sulphate exposed to 1.8 ppmv of limonene and UV-A light in different carrier gases: air (solid line), pure oxygen (dashed dotted line), and pure nitrogen (dotted line).

show that GR, the particle growth in the flow tube, increases linearly with time of exposure. These results allow us to extrapolate our laboratory experiments to smog chambers and even atmospheric time scales.

In a limited series of experiments, the uptake of limonene from the gas in the flow tube was tentatively measured using a high resolution TOF PTR-MS (Ionicon Analytika, Gmbh). This uptake was found to be below the detection limit, which would correspond to an uptake coefficient equal or lower than 10^{-4}. In spite of this, seemingly small uptake, the increase of particle mass and volume in similar experiments was shown with an Aerosol Mass Spectrometer (AMS, Aerodyne) to result exclusively from the trapping of non-condensable organic species, here the VOC.[28]

3.3 Investigating the mechanism: effect of the VOCs molecular structure

The potential selectivity of the mechanism responsible for SOA growth towards certain molecular structures was studied by performing experiments with 10 different VOCs. The results are summarized in Table 2: SOA growth was only observed with limonene, isoprene, α-pinene, β-pinene and toluene, but not with cyclohexene, n-butanol, ethanol, acetaldehyde and acetone within the residence time of our experiments.

Limonene was found to be the most efficient SOA precursor, achieving a particle growth of 27.6% for the smallest concentration (1.8 ppmv). Isoprene, α-pinene, β-pinene, and toluene resulted in particle growth ranging from 22.3 to 3.5%, but for larger concentrations than limonene (6–352 ppmv). The results in Table 2 would suggest that photo-induced SOA growth is associated with unsaturated VOCs. However, no growth is observed for cyclohexene, even when it is at concentration exceeding 500 ppmv. The results thus suggest that, more specifically, the photo-induced reaction mechanism is favoured for compounds containing unsaturated tertiary carbon atoms (isoprene, terpenes with endocyclic and/or exocyclic unsaturated tertiary sites, and toluene), over those containing secondary carbons only.

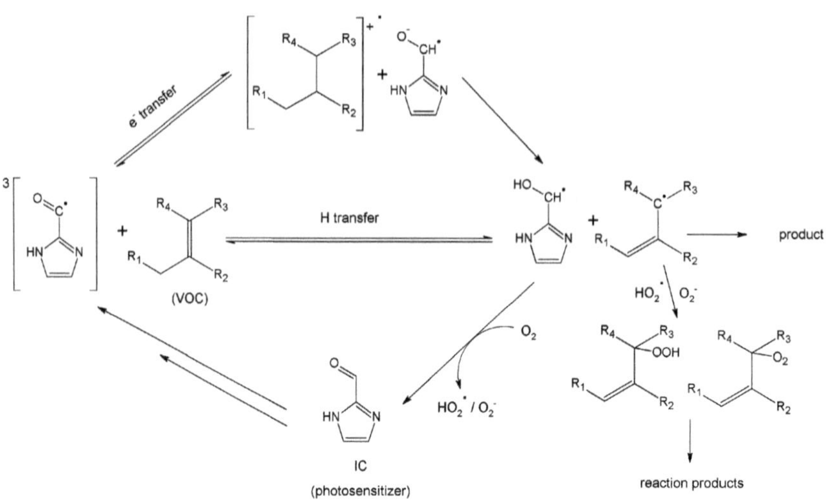

Fig. 5 Proposed reaction mechanism for the formation of active radicals in the photosensitized SOA growth.

3.4   Investigation of mechanism: role of carrier gas

To investigate the role of oxygen in the mechanism responsible for the observed SOA growth, a few experiments were performed using pure oxygen or pure nitrogen (containing $\leq$ 50 ppmv of O_2) as a carrier gas instead of air. Using IC as photosentisizers and limonene as VOC (1.8 ppmv), the SOA growth obtained was much more limited in both pure nitrogen (3.5%) and pure oxygen (6.9%) than in air (27.6%) (Fig. 4).

The very limited growth obtained in pure N_2 compared to in air indicated that O_2 plays an important role in the mechanism of SOA growth, possibly in the production of reactive species (radicals) resulting from the exchange between the photosensitizer and the VOCs. Oxygen is known to be essential in the photochemistry of dissolved organic matter with photosensitizers such as humic acids, which proceeds by the formation of singlet oxygen $(^1O_2)$[33–35] and superoxide.[36] Singlet oxygen is known to be produced by the electronic energy transfer between the excited triplet state of the photosensitizer, produced upon UV-vis light absorption, and the ground triplet state of molecular oxygen. Singlet oxygen would react with all unsaturated hydrocarbons and aromatic compounds.[37] But both singlet oxygen and superoxide could potentially be involved in the mechanisms responsible for SOA growth in this work.

The very small SOA growth observed in a pure oxygen carrier would seem to exclude the involvement of singlet oxygen as the formation of this intermediate should be favoured by large oxygen concentrations. But at high oxygen concentrations the relaxation and deactivation of the singlet oxygen by the ground triplet state might become significant[38–40] and could impede particle growth.

The involvement of singlet oxygen in the mechanism was further investigated by performing experiments where different singlet oxygen quenchers, sodium azide (NaN$_3$, Aldrich, $\geq$99.5%, 1.8 mM)[32] and hydrated nickel chloride (NiCl$_2 \cdot$6H$_2$O, MERCK, min 98%, 1 mM),[41] were added to ammonium sulfate seeds containing imidazole-2-carboxaldehyde in air carrier. The SOA growth obtained in these experiments was identical with the one obtained in their absence.

The lack of SOA growth from cyclohexene (Table 2) and the lack of effects of the singlet oxygen quenchers seem to exclude the involvement of singlet oxygen in the mechanisms. An alternative mechanism, shown in Fig. 5, could involve the direct interaction of the triplet state of the photosensitizer with the VOC[34] to produce radicals on the surface of the aerosol particle. This mechanism is similar to the one proposed by Canonica et $al.$[32] The radicals produced would then be stabilized by proton transfer and reaction with ground state oxygen. The proton transfer may also directly be produced by the abstraction of α-hydrogen to leave an allyl radical that is then stabilized by the tertiary carbon. In this redox reaction, molecular oxygen would create more reactive radicals for SOA growth, which would be limited in pure nitrogen. The excited triplet state of the photosensitizer could also potentially be quenched by ground state molecular oxygen which would account for the lower SOA growth in pure oxygen carrier.

3.5   Atmospheric relevance

Both glyoxal and ammonium sulphate containing particles are ubiquitous in the atmosphere. SOA formation from glyoxal is thus expected to occur via ionic reactions in the aerosol particles, resulting in an exceptionally large apparent

Henry's law constant[17,42] as well as *via* photosensitized reactions triggered by the condensed phase production of IC, as demonstrated by this work.

The SOA growth obtained in this work can be compared with those observed in the atmosphere, such as the one observed in the MCMA-2003 campaign in Mexico City, which was largely attributed to glyoxal reactions in the aerosols.[43] On 9 April 2003 a steady SOA growth corresponding of 3.3 $\mu g\ m^{-3}\ h^{-1}$ [12,43] was observed over 6 h of sunlight, and varying between $\sim$2 and 5.2 $\mu g\ m^{-3}\ h^{-1}$. In our experiments, a SOA mass growth of 3.2 $\mu g\ m^{-3}\ hr^{-1}$ was obtained when exposing seed particles to 1.8 ppm of limonene for 19 min. As the total concentration of VOCs expected to contribute to these processes (tertiary alkenes, branched aromatic…) in Mexico City might not exceed 100 ppbv[44,45] the new processes might account for 5 to 10% of the SOA growth observed at this location. This estimate indicates that these new processes might make a significant contribution to SOA mass in the atmosphere. But further investigation is now needed, in particular of the SOA mass produced as a function of VOC concentrations, in order to refine this estimate.

4   Conclusions

In this work, photo-induced SOA growth resulting from the interaction of glyoxal with ammonium sulfate seeds was demonstrated and investigated. Among the different light-absorbing products formed by the reactions of glyoxal in ammonium salts, imidazole-2-carboxaldehyde was found to be the most efficient photosensitizer for initiating SOA growth. SOA formation from glyoxal in the atmosphere is thus expected to be autophotocatalytic: the partitioning and reactions in ammonium-containing seeds producing photosensitizers, which, in turn, leads to further photo-induced growth of SOA.

In this study we found that SOA growth from seeds containing IC was linearly dependent on VOC concentration, and irradiation time and intensity. Different VOCs also gave very different SOA growth, those having unsaturated tertiary carbon, such as terpenes, isoprene and toluene, having the highest yields. The potential role of singlet oxygen or the photosensitizer triplet state as the key species controlling the SOA growth was investigated but needs to be studied further. The SOA mass obtained in the experiments suggests that these photo-induced processes could account for most of the SOA growth attributed to glyoxal chemistry in regions such as Mexico City. These processes need, however, to be further characterized, especially over longer time scales.

Acknowledgements

Support by the FP7 project PEGASOS (EU–FP7 project under grant agreement no. 265307) is gratefully acknowledged. C.G. thanks Olivier Piva for fruitful discussions having stimulated this work and Sumi Wren for her stylistic inputs.

References

1 M. C. Jacobson, H. C. Hansson, K. J. Noone and R. J. Charlson, *Rev. Geophys.*, 2000, **38**, 267–294.
2 M. Kanakidou, J. H. Seinfeld, S. N. Pandis, I. Barnes, F. J. Dentener, M. C. Facchini, R. Van Dingenen, B. Ervens, A. Nenes, C. J. Nielsen, E. Swietlicki, J. P. Putaud, Y. Balkanski, S. Fuzzi, J. Horth, G. K. Moortgat, R. Winterhalter, C. E. L. Myhre,

K. Tsigaridis, E. Vignati, E. G. Stephanou and J. Wilson, *Atmos. Chem. Phys.*, 2005, **5**, 1053–1123.

3 J. H. Seinfeld and J. F. Pankow, *Annu. Rev. Phys. Chem.*, 2003, **54**, 121–140.

4 Q. I. Zhang, C. O. Stanier, M. R. Canagaratna, J. T. Jayne, D. R. Worsnop, S. N. Pandis, Jimenez and J. L., *Environ. Sci. Technol.*, 2004, **38**, 4797–4809.

5 M. Prather, D. Ehhalt, F. Dentener, R. Derwent, E. Dlugokencky, E. H. I. Isaksen, J. Katima, V. Kirchhoff, P. Matson, P. Midgley and M. Wang, in *Climate Change 2001: The Scientific Basis, Contribution of WG1 to the Third Assessment report of the IPCC*, ed. J. T. Houghton, Y. Ding, D. J. Griggs, M. Noguer, P. J. Van der Linden, X. Dai, K. Maskell and C. A. Johnson, Cambridge University Press, UK, 2001.

6 V. Perraud, E. A. Bruns, M. J. Ezell, S. N. Johnson, Y. Yu, M. L. Alexander, A. Zelenyuk, D. Imre, W. L. Chang, D. Dabdub, J. F. Pankow and B. J. Finlayson-Pitts, *Proc. Natl. Acad. Sci. U. S. A.*, 2012, **109**, 2836–2841.

7 M. Hallquist, J. C. Wenger, U. Baltensperger, Y. Rudich, D. Simpson, M. Claeys, J. Dommen, N. M. Donahue, C. George, A. H. Goldstein, J. F. Hamilton, H. Herrmann, T. Hoffmann, Y. Iinuma, M. Jang, M. E. Jenkin, J. L. Jimenez, A. Kiendler-Scharr, W. Maenhaut, G. McFiggans, T. F. Mentel, A. Monod, A. S. H. Prévôt, J. H. Seinfeld, J. D. Surratt, R. Szmigielski and J. Wildt, *Atmos. Chem. Phys.*, 2009, **9**, 5155–5236.

8 J. F. Pankow, *Atmos. Environ.*, 1994, **28**, 185–188.

9 J. R. Odum, T. Hoffmann, F. Bowman, D. Collins, R. C. Flagan and J. H. Seinfeld, *Environ. Sci. Technol.*, 1996, **30**, 2580–2585.

10 A. L. Robinson, N. M. Donahue, M. K. Shrivastava, E. A. Weitkamp, A. M. Sage, A. P. Grieshop, T. E. Lane, J. R. Pierce and S. N. Pandis, *Science*, 2007, **315**, 1259–1262.

11 C. L. Heald, D. J. Jacob, R. J. Park, L. M. Russell, B. J. Huebert, J. H. Seinfeld, H. Liao and R. J. Weber, *Geophys. Res. Lett.*, 2005, **32**, L18809.

12 R. Volkamer, J. L. Jimenez, F. San Martini, K. Dzepina, Q. Zhang, D. Salcedo, L. T. Molina, D. R. Worsnop and M. J. Molina, *Geophys. Res.Lett.*, 2006, **33**, L17811.

13 R. Bergström, H. A. C. Denier van der Gon, A. S. H. Prévôt, K. E. Yttri and D. Simpson, *Atmos. Chem. Phys.*, 2012, **12**, 8499–8527.

14 M. Kuwata and S. T. Martin, *Aerosol Sci. Technol.*, 2012, **46**, 937–949.

15 M. Jang, N. M. Czoschke, S. Lee and R. M. Kamens, *Science*, 2002, **298**, 814–817.

16 J. H. Kroll, N. L. Ng, S. M. Murphy, V. Varutbangkul, R. C. Flagan and J. H. Seinfeld, *J. Geophys. Res.*, 2005, **110**, D23207.

17 R. Volkamer, P. J. Ziemann and M. J. Molina, *Atmos. Chem. Phys.*, 2009, **9**, 1907–1928.

18 J. L. Duncan, L. R. Schindler and J. T. Roberts, *Geophys. Res. Lett.*, 1998, **25**, 631–634.

19 B. Noziere and A. Cordova, *J. Phys. Chem. A*, 2008, **112**, 2827–2837.

20 B. Nozière, P. Dziedzic and A. Córdova, *J. Phys. Chem. A*, 2009, **113**, 231–237.

21 S. Myriokefalitakis, M. Vrekoussis, K. Tsigaridis, F. Wittrock, A. Richter, C. Brühl, R. Volkamer, J. P. Burrows and M. Kanakidou, *Atmos. Chem. Phys.*, 2008, **8**, 4965–4981.

22 P. Lavvas, R. V. Yelle, T. Koskinen, A. Bazin, V. Vuitton, E. Vigren, M. Galand, A. Wellbrock, A. J. Coates, J.-E. Wahlund, F. J. Crary and D. Snowden, *Proc. Natl. Acad. Sci. U. S. A.*, 2013, **110**, 2729–2734.

23 S. M. Saunders, M. E. Jenkin, R. G. Derwent and M. J. Pilling, *Atmos. Chem. Phys.*, 2003, **3**, 161–180.

24 M. M. Galloway, P. S. Chhabra, A. W. H. Chan, J. D. Surratt, R. C. Flagan, J. H. Seinfeld and F. N. Keutsch, *Atmos. Chem. Phys.*, 2009, **9**, 3331–3345.

25 C. J. Kampf, R. Jakob and T. Hoffmann, *Atmos. Chem. Phys.*, 2012, **12**, 6323–6333.

26 E. L. Shapiro, J. Szprengiel, N. Sareen, C. N. Jen, M. R. Giordano and V. F. McNeill, *Atmos. Chem. Phys.*, 2009, **9**, 2289–2300.

27 G. Yu, A. R. Bayer, M. M. Galloway, K. J. Korshavn, C. G. Fry and F. N. Keutsch, *Environ. Sci. Technol.*, 2011, **45**, 6336–6342.

28 M. E. Monge, T. Rosenørn, O. Favez, M. Müller, G. Adler, A. Abo Riziq, Y. Rudich, H. Herrmann, C. George and B. D'Anna, *Proc. Natl. Acad. Sci. U. S. A.*, 2012, **109**, 6840–6844.

29 M. Trainic, A. Abo Riziq, A. Lavi and Y. Rudich, *J. Phys. Chem. A*, 2012, **116**, 5948–5957.

30 M. Trainic, A. A. Riziq, A. Lavi, J. M. Flores and Y. Rudich, *Atmos. Chem. Phys.*, 2011, **11**, 9697–9707.

31 H. Debus, *Justus Liebigs Ann. Chem.*, 1858, **107**, 199–208.

32 S. Canonica, U. Jans, K. Stemmler and J. Hoigne, *Environ. Sci. Technol.*, 1995, **29**, 1822–1831.

33 J. P. Aguer and C. Richard, *J. Photochem. Photobiol., A*, 1996, **93**, 193–198.

34 K. You, D. Yin, L. Mao, P. Liu and H. a. Luo, *J. Photochem. Photobiol., A*, 2011, **217**, 321–325.

35 J. Pospíšil, S. Nešpůrek and J. Pilař, *Polym. Degrad. Stab.*, 2008, **93**, 1681–1688.

36 R. M. Baxter and J. H. Carey, *Nature*, 1983, **306**, 575–576.

37 E. L. Clennan, *Tetrahedron*, 2000, **56**, 9151–9179.
38 E. Afshari and R. Schmidt, *Chem. Phys. Lett.*, 1991, **184**, 128–132.
39 J. Piötz and M. Maier, *Chem. Phys. Lett.*, 1987, **138**, 419–424.
40 E. Wild, H. Klingshirn and M. Maier, *J. Photochem.*, 1984, **25**, 131–143.
41 D. J. Carlsson, T. Suprunchuk and D. M. Wiles, *Can. J. Chem.*, 1974, **52**, 3728–3737.
42 H. S. S. Ip, X. H. H. Huang and J. Z. Yu, *Geophys. Res. Lett.*, 2009, **36**, L01802.
43 R. Volkamer, F. San Martini, L. T. Molina, D. Salcedo, J. L. Jimenez and M. J. Molina, *Geophys. Res. Lett.*, 2007, **34**, L19807.
44 E. Velasco, B. Lamb, H. Westberg, E. Allwine, G. Sosa, J. L. Arriaga-Colina, B. T. Jobson, M. L. Alexander, P. Prazeller, W. B. Knighton, T. M. Rogers, M. Grutter, S. C. Herndon, C. E. Kolb, M. Zavala, B. de Foy, R. Volkamer, L. T. Molina and M. J. Molina, *Atmos. Chem. Phys.*, 2007, **7**, 329–353.
45 E. C. Fortner, J. Zheng, R. Zhang, W. Berk Knighton, R. M. Volkamer, P. Sheehy, L. Molina and M. André, *Atmos. Chem. Phys.*, 2009, **9**, 467–481.

Faraday Discussions

RSCPublishing

PAPER

Halogen-induced organic aerosol (XOA): a study on ultra-fine particle formation and time-resolved chemical characterization

Johannes Ofner,[*a] Katharina A. Kamilli,[b] Andreas Held,[b] Bernhard Lendl[a] and Cornelius Zetzsch[c]

Received 20th May 2013, Accepted 31st May 2013

DOI: 10.1039/c3fd00093a

The concurrent presence of high values of organic SOA precursors and reactive halogen species (RHS) at very low ozone concentrations allows the formation of halogen-induced organic aerosol, so-called XOA, in maritime areas where high concentrations of RHS are present, especially at sunrise. The present study combines aerosol smog-chamber and aerosol flow-reactor experiments for the characterization of XOA. XOA formation yields from α-pinene at low and high concentrations of chlorine as reactive halogen species (RHS) were determined using a 700 L aerosol smog-chamber with a solar simulator. The chemical transformation of the organic precursor during the aerosol formation process and chemical aging was studied using an aerosol flow-reactor coupled to an FTIR spectrometer. The FTIR dataset was analysed using 2D correlation spectroscopy. Chlorine induced homogeneous XOA formation takes place at even 2.5 ppb of molecular chlorine, which was photolysed by the solar simulator. The chemical pathway of XOA formation is characterized by the addition of chlorine and abstraction of hydrogen atoms, causing simultaneous carbon–chlorine bond formation. During further steps of the formation process, carboxylic acids are formed, which cause a SOA-like appearance of XOA. During the ozone-free formation of secondary organic aerosol with RHS a special kind of particulate matter (XOA) is formed, which is afterwards transformed to SOA by atmospheric aging or degradation pathways.

1   Introduction

Secondary organic aerosol (SOA) formation is currently mainly discussed pertaining to reactions of common atmospheric trace-gases like OH, O_3 and NO_x with a large variety of organic precursors. A review by Kroll and Seinfeld[1] summarizes

[a]Institute of Chemical Technologies and Analytics, Environmental and Process Analytics Division, Vienna University of Technology, Getreidemarkt 9, 1060 Vienna, Austria. E-mail: johannes.ofner@tuwien.ac.at

[b]Junior Professorship in Atmospheric Chemistry, University of Bayreuth, Dr Hans-Frisch-Strasse 1-3, 95448 Bayreuth, Germany

[c]Atmospheric Chemistry Research Laboratory, University of Bayreuth, Dr Hans-Frisch-Strasse 1-3, 95448 Bayreuth, Germany

this common understanding of SOA formation initiated by gas-phase oxidation and followed by particle-phase reactions involving the above mentioned atmospheric reactants. The evolution of organic aerosols in the atmosphere and the chemical transformation to aged organic aerosol is generally related to the presence of OH radicals or ozone.[2] Recently, Riipinen et al.[3] proposed that the condensation of low-volatility vapours and the formation of organic salts dominate the first steps in growth of ultra-fine organic particles. Over the open ocean and in marine environments, organic precursor gases, such as volatile halocarbons, are thought to contribute to marine aerosol formation.[4,5]

Apart from this general approach towards SOA formation with a major contribution of ozone and OH radicals, a few publications cover the topic of halogen-induced formation of SOA. The kinetics of the gas-phase reaction of monoterpenes with atomic chlorine have been studied e.g. by Timerghazin and Ariya,[6] implying higher reaction rates of Cl with selected monoterpenes compared to the OH-initiated reaction. They called for further research of the possible impact on the chlorine-initiated oxidation related to aerosol formation. For chlorine-induced aerosol formation from toluene with chlorine and toluene concentrations in the range of several 10^{16} molecules cm^{-3}, particle concentrations up to 3.5×10^8 cm^{-3} are reported.[7] Cai and Griffin[8] report the formation of halogen-induced SOA from different monoterpenes by chlorine atoms using an aerosol smog-chamber. These authors state that the chlorine-induced SOA formation is generally comparable to other scenarios and conclude that the chlorine-initiated oxidation of monoterpenes could be a source of SOA at sunrise. A similar study by these authors arrives at the same conclusion for the chlorine-initiated oxidation of toluene.[9] Kroll and Seinfeld[1] mention the chlorine-initiated SOA formation but conclude that it has a smaller effect on the vapour pressure than caused by OH radicals. A physico-chemical characterization of organic aerosol model substances, processed by reactive halogen species (RHS), considering the difference between halogen-induced organic aerosol and pre-existing SOA, which is processed by RHS during the period of photochemical aging is given by Ofner et al.[10] This study concludes a similar way of interaction of halogens with organic precursors as found by Kroll et al.[11] Other halogen species containing bromine and iodine are also suspected to significantly contribute to aerosol formation and processing.[12–14] Lifetimes of monoterpenes can be significantly influenced by halogens at dawn, where the reaction of chlorine atoms becomes comparable to the reaction with OH radicals.[15]

Halogen-induced organic aerosol formation is strongly dependent on the available concentration of RHS. RHS represent an abundant class of atmospheric oxidants in maritime areas. The importance of chlorine atoms as a tropospheric oxidant in the marine boundary layer was discussed in 1993,[16] concluding that the photochemical oxidation of organics by chlorine could exceed the importance of OH under special conditions. Unexpectedly high chlorine concentrations are reported by Spicer et al.[17] These authors report Cl_2 mixing ratios between 10 and 150 ppt and conclude that an unknown chlorine source must exist, which produces up to 330 ppt Cl_2 per day. A recent review summarizes very high chlorine mixing ratios:[18] 200 ppt for Cl_2, up to 421 ppt for Cl_2 and HOCl, 5.6 ppb for HCl and up to 2.1 ppb for $ClNO_2$. Different sources of RHS are reported in the scientific literature. Early smog-chamber experiments expected a possible release of molecular chlorine by homogeneous and heterogeneous reactions of different halogen sources like salts or acids.[19] The heterogeneous tropospheric chemistry of sea salt aerosol is known as an important source of RHS to the marine boundary layer.[20] The formation of $ClNO_2$ in the night-time and photolysis to Cl radicals and NO_2 during sunrise is reported as a large atomic chlorine source.[21]

The abundance of RHS in the marine boundary layer could contribute significantly to organic aerosol formation, especially at dawn and sunrise, when night-time reservoir substances are photolysed and ozone levels are still low. The present study reports aerosol-formation experiments from α-pinene with chlorine atoms. Furthermore, the chemical transformation of the organic precursor was studied using an aerosol-flow reactor coupled to an FTIR spectrometer and interpreted using 2D correlation spectroscopy[22] to obtain synchronous and asynchronous sequential mechanistic information.

2 Methods

Aerosol smog-chamber setup and experiments

Aerosol formation experiments were performed using a 700 L aerosol smog-chamber. Details on this chamber can be found at Ofner *et al.*[23] The smog chamber is equipped with a solar simulator (Osram, HMI, 4000 W). NO_2 actinometry, as reported by Bohn *et al.*,[24] was used to determine the spectral actinic flux, which is shown in Fig. 1 and compared to the solar actinic flux at the Mediterranean Sea in summer. This solar actinic flux was calculated using the STARsci v2.1 software package.[25] The solar simulator allows simulation of a solar spectrum comparable to that above the Mediterranean Sea in summer (Fig. 1). Based on the measurement of the spectral actinic flux, the photolysis of molecular chlorine was calculated inside the aerosol smog-chamber with a photolysis rate of $j_{Cl_2} = 1.89 \times 10^{-3}$ s^{-1}.

Before the experiments, the chamber was flushed with purified zero-air to avoid contamination of the chamber with aerosols or organic trace gases.[23] Zero

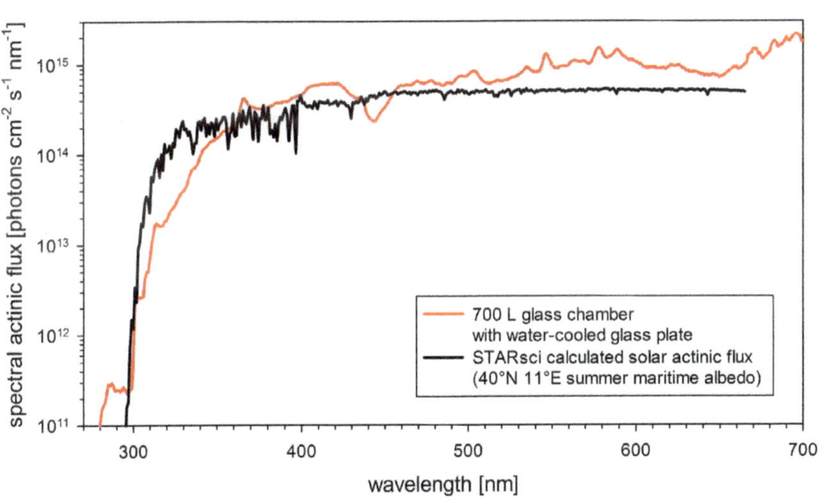

Fig. 1 The spectral actinic flux of the solar simulator of the aerosol smog-chamber compared to the solar actinic flux at the Mediterranean Sea in summer (calculated using STARsci v2.1[25]).

air was also used to keep the pressure inside the chamber above ambient conditions during the experiments. Therefore, all experimental data were corrected for dilution and wall-loss. Background measurements determined a dilution rate for the organic precursor of 3.2×10^{-3} min^{-1}. Ozone concentrations were measured using a chemiluminescence ozone analyser (Bendix-UPK8002; Bad Nauheim, Germany), exhibiting photolytically formed ozone concentrations below 10 ppb. NO$_2$ and NO concentrations were monitored using a CLD-700-Al analyser (Ecophysics, Dürnten, Switzerland) to ensure low-NO$_x$ conditions (below 10 ppb). A GC-FID system with an RTX resin column and a homemade cryostatic preconcentrator was applied to measure the decay of the organic precursor with a temporal resolution of 10 min. The aerosol number and size distribution was measured using a custom-built SMPS (IfT Leipzig, Germany) coupled to a TSI 3772 condensation particle counter (TSI, Shoreview, MN, USA). The aerosol mass formed at time t, M_t, was calculated from the measured aerosol size distributions assuming spherical particles and a particle density of 1.25 g cm^{-3}. Aerosol size distributions were corrected for wall-loss according to Pathak *et al.*[26] assuming a first-order wall-loss rate determined from plotting $\ln(M_t)$ against time after aerosol mass production has stopped. For size-dependent wall-loss corrections, we adjusted the formula of Crump and Seinfeld[27] for an arbitrary vessel to our cylindrically shaped smog chamber, and chose the turbulent energy dissipation parameter k_e in a way that the total aerosol mass derived from the size-dependent wall-loss correction was consistent with the integral wall-loss correction. Fig. 2a demonstrates a typical evolution of the aerosol size distribution from a chamber experiment related to XOA formation. The temporal evolution of aerosol mass before and after application of the wall-loss correction is shown in Fig. 2b.

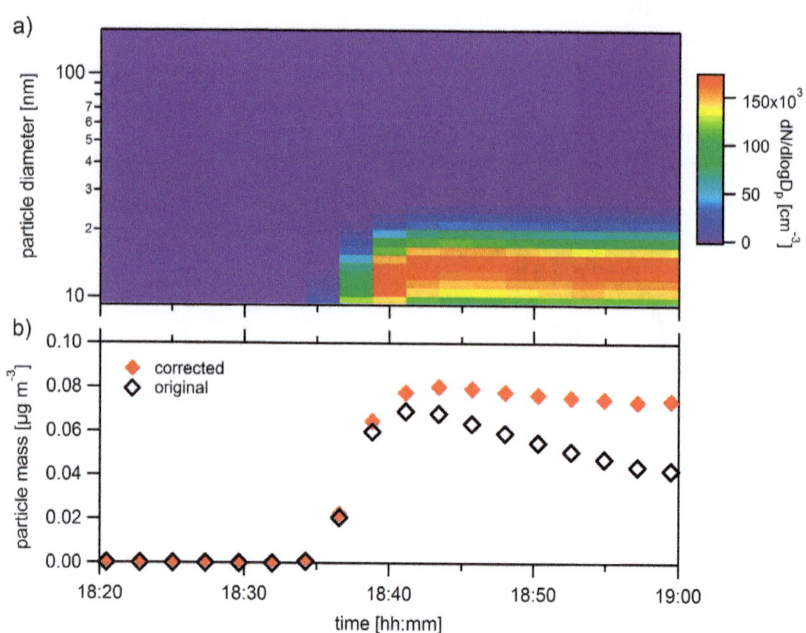

Fig. 2   a) Aerosol size distribution for a chamber experiment with α-pinene and photolyzed molecular chlorine; b) aerosol mass concentration resulting from the original and corrected SMPS data.

Table 1 Initial conditions, aerosol mass ΔM_0 and aerosol mass yield Y of smog-chamber runs

Experiment	Cl_2 [ppb]	α-pinene [ppb]	ΔM_0 [μg m^{-3}]	Y
Cl-1	2.5	10	1.3×10^{-3}	3.7×10^{-5}
Cl-2	5	10	6.0×10^{-2}	2.0×10^{-3}
Cl-3	5	10	6.4×10^{-2}	9.8×10^{-3}
Cl-4	20	20	23.7	2.4×10^{-1}
Cl-5	100	20	12.4	1.1×10^{-1}
Cl-6	200	50	175.7	6.2×10^{-1}

Aerosol formation experiments were started by switching on the solar simulator at zero-air conditions. After establishing a constant spectral actinic flux, α-pinene (Sigma-Aldrich, purity of 98%) was injected into the chamber. The concentration of the organic precursor was adjusted by injecting a known volume of the saturated vapour from the organic species. Two GC background measurements (20 min) were performed to monitor the homogeneous distribution of the precursor inside the aerosol smog-chamber and to obviate self-initiated aerosol formation. The aerosol formation was initiated by injection of a known volume of 1% molecular chlorine, diluted in N_2 (Linde, Germany) into the chamber. Table 1 summarizes the experimental conditions of the aerosol smog-chamber runs.

Aerosol flow-reactor setup and experiments

To study the chemistry of the halogen-induced organic aerosol formation, an aerosol flow-reactor was set up and coupled to a Bruker IFS 113v FTIR spectrometer, which is equipped with a normal pressure sample compartment. The sample compartment is separated from the vacuum system of the FTIR spectrometer using two KBr windows. A circular multi-reflection cell[28] with path lengths up to 105 cm was installed inside the sample compartment and coupled to the flow reactor. The flow reactor itself is described in detail by Ofner *et al.*, 2010.[29] Additionally, the lamps of a common UV face tanner (Philips UVA Typ HP 3151/A), with an emission characteristic between 270 and 450 nm and a maximum power of 1 W m^{-2} nm^{-1}, were placed around the flow-reactor for photolysis of molecular chlorine. Three mass-flow controllers (MFC) were used to control the flow conditions within the flow reactor. The sheath and reactive gas-flow were mixed using two 2000 sccm MFCs from zero-air and 1% Cl_2 in N_2 and fed to the top of the aerosol flow-reactor itself. Zero-air was saturated with the organic precursor (α-pinene – 500 Pa vapour pressure) and injected into the moveable inlet of the flow reactor. The XOA formation was initiated depending on

Table 2 Flow conditions and mixing ratios within the aerosol flow-reactor

Condition	Cl_2 [Vol%]	α-pinene [ppm]	total flow [sccm]	max. reaction time [s]
A	0.44	85	600	8.5
B	0.35	67	750	6.7
C	0.60	116	445	11.5
D	0.84	161	316	16.0

the position of the moveable inlet, where mixing of the organic vapour and the RHS occurred. The flow conditions and resulting available reaction times are shown in Table 2. The flow reactor was operated at slightly reduced pressure to avoid contamination of the multi-reflection cell and the reactor itself with low-volatile species or aerosol particles. The applied flow rates allowed an operation of the reactor under laminar conditions.

For FTIR spectroscopy, 256 scans with a resolution of 4 cm^{-1} in the range 4000–600 cm^{-1} were recorded. Spectra were recorded at 5, 10, 20, 30, and 40 cm distance of the inlet to the centre of the multi-reflection cell. Background measurements were performed when the flow reactor was flushed with zero air. The raw spectra were background corrected using the Bruker Opus 7.0 software package because of a strong contribution of Mie scattering of aerosol particles to the absorbance spectra. Further, absorptions of CO_2 at 2349 and 667 cm^{-1} were corrected to zero at the lowest stage of the reactor for series A. To allow a clear demonstration of the obtained spectral time-series, 2D correlation spectros-copy[30–32] was applied. The software package 2Dshige v1.3 (2Dshige © Shigeaki Morita, Kwansei-Gakuin University, 2004–2005) was used to calculate the synchronous and asynchronous 2D correlation plots.

3   Results and discussion

Halogen-induced ultra-fine particle formation

Aerosol formation experiments were conducted for various concentrations of α-pinene and chlorine atoms (*cf.* Table 1). After injecting molecular chlorine into

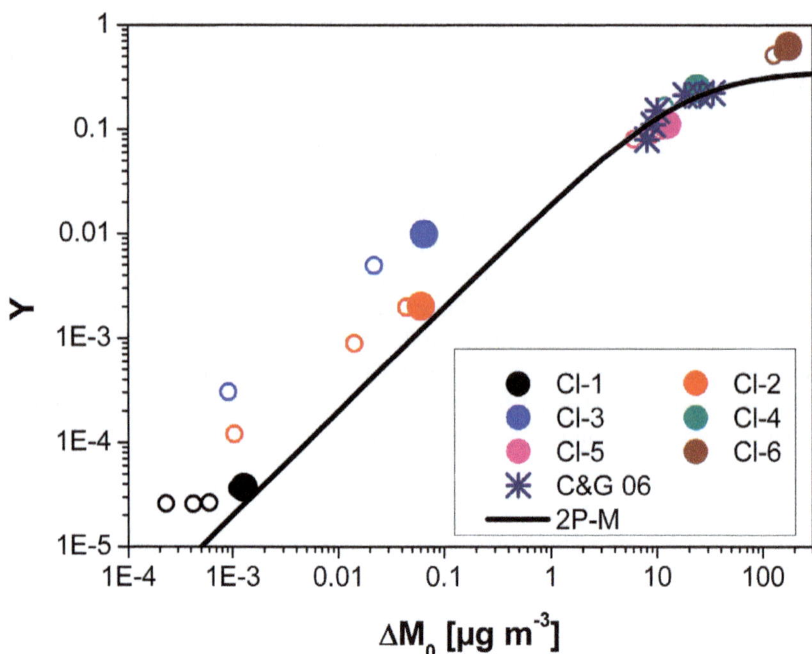

Fig. 3   Yield curves of α-pinene and chlorine from six smog-chamber experiments, and yields of six experiments and the yield curve from a two-product model by Cai and Griffin.[8] For clarity, the lowest yield from Cl-3 is not shown.

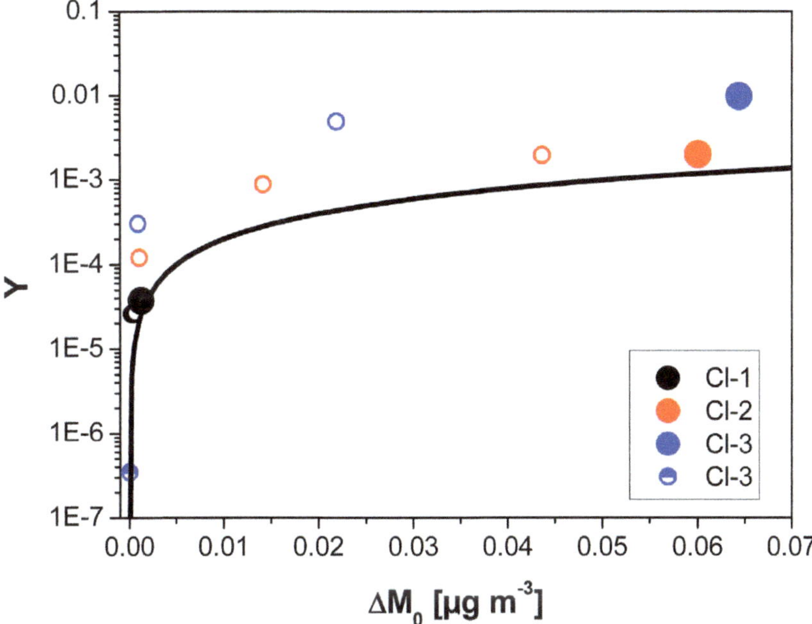

Fig. 4 Closeup of the low concentration experiments Cl-1 to Cl-3.

the illuminated aerosol smog chamber it took about 2 min until the maximum of the particle concentration was reached at low concentrations of the organic precursor and chlorine (Cl-1 to Cl-3, Table 1). In the high concentration experiments (Cl-4 to Cl-6, Table 1), particle formation started immediately after injection. Chlorine mixing ratios as low as 2.5 ppb were sufficient to induce particle formation. The aerosol yield Y was computed as the mass concentration ΔM_0 ($\mu g\ m^{-3}$) of SOA formed after the organic precursor is completely reacted, divided by the concentration of the reacted organic precursor ΔVOC ($\mu g\ m^{-3}$),

$$Y = \frac{\Delta M_0}{\Delta VOC}. \tag{1}$$

Following Pandis *et al.*,[33] in most aerosol formation experiments, Y is evaluated after the organic precursor is completely reacted, gas–particle equilibrium is established, and aerosol growth has stopped. However, several studies have investigated time-dependent aerosol formation,[34,35] *i.e.* ΔM_0 and ΔVOC are evaluated frequently in one individual experiment.

In Fig. 3 and 4, we present both the traditional aerosol yield derived from eqn (1) after α-pinene is completely reacted (large symbols), and the time-dependent yield curves (small symbols). For comparison, the yields from Cai and Griffin[8] resulting from six α-pinene and chlorine experiments are also included in Fig. 3. Finally, we compare our yields with the yield curve from a two-product model (2P-M) after Odum *et al.*[36] parameterized by Cai and Griffin.[8]

We consider the precursor mixing ratios of several ppb of α-pinene and chlorine in the experiments Cl-1 to Cl-3 to be within one or two orders of magnitude of typical ambient conditions in coastal environments with biogenic

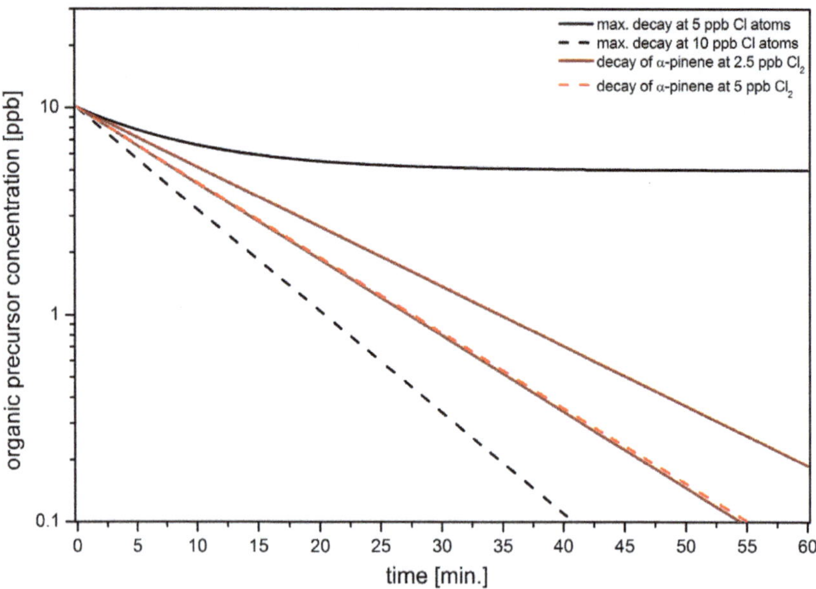

Fig. 5 Calculated maximum and measured decays of the organic precursor based on the available chlorine concentration.

emissions of volatile organic compounds. In any case, these experiments extend previous studies to aerosol mass concentrations almost four orders of magnitude lower than the work by Cai and Griffin.[8] The yields ranging from 0.0013 to 0.065 are in reasonable agreement with the two-product model shown in Fig. 3 and 4. Precursor mixing ratios in experiments Cl-4 and Cl-5 are similar to the concentration range covered by Cai and Griffin.[8] Our experiments confirm their results. Higher mixing ratios of α-pinene and chlorine in experiment Cl-6 give an aerosol yield higher than expected from the two-product model by Cai and Griffin.[8]

Overall, the observed yields are lower than the yields reported for ozone oxidation of α-pinene but larger than yields reported for photooxidation of α-pinene (*e.g.* Griffin *et al.*[37]). Therefore, the chlorine oxidation mechanism cannot be neglected in environments with high ambient chlorine concentrations.

Halogen consumption and organic precursor decays

By calculating the decay of available molecular chlorine (based on the chamber actinometry and calculation of the photolysis rate) and subsequently the availability of chlorine atoms, the maximum decay of the organic precursor was calculated. The calculation is based on the simple assumption that every chlorine atom promptly reacts with a molecule of the organic precursor and is not recovered to react with other unprocessed precursor molecules. This maximum decay is compared to the real observed decays of α-pinene, obtained from gas chromatography, for the low-concentration experiments (Fig. 5).

As indicated in Fig. 5, the maximum decay consuming all chlorine atoms for the experiments with 2.5 ppb of molecular chlorine would only allow decays from 10 down to 5 ppb. The real measured decays are much higher. Tens to hundreds

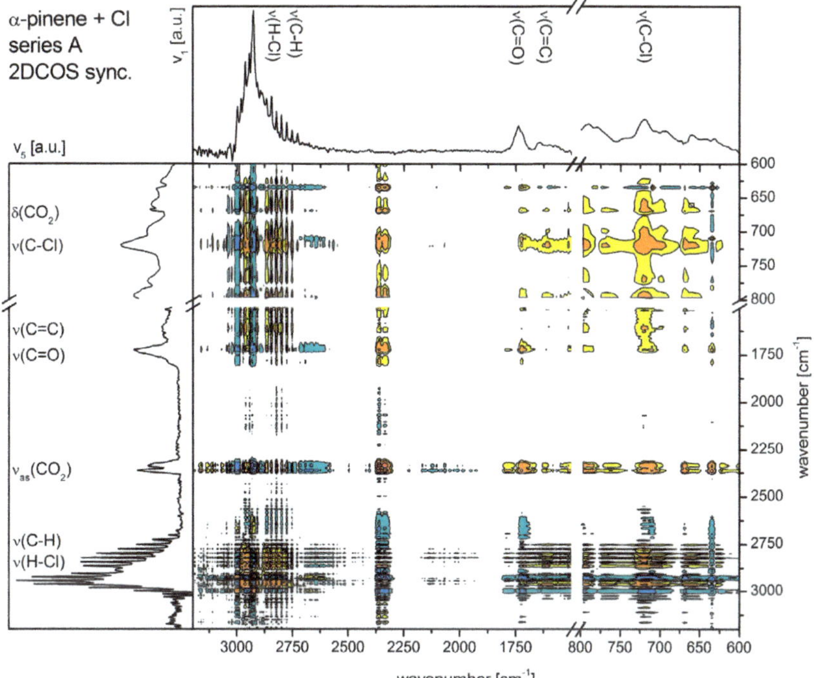

Fig. 6 Synchronous 2D correlation spectroscopy plot of photochemical XOA formation from α-pinene with chlorine at condition A, where ν_1 is the first obtained spectrum at 1.1 s and ν_5 is the last obtained spectrum at 8.5 s – yellow to red contours indicate a synchronized change of the signals in the same direction, whereas blue contours indicate a synchronized change of the signals in opposite directions.

of ppt are reached after 60 min, thus, indicating that the basic assumption that chlorine atoms are fully consumed by the same equivalent of organic precursor and not regenerated is inconsistent with the observed reaction behaviour.

For the aerosol smog-chamber experiments, the best agreement of the calculated decay and the measured decay is at 5 ppb of molecular chlorine and 10 ppb of organic precursor. At higher concentrations, the diffusion of the organic precursor and the RHS becomes limiting. XOA formation at low chlorine concentrations appears to be driven by a radical-chain-like reaction mechanism. This is consistent with an initial Cl addition pathway to alpha-pinene and subsequent opening of the ring leading to pinonaldehyde and release of Cl as proposed by Cai and Griffin.[8] A major limiting factor is the diffusivity of the gaseous species.

Chemical transformation of XOA during the formation and aging process

Chemical details on XOA formation were obtained from the aerosol flow-reactor experiments, where each series is represented by 5 single experiments at different times of the reaction. The general interpretation of the FTIR spectra and 2D correlation plots is related to the FTIR spectrum of pure α-pinene.[38] Spectra of the aerosol flow-reactor conditions A and C (Table 2) represent the highest spectral changes, thus these two time series were interpreted in detail. Condition A allows

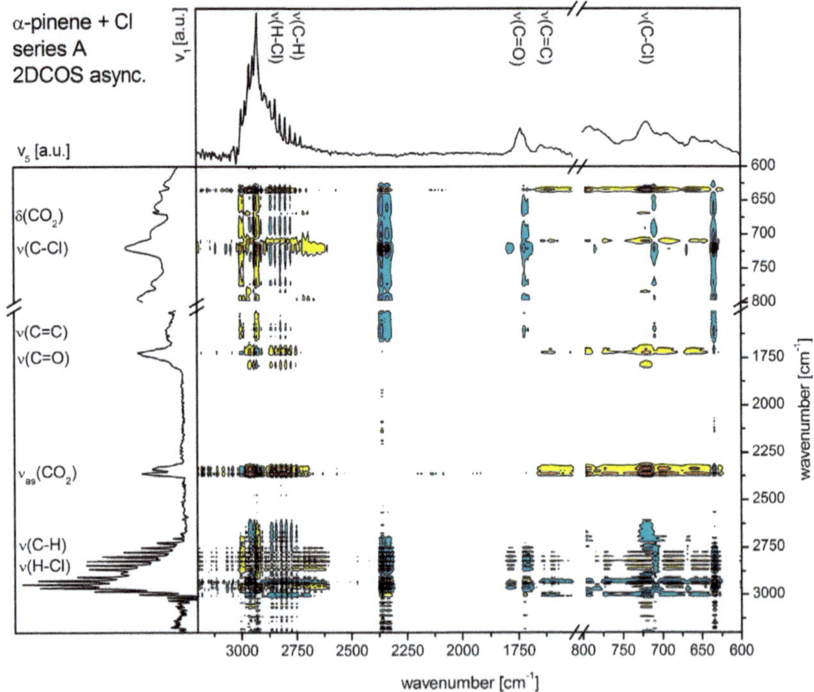

Fig. 7 Asynchronous 2D correlation spectroscopy plot of XOA formation from photochemical reactions of α-pinene with chlorine at condition A, where ν_1 is the first obtained spectrum at 1.1 s and ν_5 is the last obtained spectrum at 8.5 s – yellow to red contours indicate a change of absorption in ν_1 before ν_5, whereas blue contours indicate the other way round.

temporal steps of 1.1, 2.1, 4.2, 6.3 and 8.5 s (related to the distance of the inlet to the centre of the multi-reflection cell – see section 2 Methods), condition C of 1.4, 2.9, 5.8, 8.7 and 11.5 s. The 2D correlation of the fast condition A is more sensitive towards changes related to the gaseous species like HCl and CO_2. The slower condition C exhibits significant changes related to the aerosol particle phase like changes of the ν(C=O), ν(C=C), ν(C–H) and the ν(C–Cl) stretch vibrations. While the temporal starting point of both series should be rather similar, significant differences (*e.g.* in the ν(C=O) spectral region) are visible. These differences can be explained by changing flow conditions, different concentration profiles and a different behaviour of the laminar flow profile and the related mixing inside the flow reactor.

As mentioned above, the 2D correlation spectroscopy plots (Fig. 6 and 7) of the photochemical XOA formation from α-pinene with chlorine at aerosol flow-reactor condition A appear to be more sensitive to the gaseous species. The synchronous plot (Fig. 6) indicates a coupling of CO_2 (ν_{as}(C=O) at 2349 cm^{-1}) and HCl (narrow absorptions between 2700–3100 cm^{-1}) formation. The aerosol formation itself is also correlated with the formation of carbon chlorine bonds (ν(C–Cl) at 720 cm^{-1}), by addition of chlorine to the C=C double bond or to a radical site, and of carbonyls and/or carboxylic acids (ν(C=O) at 1700–1750 cm^{-1}). The synchronous plot exhibits an anti-correlation of the aliphatic ν(C–H) of the –CH$_3$ groups of α-pinene[38] at 2995 cm^{-1} and 2925 cm^{-1} with the

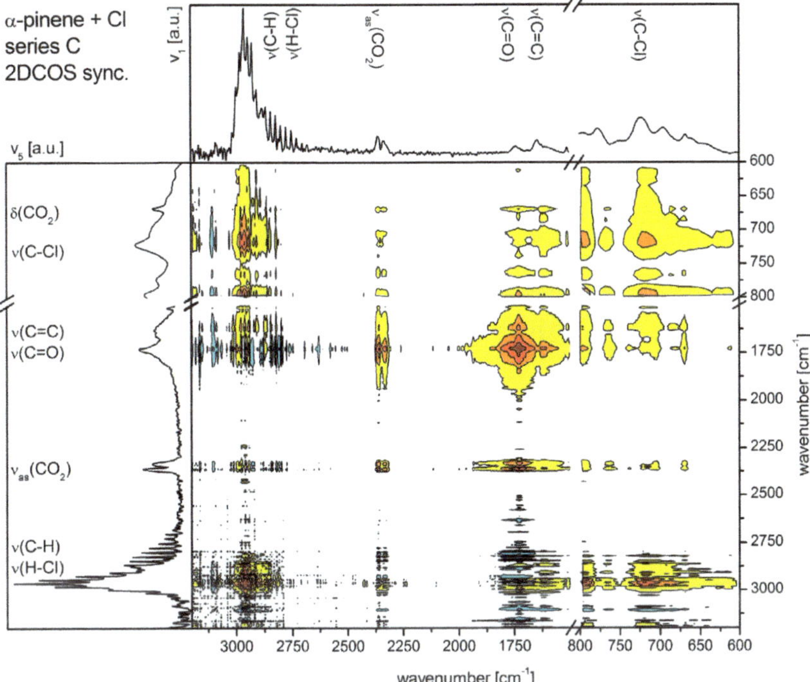

Fig. 8 A synchronous 2D correlation spectroscopy plot of photochemical XOA formation from α-pinene with chlorine at condition C; ν_1 at 1.4 s and ν_5 at 11.5 s.

overall aerosol formation (which is caused by a halogen-induced abstraction of hydrogen atoms).

Details on the sequence of transformation of single absorptions and species are given by the 2D asynchronous plot (Fig. 7). The sequential order was interpreted according to the so-called "Noda" rules.[22] No asynchronous cross-peaks exist in case of the absorptions of the C–Cl bond formation and the HCl release (which is indicated in the synchronous plot by positive cross-peaks). The formation of halogenated species in the aerosol phase is synchronous and thus coupled to the release of HCl to the gas phase. Due to the fact that the synchronous cross-peaks between CO_2 and the C–Cl formation are positive and asynchronous cross-peaks exist, the CO_2 release is following the C–Cl formation and HCl release. Missing asynchronous cross-peaks between the $\nu(C{=}O)$ and the release of CO_2 indicate a synchronous coupling of the formation of carbonyls or carboxylic acids and the release of CO_2. The sequence of $\nu(C{-}Cl)$ formation with HCl release and following C=O bond formation is also indicated by the asynchronous cross-peaks between HCl and $\nu(C{=}O)$ and $\nu(C{-}Cl)$ and $\nu(C{=}O)$.

In contrast to series A, plots from series C appear to be more sensitive towards changes of the formed aerosol phase. The synchronous plot of 2D correlation spectroscopy in Fig. 8 exhibits strong auto-correlation peaks related to the formation of carbonyls and/or carboxylic acids ($\nu(C{=}O)$ at 1700–1750 cm^{-1}) and to the formation of carbon chlorine bonds at 720 cm^{-1}. Further changes are visible in case of the olefinic $\nu(C{=}C)$ stretch vibration, which appears to broaden, caused by the formation of oligomers and more complex species. While the

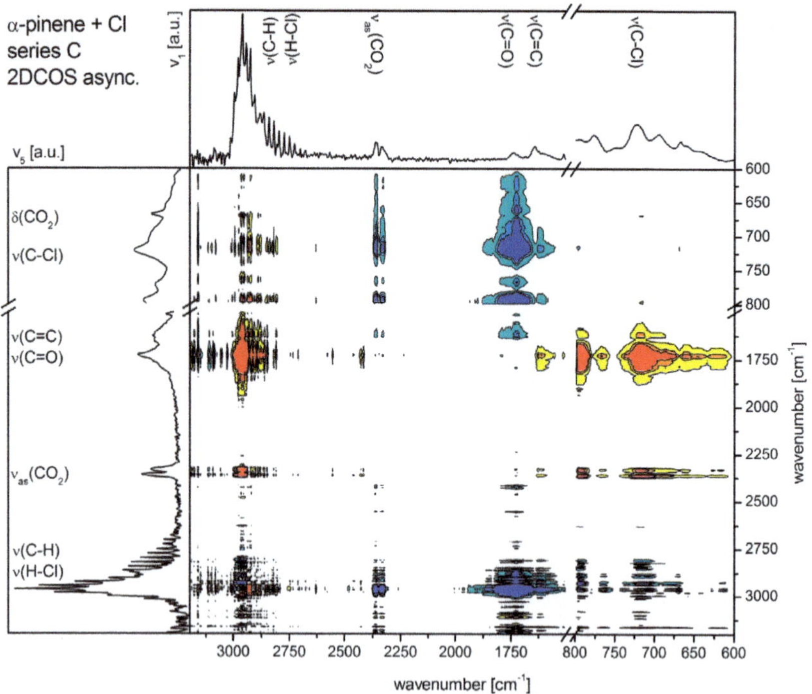

Fig. 9 Asynchronous 2D correlation spectroscopy plot of photochemical XOA formation from α-pinene with chlorine at condition C; ν_1 at 1.4 s and ν_5 at 11.5 s.

degradation of the well-defined ν(C–H) stretch vibration in Fig. 8 is not visible like in Fig. 6, a general broadening of the aliphatic carbon hydrogen bond regime is observed. This broadening, which appears like carbon-hydrogen bond formation, can be related to the formation and aging of particulate matter.

The sequence of the gaseous species is similarly demonstrated in the asynchronous plot of series C (Fig. 9). The formation of carboxylic acids or carbonyls follows after C–Cl formation. Also the transformation of the sharp ν(C=C) stretch vibration of the organic precursor (around 1650 cm^{-1}) to the broad absorption of various olefinic structures within the particulate phase takes place before ν(C=O) formation. The broad asynchronous cross-peaks between 600 and 800 cm^{-1} indicate the formation of a broad finger-print region. This can be related to the formation of oligomers and high-molecular structures or a large variety of different organic species.

The general roadmap of XOA formation, initiated by addition of chlorine and H atom abstraction, forming HCl and chlorine carbon bonds is in good agreement with the studies by Cai and Griffin[8] and Karlsson et al.[7] Also, the subsequent addition of oxygen to the so formed radical sites of the organic structure and the coupled formation of carbonyls and carboxylic acids, as well as the concurrent release of CO_2, is in agreement with the suggestions made by these authors. During further processing, the freshly formed XOA appears to degrade and becomes more SOA-like, losing the halogen-induced characteristics and appearing more "common".

Conclusions

Reactive halogen species releasing Cl radicals can induce organic aerosol formation by reaction with organic precursor gases, such as α-pinene. From the experiments in this work evidence is found both for a Cl addition pathway to α-pinene as well as an H-atom abstraction pathway. On one hand, the formation of C–Cl bonds in the FTIR spectra as well as indirect evidence for a radical-chain mechanism (*cf.* Fig. 5) is consistent with Cl addition. On the other hand, the formation of HCl observed in the flow-reactor experiments is consistent with H-atom abstraction in the reaction between Cl and α-pinene.[15] However, it is difficult to quantify the respective contributions of these pathways to the overall reaction.

The aerosol yield found in low and high concentration experiments is in general agreement with previous experimental data by Cai and Griffin[8] as well as a parameterization of the yield curve presented by the same authors. Even at precursor concentrations almost four orders of magnitude lower than the previous experiments, a reasonable agreement with the two-product parameterization is found. With a broad range of mixing ratios of α-pinene and chlorine in six experiments we could extend the data from these authors and obtain aerosol formation yields even at close-to-ambient conditions.

The chemical formation of this type of organic aerosol is halogen driven as demonstrated by the aerosol flow-reactor experiments. The formation process occurs in the absence of ozone but also even without oxygen. Thus, the formation of halogen-induced organic aerosol represents an aerosol formation process apart from the commonly considered formation pathways. The further processing of XOA with oxygen and oxygen-containing reactants causes a transformation of XOA to the common SOA which is dominated by oxygen-containing functional groups. This XOA-born SOA still exhibits carbon–chlorine bonds, thus indicating the XOA-like source of the organic aerosol. With further chemical aging and processing by atmospheric reactive species, the XOA-like character of the organic aerosol from monoterpenes and RHS is lost.

The formation of XOA from monoterpenes or other organic volatile species with RHS must not be neglected for *e.g.* maritime areas, where high RHS and VOC concentrations occur at dawn or sunset simultaneously. Although XOA appears to be transformed to SOA by aging and processing, this special type of organic aerosol might exhibit different features not commonly related to SOA or the organic aerosol in general. The halogen-driven formation and the coupled formation of solid halogen species in the particle phase significantly influences physico-chemical parameters like water-solubility, the potential to act as cloud condensation or ice nuclei, the adsorption behaviour with respect to gas-phase species as well as the interaction with sunlight (or the UV/VIS absorption spectrum[10]). This changes the influence on radiative forcing and thus modifies the influence of XOA on global warming compared to commonly considered SOA.

Acknowledgements

The authors would like to thank the German research foundation for funding within the research unit HaloProc (HE5214/5-1 and ZE792/5-2). Further the authors thank Georg Ramer from the Institute of Chemical Technologies and Analytics, Vienna University of Technology for his assistance in 2D correlation spectroscopy.

References

1 J. H. Kroll and J. H. Seinfeld, *Atmos. Environ.*, 2008, **42**, 3593–3624.

2 J. L. Jimenez, M. R. Canagaratna, N. M. Donahue, A. S. H. Prevot, Q. Zhang, J. H. Kroll, P. F. DeCarlo, J. D. Allan, H. Coe, N. L. Ng, A. C. Aiken, K. S. Docherty, I. M. Ulbrich, A. P. Grieshop, A. L. Robinson, J. Duplissy, J. D. Smith, K. R. Wilson, V. A. Lanz, C. Hueglin, Y. L. Sun, J. Tian, A. Laaksonen, T. Raatikainen, J. Rautiainen, P. Vaattovaara, M. Ehn, M. Kulmala, J. M. Tomlinson, D. R. Collins, M. J. Cubison, E. J. Dunlea, J. A. Huffman, T. B. Onasch, M. R. Alfarra, P. I. Williams, K. Bower, Y. Kondo, J. Schneider, F. Drewnick, S. Borrmann, S. Weimer, K. Demerjian, D. Salcedo, L. Cottrell, R. Griffin, A. Takami, T. Miyoshi, S. Hatakeyama, A. Shimono, J. Y. Sun, Y. M. Zhang, K. Dzepina, J. R. Kimmel, D. Sueper, J. T. Jayne, S. C. Herndon, A. M. Trimborn, L. R. Williams, E. C. Wood, A. M. Middlebrook, C. E. Kolb, U. Baltensperger and D. R. Worsnop, *Science*, 2009, **326**, 1525–9.

3 I. Riipinen, T. Yli-Juuti, J. R. Pierce, T. Petäjä, D. R. Worsnop, M. Kulmala and N. M. Donahue, *Nat. Geosci.*, 2012, **5**, 453–458.

4 C. O'Dowd, M. Facchini, F. Cavalli, D. Ceburnis, M. Mircea, S. Decesari, S. Fuzzi, Y. J. Yoon and J.-P. Putaud, *Nature*, 2004, **431**, 676–680.

5 L. J. Carpenter, S. D. Archer and R. Beale, *Chem. Soc. Rev.*, 2012, **41**, 6473–506.

6 Q. K. Timerghazin and P. A. Ariya, *Phys. Chem. Chem. Phys.*, 2001, **3**, 3981–3986.

7 R. S. Karlsson, J. J. Szente, J. C. Ball and M. M. Maricq, *J. Phys. Chem. A*, 2001, **105**, 82–96.

8 X. Cai and R. J. Griffin, *J. Geophys. Res.*, 2006, **111**, D14206.

9 X. Cai, L. D. Ziemba and R. J. Griffin, *Atmos. Environ.*, 2008, **42**, 7348–7359.

10 J. Ofner, N. Balzer, J. Buxmann, H. Grothe, P. Schmitt-Kopplin, U. Platt and C. Zetzsch, *Atmos. Chem. Phys.*, 2012, **12**, 5787–5806.

11 J. H. Kroll, N. M. Donahue, J. L. Jimenez, S. H. Kessler, M. R. Canagaratna, K. R. Wilson, K. E. Altieri, L. R. Mazzoleni, A. S. Wozniak, H. Bluhm, E. R. Mysak, J. D. Smith, C. E. Kolb and D. R. Worsnop, *Nat. Chem.*, 2011, **3**, 133–9.

12 T. Moise and Y. Rudich, *Geophys. Res. Lett.*, 2001, **28**, 4083–4086.

13 R. C. Sullivan and K. A. Prather, *Anal. Chem.*, 2005, **77**, 3861–85.

14 D. Vione, V. Maurino, S. C. Man, S. Khanra, C. Arsene, R.-I. Olariu and C. Minero, *ChemSusChem*, 2008, **1**, 197–204.

15 B. J. Finlayson-Pitts, C. J. Keoshian, B. Buehler and A. A. Ezell, *Int. J. Chem. Kinet.*, 1999, **31**, 491–499.

16 B. J. Finlayson-Pitts, *Res. Chem. Intermed.*, 1993, **19**, 235–249.

17 C. Spicer and E. Chapman, *Nature*, 1998, **1996**, 1996–1999.

18 A. Saiz-Lopez and R. von Glasow, *Chem. Soc. Rev.*, 2012, **41**, 6448–72.

19 C. Zetzsch and W. Behnke, *NATO ASI Ser., Ser. I*, 1993, **7**, 291–306.

20 B. J. Finlayson-Pitts, *Chem. Rev.*, 2003, **103**, 4801–22.

21 J. A. Thornton, J. P. Kercher, T. P. Riedel, N. L. Wagner, J. Cozic, J. S. Holloway, W. P. Dubé, G. M. Wolfe, P. K. Quinn, A. M. Middlebrook, B. Alexander and S. S. Brown, *Nature*, 2010, **464**, 271–4.

22 I. Noda, *Generalized Two-Dimensional Correlation Spectroscopy in Frontiers of Molecular Spectroscopy*, ed. Jaan Laane, Elsevier, 2009, ISBN: 978-0-444-53175.

23 J. Ofner, H.-U. Krüger, H. Grothe, P. Schmitt-Kopplin, K. Whitmore and C. Zetzsch, *Atmos. Chem. Phys.*, 2011, **11**, 1–15.

24 B. Bohn, F. Rohrer, T. Brauers and A. Wahner, *Atmos. Chem. Phys.*, 2005, **5**, 493–503.

25 A. Ruggaber, R. Dlugi and T. Nakajima, *J. Atmos. Chem.*, 1994, **18**, 171–210.

26 R. K. Pathak, C. O. Stanier, N. M. Donahue and S. N. Pandis, *J. Geophys. Res.*, 2007, **112**, D03201.

27 J. G. Crump and J. H. Seinfeld, *J. Aerosol Sci.*, 1981, **12**, 405–415.

28 J. Ofner, H.-U. Krüger and C. Zetzsch, *Appl. Opt.*, 2010, **49**, 5001.

29 J. Ofner, H.-U. Krüger and C. Zetzsch, *Z. Phys. Chem.*, 2010, **224**, 1171–1183.

30 I. Noda, *Appl. Spectrosc.*, 1993, **47**, 1329–1336.

31 I. Noda, A. Dowrey and C. Marcott, *Appl. Spectrosc.*, 2000, **54**, 236A–248A.

32 B. Muik, B. Lendl, A. Molina-Diaz, M. Valcarcel and M. J. Ayora-Cañada, *Anal. Chim. Acta*, 2007, **593**, 54–67.

33 S. N. Pandis, R. A. Harley, G. R. Cass and J. H. Seinfeld, *Atmos. Environ., Part A*, 1992, **26**, 2269–2282.

34 T. Hoffmann, J. Odum, F. Bowman, D. Collins, D. Klockow, R. C. Flagan and J. H. Seinfeld, *J. Atmos. Chem.*, 1997, **26**, 189–222.

35 N. L. Ng, J. H. Kroll, M. D. Keywood, R. Bahreini, V. Varutbangkul, R. C. Flagan, J. H. Seinfeld, A. Lee and A. H. Goldstein, *Environ. Sci. Technol.*, 2006, **40**, 2283–2297.

36 J. R. Odum, T. Hoffmann, F. Bowman, D. Collins, R. C. Flagan and J. H. Seinfeld, *Environ. Sci. Technol.*, 1996, **30**, 2580–2585.
37 R. J. Griffin, D. R. Cocker, R. C. Flagan and J. H. Seinfeld, *J. Geophys. Res.*, 1999, **104**, 3555.
38 H. W. Wilson, *Appl. Spectrosc.*, 1976, **30**, 209–212.

DISCUSSIONS

General discussion

DOI: 10.1039/C3FD90032K

Professor Wahner opened the discussion of the introductory lecture by Professor Pandis: The figure which showed the model results for the aerosol volatilities at different emissions. Why does the intermediate fraction not change, does it depend on the chosen scaling?

Professor Pandis replied: The contribution of the mid-range sources to the concentration of organics in Paris during the summer varied from 51 to 68 percent for the compounds simulated by the VBS. The small variation of the fractional contribution of this source area was a consequence of our choice of its boundaries (50 to 500 km from the city center). We have performed these source apportionment calculations using other distances (*e.g.*, 50 to 100 km from the city center) and the corresponding fractions varied a lot more.

Dr Allan commented: The changes in oxygen to carbon ratio reported at higher thermal denuder temperatures are interesting. Would you care to speculate whether these are due to condensation of reacting gases or processes within the particles?

Professor Pandis answered: We think that both of these are probably happening. We did not observe a significant change in particle mass as we heated the particles from 150 to 300 C, while the O : C increased significantly. This would support the occurrence of reactions in the particulate phase. Oxidation in the gas phase followed by condensation would probably cause a noticeable increase in the particle mass. However, this could be balanced by evaporation of unreacted organic compounds. We are planning to test the hypothesis of gas-phase reactions by passing the aerosol sample through an activated carbon denuder before it enters the thermodenuder.

Professor McFiggans commented: Could you comment on the sensitivity of the contribution of SOA to PM loading and its spatial distribution to the method used to invert dilution and thermal denuder data to retrieve volatility distributions. Can this provide a constraint on the accommodation coefficient of organic vapours on growing particles?

Professor Pandis replied: The secondary OA in our model has a number of sources: oxidation of biogenic and anthropogenic VOCs, oxidation of IVOCs, evaporation and subsequent oxidation in the gas phase of SVOCs and long range transport from outside the modeling domain. These secondary components are responsible for most of the predicted OA in all areas in all seasons, with the exception of urban areas during the winter. PMCAMx predicts that OA dominates the submicrometer PM composition everywhere (the Eastern Mediterranean is the only exception here).

The method proposed for the inversion of the thermodenuder and isothermal dilution data can provide information about potential mass transfer delays during the evaporation of the particles. These delays may be due to transport of the evaporating species inside the particle, at the interface, *etc.* All of these potential resistances are currently parameterized to a zeroth degree using the mass accommodation coefficient.

Professor Abbatt asked: How much of the challenge in correctly modeling aerosol nitrate is associated with correctly knowing the sources of ammonia?

Professor Pandis responded: The availability of ammonia often controls the formation of ammonium nitrate. This is the case especially in environments that are characterized by high levels of nitric acid and low or moderate levels of ammonia. Past work (for example S., Takahama, *et al.*, *J. Geophys. Res.*, 2004, **109**, D16S06) has shown that if the total (gas and particulate phase) nitric acid and ammonia levels are known then the available aerosol thermodynamic models can reproduce quite accurately the ammonium nitrate concentrations. Therefore, errors in predictions of ammonium nitrate by chemical transport models are often associated with either errors in simulating the formation and removal of nitric acid or the emissions and removal of ammonia.

Professor Freedman opened the discussion of the paper by Professor Johnston: Could you comment on how the O : C ratios observed on new particle formation days and on non-new particle formation days in Lewes compare to those you have measured elsewhere?

Professor Johnston answered: The O : C ratios on new particle formation days in Lewes in summer were similar to those we have measured in a boreal forest in the springtime (M. R. Pennington *et al.*, *Atmos. Chem. Phys.*, 2013, **13**, 10215–10225). In Lewes, the estimated ratios were in the order of 0.6 whereas an the boreal forest location they were in the order of 0.5 in early spring and closer to 0.6 in late spring.

Professor Freedman asked: Could you comment on why the O : C ratios on new particle formation days are so different than the values on non-new particle formation days as shown in Figure 4 of your paper?

Professor Johnston replied: On the particle formation days (12th and 13th August), the wind was from the northwest where there are few anthropogenic sources of carbonaceous aerosol within 100 km of the measurement site. On the other days (11th and 14th August), the wind was from the south, where significant combustion sources exist including a coal-fired power plant and motor vehicle emissions. The lower O : C ratios on non particle formation days most likely reflect the unoxidized character of the carbonaceous matter from these combustion sources. The higher O : C ratios on particle formation days suggest that most of the carbonaceous matter is fresh secondary organic aerosol.

Professor Pandis asked: Could the amines explain the excess nitrogen that was observed in the nanoparticles during the measurement? What was their concentration?

Professor Johnston answered: Amines are not likely candidates for excess nitrogen because their concentrations were low during the event periods (trimethylamine less than about 40 pptv; dimethylamine less than about 20 pptv).

Professor Harrison commented: It is not clear from your paper what you regard as the source of silicon in your samples. Can you please clarify?

Professor Johnston responded: We think the source of silicon in these measurements is local, possibly arising from plastic materials and/or coatings, for example containers and other items around the site that were not used as particle sampling lines, but nonetheless could have outgassed siloxanes. A local source is indicated by the time dependence of the silicon mole fraction in Figure 1b of our paper. The silicon mole fraction increased during time period 1-2, when photochemically generated sulphuric acid was condensed onto preexisting particles at the site, but then dropped when particles from the NPF event (formed remotely from the site) arrived during time period 2-3. We have observed relatively large amounts of silicon in ambient nanoparticles at an urban location (M. R. Pennington *et al.*, *J. Geophys. Res.*, 2012, **117**, D00V10) where the time dependence suggested a photochemical source, which would be consistent with photo oxidation of siloxanes. In contrast, we did not observe much silicon in ambient nanoparticles from a boreal forest (M. R. Pennington *et al.*, *Atmos. Chem. Phys.*, 2013, **13**, 10215–10225) where siloxane emissions would be expected to be low.

Professor Wahner asked: In your experimental observations you showed the detection of silicon. What is the role of silicon in the nucleation, even if it is an artifact from sampling lines *etc.*?

Professor Johnston responded: The impact of silicon on nucleation is not known, though it is likely to be small for the two particle formation days on 12th and 13th August. The particle size distributions in Figure 1a of the paper suggest that nucleation occurred away from the site, while the time dependence of the particulate silicon concentration in Figure 1b suggests a local source.

Professor Donahue remarked: The new-particle formation events shown, for example, in Figure 1 of your paper appear to be "detached". It is thus not obvious that the vapours condensing on 20 nm particles during the period 1-2 are also causing nucleation and growth at smaller sizes, as asserted in the paper. Do nano-SMPS data confirm that nucleation is indeed occurring at the site during this interval?

Professor Johnston responded: Nano-SMPS measurements (3 nm and above) for 12th August show very few particles below about 8 nm. Based on the available information, it does appear that nucleation occurred away from the site, and therefore the change in composition between time points 1 and 2 for this day does not directly indicate the chemical species associated with nucleation. However, the particle growth rate indicates that nucleation did occur during this time period. We have made similar measurements for new particle formation events at another location (M. R. Pennington et al., Atmos. Chem. Phys., 2013, 13, 10215–10225). The composition changes early in these events (equivalent to time points 1 and 2) are similar to what we report here, and did not depend on whether or not nucleation occurred directly at the site.

Professor Donahue opened the discussion of the paper by Professor Percival: The role of Criegee Intermediate stabilization is introduced in section 1.3.2 and the flowchart of Figure 1 in your paper but then more or less vanishes. Equation VIII omits the stabilization term entirely and describes a "CI" steady state as opposed to an "SCI" steady state. A substantial amount of work has been published in recent years addressing the stabilization fraction as a function of pressure, for example in Drozd and Donahue (G. T. Drozd et al., J. Phys. Chem. A, 2011, **115**, 161–166; N. M. Donahue et al., Phys. Chem. Chem. Phys., 2011, **13**, 10848–10857 and T. Berndt et al., J. Phys. Chem. Lett., 2012, **3**, 2892–2896). SCI yields can be quite low, especially for endocyclic alkenes such as a-pinene (with a yield of roughly 15% at 1 atm) and small alkenes such as isoprene. Would consideration of this not have a substantial effect on your results?

Professor Percival responded: The "stabilized fraction" is not always exactly well-known. Hasson et al.[1] suggest a value of 0.1 for 2,3-dimetrhyl-2-butene, 0.39 for ethene and Drozd and Donahue[2] show a stabilization fraction of the Criegee intermediate from tetramethylethene (0.65) that is almost a factor of seven higher than the determination of Hasson et al.,[1] and demonstrates a physically reasonable dependence on carbon number. In the model we have applied a "stabilization factor" of 0.5, which is an intermediate between that of ethene ozonolysis and that of ozonolysis of larger linear alkenes. The variation in SCI yield is such that a value of 0.5 is a realistic approximation; even if for some species it is much lower, the overall conclusions of the paper are not effected by this assumption as we know that we are underestimating the level of alkene quite considerably.

1 A. S. Hasson, et al., J. Geophys. Res., 2001, **106**(D24), 34 131–34 142
2 G. T. Drozd and N. M. Donahue, J. Phys. Chem. A, 2011, **115**(17), 4381–4387.

Professor Heard said: For the rate coefficient for the reaction of the Criegee intermediates with SO_2, and also for the yield of SO_3 from these reactions, (i) do you expect there to be a pressure dependence, and (ii) if so, what have you assumed in your model calculations?

Professor Percival responded: Recent theoretical studies[1] have suggested that the products of the reaction of SO_2 with CI at high pressure could be a secondary ozonide. First, if the rate coefficients for small CIs reported recently at 4 Torr[2,3] are not at the high pressure limit, then the rate coefficients will be larger at 760 Torr. Second, although the stabilisation of the potential adduct may be possible at 760 Torr, rather than decomposing to form a carbonyl and SO_3 as seen at 4 Torr, the likely fate of the adduct is the formation of H_2SO_4 via a reaction with water, in an analogous way to SO_3.

1 L. Vreecken *et al.*, *Phys. Chem. Chem. Phys.*, 2012, **14**, 14682–14695.
2 O. Welz, *Science*, 2012, **335**, 204–207.
3 C. A. Taatjes *et al.*, *Science*, 2013, **340**, 177–180.

Dr Cox commented: Although the role of Criegee intermediates as atmospheric oxidants was first proposed over 40 years ago in our work[1] cited by Prof. Percival, relative rate measurements for reactions of Criegee intermediates with H_2O, NO_2, SO_2 , RCHO and RC(O)OH have been the only experimental basis on which to assess the rates of these reactions. Recent direct measurements of rate coefficients for CH_2OO and CH_3CHOO reactions reported in Prof. Percival's paper show that the reactions are much faster than expected and that the steady state assumption for Criegee intermediates, used in the earlier analysis, is justified. There remains considerable uncertainty on the rates of uni-molecular loss (a value of $k_4 = 200$ s^{-1} is assigned in the paper – what is the basis for this?), and also the reaction with H_2O, which are both potentially important influences on SO_2 oxidation and aerosol formation in the atmosphere. In the light of the new knowledge, what is the prospect of designing novel chamber/relative rate experiments to reduce uncertainty?

1 R. A. Cox and S. A. Penkett, *J. Chem. Soc., Faraday Trans. 1*, 1972, **68**, 1735.

Professor Percival responded: It is correct that there is considerable uncertainty in the decomposition decay and it is reported to vary from[1] 0.3–250 s^{-1}; in this work we used the median value. The reaction with water can be important (depending on the rate coefficient), particularly in the tropics at the surface, but recent work has shown that for both carbonyl oxide and acetaldehyde oxide, no loss can be observed for the most stable conformer, *i.e.* loss *via* reaction with H_2O is reasonably represented here. Indeed, our recent work suggests that these rate coefficients are much smaller than 1×10^{-16} cm^3 molecule^{-1} s^{-1} and that the value used here is probably at the high end of the estimations. The STOCHEM-CRI estimations suggest that over 50% of the CI formed are CH_2OO and *syn*-CH_3CHOO.

1 J. D. Fenske, A. S. Hasson, A. W. Ho, S. E. Paulson, *J. Phys. Chem. A*, 2000, **104**, 9921.

Professor Carslaw asked: Have you examined the extent to which this mechanism might explain temporal and spatial variability in particle formation in the boundary layer? There is a lot of unexplained variability in nucleation rate.

Professor Percival responded: We have not examined spatial variability in nucleation rate in global terms. It is correct that it will vary a lot spatially as indicated by our WRF Chem model.

Professor Jacobson questioned: Have you compared the rate of the gas phase formation of sulfuric acid by your mechanism to the aqueous phase formation rate?

Professor Heard commented: In a recent paper by Mauldin *et al.*,[1] an interference in their OH CIMS instrument was noticed (when a scavenger was added to remove ambient OH there was still a signal), which they ascribed to Criegee intermediates, which react with SO_2 to form SO_3, and which is then converted to H_2SO_4 and detected by the CIMS. This background signal was used to generate a time-series for the concentration of the Criegee intermediates (strictly as an OH equivalent concentration), which was reported in the paper for the field site in Finland. In principle, your model could be used to calculate the concentration of the Criegee intermediates for comparison with this time series. Did you compare your steady–state Criegee concentrations calculated from equation VIII for the conditions in Finland with the time series reported by Mauldin *et al.*? This would provide an independent constraint on the accuracy of the rate coefficients you have used in your model, but using actual field data.

1 R. L. Mauldin III *et al.*, *Nature*, 2012, **488**, 193–196.

Professor Percival answered: Mauldin *et al.*,[1] suggest that CI + SO_2 chemistry can be used to explain ambient field observations and may contribute up to 50% of the H_2SO_4, which is entirely consistent with our results were we show H_2SO_4 increases from 0.01 to 200%. However, the resolution of the global model does not allow a direct comparison.

1 R. L. Mauldin III *et al.*, *Nature*, 2012, **488**, 193–196.

Professor McFiggans opened the discussion of the paper by Miss Olenius: Line 15, page 9 of your paper: "The discrepancy could be due to uncertainties in the quantum chemical formation free energies" has been used to explain figure 4 discrepancies. Have ACDC calculations been carried out to evaluate whether reasonable uncertainties in quantum chemical predicted free energies can resolve the discrepancies? Is there any/sufficient constraint on these values from measurements?

Miss Olenius answered: We have estimated the uncertainties in the calculated Gibbs free energies to be approximately ± 1 kcal mol^{-1}, and run test simulations using these values (as stated in Section 3.3 of the paper). While de-stabilizing the clusters by 1 kcal mol^{-1} results in an increase in the fraction of negatively charged tetramers containing less than two ammonias, the effect on the ammonia content of negative pentamers is not significant, that is, the major fraction of the pentamers still contain three ammonia molecules. To better assess the uncertainties in the energies that could explain the discrepancies, MCMC (Markov Chain Monte Carlo) studies are included in our future work. Naturally, also other factors may contribute to the discrepancies (see the answers to questions later in the discussion of this paper).

To our knowledge, experimental data on the thermodynamic parameters of negatively charged sulfuric acid-ammonia clusters is not available. However, in our previous work[1] we have shown a good agreement between the calculated energies of positive acid–ammonia clusters and the values measured by Froyd and Lovejoy.[2] Also, the calculated energies of pure negatively charged sulfuric acid clusters compare well with experimental values presented by Froyd and Lovejoy[3] except for the acid dimer, which, however, does not have any effect in terms of cluster kinetics, as the evaporation rate of the dimer is extremely low in any case.

1 O. Kupiainen *et al.*, *Atmos. Chem. Phys.*, 2012, **12**, 3591–3599.

2 K. D. Froyd and E. R. Lovejoy, *J. Phys. Chem. A*, 2012, **116**, 5886–5899.
3 K. D. Froyd and E. R. Lovejoy, *J. Phys. Chem. A*, 2003, **107**, 9812–9824.

Professor Johnston commented: Recently, our group published a study (B. R. Bzdek *et al.*, *J. Am. Chem. Soc.*, 2013, **135**, 3276) suggesting that there may be an activation barrier for ammonia incorporation into ammonium bisulfate clusters. Could such a barrier explain the discrepancy between the modeled and experimental cluster distributions in Figure 4 of your paper?

Miss Olenius answered: It is very possible that such a barrier, if it is high enough, may be able to explain the discrepancies. Another explanation could be the cluster fragmentation inside the instrument as stated in the paper, or both effects may play a role. The activation barriers can be easily incorporated in the model, but with the lack of reasonably accurate estimates of the barrier heights we consider it to be beyond the scope of this study.

Professor Johnston commented: A suggestion for the discrepancy between modeled and experimental cluster distributions (*e.g.* Fig. 4) is that ammonia evaporates from the clusters as they are drawn into the mass spectrometer for analysis. What are the structures of the clusters that are modeled in this work – are all of the ammonia molecules ionized (ammonium)? If so, wouldn't that inhibit evaporation of ammonia from the clusters as they are drawn into the mass spectrometer?

Miss Olenius responded: The ammonia molecules are indeed ionized, but in terms of evaporation rates ammonia is still the most loosely bound molecule in almost all the negatively charged clusters. The clusters can also fragment due to energetic collisions with gas molecules as they are accelerated inside the instrument, and one or more molecules might be knocked off especially from the smallest clusters. It must be noted though that the speculations of the effect of possible fragmentation are tentative as the phenomenon is still poorly understood, and there are other possible explanations for the discrepancies in the ammonia content (see my previous answer and the answers to later questions).

Professor Jacobson said: Have you considered using van der Waals forces in your calculation of coagulation among particles < 15 nm in diameter?

Professor Vehkamäki communicated: We have not considered van der Waals forces in this work, as we believe that the attractive forces between permanent or induced dipoles will not change the collision probablity between two neutral clusters or molecules significantly.

This issue was studied in a paper by Hanna Arstila (presently Vehkamäki) (H. Arstila, O. V. Vasl'ev and M. Kulmala, *J. Chem. Phys.*, 1997, **107**, 544), and the conclusion was that the enhancement in formation rates due to dipole–dipole interactions was less than a factor of 2, which is insignificant compared to the uncertainties of the evaporation coefficients.

Professor Nizkorodov asked: You state in the paper that "kinetic modeling of the full multicomponent acid–base–water system is impossible" because of the large disparity between the collision rates of clusters with water molecules compared to those with acid and base molecules. Would assuming that water–cluster equilibria are infinitely fast make the problem more traceable?

Miss Olenius replied: The assumption of equilibrium with respect to water is indeed justifiable and can be used to include water implicitly in the model, as stated in the paper ("Water can, nevertheless, be implemented in the model implicitly by calculating the effective collision and evaporation rates of the hydrated clusters."). However, in order to obtain the hydrate distribution of each cluster (or of the most important clusters along the growth pathways), the Gibbs free energies of hydration need to be calculated, which requires vast computational effort. As the size of the hydrate (*i.e.* the number of water molecules) increases, both the configurational sampling becomes much more complex and the time required by the quantum chemical calculations increases rapidly. We are nevertheless in the process of calculating the hydration energies, and assessing the effect of water is a topic of future work.

Professor Kiendler-Scharr said: Considering the large concentration of water in the atmosphere, it seems unlikely that clusters formed at early stages of new particle formation have no water in them. While you do describe in your paper that some water may be lost from the clusters when sampling them into the instruments for detection, I am wondering how this should lead to zero water contribution measured. Could you comment on this and give us an opinion on how the inclusion of water in the model would change your results?

Miss Olenius answered: The APi-TOF mass spectrometer includes a pressure interface section consisting of several chambers at successively lower pressures, adding to the time that a cluster spends in the instrument before it reaches the detector. In the APi-TOF, the time between a cluster entering the instrument and its detection is a few microseconds, which is enough for water molecules to evaporate from the cluster. Therefore no water is observed in the clusters, although before the sampling they are very likely to be more or less hydrated at the relative humidity of 38% at which the experiments were carried out. From the modeling point of view, water may stabilize both acid–ammonia and pure acid clusters. If the pure acid clusters are stabilized more than the acid–ammonia clusters, hydration might increase the evaporation rate of ammonia molecules. This could explain the ammonia content of the modeled (unhydrated) clusters being higher than what is observed.

Professor Cappelletti asked: Regarding the structure of the electrically neutral molecular clusters involving ammonia and sulphuric acid that you have characterized by *ab initio* calculations, did you have any evidence of proton, or more in general charge transfer and any possible dependence on the cluster size ?

Miss Olenius answered: The proton transfer from an acid to an ammonia molecule occurs as described in our previous work on the stability of electrically neutral acid–base clusters (I. K. Ortega *et al.*, *Atmos. Chem. Phys.*, 2012, **12**, 225–235). To summarize, the number of proton transfers increases with the size of the cluster, but the formation of fully deprotonated acid SO_4^{2-} in the clusters containing more ammonia than acid molecules does not occur until the cluster contains at least three acid molecules.

Professor Herrmann opened the discussion of the paper by Professor Johnston: Could you please clarify what the 'excess nitrogen' could be ?

Professor Johnston replied: Let's consider three nitrogen-containing species often discussed in the literature – ammonium nitrate, amine salts of organic acids, and organonitrates. The excess nitrogen is not likely to be ammonium nitrate since the gas phase ammonia concentration was low (less than 200 pptv), which would require a very high gas phase nitric acid concentration, on the order of 10-40 ppbv based on the Aerosol Inorganics Model, to form particulate ammonium nitrate. Amine salts are also unlikely contributors to excess nitrogen owing to low gas phase amine concentrations (40 pptv or less). While organonitrates can be produced during the day and might explain the rise in excess nitrogen during the particle formation events, several studies suggest that particle phase organonitrate concentrations are greatest at night owing to more favorable conditions for partitioning. The higher amount of excess nitrogen during the day *vs.* night argues against a partitioning mechanism for organonitrate incorporation into the particles. Therefore, none of these candidates stand out above the rest. Clearly, this is a topic that requires additional study.

Dr Murphy asked: The silicon contamination in ultrafine particle formation might have affected a number of laboratories. What did you have to do to eliminate it in your experiments?

Professor Johnston replied: In the laboratory, we often detect silicon in particles when the experimental system contains silicone-based materials and/or high vacuum grease. However, we usually do not detect silicon in particles when the apparatus is constrained to metals and Teflon-like materials.

Dr Cox remarked: During the discussion of the origin of the Si present in the nanoparticles sampled in field campaigns addressing nucleation of new particles, it was suggested that contamination from silicone tubing might be responsible. This could obviously be avoided in hindsight by using tubing fabricated from a different material. It does however highlight the potential pitfall of contamination when using novel ultrasensitive analysis techniques. We had a recent experience in our laboratory[1] using CIMS to measure hydrogen peroxide downstream of a flow tube coated with halocarbon wax. It worked beautifully until we tried a photolysis experiment, when photoproducts from the coating contaminated the mass spectrum so badly that it was unusable for H_2O_2 detection.

1 M. Pradhan, M. Kalberer, P. T. Griffiths, C. F. Braban, F. D. Pope, R. A. Cox and R. M. Lambert, *Environ. Sci. Technol.*, 2010, **44**, 1360–1365.

Professor McNeill remarked: Stanier and coworkers (Yucuis *et al.*, *Chemosphere*, 2013, **92**(8), 905–910) have observed siloxane species from personal products in the ambient atmosphere. This could possibly be a source of Si in your observations.

Professor Johnston responded: Yes, cyclic siloxanes are a potential source of silicon in the ambient particles we have analyzed. The same group you cite has modeled atmospheric oxidation and heterogeneous uptake of these species (Yucuis *et al.*, *Atmos. Environ.*, 2011, **45**, 3181–3191).

Dr Allan asked: Based on attempts to study Palas soot with an SP-AMS, I would agree how easy it is to pick up siloxane contamination from silicone tubing. I would also add that we also found it is similarly easy to get pthalate contamination from PVC tubing.

Professor Lewis opened the discussion of the paper by Professor Percival: How significant are the Criegee formation processes in high VOC, but low ozone environments such as tropical forests? Canopy and lower BL regions in these locations have the highest global alkene abundances, but often with very low ozone, with loss of many reactive VOCs by OH instead.

Professor Percival replied: Ozone will dominate loss for a range of alkenes, particularly terpenes, in these environments. Given the high shading on the forest floor and even below the canopy, the production of OH from Criegee chemistry levels will be very important and so ozonolysis will have a significant role. However, the global model study suggests that large impacts occur during outflow when high alkene levels mix with background ozone, which is elevated relative to the forest.

Professor Wahner commented: In your paper you show the possible impact of SO_2 oxidation by CI. 1. Did you include SO_2 oxidation in aerosols and clouds, which are assumed to be responsible for about half the SO_3/H_2SO_4 production? 2. By which measurements would it be possible to test your model results with respect to the importance of SO_2 oxidation by CI?

Professor Percival answered: The model does contain a very simple aerosol oxidation scheme as outlined by Stevenson *et al.*[1]

1 Stevenson *et al.*, *The tropospheric sulphur cycle and the role of volcanic SO_2*, 2003, Geological Society, London, pp. 295–305

Professor Donahue commented: Is the modeling consistent with the findings of Mauldin *et al.*[1] regarding the fraction of gas-phase H_2SO_4 produced *via* some process other than OH + SO_2?

1 R. L. Mauldin III *et al.*, *Nature*, 2012, **488**, 193–196.

Professor Percival replied: As we stated, Mauldin *et al.*,[1] suggest that CI+SO_2 chemistry can be used to explain ambient field observations and may contribute up to 50% of the H_2SO_4 which is entirely consistent with our results of up to 200% increase in H_2SO_4.

1 R. L. Mauldin III *et al.*, *Nature*, 2012, **488**, 193–196

Professor McFiggans addressed Professor Jacobson, Professor Pandis and Professor Carslaw: Is our understanding of the sulphate budget in terms of the gaseous and aqueous formation processes from large-scale models (and based on our process understanding of the oxidation timescales) likely to be substantially affected by the new Criegee chemistry? Is the rule of thumb that nitrate originates from gas phase processes and sulphate (largely) from aqueous processes under question?

Professor Carslaw responded: It is possible that the new Criegee chemistry could alter our understanding of sulphate production. Sulphate production is already surprisingly uncertain in global models, due primarily to uncertainties in aqueous phase processing. Our results show this very well. With such existing uncertainties, we might struggle to detect the need for new chemistry based on evaluation against measurements. This is the danger of not exploring uncertainty: it is possible that a single model run could differ from observations and could be improved by adding new chemistry. But we need to consider whether the uncertainty range of the existing model spans the measurements.

Professor Pandis replied: The recent studies of Li *et al.*,[1] Pierce *et al.*[2] and Sarwar *et al.*[3] have all estimated that the contributions of the new Criegee chemistry to the sulphate formation is on average a few percent or less. There are still significant uncertainties though in the corresponding kinetic data and the reaction could be quite important in specific locations and periods. Both the gas and aqueous-phase pathways can be important for the production of sulphate depending on the environment of interest (*e.g.*, existence of clouds, sunlight intensity, *etc.*). On the contrary, the gas-phase pathway dominates the formation of nitric acid in most cases.

1 J. Li *et al.*, *Atmos. Environ.*, 2013, **79**, 442–447
2 J. R. Pierce *et al.*, *Atmos. Chem. Phys.*, 2013, **13**, 3163–3176
3 G. Sarwar *et al.*, *Atmos. Environ.*, 2013, **68**, 186–197

Professor Herrmann remarked: It would be good to clarify global pathways of sulfate production. How much sulfate is being produced by homogeneous gas phase chemistry classically and by CI chemistry? How does this relate to aqueous phase production which globally is thought to dominate over gas phase production.

Professor Carslaw commented: We need to bear in mind that cloud processing (in-cloud aqueous phase production of sulphate aerosol mass) and nucleation tend to occur in different environments. Cloudy conditions, in which aqueous phase sulphate production is rapid, tend not to have much nucleation. In sunny conditions with a lot of nucleation, aqueous chemistry will not compete with this gas phase sulphuric acid production mechanism.

Professor McFiggans addressed Professor Pandis and Professor Jacobson: Whilst there may be a contribution of "primary sulphate" directly included in power station emissions to account for moist plume chemistry, in large-scale models there is frequently a requirement to consider a fraction of SO_2 emissions as primary sulphate in order to account for sub-grid sulphate production. If there is an additional process for oxidation (through Criegee chemistry), how does that affect the primary sulfate term in models? Is there room for Criegee chemistry in model-measurement sulphate discrepancies?

Professor Pandis responded: There is always some sulphate produced inside a stack or a tailpipe because of the oxidation of some of the emitted sulphur dioxide under the high temperature conditions of the exhaust. This "primary" sulphate is included in the emission inventories used by most regional chemical transport models like PMCAMx. Pierce *et al.* (J. R. Pierce *et al.*, *Atmos. Chem. Phys.*, 2013, **13**, 3163–3176) estimated that the sulphuric acid produced through Criegee chemistry globally is less than 4 percent of the total. It will be very difficult to detect such small contributions through routine long-term model-measurement comparisons. The role of Criegee chemistry as a source of sulphate could be detectable in specific sites (*e.g.*, forested areas during periods of high biogenic emissions) and periods (*e.g.*, low background sulphate levels from long-range transport).

Professor Johnston opened the discussion of the paper by Professor Pandis: In ambient measurements of nanoparticle composition during new particle formation at a given location, we have found that while the absolute growth rates of particles can change substantially from one event to the next, the chemical composition of the particles hardly changes at all. When you model new particle formation in a specific location, do you find that the chemical composition changes with growth rate or not?

Professor Pandis answered: This is an interesting observation that could help in the evaluation of ultrafine particle formation and growth models. Our simulations so far have focused on the sensitivity of the growth rates during a few specific representative nucleation events in different areas to the formation and chemical aging of secondary organics. We will perform the recommended comparisons in the next stage of the work, when we will try to simulate a range of different events in a single location.

Professor Harrison remarked: In the United Kingdom, the funding pathways for strategic and fundamental (blue skies) environmental research are different. In the past, work on nucleation processes would have been considered appropriate under both funding streams, but given that you have shown that a fairly simple parameterisation of the nucleation process is sufficient to model atmospheric nuclei formation, should we regard nucleation as a topic for consideration only under the fundamental research pathway in future?

Professor Pandis replied: Research focusing on atmospheric nucleation clearly has fundamental scientific value, but I think that it also has major practical implications. The parameterizations used have been based on hundreds of person-years of research. The fact that they do a reasonable job in reproducing existing observations in Europe, does not guarantee that these simple expressions will reproduce nucleation rates in the pre-industrial period or in future environments. The reasonable performance of some chemical transport models is encouraging, but both fundamental and strategic research is needed if we want to be able to simulate the formation of new particles in all environments in the past, present and future.

Professor Carslaw remarked: In addition to producing a model that agrees with the measured state of aerosols in the atmosphere, we also need to consider how well it captures responses to perturbations. It is possible, for example, to tune a model nucleation scheme to get approximately the correct CCN distribution. But if the model nucleation neglects the dependence on organic compounds, then it will not be useful in predicting changes in aerosol, which is the key requirement when it comes to forcing.

Professor Pandis responded: This is clearly a valid point. Our hope is that the evaluation of models in multiple locations, different chemical environments and conditions, and using not just CCN concentrations but the complete aerosol number distribution (including the total aerosol number concentration, number above 50 nm, *etc.*), particle composition, *etc.*, will reveal the model's weaknesses and strengths. This reproduction of the present, for example in environments with different concentrations of organic compounds, can play to at least some extent the same role as testing in one environment for a change in organic levels.

Professor McFiggans opened the discussion of the paper by Miss Olenius: Does the mechanistic interpretation of the comparison between the simulations and the CLOUD measurements provide us with clues about the details that will have most impact on nucleation rate and the uncertainties to which the rate will be most sensitive?

Miss Olenius answered: The comparison presented in the article demonstrates that our model is able to reproduce the main features of the experimental charged cluster distributions reasonably well, and can thus be considered a valid tool for cluster kinetics studies. The distribution comparison alone does not give information on nucleation mechanisms, but we have also compared the modeled nucleation rates (the rate at which clusters grow out of the simulation box to larger sizes) to CLOUD measurements, and found a good agreement (J. Almeida *et al.*, *Nature*, 2013, **502**, 359–363). Naturally, and also according to our results, to predict particle formation accurately, we need to know concentrations of sulfuric acid as well as compounds that are able to enhance the clustering of acid, such as ions and bases (and back-ground-level contaminants that may not be detected at all). The results are also fairly sensitive to the loss of vapors and clusters by coagulation to larger particles or chamber walls, and thus it is important to have sound experimental values for these loss rates. As for the theoretically based parameters, in the acid–ammonia system, the particle formation rate is more sensitive to uncertainties in the formation free energies than to the sticking probabilities in collisions, since a typical atmospheric ammonia concentration, and consequently the collision rate of ammonia molecules, is rather high in any case, but the evaporation rates – that are exponentially dependent on the formation free energies – of acid–ammonia clusters are non-negligible. On the other hand, the formation rate is affected also by uncertainties in the collision rates in the case that there exist kinetic barriers in the incorporation of ammonia into the clusters (see the answer to previous questions), that decrease the rate constants more than mere sticking factors.

Professor McFiggans addressed Miss Olenius: Do the interpretative tools require more complexity in order to represent the cluster distributions? Is it reasonable to exclude water from the simulations on the grounds of governing equation stiffness and draw firm mechanistic conclusions?

Miss Olenius responded: Additional factors, such as cluster hydration and kinetic barriers (see previous answers), may indeed affect the distributions, and these effects remain to be studied. As for the role of water, hydration may somewhat alter the ammonia content of the clusters, as stated previously; however, the main interaction in the studied clusters is that between acid and ammonia, which is a stronger base than water, and thus water is unlikely to have a significant effect on the qualitative distributions (in terms of, for example, the composition of the most abundant clusters). Also, we are currently working on including water in the simulations. Another level of complexity would be to model what happens to the clusters inside the instrument (evaporation of molecules and fragmentation). Then we could compare not only the modeled distribution in the chamber to the experimental distribution, but the modeled "measurable" distribution to the actually measured distribution.

Dr Ortega Colomer opened the discussion of the paper by Professor Donahue: Right now MBTCA is the compound with the highest O : C ratio that I have studied. The next one is a di-keto dicarboxylic acid ($C_{10}O_6H_{16}$). I am planing to include some of the compounds proposed by Josef Domen in his ICNAA talk (mainly carboxylic acids with few OH groups). This can be quite interesting, since one of the most interesting results that I found is that pinanediol (PD) behaves like a base, this is not too relevant for the formation of the heterodimer, but when you reach the 2,2 cluster (2 sulfuric acid, 2 PD) there is a proton transfer, and that stabilizes the clusters hugely. If any of the ELVOCS proposed by Josef are able to produce a proton transfer already in the heterodimer, that might be the key for new particle formation. On the other hand, those compounds are probably only relevant for CLOUD experiments where they used PD as a model compound. I do not think that diols are too abundant in the real atmosphere. Another of my aims is peroxides, the first try with a pinonic acid peroxide (O : C 0.4) has been quite promising, so I will try some other peroxides with a higher O : C ratio and see how it works.

If you have any suggestions for some interesting compounds to test, I will be happy to run the calculations.

Professor McNeill commented: Have you performed quantum chemistry cluster calculations for other ELVOCs besides MBTCA? How general is your observation of strong binding between MBTCA and H_2SO_4?

Professor Donahue answered: This does generalize. A forthcoming publication will describe results for a series of ELVOCS and other VOCS.

Professor McFiggans said: The study comprises both empirical and heavily theoretical elements and the overall approach can rightly be described as semi-empirical. There is a statement in the paper that, since there is limited evidence for a strong RH dependence of SOA formation "... because the treatment here is semi-empirical, we shall overlook the role of water vapor until there is empirical evidence of an important role". This approach appears rather unbalanced given the weight of theoretical considerations in the paper for which there is a similarly missing empirical evidence base. Many of the arguments for the participating nucleators are based on the fact that there will be a hydrogen-bonded network. Can the authors i) elaborate on the justification for the omission of a consideration of water, ii) comment on the likelihood of water not being involved in a hydrogen bonded network such as that proposed in nucleation and the early stages of growth, and iii) postulate on possible impacts of its participation.

Professor Donahue replied: We are not asserting that water is not involved in this hydrogen bonding network. It almost certainly is. However, the real issue is the free-energy difference between clusters and vapors at a given RH, for very, very small clusters (of an order of 4 molecules other than water). Our focus, being semi-empirical, is colored by the lack of any strong evidence for a significant RH dependence of new-particle formation in the atmosphere. Where there isn't smoke, there is probably not a fire.

Dr Krieger asked: I caution to conclude from not seeing a difference at dry conditions to conditions of higher relative humidity that water molecules are not incorporated into the clusters. Even at very dry conditions water molecules are available in much higher abundance than, for example, sulfuric acid molecules.

Professor Donahue replied: We are not suggesting that water is absent from these clusters. There is, however, considerable empirical evidence that new particle formation is relatively insensitive to relative humidity, with the exception of a suppression at the highest RH that is most probably caused by cloudiness and not RH. We thus focus on the interaction of organics and sulfuric acid, hypothesizing that those interactions govern the behavior.

Dr Colussi said: I am concerned that you ignored that after isoprene binds to protonated water clusters *via* van der Waals (not hydrogen bonding) interactions, it is subsequently protonated (*via* proton transfer) over low kinetic barriers.

Professor Donahue replied: This paper is about gas phase oxidation of alpha-pinene.

Professor Kiendler-Scharr asked: You mention in your paper that it is unlikely that certain special compounds drive the process, but rather you expect that a large body of oxidised ELVOC compounds are formed *via* gas-phase oxidation chemistry at a sufficient rate to drive new particle formation. If this is the case and we want to develop a predictive understanding of new particle formation under variable climate related parameters, to what level of detail do we need to understand the oxidation processes leading to these species and which properties of the ELVOCS – other than OSc and O : C ratio - would be the most informative ones?

E.g. would information on mean molecular weight, functional group distribution, or reactivity be important parameters to focus on in future studies?

Professor Donahue responded: Knowing the functional group distribution of at least a good selection of compounds would seem to be a good thing. Obviously, even knowing the exact structures would be better, as that would enable direct comparison with computational results. However, the computational chemistry can be used to test the hypothesis that many different functional-group distributions with about the same carbon number and O : C end up having similar excess free energies when clustered with sulfuric acid.

Professor McFiggans remarked: Thinking about useful metrics for evaluating the production rate of "nucleators" or "early growers", would a metric similar to δ(dipole moment)/δt be useful for evaluating the likely participation in a hydrogen-bonded network (and would it be accessible)?

Professor Donahue answered: Can we measure that?

Professor Nizkorodov opened the discussion of the paper by Professor Aumont: I was struck by the significant fraction of products containing multiple $-ONO_2$ groups predicted by your simulation (such as NNN, NNO, NNKK, NNKO, NNOO, *etc.*, shown in Figure 6). To the best of my knowledge, field and smog chamber observations of SOA compounds containing more than one $-ONO_2$ are rare, and may be limited to experiments with artificially high NO_x levels. Can you comment on possible reasons for unrealistically high yields of organonitrates in the model?

Professor Aumont replied: As stated in the paper, the simulated fraction of organonitrates in the particle phase appears surprisingly high, especially for the oxidation of the C10-C14 alkanes. Some reasons for this nitrate yield over-estimation are discussed in section 3.5 of the paper and include, among other, the hydrolysis of the nitrate moiety in the condensed phase and/or the possible misunderstanding of the gas-phase chemistry of multifunctional nitrates. The SOA composition given in the paper should be seen as the predicted composition based on our understanding of the atmospheric chemistry, as long as high NO_x can be sustained for a few oxidation steps. This high NO_x approximation is a severe approximation and explains the production of some species with unrealistically high yields. For example, the production of a species bearing 3 nitrate moieties (NNN) requires 3 successive oxidation steps and fairly long time scales. For a C14 species, the branching ratio for the $RO_2 + NO \rightarrow RONO_2$ reaction is ≈ 0.27. The tri-nitrate species is therefore produced with a yield of about 2% $(\approx 0.27^3)$ and appears here as a significant SOA contributor. Oxidation under realistic atmospheric conditions would substantially decrease the production of these tri-nitrate organic species. Assuming conditions representative of the transition regime between high NO_x and low NO_x conditions (*i.e.* only half of the RO_2 reacts with NO), the nitrate yield of each generation will decrease by a factor of 2, and the production of NNN therefore decreases by almost an order of magnitude $(\approx 0.5^3)$. However, even under more realistic NO_x conditions, a large fraction of organonitrates is simulated in the condensed phase with GECKO-A. As stated in the paper, this has been recently exemplified in a modeling study of the Mexico City plume, where multifunctional organic nitrates were identified as major SOA contributors as well.

Professor McFiggans asked: Is it possible to quantify the amount of the excess nitrate contribution that is caused by the high NO_x simulation regime?

Professor Aumont replied: The quantification of the nitrate moiety "excess" requires the generation of chemical schemes including peroxy + peroxy reactions. As stated in the paper, the explicit description of a C10 hydrocarbon is expected to lead to a mechanism including about 10^8 species, one order of magnitude above the size of chemical schemes that can be both generated and solved. A straight-forward quantification is therefore not reasonably practicable. In the usual application of the GECKO-A tool, simplifications are performed based on the lumping of structural isomers and generated schemes include a full description of the peroxy chemistry. Such simplified schemes would allow an estimation of the bias introduced by assuming high NO_x chemistry only, and sensitivity tests will be conducted in the future to address this point. Note, however, that such lumping leads to a partial loss of the molecular structures – key information to examine oxidative trajectories along fragmentation or functionalization routes (which was the focus of the present study).

Professor McFiggans commented: Given the recent evaluations and developments in property estimation techniques, how sensitive are the conclusions of this work to the chosen estimation method for vapour pressure and have alternative methods been evaluated?

Professor Aumont answered: In this study, vapor pressures were estimated using the Nannoolal *et al.* method.[1] Two additional structure activity relationships (SAR) are implemented in the GECKO-A modeling tool: the Simpol method[2] and the Myrdal and Yalkowsky method.[3] The vapor pressure values estimated with these 3 SARs were compared in a previous paper[4] for the set of secondary species generated by GECKO-A to describe alpha–pinene oxidation. The Nannoolal *et al.*[1] method was found to provide the best agreement of the simulated SOA yields with smog chamber observations. Note that in the Valorso *et al.* paper,[4] the simulated SOA composition appeared to be rather insensitive to the vapor pressure estimation method. The sensitivity of the SOA and OOA yields to the selected SAR was not systematically examined in this study. However, preliminary tests also showed a weak sensitivity of SOA composition to the selected SAR.

1 Y. Nannoolal *et al.*, *Fluid Phase Equilib.*, 2008, **226**, 45–63
2 J. F. Pankow and W. E. Asher, *Atmos. Chem. Phys.*, 2008, **8**, 2773–2796
3 P. B. Myrdal and S. H. Yalkowsky, *Ind. Eng. Chem. Res.*, 1997, **36**, 2494–2499
4 R. Valorso *et al.*, *Atmos. Chem. Phys.*, 2011, **11**, 6895–6910

Professor Kroll asked: You showed that nearly all the components of the SOA have the same number of carbon atoms as the original precursor. What about the gas-phase molecules – the ones that didn't condense? Are these all smaller molecules, resulting from fragmentation reactions? Similarly, most of the SOA components have at least one nitrate group – what fraction of the molecules in the gas phase also contain nitrogen?

Professor Aumont replied: Most species in the gas-phase have a smaller carbon backbone than the original precursor, *i.e.* are produced from fragmentation reactions. For an oxidation time corresponding to 5 lifetime numbers (LTN) of the parent compound, the fraction of the gas-phase carbon atoms in fragmented species to the total gas-phase carbon atoms decreases from 0.87 to 0.70 for C10 to C22 linear alkanes and is above 0.9 for alkane structures having 3 vicinal methyl groups. CO and CO_2 are the major gas-phase fragments at 5 LTN. For the oxidation of linear alkanes, these species contribute from 74% (C10) to 40% (C22) of the gas-phase carbon budget. For the 3 methyl branched alkanes, these ratios range from 42% (C10) to 30% (C22). Additional information concerning the time evolution of the fragmented *vs.* functionalized species can be found in Aumont *et al.* (B. Aumont *et al.*, *Atmos. Chem. Phys.*, 2012, **12**, 7577–7589) for *n*-alkane oxidation.

The fraction of gas-phase organic species bearing a nitrate moiety grows with oxidation time. For 5 LTN, this fraction ranges from 0.4 to 0.6 for the *n*-alkanes and from 0.2 to 0.4 for the 3 methyl branched alkanes examined in this study. Note, however, that the simulated concentration of gas-phase organic nitrates (and other secondary organic species) typically reaches a maximum for 2–3 LTN and then decreases as oxidation proceeds and progressively breaks the carbon skeleton to ultimately produce CO_2.

Professor Herrmann asked: Could you comment on the large production of organic nitrates, RNO_3. Maybe the implemented alkoxy radical chemistry needs to be re-visited?

Professor Aumont responded: As described in the paper, two recent Structure Activity Relationships (SAR) are currently implemented in GECKO-A to describe alkoxy decomposition: the SAR developed by Atkinson (R. Atkinson, *Atmos. Environ.*, 2007, **41**, 8468–8485) and the SAR developed by Vereecken and Peeters (L. Vereecken and J. Peeters, *Phys. Chem. Chem. Phys*, 2009, **11**, 9062–9074). Schemes generated with both SARs lead to a very similar SOA composition, without substantial changes in organic nitrate yields. We do not expect the organic nitrate yields to be highly sensitive to alkoxy radical chemistry. However, the gas-phase chemistry of multifunctional nitrates implemented in GECKO-A might need to be revisited in the future. As stated in the paper as an example, it was recently observed that carbonyl nitrates undergo fast photolysis and that the Kwok and Atkinson SAR used to estimate the VOC + OH reaction rates fell short in describing the OH reactivity of these multifunctional nitrates. Additional experimental studies are clearly required to improve our understanding of the behavior of multifunctional organic nitrates in the gas and condensed phases.

Professor Lewis questioned: What would be a suggested strategy to advance from predicting the aerosol products of individual higher hydrocarbons, to highly complex real-world emissions, such as the VOCs from diesel or kerosene, which may contain 10 000 or more individual components?

Professor Aumont responded: This is a challenging issue. In this paper, we explored the influence of hydrocarbon molecular structure on SOA composition and yields. For the alkanes examined here, simulation results show that branching seems to be of secondary importance for the particle oxidation state and composition. This suggests that a lumping approach using a few surrogate species might be appropriate and efficient, similar to the strategy adopted to describe ozone formation from the large number of VOCs emitted to the atmosphere. Additional tests are required to see if that behavior can be generalized to other conditions (*e.g.* low NO_x conditions) and other structures (*e.g.* cyclic species) or other families of compounds (aromatic and alkenes especially).

Professor McFiggans remarked: Does the degree of OOA production from the species in simulations within a reasonable atmospheric lifetime place any constraint on the likely magnitude of the contribution to ambient SOA from these species?

Professor Aumont answered: Simulation results rule out alkanes as a major source of OOA. The simulated SOA yields are nevertheless large for long chain alkanes, even in the case of heavily branched carbon backbones. Most SOA contributors are found to be reduced enough to be categorized as HOA (hydrocarbon like aerosols), an often-reported component in atmospheric observations based on aerosol mass spectrometer measurements. HOA is commonly assimilated as primary organic aerosols (POA). The simulation results suggest that HOA may also include SOA produced by the gas-phase oxidation of long carbon chain hydrocarbons. For the *n*-alkane oxidation, the OOA *vs.* HOA distribution of the

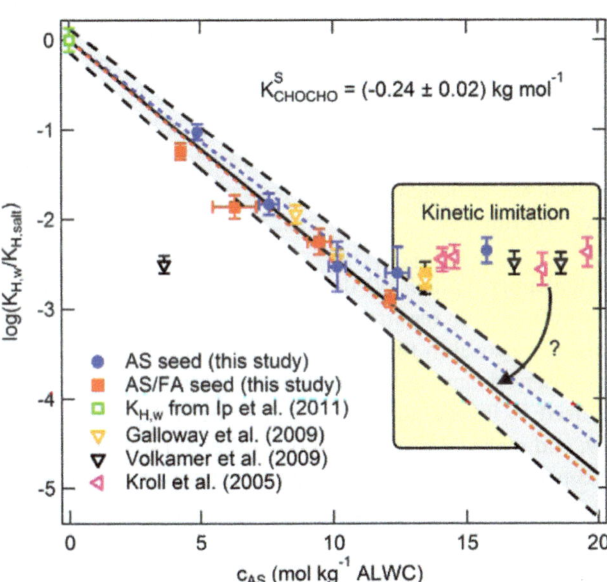

Fig. 1 Glyoxal uptake to ammonium sulfate (AS) and mixtures with fulvic acid (AS–FA) is quite consistent between literature studies independent of whether C_2H_2 was used as a source for glyoxal or not. Adopted from Kampf *et al.*, *Environ. Sci. Technol.*, 2013, **47**(9), 4236–4244. See that study for details.

SOA contributors is discussed in Aumont *et al.* (B. Aumont *et al.*, *Atmos. Chem. Phys.*, 2012, **12**, 7577–7589).

Professor Kiendler-Scharr asked: You state in your paper that "within a given homologous series, the number of generations required for the production of low volatility species must therefore grow when the carbon skeleton of the parent alkane decreases in length." This should imply that higher loss rates through, for example, deposition of the intermediate species and thereby an effect on the particle mass formed should exist. How much do your results depend on assumptions made regarding such loss processes between generations?

Professor Aumont answered: The partitioning of the secondary organic compounds between gaseous and aqueous phases has been recently examined with the GECKO-A modeling tool for a liquid water content corresponding to clouds and deliquescent aerosols (C. Mouchel-Vallon *et al.*, *Atmos. Chem. Phys.*, 2013, **13**, 1023–1037). It was found that most secondary organic species produced during the multigenerational oxidation of alkanes should dissolve in the aqueous phase during cloud events. Removal of intermediate organics by wet deposition or aqueous phase chemistry may therefore be substantial and likely play a key role in the SOA budget. The representation of these processes into GECKO-A is the object of ongoing studies.

The dry deposition of volatile organic compounds (VOCs) and its impact on secondary organic aerosols (SOA) were also recently investigated with GECKO-A in the Mexico City plume (A. Hodzic *et al.*, *Geophys. Res. Let.*, 2013, **40**, 3302–3307). The simulations showed that the dry deposition of oxidized gases is not an efficient sink for SOA and that it can reasonably be neglected in models, as it removes ≤5% of SOA within the city's boundary layer and ≈15% downwind.

Professor McNeill opened the discussion of the paper by Dr Nozière: Is the formation of SOA *via* glyoxal uptake truly autophotocatalytic? That is, in your experiments, is another VOC besides glyoxal required for photosensitized SOA formation in the glyoxal-SOA system? The results of Volkamer *et al.* (R. Volkamer *et al.*, *Atmos. Chem. Phys.*, 2009, **9**, 1907–1928) suggest a glyoxal-only autophotocatalytic mechanism, although acetylene was also present in the gas phase in their chamber experiments.

Professor Volkamer replied: We do not believe that acetylene has contributed to the rapid SOA formation reported by Volkamer *et al.*.[1] In fact, in a recent study[1] we have measured glyoxal in the gas-phase and inside aerosols, and shown that the amounts observed can explain the volume growth rates in Volkamer *et al.*[1] without the need to invoke additional SOA sources. Fig. 1 in Kampf *et al.*[2] is shown below, and makes it transparent that other studies that are conducted in absence of acetylene are actually quite consistent with our earlier data once the effect of 'salting-in' is considered. See Kampf *et al.*[2] for further discussion.

1 R. Volkamer *et al.*, *Atmos. Chem. Phys.*, 2009, **9**, 1907–1928
2 C. J. Kampf, E. M. Waxman, J. G. Slowik, J. Dommen, L. Pfaffenberger, A. P. Praplan, A. S. H. Prevot, U. Baltensperger, T. Hoffmann, and R. Volkamer, *Environ. Sci. Technol.*, 2013, 47(9), 4236–4244.

Dr Nozière answered: Our definition of autophotocatalytic only meant that VOCs such as glyoxal are able to produce chromophores ("brown carbon") in the particle phase that lead to photosensitized growth. Indeed in our experiments glyoxal was not present in the gas and other compounds were used (*e.g.*, limonene) to make the particles grow.

Dr Colussi remarked: Were your experiments done in the presence of oxygen?

Dr Nozière answered: Yes, most of the experiments presented here were performed in synthetic air. A limited series of experiments were also performed in pure oxygen (section 3.4) to check for the potential formation of singlet oxygen in the mechanism. The mechanism seems to be de-activated in pure oxygen and pure N_2, and only active in air.

Dr Colussi commented: Do you have estimates of the lifetimes of the putative triplet states required in your experiments?

Dr Nozière replied: We have recently performed laser photolysis experiments investigating this triplet state chemistry in water (in preparation). We observed the formation of a triplet state whose lifetime depends on the ionic and organic environment. This lifetime was found to be of a few microseconds, and quenched by limonene as suggested here.

Professor Abbatt asked: How does the phase of the particle affect the mechanism? In particular, given that the relative humidity is low, could the reactive species (or radicals) be confined to the surface?

Dr Nozière replied: Experiments were performed at different humidities, and with different inorganic salts mixed with succinic acid, with these mixtures exhibiting different efflorescence points. So, we certainly probed the chemistry in the liquid (with ammonium nitrate) as well as in the solid (with ammonium sulfate) without observing significant changes in the results. This suggests a surface-mediated process or a phase change during the growth of the particles due to the uptake of organics, leading to similar phases in all experiments.

Professor Heard said: Could you test this radical mechanism by trying to detect the HO_2 radicals that are generated *via* the cycling shown in Figure 5 of your paper? Each cycle has steps that generate HO_2, but there is also a reaction of HO_2, so it is the net HO_2 produced per cycle that is important. Using the concentration of precursor and photosensitizer, how much HO_2 is generated and what HO_2 concentration would be partitioned in the gas phase? It could be possible, with a very sensitive laser-induced fluorescence detection system (detection limit in the range of 10^6 molecule cm^{-3}) placed at the end of the flowtube, to measure any HO_2 produced to help confirm the mechanism.

Dr Nozière responded: We can give an upper limit for the HO_2 produced but not a more accurate answer because there are still many unknowns in the system and because the importance of HO_2-consuming reactions in the flow tube would depend on the VOC anyway (see below). The concentration of HO_2 produced can be estimated from the number of cycles performed (not the absolute concentration of photosensitizer as the latter is regenerated) and from the difference between the rates of the HO_2-producing reaction and of the HO_2-consuming ones. In addition, the fraction of HO_2 transferring back to the gas would largely depend on whether these processes occur at surfaces or in the bulk, which is unknown. Our upper estimate will thus assume that all the HO_2 produced are released into the gas (more likely with surface processes and VOCs partitioning poorly into the particles).

According to the mechanism proposed, the number of cycles performed is equal to the number of VOC molecules transferred into the particles, in turn given by their increase in volume. At the end of the flow tube the particles have typically grown from 50 to 70 nm in diameter, corresponding to a volume change of 9×10^{-16} cm^3. Assuming a density of 1 g cm^{-3} and an average molecular weight for the organic products of 150 g mol^{-1}, this corresponds to 3.7×10^6 cycles/particle at the end of the flow tube, with some uncertainties. Thus for 5000 particle cm^{-3}, the number of cycles performed at the exit of the flow tube is about 2×10^{10} cm^{-3}. This number would also be the concentration of HO_2 produced if there was no significant consumption reaction. But we do suggest in our mechanism possible reactions of HO_2 with organic radicals from the VOC, in particular peroxy radicals. These recombination reactions are, however, typically several orders of magnitudes slower than the decomposition of hydroxy–alkoxy radicals proposed as the main source for HO_2 here. Series of experiments in this paper also show that the photosensitized processes occur only in the presence of O_2, thus under conditions where HO_2 production is fast. Thus, while the exact reaction rates would have to be measured, it is probably realistic to assume that, in this system, the production of HO_2 would be much faster than its consumption and that the concentration estimated above, $\sim 2 \times 10^{10}$, is close to the net production of HO_2. An instrument detecting HO_2 down to 10^6 cm^3 should thus easily detect those produced by such a system, even if only a fraction of them transfer back into the gas.

Professor Nizkorodov said: According to the study by Sareen *et al.* (N. Sareen, S. G. Moussa, and V. F. McNeill, *J. Phys. Chem. A*, 2013, **117**, 2987–2996) light-absorbing compounds produced in the methylglyoxal + ammonium sulfate reaction are efficiently degraded by solar radiation. We carried out photodegradation measurements on the same system that support the poor photochemical stability of the absorbers (unpublished results). How would the main conclusion of this study be affected if the catalytic cycle shown in Figure 5 can only run a few times or just once?

Dr Nozière responded: IC is unstable under light and, with time, will be degraded by solar irradation. Nevertheless, we show in these experiments that this photosensitizer is still active after 20 min of exposure to UV light and, in more recent experiments (not published yet), even for up to one hour of exposure. It is thus possible that the secondary chemistry suggested here also produces absorbing species (hereby contributing to the production of brown carbon in the troposphere) leading to a kind of steady-state where the photosensitized properties are maintained over time with potentially a changing nature of the actual photosensitizer.

Professor Rudich remarked: What is the time scale needed to reach concentrations of the photosynthesised molecule so it will be active under atmospheric conditions?

Dr Noziére replied: This timescale is illustrated in Fig 1. for $(NH_4)_2SO_4$ 1mM and pH $= 9$ (<120 min). It would vary exponentially with pH and ammonium salt concentration, and linearly with glyoxal concentrations, as does the rate of imidazole formation from glyoxal (B. Noziére and A. Cordova, *J. Phys. Chem., A*, 2009, **113**, 231–237; G. Yu *et al.*, *Environ. Sci. Technol.*, 2011, **45**, 6336–6342; N. Sedehi *et al.*, *Atmos. Environ.*, 2013, **77**, 656–663). Assuming atmospheric particles containing 0.1 M of glyoxal (as in our experiments), 4 M of ammonium sulfate (equilibrium with 50% RH) instead of 1 mM, and of a pH of 5 instead of 9, the rate constants in the literature predict a slight decrease of the rate, thus an increase of the timescale, by about a factor 6 (lowering pH and increasing ammonium concentration having opposite effects). The timescale of imidazole formation would thus be of the order of 12 h, still relevant over the limetime of aerosols in the troposphere.

Dr Knopf asked: The applied VOC concentrations are very high compared to atmospheric conditions. Does the application of these high concentrations affect the particle growth process? How is particle growth sustained assuming rapid coverage of the excited particle surface by the adsorbing VOCs?

Dr Nozière responded: The use of high concentrations is a clear limitation of flow tube studies. So far in these and more recent experiments we have observed that the growth of the particles (in radius) increases linearly with exposure time, up to 50 min. Thus there does not seem to be any limitation due to surface coverage by the VOC, at least over this timescale. Inversely, we have recently observed that much lower concentrations of VOCs combined with longer exposure times resulted in the same effect as presented in this paper. Thus, as long as saturation is not observed, exposure to high concentrations can be scaled down (extrapolated) to lower concentrations with some confidence that the VOC dose exposure is an appropriate metric.

Dr Allan opened the discussion of the paper by Dr Ofner:†Regarding the point that these species have not yet been reported in ambient data, they should in theory be detectable using AMS, even if they exist solely in a transitory state. If we were to look for such compounds in the atmosphere, in which environments should we expect to find them?

Mr Held replied: Generally speaking, we expect halogen-induced organic aerosol formation in environments where high concentrations of organic precursor gases and reactive halogen species coexist. Reactive halogen species are abundant in marine environments (both open ocean and coastal areas), and in specific terrestrial environments such as salt lakes or volcanic areas. Specific examples include tropical forests in the coastal areas of South America where

---

† Dr Ofner's paper was presented by Mr Held, University of Bayreuth, Bayreuth, Germany.

organic precursor gases are strongly emitted or salt lakes in Western Australia with biogenic emissions from Eucalyptus vegetation and organic soil layers. High concentrations of reactive halogen-species cause ozone depletion and thus increase the probability of the XOA formation pathway or halogenation of pre-existing SOA (see also J. Buxmann *et al.*, *Int. J. Chem. Kinet.*, 2012, **44**, 312–326 and J. Ofner *et al.*, *Atmos. Chem. Phys.*, 2012, **12**, 5787–5806).

Professor Donahue commented: It is an error to describe an SOA chamber formation experiment conducted at an atmospheric concentration of the precursor as "atmospherically relevant". In that context the "Δ" in "ΔM" on the *x*-axis label of Figure 3 (introduced by J. R. Odum *et al.*, *Environ. Sci. Technol.*, 1996, 30(8), 2580–2585 long ago) is unfortunate. The appropriate *x*-axis is "M" alone, whereas the SOA mass yield is indeed "$\Delta M/\Delta$Precursor". Thus, to be "atmospherically relevant" an experiment ought to have organic aerosol levels ("M") of 1–10 $\mu g\ m^{-3}$. Specifically, it is an error to assume that the mass yields observed at very low precursor concentrations (and 0.01 $\mu g\ m^{-3}$) can be related directly to the atmosphere.

Mr Held responded: We agree that our chamber experiments deviate considerably from atmospherically relevant conditions with respect to the initial organic mass M_o even though the precursor concentrations are close-to-ambient. However, we still find it instructive to compare the presented yields from our chlorine oxidation experiments with similar studies using halogens or other oxidants such as ozone or OH. From these comparisons, we find that chlorine can contribute substantially to the oxidation of organic precursor gases. We do not want to suggest that the presented mass yields can be related directly to the atmosphere. We appreciate the suggestion to conduct similar experiments using organic aerosol levels of 1-10 $\mu g\ m^{-3}$.

Dr Colussi commented: These results may not apply at low tropospheric VOC concentrations

Mr Held replied: We agree. Due to the fact, that the XOA formation experiments were done under homogeneous nucleation conditions, under heterogeneous conditions (with preexisting nuclei) the XOA formation is expected to be more efficient than reported in our work.

Professor Cappelletti commented: Halogenated organic molecules can interact in the gas phase with other molecules, even with themselves, through the so called "halogen bond".[1]
Halogen bonding (XB) is used analogously with the better-known hydrogen bonding, with which halogen bonding shares numerous properties.
Specifically, the high directionality and the strength of XB, which may range between 5 and 180 kJ mol^{-1}, are of great relevance in molecular recognition processes and crystal packing. For these reasons, similar to hydrogen bonding,[2] XB can also generate very stable intermolecular clusters in the gas phase, even when simple halocarbons are involved[3] and therefore may be at the basis of a possible, not considered until now, new particle formation mechanism.

1 O. Hassel, *Science*, 1970, **17**, 497.
2 D. Cappelletti, E. Ronca, L. Belpassi, F. Tarantelli and F. Pirani, *Acc. Chem. Res.*, 2012, **45**, 1571–1580.
3 D. Cappelletti, P. Candori, F. Pirani, L. Belpassi, and F. Tarantelli, *Cryst. Growth Des.*, 2011, **11**, 4279–4283.

Mr Held responded: We appreciate this comment and agree that halogen bonding may be an important mechanism in new particle formation. More detailed flow-reactor experiments could allow identifying the contribution of halogen bonding to the XOA formation process. The spectroscopic characterization of the transformation from gaseous species to molecular clusters and finally homogeneous formed nuclei using vibrational spectroscopy related to hydrogen or halogen bonding could be very important to clarify the entire nucleation process from a spectroscopic point of view.

Professor Heard remarked: I would like to follow up on the "atmospheric relevance" theme. Using the value of $J(Cl_2)$ and the lowest $[Cl_2]$ reported in the paper, I did a quick calculation and estimated that the concentration of Cl atoms was about 2×10^8 atoms cm^{-3}. In the atmosphere, [Cl] is in the range 10^3 to 10^5 (even in the Arctic where the highest concentrations of Cl_2 have been observed). So although the product $[Cl] \times t$ might be the same in the atmosphere compared to these experiments, here [Cl] is much higher and the time (t) much lower. Given that in your experiments it is a radical chain mechanism, the yield of XOA may be a non-linear function of [Cl]. Also, chlorinated intermediates may react with OH, O_3, *etc.* in the atmosphere, and not further with Cl atoms. Cl atom reactions of intermediates are forced in this experiment owing to the high concentration of Cl atoms. In previous studies, yields of SOA in chambers have been found to be a function of the [OH] used to oxidise the parent VOCs (even though the product $[OH] \times t$ was the same as in the atmosphere, the oxidation was achieved with a higher [OH] and lower t than in the atmosphere). Could there be a similar situation here?

Mr Held responded: Indeed, the time (t) in our experiments is much shorter than what we expect or suggest to find in the atmosphere, *i.e.* the reactions occur faster due to higher-than-ambient concentrations of the reactants. We acknowledge that the XOA yield may be very sensitive to the concentration of chlorine atoms, and therefore, yields from our chamber experiments cannot be related directly to the atmosphere. Further, we acknowledge that at lower Cl concentrations, intermediates may also react with other oxidants such as OH, O_3, *etc.* However, we suggest that chlorine may contribute substantially to the oxidation of organic precursor gases in the atmosphere.

Professor Kroll asked: The aging processes that convert halogenated SOA to "standard" SOA presumably involve the conversion of C–Cl bonds to C–O bonds. Could you speculate on what types of reactions might cause such a transformation?

Mr Held replied: A detailed discussion on chemical mechanisms possibly related to the halogenation of SOA or the XOA formation process can be found in J. Ofner *et al.*, *Atmos. Chem. Phys.*, 2012, **12**, 5787–5806.

Professor Harrison asked: Have you conducted experiments specifically to establish whether these reactions lead to nucleation, or just low-volatile products?

Mr Held replied: Conducting the chamber experiments without seed aerosols, we were specifically interested in the nucleation process.

Dr Wenger said: The elimination of gaseous halogen species (*e.g.* ClO, HCl) during the ageing of XOA might provide useful information on the processes involved. Were any gaseous halogen species detected during the experiments?

Mr Held answered: Significant amounts of HCl were observed during the aerosol flow-reactor experiments. Under the chosen conditions, HCl and probably other inorganic halogen species were formed during the reaction of Cl atoms with the organic precursor. The observable timescale was too short to monitor a consumption of the formed HCl. Nevertheless, we will check the longest dataset at condition D if any hints on HCl consumption are given. A detailed discussion about the interaction of different halogen species with organic precursors and SOA particles can be found in J. Ofner *et al.*, *Atmos. Chem. Phys.*, 2012, **12**, 5787–5806.

Professor Volkamer asked: You suggested salt lakes might be a good place to look for chlorine signatures in aerosols. Do you have any evidence from your chamber work to suggest specific signatures that provide means to test the relevance of your findings in the real world?

Mr Held replied: In aerosol samples collected on ATR crystals during aerosol chamber experiments we find indications for the formation of carbon chlorine bonds using infrared spectroscopy (J. Ofner *et al.*, *Atmos. Chem. Phys.*, 2012, **12**, 5787–5806). This paper describes two effects: the effect of halogen-bond formation, caused by the interaction of reactive halogen species (RHS) with preexisting organic aerosol, and the effect of SOA on the halogen release by bromine explosion from sea-salt aerosol. We found significantly lower amounts of free RHS in the presence of SOA and explained the difference by reaction of the RHS with SOA forming halogen–carbon bonds. Buxmann *et al.* (J. Buxmann *et al.*, *Int. J. Chem. Kinet.*, 2012, **44**, 312–326) report a strong release of RHS by simulated salt pans –thus, suggesting interactions of these RHS similar to RHS from bromine explosions. In both studies, halogenated organic aerosols (halo–SOA and XOA) were analyzed using ATR FTIR spectroscopy identifying the C–Cl or C–Br stretch vibrations. Also, high resolution ICR-FT-MS and temperature-programmed-desorption MS were used. We also found significant changes related to the UV-vis spectra of our model aerosols.

Professor Villenave said: To comment about the products of Cl-initiated oxidation of organic compounds. We have studied in Bordeaux and Cork (with John Wenger) the reactions of Cl atoms with some polycyclic aromatic compounds (see the related poster presented here) and they also do not immediately form polychlorinated polymers as it is suggested in the audience, but most are simple mono- or dichlorinated compounds that you may both find in the gaseous and particulate phases (by PTR–TOF-MS and ATOF-MS, respectively).

This is important particularly if you further consider the presence of chlorinated organics in the atmosphere and possible candidates for tracers of SOA formation from Cl reactions.

Dr Colussi opened the discussion of the paper by Dr Vehkamäki: Interfacial hydronium ions can protonate most non-alkanes on surfaces. We found that gas-phase isoprene and monoterpenes are protonated on the surface of pH < 4 water.

Professor McFiggans opened the discussion of the paper by Professor Aumont: All process level modelling studies must employ study-specific assumptions which will influence the quantitative conclusions. For example, the paper by Donahue *et al.* has used a single illustrative surface tension to bind the volatility of participating "nucleators" and "growers" and the paper by Aumont *et al.* allows oxidation to proceed in the absence of competition for all of the species by (wet or dry) deposition processes. If the value of either parameter (surface tension or deposition) were changed, then the quantitative conclusions would be influenced. It is obviously necessary to assume some value. However, given the massive complexity of the atmosphere and the breadth of the parameter space, do we have a rational methodology for deciding how much atmospheric/physical realism to include in process level models to evaluate whether a mechanism/approach should be incorporated into larger-scale models?

Professor Herrmann opened the discussion of the paper by Professor Donahue: How can the mechanistic suggestions of this block of presentations be tested?

Professor Donahue replied: One test is the computational chemistry. Another is empirical – we predict that essentially any organic vapors with C* below and O : C above the range shown in Fig. 2 will be a "good" nucleator. Methods are just now becoming available to measure the composition of small clusters involved in nucleation. If the clusters are charged, the composition (nC, nH, nO) of the organics can be measured now; neutral cluster measurements are just at the edge of current methodology.

Professor Johnston asked: When moving through the two dimensional volatility–oxidation basis set to produce ELVOC (*e.g.* Fig. 2), is it possible that the organic matter might transform in part through the formation of gas phase oligomers, for example formation of a hydrogen-bonded dimer of organic acids? Such a route might involve moving right to left along the bottom of the figure to produce ELVOC. Oxidation could subsequently occur (moving from bottom to top in the ELVOC region), but is not required. Some background: we have performed calculations (J. W. DePalma *et al.*, *Phys. Chem. Chem. Phys.*, 2013, **15**, 6935) suggesting that the hydrogen bonded dimer of terpenylic acid can exist under ambient conditions in concentrations greater than 10^7 per cc, which is large enough to play a significant role in particle formation and growth.

Professor Donahue answered: The framework itself is thermodynamic and thus path independent, but obviously the kinetics of product formation and the dynamics of condensation are of great interest. Hydrogen bonded dimers would seem to be monomers from the perspective of the condensed phase – once solvated I would expect them to dissociate. However, dimer formation should be taken into account if it affects the condensation dynamics – if dimer condensation competes with monomer condensation. Covalently bound dimers would be another story.

Dr Colussi asked: Not all known chemistry is being considered. By ignoring adsorption, schemes based on gas phase chemistry and partitioning of gases into bulk phases are incomplete.

Professor Donahue replied: In general we make no claim that the chemistry represented in this scheme is complete. We have represented gas phase oxidation and heterogeneous OH uptake in previous work and simply ask what portion of the observations those processes can represent. However, the present work addresses new particle formation and specifically interaction of very low vapor pressure organics with sulfuric acid vapor. The organics discussed here MUST have been formed *via* gas-phase oxidation, so in this case the representation is for all practical purposes complete.

Professor McFiggans opened the discussion of all the preceding papers: Based on the first 8 papers, it is clear that process modelling approaches abound, ranging in breadth of processes represented, number of components treated, complexity of kinetic and thermodynamic treatment and frameworks within which they are evaluated and/or within which they can be subsequently incorporated. In paper 22 (Carslaw *et al.*) we come across an attempt to evaluate and prioritise processes according to parametric uncertainties in large-scale model treatments. However, other than the most obvious and well-understood physical processes, it appears that there is no clear agreement on the degree of structural complexity in large-scale models necessary to investigate the processes determining the atmospheric impacts of aerosol. This level of complexity cannot be arbitrarily chosen since, without an ability to adequately consider a process, a model will be unable to evaluate the parametric uncertainty to this process. As a community, we obviously have strong belief that the processes that are being investigated are important. Are there any thoughts on how structural capabilities of models should be prioritised and recommended, or are the existing frameworks capable of reasonably and fairly evaluating the parameter space?

Professor Cappelletti commented: Aerosol profile measurements have been conducted in the Arctic Boundary Layer (Ny Alesund, Svalbard) exploiting a tethered balloon able to lift aerosol instrumentations up to 1 km with a vertical resolution of 10 m approx.

More than 300 profiles of size distributions, black carbon and ozone concentration have been recorded in spring and summertime 2011 and 2012. Moreover, aerosols have been sampled on filters at various altitudes for successive off line chemical and electron microscopy analysis.

Nucleation events were recorded in spring and summer at a ground based station by SMPS, and parallel vertical profile measurements suggest that some of the early events (in spring) were confined to ground (first 200 m) within the ABL by thermal inversions. These events were characterised by low ozone concentrations and high abundance of sea salt particles rich in chlorine, which may be related to the formation of halogen induced organic aerosols, presented in the paper by J. Ofner *et al.*

Also, connected to the general discussion on new particle formation, an interesting issue is the spatial extension of nucleation events which, when stable boundary layers are present[1,2] may be confined to ground in a wide variety of basin valleys.

1 L. Ferrero, D. Cappelletti, B. Moroni, G. Sangiorgi, M.G. Perrone, S. Crocchianti and E. Bolzacchini, *Atm. Env.*, 2012, **56**, 143–153.
2. B. Moroni, D. Cappelletti, F. Marmottini, F. Scardazza, L. Ferrero and E. Bolzacchini, *Atm. Env.*, 2012, **50**, 267–277.

Mr Held replied: The release of reactive halogen species by the activation of sea salt or from the surface of salt lakes (which are sometimes basin valley like) causes a strong depletion of ozone (J. Buxmann *et al.*, *Int. J. Chem. Kinet.*, 2012, **44**, 312–326). There is a strong influence of organic species or aerosols on the halogen profiles in these release mechanisms, as can be found in Ofner *et al.* (J. Ofner *et al.*, *Atmos. Chem. Phys.*, 2012, **12**, 5787–5806). We agree that the combination of a strong halogen source with VOC release at stable boundary layer conditions supports the formation of XOA.

Faraday Discussions

RSCPublishing

PAPER

Average chemical properties and potential formation pathways of highly oxidized organic aerosol

Kelly E. Daumit,[a] Sean H. Kessler[b] and Jesse H. Kroll[*ab]

Received 20th March 2013, Accepted 30th April 2013

DOI: 10.1039/c3fd00045a

Measurements of ambient organic aerosol indicate that a substantial fraction is highly oxidized and low in volatility, but this fraction is generally not reproduced well in either laboratory studies or models. Here we describe a new approach for constraining the viable precursors and formation pathways of highly oxidized organic aerosol, by starting with the oxidized product and considering the possible reverse reactions, using a set of simple chemical rules. The focus of this work is low-volatility oxidized organic aerosol (LV-OOA), determined from factor analysis of aerosol mass spectrometer data. The elemental composition and volatility of the aerosol enable the determination of its position in a three-dimensional chemical space (defined by H/C, O/C, and carbon number) and thus its average chemical formula. Consideration of possible back-reactions then defines the movement taken through this chemical space, constraining potential reaction pathways and precursors. This approach is taken for two highly oxidized aerosol types, an average of LV-OOA factors from ten field campaigns (average formula $C_{10.5}H_{13.4}O_{7.3}$), and extremely oxidized LV-OOA (from Mexico City, average formula $C_{10}H_{12.1}O_{8.4}$). Results suggest that potential formation pathways include functionalization reactions that add multiple functional groups per oxidation step, oligomerization of highly oxidized precursors, and, in some cases, fragmentation reactions that involve the loss of small, reduced fragments.

1 Introduction

Organic aerosol (OA) constitutes a substantial fraction (20%–90% by mass[1]) of atmospheric fine particulate matter, and thus a thorough understanding of its formation and evolution is necessary in order to better evaluate the effects of particulate matter on climate and human health. However, despite substantial research efforts, there remains a great deal unknown about the chemical mechanisms important to OA. Especially uncertain are the sources and chemistry of the

[a]MIT Department of Civil and Environmental Engineering, Cambridge, MA, United States. E-mail: jhkroll@mit.edu; Tel: +617 253 2409

[b]MIT Department of Chemical Engineering, Cambridge, MA, United States

most oxidized fraction of OA, sometimes termed "humic-like substances" (HULIS) (when measured using offline techniques of fractionated filter samples),[2] or low-volatility oxidized organic aerosol (LV-OOA) (when determined from factor analysis of online aerosol mass spectrometer data).[3] This class of atmospheric organic matter is ubiquitous in the atmosphere,[1,3] highly oxidized (with an average carbon oxidation state of approximately 0–1),[4] exceedingly low in volatility (saturation concentration $c^* \sim 0.1$ to $<10^{-7}$ µg m^{-3}),[5] and generated relatively quickly (over time scales of $\sim$1–3 days).[3,6]

While this highly oxidized OA is presumed to be secondary in nature (formed from chemical transformations of gas-phase organics that subsequently condense as particulate matter), its detailed formation pathways are poorly understood and have not been adequately reproduced in the laboratory. Laboratory chamber experiments of secondary organic aerosol (SOA) formed from various classes of precursors tend to produce aerosol that is relatively unoxidized (average carbon oxidation state $(OS_C) \sim -1.1$ to 0.1).[4,7] Several efforts to "age" OA in laboratory chambers have also fallen short of reaching oxidation levels as high as those measured for ambient aerosol.[8,9] Those studies that have succeeded in generating highly oxidized SOA have generally been able to do so only with a select few precursors[10,11] or at OH exposures substantially higher than those relevant to the formation of oxidized aerosol in the atmosphere.[12–14] Alternative routes for aging aerosol, such as aqueous-phase oxidation[15] or photosensitized reactions[16] have received recent attention, but there is still much unknown about their chemistry and role in OA formation.

Recent modeling efforts have also attempted to reproduce the amount and degree of oxidation of organic aerosol as measured in field studies, by introducing additional aging chemistry to simulate oxidation beyond the initial SOA formation.[17–19] However, such models, whether simulating a specific environment (*e.g.*, Mexico City[17,20,21]), or SOA formation from a specific class of precursors,[18,22] still tend to be unsuccessful in reproducing the formation of highly oxidized aerosol. They either underpredict OS_C at a given loading[17,20,21] or predict the formation of highly oxidized OA over longer time scales than what is observed.[23] These model-measurement discrepancies suggest major gaps in our understanding of the chemistry underlying the formation of highly oxidized organic aerosol. Unfortunately at present we have very little information as to whether these gaps relate to uncertainties in reaction mechanisms, SOA precursors, or both.

Here we present a new approach for constraining the formation of highly oxidized aerosol, involving the use of chemical properties of the aerosol to better understand its possible formation pathways and precursors. In contrast to most modeling approaches that start with potential precursors and simulate their forward evolution in an attempt to generate products chemically similar to those measured in the atmosphere, we start with the known products (highly oxidized OA components) and work backwards toward reactants (SOA precursors). This approach is similar to the retrosynthetic approach of Pun *et al.* for assessing formation pathways of atmospheric water-soluble organic compounds.[24] That approach was highly molecule- and reaction-specific, involving the identification of detailed chemical reactions that link individual precursors to individual aerosol components. Here we take a more generalized view of the chemistry, assessing classes of reactions that could ultimately lead to the formation of highly oxidized, low-volatility organics. Such an approach requires knowledge of the chemical properties of the "targets" (molecules representing the oxidized fraction of the aerosol); we define these in terms of the average chemical formulas of their constituent molecules, as determined from properties measured in field studies as well as structure–activity relationships. From these targets, a set of simple, general rules governing possible atmospheric reactions lets us work backwards, allowing for identification of viable reaction pathways and aerosol precursors (which are also defined in terms of their average chemical properties). By focusing on generic processes rather than detailed chemical structures and mechanisms, we can draw general conclusions about the types of processes and precursors that might yield highly oxidized organic species. These can be used to provide guidance in identifying relevant reaction conditions and precursors in future studies.

In the following section we describe this methodology in detail, including the chemical space we use to describe the organics, the characterization of our targets, and the types and characteristics of reverse pathways considered. Results – the chemical characteristics of our targets (constrained by AMS measurements of LV-OOA components) and the possible range of aerosol formation pathways – are then described. We conclude with discussions of implications of this work for our understanding of highly oxidized organic aerosol, as well as potential areas of future work.

2   Methodology

We describe molecules in terms of their position in a three-dimensional chemical space, defined by their carbon number (n_C), oxygen-to-carbon ratio (O/C), and hydrogen-to-carbon ratio (H/C). O/C and H/C have already seen substantial use in describing the average properties of atmospheric organic aerosol using the two-dimensional van Krevelen diagram;[7,25] however, this provides no information on the size of the molecule (a crucial factor governing volatility). By considering n_C as a third dimension to the van Krevelen space, we can represent any molecular formula $C_xH_yO_z$ as a single, unique point, and can relate its position in chemical space to volatility.[12,13] This space is equivalent to the three-dimensional space defined by carbon number, hydrogen number, and oxygen number (n_C vs. n_H vs. n_O); we choose to use O/C and H/C here since they are also used in the van Krevelen diagram, and are more amenable to visualization and comparison of elemental composition (hydrogen and oxygen content) of molecules of different sizes. A previously proposed alternative three-dimensional space uses molecular weight, heteroatom mass, and double bond equivalents.[26] While this has advantages for describing certain atmospheric systems, especially those with nitrogen or sulfur containing species, the space proposed here is more conducive to the description of atmospheric reactions in terms of changes to carbon, hydrogen, and oxygen content of the molecules.

The three-dimensional space we use here is closely related to many of the two-dimensional frameworks that have recently been used to describe and/or model organic aerosol. These include the van Krevelen plot,[7] polarity vs. n_C,[27] OS_C vs. saturation vapor concentration (c^*),[3,19] the f44 vs. f43 space for plotting AMS data,[25,28] OS_C vs. n_C,[4] and n_O vs. n_C.[18] Although adding a third dimension introduces complexity to any of these descriptions, it allows for more chemical information to be represented. It explicitly includes the effects of carbon number, which are missing from functional-group-based frameworks (e.g., van Krevelen and f44–f43), and it requires fewer assumptions about the identity or distributions of functional groups than frameworks that reduce the descriptions of functional groups to a single dimension (polarity–n_C, OS_C–n_C, OS_C–c^*, n_O–n_C). We note that isomeric species of a given formula $C_xH_yO_z$ are not distinguished even in this 3-D space, which can lead to errors in the calculation of c^* when the identity of the functional groups changes, as discussed below. Nonetheless, chemical transformations and properties (*e.g.*, c^*) can be represented in this space in a straightforward manner.

I. Characterization of the target

Our general approach is to start with the product (molecules representative of highly oxidized organic aerosol) and work backwards. We refer to this species as the "target" and the backwards reactions as "transforms" following the conventions of retrosynthetic analysis.[29] Key to this approach is the accurate determination of the chemical formula (position in O/C–H/C–n_C space) of the target.

Since the target in this case (highly oxidized organic aerosol) is not a single molecule, but rather a complex mixture, we define it in terms of its average properties and average chemical formula. In the present study we determine this from field studies using the Aerodyne High Resolution-Time of Flight-Aerosol Mass Spectrometer (HR-ToF-AMS, or simply AMS),[30,31] which provides measurements of the elemental ratios of OA.[32,33] Positive matrix factorization (PMF) of AMS data[34] typically yields several factors, of which LV-OOA is the most highly oxidized. We take this to be representative of the most oxidized organic aerosol, and use measurements of LV-OOA elemental ratios to define the location of our target in van Krevelen space.

Because the AMS provides no information regarding carbon number (the third dimension of our chemical space), we determine n_C from field measurements of aerosol volatility (c^*).[5,36,38] The carbon number can be estimated from c^* using structure–activity relationships (SARs) for determining vapor pressure. In this work we use SIMPOL,[39] allowing for the direct determination of c^* from n_C and the functional group abundance, as described below. We focus on partitioning only into the condensed organic phase, as described by c^*. Partitioning into liquid water could also be included using this general approach, but it would require use of a SAR for estimating the volatility over water (Henry's law constant), which is beyond the scope of this work.

Key inputs for SIMPOL (and most other SARs for estimating vapor pressures) are the abundances of different functional groups in the molecule.[39] In order to determine these from our values for n_C, H/C, and O/C, we make two assumptions. The first is that functional groups in the molecule are limited to carbonyls, hydroxyls, or some combination of the two (*e.g.*, carboxylic acids). Since each functional group contains only one oxygen atom, their abundances can be related to the oxygen number of the molecule:

$$n_O = n_C O/C = n_{-OH} + n_{=O} \qquad (1)$$

where n_O, n_C, and O/C are as defined above, and n_{-OH} and $n_{=O}$ are the number of hydroxyl groups and carbonyl groups in the molecule, respectively. While other functional groups are also likely to be present in organic aerosol, several groups can be approximated using this simple treatment, as discussed below.

The second assumption is that all sites of unsaturation (double bond equivalents, DBE) in the molecule arise from carbonyl groups:

$$n_{=O} = \text{DBE} = 1 + n_C\left(1 - \frac{1}{2}\text{H/C}\right) \tag{2}$$

We therefore assume that our target contains no carbon–carbon double bonds or rings. While this is reasonable for aliphatic C=C bonds, which are highly susceptible to oxidation and unlikely to survive significant atmospheric processing, cyclic (or aromatic) structures may be present in highly oxidized OA. The effect of neglecting any rings present is to overestimate $n_{=O}$ and underestimate n_{-OH}, thereby overestimating c^* somewhat.

The two assumptions above (eqn (1) and (2)) allow for the straightforward derivation of expressions relating elemental ratios, functional groups, volatility, and carbon number. The number of hydroxyl groups (n_{-OH}) can be determined by combining eqn (1) and (2):

$$n_{-OH} = n_O - n_{=O} = n_C\text{O/C} - \left(1 + n_C\left(1 - \frac{1}{2}\text{H/C}\right)\right)$$

$$= -1 + n_C\left(\text{O/C} + \frac{1}{2}\text{H/C} - 1\right) \tag{3}$$

The carbon number (n_C) and functional group abundances (n_{-OH} and $n_{=O}$) allow for the calculation of c^* (μg m^{-3}) using SIMPOL:

$$\log_{10}c^* = \log_{10}(\alpha P_{\text{sat}}) = \log_{10}\alpha + b_0 + b_C n_C + b_{=O}n_{=O} + b_{-OH}n_{-OH}$$

$$= \log_{10}\alpha + b_0 + b_C n_C + b_{=O}\left(1 + n_C\left(1 - \frac{1}{2}\text{H/C}\right)\right) \tag{4}$$

$$+ b_{-OH}\left(-1 + n_C\left(\text{O/C} + \frac{1}{2}\text{H/C} - 1\right)\right)$$

where P_{sat} is the vapor pressure in atm, α is the conversion from P_{sat} to c^*, and the b terms are the different group contribution terms for quantifying the contribution of each chemical moiety to vapor pressure: b_0 is the zero order term, b_C is the carbon number term, $b_{=O}$ is the carbonyl group term, and b_{-OH} is the hydroxyl group term (for MW = 200 g mol^{-1} and T = 293 K, equal to 1.79, −0.438, −0.935, and −2.23, respectively).[39] The conversion factor α is given by $\alpha = 10^6(\text{MW})/RT$, or 8.314×10^9 at 293 K, where MW is the average molecular weight (g mol^{-1}) of the molecules making up the absorbing phase, R is the ideal gas constant (8.21×10^{-5} atm m^3 mol^{-1} K^{-1}), and T is the temperature in K. For these calculations we use an assumed MW value of 200 g mol^{-1}; the actual value used has little effect on the results. Rearranging eqn (4) to solve for carbon number, and substituting in values of $n_{=O}$ and n_{-OH} from eqn (2) and (3), we obtain

$$n_C = \frac{\log_{10}c^* - \log_{10}\alpha - b_0 - b_{=O} + b_{-OH}}{b_C + b_{=O}\left(1 - \frac{1}{2}\text{H/C}\right) + b_{-OH}\left(\text{O/C} + \frac{1}{2}\text{H/C} - 1\right)} \tag{5}$$

Eqn (5) allows for the determination of carbon number from elemental (H/C, O/C) and volatility (c^*) data. This approach, used in previous descriptions of heterogeneous oxidation systems,[12,13] is similar to that of Donahue et al.[19] for estimating carbon number from OS_C (or O/C) and c^* alone; however that approach required making assumptions about functional group distribution (specifically that $n_{-OH} = n_{=O}$). With the explicit inclusion of H/C, we can directly calculate n_{-OH} and $n_{=O}$, which allows for a more accurate estimation of n_C.

Eqn (5) relies critically on the assumption that target molecules contain only hydroxyl and carbonyl groups. While molecules containing other functional groups will be less accurately represented, many such functional groups are reasonably approximated as carbonyls, alcohols, or some combination of the two. Because the vapor pressure effect of an acid group ($b_{(=O)OH} = -3.58$) is similar to that of a hydroxyl plus a carbonyl ($b_{-OH} + b_{=O} = -3.165$), treating an acid as the sum of the two introduces only a minor error. Likewise, epoxides are very similar to carbonyls in terms of both bonding and effect on volatility: like carbonyls, they contribute one DBE, and the vapor pressure effect of an ether ($b_{-O-} = -0.718$) is similar to that of a carbonyl. Similarly, nitrate and hydroperoxyl groups are connected to the carbon skeleton *via* a single C–O bond and so have similar bonding as a hydroxyl, and also have roughly the same vapor pressure effect ($b_{-NO_3} = -2.23$, $b_{-OOH} = -2.48$). These functional groups have additional oxygen atoms, but in the AMS these are likely lost as NO_2 or OH fragments[40] and so it is likely that only one oxygen atom is measured (as is the case for a hydroxyl group). Since these moieties contribute one fewer hydrogen than a hydroxyl group, the measured H/C may not be exactly the same as for the corresponding alcohol, but this will have only a minor effect on results. Unfortunately, not all functional groups are as well-represented in the above treatment. Acyclic ethers have the same bonding (and contribution to O/C and H/C) as hydroxyl groups, but have a substantially smaller effect on vapor pressure ($b_{-O-} = -0.718$ *vs.* $b_{-OH} = -2.23$). Similarly, esters, like acids, will be treated as a carbonyl plus a hydroxyl group,

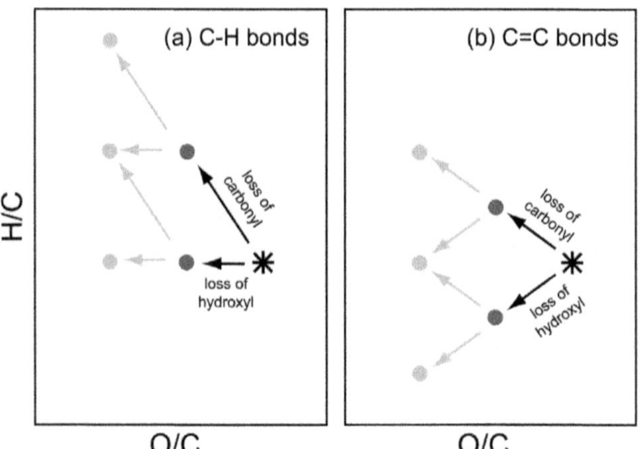

Fig. 1 Possible transforms (back-reactions) available to a target molecule, assuming functionalization reactions only (with no change to carbon number). Loss of functional groups is illustrated in terms of movement in van Krevelen space, depicting changes in H/C and O/C. The exact trajectory depends on the identity both of the functional group and the bond being replaced. Panel a: functionalization of C–H bonds (saturated case), with slopes of −2 for carbonyl groups and 0 for hydroxyl groups. Panel b: functionalization of C=C bonds or rings (unsaturated case), with slopes of −1 for carbonyl groups and +1 for hydroxyl groups. In either case, changes to more than one functional group (light grey arrows and points) can be described by simple vector addition.

but their actual effect on vapor pressure is far smaller ($b_{(=O)O^-} = -1.20$ *vs.* $b_{-OH} + b_{=O} = -3.165$). There is no good way to account for these errors using elemental ratios alone; direct measurements of functional group abundances in OA (*e.g.*, using FTIR[41]) would thus greatly aid in this general approach for determining the chemical formula and properties of the target.

We note there are ways to determine the chemical formula (position in a given chemical space) other than the elemental- and volatility-based approach described above. One widely-used technique is ultrahigh-resolution electrospray ionization mass spectrometry (UHR-ESI-MS) of filter samples to directly determine exact chemical formulas.[42,43] There are currently far fewer measurements of ambient OA by UHR-ESI-MS than by AMS, but this is another promising technique for characterizing the target. While concerns such as variations in sensitivity toward different compounds, and the possibility of fractionation arising from dissolution in a given solvent, still need to be addressed, estimates of OS_C and n_C from AMS and ESI measurements are generally in good agreement.[4]

II. Transforms

Once the target is well-defined, transforms (the reverse of atmospheric reactions) allow for the formation chemistry to be traced backwards toward potential precursors. Again, the aim is not to identify specific formation pathways and precursor molecules, but rather to examine the key features of viable ones. Here, we consider three reaction classes—functionalization, fragmentation, and oligomerization—each of which is discussed below. Simple chemical rules for each reaction class help constrain their potential role and possible precursors.

a) Functionalization. Functionalization reactions are those that add (or interconvert) functional groups without other changes to the carbon skeleton. They are easily represented in a 2-D van Krevelen plot since they involve no change in n_C. Movement through this space has been used previously to describe different functional group additions.[7,25] Reverse functionalization reactions connecting target molecules back to their precursors involve the loss of carbonyl or hydroxyl groups, as shown in Fig. 1. The exact direction in van Krevelen space depends not only on the functional group involved but also the nature of the precursor, specifically the bond (C–H *vs.* C=C) being replaced by the functional group. Fig. 1a shows the "saturated" case, in which the functional groups replace C–H bonds, which has been considered previously.[7] Removing a single functional group decreases O/C by $1/n_C$, with the effect on H/C depending on the identity of the group. Loss of a carbonyl decreases H/C by $2/n_C$ (following a slope of -2 on the van Krevelen plot), while removing a hydroxyl has no effect on the H/C ratio (following a slope of 0). These steps are fully additive, such that the loss of multiple functional groups can be described simply by the sum of the vectors associated with the individual groups. For example, removing a hydroxycarbonyl or carboxylic acid (the equivalent of removing one hydroxyl and one carbonyl) decreases O/C by $2/n_C$ and increases H/C by $2/n_C$ (thereby following a slope of -1). Fig. 1b shows the "unsaturated" case, in which the functional group replaces a carbon-carbon double bond (or C–C single bond within a ring). Loss of a functional group still decreases O/C by $1/n_C$, but now removing a carbonyl increases H/C by only $1/n_C$ (slope $= -1$), whereas removing a hydroxyl actually decreases H/C, by $1/n_C$ (slope $= +1$). Note that the oxidation of C=C bonds typically involves changes to two functional groups, so movement through this space can be quite rapid. In either case (saturated or unsaturated), for a given n_C, each point in van Krevelen space corresponds to a value of c^*, allowing the change in volatility associated with any transform to be easily computed.

We can define the range of potential precursors of our target by providing two fundamental limits on what reverse reactions are possible: (1) a molecule cannot lose more of a given functional group than it actually has; and (2) going backwards, the average carbon oxidation state cannot increase, since reduction reactions are thermodynamically unfavorable. This can significantly narrow the area in chemical space that defines the potential precursors, as discussed in the Results section.

b) Fragmentation. Fragmentation reactions decrease the carbon number of the molecule *via* the cleavage of carbon–carbon bonds. Since these reactions are oxidative, they will involve the addition of functional groups at the site of the bond breakage. (The scission of a single bond within a ring is therefore not considered a fragmentation reaction, and is instead a functionalization reaction, involving the addition of multiple functional groups.) Fragmentation reactions can result in rapid oxidation of the carbon, since they can lead to a high abundance of functional groups on relatively small molecules. However, the decrease in carbon number can also lead to a substantial increase in volatility. This tradeoff likely limits the importance of fragmentation pathways for the formation of very oxidized, low-volatility species.

Because n_C changes during fragmentation, simple 2-D van Krevelen plots are not sufficient to accurately capture these transformations. Furthermore, because of the range of possible precursors (with any number of carbon, oxygen, or hydrogen atoms), there are no simple generic transforms to describe this change. Nonetheless, we can apply constraints that considerably narrow the chemical space associated with the possible transforms and precursors. One approach is to assume that functional groups are evenly distributed across all the carbon in a molecule, and that this distribution is conserved during the fragmentation process (with the exception of any functional groups added during fragmentation). This statistical treatment of fragmentation is similar to that used in the recently-developed Statistical Oxidation Model.[18] For the present study, we assume that two functional groups (one hydroxyl and one carbonyl) are added to a given fragment, but this distribution can be altered as necessary. Such assumptions (plus the requirement that n_C of the precursor must exceed that of the target), give the following constraints: $n_{C,p} > n_{C,t}$, $n_{-OH,p} = (n_{-OH} - 1)(n_{C,p}/n_{C,t})$, and $n_{=O,p} = (n_{=O,t} - 1)(n_{C,p}/n_{C,t})$, where subscript "t" refers to the target and subscript "p" refers to the precursor. The subtraction of 1 from $n_{-OH,t}$ and $n_{=O,t}$ accounts for the addition of the two functional groups upon fragmentation.

An alternative, more exact approach is to allow the functional groups to be localized somewhere on a molecule, rather than distributed evenly across it. This allows for a broader range of potential precursors, since fragments can have varying degrees of oxidation (*i.e.*, one that is more oxidized than the other). In this case, the number of functional groups in the target provides constraints on precursors: after taking into account any new functional groups, the precursor cannot have fewer hydroxyl or carbonyl groups than the target. Again, if we assume that two groups (one hydroxyl and one carbonyl) are added to the fragment, thereby increasing $n_{=O}$ and n_{-OH} each by one, we obtain the following constraints: $n_{C,p} > n_{C,t}$, $n_{-OH,p} \geq (n_{-OH,t} - 1)$, and $n_{=O,p} \geq (n_{=O,t} - 1)$.

Table 1 HR-ToF-AMS data from measurements at 10 field locations. Elemental ratios are reported for the average LV-OOA PMF factor at each site and are used to calculate each average LV-OOA OS_C ($\sim 2O/C - H/C$)

Measurement Location	O/C	H/C	OS_C	Reference
Riverside, CA (2005)	0.72	1.27	0.17	Docherty 2011[45]
Mexico City Aircraft (2006)	1.02	1.12	0.92	DeCarlo 2010[35]
Mexico City Ground (2006)	0.84	1.21	0.47	Aiken 2008[32]
Kaiping, China (2008)	0.64	1.30	−0.02	Huang 2011[37]
Barcelona, Spain (2009)	0.75	1.18	0.32	Mohr 2012[46]
Paris, France (2009)	0.73	1.33	0.13	Crippa 2013[47]
New York City, NY (2009)	0.63	1.29	−0.03	Sun 2011[48]
Hong Kong, China (2009)	0.59	1.26	−0.08	He 2011[49]
Shanghai, China (2010)	0.65	1.49	−0.19	Huang 2012[50]
Sacramento, CA (2010)	0.54	1.32	−0.24	Setyan 2012[51]
Heshan, China (2010)	0.55	1.30	−0.20	Gong 2012[52]

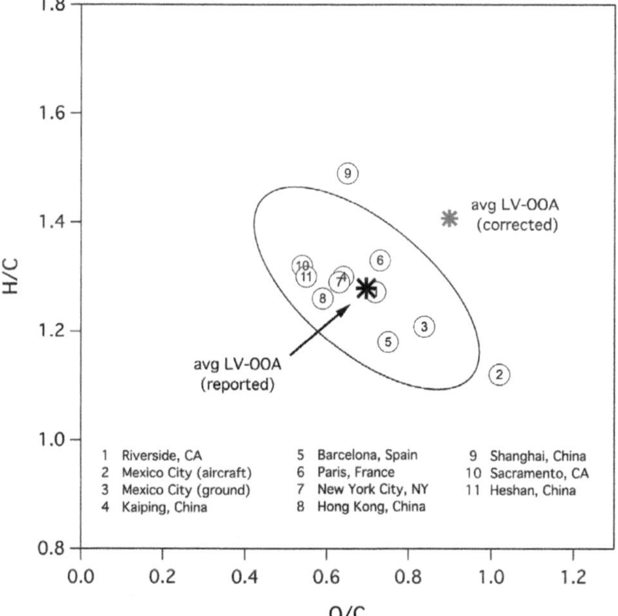

Fig. 2 Van Krevelen diagram showing the reported elemental ratios of the LV-OOA factors from HR-AMS field campaigns (see Table 1). The black star is the average reported O/C and H/C from all measurements, with the ellipse indicating the covariance of the data within 1.96 standard deviations of the uncorrected mean value. The grey star is the average elemental composition after applying the recommended correction of 1.3 and 1.1 to O/C and H/C, respectively.[53]

c) Oligomerization. Oligomerization reactions occur when two molecular species combine to form a larger one. Because of the associated increase in carbon number, oligomerization reactions are an efficient way to lower vapor pressure. However, this type of reaction is typically non-oxidative[44] so is not accompanied by an overall change in OS_C. Therefore the carbon atoms in the precursors must, on average, be as oxidized as those in the target.

As with fragmentation, oligomerization transforms are more difficult to describe than those for functionalization, since any number of changes to n_C, n_H, and n_O are possible. However, we can still apply basic constraints related to the allowable number of carbons and functional groups to narrow the range of potential precursors. The same statistical method described for fragmentation reactions can be used for oligomerization to constrain the precursor molecules. Here, we require the n_C of the precursor to be smaller than that of the target, and we assume that the precursor molecules have the same functional group distribution as the target, giving the following constraints: $n_{C,p} < n_{C,t}$, $n_{-OH,p} = n_{C,p}(n_{-OH,t}/n_{C,t})$, and $n_{=O,p} = n_{C,p}(n_{=O,t}/n_{C,t})$.

As with fragmentation, we can expand the range of precursors further by allowing the functional groups to be unevenly distributed between the two precursors. This allows one of the oligomerization reactants to be more functionalized that the other. While a large number of different oligomerization transformations are possible, we assume that for any oligomerization reaction each precursor must have at least one functional group. We also allow the total oxygen and hydrogen content to change somewhat by the gain or loss of a water molecule, or by the conversion of a carbonyl to an alcohol (as in a hemiacetal formation). These assumptions, plus the requirement that the precursor n_C be lower than the target n_C, yield the following simple rules for the number of oxygen atoms and carbonyl and hydroxyl groups: $n_{O,p} \leq n_{O,t}$, $n_{=O,p} \leq (n_{=O,t} + 1)$, and $n_{-OH,p} \leq (n_{-OH,t} + 1)$. It should be noted that if ester or ether linkages are formed during the oligomerization reaction, they will introduce an error in the c^* calculation. However, these effects are likely to be relatively small since the change in c^* is governed mostly by changes to the carbon number, rather than by interconversions of functional groups.

3   Results

I. LV-OOA target

Properties of the highly oxidized OA target are determined from HR-ToF-AMS field measurements. Table 1 summarizes AMS elemental ratios for studies that report H/C and O/C for the LV-OOA PMF factor (or, more generally, the most oxidized of multiple OOA factors, which includes factors such as OOA-I[32] or MO-OOA[51]). Fig. 2 shows the location of these factors in van Krevelen space. The degree of oxidation varies widely from study to study, indicating that LV-OOA is not one class of compounds but rather an aged form of SOA with a composition that varies as a function of location, emissions profile, and/or oxidant exposure. For example, LV-OOA from the MILAGRO field campaign in the Mexico City Metropolitan Area (MCMA)[32,35] is substantially more oxidized than that measured at the other field sites. Nonetheless, we use the average values of LV-OOA H/C and O/C (1.28 and 0.696, respectively) from these studies as our generic highly-oxidized OA target. This is shown as a black star in Fig. 2. The ellipse represents the covariance of the data within 1.96 standard deviations of the uncorrected mean value. AMS elemental ratios for ambient organic aerosol are typically corrected using empirical factors to account for biases in ion fragmentation.[32] Recent calibration studies indicate that the recommended factors result in an underestimation of O/C and H/C values for a range of multi-functional oxidized organic standards, indicating that AMS elemental ratios reported for ambient environments will

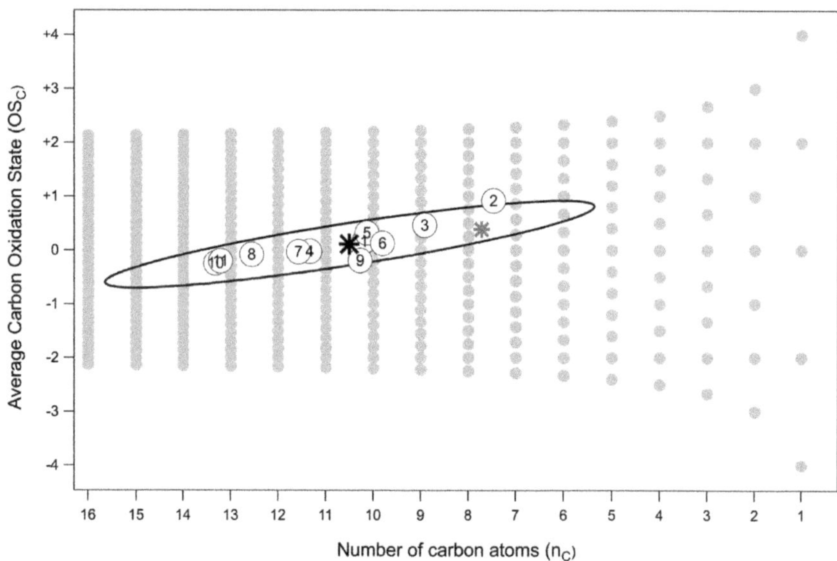

Fig. 3 Locations of the different LV-OOA factors in OS_C–n_C space. Grey circles indicate the possible combinations of average carbon oxidation state and carbon number.[4] Numbered points denote LV-OOA from all ten field campaigns (numbers are the same as in Fig. 2); OS_C is determined from elemental ratios, and n_C from eqn (5), using a c^* of 10^{-3} μg m^{-3}. The black star shows the average values for all the factors, with the ellipse depicting the uncertainty in its n_C and OS_C values (see text for details). The grey star shows the position of the average LV-OOA after applying the recommended correction of 1.3 and 1.1 to O/C and H/C, respectively.[53]

need to be modified by new correction factors (effectively 1.3 for O/C and 1.1 for H/C).[53] The average ratios calculated after using these new proposed correction factors are also shown in Fig. 2.

Volatility of the aerosol, as measured with thermodenuders, has to our knowledge been reported for only a limited number of field studies,[38] and values for c^* have been determined only for the MCMA ground-based data.[5] The decrease in the LV-OOA factor upon heating in the thermodenuder is consistent with LV-OOA volatilities spanning a range of c^* values between 10^{-7} and 10^{-1} μg m^{-3}.[5] However, since the LV-OOA measured in MCMA appears not to be representative of LV-OOA everywhere (Fig. 2), these volatility ranges may not be widely applicable to other locations. On the other hand, the elemental composition of Riverside LV-OOA is close to that of the overall LV-OOA average (Fig. 2), suggesting its volatility, which is substantially higher than that of LV-OOA in MCMA,[38] may be more representative of typical LV-OOA. We choose c^* in the range of 10^{-5} to 10^{-1} μg m^{-3} as reasonable values based on the reported c^* distribution.[5] From these c^* values, carbon numbers can be calculated (eqn (5)): for $\log_{10}(c^*)$ of -1, -3, and -5, the corresponding chemical formulas are $C_{9.2}H_{11.8}O_{6.4}$, $C_{10.5}H_{13.4}O_{7.3}$, and $C_{11.8}H_{15.1}O_{8.2}$, respectively, for the reported average, and $C_{6.8}H_{9.5}O_{6.1}$, $C_{7.7}H_{10.9}O_7$, and $C_{8.7}H_{12.2}O_{7.9}$, respectively, for the corrected average.[53] These formulas are not meant to define exact chemical species or specific molecules, but instead represent averages of all the compounds in LV-OOA.

The sources of uncertainty considered in evaluating the average number of carbon atoms in the LV-OOA mixture arise from estimated errors in the elemental

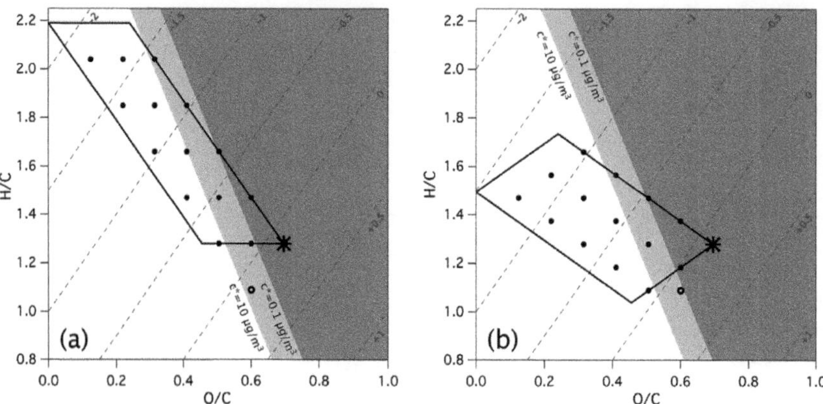

Fig. 4 Functionalization transforms available to the LV-OOA target (black star). These are calculated for $n_C = 10.5$ and $c^* = 10^{-3}$ µg m^{-3}, for the fully saturated case (panel a) and the fully unsaturated case (panel b). Potential oxidative transforms and precursors are bounded by the parallelograms, whose dimensions are defined by the number of hydroxyl and carbonyl functional groups in the target. Each black circle depicts an intermediate precursor, formed from the removal of an integer number of functional groups; the leftmost corner of each parallelogram (y-intercept) gives the limiting H/C value of the unfunctionalized hydrocarbon precursor. OS$_C$ contours are shown as dashed grey lines, and shading indicates the predominant phase of the species: gas phase (white, $c^* > 10$ µg m^{-3}), semi-volatile (light grey, c^* 0.1 to 10 µg m^{-3}), or condensed-phase (dark grey, $c^* < 0.1$ µg m^{-3}). Multiple functional groups (including at least one hydroxyl group) must be lost for the immediate precursor of the target to be in the gas phase. Nonoxidative processes (*e.g.* hydrolysis) may also be important in SOA formation; the open circles indicate immediate hydrolysis precursors.

analysis calculation, the distribution of vapor pressures of the compounds within the mixture, and variation in the elemental ratios measured in the different field studies. The measurement error arising from elemental analysis is estimated to be ±30% of the reported O/C and ±10% for H/C.[32,33] Because these errors were reported for individual compounds, we expect the measurement error of a large mixture of compounds to be substantially smaller, so that the values given represent the most conservative case. The uncertainty in c^* for the average of all OOA components is taken to be $10^{-3\pm2}$ µg m^{-3}, as approximated from the distribution of values reported.[5] The variance in elemental ratios for the mixtures is simply the variance of the data presented in Table 1, multiplied by 1.96^2 in order to reflect the range in which 95% of the data is expected to fall. These errors are propagated forward using the sensitivity of n_C to O/C, H/C, and $\log_{10}(c^*)$ as described by eqn (5), to give three separate values of the uncertainty in the carbon number, δ_{NC}. The contributions from elemental analysis measurement error, variation in the volatility, and variance are 3.3, 1.3, and 3.8, respectively. An overall uncertainty, $\delta_{NC,tot}$, is calculated as the square root of sum of squares of each separate uncertainty, to yield a final estimate of 10.5 ± 5.2 for the LV-OOA carbon number, with the greatest source of uncertainty being the variance in elemental ratios reported for each field study discussed here.

Fig. 3 (OS$_C$ *vs.* n_C) shows the location of the different LV-OOA factors in carbon oxidation state *vs.* carbon number space.[4] Also shown is the average LV-OOA target, with the ellipse depicting the uncertainty associated with its n_C and OS$_C$ values. The rotated ellipse arises from the covariance between the outputs and reflects the fact that more highly oxidized compounds require a smaller carbon number for a given value of c^*. The carbon numbers determined here ($n_C \sim 6\text{–}15$) are slightly larger than those estimated by Donahue *et al.* ($n_C \sim 5\text{–}10$) using a similar technique and the same general c^* range.[54] This difference in part results from the lower value of OS_C that we use, and our direct determination of the abundance of hydroxyl and carbonyl functional groups individually, rather than assuming the two are equally abundant.

Functionalization. Taking the LV-OOA target determined using $c^* = 10^{-3}\ \mu g\ m^{-3}$ ($C_{10.5}H_{13.4}O_{7.3}$), and assuming it was formed *via* functionalization reactions only (with changes to H/C and O/C but not carbon number), allows us to work backwards toward precursors in van Krevelen space, using the simple rules described in the Methodology section. Fig. 4 shows the potential formation pathways and precursors, but is different than most other van Krevelen plots used to represent atmospheric OA since it includes gas-phase as well as particulate organics. The black star denotes the target, and the black circles depict the possible transforms associated with functionalization (as depicted in Fig. 1), with each point corresponding to the loss of an integer number of functional groups. Fig. 4a shows the fully saturated case in which functional groups replace C–H bonds only, and Fig. 4b shows the unsaturated case in which they replace only C=C bonds.

The parallelograms represent the range of possible transforms and precursors, and are bounded by the requirement that the target cannot lose more hydroxyl or carbonyl groups than it has (so that the dimensions of the parallelogram are given by the number of hydroxyl and carbonyl groups). The left corner of the parallelogram (*y*-intercept in the van Krevelen diagram) corresponds to the fully de-functionalized hydrocarbon precursor. For the saturated case, this gives a precursor with a formula consistent with an acyclic alkane. For the unsaturated case, it gives an alkene containing $n_O/2$ C=C bonds (or 3.7 for our target). This corresponds to a precursor of formula $C_{10.5}H_{15.7}$, which interestingly is similar to that of monoterpenes, a well-studied class of SOA precursors. These two limiting cases bound the H/C of the hydrocarbon precursor to between 1.5 and 2.2.

The areas defined by the parallelograms in Fig. 4, and movement within them, are not necessarily tied to specific chemistry, but rather describe the range of possible precursors. The dashed grey lines are carbon oxidation state contours ($OS_C \approx 2O/C - H/C$). From these, it can be seen that replacing C–H bonds with functional groups (Fig. 4a) moves more directly along the OS_C gradient than replacing C=C bonds (Fig. 4b) does, and results in a larger increase in OS_C per functional group. At the same time, the oxidation of C=C bonds usually involves the addition of two functional groups per oxidation step. It should be noted that the areas bounded by the parallelograms are for oxidation reactions only; non-oxidative transformations (involving movement along lines of constant OS_C) could lead to a wider range of possible precursors. One example is the dehydration of the target molecule (open circles in Fig. 4), whose corresponding forward reaction is the hydrolysis of a carbonyl or epoxide group to form two hydroxyl groups, believed to be an important reaction in the formation of isoprene SOA.[55]

The shading in Fig. 4 indicates the volatility (and therefore predominant phase) as a function of location in van Krevelen space: volatile/gas-phase (white, $c^* > 10\ \mu g\ m^{-3}$), semi-volatile (light grey, $0.1 < c^* < 10\ \mu g\ m^{-3}$), or low-volatility/particle-phase (dark grey, $c^* < 0.1\ \mu g\ m^{-3}$). This is similar to the work of Kessler *et al.* in which c^* cutoffs were calculated as a function of carbon number.[12,13] While the exact cutoffs will vary with organic aerosol loading, these are reasonable values for most ambient conditions. Values for c^* were calculated using eqn (4), though for Fig. 4b it was modified to allow for one DBE to be from a C=C bond:

$$\log_{10}c^* =$$

$$\log_{10}\alpha + b_0 + b_C n_C + b_{=O}\left(n_C\left(1 - \frac{1}{2}H/C\right)\right) + b_{-OH}\left(n_C\left(O/C + \frac{1}{2}H/C - 1\right)\right)$$

$$(6)$$

While the timescales of formation of highly oxidized OA are highly uncertain, there is evidence (based on Mexico City data) that the process is relatively fast, on the order of a day.[3,6] This is faster than the rate of heterogeneous oxidation (which occurs on the order of several days),[12,13,56,57] suggesting that the immediate precursor to highly oxidized OA is likely not in the particle phase. The production of low-volatility products by gas-phase oxidation leads to a "trapping" in the particle phase where the rate of oxidation slows dramatically.[58,59] Thus the rapid formation of highly oxidized OA *via* functionalization reactions would seem to require gas-phase precursors. Semi-volatile compounds (light grey area in Fig. 4) may serve as gas-phase precursors, but their oxidation reactions tend to be slow, since they spend some fraction of their time in the particle phase; in addition they are highly susceptible to depositional losses to other environmental surfaces. Thus, in the absence of fast particle phase oxidation reactions, the most likely

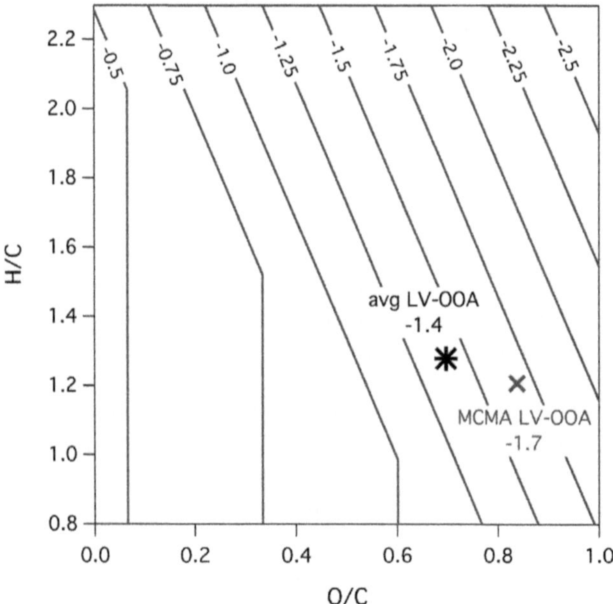

Fig. 5 Contour plot showing the influence of a single carbon atom in a molecule, including its associated functional groups, on the molecule's volatility ($\partial(\log_{10} c^*)/\partial n_C$), as calculated from eqn (4). These are calculated for $n_C = 10.5$, but do not vary significantly with carbon number above $n_C = 7$. Instead H/C and O/C have the strongest influence on "per-carbon volatility" which is greatest when both ratios are high (top right corner, when hydroxyl groups are abundant).[12,13] The discontinuities near the left of the figure reflect the fact that eqn (4) does not always hold at low O/C, where unsaturation arises from C=C double bonds or rings rather than carbonyl groups; thus when DBE > n_O, eqn (4) is modified to allow for C=C double bonds. LV-OOA targets used in this study are denoted by markers.

immediate gas-phase precursors to highly oxidized aerosol are those that are fully in the gas phase; as shown in Fig. 4, this requires the precursor to have substantially fewer functional groups than the target (at least two fewer hydroxyl groups or three fewer mixed hydroxyl/carbonyl groups). Known routes for the addition of multiple functional groups in a single oxidation step include the oxidation of carbon–carbon double bonds or oxidative ring-opening reactions, both of which add at least two (and possibly more) functional groups to the molecule. In addition, some alkoxy isomerization reactions can add three or more functional groups at a time;[60] plus there is evidence for other, poorly-understood mechanisms that involve the rapid addition of multiple functional groups.[61]

The above discussion assumes that the formation of low-volatility organics is rapid compared to the rate of heterogeneous oxidation, and also that gas-particle partitioning is essentially instantaneous. In general, the formation kinetics of the most oxidized fraction of OA are not well constrained; if they are substantially slower in most environments than in MCMA, heterogeneous oxidation could play an important role, allowing the immediate oxidative precursors to be in the condensed phase. This uncertainty underscores the need for additional ambient studies of OA at a variety of photochemical ages. Additionally, recent studies of partitioning kinetics suggest that condensable organics may remain in the gas phase longer than predicted by assumptions of thermodynamic equilibrium.[62] This could potentially allow for additional gas-phase oxidation even when the precursors have volatilities below the semi-volatile/condensed-phase cutoff. On the other hand, several experimental studies suggest a mass transfer limitation to

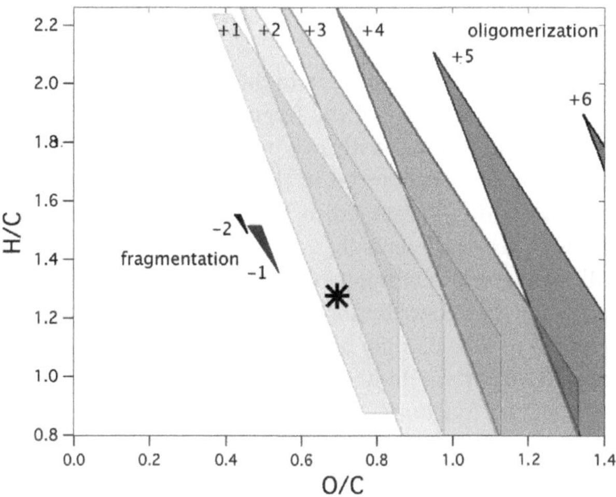

Fig. 6 Ranges of possible precursors of fragmentation (solid-shaded areas) and oligomerization (lightly-shaded areas) reactions that can form the LV-OOA target (black star). Three-dimensional chemical space is depicted by showing the range in van Krevelen space available to precursors of different carbon numbers. The numbers correspond to the carbon number change of the reaction; for example "−2" refers to a fragmentation reaction in which the target has two fewer carbon atoms than the precursor. The target can be formed *via* fragmentation of a gas-phase precursor only if a small (C_1–C_2) reduced fragment is lost; oligomerization is a more viable channel, though it requires the precursors to be low in volatility and thus extremely oxidized. In both cases, the range of possible precursors is considerably narrower if it is assumed that functional groups are evenly distributed on the target and precursor(s).

the evaporation of particulate organics;[63–65] this would have the opposite effect, enhancing the "trapping" effect and causing the semi-volatile precursors to behave more like the least-volatile species.

The potential formation pathways of highly oxidized aerosol also depend on our knowledge of the physicochemical properties (elemental ratios and $c*$), which can be uncertain. While the volatility ($c*$) chosen for our target has a relatively minor effect on calculated carbon number, it strongly influences the potential formation processes, most importantly by affecting the phase of the precursors. Decreasing the $c*$ of the target (for example, from 10^{-3} to 10^{-5} $\mu g\ m^{-3}$, increasing the calculated n_C from 10.5 to 11.8), increases the number of functional groups that must be removed in order for the immediate precursor to be in the gas phase. Therefore targets with low $c*$ are less likely to be formed from functionalization reactions, whereas targets with higher $c*$ can be easily formed from reactions that add only 1–2 functional groups.

The elemental ratios used for the target in Fig. 4 represent the average from a number of studies (Table 1), yielding an average chemical formula of $C_{10.5}H_{13.4}O_{7.3}$ (with 2.5 hydroxyl groups and 4.8 carbonyl groups). If we use the corrected values for the elemental ratios[53] (grey star in Fig. 2), but the same $c*$ as in the base case (10^{-3} $\mu g\ m^{-3}$), we obtain a target with a chemical formula of $C_{7.7}H_{10.9}O_7$ (3.7 hydroxyl groups and 3.3 carbonyl groups). The higher $n_{-OH}/n_{=O}$ ratio implies a lower carbon number at a given $c*$, since hydroxyl groups have a greater effect on $c*$ than do carbonyl groups. While this has little effect on the number of functional groups separating the target and a gas-phase precursor (since the separation is governed primarily by the volatility of the target), it can affect the overall pathways connecting the oxidized aerosol to hydrocarbon precursors; additional oxidation steps will be required to form a target with a higher number of functional groups.

Fragmentation reactions. In addition to the chemical constraints listed in the Methodology section, the phase of the fragmentation precursor also limits the possible transforms associated with this reaction. As in the functionalization case, we require the precursor to be in the gas phase ($c* > 0.1$ $\mu g\ m^{-3}$) since heterogeneous oxidation is assumed to be too slow to account for the rate of formation of highly oxidized aerosol.

The overall, average effect of fragmentation reactions on volatility is illustrated in Fig. 5, which shows the contribution of a single carbon atom (and its associated functional groups) to the volatility of a molecule of $n_C = 10.5$. (Results are only weakly dependent on carbon number above $n_C = 7$.) This is calculated from eqn (4), assuming that the functional groups are evenly distributed throughout the molecule (i.e., the number of groups on the carbon atom is equal to the number of groups on the entire molecule divided by n_C). For our target (black star in Fig. 5), each carbon atom decreases the vapor pressure by 1.4 orders of magnitude. This is a full order of magnitude larger than the effect of a non-functionalized carbon (e.g., a CH_2 group). This simple view of "per-carbon volatility" helps guide assessments of the viability of fragmentation reactions as steps in the formation of highly oxidized organic aerosol; however, as discussed in the Methodology section, it does not accurately represent all fragmentation reactions, since functional groups generally are not evenly distributed throughout a given molecule. Here we consider potential fragmentation precursors both with and without evenly-distributed functional groups.

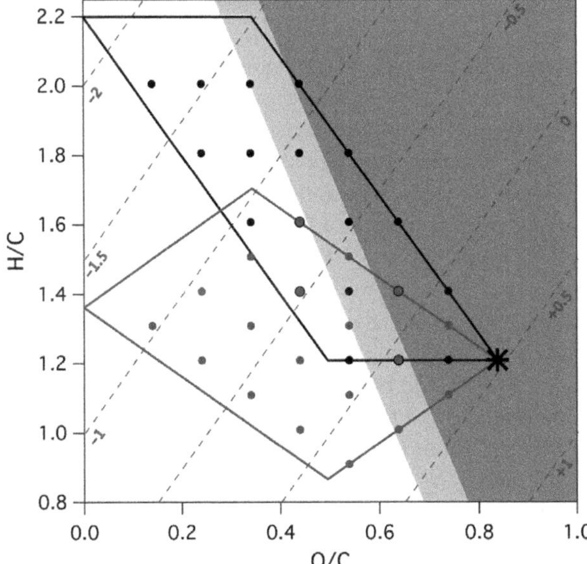

Fig. 7 Functionalization transforms available to the LV-OOA target from the MILAGRO ground site,[31,37] using $n_C = 10$ and $c^* = 10^{-5}$ μg m^{-3}. (These are different than the values in Fig. 3–4 which used c^* of 10^{-3} μg m^{-3}.) This plot is similar to Fig. 4, except that the saturated and unsaturated cases are shown together, as black and grey parallelograms, respectively (the c^* cutoffs are given for the saturated case only). Because of the exceedingly low volatility of these organics,[5,37] their formation from gas-phase precursors can occur only if a large number of functional groups (three or more) are added to the precursor in a single oxidation step.

Assuming the functional groups are distributed evenly, fragmentation of a gas-phase molecule cannot form highly oxidized, low-volatility OA. This is because the decrease in volatility from the additional functional groups (in this case, one hydroxyl group and one carbonyl group, lowering c^* by 3.2 orders of magnitude) is largely if not completely offset by the volatility increase from the carbon loss (1.4 orders of magnitude per carbon atom). As a result, the large decrease in volatility needed to form a low-volatility target (2 orders of magnitude) is not possible, a conclusion broadly consistent with the findings of Cappa and Wilson.[18] However, when functional groups are not evenly distributed, and reduced carbon can be lost during fragmentation, the formation of the target *via* fragmentation is possible in some cases. The carbon numbers and elemental ratios of viable precursors are depicted in Fig. 6 (solid-shaded areas). The allowed precursors are only slightly larger than the target, and are significantly less oxidized, indicating that fragmentation can produce highly oxidized organic aerosol only when the precursor loses a small (C_1 or C_2) and essentially unfunctionalized fragment. (This exact requirement depends on the identity and number of functional groups added to the fragment). However, molecules that are more volatile than our target are more likely to form *via* fragmentation, since they may be more easily formed from gas-phase precursors. Similarly, less-oxidized species can also be formed from fragmentation reactions, since the increase in volatility from losing carbon atoms is not as great (top left corner of Fig. 5). Thus, while fragmentation reactions probably do not lead to the immediate formation of highly oxidized OA, they might occur as part of the overall reaction mechanism, primarily in the earlier oxidation steps.

Oligomerization reactions. Because oligomerization reactions decrease volatility, they are generally more viable steps in the formation of oxidized OA components than fragmentation reactions. As with fragmentation reactions, the phase of the precursor introduces constraints beyond those described in the Methodology section: since the probability of two gas-phase precursors reacting is quite low under most conditions, at least one precursor must be present in the particle phase ($c^* < 10$ µg m^{-3}).

Since oligomerization reactions involve changes to n_C, their effects can be visualized similarly to the effects of fragmentation reactions, by treating all carbon as evenly functionalized (Fig. 5). Assuming that functional groups are evenly-distributed in both the precursors and the targets, the condensed-phase precursor can be smaller than the target by one or two carbon atoms. Precursors with even fewer carbon atoms will be so volatile that they will be present in the gas phase only, and therefore cannot participate in oligomerization reactions. However, a wider range of precursors is possible if the functional groups are not evenly distributed, and instead the two precursor molecules have different levels of functionalization. The range of possible oligomerization precursors are shown as lightly-shaded regions in Fig. 6; lower carbon numbers are possible when the majority of the functional groups are on the condensed-phase precursor. Thus oxidized organic aerosol could form from the oligomerization of smaller, highly-oxidized molecular species, which themselves may be formed by previous functionalization or fragmentation reactions. This is broadly consistent with the view of formation of SOA from aqueous chemistry;[15] even though we do not explicitly model partitioning into the aqueous phase, this work provides some constraints into the chemical nature (H/C and O/C ratios) of these smaller species. On the other hand, the target itself (of formula $C_{10.5}H_{13.4}O_{7.3}$) is not consistent with the oligomeric species measured in SOA, such as those from monoterpene oxidation (which typically have twenty or more carbon atoms and relatively low OS_C values[66]), isoprene photooxidation, or glyoxal oligomerization (which typically have higher H/C values[11]). Thus, even though these oligomerization reactions can occur under atmospheric conditions, they do not appear to dominate formation of the most oxidized fraction of the aerosol.

MCMA case study. Although LV-OOA in Mexico City is significantly more oxidized than that measured in other locations (Table 1 and Fig. 2), it has received substantial study and to our knowledge is the only LV-OOA factor whose volatility and elemental ratios have both been reported. Here, we examine LV-OOA at the T0 ground site at MILAGRO[32,38] as a case study, in order to gain insight into potential formation pathways for this exceedingly highly oxidized organic aerosol. Using a c^* of $10^{-5\pm2}$ µg m^{-3} gives a target carbon number estimate of 10.0 ± 3.4 with the chemical formula $C_{10}H_{12.1}O_{8.4}$, (3.4 hydroxyl groups and 5 carbonyl groups). The sources of uncertainty studied in this case for the carbon number are the error in computing elemental ratios and the deviations in c^*, which give values of δ_{NC} of 3.2 and 1.1, respectively. As with the average LV-OOA, the overall error is computed from the sum of the square of the errors. The potential functionalization precursors are shown in Fig. 7 for the saturated and unsaturated cases. Because the target in this case is so low in volatility, the functionalization transforms involve the loss of at least two hydroxyl groups, or all five carbonyl groups, to form precursors that are semi-volatile, and even more for the precursor to be present exclusively in the gas phase. This suggests that gas-phase functionalization reactions would have to be exceedingly rapid to lead to the formation of organic material that is so high in oxidation state and low in volatility. Similarly, this class of organics cannot be formed directly from fragmentation reactions (even if the loss of small, reduced fragments are allowed). Again, some fragmentation reactions might occur earlier in the overall oxidation mechanism, but they do not play a role in the last step that forms this target. As in our average LV-OOA target (Fig. 6), oligomerization could be a viable step in the formation of these compounds, provided highly oxidized organic precursor species are present in the condensed phase.

4   Conclusions

We have presented a new general approach for assessing the potential pathways and precursors involved in the formation of highly oxidized organic aerosol, which involves starting from the product or "target" and working backwards. We define this target based on measurements of volatility and average elemental ratios of the AMS LV-OOA factor, yielding an average approximate chemical formula of $C_{10.5}H_{13.4}O_{7.3}$. The utility of our retrosynthetic method for constraining its formation chemistry has been demonstrated for this average LV-OOA target, as well as an exceedingly oxidized case, LV-OOA measured at the MILAGRO ground site. Our results suggest that the most probable LV-OOA-forming reactions are functionalization reactions that add multiple functional groups (including at least one hydroxyl group) at a time, as well as oligomerization reactions of smaller, highly oxidized precursors. Fragmentation reactions are only possible as the final step if they involve the loss of a small number of relatively reduced carbon atoms. While the specific aim of this study was to apply this technique to LV-OOA, this approach could be useful for understanding a wider range of targets, including other fractions of the aerosol (*e.g.*, SV-OOA[3]), as well as SOA generated in laboratory studies. However, this requires some information about the volatility and/or carbon number of the aerosol of interest.

In the present work, we limited the partitioning of the organic species to only two phases, the gas phase and the condensed-organic phase. Recent work has suggested the importance of the liquid-water phase as a medium for oxidation reactions,[15] which could also be included in this treatment. This would require the determination of the volatilities of organics over water (Henry's Law constant) from chemical properties (elemental ratios, functional group distributions, *etc.*). Otherwise the general approach, including the transforms that describe the chemical reactions, is the same as described in this work.

This retrosynthetic approach relies critically on the accurate characterization of the target molecule and key characteristics of its formation. This present work highlights the need for accurate measurements of the elemental ratios and volatility of organic aerosol. Field measurements of c^* are limited, hindering accurate estimates of carbon number. Better constraints on c^* of the oxidized aerosol could significantly improve the accuracy of our target carbon number. There is also a need for better constraints on the kinetics of the formation of highly oxidized aerosol. This has been measured only in the outflow from MCMA, which might not be widely representative. This rate is a key constraint on the formation pathway, as it limits the number of oxidation steps that are possible as well as the phase in which those reactions can happen. Additionally, characterization of highly-oxidized OA using techniques other than AMS would be useful. The direct measurement of functional group distributions (*e.g.*, using FTIR[41]), can provide an alternative, and potentially more comprehensive, description of the relationships between chemical structure, volatility, and formation chemistry. Similarly, direct measurements of n_C (*e.g.*, using UHR-ESI-MS) would improve the accuracy of the target chemical formula and could provide more detailed information on the formation pathways of a wide range of molecular compounds.

We have made an effort to demonstrate the utility of this retrosynthetic approach for exploring the possible pathways by which highly oxidized aerosol is formed. By starting at the known endpoint/target, we improve our ability to identify pathways that can actually result in the formation of low-volatility, highly-oxidized organic compounds. This is in contrast to standard forward approaches that require some guesswork as to the starting molecules and pathways that may result in the formation of the target. This general approach could be particularly powerful when applied to rigorous models of oxidation mechanisms, gas-particle partitioning, and aerosol chemistry.[18,22,54,67] Running such models backwards from well-defined targets may help identify key classes (or even structures) of organic compounds, as well as reaction conditions, that are most likely to lead to the formation of this important class of organic aerosol.

Acknowledgements

This work was supported by the National Science Foundation, under grant numbers CHE-1012809 and AGS-1056225. We thank Manjula Canagaratna for sharing new AMS elemental analysis correction factors.

References

1 Q. Zhang, J. L. Jimenez, M. R. Canagaratna, J. D. Allan, H. Coe, I. Ulbrich, M. R. Alfarra, A. Takami, A. M. Middlebrook, Y. L. Sun, K. Dzepina, E. Dunlea, K. Docherty, P. F. DeCarlo, D. Salcedo, T. Onasch, J. T. Jayne, T. Miyoshi, A. Shimono, S. Hatakeyama, N. Takegawa, Y. Kondo, J. Schneider, F. Drewnick, S. Borrmann, S. Weimer, K. Demerjian, P. Williams, K. Bower, R. Bahreini, L. Cottrell, R. J. Griffin, J. Rautiainen, J. Y. Sun, Y. M. Zhang and D. R. Worsnop, *Geophys. Res. Lett.*, 2007, **34**, L13801.
2 E. R. Graber and Y. Rudich, *Atmos. Chem. Phys.*, 2006, **6**, 729–753.
3 J. L. Jimenez, M. R. Canagaratna, N. M. Donahue, A. S. H. Prevot, Q. Zhang, J. H. Kroll, P. F. DeCarlo, J. D. Allan, H. Coe, N. L. Ng, A. C. Aiken, K. S. Docherty, I. M. Ulbrich, A. P. Grieshop, A. L. Robinson, J. Duplissy, J. D. Smith, K. R. Wilson, V. A. Lanz, C. Hueglin, Y. L. Sun, J. Tian, A. Laaksonen, T. Raatikainen, J. Rautiainen, P. Vaattovaara, M. Ehn, M. Kulmala, J. M. Tomlinson, D. R. Collins, M. J. Cubison, J. Dunlea, J. A. Huffman, T. B. Onasch, M. R. Alfarra, P. I. Williams, K. Bower, Y. Kondo, J. Schneider, F. Drewnick, S. Borrmann, S. Weimer, K. Demerjian, D. Salcedo, L. Cottrell, R. Griffin, A. Takami, T. Miyoshi, S. Hatakeyama, A. Shimono, J. Y. Sun, Y. M. Zhang, K. Dzepina, J. R. Kimmel, D. Sueper, J. T. Jayne, S. C. Herndon, A. M. Trimborn, L. R. Williams, E. C. Wood, A. M. Middlebrook, C. E. Kolb, U. Baltensperger and D. R. Worsnop, *Science*, 2009, **326**, 1525–1529.
4 J. H. Kroll, N. M. Donahue, J. L. Jimenez, S. H. Kessler, M. R. Canagaratna, K. R. Wilson, K. E. Altieri, L. R. Mazzoleni, A. S. Wozniak, H. Bluhm, E. R. Mysak, J. D. Smith, C. E. Kolb and D. R. Worsnop, *Nat. Chem.*, 2011, **3**, 133–139.
5 C. D. Cappa and J. L. Jimenez, *Atmos. Chem. Phys.*, 2010, **10**, 5409–5424.
6 R. Volkamer, J. L. Jimenez, F. San Martini, K. Dzepina, Q. Zhang, D. Salcedo, L. T. Molina, D. R. Worsnop and M. J. Molina, *Geophys. Res. Lett.*, 2006, **33**, L17811.

7 C. L. Heald, J. H. Kroll, J. L. Jimenez, K. S. Docherty, P. F. DeCarlo, A. C. Aiken, Q. Chen, S. T. Martin, D. K. Farmer and P. Artaxo, *Geophys Res. Lett.*, 2010, **37**, L08803.
8 N. M. Donahue, K. M. Henry, T. F. Mentel, A. Kiendler-Scharr, C. Spindler, B. Bohn, T. Brauers, H. P. Dorn, H. Fuchs, R. Tillmann, A. Wahner, H. Saathoff, K. H. Naumann, O. Mohler, T. Leisner, L. Muller, M. C. Reinnig, T. Hoffmann, K. Salo, M. Hallquist, M. Frosch, M. Bilde, T. Tritscher, P. Barmet, A. P. Praplan, P. F. DeCarlo, J. Dommen, A. S. H. Prevot and U. Baltensperger, *Proc. Natl. Acad. Sci. U. S. A.*, 2012, **109**, 13503–13508.
9 L. Qi, S. Nakao, Q. Malloy, B. Warren and D. R. Cocker III, *Atmos. Environ.*, 2010, **44**, 2990–2996.
10 R. Bahreini, M. D. Keywood, N. L. Ng, V. Varutbangkul, S. Gao, R. C. Flagan, J. H. Seinfeld, D. R. Worsnop and J. L. Jimenez, *Environ. Sci. Technol.*, 2005, **39**, 5674–5688.
11 P. S. Chhabra, N. L. Ng, M. R. Canagaratna, A. L. Corrigan, L. M. Russell, D. R. Worsnop, R. C. Flagan and J. H. Seinfeld, *Atmos. Chem. Phys.*, 2011, **11**, 8827–8845.
12 S. H. Kessler, J. D. Smith, D. L. Che, D. R. Worsnop, K. R. Wilson and J. H. Kroll, *Environ. Sci. Technol.*, 2010, **44**, 7005–7010.
13 S. H. Kessler, T. Nah, K. E. Daumit, J. D. Smith, S. R. Leone, C. E. Kolb, D. R. Worsnop, K. R. Wilson and J. H. Kroll, *J. Phys. Chem. A*, 2012, **116**, 6358–6365.
14 A. T. Lambe, T. B. Onasch, D. R. Croasdale, J. P. Wright, A. T. Martin, J. P. Franklin, P. Massoli, J. H. Kroll, M. R. Canagaratna, W. H. Brune, D. R. Worsnop and P. Davidovits, *Environ. Sci. Technol.*, 2012, **46**, 5430–5437.
15 B. Ervens, B. J. Turpin and R. J. Weber, *Atmos. Chem. Phys.*, 2011, **11**, 11069–11102.
16 E. Gómez Alvarez, H. Wortham, R. Strekowski, C. Zetzsch and S. Gligorovski, *Environ. Sci. Technol.*, 2012, **46**, 1955–1963.
17 K. Dzepina, C. D. Cappa, R. M. Volkamer, S. Madronich, P. F. DeCarlo, R. A. Zaveri and J. L. Jimenez, *Environ. Sci. Technol.*, 2011, **45**, 3496–3503.
18 C. D. Cappa and K. R. Wilson, *Atmos. Chem. Phys.*, 2012, **12**, 9505–9528.
19 N. M. Donahue, S. A. Epstein, S. N. Pandis and A. L. Robinson, *Atmos. Chem. Phys.*, 2011, **11**, 3303–3318.
20 K. Dzepina, R. M. Volkamer, S. Madronich, P. Tulet, I. M. Ulbrich, Q. Zhang, C. D. Cappa, P. J. Ziemann and J. L. Jimenez, *Atmos. Chem. Phys.*, 2009, **9**, 5681–5709.
21 A. Hodzic, J. L. Jimenez, S. Madronich, M. R. Canagaratna, P. F. DeCarlo, L. Kleinman and J. Fast, *Atmos. Chem. Phys.*, 2010, **10**, 5491–5514.
22 B. Aumont, R. Valorso, C. Mouchel-Vallon, M. Camredon, J. Lee-Taylor and S. Madronich, *Atmos. Chem. Phys.*, 2012, **12**, 7577–7589.
23 J. Lee-Taylor, S. Madronich, B. Aumont, A. Baker, M. Camredon, A. Hodzic, G. S. Tyndall, E. Apel and R. A. Zaveri, *Atmos. Chem. Phys.*, 2011, **11**, 13219–13241.
24 B. K. Pun, C. Seigneur, D. Grosjean and P. Saxena, *J. Atmos. Chem.*, 2000, **35**, 199–223.
25 N. L. Ng, M. R. Canagaratna, J. L. Jimenez, P. S. Chhabra, J. H. Seinfeld and D. R. Worsnop, *Atmos. Chem. Phys.*, 2011, **11**, 6465–6474.
26 Y. Wei, T. Cao and J. E. Thompson, *Atmos. Environ.*, 2012, **62**, 199–207.
27 J. F. Pankow and K. C. Barsanti, *Atmos. Environ.*, 2009, **43**, 2829–2835.
28 N. L. Ng, M. R. Canagaratna, Q. Zhang, J. L. Jimenez, J. Tian, I. M. Ulbrich, J. H. Kroll, K. S. Docherty, P. S. Chhabra, R. Bahreini, S. M. Murphy, J. H. Seinfeld, L. Hildebrandt, N. M. Donahue, P. F. DeCarlo, V. A. Lanz, A. S. H. Prévôt, E. Dinar, Y. Rudich and D. R. Worsnop, *Atmos. Chem. Phys.*, 2010, **10**, 4625–4641.
29 E. J. Corey and X.-M. Cheng, *The Logic of Chemical Synthesis*, Wiley, New York, 1995.
30 P. F. DeCarlo, J. R. Kimmel, A. Trimborn, M. J. Northway, J. T. Jayne, A. C. Aiken, M. Gonin, K. Fuhrer, T. Horvath, K. S. Docherty, D. R. Worsnop and J. L. Jimenez, *Anal. Chem.*, 2006, **78**, 8281–8289.
31 M. R. Canagaratna, J. T. Jayne, J. L. Jimenez, J. D. Allan, M. R. Alfarra, Q. Zhang, T. B. Onasch, F. Drewnick, H. Coe, A. Middlebrook, A. Delia, L. R. Williams, A. M. Trimborn, M. J. Northway, P. F. DeCarlo, C. E. Kolb, P. Davidovits and D. R. Worsnop, *Mass Spectrom. Rev.*, 2007, **26**, 185–222.
32 A. C. Aiken, P. F. DeCarlo, J. H. Kroll, D. R. Worsnop, J. A. Huffman, K. S. Docherty, I. M. Ulbrich, C. Mohr, J. R. Kimmel, D. Sueper, Y. Sun, Q. Zhang, A. Trimborn, M. Northway, P. J. Ziemann, M. R. Canagaratna, T. B. Onasch, M. R. Alfarra, A. S. H. Prevot, J. Dommen, J. Duplissy, A. Metzger, U. Baltensperger and J. L. Jimenez, *Environ. Sci. Technol.*, 2008, **42**, 4478–4485.
33 A. C. Aiken, P. F. DeCarlo and J. L. Jimenez, *Anal. Chem.*, 2007, **79**, 8350–8358.
34 I. M. Ulbrich, M. R. Canagaratna, Q. Zhang, D. R. Worsnop and J. L. Jimenez, *Atmos. Chem. Phys.*, 2009, **9**, 2891–2918.
35 P. F. DeCarlo, I. M. Ulbrich, J. Crounse, B. de Foy, E. J. Dunlea, A. C. Aiken, D. Knapp, A. J. Weinheimer, T. Campos, P. O. Wennberg and J. L. Jimenez, *Atmos. Chem. Phys.*, 2010, **10**, 5257–5280.

36 C. D. Cappa, *Atmos. Meas. Tech.*, 2010, **3**, 579–592.
37 X. F. Huang, L. Y. He, M. Hu, M. R. Canagaratna, J. H. Kroll, N. L. Ng, Y.-H. Zhang, Y. Lin, L. Xue, T. L. Sun, X. G. Liu, M. Shao, J. T. Jayne and D. R. Worsnop, *Atmos. Chem. Phys.*, 2011, **11**, 1865–1877.
38 J. A. Huffman, K. S. Docherty, A. C. Aiken, M. J. Cubison, I. M. Ulbrich, P. F. DeCarlo, D. Sueper, J. T. Jayne, D. R. Worsnop, P. J. Ziemann and J. L. Jimenez, *Atmos. Chem. Phys.*, 2009, **9**, 7161–7182.
39 J. F. Pankow and W. E. Asher, *Atmos. Chem. Phys.*, 2008, **8**, 2773–2796.
40 D. K. Farmer, A. Matsunaga, K. S. Docherty, J. D. Surratt, J. H. Seinfeld, P. J. Ziemann and J. L. Jimenez, *Proc. Natl. Acad. Sci. U. S. A.*, 2010, **107**, 6670–6675.
41 S. F. Maria, L. M. Russell, B. J. Turpin and R. J. Porcja, *Atmos. Environ.*, 2002, **36**, 5185–5196.
42 P. Lin, A. G. Rincon, M. Kalberer and J. Z. Yu, *Environ. Sci. Technol.*, 2012, **46**, 7454–7462.
43 A. S. Wozniak, J. E. Bauer, R. L. Sleighter, R. M. Dickhut and P. G. Hatcher, *Atmos. Chem. Phys.*, 2008, **8**, 5099–5111.
44 J. H. Kroll and J. H. Seinfeld, *Atmos. Environ.*, 2008, **42**, 3593–3624.
45 K. S. Docherty, A. C. Aiken, J. A. Huffman, I. M. Ulbrich, P. F. DeCarlo, D. Sueper, D. R. Worsnop, D. C. Snyder, R. E. Peltier, R. J. Weber, B. D. Grover, D. J. Eatough, B. J. Williams, A. H. Goldstein, P. J. Ziemann and J. L. Jimenez, *Atmos. Chem. Phys.*, 2011, **11**, 12387–12420.
46 C. Mohr, P. F. DeCarlo, M. F. Heringa, R. Chirico, J. G. Slowick, R. Richter, C. Reche, A. Alastuey, X. Querol, R. Seco, J. Penuelas, J. L. Jimenez, M. Crippa, R. Zimmermann, U. Baltensperger and A. S. H. Prevot, *Atmos. Chem. Phys.*, 2012, **12**, 1649–1665.
47 M. Crippa, I. El Haddad, J. G. Slowik, P. F. DeCarlo, C. Mohr, M. R. Heringa, R. Chirico, N. Marchand, J. Sciare, U. Baltensperger and A. S. H. Prevot, *J. Geophys. Res.: Atmos.*, 2013, **118**, 1950–1963.
48 Y.-L. Sun, Q. Zhang, J. J. Schwab, K. L. Demerjian, W.-N. Chen, M.-S. Bae, H.-M. Hung, O. Hogrefe, B. Frank, O. V. Rattigan and Y.-C. Lin, *Atmos. Chem. Phys.*, 2011, **11**, 1581–1602.
49 L.-Y. He, X.-F. Huang, L. Xue, M. Hu, Y. Lin, J. Zheng, R. Zhang and Y.-H. Zhang, *J. Geophys. Res.*, 2011, **116**, D12304.
50 X.-F. Huang, L.-Y. He, L. Xue, T.-L. Sun, L.-W. Zeng, Z.-H. Gong, M. Hu and T. Zhu, *Atmos. Chem. Phys.*, 2012, **12**, 4897–4907.
51 A. Setyan, Q. Zhang, M. Merkel, W. B. Knighton, Y. Sun, C. Song, J. E. Shilling, T. B. Onasch, S. C. Herndon, D. R. Worsnop, J. D. Fast, R. A. Zaveri, L. K. Berg, A. Wiedensohler, B. A. Flowers, M. K. Dubey and R. Subramanian, *Atmos. Chem. Phys.*, 2012, **12**, 8131–8156.
52 Z. Gong, Z. Lan, L. Xue, L. Zeng, L. He and X. Huang, *Front. Environ. Sci. Eng.*, 2012, **6**, 725–733.
53 M. R. Canagaratna, P. Massoli, J. L. Jimenez, S. Kessler, Q. Chen, L. Hildebrandt, E. Fortner, L. Williams, K. Wilson, J. Surratt, N. Donahue, J. Kroll, J. T. Jayne and D. R. Worsnop, *Improved calibration of O/C and H/C Ratios obtained by Aerosol Mass Spectrometry of Organic Species*, manuscript in preparation.
54 N. M. Donahue, J. H. Kroll, S. N. Pandis and A. L. Robinson, *Atmos. Chem. Phys.*, 2012, **12**, 615–634.
55 F. Paulot, J. D. Crounse, H. G. Kjaergaard, A. Kurten, J. M. St. Clair, J. H. Seinfeld and P. O. Wennberg, *Science*, 2009, **325**, 730–733.
56 I. J. George and J. P. D. Abbatt, *Nat. Chem.*, 2010, **2**, 713–722.
57 A. L. Robinson, N. M. Donahue and W. F. Rogge, *J. Geophys. Res.*, 2006, **111**, D03302.
58 J. D. Hearn and G. D. Smith, *Int. J. Mass Spectrom.*, 2006, **258**, 95–103.
59 E. A. Weitkamp, K. E. H. Hartz, A. M. Sage, N. M. Donahue and A. L. Robinson, *Environ. Sci. Technol.*, 2008, **42**, 5177–5182.
60 P. J. Ziemann and R. Atkinson, *Chem. Soc. Rev.*, 2012, **41**, 6582–6605.
61 M. Ehn, E. Kleist, H. Junninen, T. Petäjä, G. Lönn, S. Schobesberger, M. Dal Maso, A. Trimborn, M. Kulmala, D. R. Worsnop, A. Wahner, J. Wildt and T. F. Mentel, *Atmos. Chem. Phys.*, 2012, **12**, 5113–5127.
62 M. Shiraiwa and J. H. Seinfeld, *Geophys. Res. Lett.*, 2012, **39**, L24801.
63 T. D. Vaden, D. Imre, J. Beránek, M. Shrivastava and A. Zelenyuk, *Proc. Natl. Acad. Sci. U. S. A.*, 2011, **108**, 2190–2195.
64 V. Perraud, E. A. Bruns, M. J. Ezell, S. N. Johnson, Y. Yu, M. L. Alexander, A. Zelenyuk, D. Imre, W. L. Chang, D. Dabdub, J. F. Pankow and B. J. Finlayson-Pitts, *Proc. Natl. Acad. Sci. U. S. A.*, 2012, **109**, 2836–2841.
65 A. P. Grieshop, N. M. Donahue and A. L. Robinson, *Geophys. Res. Lett.*, 2007, **34**, L14810.
66 M. P. Tolocka, M. Jang, J. M. Ginter, F. J. Cox, R. M. Kamens and M. V. Johnston, *Environ. Sci. Technol.*, 2004, **38**, 1428–1434.
67 X. Zhang and J. H. Seinfeld, *Atmos. Chem. Phys.*, 2013, **13**, 5907–5926.

PAPER

Atmospheric aerosols in Amazonia and land use change: from natural biogenic to biomass burning conditions

Paulo Artaxo,[*a] Luciana V. Rizzo,[b] Joel F. Brito,[a] Henrique M. J. Barbosa,[a] Andrea Arana,[a] Elisa T. Sena,[a] Glauber G. Cirino,[c] Wanderlei Bastos,[d] Scot T. Martin[e] and Meinrat O. Andreae[f]

Received 9th April 2013, Accepted 21st May 2013
DOI: 10.1039/c3fd00052d

In the wet season, a large portion of the Amazon region constitutes one of the most pristine continental areas, with very low concentrations of atmospheric trace gases and aerosol particles. However, land use change modifies the biosphere–atmosphere interactions in such a way that key processes that maintain the functioning of Amazonia are substantially altered. This study presents a comparison between aerosol properties observed at a preserved forest site in Central Amazonia (TT34 North of Manaus) and at a heavily biomass burning impacted site in south-western Amazonia (PVH, close to Porto Velho). Amazonian aerosols were characterized in detail, including aerosol size distributions, aerosol light absorption and scattering, optical depth and aerosol inorganic and organic composition, among other properties. The central Amazonia site (TT34) showed low aerosol concentrations ($PM_{2.5}$ of 1.3 ± 0.7 μg m^{-3} and 3.4 ± 2.0 μg m^{-3} in the wet and dry seasons, respectively), with a median particle number concentration of 220 cm^{-3} in the wet season and 2200 cm^{-3} in the dry season. At the impacted site (PVH), aerosol loadings were one order of magnitude higher ($PM_{2.5}$ of 10.2 ± 9.0 μg m^{-3} and 33.0 ± 36.0 μg m^{-3} in the wet and dry seasons, respectively). The aerosol number concentration at the impacted site ranged from 680 cm^{-3} in the wet season up to $20\,000$ cm^{-3} in the dry season. An aerosol chemical speciation monitor (ACSM) was deployed in 2013 at both sites, and it shows that

[a]Institute of Physics, University of São Paulo, Rua do Matão, Travessa R, 187. CEP 05508-090, São Paulo, S.P., Brazil. E-mail: artaxo@if.usp.br

[b]Department of Earth and Exact Sciences, Institute of Environmental, Chemical and Pharmaceutics Sciences, Federal University of São Paulo, UNIFESP - Campus Diadema, Rua Prof. Artur Riedel, 275, CEP 09972-270, Diadema – São Paulo, Brazil

[c]INPA - Instituto Nacional de Pesquisas da Amazônia, Av. André Araújo, 2.936 – CEP 69067-375, Manaus, Brazil

[d]Laboratório de Biogeoquímica Ambiental Wolfgang C. Pfeiffer, Universidade Federal de Rondônia - UNIR, Rondônia, Brazil

[e]School of Engineering and Applied Sciences and Department of Earth and Planetary Sciences, Harvard University, 29 Oxford St., Pierce Hall, Cambridge, Massachusetts, 02138, USA

[f]Biogeochemistry Department, Max Planck Institute for Chemistry, P.O. Box 3060, 55020 Mainz, Germany

organic aerosol account to 81% to the non-refractory PM1 aerosol loading at TT34, while biomass burning aerosols at PVH shows a 93% content of organic particles. Three years of filter-based elemental composition measurements shows that sulphate at the impacted site decreases, on average, from 12% of $PM_{2.5}$ mass during the wet season to 5% in the dry season. This result corroborates the ACSM finding that the biomass burning contributed overwhelmingly to the organic fine mode aerosol during the dry season in this region. Aerosol light scattering and absorption coefficients at the TT34 site were low during the wet season, increasing by a factor of 5, approximately, in the dry season due to long range transport of biomass burning aerosols reaching the forest site in the dry season. Aerosol single scattering albedo (SSA) ranged from 0.84 in the wet season up to 0.91 in the dry. At the PVH site, aerosol scattering coefficients were 3–5 times higher in comparison to the TT34 site, an indication of strong regional background pollution, even in the wet season. Aerosol absorption coefficients at PVH were about 1.4 times higher than at the forest site. Ground-based SSA at PVH was around 0.92 year round, showing the dominance of scattering aerosol particles over absorption, even for biomass burning aerosols. Remote sensing observations from six AERONET sites and from MODIS since 1999, provide a regional and temporal overview. Aerosol Optical Depth (AOD) at 550 nm of less than 0.1 is characteristic of natural conditions over Amazonia. At the perturbed PVH site, AOD_{550} values greater than 4 were frequently observed in the dry season. Combined analysis of MODIS and CERES showed that the mean direct radiative forcing of aerosols at the top of the atmosphere (TOA) during the biomass burning season was -5.6 ± 1.7 W m^{-2}, averaged over whole Amazon Basin. For high AOD (larger than 1) the maximum daily direct aerosol radiative forcing at the TOA was as high as -20 W m^{-2} locally. This change in the radiation balance caused increases in the diffuse radiation flux, with an increase of Net Ecosystem Exchange (NEE) of 18–29% for high AOD. From this analysis, it is clear that land use change in Amazonia shows alterations of many atmospheric properties, and these changes are affecting the functioning of the Amazonian ecosystem in significant ways.

1   Introduction

Amazonia is an excellent laboratory to study atmospheric processes that are characteristic of natural conditions, as they existed prior to the impact of industrialization on the regional and global atmosphere.[1] The strong coupling between the atmosphere and the forest can be seen as characteristic of conditions before large-scale deforestation changed land use in Europe, North America, and other regions. Furthermore, the region sustained a strong hydrological cycle that is maintained by large water vapour emissions from the forest as well as cloud condensation nuclei (CCN) produced from forest emissions,[2,3] and one important location of deep tropical convection. However, the vast forest–river system of Amazonia is changing due to expansion and intensification of agriculture, logging, and urban footprints.[4] Indications that the hydrological cycle in Amazonia is being intensified in the last two decades add a key issue in the changes in Amazonia.[5] Recently two strong droughts in 2005 and 2010 have received attention,[6,7] as potential indicators of increase in climate extremes in Amazonia that feeds back into the forest carbon processing.[8]

The Brazilian Amazon extends over about 5.5 million km^2, corresponding to 61% of the area of the country of Brazil. Deforestation has changed about 18% of the original forest area, mostly in southern and western Amazonia.[9] The forests and soils of the Amazon basin also store a large amount of organic carbon (around 200 Pg C), which may potentially be released to the atmosphere through forest to pasture conversion or logging or because of biome changes.[10] The river system is responsible for about 20% of the world's freshwater discharge, and this forest–river system is vulnerable to climate change.[11] In addition, responses and feedbacks of this biome to changes in climate and land-use could affect regional and global climate.[12,13]

Human activities in Amazonia over the last 50 years have had a significant impact on a considerable part of the region, especially along the southern perimeter.[14] A recent steep decline in annual deforestation rates in the Brazilian Amazon from 27 800 km² yr⁻¹ in 2004 to 4660 km² yr⁻¹ in 2012 is recorded in the time series shown in Fig. 1, as measured by the PRODES (Projeto de Monitoramento do Desflorestamento na Amazônia Legal) program from INPE (The Brazilian National Institute for Space Research) for the Brazilian Amazon. The reduction from 2004 to 2012 observed in Fig. 1 is an impressive achievement, but there are questions if these relatively recent low deforestation rates can be maintained over the next decades,[12] because of socio-economic pressures as well as a result of a changing global climate.[13,15] The dominant factors of public policies, climate, economic issues, and so forth that successfully contributed to the reduction of deforestation rates in recent years have not been fully dissected. There is also increased uncertainty over the sensitivity of carbon storage in the tropics to climate warming, increasing atmospheric CO_2 concentrations, and tropical temperature anomalies.[16]

The large carbon stock of Amazonia is sensitive to processes that alter precipitation and radiation. The two large droughts of 2005 and 2010[6,7] released large amounts of carbon to the atmosphere, indicating a high sensitivity of this system to altered conditions. High concentrations of aerosol particles in the atmosphere due to biomass burning decrease the amount of photosynthetically radiation to varying canopy levels, affecting sensible and latent heat fluxes at the

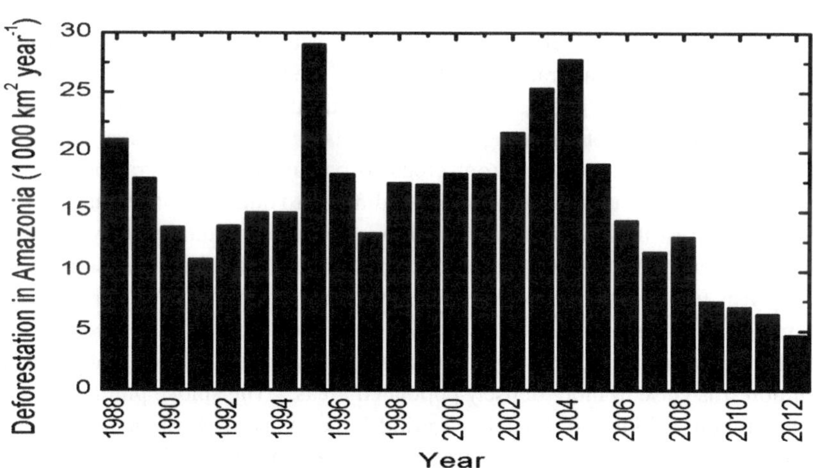

Fig. 1 Annual deforestation rates in the Brazilian Amazonia from 1977 to 2012 measured by the PRODES (Projeto de Monitoramento do Desflorestamento na Amazônia Legal) program from INPE (The Brazilian National Institute for Space Research).

surface.[17] The changes in radiation flux due to aerosol and clouds and in the ratio of diffuse to direct radiation have a large impact on Net Ecosystem Exchange (NEE), with an enhancement in carbon uptake of 18–29%[18] for AOD (Aerosol Optical Depth) changes from 0.1 to 1. At the Tapajos National Forest (FLONA-Tapajos, Santarem-PA), an increase in carbon uptake was mostly attributed to the increased diffuse radiation in the sub-canopy layer.[19] On the other hand, for very large aerosol loadings (AOD > 2 at 550 nm), NEE decreases significantly, indicating the large impact of aerosols on ecosystem functioning.[18] Studies of effects of aerosol particles on changes in cloud properties and precipitation were the focus of several previous studies in Amazonia.[20–25]

The Large-scale Biosphere-Atmosphere Experiment in Amazonia (LBA) is a long term multinational, interdisciplinary research program, led by Brazil through its Ministry of Science and Technology and operated by INPA (The Brazilian National Institute for Amazonian Research.[8] The focus of LBA is to understand how Amazonia functions as a regional entity in the Earth system and how on-going changes in land use and climate are affecting the Amazonian ecosystem.[11] LBA seeks to supply a scientific basis for addressing the sustainability of development in the region through an interdisciplinary scientific agenda, which spans the fields of physical climate, hydrology, biogeochemistry, ecology, economics, and human dimensions of land-use change. The integrated understanding of these complex issues requires understanding its underlying spatial and temporal heterogeneity, as related to climate, rivers, soils, vegetation, and land-use drivers across the Basin. The Amazon forest is heterogeneous and complex, and it is essential to have a comprehensive scientific understanding of the feedbacks among climate, atmospheric composition, land-use, fire, and the socio-economic drivers.[12] Atmospheric studies over the Amazonian Basin started in the 1980's with the Brushfire experiment,[26] and the ABLE-2A (Amazon Boundary Layer Experiment 2A) and 2B campaigns,[27–29] followed by several integrated experiments, such as LBA-EUSTACH,[30] LBA-CLAIRE,[31] AMAZE-08,[32] and others. In general, these studies addressed one or both topics: (1) biomass burning emissions and effects and (2) emissions and reactivity of natural trace gases and biogenic particles.

In the Amazon, fires are almost exclusively caused by humans, and occurrences of natural fires are rare events because of the high precipitation rates even in the dry season.[33,34] Fires have been routinely used as a clearing tool during the dry season, in preparation for agricultural fields after a forested patch has been slashed down, or conversion from crops to pasture for cattle grazing. As the amount of precipitation is high, making it difficult to burn the forest, farmers cut down the forest in the end of the wet season (May–June) and let the wood and slash dry until September in order to clear the land for agricultural or pasture use. Every year from September to November, large amounts of smoke can be easily observed using remote sensing observations,[35] mainly in the region called the "arc of deforestation" in the southern part of Amazonia, a region where the tropical rainforest is close to more densely populated areas.[36] The smoke plume extends over millions of km^2, ultimately covering large areas of South America, with significant impacts extending far from the Amazonian region.[37–39] In particular, the effects of biomass-burning aerosol particles on human health, such as increased incidences of morbidity, mortality, and asthma, are significant in the region of the so-called deforestation arc.[40] Many studies in the last few years have analysed the role of aerosol particles in cloud development and suppression in Amazonia,[21,22,41] showing that as pristine conditions have quite low CCN concentrations and high water vapour, any increase in CCN numbers can significantly affect cloud development and invigoration.[42]

Aerosol mass spectrometry is a powerful tool for characterizing aerosols and their atmospheric processing. In AMAZE-08 (Amazonian Aerosol Characterization Experiment-2008), an Aerodyne high-resolution time-of-flight aerosol mass spectrometer (HR-ToF-AMS) was used for the first time in South America; it was applied to the characterization of Amazonian organic aerosols at a time resolution of better than 5 min.[24,32,43] Prior to the AMAZE-08 campaign, the contribution of primary biogenic aerosols to the accumulation mode was less well characterized than that to the coarse mode.[44,45] Given that Amazonian aerosols have primary and secondary contributions, marker signals were sought in the collected AMS data. The results showed that primary aerosols in Amazonia consist of carbohydrates, waxes, *etc.* from biogenic debris, whereas secondary aerosols are oxidation products of isoprene, terpenes, and other VOCs. Furthermore, the mass spectral data were also able to identify time periods influenced by African aerosol advection.[46,47] During clean periods, the mass spectra showed the submicron aerosol composition to be dominated by 90% organic matter and 10% sulphate, at average mass concentrations of 0.6 $\mu g\ m^{-3}$. The oxygen-to-carbon (O : C) ratio was 0.42. When influenced by African emissions, the average composition changed to 75% organic and 25% sulphate, with mass concentration of 0.9 $\mu g\ m^{-3}$ and O : C of 0.49. The study concluded that submicron aerosol in the wet season of Amazonia was largely dominated by secondary organic aerosol related to gaseous BVOC emissions from the forest flora. Furthermore, Pöhlker *et al.*,[48] added the finding that potassium is present in more than 90% of particles larger than 20 nm, again showing the importance of biogenic forest emissions on SOA formation in the Amazonian atmosphere, and the strong coupling between the biology of the forest and atmospheric composition.[24]

In pristine regions of Amazonia, the biology of the forest has close links with atmospheric aerosol and trace gases.[24,32,44,48,49] Ground based microorganisms could play a role in cloud process in Amazonia,[50] since strong convection can transport biogenic particles to the altitudes of cloud formation. Fungi and other primary biological particles also play a role in aerosol organic components in Amazonia.[45] Biomass burning and land use change interfere with the natural cycles and feedbacks. To better understand the effects of these changes, it is important to physically and chemically characterize the aerosol population in Amazonia, aiming to identify trends over time and landscape. Forest VOCs emissions[51–53] through atmospheric photo-oxidation chemistry produce most of the secondary organic aerosol (SOA) in the fine mode over Amazonia.[54–56] Primary biogenic aerosols dominate the coarse mode particles showing the very close relationship between emissions from the vegetation and aerosol concentrations and composition in Amazonia.[49,57–59]

This paper presents a perspective and analysis of the effects of land use change on atmospheric properties in Amazonia on a long-term basis. Sampling of trace gases and aerosols was performed for 3 years continuously in two locations: a preserved forest site in central Amazonia (TT34) and a biomass burning impacted site in Southwestern Amazonia (PVH), where land use change has been a strong process since the 1980s. Trace gases and aerosol properties were analysed at these

Fig. 2 Map of Amazonia showing the locations of the two atmospheric sampling sites at Porto Velho (PVH), and Central Amazonia (TT34) (yellow markers).

two sites from 2008 to 2012. Large inter annual variability in climatic conditions in Amazonia makes long term studies necessary. Also, the effects of changing aerosol loading in the atmosphere were analysed in two aspects: changes in radiative forcing as well as changes in carbon uptake due to increases in diffuse radiation associated with increased aerosol loading.

2 Experimental

2.1 Sampling sites

Amazonia is a continental scale region, with most of the land use change occurring in the southern part of the Basin (Fig. 2). The *in situ* trace gas and aerosol measurements reported here were taken at two sites: one at a preserved forest area in central Amazonia (named TT34 site hereafter), and the other at a biomass burning impacted area in south western Amazonia (named PVH hereafter).

The first site (TT34) is representative of near-pristine conditions during the wet season. In this area, from the point of view of aerosol properties, the wet season corresponds to the period Jan–Jun, whereas the dry season corresponds to Jul–Dec.[60] This forest site is located 60 km northwest to Manaus urban area, a fast developing city with a population of about 2 million people. Observations were taken at the TT34 tower located at the ZF2 Ecological Station (2.59°S, 60.21°W, 110 m asl), where aerosols and trace gases were measured almost continuously from February 2008 to June 2011. This site is within the INPA (Brazilian National Institute for Amazonian Research) Cuieiras forest reserve. No biomass burning occurs in the reservation or close to the site. Most of the time, the prevailing trade winds blow over vast expanses of intact tropical forest before reaching the measurement tower. However, as will be further discussed, the site was affected by regional transport of pollutants, most of it from long-range transported biomass burning emissions. All measurements were taken under dry conditions (RH 30–40%) by use of an automatic diffusion dryer in the sampling line.[61] An inlet with 50% aerodynamic cut-off of 7 μm was used for sampling. Inlet lines ran from the measurement level (39 m agl, about 10 m above the canopy height) to an air-conditioned container at ground level. Housing for the researchers and a diesel generator that provided the power supply were located respectively 0.33 km and 0.72 km to the west of the sampling site (downwind). A detailed description of the ZF2 TT34 tower measurement site and surrounding area can be found in Martin *et al.*[32]

The second site (PVH) is located near Porto Velho city, the capital of the state of Rondônia. The sampling site was located in an area of mixed forest and open vegetation in an ecological reservation about 5 km NE (upwind) from the city (8.69°S, 63.87°W). The whole region of Rondônia has been under the effect of land use change since the 1980s. Most of the measurements were taken under relatively dry conditions (RH < 50%) by use of diffusion dryers. An inlet having a 50% aerodynamic cut-off of 10 μm was used for sampling. Inlet lines ran from the measurement level (5 m agl) to an air-conditioned house. This site represents large land use changes and associated regional biomass burning as characteristic of Amazonia in contact with human interferences. At the PVH site, the dry season is from June to December, and the wet season from January to May.

Adding to the ground based *in situ* measurements, remote sense data from the NASA/AERONET sunphotometers network and from the MODIS satellite sensor were also analysed. Reported AOD observations from AERONET are level 2.0, and data were taken from two sites in southern Amazonia (same region as the PVH site): Abracos Hill (10S; 62W) and Ji-Parana-SE (10S; 61W), and from two sites in central Amazonia (same forest reservation as the TT34 site): Balbina (1S; 59W), and Manaus-EMBRAPA (2S; 59W). AOD observations from MODIS were integrated for an area with 40 km radius around the AERONET measurement sites. MODIS and AERONET data between 1999 and 2012 are analysed in this paper.

2.2 Instrumentation

Measurements of aerosol particle number size distribution, number concentration, mass and elemental composition were made. Two fine mode mobility particle size spectrometers (10–500 nm) were used interchangeably: a TSI-3080 SMPS (Scanning Mobility Particle Sizer) and a custom-made SMPS designed at Lund University according to EUSAAR (European Supersites for Atmospheric Aerosol Research) standards.[62] Particle number concentrations were measured using condensation particle counters (TSI CPC models 3010, 3785, 3772). Stacked Filter Units (SFU) fitted with PM_{10} inlets were used to collect fine mode ($D_p < 2.0$ μm) and coarse mode ($2.0 < D_p < 10.0$ μm) aerosols, with integrating periods ranging from 2 to 5 days depending on the aerosol loading. Fine and coarse mode Nuclepore filters were analysed for particulate mass, following the measurement

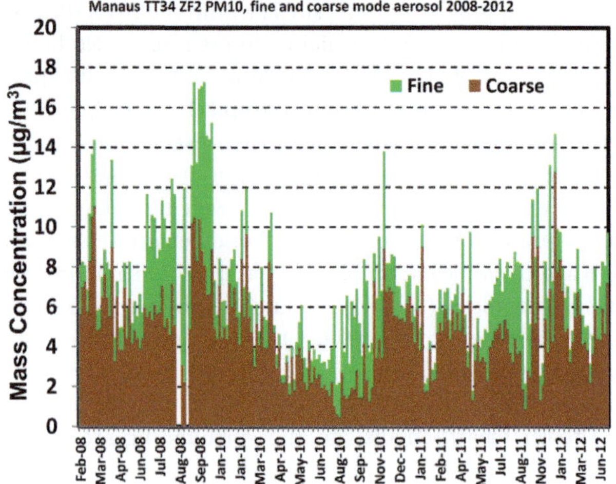

Fig. 3   Time series of fine ($PM_{2.5}$) and coarse mode aerosol mass concentrations at the central Amazonia TT34 forest site from 2008 to 2012.

protocol of the US Environmental Protection Agency for weighing filters, in a controlled atmosphere at 35% RH and 20 °C. Equivalent black carbon concentration (BC_e) was measured in the fine and coarse mode from Nuclepore filters, using an optical reflectance method calibrated with Monarch black carbon standards. About 25 trace elements were measured using X-ray fluorescence analysis with a Pan Analytical Epsilon 5 X-ray spectrometer. Precision for elemental composition was 10% for most of the measured elements, increasing to 20% for elements close to the detection limits. Ozone mixing ratios were

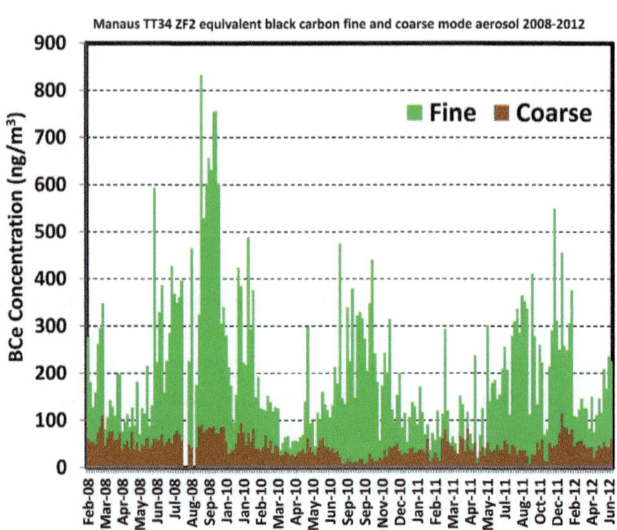

Fig. 4   Time series of equivalent black carbon in fine and coarse mode aerosols at the TT34 forest site from 2008 to 2012. The increase in fine mode BC concentrations in the dry season is due to long-range transport of biomass burning aerosol.

measured with Thermo Environment 49i ozone monitors, while CO was measured with a Picarro analyser model G2301. The zero point of the instruments was checked on a weekly basis.

An Aerodyne Research Aerosol Chemical Speciation Monitor (ACSM)[63] was deployed at the TT34 (central Amazonia) and PVH (Porto Velho) sites. The ACSM is a compact version of the Aerodyne's Aerosol Mass Spectrometer,[64] designed to characterize and monitor, under routine stable operation, the mass and chemical composition of non-refractory submicron particulate matter. Under ambient conditions, mass concentrations of particulate organics, sulphate, nitrate, ammonium, and chloride were obtained with a detection limit < 0.2 $\mu g\ m^{-3}$ for 30 min of signal averaging. Due to the high concentration of organics relative to sulphate, nitrate, chloride and ammonium, especially when sampling strongly biomass burning impacted air masses, corrections to the instrument chemical assignment were performed.[65,66]

Aerosol particle scattering coefficients were measured using three-wavelength integrating nephelometers (TSI-3563 and Ecotech Aurora 3000).[67] The instruments were calibrated periodically using filtered air and CO_2. Data were corrected

Table 1 Average aerosol elemental concentrations (2008–2012) at the TT34 forest site for dry and wet seasons. The table includes averages for particulate matter (PM), equivalent black carbon (BC$_e$) and trace elements. The concentrations are ng m^{-3}. The uncertainty is the standard deviation and N is the number of samples that the element was measured above detection limits. Fine mode refers to aerosol aerodynamic diameters up to 2.0 μm and coarse mode from 2.0 μm to 10 μm

	Dry Season				Wet Season			
	Fine Mode	N	Coarse Mode	N	Fine Mode	N	Coarse Mode	N
PM	3400 ± 2000	94	4425 ± 2429	94	1300 ± 700	104	5000 ± 2000	104
BC$_e$	235 ± 156	94	41 ± 22	94	98 ± 83	104	47 ± 20	104
Na	20 ± 18	64	$28. \pm 25$	66	10.0 ± 8.3	65	26 ± 24	79
Mg	4.6 ± 6.2	84	8.1 ± 8.8	92	10 ± 13	87	13 ± 14	94
Al	18 ± 20	96	25 ± 25	97	42 ± 64	106	44 ± 63	100
Si	27 ± 34	95	37 ± 37	97	77 ± 116	107	85 ± 122	103
P	4.8 ± 3.3	97	17 ± 12	96	3.0 ± 1.9	110	23.0 ± 8.0	110
S	175 ± 114	97	38 ± 24	97	74 ± 45	110	35 ± 16	110
Cl	2.0 ± 1.9	46	20 ± 25	96	1.7 ± 1.7	70	46 ± 57	110
K	74 ± 66	97	54 ± 30	97	26 ± 21	110	68 ± 22	110
Ca	4.9 ± 7.0	97	13 ± 22	97	6.1 ± 7.6	110	14 ± 14	110
Ti	1.5 ± 1.4	82	2.7 ± 2.5	93	2.8 ± 4.1	100	3.4 ± 4.5	100
V	0.2 ± 0.3	73	0.2 ± 0.3	63	0.2 ± 0.4	73	0.1 ± 0.1	67
Cr	0.3 ± 0.4	52	0.3 ± 0.2	76	0.2 ± 0.2	67	0.2 ± 0.2	94
Mn	0.3 ± 0.3	83	0.4 ± 0.4	84	0.4 ± 0.5	92	0.6 ± 0.6	97
Fe	12 ± 12	97	21 ± 18	97	20 ± 27	105	23 ± 31	110
Ni	0.2 ± 0.2	74	0.1 ± 0.1	69	0.2 ± 0.3	88	0.1 ± 0.1	68
Cu	1.3 ± 4.5	75	0.3 ± 0.9	86	0.2 ± 0.5	89	0.2 ± 0.2	96
Zn	1.6 ± 3.6	97	1.0 ± 1.9	92	0.4 ± 0.4	103	0.6 ± 0.5	110
Br	0.7 ± 0.5	83	0.3 ± 0.3	75	0.3 ± 0.4	81	0.3 ± 0.4	84
Rb	0.2 ± 0.3	25	0.2 ± 0.2	26	0.2 ± 0.2	39	0.2 ± 0.1	52
Sr	1.3 ± 1.1	15	0.6 ± 0.8	7	1.1 ± 1.2	14	0.5 ± 0.5	14
Sb	1.5 ± 1.2	9	1.1 ± 1.1	13	1.4 ± 0.9	15	1.0 ± 0.9	14
Pb	0.4 ± 0.5	78	0.3 ± 0.5	63	0.4 ± 0.5	83	0.4 ± 0.6	75
BC$_e$/PM (%)	6.9 ± 7.8	94	0.9 ± 0.9	94	7.5 ± 11.8	104	0.9 ± 1.0	104
SO4/PM (%)	15.4 ± 17.1	94	2.6 ± 2.9	94	17.1 ± 19.3	104	2.1 ± 2.4	104

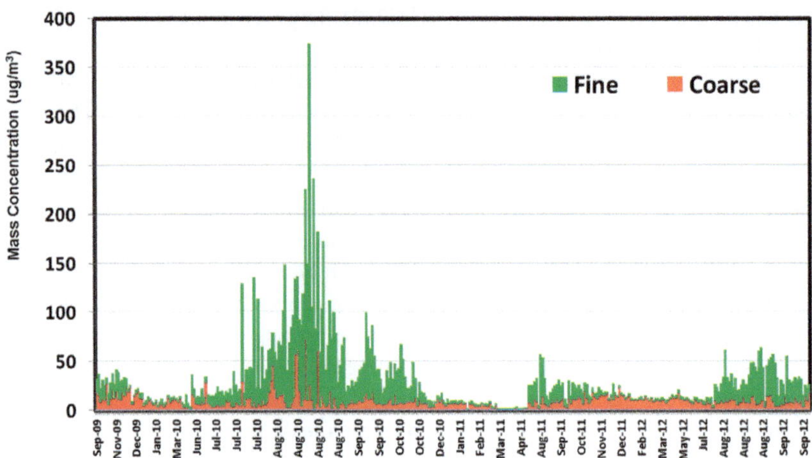

Fig. 5 Time series of fine and coarse mode aerosol mass concentrations at the PVH anthropogenic impacted site, from 2009 to 2012.

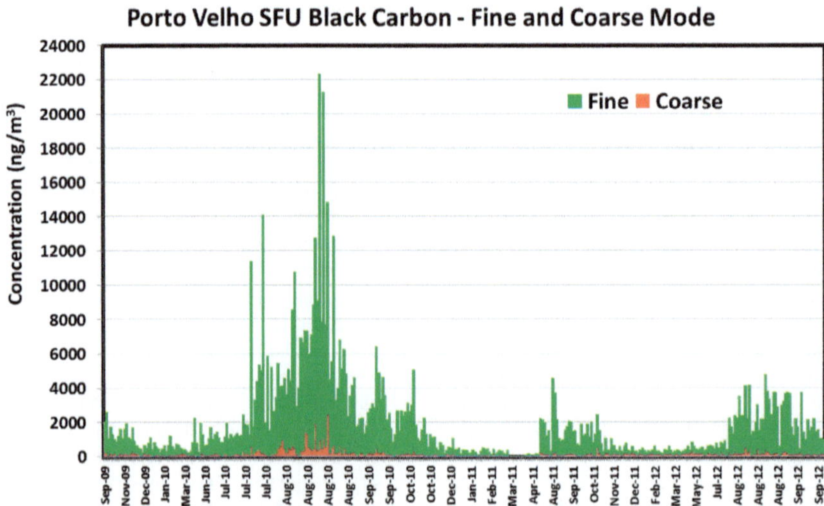

Fig. 6 Average aerosol equivalent black carbon BC_e concentrations for fine and coarse mode at the PVH impacted site from 2009 to 2012.

for truncation errors.[68] Aerosol particle absorption was measured using MAAP absorption photometers (MultiAngle Absorption Photometry – Thermo Inc., Model 5012).[69] The MAAP reports equivalent black carbon (BC_e) concentrations at 637 nm, which were converted to absorption coefficients assuming a mass absorption coefficient of 6.6 m^2 g^{-1}. A 5% correction was applied to the data to account for an adjustment of wavelength. Pressure and temperature measured inside the nephelometer were used for adjusting scattering and absorption coefficients to 1013 mbar and 0 °C.

Table 2 Average aerosol elemental concentrations (2009–2012) at the PVH site (Porto Velho, Rondônia) for dry and wet seasons. The table includes averages for particulate matter (PM), equivalent black carbon (BC_e) and trace elements. The concentrations are in ng m^{-3}. The uncertainty is the standard deviation and N is the number of samples that the element was measured above detection limits. Fine mode refers to aerosol aerodynamic diameters up to 2.0 μm and coarse mode from 2.0 μm to 10 μm

| | Dry Season | | | | Wet Season | | | |
	Fine Mode	N	Coarse Mode	N	Fine Mode	N	Coarse Mode	N
PM	33000 ± 36000	209	10200 ± 9000	209	3600 ± 4500	150	8800 ± 5300	150
BC_e	2801 ± 2922	209	222 ± 230	209	405 ± 369	150	102 ± 58	150
Na	48 ± 54	124	44 ± 47	120	14 ± 22	102	16 ± 25	106
Mg	32 ± 30	67	69 ± 144	87	6.5 ± 6.8	61	10 ± 17	81
Al	194 ± 166	209	351 ± 312	201	29 ± 49	150	63 ± 94	142
Si	198 ± 166	209	418 ± 370	207	38 ± 59	150	88 ± 118	145
P	19 ± 12	209	19 ± 17	180	7.3 ± 6.1	150	45 ± 31	150
S	568 ± 469	209	120 ± 138	163	146 ± 172	150	55 ± 40	150
Cl	13 ± 18	87	9.0 ± 10	128	1.6 ± 2.3	20	13 ± 13	147
K	506 ± 418	209	130 ± 124	209	71 ± 88	150	121 ± 70	150
Ca	17 ± 14	209	43 ± 42	204	4.5 ± 5.6	150	16 ± 23	149
Ti	15 ± 13	209	30 ± 26	205	1.9 ± 3.0	150	6.0 ± 8.3	144
V	1.5 ± 2.0	67	1.8 ± 2.6	88	0.3 ± 0.4	82	0.2 ± 0.4	84
Cr	4.0 ± 5.1	78	3.1 ± 4.6	124	0.8 ± 1.1	54	0.6 ± 0.9	91
Mn	2.7 ± 3.4	119	3.1 ± 4.0	123	0.4 ± 0.6	119	0.7 ± 0.5	127
Fe	178 ± 150	209	340 ± 286	209	27 ± 41	150	79 ± 95	150
Ni	1.4 ± 2.6	75	1.2 ± 2.0	30	0.2 ± 0.2	46	0.2 ± 0.4	27
Cu	1.8 ± 2.7	133	1.5 ± 2.2	95	0.7 ± 2.3	129	0.4 ± 0.4	128
Zn	3.3 ± 2.5	170	2.0 ± 2.0	139	1.0 ± 1.5	143	1.3 ± 1.8	147
Br	5.9 ± 7.4	180	3.2 ± 4.0	122	0.7 ± 0.9	115	0.5 ± 0.4	105
Rb	2.7 ± 2.9	133	2.2 ± 3.2	139	0.3 ± 0.5	98	0.4 ± 0.4	99
Sr	8.6 ± 14	64	10 ± 15	77	1.9 ± 2.5	47	1.6 ± 1.9	52
Sb	8.5 ± 6.0	15	20 ± 26	27	1.6 ± 1.5	28	2.2 ± 2.8	42
Pb	6.2 ± 9.4	142	5.9 ± 8.9	147	1.1 ± 1.2	105	0.9 ± 1.0	93
BC_e/PM (%)	8.5 ± 8.1	209	2.2 ± 2.6	209	11.3 ± 8.2	150	1.2 ± 1.1	150
SO4/PM (%)	5.2 ± 3.9	209	3.5 ± 4.6	163	12.2 ± 11.5	150	1.9 ± 2.3	150

3 Aerosol mass, black carbon and elemental composition in Amazonia

Long-term fine and coarse mode aerosol collection with filters allows obtaining mass concentration, black carbon and trace elements, indicating aerosol sources and processes. Fig. 3 shows the time series of fine and coarse mode aerosol mass concentrations at the TT34 forest site from 2008 to 2012. The fine mode concentration in the wet season is quite low (1.3 ± 0.7 μg m^{-3}), and even dry season values with an average of 3.4 ± 2.0 μg m^{-3} are a low concentration for PM$_{2.5}$. The mass concentration of coarse mode aerosols, essentially containing biogenic and soil dust aerosols, do not have a strong seasonal dependence, with an average mass concentration of 4.8 ± 2.3 μg m^{-3}. These values are in agreement with previous observations in a location close to TT34 (Balbina, ZF2), between 1998 and 2005.[44] No trends in aerosol mass concentrations can be observed in the preserved central Amazonia. The concentrations of equivalent black carbon at the TT34 site have a pronounced seasonal behaviour (Fig. 4), with increased concentrations from August to December due to long-range transport of biomass burning aerosol. There is also a small absorption component in the coarse mode aerosol (average 45 ± 21 ng m^{-3}) due to the absorption characteristics of natural biogenic aerosol particles. The average equivalent BC$_e$ fraction in fine mode aerosol is about 7% in central Amazonia, constant for the dry and wet seasons. Table 1 summarizes the average values for aerosol mass and BC$_e$ concentrations for fine and coarse mode aerosols for the TT34 central Amazonia site.

Fig. 5 shows the time series of fine and coarse mode aerosol mass concentrations at the PVH site from 2009 to 2012. In 2010, a strong drought in Amazonia significantly increased the smoke emissions from biomass burning (Fig. 5).[70] The increase in BC$_e$ concentrations in 2010 was likewise significant, as can be seen in Fig. 6, showing the time series of BC$_e$ concentrations at the PVH site from 2009 to 2012. BC$_e$ concentrations as high as 22 µg m^{-3} were observed in the fine mode. This plot shows the effects of strong droughts in changing atmospheric properties in Amazonia. The years of 2009, 2011, and 2012, with usual amounts of precipitation in the region, had BC$_e$ fine mode concentrations of 2–4 µg m^{-3}. As a matter of comparison, Artaxo et al.[39] reported an average fine BCe concentration of 1.8 µg m^{-3} at a nearby site (Jaru ecological reserve, in the Rondônia state) in the dry season of 1999. Table 2 shows the average fine and coarse mode mass and BC$_e$ concentrations at the PVH site. Compared to the TT34 site, we observe that the average fine mode concentration increases by a factor of 10 in the dry season and by a factor of 3 in the wet season, indicating that even in the absence of large scale biomass burning, land use change increases fine mode aerosol concentrations at the impacted southern Amazonia site.

Sources and atmospheric processes influence the aerosol elemental composition. Therefore, it is expected to observe significant differences in aerosol composition depending on land use change and season. In Amazonia, the inorganic component typically constitutes 10–20% of the fine mode mass, and about 10% of the coarse fraction mass (e.g., Martin et al.[44]). Tables 1 and 2 show the average aerosol elemental concentrations, respectively, for TT34 and PVH sites, for dry and wet seasons. In the wet season at the TT34 forest site, fine mode sulphur concentrations averaged at a low 74 ± 45 ng m^{-3}, increasing by a factor of 2 in the dry season (Table 1). At the PVH site in the dry season, fine mode sulphur is 8 times higher in comparison to the TT34 site (Table 2). Fine mode sulphur levels observed at the PVH site in the dry season agree with previously reported values for the dry season of 1999 at a nearby site in the state of Rondônia (529 ng m^{-3}, ref. 39). Increased sulphur loading in the dry season is associated with biomass burning emissions. Among inorganic ions, sulphate is the most prominent compound in biomass burning aerosols in Amazonia, according to Yamasoe et al.[71] Nevertheless, average PVH sulphate contribution to fine mode mass is greater in the wet season (11.5%) in comparison to the dry season (5.2%). This is associated with the fact that fine biomass burning aerosols show a significantly enhanced organic content, as will be discussed on Section 5.

Concentrations of elements associated with industrial and fossil fuel emissions (Cr, V, Mn, Ni, Pb, and others) were typically very low at <1 ng m^{-3} at the TT34 site. At the PVH impacted site, these elements are a bit more abundant, particularly in the dry season. Soil dust associated elements, such as Al, Si, Ti, Fe, had relatively low concentrations at the TT34 forest site (<300 ng m^{-3}). This can

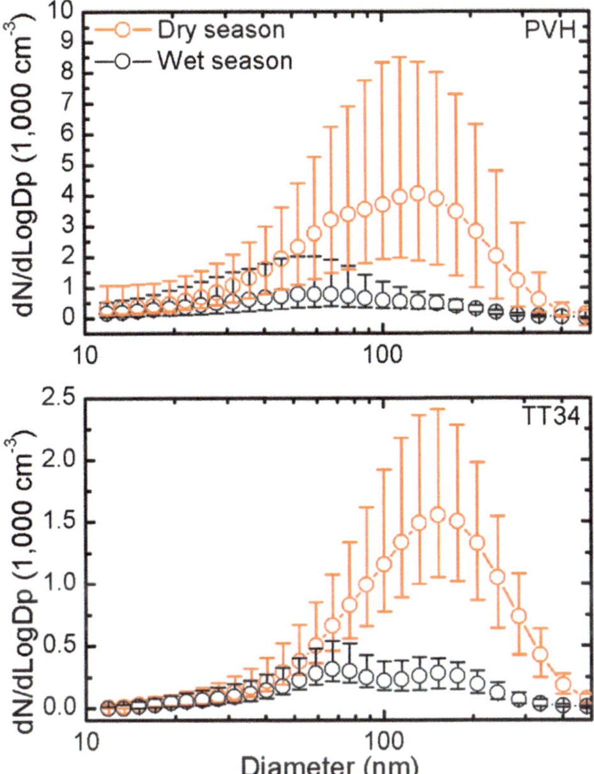

Fig. 7 Median of aerosol number size distributions measured at southern Amazonia (PVH) and central Amazonia (TT34). Circles represent median values, and bars represent 10 and 90 percentiles.

be explained by the fact that forest soil is covered with a thick layer of plant debris that suppresses local soil dust emission.[38,39] The TT34 site receives episodic Sahara dust events during January–May that significantly enhances ground based soil dust elements.[46,47] At the PVH site in the dry season, soil dust elements in the fine mode are about 8 times higher in comparison to the TT34 site. A combination of surrounding pasture landscapes, dryer climate and enhanced convection during the biomass burning season explains the increased mineral dust concentrations observed at the impacted PVH site. In the coarse mode, biological trace elements K, P and Zn[49,72] were detected in significant concentrations at both sites. No significant seasonal variation was observed in the concentration of inorganic biogenic tracers, suggesting a rather constant biogenic aerosol contribution along the year. The concentration of biogenic tracers in the coarse mode are increased by a factor of 2 at PVH in comparison to TT34, but the proportion of these elements to the coarse mode mass is similar at both sites.

As expected, differences between the two sites regarding aerosol mass and composition is most significant during the dry season, since the PVH site receives greater influence from biomass burning emissions. In the wet season, differences in aerosol properties between the sites are smaller. Even though, aerosol total mass and most element concentrations are increased by a factor of about 2 at PVH in comparison to TT34 during the wet season, indicating the presence of a

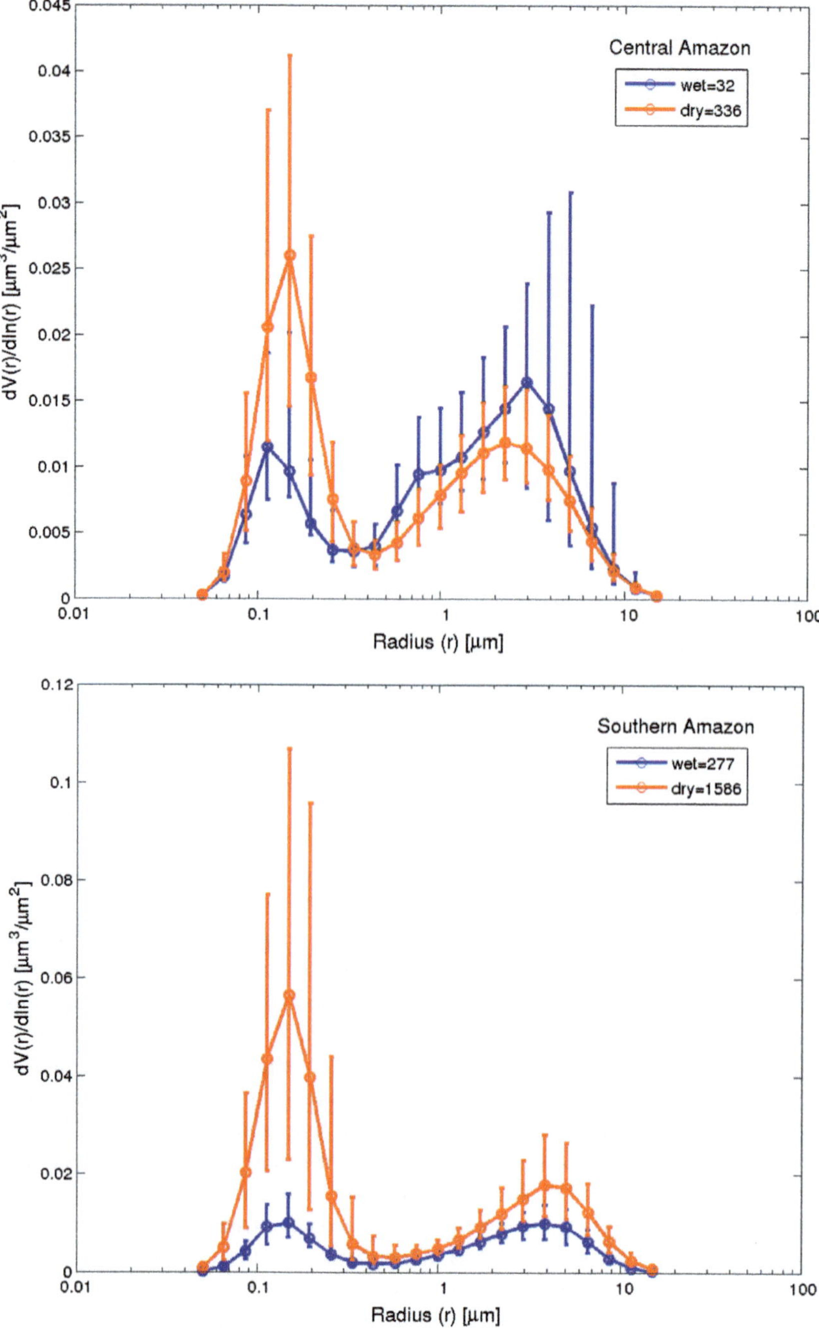

Fig. 8 Aerosol size distributions from TT34 (Central Amazonia) (top) and PVH (Southern Amazonia) (bottom) AERONET sites are shown for wet (Jul–Nov, blue) and dry (Dec–Jun, red) seasons. Error bars indicate the 25% and 75% quartiles. The legends show the number of level 2.0 almucantar retrievals.

regional pollution background in southern Amazonia, even in the absence of biomass burning fires.

4 Aerosol number concentrations and submicrometer size distributions

Aerosols occur in the atmosphere with diameters ranging through several orders of magnitude: from nanometres to tenths of micrometres. Characterization of aerosol size distributions is important because the particle size influence aerosol light scattering and absorption as well as cloud droplet activation and impacts on human health. In Fig. 7, we present medians of submicrometer aerosol number size distributions measured at PVH and TT34 sites in 2012 and 2008, respectively, for the dry and wet seasons. The integral of the number size distribution curve provides the aerosol number concentration, which increased by a factor of four at both sites from the wet to the dry season. At the TT34 preserved forest site, the median number concentration was 220 cm^{-3}, almost a factor of three lower than the wet season at the PVH site (680 cm^{-3}). This difference can be explained by the increased regional aerosol loading in southern Amazonia, in agreement with the aerosol scattering (*cf.* Section 6) and mass (*cf.* Section 3). In the dry season, aerosol number concentrations reached up to 20 000 cm^{-3} at PVH, compared to 2200 cm^{-3} at TT34.

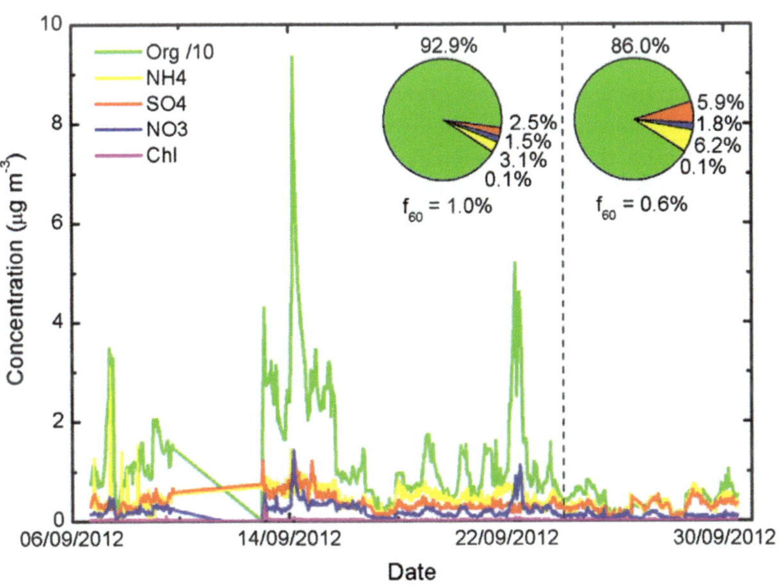

Fig. 9 Submicron non-refractory aerosol composition at the PVH site. The studied period of time comprises the end of the dry season, here chosen at 25th September and indicated using a dashed line. The parameter f_{60} is a surrogate for the contribution of biomass burning on OA (see text for details). The aerosol composition shows a significant enhancement of organics relative to other species due to biomass burning. Average concentration of monitored species prior 25th September 2012 is 13.9 µg m^{-3}, 0.46 µg m^{-3}, 0.37 µg m^{-3}, 0.22 µg m^{-3} and 0.01 µg m^{-3} for Org, NH4, SO4, NO3, and Chl, respectively. The non-refractory PM1 aerosol concentration after 25th September yields 4.0 µg m^{-3}, 0.29 µg m^{-3}, 0.27 µg m^{-3}, 0.08 µg m^{-3} and 0.005 µg m^{-3} for Org, NH4, SO4, NO3, and Chl, respectively.

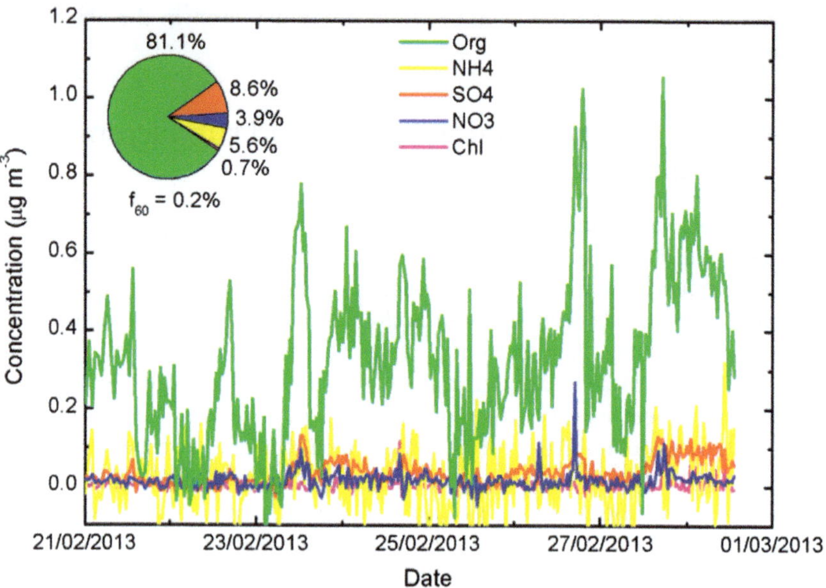

Fig. 10   Submicron non-refractory aerosol composition at the TT34 site (central Amazonia) during the wet season. The parameter f_{60} is a surrogate for the contribution of biomass burning on OA (see text for details). The non-refractory PM1 aerosol concentration after 25th September yields 0.33 µg m^{-3}, 0.02 µg m^{-3}, 0.03 µg m^{-3}, 0.02 µg m^{-3} and 0.002 µg m^{-3} for Org, NH4, SO4, NO3, and Chl, respectively.

In the dry season, the aerosol number size distribution was dominated by the accumulation mode at both sites, corresponding to aged biomass burning aerosols and having mean geometric diameters of 90 nm at PVH and 130 nm at TT34. In the wet season, the accumulation mode and Aitken mode were often separated, showing the so-called Hoppel minimum, separating particles that have been subjected to in-cloud processing from those that have not.[32,44] New particle formation and subsequent growth of ultrafine particles were rarely observed. Nevertheless, bursts of ultrafine particles with diameters in the range 10–40 nm were frequently observed at the TT34 site during the wet season, associated with low aerosol absorption coefficients (<4 Mm^{-1}), lower temperatures (<30 °C), and relative humidity above 70%. This result suggests that processes of new particle formation in Amazonia differ from those in coniferous forests of northern Europe.[73,74] The coarse mode primary biogenic particles size distribution is centred around 6 micrometres, which is a typical size for primary biological particles, such as fungal spores.[24,59]

Aerosol size distribution retrievals from AERONET CIMEL sunphotometers operated in the last 10 years in Amazonia presents a bimodal size distribution shown in Fig. 8 for both sites (PVH and TT34) and dry and wet seasons. The accumulation mode median diameter from ground based measurements agree quite well with column integrated values, possibly because strong convection enhances vertical mixing. The coarse mode AERONET size distribution retrievals peak at about 2–3 micrometres (Fig. 8). This value is consistent with fluorescent biological aerosol particles (FBAP) number size distribution measurements using UV-APS fluorescence techniques that peaks at 2.3 µm at TT34 site.[59]

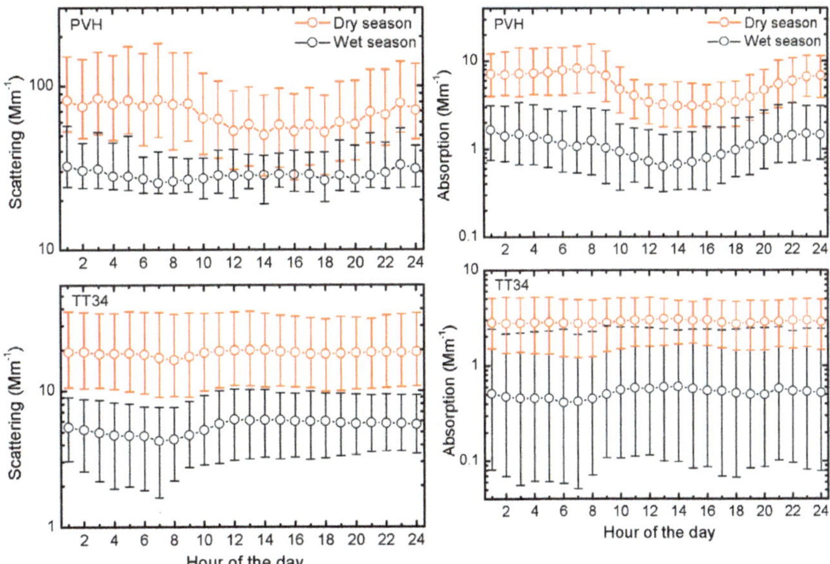

Fig. 11 **(a)** Diurnal cycle of light scattering coefficient σ_s at 637 nm in Mm^{-1} in Porto Velho (PVH – top) and Central Amazonia (TT34 – bottom) for the dry (red) and wet (black) seasons from 2009 to 2012. Circles represent median values, and bars represent 10 and 90 percentiles. **(b)** Diurnal cycle of light absorption coefficient σ_a at 637 nm in Mm^{-1} in Porto Velho (PVH – top) and central Amazonia (TT34 – bottom) for the dry (red) and wet (black) seasons from 2009 to 2012. Circles represent median values, and bars represent 10 and 90 percentiles.

5   Aerosol mass spectrometry measurements in Amazonia

Aerosol mass spectrometry (AMS) is a valuable tool to measure aerosol composition in real time. ACSMs were recently deployed at the sites in Porto Velho (PVH) and central Amazonia (TT34), with the intention that they begin a multiyear record of data. At the PVH site, the system collected data from 6 to 30 September 2012 during the SAMBBA (South American Biomass Burning Analysis) campaign. This period comprised the transition between the end of the dry season, hereafter named intense biomass burning (BB) period, and the beginning of the wet season for that year (moderate BB period). Based on meteorological data, aerosol composition and optical properties, the date of 25 September was chosen to divide the measurements into distinct periods.

An important parameter retrieved using the AMS/ACSM system is the fraction f_{60} of the organic signal at m/z 60. Given the fragmentation pattern of levoglucosan, mannosan, galactosan, and other pyrolysis by-products of cellulose in the AMS systems, the intensity of f_{60} is used as a surrogate for regional biomass burning influence.[74] An f_{60} of 0.3% is usually chosen as the threshold value to indicate air masses influenced by BB emissions.[75] The aerosol composition for the PVH site is shown in Fig. 9. The f_{60} value during the intense BB period is on average almost twice the value measured during the moderate BB period (1.0% *versus* 0.6%). During the intense BB period, the PM1 non-refractory aerosol composition is largely dominated by organic substances (92.9%), with sulphate accounting for only 2.5% of aerosol mass. During the moderate BB period, organic

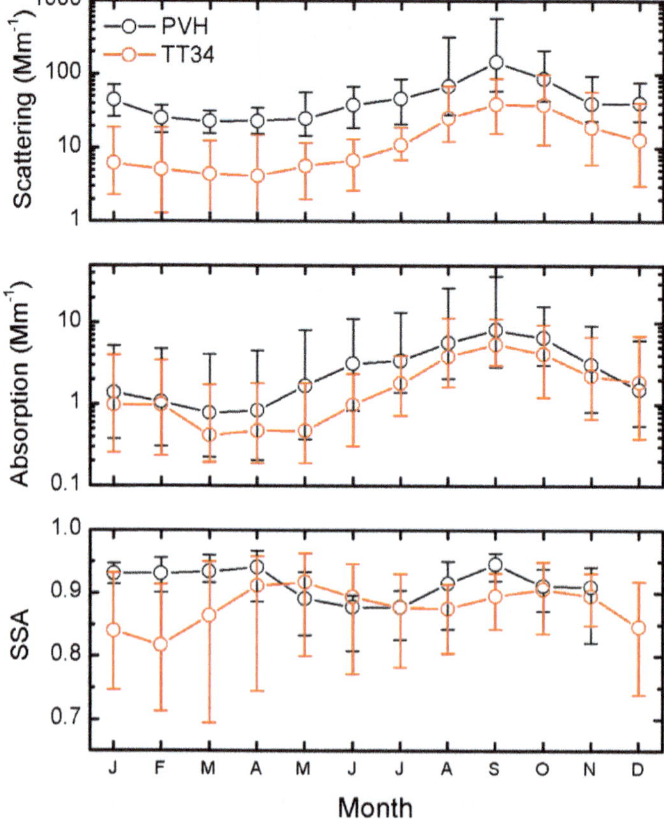

Fig. 12 Monthly statistics (2009–2012) for light scattering coefficient σ_s at 637 nm and light absorption coefficient σ_a at 637 nm in Mm^{-1} for Porto Velho (PVH, in black) and central Amazonia (TT34, in red). The bottom plot shows the monthly statistics for the single scattering albedo (SSA) at 637 nm. Circles represent median values, and bars represent 10 and 90 percentiles.

substances accounted for 86% and sulphate for 5.9%, at an average concentration of 4.7 μg m^{-3}. At the central Amazonia TT34 site, the first deployment of the ACSM collected data from 21 February to 1 March 2013 (wet season). Submicron aerosol composition at the TT34 site is shown in Fig. 10. For the studied period, organic substances comprised 81% of mass and sulphate 8.6%, with an average concentration of 0.4 μg m^{-3}, consistent with values obtained during AMAZE-08.[43] The f_{60} parameter was on average 0.2% during the studied period, indicating the absence of an effect of regional biomass burning. Chen *et al.*[43] noted that African biomass burning aerosols are extensively oxidized in long-range transport such that f_{60} no longer serves as a marker signal in this case.

Previously reported organic contributions to submicron non-refractory PM1 composition of biomass/wood burning emissions range from 97.8% (*Pinus ponderosa* smoke[76]), 85% (aircraft measurements over West Africa),[74] 83% (aircraft measurements over boreal forest in North America,[77] 60% (aircraft measurements intercepting biomass burning plumes from Russia and Kazakhstan),[77] to 42.8% (*Serenoa repens* smoke[76]). Considering only ambient measurements, organic substances range from 85% to 60% of PM$_1$, with sulphate as high as 30% in the

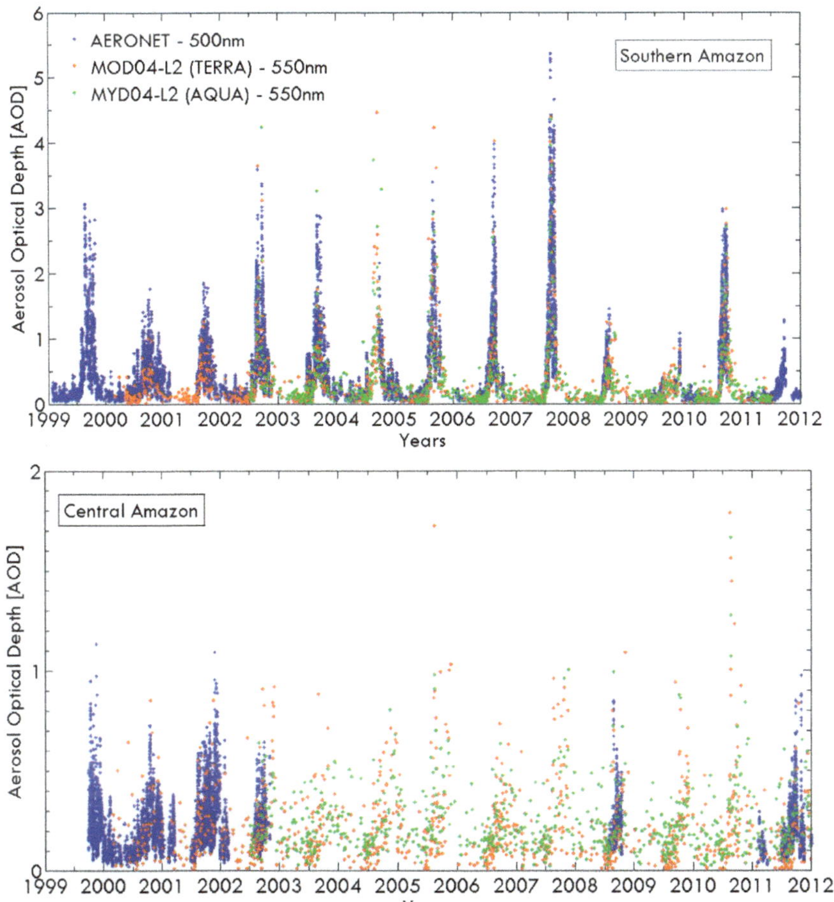

Fig. 13 **(a)** Time series of aerosol optical depth at the PVH site with MODIS (550 nm) and AERONET (500 nm) retrievals from 1999 to 2012. **(b)** Time series of aerosol optical depth from the central Amazonia TT34 site with MODIS retrievals at 550 nm from 2000 to 2012.

latter case. At the PVH site, biomass burning significantly enhanced the organic content of the fine aerosol to above 90%. The elemental composition results discussed in Section 3 corroborate this high contribution of organic substances and the corresponding relatively low contribution from sulphate for biomass burning in Amazonia. Results listed in Table 2 indicate a contribution of 11.5% by sulphate in PM2.5 during the wet season and 5.2% on average during the dry season. These changes in aerosol composition have important consequences in many aspects, including the optical properties, as discussed in the following subsection.

6   Aerosol optical properties

Knowledge of aerosol optical properties is important to quantify the direct and indirect effects of aerosols on the radiation balance.[78,79] In addition to the light scattering $\sigma_s(\lambda)$ and absorption $\sigma_a(\lambda)$ coefficients, single scattering albedo (SSA) is

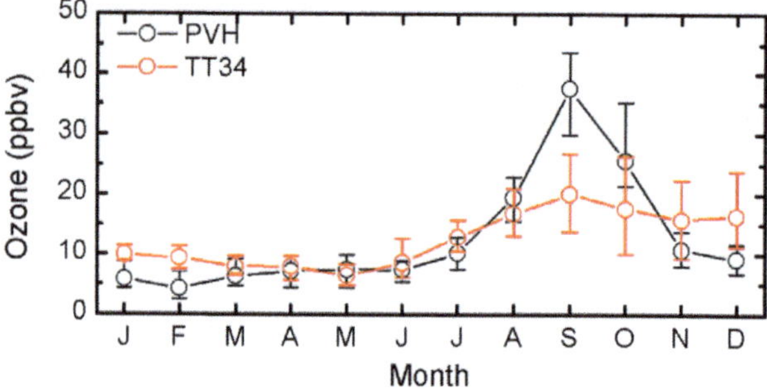

Fig. 14 Seasonal variability (2009–2012) of ozone volume mixing ratios in Porto Velho (PVH) and central Amazonia (Manaus TT34 tower). Circles represent median values, and bars represent 10 and 90 percentiles.

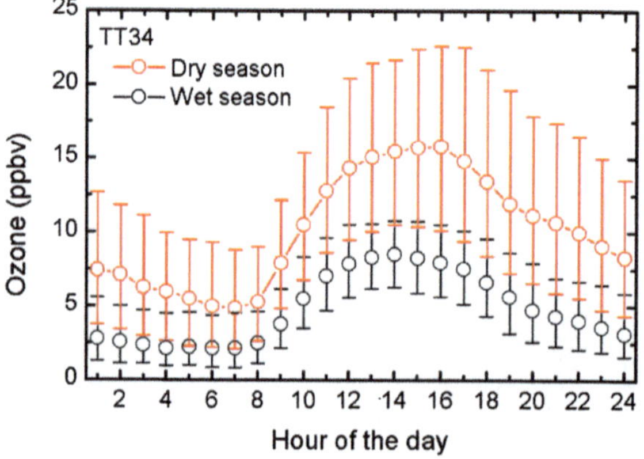

Fig. 15 Diurnal cycle of ozone volume mixing ratios in central Amazonia (TT34) for the dry (red curve) and wet seasons (black curves) from 2009 to 2012. Circles represent median values, and bars represent 10 and 90 percentiles.

a critical parameter in the aerosol effects on climate. SSA(λ) is calculated as the scattering coefficient divided by the extinction coefficient ($\sigma_e(\lambda) = \sigma_s(\lambda) + \sigma_a(\lambda)$); being an intensive aerosol property, it does not depend on particle amount. Fig. 11a and b show the median diurnal cycles of particle scattering and absorption coefficients at PVH and TT34 sites, distinguishing between dry and wet seasons. At the PVH site, the diurnal variability was dominated by the boundary layer dynamics: higher coefficients at night when the atmosphere is stable and thin and lower coefficients during the day when the atmosphere is well mixed and dispersion is favoured. This behaviour was similar in both seasons as a result of the regional pollution background observed at the PVH site all through the year. At the preserved TT34 forest site during the wet season, a different behaviour was observed: the scattering diurnal variability was modulated by the biogenic production of secondary organic aerosols during the day. As discussed in detail by Rizzo et al.,[60] a daytime 20% increase in submicrometer particle diameter was observed, which may cause an increase of 50–70% in the scattering coefficients between 9:00 and 12:00 local time. This feature cannot be seen during

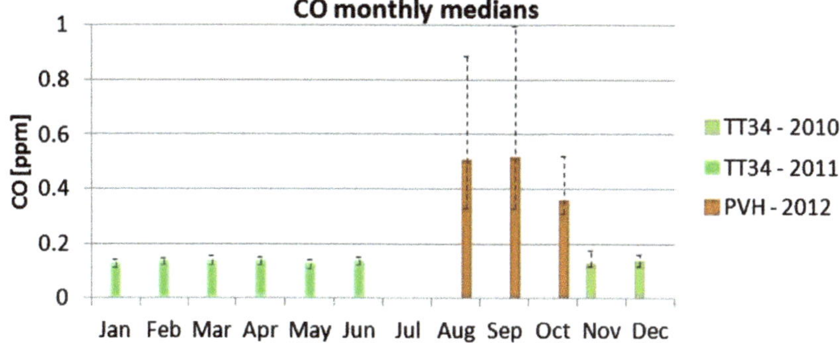

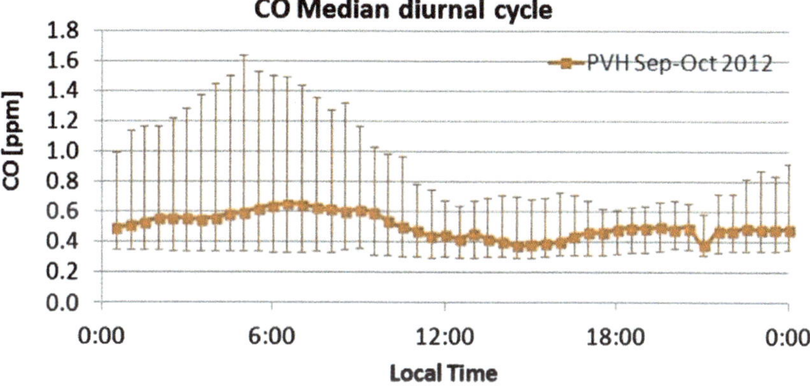

Fig. 16 Carbon monoxide monthly medians for central Amazonia (TT34) and Porto Velho (PVH), together with diurnal variability of CO in PVH during the dry season.

the dry season, since the regional transport of biomass burning particles overwhelms the biogenic processes.

Fig. 12 shows monthly statistics for aerosol light scattering (550 nm), absorption (637 nm), and SSA (637 nm). There is an enhancement by a factor of 10 in σ_s at PVH compared to TT34 for both dry and wet season. This enhancement is another indication that land use change in southern Amazonia is increasing the regional aerosol loading, even in the wet season when biomass burning is suppressed. Concerning aerosol absorption, the difference between Porto Velho and central Amazonia is much smaller throughout the year. In January and February (wet season at both sites), σ_a was around 1.0 Mm^{-1}. Since there are no biomass burning activities in this period, the observed absorption coefficients at that time are attributed to the so-called "brown carbon",[80] *i.e.*, light absorbing biogenic particles.[24,60,81] The dry season starts earlier at PVH (May) in comparison to TT34 (July), as depicted in Fig. 12 (absorption in the middle panel). In the wet season, SSA values (Fig. 12 lower panel) are significantly lower at TT34 (SSA = 0.83) compared to PVH (SSA = 0.93). This difference is a result of low aerosol scattering coefficients (driven by low aerosol loadings) combined with relatively high absorption coefficients (driven by the presence of light absorbing biogenic aerosols) at the forest site during the wet season. Low SSA values do not occur in the wet season at PVH because of the higher scattering particle loading in that region. At PVH, a decrease in SSA was observed from September to October, coincident with the decrease of the aerosol organic fraction and an increase of sulphate, as discussed in section 4. This behaviour is at first unexpected, because sulphate aerosols are efficient light scatterers, meaning that increased sulphate content should be associated with higher SSA values. In spite of the fact that some organic compounds may absorb light at shorter wavelengths (<500 nm), acting as brown carbon, they are efficient scatterers with little absorption at longer wavelengths.

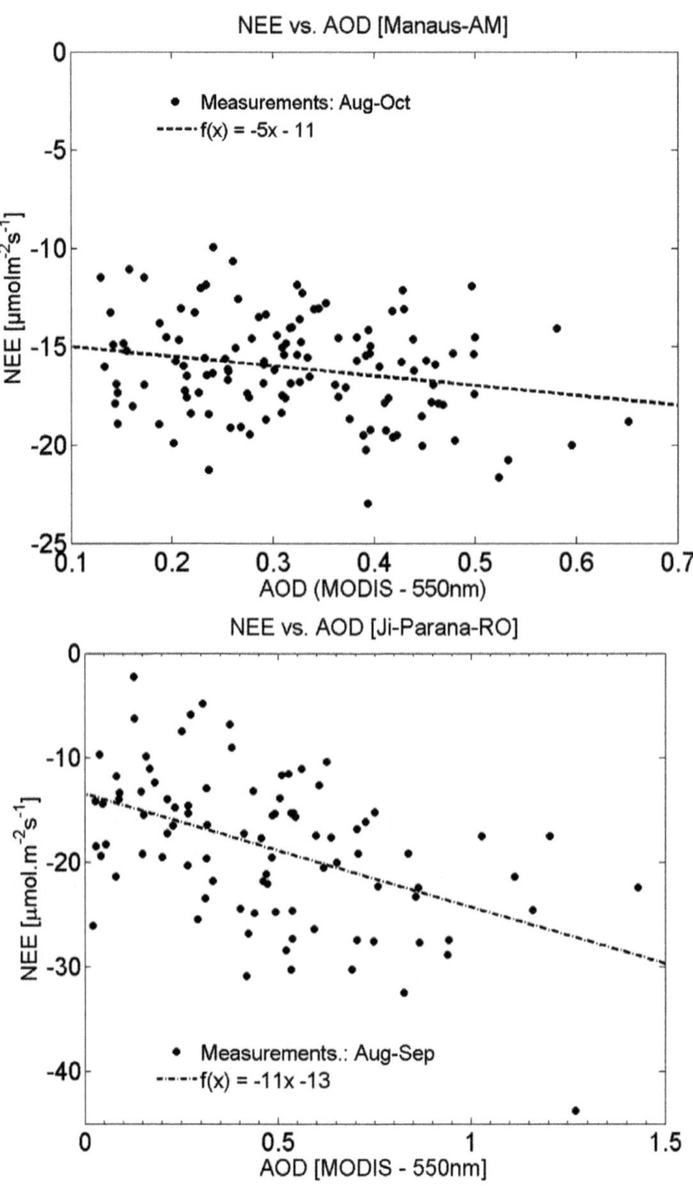

Fig. 17 Effects of aerosols on carbon uptake expressed as Net Ecosystem Exchange (NEE) at central Amazonia (nearby the TT34 forest preserved site), and at southern Amazonia (same region as the PVH impacted site). Data points refer to daily averages in the dry season (Aug–Sep), between 1999–2009 (TT34) and between 1999–2002 (PVH).

Since SSA was calculated at 637 nm, higher SSA values in September may be explained by increased light scattering due to organic aerosols released by biomass burning activities.

Fig. 13a and b shows the time series of $AOD_{500-550}$ in central Amazonia (TT34 site) and in southern Amazonia (PVH site) between 1999 and 2012, from both AERONET and MODIS observations. There is an excellent agreement between the AERONET and MODIS datasets, in spite of the different methods used in each AOD retrieval. A strong seasonal variation is evident at both regions. In the preserved central Amazonia, AOD reached up to 1.0 in the dry season, whereas in the impacted southern Amazonia it reached very high values up to 5.0 at 500 nm. These long-term time series plots give an overview of the variable aerosol loading in Amazonia, both from wet to dry seasons and also the large annual variability due to climatic conditions. It is possible to observe that after the deforestation started to get reduced (see Fig. 1) AOD at the PVH site was reduced, but at the strong drought of 2010, smoke has increase significantly compared to 2008 and 2009.

Overall, the seasonal variability of the aerosol optical properties indicates that in the dry season biomass burning particle emissions overwhelm biogenic emissions, significantly changing the physical properties of the natural aerosol population. This increase in particles affects the regional radiation budget, cloud microphysics, and CO_2 uptake by the vegetation.[18,19,82–84]

7 Land use changing trace gas concentrations

Ozone is an important trace gas in Amazonia because it can damage vegetation at high concentrations and is an indicator of photochemical activity. Fig. 14 shows the seasonal variability of median monthly ozone mixing ratios between 13:00 and 16:00 LT in central Amazonia (TT34) and PVH. Ozone mixing ratios are elevated in the dry season of both sites because of biomass burning emissions of precursors (nitrogen oxides and VOCs) and the larger radiation flux and photo-chemical activity caused by reduced cloud cover. The higher values at mid-day in PVH compared to TT34 are the effect of larger impact of ozone precursors emitted by biomass burning. Fig. 15 shows the diurnal cycle of ozone mixing ratios in Central Amazonia (TT34) for the dry and wet seasons. The ozone mixing ratio goes to low values at nighttime (3 ppb in the wet season and 7 ppb in the dry season), being consumed by vegetation uptake and reaction with NO and VOC from soils and vegetation.[85] The dry-season ozone values and diel cycles measured at TT34 are very similar to results obtained in July 1985 in a closed forest at Reserva Ducke and in July 2001 in an open area near the Balbina Reservoir, both in the region north of Manaus.[86,87]

Ozone measurements in Rondônia have been made previously at sites in a closed rainforest and on a cattle pasture about 250 km SE of Porto Velho.[30,85] Midday maxima above the canopy were about 15 ppb in the wet season and 50 ppb in the dry season at both sites. At night, ozone decreased to around 6 ppb at the forest site and 10 ppb at the pasture site. The elevated values of ozone over Rondônia as compared to those over the Central Amazon are the result of much higher pollution levels resulting from land use change in the arc of deforestation at the southern perimeter of the Amazon Basin.

These elevated pollution burdens can readily be seen by comparing the levels of carbon monoxide, which is one of the best tracers for biomass burning

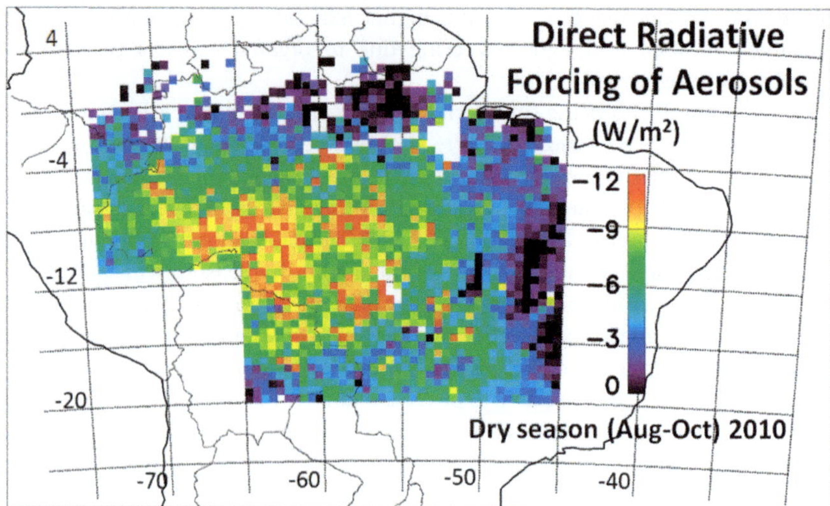

Fig. 18 Average spatial distribution of the direct radiative forcing (DRF) of biomass burning aerosols in Amazonia during the dry season (August to October) of 2010. Forcing derived from calculations using a combination of MODIS and CERES sensors data. During this three-month period, the daily-average radiative forcing of aerosols for the whole area was on average -5.3 ± 0.1 W m^{-2}.

emissions.[88] Fig. 16 shows the carbon monoxide monthly medians for central Amazonia (TT34) and Porto Velho (PVH). Unfortunately, there is no continuous CO data for the two sites. The average CO mixing ratio in central Amazonia is about 130 ppb. This can be compared to about 400 ppb CO in Porto Velho during the dry season due to biomass burning emissions. The lower plot in Fig. 16 shows the diurnal variability of CO mixing ratios at PVH, clearly dominated by the atmospheric boundary layer dynamics: CO accumulation at night due to lower boundary layer height and minimum CO mixing ratios at 15:00 LT when the boundary layer is well mixed.

8   Interactions between carbon cycle and aerosols

Amazonia is a large reservoir of carbon that can be quickly mobilized to the atmosphere through deforestation, changes in the radiation balance, in precipitation patterns, or in climate.[9] The increase in atmospheric aerosol loading increases the ratio of diffuse to direct radiation.[18,89] Forests show an increase in carbon uptake with an increase in diffuse radiation due to the larger canopy penetration of the scattered solar radiation.[17,19] In order to quantify this effect, AOD measurements from the MODIS (Moderate Resolution Imaging Spectror-adiometer) sensor (on board the TERRA and AQUA satellites), were validated with AOD measurements from the NASA/AERONET network of sun photometers, and compared to Net Ecosystem Exchange (NEE) variations. Daily averages of AOD were taken based on the MODIS MOD/MYD_04_L2 product. Two forest sites were studied during the dry season (Aug–Sep): the Jaru biological reservation site in southern Amazonia (corresponding to the same region as the PVH site), between 1999 and 2002, and the Cuieiras reservation K34 tower in central Amazonia (1 km away from the TT34 tower site), between 1999 and 2009. Carbon fluxes were measured using eddy-correlation techniques.[90] NEE was calculated according to the relationship NEE $=$ Fc $+$ Stg,[90,91] where Fc is the turbulent flux of CO_2 above the canopy, and Stg is the non-turbulent storage term, defined as the temporal variation of the CO_2 concentration measured in a vertical profile from the soil surface to the height of eddy correlation measurements.

Fig. 17 shows the effects of aerosols on carbon uptake expressed as NEE in central Amazonia and in southern Amazonia. In southern Amazonia, a 29% increase in carbon uptake (NEE) was observed when the AOD increased from 0.10 to 1.5. In central Amazonia, the effect is smaller, being approximately an increase of 20% of the NEE compared to clean conditions. The increase of 29% and 20% in NEE is attributed mainly to an increase (50%) in the PAR (photosynthetically active radiation) diffuse radiation flux. Oliveira et al.[18] observed a similar increase in NEE of about 18% at the southern Amazonian site for a reduction of 70% of clear-sky irradiance. For even higher aerosol loadings in the atmosphere (AOD > 2), leading to strong reductions of solar radiation flux, Oliveira et al.[18] reported that the photosynthetic process is almost shut down, with NEE values close to zero. Considering long-range transport of biomass burning aerosols in Amazonia, changes in NEE may occur over large areas in Amazonia, modifying the potential for forest ecosystems to absorb atmospheric CO_2. Previous studies at other sites in Amazonia[19] have shown similar NEE enhancements, but the use of MODIS to obtain AOD allows for the determination of NEE changes in regions where AERONET sun photometers measurements are not available.

9   Large scale radiative forcing of aerosols and surface albedo change

Aerosol particles scatter and absorb solar radiation directly, affecting the Earth's radiation budget and modifying the atmospheric temperature profile.[78] Because aerosol particles act as cloud condensation nuclei, aerosols and clouds are intrinsically coupled through processes that control cloud formation and development, spatial coverage, lifetime, and precipitation efficiency.[92–95] Anthropogenic increases in aerosol particles modify cloud microphysical properties by decreasing the average cloud droplet size, because more particles are sharing the same amount of liquid water. This reduction affects the atmospheric energy balance since the larger number of smaller droplets reflects more incoming solar radiation back to space than fewer, larger droplets.[96] The radiative forcing due to the indirect effect of aerosols on cloud albedo presents the highest uncertainty of all climate-forcing mechanisms reported in the Fourth IPCC assessment.[78] Furthermore, the impact of aerosol particles on the precipitation rate remains a topic of much debate. Some studies suggest that increases in aerosol loading decrease the rain rate from convective clouds,[94,97,98] while others claim the opposite.[84,99,100] The actual response of convective clouds to increasing aerosol loadings is most likely a non-monotonic function with a maximum at intermediate aerosol concentrations.[95]

The presence of absorbing aerosols in the atmosphere can enhance dissipation of cloud droplets, decreasing cloud cover, an effect called the semi-indirect effect of aerosols. It has been observed[21,23] and simulated[41] for clouds over the Amazon forest in the presence of smoke aerosols. Koren et al.[23] showed that there is a competing effect between the microphysical effects and aerosol absorption in the

Amazonia: smoke aerosols can either invigorate clouds, increasing cloud fraction and height due to an increase in the cloud droplet concentration, or alternatively inhibit cloud development due to aerosol absorption within the cloud.

The direct radiative forcing (DRF) of biomass burning aerosols over Amazonia has been assessed using ground-based instruments, airborne measurements or satellite remote sensing.[79,101,102] Biomass burning aerosols can be transported over long distances and therefore have a large-scale impact on the radiative balance over the Amazon Basin. Simultaneous retrievals of the flux at the top of the atmosphere (TOA) and AOD, obtained by the CERES (Clouds and the Earth's Radiant Energy System) and MODIS (Moderate Resolution Imaging Spectro Radiometer) sensors aboard the Terra and Aqua satellites, have been used to assess the DRF at the TOA with high spatial coverage.[79,102] A similar approach was used to evaluate the spatial distribution of the 24 h-average direct radiative forcing (DRF) of biomass burning aerosols at the TOA during the Amazonian dry season (August to October) of 2010, a year during which a severe drought affected the whole region (Fig. 18).

Fig. 18 indicates that during the dry season the aerosol DRF presents high spatial variability, with significant impacts over central and south-western Amazonia, where the aerosol DRF reached average values down to -12 W m^{-2}. In the region of the PVH site, the DRF was on average -10 W m^{-2}. Central Amazonia shows much lower DRF absolute values, varying from about 0 to -3 W m^{-2} near the TT34 forest site. Several factors contribute to these heterogeneities in the DRF spatial distribution, such as the location of fire sources in the arc of deforestation (southern Amazonia), the predominantly east–west wind pattern during the period, and the high cloudiness and precipitation rate in central Amazonia. During this three-month period, the daily-average radiative forcing of aerosols for the whole area was on average -5.3 ± 0.1 W m^{-2}.

The replacement of tropical forests by croplands and pastures also increases the broadband surface albedo, modifying the radiative balance over the deforested area. In a recent paper, Sena *et al.*[79] reported the first estimate for the clear-sky land-use change radiative forcing (LURF) due to deforestation in Amazonia. Using remote sensing, the authors obtained a value of -7.3 ± 0.9 W m^{-2} for the annual average LURF over Rondônia, the state in which the PVH site is located. This finding is very important because changes in surface albedo affect the radiative budget throughout the year, while the impact of aerosols in Amazonia is most significant only during the biomass-burning season, which lasts about two to three months. Therefore, the impact of surface albedo change is about 6–7 times higher than the impact of aerosols in the annual energy balance in Amazonia. Deforestation also reduces evapotranspiration, decreasing the column water vapour (CWV) by about 6–10%.[79,103] As water vapour is a greenhouse gas, this decrease in CWV over the deforested area further contributes to cooling, with a reduction at the radiative flux at the TOA of the order of -0.4 to -1.2 W m^{-2}.[79]

10   Conclusions

Aerosol physical and chemical properties were monitored along 4 years of measurements in two different regions in the Amazon Basin: central Amazonia, enclosing the TT34 preserved forest site, and southern Amazonia, enclosing the PVH site highly impacted by biomass burning and land use change. Significant seasonal variations were observed on aerosol mass and number concentrations at both regions, as well as in black carbon equivalent concentration. Seasonal variations were attributed to biomass burning emissions, which are long range transported and widespread in the Amazon Basin during the dry season. Reduction on precipitation and on aerosol deposition rates during the dry season also plays an important role on changing the atmospheric properties across Amazonia.

Significant differences were observed between the TT34 and PVH sites. Central Amazonia during the wet season showed to be one of the cleanest continental areas worldwide, with aerosol number concentrations in the range of 300–500 cm^{-3}. In contrast, we observed that the southern Amazonian region shows important changes in atmospheric composition and aerosol loading even in the wet season, indicating the existence of a regional pollution haze associated with land use changes that have been taking place in the last decades.

New findings were provided by the recently deployed ACSM (Aerosol Chemical Speciation Monitor), indicating that biomass burning emissions significantly enhanced the organic content of the fine aerosol to above 90%, reducing the contribution of inorganic species like sulphate to the fine mode aerosol mass (from 11.5% in the wet season to 5.2% in the dry season at the PVH site). Long term measurements of aerosol mass spectra in high time resolution are planned for the next years in Amazonia, and will help to fill gaps related to sources and dynamical processes of aerosols in the Amazon Basin.

Remotely sensed aerosol properties in the Amazon Basin indicated that land use change in Amazonia significantly alters aerosol loading as well as radiation balance and carbon uptake by the forest. The observed changes in the radiation balance were significant, with radiation flux reductions of down to -20 W m^{-2} over large regions. The increase in diffuse radiation resulting from moderate increases in aerosol loading (AOD from 0.1 to 1.2 at 550 nm) affects the carbon balance in tropical regions by increasing NEE. A further increase (AOD > 1.3) produces a sharp decrease in carbon uptake due to the reduction in solar radiation flux at the surface, according to Oliveira et al.[18]

In addition to biomass burning, urbanization is on the rise in Amazonia. The GoAmazon 2014–2015 experiment will analyse the impacts of the urban plume of Manaus on the atmospheric chemistry downwind of the city.[104] Changes in cloud properties and precipitation will also be analysed in the GoAmazon experiment. A very clean site, the Amazon Tall Tower Observatory (ATTO) is being constructed north of Manaus in a very remote region, and will be a long-term observatory for trace gases and aerosols, as well as a platform to study biosphere–atmosphere interactions in pristine central Amazonia.

Acknowledgements

We would like to thank the INPA LBA central office in Manaus for logistical support for the operation of the TT34 site. This research was funded by the FAPESP projects 2008/58100-2, 2010/52658-1, 2011/50170-4, 2012/14437-9 and CNPq project 475735-2012-9, INCT Amazonia, the FP6 project EUCAARI (Contract 34684), and the Max Planck Society. We thank Eric Swietlicki, Alfred Wiedensohler and Markku Kulmala for support at the central Amazonia ZF2 site. We thank Sandra Hacon (FIOCRUZ), Fernando Morais (IFUSP), Alcides C. Ribeiro (IFUSP), Ana L. Loureiro (IFUSP), Fábio de Oliveira Jorge (IFUSP), Lívia Oliveira

(INPA) and Joiada Silva (UNIR-Porto Velho) for technical support and help in keeping the sites fully operational.

References

1 M. O. Andreae, Aerosols before pollution, *Science*, 2007, **315**, 50–51.
2 M. A. F. Silva Dias, S. Rutledge, P. Kabat, P. L. Silva Dias, C. Nobre, G. Fisch, A. J. Dolman, E. Zipser, M. Garstang, A. Manzi, J. D. Fuentes, H. Rocha, J. Marengo, A. Plana-Fattori, L. Sá, R. Alvalá, M. O. Andreae, P. Artaxo, R. Gielow and L. V. Gatti, Clouds and rain processes in a biosphere atmosphere interaction context in the Amazon Region, *J. Geophys. Res.*, 2002, **107**(D20), 8072–8092.
3 R. A. Betts, P. M. Cox, M. Collins, P. P. Harris, C. Huntingford and C. D. Jones, The role of ecosystem-atmosphere interactions in simulated Amazonian precipitation decrease and forest dieback under global climate warming, *Theo. Appl. Clim.*, 2004, **78**, 157–175.
4 B. S. Soares-Filho, D. C. Nepstad, L. M. Curran, G. C. Cerqueira, R. A. Garcia, C. A. Ramos, E. Voll, A. McDonald, P. Lefebvre and P. Schlesinger, Modeling conservation in the Amazon Basin, *Nature*, 2006, **440**, 520–523.
5 M. Gloor, R. J. W. Brienen, D. Galbraith, T. R. Feldpausch, J. Schöngart, J.-L. Guyot, J. C. Espinoza, J. Lloyd and O. L. Phillips, Intensification of the Amazon hydrological cycle over the last two decades, *Geophys. Res. Lett.*, 2013, **40**, 1–5, DOI: 10.1002/grl.50377.
6 J. A. Marengo, C. A. Nobre, J. Tomasella, M. F. Cardoso and M. D. Oyama, Hydroclimatic and ecological behaviour of the drought of Amazonia in 2005, *Philos. Trans. R. Soc. London, Ser. B*, 2008, **363**, 1773–1778.
7 J. A. Marengo, J. Tomasella, L. M. Alves, W. R. Soares and D. A. Rodriguez, The drought of 2010 in the context of historical droughts in the Amazon region, *Geophys. Res. Lett.*, 2011, **38**, L12703.
8 E. A. Davidson and P. Artaxo, Globally significant changes in biological processes of the Amazon Basin: results of the Large-scale Biosphere-Atmosphere Experiment, *Global Change Biol.*, 2004, **10**, 519–529.
9 J. P. Ometto, A. D. Nobre, H. R. Rocha, P. Artaxo and L. A. Martinelli, Amazonia and the modern carbon cycle: lessons learned, *Oecologia*, 2005, **143**(4), 483–500.
10 Y. Malhi, J. T. Roberts, R. A. Betts, T. J. Killeen, W. H. Li and C. A. Nobre, Climate change, deforestation, and the fate of the Amazon, *Science*, 2008, **319**, 169–172.
11 C. Nobre, J. Marengo and P. Artaxo, Understanding the climate of Amazonia: Progress from LBA, in: *Amazonia and Global Change*, ed. M. Keller, M. Bustamante, J. Gash and P. S. Dias, American Geophysical Union, Geophysical Monograph, 2009, vol. 186, pp. 147–150, ISBN: 978-0-87590-449-8.
12 E. A. Davidson, A. C. de Araújo, P. Artaxo, J. K. Balch, I. F. Brown, M. M. da C. Bustamante, M. T. Coe, R. S. DeFries, M. Keller, M. Longo, J. W. Munger, W. Schroeder, B. S. Soares-Filho, C. M. Souza Jr. and S. C. Wofsy, The Amazon Basin in Transition, *Nature*, 2012, **481**, 321–328.
13 C. Huntingford, P. Zelazowski, D. Galbraith, L. M. Mercado, S. Sitch, R. Fisher, M. Lomas, A. P. Walker, C. D. Jones, B. B. B. Booth, Y. Malhi, D. Hemming, G. Kay, P. Good, S. L. Lewis, O. L. Phillips, O. K. Atkin, J. Lloyd, E. Gloor, J. Zaragoza-Castells, P. Meir, R. Betts, P. P. Harris, C. Nobre, J. Marengo and P. M. Cox, Simulated resilience of tropical rainforests to CO2-induced climate change, *Nat. Geosci.*, 2013, **6**, 268–273.
14 P. Artaxo, W. Maenhaut, H. Storms and R. Van Grieken, Aerosol characteristics and sources for the Amazon basin during the wet season, *J. Geophys. Res.*, 1990, **95**(D10), 16971–16985.
15 P. Artaxo, Reductions in deforestation rates in Amazonia, *Global Atmospheric Pollution Forum Newsletter*, 2010, **8**, 2–3.
16 P. M. Cox, D. Pearson, B. B. Booth, P. Friedlingstein, C. Huntingford, C. D. Jones and C. M. Luke, Sensitivity of tropical carbon to climate change constrained by carbon dioxide variability, *Nature*, 2013, **494**, 341–344.
17 M. A. Yamasoe, C. von Randow, A. O. Manzi, J. S. Schafer, T. F. Eck and B. N. Holben, Effect of smoke and clouds on the transmissivity of photosynthetically active radiation inside the canopy, *Atmos. Chem. Phys.*, 2006, **6**, 1645–1656.
18 P. H. F. Oliveira, P. Artaxo, C. Pires Jr, S. de Lucca, A. Procópio, B. Holben, J. Schafer, L. F. Cardoso, S. C. Wofsy and H. R. Rocha, The effects of biomass burning aerosols and clouds on the CO2 flux in Amazonia, *Tellus, Ser. B*, 2007, **59B**(3), 338–349.
19 C. E. Doughty, M. G. Flanner and M. L. Goulden, Effect of smoke on subcanopy shaded light, canopy temperature, and carbon dioxide uptake in an Amazon rainforest, *Global Biogeochem. Cycles*, 2010, **24**, GB3015.

20 M. O. Andreae, D. Rosenfeld, P. Artaxo, A. A. Costa, G. P. Frank, K. M. Longo and M. A. F. Silva-Dias, Smoking rain clouds over the Amazon, *Science*, 2004, **303**(5662), 1337–1342.

21 I. Koren, Y. J. Kaufman, L. A. Remer and J. V. Martins, Measurement of the effect of Amazon smoke on inhibition of cloud formation, *Science*, 2004, **303**, 1342–1345.

22 I. Koren, Y. J. Kaufman, D. Rosenfeld, L. A. Remer and Y. Rudich, Aerosol invigoration and restructuring of Atlantic convective clouds, *Geophys. Res. Lett.*, 2005, **32**, L14828.

23 I. Koren, J. V. Martins, L. A. Remer and H. Afargan, Smoke invigoration *versus* inhibition of clouds over the Amazon, *Science*, 2008, **321**, 946–949.

24 U. Pöschl, S. T. Martin, B. Sinha, Q. Chen, S. S. Gunthe, J. A. Huffman, S. Borrmann, D. K. Farmer, R. M. Garland, G. Helas, J. L. Jimenez, S. M. King, A. Manzi, E. Mikhailov, T. Pauliquevis, M. D. Petters, A. J. Prenni, P. Roldin, D. Rose, J. Schneider, H. Su, S. R. Zorn, P. Artaxo and M. O. Andreae, Rainforest aerosols as biogenic nuclei of clouds and precipitation in the Amazon, *Science*, 2010, **329**, 1513–1516.

25 A. J. Prenni, M. D. Petters, S. M. Kreidenweis, C. L. Heald, S. T. Martin, P. Artaxo, R. M. Garland, A. G. Wollny and U. Poeschl, Relative roles of biogenic emissions and Saharan dust as ice nuclei in the Amazon basin, *Nat. Geosci.*, 2009, **2**, 402–405.

26 P. J. Crutzen, A. C. Delany, J. P. Greenberg, P. Haagenson, L. Heidt, R. Lueb, W. Pollock, W. Seiler, A. F. Wartburg and P. R. Zimmerman, Tropospheric chemical composition measurements in Brazil during the dry season, *J. Atmos. Chem.*, 1985, **2**, 233–256.

27 R. C. Harriss, S. C. Wofsy, M. Garstang, E. V. Browell, L. C. B. Molion, R. J. McNeal, J. M. Hoell, R. J. Bendura, S. M. Beck, R. L. Navarro, J. T. Riley and R. L. Snell, The Amazon Boundary-Layer Experiment (ABLE-2A) - dry season 1985, *J. Geophys. Res.*, 1988, **93**, 1351–1360.

28 R. C. Harriss, M. Garstang, S. C. Wofsy, S. M. Beck, R. J. Bendura, J. R. B. Coelho, J. W. Drewry, J. M. Hoell, P. A. Matson, R. J. McNeal, L. C. B. Molion, R. L. Navarro, V. Rabine and R. L. Snell, The Amazon Boundary-Layer Experiment - Wet Season 1987, *J. Geophys. Res.*, 1990, **95**, 16721–16736.

29 M. O. Andreae, E. V. Browell, M. Garstang, G. L. Gregory, R. C. Harriss, G. F. Hill, D. J. Jacob, M. C. Pereira, G. W. Sachse, A. W. Setzer, P. L. S. Dias, R. W. Talbot, A. L. Torres and S. C. Wofsy, Biomass-burning emissions and associated haze layers over Amazonia, *J. Geophys. Res*, 1988, **93**, 1509–1527.

30 M. O. Andreae, S. S. de Almeida, P. Artaxo, C. Brandão, F. E. Carswell, P. Ciccioli, A. Culf, J. L. Esteves, J. Gash, J. Grace, P. Kabat, J. Lelieveld, Y. Malhi, A. O. Manzi, F. X. Meixner, A. Nobre, C. Nobre, M. A. L. Ruivo, M. A. Silva-Dias, P. Stefani, R. Valentini, J. von Jouanne and M. Waterloo, Biogeochemical cycling of carbon, water, energy, trace gases and aerosols in Amazonia: The LBA-EUSTACH experiments, *J. Geophys. Res.*, 2002, **107**(D20), 8066–8091.

31 P. Formenti, M. O. Andreae, L. Lange, G. Roberts, J. Cafmeyer, I. Rajta, W. Maenhaut, B. N. Holben, P. Artaxo and J. Lelieveld, Saharan dust in Brazil and Suriname during the Large-Scale Biosphere-Atmosphere Experiment in Amazonia (LBA) - Cooperative LBA Regional Experiment (CLAIRE) in March 1998, *J. Geophys. Res.*, 2001, **106**, 14919–14934.

32 S. T. Martin, M. O. Andreae, D. Althausen, P. Artaxo, H. Baars, S. Borrmann, Q. Chen, D. K. Farmer, A. Guenther, S. S. Gunthe, J. L. Jimenez, T. Karl, K. Longo, A. Manzi, T. Pauliquevis, M. D. Petters, A. J. Prenni, U. Pöschl, L. V. Rizzo, J. Schneider, J. N. Smith, E. Swietlicki, J. Tota, J. Wang, A. Wiedensohler and S. R. Zorn, An overview of the Amazonian Aerosol Characterization Experiment 2008 (AMAZE-08), *Atmos. Chem. Phys. Discuss.*, 2010b, **10**, 18139–18195.

33 P. J. Crutzen and M. O. Andreae, Biomass burning in the tropics: Impact on atmospheric chemistry and biogeochemical cycles, *Science*, 1990, **250**, 1669–1678.

34 D. Bowman, J. Balch, P. Artaxo, W. Bond, M. Cochrane, C. D'Antonio, R. DeFries, F. Johnston, J. Keeley, M. Krawchuk, C. Kull, M. Mack, M. Moritz, S. Pyne, C. Roos, A. Scott, N. Sodhi and T. Swetnam, The human dimension of fire regimes on Earth, *J. Biogeogr.*, 2011, **38**, 2223–2236.

35 M. O. Andreae, Global distribution of fires seen from space, *EOS, Trans. Am. Geophys. Union*, 1993, **74**(12), 129–144.

36 S. Fuzzi, S. Decesari, M. C. Facchini, F. Cavalli, L. Emblico, M. Mircea, M. O. Andreae, I. Trebs, A. Hoffer, P. Guyon, P. Artaxo, L. V. Rizzo, L. L. Lara, T. Pauliquevis, W. Maenhaut, N. Raes, X. G. Chi, O. L. Mayol-Bracero, L. L. Soto-Garcia, M. Claeys, I. Kourtchev, J. Rissler, E. Swietlicki, E. Tagliavini, G. Schkolnik, A. H. Falkovich, Y. Rudich, G. Fisch and L. V. Gatti, Overview of the inorganic and organic composition of size-segregated aerosol in Rondonia, Brazil, from the biomass-burning period to the onset of the wet season, *J. Geophys. Res.*, 2007, **112**, D01201.

37 M. O. Andreae, P. Artaxo, H. Fischer, S. R. Freitas, J. M. Gregoire, A. Hansel, P. Hoor, R. Kormann, R. Krejci, L. Lange, J. Lelieveld, W. Lindinger, K. Longo, W. Peters, M. de Reus, B. Scheeren, M. Dias, J. Strom, P. F. J. van Velthoven and J. Williams, Transport of biomass burning smoke to the upper troposphere by deep convection in the equatorial region, *Geophys. Res. Lett.*, 2001, **28**, 951–954.

38 P. Artaxo, H. Storms, F. Bruynseels, R. Van Grieken and W. Maenhaut, Composition and sources of aerosols from the Amazon Basin, *J. Geophys. Res.*, 1988, **93**, 1605–1615.

39 P. Artaxo, J. V. Martins, M. A. Yamasoe, A. S. Procópio, T. M. Pauliquevis, M. O. Andreae, P. Guyon, L. V. Gatti and A. M. C. Leal, Physical and chemical properties of aerosols in the wet and dry season in Rondônia, Amazonia, *J. Geophys. Res.*, 2002, **107**(D20), 8081–8095.

40 E. Ignotti, J. G. Valente, K. M. Longo, S. R. Freitas, S. S. Hacon and P. Artaxo, Impact on human health of particulate matter emitted from burnings in the Brazilian Amazon region, *Rev. Saúde Pública*, 2010, **44**(1), 121–130.

41 G. Feingold, H. L. Jiang and J. Y. Harrington, On smoke suppression of clouds in Amazonia, *Geophys. Res. Lett.*, 2005, **32**, L02804.

42 G. McFiggans, P. Artaxo, U. Baltensperger, H. Coe, C. Facchini, G. Feingold, S. Fuzzi, M. Gysel, A. Laaksonen, U. Lohmann, T. Mentel, D. Murphy, C. O'Dowd and J. Snider, The Effect of Physical & Chemical Aerosol Properties on Warm Cloud Droplet Activation, *Atmos Chem Phys*, 2006, **6**, 2593–2649.

43 Q. Chen, D. K. Farmer, J. Schneider, S. R. Zorn, C. L. Heald, T. G. Karl, A. Guenther, J. D. Allan, N. Robinson, H. Coe, J. R. Kimmel, T. Pauliquevis, S. Borrmann, U. Poschl, M. O. Andreae, P. Artaxo, J. L. Jimenez and S. T. Martin, Mass spectral characterization of submicron biogenic organic particles in the Amazon Basin, *Geophys. Res. Lett.*, 2009, **36**(1–5), L20806.

44 S. T. Martin, M. O. Andreae, P. Artaxo, D. Baumgardner, Qi Chen, A. H. Goldstein, A. B. Guenther, C. L. Heald, O. L. Mayol-Bracero, P. H. McMurry, T. Pauliquevis, U. Pöschl, K. A. Prather, G. C. Roberts, S. R. Saleska, M. A. Silva Dias, D. V. Spracklen, E. Swietlicki and I. Trebs, Sources and Properties of Amazonian Aerosol Particles, *Rev. Geophys.*, 2010, **48**, RG2002.

45 W. Elbert, P. E. Taylor, M. O. Andreae and U. Pöschl, Contribution of fungi to primary biogenic aerosols in the atmosphere: wet and dry discharged spores, carbohydrates, and inorganic ions, *Atmos. Chem. Phys.*, 2007, **7**, 4569–4588.

46 H. Baars, A. Ansmann, D. Althausen, R. Engelmann, P. Artaxo, T. Pauliquevis and R. Souza, Further evidence for significant smoke transport from Africa to Amazonia, *Geophys. Res. Lett.*, 2011, **38**, 1–6, DOI: 10.1029/2011GL049200.

47 A. Ansmann, H. Baars, M. Tesche, D. Muller, D. Althausen, R. Engelmann, T. Pauliquevis and P. Artaxo, Dust and smoke transport from Africa to South America: Lidar profiling over Cape Verde and the Amazon rainforest, *Geophys. Res. Lett.*, 2009, **36**, L11802.

48 C. Pöhlker, K. T. Wiedemann, B. Sinha, M. Shiraiwa, S. S. Gunthe, M. Smith, H. Su, P. Artaxo, Q. Chen, Y. Cheng, W. Elbert, M. K. Gilles, A. L. D. Kilcoyne, R. C. Moffet, M. Weigand, S. T. Martin, U. Pöschl and M. O. Andreae, Biogenic potassium salt particles as seeds for secondary organic aerosol in the Amazon, *Science*, 2012, **337**, 1075–1078.

49 P. Artaxo and H.-C. Hansson, Size distribution of biogenic aerosol particles from the Amazon basin, *Atmos. Environ.*, 1995, **29**(3), 393–402.

50 S. Ekström, B. Nozière, M. Hultberg, T. Alsberg, J. Magnér, E. D. Nilsson and P. Artaxo, A possible role of ground-based microorganisms on cloud formation in the atmosphere, *Biogeosciences*, 2010, **7**, 387–394.

51 M. Claeys, B. Graham, G. Vas, W. Wang, R. Vermeylen, V. Pashynska, J. Cafmeyer, P. Guyon, M. O. Andreae, P. Artaxo and W. Maenhaut, Formation of secondary organic aerosols through photooxidation of isoprene, *Science*, 2004, **303**, 1173–1176.

52 T. Karl, A. B. Guenther, A. Turnipseed, G. Tyndall, P. Artaxo and S. T. Martin, Rapid Formation of Isoprene Photo-oxidation Products Observed in Amazonia, *Atmos. Chem. Phys.*, 2009, **9**, 7753–7767.

53 T. Karl, A. Guenther, R. J. Yokelson, J. Greenberg, M. Potosnak, D. R. Blake and P. Artaxo, The tropical forest and fire emissions experiment: Emission, chemistry, and transport of biogenic volatile organic compounds in the lower atmosphere over Amazonia, *J. Geophys. Res.*, 2007, **112**, D18302.

54 U. Kuhn, M. O. Andreae, C. Ammann, A. C. Araujo, E. Brancaleoni, P. Ciccioli, T. Dindorf, M. Frattoni, L. V. Gatti, L. Ganzeveld, B. Kruijt, J. Lelieveld, J. Lloyd, F. X. Meixner, A. D. Nobre, U. Poschl, C. Spirig, P. Stefani, A. Thielmann, R. Valentini and J. Kesselmeier, Isoprene and monoterpene fluxes from Central Amazonian rainforest inferred from tower-based and airborne measurements, and implications

on the atmospheric chemistry and the local carbon budget, *Atmos. Chem. Phys.*, 2007, **7**, 2855–2879.

55 C. J. Ebben, I. S. Martinez, M. Shrestha, A. M. Buchbinder, A. L. Corrigan, A. Guenther, T. Karl, T. Petaja, W. W. Song, S. R. Zorn, P. Artaxo, M. Kulmala, S. T. Martin, L. M. Russell, J. Williams and F. M. Geiger, Contrasting organic aerosol particles from boreal and tropical forests during HUMPPA-COPEC-2010 and AMAZE-08 using coherent vibrational spectroscopy, *Atmos. Chem. Phys.*, 2011, **11**, 10327–10329.

56 C. J. Ebben, M. Shrestha, I. S. Martinez, A. L. Corrigan, A. A. Frossard, W. W. Song, D. R. Worton, T. Petäjä, J. Williams, L. M. Russell, M. Kulmala, A. H. Goldstein, P. Artaxo, S. T. Martin, R. J. Thomson and F. M. Geiger, Secondary Organic Aerosol Particles from Southern Finland, Amazonia, and California Studied by Coherent Vibrational Spectroscopy, *J. Phys. Chem. A*, 2012, **116**, 8271–8290.

57 P. Artaxo, E. T. Fernandes, J. V. Martins, M. A. Yamasoe, P. V. Hobbs, W. Maenhaut, K. M. Longo and A. Castanho, Large Scale Aerosol Source Apportionment in Amazonia, *J. Geophys. Res.*, 1998, **103**(D24), 31837–31848.

58 P. Artaxo, F. Gerab, M. A. Yamasoe and J. V. Martins, Fine Mode Aerosol Composition in Three Long Term Atmospheric Monitoring Sampling Stations in the Amazon Basin, *J. Geophys. Res.*, 1994, **99**(D11), 22.857–22.868.

59 J. A. Huffman, B. Sinha, R. M. Garland, A. Snee-Pollmann, S. S. Gunthe, P. Artaxo, S. T. Martin, M. O. Andreae and U. Pöschl, Size distributions and temporal variations of biological aerosol particles in the Amazon rainforest characterized by microscopy and real-time UV-APS fluorescence techniques during AMAZE-08, *Atmos. Chem. Phys.*, 2012, **12**, 11997–12019.

60 L. V. Rizzo, P. Artaxo, T. Muller, A. Wiedensohler, M. Paixão, G. G. Cirino, A. Arana, E. Swietlicki, P. Roldin, E. O. Fors, K. T. Wiedemann, L. S. M. Leal and M. Kulmala, Long term measurements of aerosol optical properties at a primary forest site in Amazonia, *Atmos. Chem. Phys.*, 2013, **13**, 1–23.

61 T. M. Tuch, A. Haudek, T. Muller, A. Nowak, H. Wex and A. Wiedensohler, Design and performance of an automatic regenerating adsorption aerosol dryer for continuous operation at monitoring sites, *Atmos. Meas. Tech.*, 2009, **2**, 417–422.

62 A. Wiedensohler, W. Birmili, A. Nowak, A. Sonntag, K. Weinhold, M. Merkel, B. Wehner, T. Tuch, S. Pfeifer, M. Fiebig, A. M. Fjäraa, E. Asmi, K. Sellegri, R. Depuy, H. Venzac, P. Villani, P. Laj, P. Aalto, J. A. Ogren, E. Swietlicki, P. Williams, P. Roldin, P. Quincey, C. Hüglin, R. Fierz-Schmidhauser, M. Gysel, E. Weingartner, F. Riccobono, S. Santos, C. Grüning, K. Faloon, D. Beddows, R. Harrison, C. Monahan, S. G. Jennings, C. D. O'Dowd, A. Marinoni, H.-G. Horn, L. Keck, J. Jiang, J. Scheckman, P. H. McMurry, Z. Deng, C. S. Zhao, M. Moerman, B. Henzing, G. de Leeuw, G. Löschau and S. Bastian, Mobility particle size spectrometers: harmonization of technical standards and data structure to facilitate high quality long-term observations of atmospheric particle number size distributions, *Atmos. Meas. Tech.*, 2012, **5**, 657–685.

63 N. L. Ng, S. C. Herndon, A. Trimborn, M. R. Canagaratna, P. L. Croteau, T. B. Onasch, D. Sueper, D. R. Worsnop, Q. Zhang, Y. L. Sun and J. T. Jayne, An Aerosol Chemical Speciation Monitor (ACSM) for Routine Monitoring of the Composition and Mass Concentrations of Ambient Aerosol, *Aerosol Sci. Technol.*, 2011, **45**, 780–794.

64 J. T. Jayne, D. C. Leard, X. Zhang, P. Davidovits, K. A. Smith, C. E. Kolb and D. R. Worsnop, Development of an Aerosol Mass Spectrometer for Size and Composition Analysis of Submicron Particles, *Aerosol Sci. Technol.*, 2000, **33**, 49–70.

65 M.-S. Bae, J. J. Schwab, Q. Zhang, O. Hogrefe, K. L. Demerjian, S. Weimer, K. Rhoads, D. Orsini, P. Venkatachari and P. K. Hopke, Interference of organic signals in highly time resolved nitrate measurements by low mass resolution aerosol mass spectrometry, *J. Geophys. Res.*, 2007, **112**, D22305.

66 S. K. Akagi, J. S. Craven, J. W. Taylor, G. R. McMeeking, R. J. Yokelson, I. R. Burling, S. P. Urbanski, C. E. Wold, J. H. Seinfeld, H. Coe, M. J. Alvarado and D. R. Weise, Evolution of trace gases and particles emitted by a chaparral fire in California, *Atmos. Chem. Phys.*, 2012, **12**, 1397–1421.

67 T. L. Anderson, D. S. Covert, S. F. Marshall, M. L. Laucks, R. J. Charlson, A. P. Waggoner, J. A. Ogren, R. Caldow, R. L. Holm, F. R. Quant, G. J. Sem, A. Wiedensohler, N. A. Ahlquist and T. S. Bates, Performance characteristics of a high-sensitivity, three-wavelength, total scatter/backscatter nephelometer, *Am. Meteorol. Soc.*, 1996, **13**, 967–986.

68 T. L. Anderson and J. A. Ogren, Determining Aerosol Radiative Properties Using the TSI 3563 Integrating Nephelometer, *Aerosol Sci. Technol.*, 1998, **29**, 57–69.

69 A. Petzold, H. Schloesser, P. Sheridan, W. P. Arnott, J. Ogren and A. Virkkula, Evaluation of Multiangle Absorption Photometry for Measuring Aerosol Light Absorption, *Aerosol Sci. Technol.*, 2005, **39**, 40–51.

70 S. L. Lewis, P. M. Brando, O. L. Phillips, G. M. F. van der Heijden and D. Nepstad, The 2010 Amazon Drought, *Science Brevia*, 2011, **331**(6017), 554, DOI: 10.1126/science.1200807.

71 M. A. Yamasoe, P. Artaxo, A. H. Miguel and A. G. Allen, Chemical composition of aerosol particles from direct emissions of biomass burning in the Amazon Basin: Water-soluble species and trace elements, *Atmos. Environ.*, 2000, **34**, 1641–1652.

72 L. V. Rizzo, P. Artaxo, T. Karl, A. B. Guenther and J. P. Greenberg, Aerosol properties, in-canopy gradients, turbulent fluxes and VOC concentrations at a pristine forest site in Amazonia, *Atmos. Environ.*, 2010, **44**(4), 503–511.

73 M. Kulmala, *et al.*, General overview: European Integrated project on Aerosol Cloud Climate and Air Quality interactions (EUCAARI) – integrating aerosol research from nano to global scales, *Atmos. Chem. Phys.*, 2011, **11**, 13061–13143.

74 G. Capes, B. Johnson, G. McFiggans, P. I. Williams, J. Haywood and H. Coe, Aging of biomass burning aerosols over West Africa: Aircraft measurements of chemical composition, microphysical properties, and emission ratios, *J. Geophys. Res.*, 2008, **113**, D00C15; M. Kulmala, H. Vehkamaki, T. Petaja, M. Dal Maso, A. Lauria, V.-M. Kerminen, W. Birmili and P. H. McMurry, Formation and growth rates of ultrafineatmospheric particles: a review of observations, *J. Aerosol Sci.*, 2004, **35**, 143–176.

75 M. J. Cubison, A. M. Ortega, P. L. Hayes, D. K. Farmer, D. Day, M. J. Lechner, W. H. Brune, E. Apel, G. S. Diskin, J. A. Fisher, H. E. Fuelberg, A. Hecobian, D. J. Knapp, T. Mikoviny, D. Riemer, G. W. Sachse, W. Sessions, R. J. Weber, A. J. Weinheimer, A. Wisthaler and J. L. Jimenez, Effects of aging on organic aerosol from open biomass burning smoke in aircraft and laboratory studies, *Atmos. Chem. Phys.*, 2011, **11**, 12049–12064.

76 K. A. Lewis, W. P. Arnott, H. Moosmüller, R. K. Chakrabarty, C. M. Carrico, S. M. Kreidenweis, D. E. Day, W. C. Malm, A. Laskin, J. L. Jimenez, I. M. Ulbrich, J. A. Huffman, T. B. Onasch, A. Trimborn, L. Liu and M. I. Mishchenko, Reduction in biomass burning aerosol light absorption upon humidification: roles of inorganically-induced hygroscopicity, particle collapse, and photoacoustic heat and mass transfer, *Atmos. Chem. Phys.*, 2009, **9**, 8949–8966.

77 Y. Kondo, H. Matsui, N. Moteki, L. Sahu, N. Takegawa, M. Kajino, Y. Zhao, M. J. Cubison, J. L. Jimenez, S. Vay, G. S. Diskin, B. Anderson, A. Wisthaler, T. Mikoviny, H. E. Fuelberg, D. R. Blake, G. Huey, A. J. Weinheimer, D. J. Knapp and W. H. Brune, Emissions of black carbon, organic, and inorganic aerosols from biomass burning in North America and Asia in 2008, *J. Geophys. Res.*, 2011, **116**, D08204.

78 P. Forster, V. Ramaswamy, P. Artaxo, T. Berntsen, R. Betts, D. W. Fahey, J. Haywood, J. Lean, D. C. Lowe, G. Myhre, J. Nganga, R. Prinn, G. Raga, M. Schulz and R. Van Dorland, 2007: Changes in Atmospheric Constituents and Radiative Forcing, in: *Climate Change 2007: The Physical Science Basis*, Contribution of Working Group I to the Fourth Assessment Report of the Intergovernmental Panel on Climate Change [S. Solomon, D. Qin, M. Manning, Z. Chen, M. Marquis, K. B. Averyt, M. Tignor and H. L. Miller (eds.)], Cambridge University Press, Cambridge, UK and New York, NY, USA.

79 E. T. Sena, P. Artaxo and A. L. Correia, Spatial variability of the direct radiative forcing of biomass burning aerosols and the effects of land use change in Amazonia, *Atmos. Chem. Phys.*, 2013, **13**, 1261–1275.

80 M. O. Andreae and A. Gelencser, Black carbon or brown carbon? The nature of light-absorbing carbonaceous aerosols, *Atmos. Chem. Phys.*, 2006, **6**, 3131–3148.

81 L. V. Rizzo, A. L. Correia, P. Artaxo, A. S. Procópio and M. O. Andreae, Spectral dependence of aerosol light absorption over the Amazon Basin, *Atmos. Chem. Phys.*, 2011, **11**, 8899–8912.

82 A. S. Procópio, L. A. Remer, P. Artaxo, Y. J. Kaufman and B. N. Holben, Modeled spectral optical properties for smoke aerosols in Amazonia, *Geophys. Res. Lett.*, 2003, **30**, 2265–2270.

83 A. S. Procópio, P. Artaxo, Y. J. Kaufman, L. A. Remer, J. S. Schafer and B. N. Holben, Multiyear analysis of Amazonian biomass burning smoke radiative forcing of climate, *Geophys. Res. Lett.*, 2004, **31**, L03108–L03112.

84 I. Koren, O. Altaratz, L. A. Remer, G. Feingold, J. V. Martins and R. H. Heiblum, Aerosol-induced intensification of rain from the tropics to mid-latitudes, *Nat. Geosci.*, 2012, **5**, 118–122.

85 U. Rummel, C. Ammann, G. A. Kirkman, M. A. L. Moura, T. Foken, M. O. Andreae and F. X. Meixner, Seasonal variation of ozone deposition to a tropical rainforest in southwest Amazonia, *Atmos. Chem. Phys.*, 2007, **7**, 5415–5435.

86 V. W. J. H. Kirchhoff, Surface ozone measurements in Amazonia, *J. Geophys. Res.*, 1988, **93**, 1469–1476.

87 M. A. L. Moura, F. X. Meixner, I. Trebs, L. C. B. Molion and M. F. do Nascimento, Filho, Medições de NO–NO2–O3 na Amazônia Central durante o experimento LBA/CLAIRE-2001, *Rev. Bras. Meteorol.*, 2004, **19**, 49–58.

88 M. O. Andreae, P. Artaxo, V. Beck, M. Bela, C. Gerbig, K. Longo, J. W. Munger, K. T. Wiedemann and S. C. Wofsy, Carbon monoxide and related trace gases and aerosols over the Amazon Basin during the wet and dry seasons, *Atmos. Chem. Phys.*, 2012, **12**, 6041–6065.

89 L. M. Mercado, N. Bellouin, S. Sitch, O. Boucher, C. Huntingford, M. Wild and P. M. Cox, Impact of changes in diffuse radiation on the global land carbon sink, *Nature*, 2009, **458**, 1014–1018.

90 A. C. Araujo, A. J. Dolman, M. J. Waterloo, J. H. C. Gash, B. Kruijt, F. B. Zanchi, J. M. E. De Lange, R. Stoevelaar, A. O. Manzi, A. D. Nobre, R. N. Lootens and J. Backer, The spatial variability of CO2 storage and the interpretation of eddy covariance fluxes in central Amazonia, *Agric. Forest Meteorol.*, 2010, **150**, 226–237.

91 C. Von Randow, A. O. Manzi, B. Kruijt, P. J. Oliveira, F. B. Zanchi, R. L. Silva, M. G. Hodnett, J. H. C. Gash, J. A. Elbers, M. J. Waterloo, F. L. Cardoso and P. Kabat, Comparative measurements and seasonal variations in energy and carbon exchange over forest and pasture in Southwest Amazonia, *Theor. Appl. Climatol.*, 2004, **78**, 5–26.

92 J. Haywood, Estimates of the direct and indirect radiative forcing due to tropospheric aerosols: A review, *Rev. Geophys.*, 2000, **38**, 513–543.

93 D. Rosenfeld and I. M. Lensky, Satellite-based insights into precipitation formation processes in continental and maritime convective clouds, *Bull. Am. Meteorol. Soc.*, 1998, **79**, 2457–2476.

94 D. Rosenfeld, Suppression of rain and snow by urban and industrial air pollution, *Science*, 2000, **287**, 1793–1796.

95 D. Rosenfeld, U. Lohmann, G. B. Raga, C. D. O'Dowd, M. Kulmala, S. Fuzzi, A. Reissell and M. O. Andreae, Flood or drought: How do aerosols affect precipitation?, *Science*, 2008, **321**, 1309–1313.

96 S. Twomey, Pollution and the planetary albedo, *Atmos. Environ.*, 1974, **8**, 1251–1256.

97 J. Huang, C. Zhang and J. M. Prospero, Large-scale effect of aerosols on precipitation in the West African Monsoon region, *Q. J. R. Meteorol. Soc.*, 2009, **135**, 581–594.

98 J. H. Iang, H. Su, M. R. Schoeberl, S. T. Massie, P. Colarco, S. Platnick and N. J. Livesey, Clean and polluted clouds: Relationships among pollution, ice clouds, and precipitation in South America, *Geophys. Res. Lett.*, 2008, **35**, L14804.

99 J. C. Lin, T. Matsui, R. A. Pielke Sr and C. Kummerow, Effects of biomass-burning-derived aerosols on precipitation and clouds in the Amazon Basin: A satellite-based empirical study, *J. Geophys. Res.*, 2006, **111**, D19204.

100 J. A. Martins, M. A. F. Silva Dias and F. L. T. Gonçalves, Impact of biomass burning aerosols on precipitation in the Amazon: A modelling case study, *J. Geophys. Res.*, 2009, **114**, D02207.

101 S. A. Christopher, X. Li, R. M. Welch, J. S. Reid, P. V. Hobbs, T. F. Eck and B. Holben, Estimation of surface and top-of-atmosphere shortwave irradiance in biomass-burning regions during SCAR-B, *J. Appl. Meteorol.*, 2000, **39**, 1742–1753.

102 F. Patadia, P. Gupta, S. A. Christopher and J. S. Reid, A multisensor satellite-based assessment of biomass burning aerosol radiative impact over Amazonia, *J. Geophys. Res.*, 2008, **113**, D12214.

103 J. E. Ten Hoeve, L. A. Remer and M. Z. Jacobson, Microphysical and radiative effects of aerosols on warm clouds during the Amazon biomass burning season as observed by MODIS: impacts of water vapour and land cover, *Atmos. Chem. Phys.*, 2011, **11**, 3021–3036.

104 U. Kuhn, L. Ganzeveld, A. Thielmann, T. Dindorf, G. Schebeske, M. Welling, J. Sciare, G. Roberts, F. X. Meixner, J. Kesselmeier, J. Lelieveld, O. Kolle, P. Ciccioli, J. Lloyd, J. Trentmann, P. Artaxo and M. O. Andreae, Impact of Manaus City on the Amazon Green Ocean atmosphere: ozone production, precursor sensitivity and aerosol load, *Atmos. Chem. Phys.*, 2010, **10**, 9251–9282.

Faraday Discussions

RSCPublishing

PAPER

Sulfate radical-initiated formation of isoprene-derived organosulfates in atmospheric aerosols

J. Schindelka, Y. Iinuma, D. Hoffmann and H. Herrmann*

Received 19th March 2013, Accepted 16th April 2013

DOI: 10.1039/c3fd00042g

Recent studies show that isoprene-derived organosulfates are an important fraction of ambient secondary organic aerosol (SOA), adding up to 20% to the organic mass. Organosulfates with m/z of 199 and 183 relating to C_4 compounds are found in ambient and laboratory generated SOA and a sulfate radical induced oxidation of methacrolein (MACR) and methyl vinyl ketone (MVK) has been shown to be a possible formation mechanism. In the present study, experiments on the sulfate radical-induced oxidation of methacrolein and methyl vinyl ketone were performed in bulk aqueous phase, as well as in an aerosol chamber, and finally compared with ambient PM_{10} samples collected at a rural East German village during the summer 2008, to investigate their relevance in aqueous phase SOA formation. Samples from aqueous phase experiments and extracts from filters were analysed with UPLC/(−)ESI-IMS-QTOFMS. All the samples showed the abundance of highly oxidised organosulfates with m/z 153, 155, 167, 183 and 199 corresponding to the species found in ambient particle samples. In the bulk phase studies with laser-induced sulfate radical formation, the signal intensities increased with increasing number of laser pulses, indicating the sulfate radical-induced formation of these organosulfates. Additionally, the chamber experiments showed a particle mass growth of about 10 μg m^{-3} and 4 μg m^{-3} for experiments on the reactive uptake of MACR and MVK with a sulfate radical precursor ($K_2S_2O_8$) in the seed particles. Correlations of the C_2 to C_5 organosulfate species (including the m/z 215, $C_5H_{11}O_7S^-$), detected in the ambient samples were found to be very strong ($r > 0.8$), indicating that these compounds are formed from similar mechanisms and under equal environmental conditions. This study shows that sulfate radical-induced oxidation in the aqueous particle phase provides a reasonable explanation for the formation of these organosulfates from methacrolein and methyl vinyl ketone.

1 Introduction

In the past decade, evidence from laboratory and field studies has shown that isoprene is an important precursor compound for the formation of secondary

Leibniz-Institute for Tropospheric Research, Permoserstr. 15, D-04318 Leipzig, Germany. E-mail: hartmut. herrmann@tropos.de organic aerosol (SOA) mass, and isoprene-derived compounds were found to add substantial mass to the ambient organic aerosol.[1-4] A large amount of isoprene is emitted globally (660 TgC y^{-1}), mainly from regions with broad-leaf trees, such as tropical rain forests.[5] Due to its two double bonds isoprene is highly reactive towards atmospheric oxidants like OH, NO_3 radicals and, to a lesser extent, ozone and is rapidly degraded to form semi-volatile organic compounds (SVOC). These SVOCs can then partition into the particle phase and/or react further to form low volatile organic compounds leading to SOA formation.

Recent field studies on ambient aerosols sampled in different ground-based studies in the US,[6-8] the Southeast Pacific,[9] Hungary,[10-12] Germany,[13] and rain and fog water[14,15] showed organosulfates as an important fraction of ambient secondary organic aerosol, and estimated a contribution of up to 30% to the organic mass.[7,11,16] An isoprene derived organosulfate with m/z 215 ($C_5H_{11}O_7S^-$) is often reported to be the most abundant organosulfate compound in ambient aerosols[6-8,10,11,14,17] and was found in up to 80% of the particles.[17] Froyd et $al.$[17] have shown that the amount of the isoprene-derived organosulfates correlates well with the particulate acidic sulfate that was mainly found in the free troposphere over tropical regions, and up to 20% of the organic mass of particles sampled in altitudes of 5–7 km are estimated to originate from isoprene epoxydiol (IEPOX)-originating organosulfates.

The formation of the isoprene-derived organosulfate ($C_5H_{11}O_7S^-$) with m/z 215 is currently explained by the acid-catalysed sulfation of IEPOX that is suggested to form from the OH radical induced oxidation of isoprene under NO_x-free conditions.[18,19] It is noted that there is an ongoing discussion about the exact mechanisms of the gas phase isoprene oxidation for low NO_x mixing ratios.[18-23] IEPOX can partition into the particle-phase via reactive uptake and reacts in the presence of acidic sulfate particles to form 2-methyltetrols (2-methylthreitol and 2-methylerythritol, $C_5H_{12}O_4$) and their organosulfates ($C_5H_{12}O_7S$).[19,24-27] An alternative formation pathway is the nucleophilic substitution of tertiary organonitrates.[26,27] This pathway is likely important in high-NO_x regimes in which the oxidation of isoprene with OH radicals partly results in the formation of organonitrates that subsequently partition into the particle phase. Darer et $al.$[26] have reported tertiary organonitrates to be thermodynamically unstable and the nitrate group exchanges rapidly with sulfate or water to form tertiary organosulfates or alcohols. The esterification of alcohols by acidic sulfate has been suggested as a major organosulfate formation pathway in the past but it was found to be insignificant because the equilibrium constants of these reactions are too small at the typical acidity of ambient SOA.[28,29]

Apart from the IEPOX-originating organosulfates ($C_5H_{11}O_7S^-$), smaller C_2 to C_4 organosulfates have been reported from ambient aerosol samples as well as laboratory isoprene oxidation studies.[6-8,10-12,14,17,19,30] From their high-resolution mass spectrometric data, the compounds with m/z 139, 153, 155 and 169 were attributed to the organosulfates of glycolaldehyde, hydroxyacetone, glycolic acid and lactic acid.[6,31,32] Furthermore, organosulfates with m/z of 199 and 183 originating from methylglyceric acid and methacrolein or methyl vinyl ketone have been found in ambient aerosol samples.[10,12,17,30]

Recently, organosulfate formation under irradiation has been reported from an aerosol chamber study about glyoxal uptake onto ammonium sulfate seed particles, indicating that a sulfate radical may provide an additional organosulfate formation pathway.[31] Natural sources of sulfate radicals in the atmospheric aqueous-phase are the oxidation of S(IV)-species in the presence of transition metals, such as manganese, or oxidants like hydrogen peroxide.[33–37] Another source is the reaction of sulfate ions with OH radicals.[35,38] Indeed it has been shown in the study by Perri *et al.*[39] that organosulfates are formed in an aqueous solution of glycolaldehyde and sulfuric acid in the presence of OH radicals. At the same time Nozière *et al.*[40] have performed the irradiation of solutions containing isoprene or its oxidation products (methacrolein and methyl vinyl ketone) and sulfate radical precursors (ammonium sulfate, sodium sulfate or potassium peroxodisulfate), and provided evidence that sulfate radicals indeed play a role in the formation of isoprene originating organosulfates. Estimated time constants for the formation through the radical pathway are several hours, indicating that the sulfate radical reactions, especially with unsaturated compounds, can reasonably explain the formation of these smaller C_2 to C_4 organosulfates.[40]

In the present study, the sulfate radical-induced oxidation of methyl vinyl ketone (MVK) and methacrolein (MACR) in the bulk aqueous-phase and in an aerosol chamber was investigated to elucidate the mechanisms leading to C_2–C_4 organosulfates that are also found in ambient aerosol samples. The samples from bulk aqueous phase reactions, aerosol chamber experiments and field studies in Germany were compared using Ultra Performance Liquid Chromatography coupled to Electrospray Ionisation Time-Of-Flight Mass Spectrometry (UPLC/ESI-TOFMS) to obtain insights into the mechanisms leading to the formation of the C_2 to C_4 organosulfates. Finally, the potential importance of this mechanism to SOA formation is discussed based on the results obtained from this study.

2   Experimental

Laser-induced oxidation experiments in the bulk aqueous phase

An aqueous solution of 5 mM methacrolein (MACR, 95%, Sigma-Aldrich, St. Louis, MI, USA) or methyl vinyl ketone (MVK, 99%, Sigma-Aldrich, St. Louis, MI, USA) with 30 mM of potassium peroxodisulfate (Fluka, puriss., ≥99.0%), either acidified with sulfuric acid (Fluka, puriss., 98%) or not (see Table 1) was irradiated with laser pulses from an excimer laser. With a gas mixture of KrF, the laser produces light at a wavelength of 248 nm. A quartz cuvette with a volume of 3 mL was used as a reaction vessel. The number of pulses used was 5, 10 and 20 with energies of 190 mJ per laser pulse.

Chamber experiments

Six sets of chamber experiments (see Table 2) were performed on the formation of organosulfates from MACR and MVK in the indoor aerosol chamber LEAK. A brief

Table 1   Experimental conditions of the bulk phase studies

Reactant	Conc./mM	Oxidant	Conc./mM	pH	Number of laser pulses
MACR	5	$K_2S_2O_8$	30	3.8	0, 5, 10, 20
		$K_2S_2O_8/H_2SO_4$	30/10	1.2	0, 5, 10, 20
MVK	5	$K_2S_2O_8$	30	3.8	0, 5, 10, 20
		$K_2S_2O_8/H_2SO_4$	30/10	1.2	0, 5, 10, 20

Table 2 Experimental conditions of the chamber experiments

VOC	Mixing ratio/ppb	Seed particle	Conc. of atomising solutions/mM	Light	Initial seed particle volume $\times 10^{-6}$/cm^3 m^{-3}
MACR	1000	$K_2S_2O_8$/ H_2SO_4	30/50	UV/Vis	32.7
		H_2SO_4	50	UV/Vis	15.4
		H_2SO_4	50	dark	15.5
MVK	1000	$K_2S_2O_8$/ H_2SO_4	30/50	UV/Vis	29.7
		H_2SO_4	50	UV/Vis	15.8
		H_2SO_4	50	dark	15.8

description of the chamber setup can be found elsewhere.[41] Before starting the experiment the chamber was flushed for 18 h with clean dry synthetic air at a flow rate of 200 L min^{-1} and thermostated at 19 °C. To achieve the relative humidity (RH) used for the experiments the airflow was humidified in the morning prior to the experiment until it reached 75% RH. Temperature as well as humidity was recorded using a combined sensor connected to a PC (Vaisala HMP 238, Helsinki, Finland). Seed particles were introduced by nebulising a solution containing either only sulfuric acid or a mixture of potassium peroxodisulfate and sulfuric acid, as summarised in Table 1, and the particle size distributions were monitored using a scanning mobility particle sizer (SMPS) system (8–850 nm, IFT-SMPS, Leipzig, Germany). From an injection port of the LEAK chamber, MACR or MVK were injected using a syringe at 200 L min^{-1} of air for 2 min. A proton-transfer-reaction mass spectrometer (IONICON, Innsbruck, Austria) was used for monitoring the mixing ratios of MACR and MVK. Four Hg lamps with a wavelength of 254 nm as well as 16 UV lamps (100 W Eversun Super, Phillips, Netherlands) were used to initiate the sulfate radical production in the particles. After a reaction time of 2 h particles were sampled on a borosilicate glass fibre filter coated with fluorocarbon (47 mm diameter, PALLFLEX T60A20, Pall, NY, USA) with a total sampling volume of 1.8 m^3. To avoid artefacts from the absorption of gas phase products an annular denuder (5 channel, 400 mm length, URG, Chapel Hill, NC, USA) coated with XAD-4 resin (Sigma Aldrich) was used in front of the filter holder. One half of the chamber sample filter was cut into small pieces using ceramic scissors and inserted into a syringe ($V = 1$ mL). The SOA products were extracted with a solution of 250 µL methanol (LC-MS grade) from the filter in the syringe and filtered through a Teflon syringe filter (0.2 µm pore size, Acrodisc, Pall, NY, USA) to remove the insoluble material.

Ambient samples

Twenty 24 h ambient summertime Hi-Vol PM$_{10}$ filter samples were collected between 15th August and 4th September 2008 in the rural village of Seiffen that lies near the border of Germany and the Czech Republic (50°38′50″ N, 13°27′08″ E, 647m above sea level). The sampling site is subjected to emissions from surrounding coniferous trees and regional industries. A small portion of the sample filter (approximately 6% of the total filter area) was cut into small pieces and extracted with 1mL of methanol (LC-MS Chromasolv, 99.9%, Sigma-Aldrich, St. Louis, MI, USA). Insoluble materials in the extract were removed by a Teflon

Table 3 Elemental composition with exact masses, number of isomers and suggested structures for detected organosulfates from bulk aqueous phase, chamber and field studies

Reactant	m/z [M − H]	Error [mDa]	iFit^a value	Formula	No. of isomers	Suggested structure	Bulk phase	Chamber	Ambient
MACR	152.9863	−2.2	0.1	$C_3H_5O_5S^-$	1		X	X	X
	167.0020	−0.1	0.0	$C_4H_7O_5S^-$	1		X	X	X
	180.9812	−0.5	3.3	$C_4H_5O_6S^-$	1		X	—	—
	182.9969	−1.1	0.5	$C_4H_7O_6S^-$	6		X	X	X
	198.9918	2.3	1.6	$C_4H_7O_7S^-$	1		X	X	X
	221.0125	2.1	5.0	$C_7H_9O_6S^-$	1		X	—	—
	251.0231			$C_8H_{11}O_7S^-$	1		X	X	—
	253.0387	4.0	0.1	$C_8H_{13}O_7S^-$	4		X	X	—

Table 3 (Contd.)

Reactant	m/z [M − H]	Error [mDa]	iFit[a] value	Formula	No. of isomers	Suggested structure	Bulk phase	Chamber	Ambient
	269.0337	4.8	0.7	$C_8H_{13}O_8S^-$	1		X	—	—
MVK	152.9863	2.1	1.3	$C_3H_5O_5S^-$	1		X	X	X
	154.9656	1.9	0.7	$C_2H_3O_6S^-$	1		X	X	X
	167.0020	−0.3	0.0	$C_4H_7O_5S^-$	2		X	X	X
	182.9969	−1.7	0.0	$C_4H_7O_6S^-$	1		X	—	X
	198.9918	−2.6	3.3	$C_4H_7O_7S^-$	1		X	X	X
	225.0074	0.5	0.0	$C_6H_9O_7S^-$	3		X	—	—
	235.0282	2.0	0.0	$C_8H_{11}O_6S^-$	5		X	—	—
	237.0438	2.1	0.0	$C_8H_{13}O_6S^-$	2		X	—	—
	251.0231	−0.3	3.3	$C_8H_{11}O_7S^-$	2		X	—	—

Table 3 (*Contd.*)

Reactant	m/z [M − H]	Error [mDa]	iFit[a] value	Formula	No. of isomers	Suggested structure	Bulk phase	Chamber	Ambient
	253.0387	0.3	0.0	$C_8H_{13}O_7S^-$	2	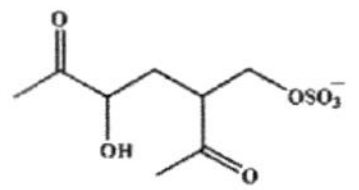	X	—	—
	321.0650	1.3	3.0	$C_{12}H_{17}O_8S^-$	1	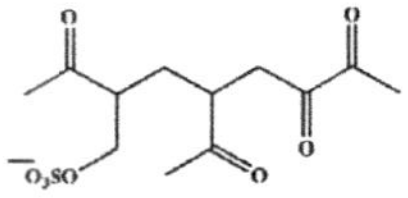	X	—	—

[a] iFit: The likelihood that the isotopic pattern of the elemental composition matches a cluster of peaks in the spectrum. An iFit value close to zero indicates a better match between the theoretical and measured data.

syringe filter (0.2 μm pore size, Acrodisc, Pall, NY, USA), and the resulting extract solution was dried under a gentle stream of nitrogen gas at 10 °C. The dry residue of the extract was recovered in a solution of 200 μL of methanol/water (50/50, v/v) for the UPLC/ESI-TOFMS analysis.

UPLC/(−)ESI-TOFMS (Ultra Performance Liquid Chromatography coupled to Electrospray Ionisation Time-of-Flight Mass Spectrometry)

Samples obtained from the experiments performed in the bulk aqueous solutions and the aerosol chamber were analysed using an Acquity UPLC system coupled to Synapt HD TOFMS (UPLC/(−)ESI-TOFMS, Waters, Milford, MA, USA). The analytes were separated on an Acquity UPLC HSS T3 column (2.1 × 100 mm, 1.8 μm) with the following eluent gradient programme: held 99% A (0.1% acetic acid in ultrapure water) and 1% B (methanol, UHPLC gradient grade, Fischer Scientific UK, Loughborough) constant for 1.25 min, increased B to 100% in 6.75 min, held constant for 2 min, returned to the initial eluent composition and re-equilibrated for 2 min. The column temperature was set to 40 °C and the flow rate of the eluent was 0.5 mL min^{-1}. The injection volumes for the bulk aqueous solution and aerosol chamber samples were 1 μL and 5 μL, respectively. Details about the operational conditions of Synapt HD TOFMS can be found elsewhere.[41]

3   Results and discussion

3.1   Organosulfate formation in bulk aqueous-phase

3.1.1   **Detection of organosulfates.** A number of different organosulfate species were found to form from the oxidation of MACR and MVK with sulfate

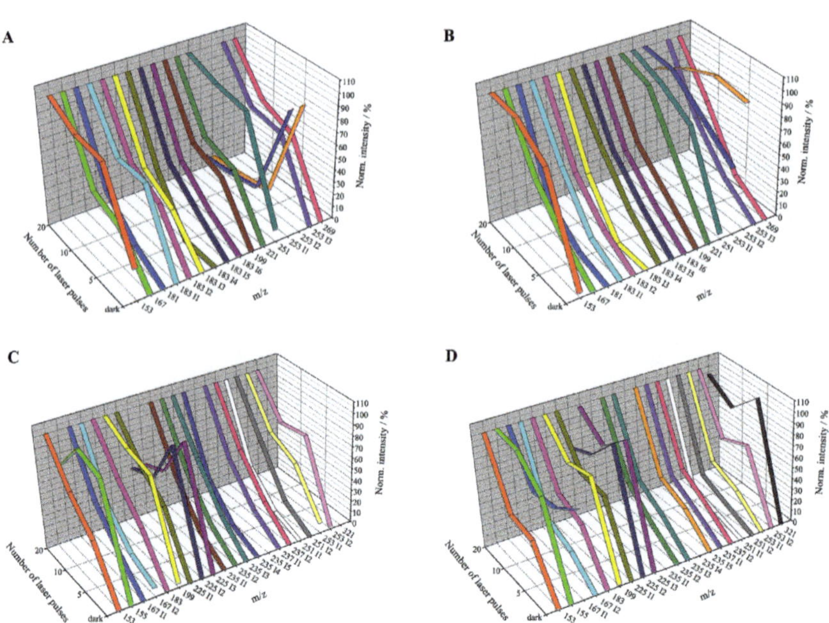

Fig. 1   Normalised signal intensities for the oxidation of MACR under non-acidified (A) and acidified (B) conditions, and for the oxidation of MVK under non-acidified (C) and acidified (D) conditions.

radicals. The high-resolution m/z values and elemental compositions determined from their TOFMS data as well as the number of isomers and suggested structures are summarised in Table 3.

The normalised intensities for all the detected compounds as a function of a number of laser pulses are depicted in Fig. 1. In general, the intensities for most of the signals increase with an increase in number of laser pulses, indicating that their formation involves a sulfate radical induced oxidation. Methacrolein and methyl vinyl ketone show both the formation of the hydroxyacetone sulfate ester (m/z 153), whereas the glycolic acid sulfate ester with m/z 155 was only detected in the oxidation of MVK, consistent with the results from Nozière et al.[40]

Besides these small C_2–C_3 compounds, several C_4 compounds with a different degree of oxidation were detected for both analytes. In particular the detection of species with m/z 183 and 199 is of interest as evidence for their existence was shown by studies of ambient aerosols.[6–8,10,12,17] Six isomers were found for the MACR originating organosulfates with m/z 183 (Fig. 1A and B), whereas just one isomer was detected for MVK (Fig.1C and D). Very recently Safi Shalamzari et al.[12] proposed that crotonaldehyde or MVK can be a source for these organosulfates with m/z 183. The obtained MS^2 data from a peak in the K-puszta aerosol was correlating with a synthesised standard of the sulfate ester of 3,4-dihydroxybutan-2-one with the carbon backbone of MVK.

Methacrolein and methyl vinyl ketone also form a product with m/z 199. Several peaks have been detected with m/z 199 in the K-puszta aerosol and one species is suggested to be the organosulfate of methylglyceric acid, an oxidation product of MACR.[10,12] Gómez-González et al.[10] suggested that the second, more intense peak originated from 2-hydroxy-1,4-butanedial due to its MS fragmentation. The results of the present study indicate that formation from the reaction of MVK and sulfate radicals is another source of a compound with this m/z value and could explain the second peak in the K-puszta aerosol samples because this organosulfate likely gives the same fragmentation pattern.

Comparing the reaction products of both analytes, methyl vinyl ketone produces more oligomeric compounds (Fig. 1C and D). Compounds with 6, 8 and 12 carbon atoms were found in this study and one with 16 carbon atoms was reported by Nozière et al.[40] The generation of dimeric C_8-structures is common for both analytes. Within the present study, three different species for MACR and

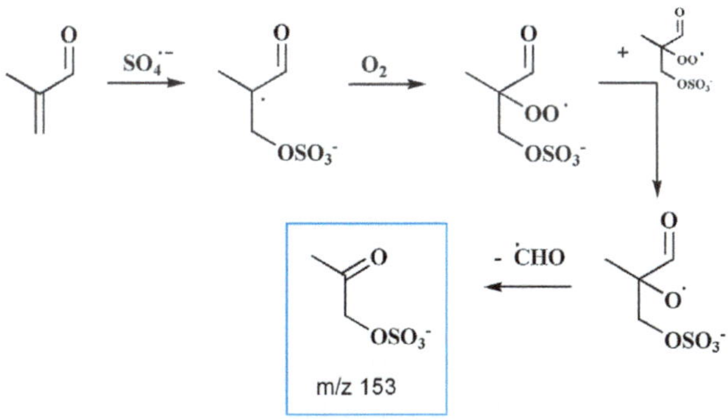

Fig. 2 A possible reaction mechanism for the formation of the hydroxyacetone sulfate ester m/z 153 from MACR.

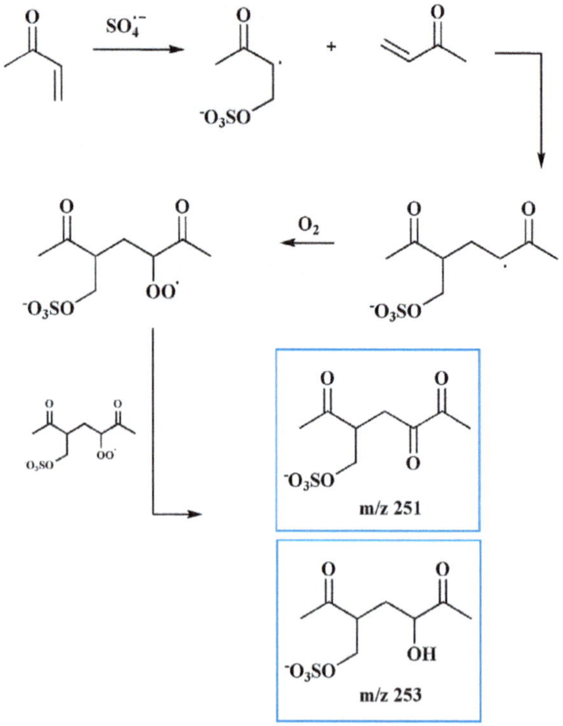

Fig. 3 A possible reaction mechanism for the formation of the dimeric organosulfates m/z 251 and 253 from MVK.

four for MVK have been detected. Some dimers from MACR and MVK show a decrease in intensity with increasing number of laser pulses. Two isomers with m/z 253 from MACR (Fig. 1A) show the highest intensities in the sample without irradiation, probably formed due to the thermal decomposition of peroxodisulfate.[42] The following decrease in intensity is either due to a photo-instability or a radical-induced degradation of these dimers.

Comparisons of experiments with acidified and non-acidified peroxodisulfate solutions show differences in the production of oligomers. Some products are only observed under either non-acidified or acidified conditions, such as the C_{12} compound with m/z 321 (Fig. 1D). Additionally, the compound with m/z 167 from MACR is preferentially formed under acidified conditions, as the signal intensity is considerably increased when sulfuric acid was added to the solution. This indicates that other acid-catalysed reaction mechanisms, such as hydration or polymerisation, are also involved under acidic conditions.

Overall, the results of the present product study agree well with the former study of Nozière *et al.*,[40] though there are also some small differences. In the present study, several smaller oxidation products of MACR were found, such as the hydroxyacetone originating organosulfate, organosulfates with m/z 167 and 181 and different species with m/z 183. Nozière *et al.*[40] only reported the formation of the compounds with m/z 199 for MACR and MVK. Additionally, more dimeric structures were detected from the oxidation of both analytes compared to the study of Nozière *et al.*[40]

3.1.2 Reaction mechanisms. Fig. 2 and 3 show suggested structures and reaction pathways for the formation of the compounds with m/z 153 from MACR and m/z 251 and 253 from MVK. The figures summarise known mechanisms for radical reactions in the aqueous phase leading to the formation of oligomeric structures from chain extending reactions and the production of peroxy radicals and their decay *via* unimolecular and bimolecular reactions.

A sulfate radical abstracts an H atom from an organic molecule or adds itself to a C–C double bond. In both cases a carbon-centred radical forms, and it undergoes different successive reactions depending on its structure. For example, the formed radical may react with another unsaturated compound to extend the carbon chain to form a dimeric carbon-centred radical. The other important reaction is the formation of peroxyl radicals from the addition of oxygen.

Von Sonntag and Schuchmann[43] compiled the reaction mechanisms of peroxyl radicals in the aqueous phase, where, in the case of α-hydroxy-peroxyradicals, the HO_2 elimination and the recombination between two RO_2 radicals followed by disproportionation are the most important reactions. If the unimolecular HO_2 elimination reaction is not possible or very slow, peroxyl radicals can recombine to form tetroxides as intermediate species.[44,45] The latter then decomposes in different pathways, producing compounds with carbonyl or alcohol functional groups or alkoxy radicals. Alkoxy radicals are unstable and undergo isomerisation reactions or decompose, producing smaller compounds and radicals by bond breakage.[46–48] However, not much is known about these processes in aqueous solution compared to alkoxy radical gas phase reactions.[49–51] The formation of alkoxy radicals is especially important in the decomposition of tetroxides formed from two tertiary peroxy radicals as, in this case, it is the only possible reaction mechanism due to the lack of α-H-atoms.[52]

The addition of the sulfate radical to the double bond of MACR (Fig. 2) or MVK (Fig. 3) should mainly take place at the terminal carbon atoms. These positions are favoured as they are less sterically hindered and lead to the production of a tertiary alkyl radical for MACR and a secondary one for MVK, which are more stable than primary alkyl radicals. Both isomers of the alkyl radicals can undergo different successive reactions, as mentioned above. For example, the reaction of the MACR originating tertiary alkyl radical with oxygen is shown in Fig. 2. This reaction results in the formation of a tertiary peroxyl radical. The latter reacts under recombination due to the lack of an α-hydroxy group for an HO_2 elimination. As described above, a recombination and further decomposition of two tertiary peroxyl radicals results in the formation of two tertiary alkoxy radicals. They can react further *via* elimination of a formyl radical whereby the hydroxyacetone sulfate ester (m/z 153) is formed.

Tetroxides that are formed from the secondary peroxyl radical of MVK can react in the same manner as the pathways described above to produce compounds with a carbonyl or alcohol functional group. This is shown in the last reaction step in the formation of dimers of MVK (Fig. 3). The first radical chain reaction step likely proceeds *via* addition of the secondary alkyl radical to the terminal carbon atom of another MVK molecule. A dimeric secondary alkyl radical is formed, which can react again with a MVK molecule producing a trimeric alkyl radical or form a secondary peroxyl radical with oxygen. As shown in Fig. 3, the reaction with oxygen and the following tetroxide formation and decomposition produces a compound with a carbonyl functional group and one with an alcohol functional group corresponding to the compounds with m/z 251 and 253.

The other organosulfate compounds from MACR and MVK are expected to form from similar reactions but cannot be completely explained with these mechanisms. Nozière *et al.*[40] considered the recombination of a sulfate alkyl radical with an OH radical as an additional mechanism and proposed therewith a slightly different mechanism for the formation of the compound m/z 253 from MACR or MVK. As shown in Fig. 3, we agree with the chain reaction of the alkyl radical and the double bond of another MVK molecule. Following the chain reaction Nozière *et al.*[40] suggested that the dimer with m/z 253 formed from a recombination of the dimeric alkyl radical with an OH radical. Assuming the background OH radical concentration to be very low ($\sim 10^{-14}$ M) and the mean concentration of oxygen to be $\sim 10^{-4}$ M, such a reaction cannot be favoured under these conditions, even when we suggest a rate constant of about 10^9 M^{-1} s^{-1} for the recombination and a rate constant as low as 10^6 M^{-1} s^{-1} for the reaction with oxygen.[53,54]

3.1.3 Kinetics of radical-induced organosulfate formation. Kinetics of sulfate radical reactions with organic species in the aqueous-phase are commonly determined by applying laser flash photolysis or pulse radiolysis. A less common method is the production of sulfate radicals by thermal decomposition, as it can only be used for studies on longer time scales. The peroxodisulfate anion is often used as a precursor because it decomposes to form two sulfate radicals. The peroxodisulfate anion has a similar absorption to that of hydrogen peroxide. Its extinction coefficient increases exponentially towards the wavelength of 200 nm and corresponds to the values of 20 and 27.5 M^{-1} cm^{-1} at 254 and 248 nm, respectively.[55,56] For the determination of the quantum yield of the peroxodisulfate photolysis the production of sulfate radical has to be observed directly from absorption or indirectly from reactions with scavengers such as chloride or bromide ions.[55] The UV/VIS spectrum of the sulfate radical shows a broad absorption band between 350–550 nm with a maximum extinction coefficient of 1544 M^{-1} cm^{-1} at 445 nm, which allows the direct measurement of the radicals.[54]

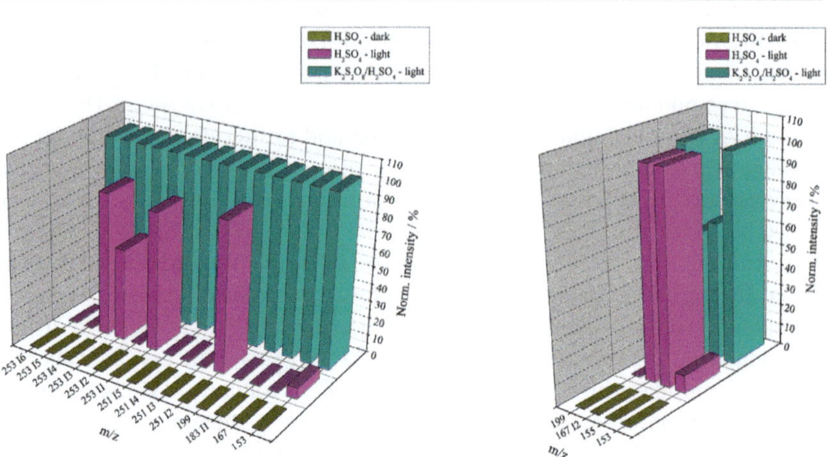

Fig. 4 Normalised intensities of observed compounds from the chamber experiments on MACR (left) and MVK (right); accounted isomers match the ones found in the bulk phase studies, whereas the isomers 2–5 and 4–6 for m/z 251 and 253 are newly detected ones from the oxidation of MACR.

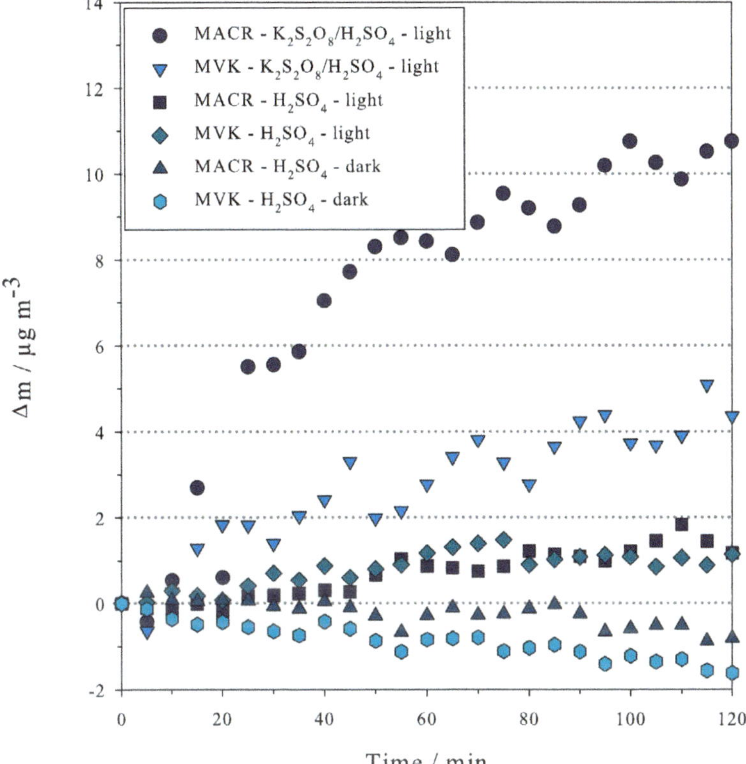

Fig. 5 Time series of particle mass increase (Δm) for all six chamber experiments.

A compilation of absorption spectra can be found in Herrmann *et al.*[54] Due to loss reactions of sulfate radicals, including self recombination and the reaction with water, the quantum yield of the sulfate radical from the peroxodisulfate anion was found to be 1.4 at $\lambda = 248$ nm.[55,56]

The sulfate radical is a strong oxidant and can therefore initiate electron transfer reactions. This fact is used in the generation of NO_3 radicals from nitrate ions.[57] The important reaction mechanisms with organic species are the H-abstraction and the addition to a C–C double bond. When the kinetics of reactions with hydroxyl and sulfate radicals are compared, they show different trends for these two mechanisms. The OH-induced H-abstraction from aliphatic alcohols occurs at a second order rate constant of $\sim 10^9$ M^{-1} s^{-1}, whereas the reaction of the same compounds with sulfate radicals is about two orders of magnitude slower.[54] Nevertheless, the difference in the reaction rates for the addition of the two radicals to a C–C double bond of an unsaturated compound can be less than one order of magnitude. Having a closer look at the reaction kinetics of methacrolein, the k_{2nd} of the reaction with OH[58–60] is 8–13×10^9 M^{-1} s^{-1} and 1.2×10^9 M^{-1} s^{-1} for the reaction with the sulfate radical.[58] This is similar for isoprene, where the k_{2nd} of the reaction with OH[60] is 1.2×10^{10} M^{-1} s^{-1} and 2.1×10^9 M^{-1} s^{-1} for the addition of a sulfate radical.[36] When the concentrations of hydroxyl and sulfate radicals are equal, the reaction of sulfate radicals with unsaturated compounds can be competitive with the reaction with hydroxyl radicals.

3.2   Chamber studies in SOA formation

The formation of organosulfate from MACR and MVK was investigated under three different reaction conditions in the TROPOS aerosol chamber. All experiments were performed at elevated relative humidity as the latest studies have shown that the increase in the relative humidity enhances the formation of organosulfates.[30,61] To promote the partitioning of MACR and MVK into the particle phase, an initial mixing ratio of 1 ppm was applied to produce a high partial pressure over the aqueous particle phase. Furthermore, sulfuric acid was added to the seed particles to increase an uptake of both OVOCs from possible acid-catalysed enolisation or hydration. Two sets of blank experiments were performed to determine whether the organosulfates are formed from a sulfation reaction under the dark conditions (*i.e.* OVOC/acidic sulfate seed particles/no UV) or whether background sulfate radical production significantly contributes to the organosulfate formation (*i.e.* OVOC/acidic sulfate seed particles/UV). The experiments with a mixture of potassium peroxodisulfate and sulfuric acid under irradiation were designed to simulate the aqueous-phase sulfate radical-induced formation of organosulfates. The PTR-MS data obtained from these experiments indicate that the loss of the OVOCs due to photolysis was less than 10% over the course of the experiment.

Fig. 4 and 5 show the normalised intensities of all detected compounds from the filter samples and the particle mass growth from the reactive uptake of MACR and MVK for the three different reaction conditions. The Δm values were estimated from the ΔV values obtained from the SMPS data with a density of 1 g cm^{-3}. Initial seed particle volumes for all the experiments are summarised in Table 2.

Both experiments performed under the dark conditions show no organic mass growth, consistent with the results of the filter analysis that showed no organosulfate species in these samples. This indicates that an acid-catalysed sulfation mechanism cannot be important under these conditions, in line with the estimations of Minerath *et al.*[28] In these experiments, only wall loss of particles reduced the overall mass of the particles dispersed.

However, under irradiation conditions, a slight increase in particle mass was observed for both MVK and MACR, resulting in an organic particle mass formation of about 1 µg m^{-3} during a chamber run of 2 h duration. This is consistent with the appearance of organosulfates in the filter samples from these experiments. Although the increases in SOA mass were similar, the SOA products were distinctively different when the experiments of MACR and MVK are compared. Both OVOCs produce the sulfate ester of hydroxyacetone (*m/z* 153) that was also found from the bulk phase studies. Dimeric compounds with *m/z* 251 and 253 were detected only from the oxidation of MACR. Methyl vinyl ketone in turn showed to produce the compound with *m/z* 167 in quite high amounts as well as the glycolic acid sulfate ester (*m/z* 155). The results from this set of experiments indicate that a small amount of sulfate radicals formed in the presence of acidic sulfate seed particles and UV light, most likely due to background OH radical production. The formation of organosulfates under similar conditions has also been reported from the experiments of Galloway *et al.*[31]

The addition of a sulfate radical source (potassium peroxodisulfate) to the seed solution led to an increase of the particle mass production and the formation of additional organosulfate species from both OVOCs. After 2 h of irradiation in the presence of acidified potassium peroxodisulfate seed particles, 10 µg m^{-3} of particle mass was produced from MACR, and 4 µg m^{-3} was produced from MVK. This is consistent with the spate of organosulfate species detected in the filter samples. In the MVK experiment, the compounds with m/z 153, 155, 167 and 199 were found at high intensities but no dimeric compounds were detected. In contrast, MACR led to the formation of a series of monomeric and dimeric species. Besides the compound with m/z 153, the compounds with m/z 167, 183 and 199 were detected in the MACR sample. In comparison to the results from the bulk phase studies, only the first isomer of the m/z 183 compounds was found from the chamber experiment. Furthermore, more dimeric compounds were

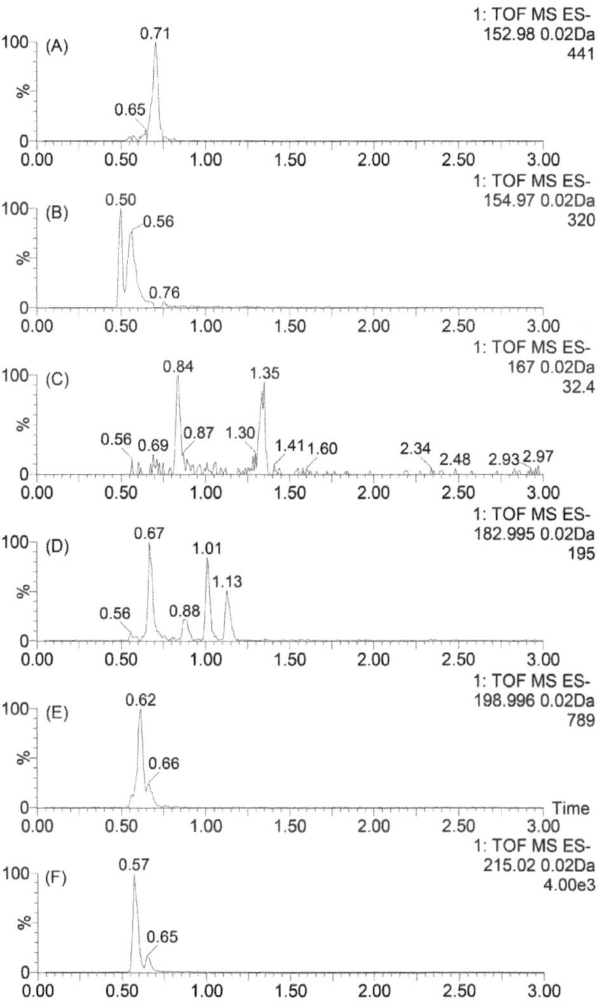

Fig. 6 Typical extracted ion chromatograms (EICs) for organosulfates of (A) hydroxyacetone (m/z 153, C$_3$H$_5$O$_5$S$^-$, 0.71 min), (B) glycolic acid (m/z 155, C$_2$H$_3$O$_6$S$^-$, 0.51 min), (C) methyl vinyl ketone (m/z 167, C$_4$H$_7$O$_5$S$^-$, 1.35 min), (D) 1,2-dihydroxy-3-butanone (m/z 183, C$_4$H$_7$O$_6$S$^-$, 1.13 min), (D) 2-methyl glyceric acid (m/z 199, C$_4$H$_7$O$_7$S$^-$, 0.62 min), and (F) 2-methyltetrols (m/z 215, C$_5$H$_{11}$O$_7$S$^-$) detected in the ambient aerosol sample collected in Seiffen, Germany during summer 2008.

found from the chamber sample. Although compounds with m/z 251 were detected in both the chamber and bulk phase samples, they were not the same compounds. For the m/z 253 compounds, an additional three isomers to the isomers 1–3 from bulk phase studies were found in the chamber study. From this higher number of products, especially of dimers, the difference in particle mass growth from MACR and MVK is comprehensible, as oligomeric species add more mass to the SOA in the MACR experiment.

Recent chamber studies on the SOA formation from the photo-oxidation of MACR suggest the reaction conditions play an important role in the SOA formation mechanisms. At low relative humidity the formation of 2-methyl-glyceric acid (2-MG) and its oligoesters significantly add to the organic mass but the formation of these compounds is inhibited with increasing relative humidity.[19,30,62] Nevertheless, particle growth was observed with increasing relative humidity as organosulfates originating from MACR increasingly add to particle mass.[30] Little is known about the SOA formation from MVK oxidation, though Kroll et al.[3] suggested that the MVK oxidation does not add significantly to the particle organic mass. However, bulk phase studies indicate that the formation of SOA from OH radical-induced oxidation of MACR and MVK can be important, due to the production of small organic acids and a possibly radical-induced oligomerisation.[63–66] For these aqueous phase studies, MACR and MVK were assumed to partition into the particle phase and react further, whereas the chamber studies mentioned before report gas phase oxidation products of the OVOCs contributing to SOA formation, e.g. MPAN. Reactive uptake of MACR and MVK is another uncertainty, which is discussed in the following.

From the Henry coefficients of 6.5 M atm^{-1} and 41 M atm^{-1} for MACR and MVK, respectively, they cannot be expected to contribute to SOA formation simply by gas-to-particle conversion or condensation to the pre-existing particles.[67] Although, high aqueous phase concentrations of up to 0.8 µmol L^{-1} and 3 µmol L^{-1} for MACR and MVK were reported from the analysis of cloud water sampled in a remote site, which are about two orders of magnitude higher than expected from their Henry coefficients.[68] As a reactive uptake is a complex process,[69] which is influenced by different particle properties, such as acidity,[70,71] ionic strength,[72] present organics[73] and bulk phase reactions, only a few studies are available about the uptake of MACR and MVK so far.[71,74] Nozière et al.[71] found an increase of the effective Henry coefficient of MVK of about two orders of magnitude with increasing fraction of sulfuric acid but in contrast, no acid dependency and reactivity for MACR. However, a recent study by Liu et al.[74] reports an increasing irreversible uptake of MACR with increasing fraction of sulfuric acid when hydrogen peroxide is present. The reaction of MACR with hydrogen peroxide is a rather slow process with a rate constant of about 0.13 M^{-1} s^{-1}.[75] It can be assumed that a reaction of radicals with MACR, which is faster than the reaction with hydrogen peroxide (cf. section 4), could increase the rate of the uptake under acidic conditions.

The series of chamber experiments in the present study were conducted to determine possible mechanisms of the sulfate radical-induced oxidation of MACR and MVK under more atmospherically relevant conditions. For this purpose, the gas-phase OVOCs were subjected to phase-transfer to react with the sulfate radical produced in the particle phase, in which the phase-transfer acts as one of the limiting factors for SOA formation, similar to the real environment. Indeed, the results of these experiments reveal different behaviours of MACR and MVK to

Table 4 Correlation coefficients (r) determined between C2–C5 organosulfates and PM10 SO_4^{2-} in the ambient samples ($n = 20$)

	m/z 153	m/z 155	m/z 167	m/z 183	m/z 199	m/z 215	PM10 SO_4^{2-}
m/z 153	1.00						
m/z 155	0.92	1.00					
m/z 167	0.88	0.84	1.00				
m/z 183	0.94	0.95	0.90	1.00			
m/z 199	0.93	0.96	0.83	0.97	1.00		
m/z 215	0.91	0.92	0.76	0.86	0.90	1.00	
PM10 SO_4^{2-}	0.70	0.44	0.81	0.66	0.42	0.53	1.00

partitioning, radical induced organosulfate formation and subsequent SOA formation.

As no particle mass was produced from the dark experiments one can infer that acid-catalysed oligomerisation reactions, such as aldol condensation and hydration/polymerisation, were unimportant under the experimental conditions used in the present study. The increase in particle growth from an increasing concentration of sulfate radicals in the seed particles in the experiments with irradiation indicates that the radical reaction was one of the determining factors for the organic mass production. Therefore, the higher SOA formation of MACR likely originates from a more pronounced partitioning of MACR than MVK, and its higher reactivity towards sulfate radicals. Although no rate constant is reported for the reaction of MVK with a sulfate radical we can assume that it reacts with a sulfate radical in a similar manner to an OH radical, in which the reaction of an OH radical with MACR is faster than that with MVK.

The differences in dimer formation from both OVOCs, *i.e.* the fact that only MACR did form dimers, confirm the assumption that MACR is taken up in higher amounts than MVK. Contrary to MVK, the amount of unreacted MACR was high enough that it could have acted as a reaction partner for produced sulfate alkyl radicals. Thus, this chain reaction (*cf.* section 3.1.2) was competitive with the reaction with oxygen. Comparing the results of the bulk phase and chamber studies, dimer formation seems to be not favoured when aqueous phase concentration of the OVOC becomes too low, as seems to be the case with MVK in these chamber experiments, and can therefore be of minor importance in the ambient environment. Indeed, no signals for compounds with m/z 251 and 253 related to isoprene-derived organosulfates were found in ambient aerosol samples so far. This seems reasonable as reported cloud droplet concentrations of MACR and MVK were in the range of 10^{-6} M.[68] The reaction of generated sulfate alkyl radicals with the unreacted OVOC cannot be competitive as long as the concentration of oxygen[53] is in the range of 10^{-4} M or OVOC concentrations are not increased. More likely, the smaller C_2–C_4 organosulfates are formed from the described peroxy radical chemistry.

Besides the production of dimers, glycolic acid- and hydroxyacetone-derived organosulfates (m/z 155 and 153) were formed in the present study. This proves that these smaller organosulfates are not only produced by the reactive uptake of C_2 and C_3 compounds, *e.g.* the production of the m/z 155 from the reactive uptake of glyoxal under irradiation,[31] but the oxidation of C_4 compounds can be another source for their formation in atmospheric particles.

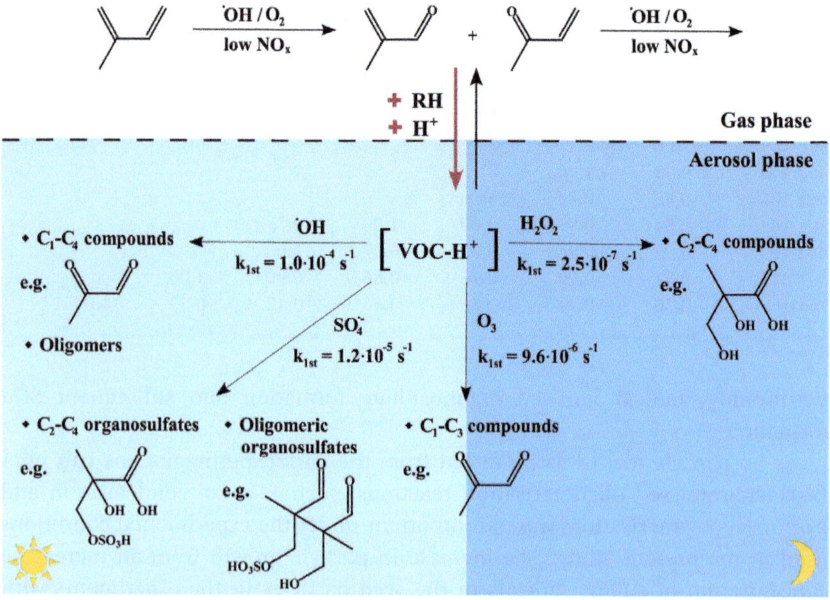

Fig. 7 Scheme of known reactions and products of MACR and MVK in the aqueous particle phase; first-order rate constants belong to the reactions of MACR with the oxidants.

3.3 Atmospheric evidence

A series of C_2 to C_4 organosulfates that are observed in the bulk-phase and chamber samples are detected in the ambient PM_{10} samples collected at a rural East German village during the summer 2008. Fig. 6A–E shows typical extracted ion chromatograms (EICs) obtained from the analysis of an ambient aerosol sample collected at the sampling site. They are attributed to the organosulfates of hydroxyacetone (m/z 153, $C_3H_5O_5S^-$, 0.71 min), glycolic acid (m/z 155, $C_2H_3O_6S^-$, 0.51 min), methyl vinyl ketone (m/z 167, $C_4H_7O_5S^-$, 1.35 min), 1,2-dihydroxy-3-butanone (m/z 183, $C_4H_7O_6S^-$, 1.13 min) and 2-methyl glyceric acid (m/z 199, $C_4H_7O_7S^-$, 0.62 min) based on the results obtained from the bulk phase and aerosol chamber experiments and data presented in the literature.[6,10,12,30,32] Atmospheric evidence for the existence of these compounds in aerosols and rainwater has been shown in the past.[10,12,14,17,30,32] Owing to the lack of authentic standard compounds, quantification for these compounds was not performed in the present study and only their intensities are compared for the discussion.

Among these organosulfates, the intensity of the m/z 199 compound was typically the highest followed by the m/z 183, m/z 155, m/z 153 and m/z 167. In addition to these, a peak corresponding to organosulfates of 2-methyltetrols (m/z 215, $C_5H_{11}O_7S^-$)[6,10,11,19] was detected in the ambient samples at much higher intensity than these C_2–C_4 organosulfates (Fig. 6). Relatively high abundances of m/z 215 and m/z 199 compounds are consistent with the above-mentioned studies, which reported the same trend in the ambient aerosol samples. It is noted that lactic acid organosulfate (m/z 169, $C_3H_5O_6S^-$, not shown) originating from the further oxidation of methylglyoxal[12,32] was also detected in these ambient samples, though it was not detected in the bulk-phase and aerosol chamber samples. It is noted that dimeric organosulfates that were observed in the bulk and aerosol chamber samples were not detected in the ambient samples. It is likely that the formation of these dimeric compounds were not favoured in the ambient aerosols because the organosulfate precursor compounds were present at much lower concentrations than the laboratory samples and when the dimerisation reactions were initiated it is likely that the reactions yield higher molecular weight compounds (HMWC) that are hetero-dimers and -oligomers as other organic reaction partners are abundant in the ambient particles. These hetero-dimeric and -oligomeric organosulfates are likely constituents of multi-functional HMWC that are known to be an important fraction of water-soluble organic molecules present in the aerosols (*e.g.* Nizkorodov *et al.*[76] and references therein), fog[15] and rain water.[14]

All of the above-mentioned C_2 to C_5 organosulfate species detected in the ambient samples correlate very well with each other (Table 4), indicating that similar environmental conditions and formation mechanisms are involved to form these compounds. In addition, statistically significant correlations were found between PM_{10} sulfate and these organosulfates though hydroxyacetone organosulfate (*m/z* 153) and 2-methyl glyceric acid organosulfate (*m/z* 199) showed somewhat smaller correlation coefficients ($r \approx 0.4$). The good correlations between anthropogenic PM_{10} sulfate and biogenic organosulfates may be caused by enhanced formation of biogenic SOA in polluted air mass,[77] and subsequent reactions of these compounds with the particle-phase sulfate.

4   Atmospheric implications

Fig. 7 shows a compilation of known oxidants, reaction kinetics and types of reaction products for MACR and MVK in the aqueous particles. As their Henry coefficients are small, aqueous phase reactions were not expected to occur and add significantly to the organic mass. Therefore, only a few studies dealt with the aqueous phase oxidation products from both OVOCs, whereby rate constants for the reactions with all four oxidants were only determined for the oxidation of MACR so far. In the oxidation of MACR and MVK OH radicals produce smaller oxidation products, such as methylglyoxal and oxalic acid, but also higher molecular weight compounds are formed.[64–66] The reaction of hydrogen peroxide with MACR was found to form 2-MG in the presence of acidity[1,74] and the ozonolysis of MACR and MVK leads to the formation of smaller oxidation products, such as methylglyoxal and formaldehyde.[78] Comparing the first-order rate constants (Fig. 7) that are calculated from the second-order rate constants and particle phase concentrations of the oxidants, the reaction with OH radicals is the main sink for MACR.[54,60] Competitive sinks are the reactions with sulfate radicals[54,58] and ozone[75] as about 10% of MACR can react with each oxidant. The reaction with hydrogen peroxide can be important for the nighttime chemistry as it is a slow reaction,[74,75] provided hydrogen peroxide is still available in the atmospheric system. It should be noted that this first-order constant can only be an upper value as hydrogen peroxide is photolysed to form OH radicals during the daytime and therefore the aqueous particle phase concentration can be much lower during nighttime.

In this study we found evidence for the formation of organosulfates following reactive uptake of MACR and MVK. Contrary to the expectation based on the former study by Kroll *et al.*,[3] MVK also forms a distinct amount of particle mass. As 1 ppm of each OVOC was present in the gas-phase in the chamber experiment, a high gas phase partial pressure was produced and therefore a partitioning was promoted. We suggest that the particle phase sulfate radical chemistry can play a role in SOA formation under appropriate conditions in which these OVOCs are likely present in aqueous particles.

Clearly, in the atmosphere the most important prerequisite condition is the presence of SO_2. As over 60% of the SO_2 is of anthropogenic origin over the continents,[79] one can assume that continental rural locations where anthropogenic emissions have a strong influence provide the best conditions for organosulfate formation. It should be considered that a strong pollution with OVOCs might inhibit the formation of sulfate radicals and hence the formation of organosulfates *via* this pathway, as a high OVOC content can disturb the oxidation chain of SO_2 in the aqueous particle phase by sulfate radical scavenging.[34,36,37]

Besides a source for the emission of SO_2, lower temperatures and high relative humidity can promote the organosulfate formation. These latter conditions can prevail either under certain meteorological conditions or at higher altitudes. As the lifetimes of MACR and MVK are about 6–10 h in the troposphere[80] a transport to higher altitudes above the boundary layer can occur. The lower temperatures and a higher relative humidity promote the partitioning of OVOCs into a particle phase by the condensation on a surface or an uptake into the aqueous particles and droplets. Furthermore, the particle fraction of acidic sulfate is found to be significantly higher at heights of 5–20 km, and this increases the formation of isoprene derived organosulfates from the partitioning of IEPOX.[17] This likely also influences the reactive uptake of MACR and MVK but further studies are warranted here. Additionally, the light intensity is stronger in higher altitudes, whereby photo-oxidation processes in the aqueous phase can be enhanced. Similar conditions might occur in radiation fogs that form from the rapid cooling of the surface layer, *e.g.* due to the inflow of cold air. Radiation fog droplets are condensed mixtures of particles, water-soluble organic gas phase compounds and oxidants. Photo-oxidative processes at the upper surface of the fog layer lead to the formation of highly oxidised and multifunctional compounds, such as small organic acids, but high molecular weight compounds with sulfate functions have already been detected in fog samples.[15,81]

5 Summary

In the present study, we address a likely occurrence of radical-initiated organosulfate formation mechanisms for these C_2–C_4 organosulfates over non-radical initiated mechanisms in the atmospheric wet particles. Laboratory evidence obtained from the present study suggests that the formation of these smaller organosulfates can be better explained by the radical mechanisms than non-radical mechanisms, and we provide evidence for the atmospheric relevance of these compounds. Nevertheless, their formation mechanisms remain elusive and further studies are warranted to draw a clear conclusion about the influence of anthropogenic emissions on the enhanced biogenic SOA formation, mechanisms, SOA product distributions and likely importance of the radical and non-radical initiated organosulfate formation in ambient wet particles.

Acknowledgements

The work was supported by the European Community's Seventh Framework Programme ([FP7/2007-2013]) EU project PEGASOS under grant agreement no. 265307.

References

1 M. Claeys, W. Wang, A. C. Ion, I. Kourtchev, A. Gelencsér and W. Maenhaut, *Atmos. Environ.*, 2004, **38**, 4093–4098.
2 H. J. Lim, A. G. Carlton and B. J. Turpin, *Environ. Sci. Technol.*, 2005, **39**, 4441–4446.
3 J. H. Kroll, N. L. Ng, S. M. Murphy, R. C. Flagan and J. H. Seinfeld, *Geophys. Res. Lett.*, 2005, **32**, DOI: 10.1029/2005GL023637.
4 J. H. Kroll, N. L. Ng, S. M. Murphy, R. C. Flagan and J. H. Seinfeld, *Environ. Sci. Technol.*, 2006, **40**, 1869–1877.
5 A. Guenther, T. Karl, P. Harley, C. Wiedinmyer, P. I. Palmer and C. Geron, *Atmos. Chem. Phys.*, 2006, **6**, 3181–3210.
6 J. D. Surratt, J. H. Kroll, T. E. Kleindienst, E. O. Edney, M. Claeys, A. Sorooshian, N. L. Ng, J. H. Offenberg, M. Lewandowski, M. Jaoui, R. C. Flagan and J. H. Seinfeld, *Environ. Sci. Technol.*, 2007, **41**, 517–527.
7 L. E. Hatch, J. M. Creamean, A. P. Ault, J. D. Surratt, M. N. Chan, J. H. Seinfeld, E. S. Edgerton, Y. X. Su and K. A. Prather, *Environ. Sci. Technol.*, 2011, **45**, 5105–5111.
8 L. E. Hatch, J. M. Creamean, A. P. Ault, J. D. Surratt, M. N. Chan, J. H. Seinfeld, E. S. Edgerton, Y. X. Su and K. A. Prather, *Environ. Sci. Technol.*, 2011, **45**, 8648–8655.
9 L. N. Hawkins, L. M. Russell, D. S. Covert, P. K. Quinn and T. S. Bates, *J. Geophys. Res.-Atmos.*, 2010, **115**, DOI: 10.1029/2009JD013276.
10 Y. Gómez-González, J. D. Surratt, F. Cuyckens, R. Szmigielski, R. Vermeylen, M. Jaoui, M. Lewandowski, J. H. Offenberg, T. E. Kleindienst, E. O. Edney, F. Blockhuys, C. Van Alsenoy, W. Maenhaut and M. Claeys, *J. Mass Spectrom.*, 2008, **43**, 371–382.
11 J. D. Surratt, Y. Gómez-González, A. W. H. Chan, R. Vermeylen, M. Shahgholi, T. E. Kleindienst, E. O. Edney, J. H. Offenberg, M. Lewandowski, M. Jaoui, W. Maenhaut, M. Claeys, R. C. Flagan and J. H. Seinfeld, *J. Phys. Chem. A*, 2008, **112**, 8345–8378.
12 M. Safi Shalamzari, O. Ryabtsova, A. Kahnt, R. Vermeylen, M.-F. Hérent, J. Quetin-Leclercq, P. Van der Veken, W. Maenhaut and M. Claeys, *Rapid Commun. Mass Spectrom.*, 2013, **27**, 784–794.
13 Y. Iinuma, C. Müller, T. Berndt, O. Böge, M. Claeys and H. Herrmann, *Environ. Sci. Technol.*, 2007, **41**, 6678–6683.
14 K. E. Altieri, B. J. Turpin and S. P. Seitzinger, *Atmos. Chem. Phys.*, 2009, **9**, 2533–2542.
15 L. R. Mazzoleni, B. M. Ehrmann, X. H. Shen, A. G. Marshall and J. L. Collett, *Environ. Sci. Technol.*, 2010, **44**, 3690–3697.
16 M. P. Tolocka and B. Turpin, *Environ. Sci. Technol.*, 2012, **46**, 7978–7983.
17 K. D. Froyd, S. M. Murphy, D. M. Murphy, J. A. de Gouw, N. C. Eddingsaas and P. O. Wennberg, *Proc. Natl. Acad. Sci. U. S. A.*, 2010, **107**, 21360–21365.
18 F. Paulot, J. D. Crounse, H. G. Kjaergaard, A. Kurten, J. M. St Clair, J. H. Seinfeld and P. O. Wennberg, *Science*, 2009, **325**, 730–733.
19 J. D. Surratt, A. W. H. Chan, N. C. Eddingsaas, M. N. Chan, C. L. Loza, A. J. Kwan, S. P. Hersey, R. C. Flagan, P. O. Wennberg and J. H. Seinfeld, *Proc. Natl. Acad. Sci. U. S. A.*, 2010, **107**, 6640–6645.
20 J. Peeters and J. F. Müller, *Phys. Chem. Chem. Phys.*, 2010, **12**, 14227–14235.
21 J. Peeters, T. L. Nguyen and L. Vereecken, *Phys. Chem. Chem. Phys.*, 2009, **11**, 5935–5939.
22 H. F. Zhang, W. Rattanavaraha, Y. Zhou, J. Bapat, E. P. Rosen, K. G. Sexton and R. M. Kamens, *Atmos. Environ.*, 2011, **45**, 4507–4521.
23 P. J. Crutzen, J. Williams, U. Pöschl, P. Hoor, H. Fischer, C. Warneke, R. Holzinger, A. Hansel, W. Lindinger, B. Scheeren and J. Lelieveld, *Atmos. Environ.*, 2000, **34**, 1161–1165.
24 E. C. Minerath and M. J. Elrod, *Environ. Sci. Technol.*, 2009, **43**, 1386–1392.
25 E. C. Minerath, M. P. Schultz and M. J. Elrod, *Environ. Sci. Technol.*, 2009, **43**, 8133–8139.
26 A. I. Darer, N. C. Cole-Filipiak, A. E. O'Connor and M. J. Elrod, *Environ. Sci. Technol.*, 2011, **45**, 1895–1902.
27 K. S. Hu, A. I. Darer and M. J. Elrod, *Atmos. Chem. Phys.*, 2011, **11**, 8307–8320.
28 E. C. Minerath, M. T. Casale and M. J. Elrod, *Environ. Sci. Technol.*, 2008, **42**, 4410–4415.

29 Q. Zhang, J. L. Jimenez, D. R. Worsnop and M. Canagaratna, *Environ. Sci. Technol.*, 2007, **41**, 3213–3219.

30 H. F. Zhang, Y. H. Lin, Z. F. Zhang, X. L. Zhang, S. L. Shaw, E. M. Knipping, R. J. Weber, A. Gold, R. M. Kamens and J. D. Surratt, *Environ. Chem.*, 2012, **9**, 247–262.

31 M. M. Galloway, P. S. Chhabra, A. W. H. Chan, J. D. Surratt, R. C. Flagan, J. H. Seinfeld and F. N. Keutsch, *Atmos. Chem. Phys.*, 2009, **9**, 3331–3345.

32 C. N. Olson, M. M. Galloway, G. Yu, C. J. Hedman, M. R. Lockett, T. Yoon, E. A. Stone, L. M. Smith and F. N. Keutsch, *Environ. Sci. Technol.*, 2011, **45**, 6468–6474.

33 M. R. Hoffmann and J. O. Edwards, *J. Phys. Chem.*, 1975, **79**, 2096–2098.

34 I. Grgić, A. Dovžan, G. Berčič and V. Hudnik, *J. Atmos. Chem.*, 1998, **29**, 315–337.

35 H. Herrmann, *Chem. Rev.*, 2003, **103**, 4691–4716.

36 K. J. Rudziński, *J. Atmos. Chem.*, 2004, **48**, 191–216.

37 K. J. Rudziński, L. Gmachowski and I. Kuznietsova, *Atmos. Chem. Phys.*, 2009, **9**, 2129–2140.

38 P. Y. Jiang, Y. Katsumura, R. Nagaishi, M. Domae, K. Ishikawa, K. Ishigure and Y. Yoshida, *J. Chem. Soc., Faraday Trans.*, 1992, **88**, 1653–1658.

39 M. J. Perri, Y. B. Lim, S. P. Seitzinger and B. J. Turpin, *Atmos. Environ.*, 2010, **44**, 2658–2664.

40 B. Nozière, S. Ekström, T. Alsberg and S. Holmström, *Geophys. Res. Lett.*, 2010, **37**, DOI: 10.1029/2009GL041683.

41 Y. Iinuma, O. Böge, A. Kahnt and H. Herrmann, *Phys. Chem. Chem. Phys.*, 2009, **11**, 7985–7997.

42 I. M. Kolthoff and I. K. Miller, *J. Am. Chem. Soc.*, 1951, **73**, 3055–3059.

43 C. von Sonntag and H. P. Schuchmann, *Angew. Chem., Int. Ed. Engl.*, 1991, **30**, 1229–1253.

44 G. A. Russell, *J. Am. Chem. Soc.*, 1957, **79**, 3871–3877.

45 J. E. Bennett and R. Summers, *Can. J. Chem.*, 1974, **52**, 1377–1379.

46 V. M. Berdnikov, N. M. Bazhin, V. K. Fedorov and O. V. Polyakov, *Kinet. Catal.*, 1972, **13**, 986–987.

47 B. C. Gilbert, R. G. G. Holmes, H. A. H. Laue and R. O. C. Norman, *J. Chem. Soc., Perkin Trans. 2*, 1976, 1047–1052.

48 G. D. Mendenhall and E. M. Y. Quinga, *Int. J. Chem. Kinet.*, 1985, **17**, 1187–1190.

49 R. Atkinson, *Int. J. Chem. Kinet.*, 1997, **29**, 99–111.

50 R. Atkinson, *Atmos. Environ.*, 2007, **41**, 8468–8485.

51 J. J. Orlando, G. S. Tyndall and T. J. Wallington, *Chem. Rev.*, 2003, **103**, 4657–4689.

52 B. C. Gilbert, P. David, R. Marshall, R. O. C. Norman, N. Pineda and P. S. Williams, *J. Chem. Soc., Perkin Trans. 2*, 1981, 1392–1400.

53 M. I. Guzmán, A. J. Colussi and M. R. Hoffmann, *J. Phys. Chem. A*, 2006, **110**, 3619–3626.

54 H. Herrmann, D. Hoffmann, T. Schaefer, P. Bräuer and A. Tilgner, *ChemPhysChem*, 2010, **11**, 3796–3822.

55 H. Herrmann, *Phys. Chem. Chem. Phys.*, 2007, **9**, 3935–3964.

56 G. Mark, M. N. Schuchmann, H. P. Schuchmann and C. von Sonntag, *J. Photochem. Photobiol., A*, 1990, **55**, 157–168.

57 H. Herrmann, M. Exner, H. W. Jacobi, G. Raabe, A. Reese and R. Zellner, *Faraday Discuss.*, 1995, **100**, 129–153.

58 G. V. Buxton, G. A. Salmon and J. E. Williams, *J. Atmos. Chem.*, 2000, **36**, 111–134.

59 S. Gligorovski, D. Rousse, C. H. George and H. Herrmann, *Int. J. Chem. Kinet.*, 2009, **41**, 309–326.

60 D. Huang, X. Zhang, Z. M. Chen, Y. Zhao and X. L. Shen, *Atmos. Chem. Phys.*, 2011, **11**, 7399–7415.

61 H. Zhang, J. D. Surratt, Y. H. Lin, J. Bapat and R. M. Kamens, *Atmos. Chem. Phys.*, 2011, **11**, 6411–6424.

62 T. B. Nguyen, P. J. Roach, J. Laskin, A. Laskin and S. A. Nizkorodov, *Atmos. Chem. Phys.*, 2011, **11**, 6931–6944.

63 I. El Haddad, Y. Liu, L. Nieto-Gligorovski, V. Michaud, B. Temime-Roussel, E. Quivet, N. Marchand, K. Sellegri and A. Monod, *Atmos. Chem. Phys.*, 2009, **9**, 5107–5117.

64 Y. Liu, I. El Haddad, M. Scarfogliero, L. Nieto-Gligorovski, B. Temime-Roussel, E. Quivet, N. Marchand, B. Picquet-Varrault and A. Monod, *Atmos. Chem. Phys.*, 2009, **9**, 5093–5105.

65 Y. Liu, F. Siekmann, P. Renard, A. El Zein, G. Salque, I. El Haddad, B. Temime-Roussel, D. Voisin, R. Thissen and A. Monod, *Atmos. Environ.*, 2012, **49**, 123–129.

66 X. Zhang, Z. M. Chen and Y. Zhao, *Atmos. Chem. Phys.*, 2010, **10**, 9551–9561.

67 L. T. Iraci, B. M. Baker, G. S. Tyndall and J. J. Orlando, *J. Atmos. Chem.*, 1999, **33**, 321–330.

68 D. van Pinxteren, A. Plewka, D. Hofmann, K. Müller, H. Kramberger, B. Svrcina, K. Bachmann, W. Jaeschke, S. Mertes, J. L. Collett and H. Herrmann, *Atmos. Environ.*, 2005, **39**, 4305–4320.

69 J. P. D. Abbatt, A. K. Y. Lee and J. A. Thornton, *Chem. Soc. Rev.*, 2012, **41**, 6555–6581.

70 M. S. Jang, N. M. Czoschke, S. Lee and R. M. Kamens, *Science*, 2002, **298**, 814–817.
71 B. Nozière, D. Voisin, C. A. Longfellow, H. Friedli, B. E. Henry and D. R. Hanson, *J. Phys. Chem. A*, 2006, **110**, 2387–2395.
72 J. H. Kroll, N. L. Ng, S. M. Murphy, V. Varutbangkul, R. C. Flagan and J. H. Seinfeld, *J. Geophys. Res.-Atmos.*, 2005, **110**.
73 M. B. Williams, R. R. H. Michelsen, J. L. Axson and L. T. Iraci, *Atmos. Environ.*, 2010, **44**, 1145–1151.
74 Z. Liu, L. Y. Wu, T. H. Wang, M. F. Ge and W. G. Wang, *J. Phys. Chem. A*, 2012, **116**, 437–442.
75 X. Zhang, Z. M. Chen, H. L. Wang, S. Z. He and D. M. Huang, *Atmos. Environ.*, 2009, **43**, 4465–4471.
76 S. A. Nizkorodov, J. Laskin and A. Laskin, *Phys. Chem. Chem. Phys.*, 2011, **13**, 3612–3629.
77 J. A. de Gouw, A. M. Middlebrook, C. Warneke, P. D. Goldan, W. C. Kuster, J. M. Roberts, F. C. Fehsenfeld, D. R. Worsnop, M. R. Canagaratna, A. A. P. Pszenny, W. C. Keene, M. Marchewka, S. B. Bertman and T. S. Bates, *J. Geophys. Res.-Atmos.*, 2005, **110**, DOI: 10.1029/2004JD005623.
78 Z. M. Chen, H. L. Wang, L. H. Zhu, C. X. Wang, C. Y. Jie and W. Hua, *Atmos. Chem. Phys.*, 2008, **8**, 2255–2265.
79 C. Lee, R. V. Martin, A. van Donkelaar, H. Lee, R. R. Dickerson, J. C. Hains, N. Krotkov, A. Richter, K. Vinnikov and J. J. Schwab, *J. Geophys. Res.-Atmos.*, 2011, **116**, DOI: 10.1029/2010JD014758.
80 T. Gierczak, J. B. Burkholder, R. K. Talukdar, A. Mellouki, S. B. Barone and A. R. Ravishankara, *J. Photochem. Photobiol., A*, 1997, **110**, 1–10.
81 J. L. Collett, P. Herckes, S. Youngster and T. Lee, *Atmos. Res.*, 2008, **87**, 232–241.

Faraday Discussions

RSC Publishing

Formation of secondary organic aerosol marker compounds from the photooxidation of isoprene and isoprene-derived alkene diols under low-NO$_x$ conditions†

Wu Wang,[ab] Yoshiteru Iinuma,[c] Ariane Kahnt,[a] Oxana Ryabtsova,[a] Anke Mutzel,[c] Reinhilde Vermeylen,[a] Pieter Van der Veken,[a] Willy Maenhaut,[ad] Hartmut Herrmann[c] and Magda Claeys*[a]

Received 15th May 2013, Accepted 31st May 2013

DOI: 10.1039/c3fd00092c

In the present work, we have evaluated whether isomeric C$_5$-alkene diols (1,2-dihydroxy-2-methyl-3-butene, 1,2-dihydroxy-3-methyl-3-butene, and 1,4-dihydroxy-2-methyl-2-butene (*cis* + *trans*)), which have first been detected upon photooxidation of isoprene in the absence of NO and are known to be formed in the ambient atmosphere, can serve as precursors for the 2-methyltetrols, C$_5$-alkene triols, and 2-methylglyceric acid under low-NO$_x$ conditions. The C$_5$-alkene diols were prepared following published synthesis procedures. It is shown that under the applied chamber conditions the isomeric C$_5$-alkene diols give rise to 2-methyltetrols with different *threo*/*erythro* abundance ratios and that certain diols produce 2-methylglyceric acid, but that they do not form C$_5$-alkene triols. Furthermore, it is shown that the photooxidation of isoprene under the applied chamber conditions employing photolysis of H$_2$O$_2$ under dry conditions yields relatively small amounts of C$_5$-alkene triols compared to those of the 2-methyltetrols, unlike under ambient conditions. It is argued that the chamber conditions are not optimal for the formation of C$_5$-epoxydiols, which serve as gas-phase precursors for the C$_5$-alkene triols, and likely as in some previous studies favor the formation of C$_5$-alkene diols as a result of RO$_2$ + RO$_2$ reactions.

Introduction

Isoprene is the largest source of nonmethane hydrocarbons in the atmosphere, estimated at about 500 Tg C per year.[1] It is mainly emitted by broadleaf vegetation and plays a critical role in tropospheric chemistry over large regions of the globe. In

[a]Department of Pharmaceutical Sciences, University of Antwerp, Belgium. E-mail: magda.claeys@ua.ac.be
[b]Institute of Environmental Pollution and Health, Shanghai University, Shanghai, P.R. China
[c]Leibniz-Institut für Troposphärenforschung, Leipzig, Germany
[d]Department of Analytical Chemistry, Ghent University, Belgium

† Electronic supplementary information (ESI) available. See DOI: 10.1039/c3fd00092c

pristine regions, such as the Amazon rainforest, the oxidation of isoprene by the hydroxyl (OH) radical occurs in the presence of low NO_x $(NO + NO_2)$ concentrations, whereas in polluted regions the photooxidation occurs in the presence of high NO_x concentrations. The underlying chemistries leading to particle-phase products, however, are not fully understood. Major marker compounds of secondary organic aerosol (SOA) from isoprene photooxidation are the diastereoisomeric 2-methyl-tetrols (2-methylthreitol and 2-methylerythritol), the isomeric C_5-alkene triols [2-methyl-1,3,4-trihydroxy-1-butene (*cis* and *trans*) and 3-methyl-2,3,4-trihydroxy-1-butene], and 2-methylglyceric acid. The 2-methyltetrols and the C_5-alkene triols were first discovered in fine aerosol collected from the Amazon forest,[2,3] whereas 2-methylglyceric acid was first detected in fine aerosol collected from K-puszta, Hungary.[4] Laboratory experiments demonstrated that these SOA marker compounds can be formed through photooxidation of isoprene under different conditions; while the 2-methyltetrols are formed under various NO_x regimes,[5,6] the formation of the C_5-alkene triols requires low-NO_x conditions[6] in contrast to that of 2-methylglyceric acid which is enhanced under high-NO_x conditions.[6]

Isoprene SOA chemistry is tightly linked to gas-phase photooxidation processes, which yield a number of volatile gas-phase products, including methacrolein, methyl vinyl ketone, methyl furan, C_5-alkene diols and C_5-hydroxyhydroperoxides as first-generation products.[7-10] Different mechanistic pathways, involving gas- and particle-phase reactions, have been suggested to explain the formation of the 2-methyltetrols through photooxidation of isoprene under low NO_x conditions in the presence or absence of acidic or neutral ammonium sulfate seed aerosol.[2,4,9-12] Key intermediates in the pathway first reported by Claeys *et al.*[2] and later elaborated by Böge *et al.*[12] and Kleindienst *et al.*[9] are the C_5-alkene diols, whereas those in the pathway reported by Paulot *et al.*[10] are the C_5-epoxydiols. With regard to the first pathway, it has been shown by Böge *et al.*[12] that the C_5-alkene diol, 1,2-dihydroxy-2-methyl-3-butene, serves as a precursor for the 2-methyltetrols. In the present work, we have further evaluated whether isomeric C_5-alkene diols, which have first been detected upon photoox-idation of isoprene in the absence of NO[9] and are known to be formed in the ambient atmosphere,[9] can serve as precursors for the 2-methyltetrols, C_5-alkene triols, and 2-methylglyceric acid under low-NO_x conditions.

2   Experimental section

2.1.   Preparation and characterization of C_5-alkene diols

Three different positional isomers of the C_5-alkene diols were synthesized according to published procedures. A brief overview of the synthetic routes is presented in Scheme 1. 1,2-Dihydroxy-2-methyl-3-butene (diol 1) (**1**) was prepared by hydrolysis of 2-methyl-2-vinyloxiran.[8] 1,2-Dihydroxy-3-methyl-3-butene (diol 2) (**2**) was prepared from 3,4-epoxytetrahydrofuran in an alkylative double ring opening reaction with methyl lithium.[13] 1,4-Dihydroxy-2-methyl-2-butene (diol 3) (**6**) was obtained in a 3-step reaction starting from isoprene, involving bromina-tion of isoprene, replacement of the bromine groups by acetate groups, and saponification.[8] Details about the synthesis of the isomeric C_5-alkene diols and their characterization by [1]H-nuclear magnetic resonance (^{1}H-NMR) spectroscopy and gas chromatography/mass spectrometry (GC-MS) with prior trimethylsilyla-tion are provided in the ESI.†

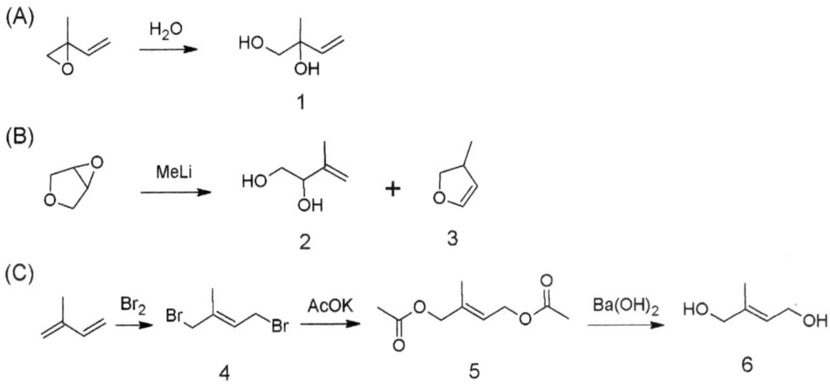

Scheme 1 Overview of the synthetic procedures leading to (A) 1,2-dihydroxy-2-methyl-3-butene (**1**; diol 1), (B) 1,2-dihydroxy-3-methyl-3-butene (**2**; diol 2), and (C) 1,4-dihydroxy-2-methyl-2-butene (*cis + trans*) (**6**; diol 3).

2.2. Chamber experiments

Chamber experiments were conducted at the Leibniz-Institut für Troposphärenforschung (TROPOS), using the LEAK chamber (volume: 19 m^3), which was described by Iinuma *et al.*[14] The aerosol chamber was flushed with clean

Table 1 Initial conditions for the formation of SOA from the photooxidation of isoprene and C_5-alkene diols using $(NH_4)_2SO_4$ seed aerosol at 20 °C under low-NO_x conditions. Abbreviation: HC, hydrocarbon

Experiment	Date (2012)	Starting compound/oxidant	Initial HC (ppb)	Duration	RH (%)
1	21 Feb	isoprene/O_3	94	3 h	30
2	20 Feb	isoprene/H_2O_2	94	3 h	dry
3[a]	23 Apr	isoprene/H_2O_2	94	3 h	dry
4	22 Feb	isoprene/H_2O_2	94	5 h	dry
5	23 Feb	diol 1/H_2O_2	116	2 h	dry
6	24 Feb	diol 1/H_2O_2	116	2.5 h	dry
7[a]	24 Apr	diol 2/H_2O_2	116	2 h	dry
8 a[a]	25 Apr	diol 2/H_2O_2	116	20 min	dry
8 b[a]				2 h	
9[a]	30 Jul	diol 2/H_2O_2	116	2 h	dry
10 a[a]	31 Jul	diol 2/H_2O_2	116	10 min	dry
10 b[a]				20 min	
10 c[a]				30 min	
10 d[a]				40 min	
10 e[a]				50 min	
10 f[a]				2 h	
11[a]	01 Aug	diol 3/H_2O_2	116	2 h	dry
12 a[a]	02 Aug	diol 3/H_2O_2	116	10 min	dry
12 b[a]				20 min	
12 c[a]				30 min	
12 d[a]				40 min	
12 e[a]				50 min	
12 f[a]				2 h	

[a] PTR-MS was used to monitor the mixing ratio of the hydrocarbon.

synthetic air for at least 16 h at a flow rate of 200 L min^{-1} prior to each experiment. The relative humidity (RH) and temperature were monitored with a sensor (HMP 238, Vaisala, Finland), the ozone concentration was monitored using an ozone monitor (ML 9812, Monitor Labs, Englewood, CO, USA), and the NO$_x$ (NO and NO2) concentrations were monitored using an NO$_x$ analyzer (Model 42S, Thermo Environmental Instruments, Franklin, MA, USA). The initial conditions for the chamber experiments are given in Table 1. In selected experiments (marked with *), the mixing ratio of isoprene and gas-phase oxidation products was monitored using a proton-transfer-reaction mass spectrometer (PTR-MS; ION-ICON, Innsbruck, Austria). The PTR-MS results reveal that isoprene is completely consumed in 1 h 45 min (Fig. S2; ESI†), whereas the C$_5$-alkene diols 2 and 3 are converted within the first 20 min (not shown) after introduction into the chamber. Seed particles were produced by nebulizing a solution containing $(NH_4)_2SO_4$ (60 mM) and injected directly into the chamber. Particle size distributions between 3–900 nm were monitored every 5 min using a differential mobility particle sizer (DMPS). Ozone was produced using an ultraviolet (UV) lamp and a stream of oxygen and injected directly into the chamber through an injection port (Exp. 1). OH radicals were generated by using UV-C light (UV-Strahler 30 W, UMEX GmbH, Germany) to photolyze H_2O_2 (30%) (Trace SELECT® Ultra, Fluka), which was introduced constantly into the chamber at 120 μL h^{-1} and carried by an air stream of 4 L min^{-1}. After introduction of seed particles and ozone or H_2O_2, a known volume of isoprene (purity > 99%, Sigma-Aldrich; 8 μL) or C$_5$-alkene diols (10 μL in 500 μL H_2O) was injected using a stream of synthetic air from an injection port (200 L min^{-1}) with a syringe within 2 or 5 min, respectively. After measuring the initial particle size distribution, the UV-lamps were switched on to initiate the reactions. The average chamber temperature was about 20 °C. The SOA mass formation was monitored using a Tandem Differential Mobility Particle Sizer (TDMPS) by measuring the volume change of the particles in the chamber and assuming a density of 1 g cm^{-3}. The SOA samples were collected on a Teflon filter (47-mm diameter, Pallflex T60A20, Pall, NY, USA) after the reaction was terminated by switching off the UV-lamps. The sampling time was 1 h with a total sampling volume of 1.8 m^3, except in the time-resolved experiments, where the sampling time was 10 min. For the final SOA collections, XAD-4 coated denuders (5 channel, 400 mm length, URG, Chapel Hill, NC, USA) were installed in front of the filter pack to remove semi-volatile compounds. The filter samples were stored in a freezer (−20 °C) until analysis. Furthermore, in selected experiments with isoprene 500 mL of the gas phase was bubbled through a solution of acidified methanol (20 mL; 1% HCOOH) kept in a glass bottle at 0 °C in an effort to trap C$_5$-epoxydiols and convert them to the 2-methyltetrol mono-methylethers through acid-catalyzed opening of the oxiran ring and reaction with methanol.

2.3. Ambient aerosol samples

Archived PM2.5 (particulate matter with an aerodynamic diameter $\leq$ 2.5 μm) aerosol samples collected on quartz fibre filters were analyzed in the same way as SOA samples. The ambient aerosol samples were collected during the BIOSOL (Formation mechanisms, marker compounds, and source apportionment for BIOgenic atmospheric aeroSOLs) campaign, which took place from 24 May to 29 June 2006 at K-puszta, Hungary, a rural site located on the Great Hungarian

Plain (46°58′N, 19°35′E, 125 m above sea level), 15 km northwest from the nearest town Kecskemét, and 80 km southeast from Budapest. For more information on this campaign, and collection and storage of the samples, see Maenhaut *et al.*[15] Only GC-MS results for a selected day of the warm period (26 June 2006) are presented here for comparison purposes.

2.4.   Gas chromatography/mass spectrometry (GC-MS)

All filters were analyzed for polar compounds by GC-MS with prior trimethylsilylation using a method that was adapted from that reported by Pashynska *et al.*[16] The sample workup consisted of spiking the filter with a suitable amount of internal recovery standard (methyl-β-D-xylopyranoside), extraction of all or a part of the filter with methanol under ultrasonic agitation, and derivatization of carboxyl and hydroxyl groups into trimethylsilyl (TMS) derivatives employing as reagent *N,O*-bis(trimethylsilyl)trifluoroacetamide (BSTFA) (Sigma-Aldrich) without trimethylsilylchloride (TMCS) catalyst in pyridine (2 : 1, v/v). The use of TMCS catalyst was avoided in order to allow the detection of labile intermediates if formed (*e.g.*, C_5-trihydroxyhydroperoxy intermediates); however, owing to the absence of TMCS catalyst in the derivatization mixture the trimethylsilylation of the 2-methyltetrols was only found to be complete after 2 days. The extract was divided into parts; one part was trimethylsilylated, whereas other parts were stored in a refrigerator at 4 °C for additional complimentary tests. GC-MS analyses were performed with a system comprising a TRACE GC2000 gas chromatograph, which was coupled to a Polaris Q ion trap mass spectrometer equipped with an external ionization source (Thermo Scientific, San Jose, CA, USA). A Heliflex AT-5MS fused silica capillary column [5% phenyl, 95% methylpolysiloxane, 0.25 μm film thickness, 30 m × 0.25 mm internal diameter (i.d.)] preceded by a deactivated fused silica precolumn (2 m × 0.25 μm i.d.) (Alltech, Deerfield, IL, USA) was used to separate the derivatized extracts. Helium was used as carrier gas at a flow rate of 1.2 mL min^{-1}. The temperature program was as follows: isothermal hold at 50 °C, temperature ramp of 3 °C min^{-1} up to 200 °C, isothermal hold at 200 °C for 2 min, temperature ramp of 30 °C min^{-1} up to 310 °C, and isothermal hold at 310 °C for 2 min. The analyses were mainly performed in the full scan mode (mass range: *m/z* 50–800), and in the electron ionization (EI) mode. The ion source was operated at an electron energy of 70 eV and a temperature of 200 °C. The temperatures of the GC injector and the GC-MS transfer were 250 °C and 280 °C, respectively. The identities of the marker compounds were confirmed by comparing their EI mass spectra with those reported in the literature.[2–4,5] In addition to the 2-methyltetrols, the C_5-alkene triols, and 2-methylglyceric acid, C_5-trihydroxymonocarboxylic (MW 150; MW TMS-derivatives 438) and C_5-dihydroxydicarboxylic acids (MW 164; MW TMS derivatives 452), each occurring as a diastereoisomeric pair (*threo* and *erythro*), could be identified in SOA samples and ambient PM2.5 samples at a significant relative abundance; therefore, these C_5-dicarboxylic acids are also taken into account in the calculation of the percentage of the mass of SOA marker compounds to that of the total identified SOA. The retention times of the trimethylsilylated derivatives of the C_5-monohydroxydicarboxylic acids were 40.74 and 41.05 min, whereas those of the C_5-dihydroxydicarboxylic acids were 42.15 and 42.63 min. The quantitation of the SOA marker compounds was based on an internal standard calibration procedure employing methyl-L-xylanopyranoside as

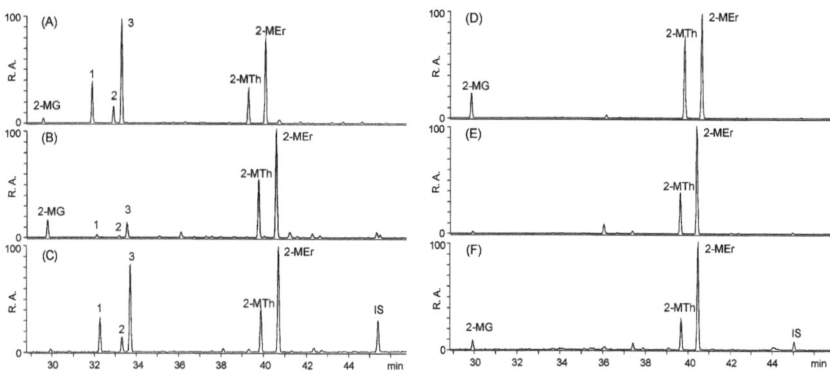

Fig. 1   GC-MS extracted ion chromatograms (*m/z* 219 + 231) obtained for SOA from ambient aerosol sample collected in K-puszta, Hungary (A), the photooxidation of isoprene with OH, exp. 1 (B), isoprene with ozone, exp. 2 (C), and 1,2-dihydroxy-2-methyl-2-butene, exp. 5 (D), 1,2-dihydroxy-3-methyl-3-butene, exp. 9 (E), and 1,4-dihydroxy-2-methyl-2-butene (*cis* + *trans*), exp. 11 (F) with OH. Abbreviations: 2-MG: 2-methylglyceric acid; **1**: *cis*-2-methyl-1,3,4-trihydroxy-1-butene; **2**: 3-methyl-2,3,4-trihydroxy-1-butene; **3**: *trans*-2-methyl-1,3,4-trihydroxy-1-butene; 2-MTh: 2-methylthreitol; 2-MEr: 2-methylerythritol; R.A., relative abundance; IS: internal standard.

internal recovery standard and erythritol (purity 99%; Sigma-Aldrich) as a surrogate standard. The peak areas of the total ion chromatograms (TICs) served as input data for the quantitative determinations.

3   Results and discussion

Fig. 1 shows extracted ion chromatograms (EICs) (*m/z* 219 + 231) obtained for ambient PM2.5 aerosol collected from K-puszta, Hungary, and selected SOA samples from the chamber (without added NO) in the presence of neutral ammonium sulfate seed aerosol. The 2-methyltetrols (*i.e.*, 2-methylthreitol and 2-methylerythritol) are detected as abundant marker compounds in the photo-oxidation of isoprene and the isomeric C_5-alkene diols, whereas the C_5-alkene triols are only detected in ambient PM2.5 aerosol and in the SOA samples from the chamber irradiation of isoprene with ozone or H_2O_2. It thus appears that under the chamber conditions the C_5-alkene diols serve as gas-phase intermediates for the 2-methyltetrols but not for the C_5-alkene triols.

Table 2 shows the concentrations of SOA tracers formed from the irradiation reactions and their percentage contributions to the sum of all tracers; the last two columns of Table 2 give the concentration of the total SOA mass, as derived from the TDMPS measurements, and the percentage contribution of the sum of all tracers to the total SOA mass. It can be seen in Table 2 that in the case of the time-resolved SOA collections for the diol 2 and 3 experiments (experiments 10 and 12) the percentage contribution of the sum of all tracers to the total SOA mass is the highest for the first collection (after 10 min), indicating a fast formation of the major tracers, *i.e.*, the 2-methyltetrols, which together account for over 80% of the sum of all tracers, and thus suggesting that the 2-methyltetrols are not stable in the course of the experiment. Table 3 lists the average percentage contributions to the sum of all tracers and average *threo/erythro* ratios per type of experiment; the average for the diol 2 experiments were calculated with the data from the final

Table 2 Concentrations of SOA tracers and total SOA mass formed from irradiation reactions and their percentage contributions

	SOA tracers in ng m^{-3} (% of $\sum$tracers)							Total SOA	$\sum$tracers % of
Exp.	2-methylglyceric acid	C$_5$-alkenetriols	2-methylthreitol	2-methylerythritol	C$_5$-acids	C$_5$-diacids	$\sum$tracers	mass (μg m^{-3})	total SOA mass
1	0.18 (0.72)	11.7 (47)	3.7 (14.8)	9.1 (36)	—	—	25	—	—
2	11.4 (9.8)	8.0 (6.9)	29 (25)	53 (46)	13.4 (11.6)	1.33 (1.15)	116	19	0.6
3	18.0 (12.7)	23 (16.3)	34 (24)	55 (39)	10.2 (7.2)	1.34 (0.94)	142	4	3.5
5	79 (16.1)	—	167 (34)	230 (47)	14.2 (2.9)	4.3 (0.86)	490	27	1.83
7	5.9 (5.3)	—	23 (21)	79 (71)	2.5 (2.3)	0.69 (0.62)	111	3	3.7
8a	59 (6.1)	46 (4.7)	250 (26)	540 (56)	73 (7.5)	—	970	8	12.1
8b	—	—	42 (20)	168 (80)	—	—	210	2	10.5
9	3.4 (1.48)	—	63 (27)	154 (67)	12.2 (5.3)	1.81(0.79)	230	8	2.9
10a	55 (2.6)	8.7 (0.41)	640 (30)	1070 (51)	310 (14.8)	—	2100	5	42
10b	147 (3.9)	15.0 (0.39)	1310 (34)	2000 (53)	290 (7.6)	—	3800	27	13.9
10c	116 (5.5)	5.5 (0.26)	650 (31)	1100 (52)	210 (10.0)	—	2100	29	7.2
10d	87 (3.1)	10.9 (0.39)	890 (32)	1420 (51)	400 (14.3)	7.3 (0.26)	2800	25	11.3
10e	81 (4.3)	—	550 (29)	1170 (63)	73 (3.9)	—	1870	20	9.4
10f	17.9 (9.0)	—	59 (30)	109 (55)	9.1 (4.6)	4.9 (2.5)	200	8	2.5
11	8.6 (16.5)	—	9.4 (18.1)	34. (65)	—		52	3	1.73
12a	86 (3.2)	13.0 (0.48)	950 (35)	1590 (59)	30 (1.11)	—	2700	4	67
12b	116 (4.8)	12.5 (0.52)	780 (33)	1410 (59)	38 (1.58)	—	2400	15	15.7
12c	109 (6.9)	7.9 (0.50)	480 (30)	930 (58)	48 (3.0)	15.2 (0.96)	1590	13	12.2
12d	145 (8.5)	7.1 (0.42)	510 (30)	1000 (58)	45 (2.6)	3.7 (0.22)	1710	10	17.1
12e	88 (8.1)	9.8 (0.91)	330 (31)	620 (57)	28 (2.6)	3.0 (0.28)	1080	8	13.5
12f	12.1 (9.7)	—	33 (26)	75 (60)	3.6 (2.9)	1.65 (1.32)	125	1	12.5

Table 3 Percentage contributions of the SOA marker compounds to the total identified SOA mass and *threo/erythro* abundance ratios of 2-methyltetrol diastereoisomers. The data listed are averages per type of experiment (see text for details)

| Experiments $(n)^a$ | % of $\sum$(SOA) | | | | | | | *threo/ erythro* ratio |
	2-MG	C_5-alkene triols	2-MTs	C_5-acids	C_5-diacids	*threo*	*erythro*	
isoprene/O$_3$ (1)	0.7	47	51	—	—	15	36	0.42
isoprene/OH (2)	11	11	67	9.4	1.1	24	42	0.57
diol 1/OH (1)	16	—	81	2.9	0.9	34	47	0.72
diol 2/OH (2)	3.9	—	93	3.0	1.0	24	68	0.35
diol 3/OH (1)	13	—	85	—	—	22	63	0.35

^{a} number of experiments.

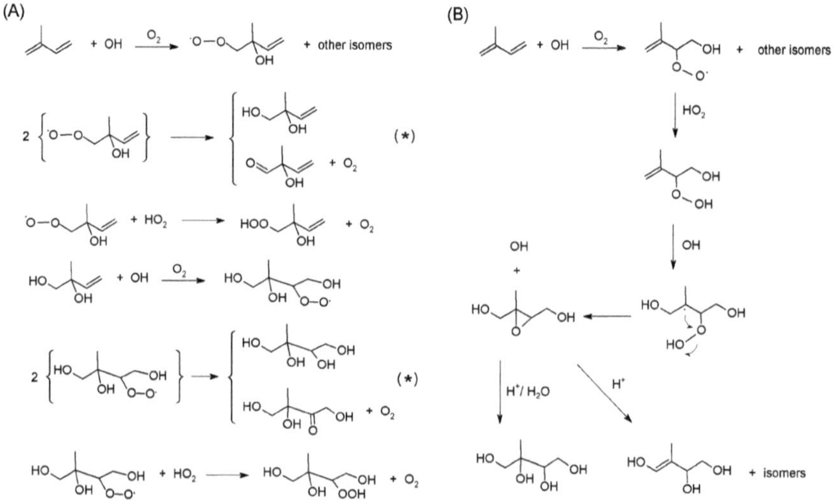

(A)

(B)

Scheme 2 Pathways reported for the formation of the 2-methyltetrols through photooxidation of isoprene: (A) pathway involving C_5-alkene diols as gas-phase intermediates (reproduced from Kleindienst *et al.*[9]); and (B) pathway involving C_5-epoxydiols as gas-phase intermediates (adapted from Surratt *et al.*[11]). The reactions marked with * are relevant to the current study.

aerosol collection (*i.e.*, experiments 7 and 9). It is worth noting in Table 3 that the *threo/erythro* ratio of the 2-methyltetrol diastereoisomers is different for the isoprene and C_5-alkene diol experiments, with the highest ratio for diol 1 (0.72; Fig. 1D) and the lowest ones for diols 2 and 3 (0.35; Fig. 1E and F), indicating that diols 2 and 3 more than diol 1 favor the formation of 2-methylerythritol. Interestingly, the overall average *threo/erythro* ratio of the 2-methyltetrols for the three isomeric C_5-alkene diols (0.48) falls between those obtained in the isoprene/O$_3$ experiment (0.42) and the photooxidation of isoprene (0.57). Table 3 further shows that the percentage contributions of the 2-methyltetrols to the sum of all tracers were above 50% for the isoprene experiments (*i.e.*, 67% for isoprene/OH and 51% for isoprene/O$_3$) and were even higher for the C_5-alkene diol experiments (*i.e.*, 81% for diol 1, 93% for diol 2, and 85% for diol 3).

Furthermore, it can be seen that 2-methylglyceric acid is not only formed from isoprene but also from certain C_5-alkene diols (Fig. 1); the percentage of the mass fraction of 2-methylglyceric acid to the sum of all tracers amounts to 11%, 16% and 13% for the isoprene/OH, the diol 1/OH, and the diol 3/OH reactions, respectively (Table 3). Relatively less 2-methylglyceric acid is formed in the diol 2/OH reaction (4%) and little upon isoprene ozonolysis (0.7%).

The question arises whether the applied chamber conditions reflect those encountered under ambient conditions, as will be discussed in this and the following paragraphs. Kleindienst et al.[9] proposed a gas-phase mechanism for the formation of the 2-methyltetrols from isoprene through C_5-alkene diols and per-oxy radicals involving trihydroxyperoxy radicals, which might react with HO_2 to form C_5-trihydroxyhydroperoxides and partition to the particle phase [Scheme 2(A)]. In the present study, however, no evidence could be found for the formation of C_5-trihydroxyhydroperoxides; if such hydroperoxide products would have been formed and would be stable under our conditions, we would expect the formation of oxo derivatives as thermal degradation products of the trimethylsilylated hydroperoxides upon GC-MS analysis, similar to the behavior of trimethylsilylated unsaturated fatty acid hydroperoxides.[17] The photooxidation experiments with isoprene (experiments 2–4) suggest that the 2-methyltetrols are directly generated from intermediate C_5-alkene diols under the chamber conditions, likely as a result of RO_2 + RO_2 reactions on the generated C_5-trihydroxyperoxy radicals. In a more recent study, Paulot et al.[10] detected gas-phase photooxidation products of isoprene with a molecular weight of 118, identified these products as isomeric C_5-epoxydiols and proposed them as gas-phase intermediates in the formation of the 2-methyltetrols and the C_5-alkene triols. A subsequent study by Surratt et al.[11] confirmed that C_5-epoxydiols indeed serve as the intermediate gas-phase precursors for the 2-methyltetrols and C_5-alkene triols in the particle phase [Scheme 2(B)], as well as for C_5-epoxydiol dimers and organosulfates. Our efforts in the present study to detect C_5-epoxydiols in the gas phase upon photooxidation of isoprene (experiments 2–4) by a trapping experiment in acidified methanol failed; a possible explanation is that under our experimental conditions they are not formed as major gas-phase intermediates. An indication that they are produced as minor gas-phase intermediates serving as precursors for the 2-methyltetrols is that the C_5-alkene triols, which have been shown to be formed from C_5-epoxydiols,[10] are only generated at a rather low relative abundance [Fig. 1(B)].

It thus appears that our experimental conditions do not favor the formation of C_5-epoxydiols but lead to the formation of C_5-alkene diols, likely as a result of RO_2 + RO_2 reactions on the corresponding hydroperoxy radicals [Scheme 2(A)]. The

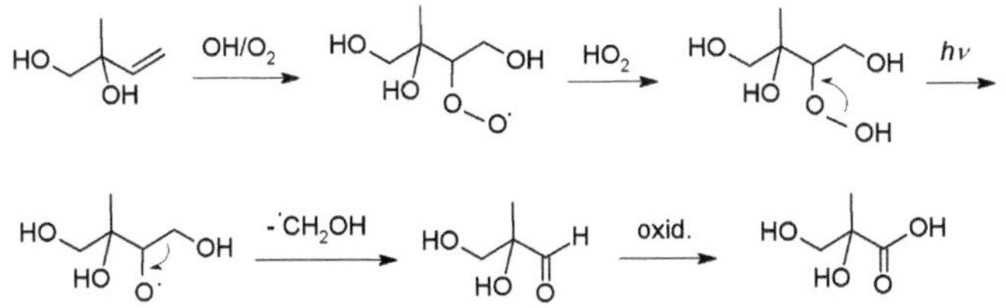

Scheme 3 A proposed pathway for the formation of 2-methylglyceric acid via the C_5-alkene diol 1,2-dihydroxy-2-methyl-3-butene (diol 1) under low-NO_x conditions.

observations, however, that the C_5-alkene diols have been detected in the ambient atmosphere[9] and can readily be converted to 2-methyltetrols, as shown in the present study, suggest that the route through C_5-alkene diol gas-phase intermediates could be a possible minor route leading to the 2-methyltetrols. On the basis of the *threo/erythro* abundance ratio of the 2-methyltetrols, it is, however, not possible to determine to which extent this route is followed; assuming that the isomeric C_5-alkene diols are produced in comparable amounts as shown by Kleindienst *et al.*[9] the expected *threo/erythro* ratio is 0.44, very close to the ratio of 0.45 observed for ambient fine aerosol (Fig. 1). It is noted that the isoprene mixing ratios applied in our study (94 ppb) is lower than the lowest mixing ratios that were applied in the studies of Ruppert and Becker[8] (*i.e.*, 1.86 ppm) and Kleindienst *et al.*[9] (*i.e.*, 8.7 ppm) for which C_5-alkene diols were detected, but somewhat higher than those of Paulot *et al.*[10] (*i.e.*, 20.9 ppb) and Surratt *et al.*[11] (*i.e.*, 40 ppb) for which C_5-epoxydiols and C_5-alkene triols, respectively, were reported. A notable difference, however, between the experimental conditions in the studies of Ruppert and Becker,[8] Kleindienst *et al.*[9] and our study and the studies of Paulot *et al.*[10] and Surratt *et al.*[11] is the way in which the H_2O_2 was introduced into the chamber; while H_2O_2 was continuously introduced in the first two studies and ours, a known mass of H_2O_2 was introduced in the beginning of the experiment (first 15–20 min) in the last two studies, likely giving rise to significant HO_2 radical levels by the $OH + H_2O_2$ reaction, which is favored at the slow chamber photolysis rate of H_2O_2. It is worth noting that the product distribution of isoprene SOA marker compounds obtained in the photolysis reaction of ozone [Fig. 1(C)] under more atmospherically relevant conditions in the presence of water vapor (RH 30%) is comparable to that obtained for ambient aerosol [Fig. 1(A)]; more specifically, the 2-methyltetrols and C_5-alkene triols show comparable relative abundances. A possible explanation is that photolysis of ozone in the presence of water vapor favors $RO_2 + HO_2$ reactions, thereby facilitating C_5-epoxydiol and subsequent 2-methyltetrol + C_5-alkene triol formation [Scheme 2(B)]. In this context, recent experiments in the gas phase by Berndt[19] on the OH radical reaction of isoprene for low-NO_x conditions employing photolysis of ozone suggested a possible effect of water on the kinetics or the reaction mechanism. However, this new water-dependent gas-phase isoprene chemistry still needs to be explored in detail.

As to the formation of 2-methylglyceric acid through photooxidation of isoprene, it has been shown that this marker compound is mainly formed in chamber experiments under high-NO_x conditions[5] and that it involves methacroylperoxynitrate (MPAN) and its subsequent oxidation product 2-methyloxirane-carboxylic acid as gas-phase intermediates.[11,18] The results obtained in the present study show that 2-methylglyceric acid is formed as a minor marker compound through photooxidation of isoprene at low-NO_x conditions, and can also be produced from the C_5-alkene diols 1 and 3 under such conditions. It thus appears that there is a minor low-NO_x pathway operating in the formation of 2-methylgyceric acid. A possible pathway leading to the formation of 2-methylglyceric acid from diol 1 involving the photolysis of 2-hydroperoxy-1,3,4-trihydroxy-3-methylbutane is presented in Scheme 3. In this respect, it is worth mentioning that low atmospheric levels of 2-methylglyceric acid (on average 1.1 ng m^{-3}) have been measured in ambient fine aerosol collected during a 2004 summer episode at a boreal forest site, *i.e.*, Hyytiälä, Finland.[20]

4   Conclusions

Additional insights have been obtained on pathways leading to the formation of isoprene SOA marker compounds and the discrepant results reported in the literature for the underlying formation mechanisms, more specifically, with regard to the formation of the 2-methyltetrols which can either be explained by acid-catalyzed hydrolysis of C_5-epoxydiols or further photooxidation of C_5-alkene diols. We show that isomeric C_5-alkene diols serve as precursors for the 2-methyltetrols, *i.e.*, 2-methylthreitol and 2-methylerythritol, under low-NO_x conditions, and rule out their involvement in the formation of the C_5-alkene triols. Furthermore, we demonstrate that there is a minor low-NO_x pathway leading to 2-methylglyceric acid, which may be relevant to ambient conditions. Our efforts to detect C_5-epoxydiols in the gas phase upon photooxidation of isoprene were not successful, likely because of the prevalence of $RO_2 + RO_2$ reactions on the first-order C_5-hydroxyhydroperoxy radicals leading to the formation of C_5-alkene diols. Further experiments are warranted to optimize chamber conditions that favor the formation of C_5-alkene triols to better simulate ambient low-NO_x conditions and allow the chemical identification of intermediate C_5-epoxydiols.

Acknowledgements

This work was supported by the Belgian Federal Science Policy Office through the network project "Biogenic Influences on Oxidants and Secondary Organic Aerosol: theoretical, laboratory and modelling investigations (BIOSOA)" and Shanghai Leading Academic Disciplines (S30109).

References

1 A. Guenther, C. N. Hewitt, D. Erickson, R. Fall, C. Geron, T. Graedel, P. Harley, L. Klinger, M. Lerdau, W. A. Mckay, T. Pierce, B. Scholes, R. Steinbrecher, R. Tallamraju, J. Taylor and P. Zimmerman, *J. Geophys. Res.*, 1995, **100**, 8873.

2 M. Claeys, B. Graham, G. Vas, W. Wang, R. Vermeylen, V. Pashynska, J. Cafmeyer, P. Guyon, M. O. Andreae, P. Artaxo and W. Maenhaut, *Science*, 2004, **303**, 1173.

3 W. Wang, I. Kourtchev, B. Graham, J. Cafmeyer, W. Maenhaut and M. Claeys, *Rapid Commun. Mass Spectrom.*, 2005, **19**, 1343.

4 M. Claeys, W. Wang, A. C. Ion, I. Kourtchev, A. Gelencsér and W. Maenhaut, *Atmos. Environ.*, 2004, **38**, 4093.

5 E. O. Edney, T. E. Kleindienst, M. Jaoui, M. Lewandowski, J. H. Offenberg, W. Wang and M. Claeys, *Atmos. Environ.*, 2005, **39**, 5281.

6 J. D. Surratt, S. M. Murphy, J. H. Kroll, N. L. Ng, L. Hildebrandt, A. Sorooshian, R. Szmigielski, R. Vermeylen, W. Maenhaut, M. Claeys, R. C. Flagan and J. H. Seinfeld, *J. Phys. Chem. A*, 2006, **110**, 9665.

7 A. Miyoshi, S. Hatakeyama and N. Washida, *J. Geophys. Res.*, 1994, **99**, 18779.

8 L. Ruppert and K. H. Becker, *Atmos. Environ.*, 2000, **34**, 1529.

9 T. E. Kleindienst, M. Lewandowski, J. H. Offenberg, M. Jaoui and E. O. Edney, *Atmos. Environ.*, 2009, **9**, 6541.

10 F. Paulot, J. D. Crounse, H. G. Kjaergaard, A. Kürten, J. M. St. Clair, J. H. Seinfeld and P. O. Wennberg, *Science*, 2009, **325**, 730.

11 J. D. Surratt, A. W. H. Chan, N. C. Eddingsaas, M. N. Chan, C. L. Loza, A. J. Kwan, S. P. Hersey, R. C. Flagan, P. O. Wennberg and J. H. Seinfeld, *Proc. Natl. Acad. Sci. U. S. A.*, 2010, **107**, 6640.

12 O. Böge, Y. Miao, A. Plewka and H. Herrmann, *Atmos. Environ.*, 2006, **40**, 2501.

13 D. M. Hodgson, M. A. H. Stent and F. X. Wilson, *Org. Lett.*, 2001, **3**, 3401.

14 Y. Iinuma, O. Boge, M. Keywood, T. Gnauk and H. Herrmann, *Environ. Sci. Technol.*, 2009, **43**, 280.

15 W. Maenhaut, N. Raes, X. Chi, J. Cafmeyer and W. Wang, *X-Ray Spectrom.*, 2008, **37**, 193.
16 V. Pashynska, R. Vermeylen, G. Vas, W. Maenhaut and M. Claeys, *J. Mass Spectrom.*, 2002, **37**, 1249.
17 A. N. Grechkin, L. S. Mukhtarova and M. Hamberg, *Chem. Phys. Lipids*, 2005, **138**, 93.
18 Y.-H. Lin, H. Zhang, H. O. T. Pye, Z. Zhang, W. J. Marth, S. Park, M. Arashiro, T. Cui, S. H. Budisulistiorini, K. G. Sexton, W. Vizuete, Y. Xie, D. J. Luecken, I. R. Piletic, E. O. Edney, L. J. Bartolotti, A. Gold and J. D. Surratt, *Proc. Natl. Acad. Sci. U. S. A.*, 2013, **110**, 6718.
19 T. Berndt, *J. Atmos. Chem.*, 2012, **69**, 253.
20 I. Kourtchev, T. Ruuskanen, W. Maenhaut, M. Kulmala and M. Claeys, *Atmos. Chem. Phys.*, 2005, **5**, 2761.

Faraday Discussions

RSC Publishing

PAPER

Including phase separation in a unified model to calculate partitioning of vapours to mixed inorganic–organic aerosol particles†

David Topping,[*a] Mark Barley[‡b] and Gordon McFiggans[b]

Received 22nd March 2013, Accepted 10th May 2013

DOI: 10.1039/c3fd00047h

A simple approach to calculate liquid–liquid phase separation has been developed based on the derivation of partitioning coefficients between multiple liquid phases and inclusion in a framework used to partition an arbitrary number of compounds between the vapour and particle phases in an atmospheric aerosol. The representation compares favourably with a more complex and expensive benchmark gas-particle thermodynamic model for simple well-constrained systems. The model has then been applied to consider liquid phase separation in multicomponent particles formed by the equilibration of organic products generated by a near-explicit model of VOC oxidation. Inclusion of phase separation decreases the predicted mass of condensed organic material by $\sim$10 to $\sim$50%, dependent on the concentration of semi-volatile components and ambient conditions in the model scenario. The current study considers only two liquid phases, but the framework can readily accommodate an arbitrary number, though this is beyond the scope of the current work. Uncertainty introduced by the omission of phase separation is far lower than existing uncertainties in pure component vapour pressures, where orders of magnitude differences in predicted mass are found, though the bias introduced when choosing a particular method for estimating the saturation vapour pressure will influence the magnitude of phase separation. The proposed technique is the first to be used to practically deal with the many hundreds or thousands of components present in the ambient atmospheric aerosol.

1 Introduction

Atmospheric aerosol particles from both anthropogenic and biogenic sources are responsible for substantial uncertainty in the Earth system. They impact on

---

[a]National Centre for Atmospheric Science (NCAS), University of Manchester, Oxford Road, Manchester, M13 9PL, England. E-mail: david.topping@manchester.ac.uk; Tel: +0161 306 2490

[b]Centre for Atmospheric Science (CAS), School of Earth, Atmospheric and Environmental Science (SEAES), University of Manchester, Oxford Road, Manchester, M13 9PL, England

† Electronic supplementary information (ESI) available. See DOI: 10.1039/c3fd00047h

‡ Dr Mark Barley currently has a visitor status with the University.

climate by directly scattering and absorbing solar radiation (the direct effect), as well as regulating the properties of clouds[1,2] (the indirect effects). On regional scales, particulate pollution is a major contributor to deteriorating air quality.[3] Both impacts are poorly quantified. There are substantial uncertainties in many fundamental parameters that are required to predict impacts of sub-micron atmospheric aerosol. Many result from their significant organic mass fraction (20–90%),[4] comprising many thousands of compounds with poorly constrained properties. A wide range of methods exist to predict the equilibrium composition in multicomponent inorganic–organic aerosol particles. The most complex rely on minimisation of the Gibbs energy of the system (*e.g.* see Topping *et al.*[5] for a brief review). These models account for the contribution of each of the phases present in the system to its total Gibbs energy, numerically searching for the global minimum. The number of phases, vapour or condensed, determines the complexity of the problem. Models that search the entire Gibbs energy surface (GFEMN[6]), those that rely on constrained nonlinear global minimisation (ADDEM,[5] E-AIM[7]), and those that employ a hybrid combination of the two (MCM-EVAPORATION-AIOMFAC[8]) have all been used to investigate atmospheric aerosol thermodynamics.

The physical state and morphology of internally mixed organic/inorganic aerosol particles is largely uncertain[9] and the wide range of concentrations and properties of organic compounds found within aerosol particles in the atmosphere suggest that liquid–liquid phase separation might occur under certain ambient conditions. The ability to predict liquid–liquid equilibria (LLE) will add complexity to a model. Within the absolute Gibbs energy minimisation approach, the concentration of each component in the additional liquid phase must be reflected in the energy summation. Using absolute concentrations, Topping *et al.*[5] and Ansari and Pandis[6] discussed the problems associated with uncertainties in fundamental parameters, such as Gibbs energies of formation, on predicting mass transfer across phase boundaries. More recently, Zuend *et al.*[10] demonstrated that if the condensed phase system is constrained to two liquid phases, an expression for the normalised Gibbs energy can be derived which mitigates this problem, similar to the method used by Ming and Russell.[11] A limiting factor in any thermodynamic equilibrium model is the applicability of the method chosen for prediction of activity coefficients in a solution. Prediction of the distribution of components in multiple phases in a broad range of mixed inorganic/organic systems has previously been limited by the narrow range of applicability of available activity coefficient methods. A review of the development, evaluation and improvement of such methods is beyond the scope of this report. Recently, Zuend *et al.*[12] extended the AIOMFAC functional group method, demonstrating its broad applicability by predicting liquid–liquid phase separation in simple mixtures and demonstrating the potential importance for secondary organic aerosol (SOA) mass loadings.[8]

Available thermodynamic methods that account for partitioning between two liquid phases are computationally expensive and the computational expense scales with each additional component.[8] Ultimately, assessing the importance of LLE for gas-particle partitioning requires the ability to account for the many numbers of organics expected to occur within the aerosol phase. Whilst empirical parameterisations based on organic : sulphate and O : C ratios have been constrained to specific systems,[13] there is no general consensus on how to account for

LLE across a broad compositional parameter space. Similarly, laboratory studies using a limited range of compounds may not be indicative of behaviour in the atmosphere. For vapour–liquid equilibria (VLE), numerous investigations have relied on bulk absorptive partitioning theory. In this approach, the equations defining chemical equilibrium are recast and all equations defining partitioning between a condensed and gaseous phase solved simultaneously rather than directly, minimising the Gibbs energy. Recently, McFiggans et al.[14] and Topping and McFiggans[15] demonstrate that this approach can be used to account for any number of condensing compounds, particle size and non-ideality through use of an iterative solver. In this study, we extend this approach to account for predictions of LLE for any number and type of organic compounds and thus derive a generic simplified method for predicting coupled VLE and LLE in atmospheric aerosol particles. Comparisons with more complex benchmark models are made along with a discussion as the wider significance of LLE in relation to additional uncertainties in other properties.

2   New LLE model methodology

Central to absorptive partitioning theory is the definition of a partitioning coefficients across two phases. For VLE these are defined as:[14]

$$\xi_i^{\text{VLE}} = \left(1 + \frac{C_i^*}{\text{COA}}\right)^{-1} \tag{1}$$

Where ξ_i^{VLE} is the partitioning coefficient of component i between the condensed and gaseous phase, C_i^* the saturation concentration (μg m^{-3}) and COA the total condensed phase concentration (μg m^{-3}). For LLE similar expressions can be derived starting from the condition that component 'i' has to have the same chemical potential in the two liquid phases (see supplementary material†). In doing so, the partitioning coefficient for component i between two liquid phases (ξ_i^{LLE}) is given by:

$$\xi_i^{\text{LLE}} = \left[1 + \left(\frac{N_{\text{T}}}{N_{\text{T}}^\alpha} - 1\right)\frac{f_i^\alpha}{f_i^\beta}\right]^{-1} \tag{2}$$

Where N_{T} is the total number of moles in the mixture, N_{T}^α the total number of moles in one liquid phase (denoted phase α), and $\dfrac{f_i^\alpha}{f_i^\beta}$ the ratio of activity coefficients of component i in phase α and β (on a mole fraction scale). The equilibrium condition can then be expressed as:

$$\sum \left[1 + \left(\frac{N_{\text{T}}}{N_{\text{T}}^\alpha} - 1\right)\frac{f_i^\alpha}{f_i^\beta}\right]^{-1} n_i^{\text{T}} - n_i^\alpha = 0 \tag{3}$$

where n_i^{T} is the total molar amount of component i in the mixture and n_i^α the total concentration of component i in phase α. Using the analogy of McFiggans,[14] one option is to solve for the value of N_{T}^α using a Newton approach. If the activity coefficients $\dfrac{f_i^\alpha}{f_i^\beta}$ are allowed to vary at each step, there are potential instability concerns and sensitivity to the initial starting point. The likelihood of such problems increases with the number of components within the mixture. In a binary system, each value of N_T^α has one unique combination of individual components, whilst for higher order systems this is no longer true. There are also potential problems with predictions of $\dfrac{f_i^\alpha}{f_i^\beta}$ in extreme conditions (*e.g.* very high ionic strengths) where values of $\dfrac{f_i^\alpha}{f_i^\beta}$ force the solver into an infinite loop. AIOMFAC has been constrained to many systems with varying organic functionality and inorganic compounds. Comprised of 'long range', 'mid range' and 'short range' contributions to the excess Gibbs energy, the mid range terms account for organic–inorganic interactions on a molality and ionic strength scale. Predicting activity coefficients in supersaturated solutions may therefore lead to highly nonlinear behaviour, as noted by Zuend *et al.*[8] Performing a full failure analysis on the Newton method would then be system dependent. As an alternative, global optimisation routines can be applied to solve an adapted form of eqn (3) given by the following:

$$\mathrm{ABS}\left(\sum \left[1 + \left(\frac{N_T}{N_T^\alpha} - 1 \right) \frac{f_i^\alpha}{f_i^\beta} \right]^{-1} - \xi_i^{\mathrm{LLE}} \right) \tag{4}$$

where the variables being optimised, the LLE partitioning coefficients ξ_i^{LLE}, vary between 0–1. Choice of global optimisation routine would require the ability to set bounds on ξ_i^{LLE} and the ability to use finite difference to calculation the Jacobian matrix. Following Topping *et al.*,[5] one option would be to use a Sequential Least SQuares Programming ('SLSQP') optimisation algorithm, though discussion of the ability of various global minimisation algorithms to ensure convergence on the global minimum is beyond the scope of this paper. In the approach of Zuend *et al.*,[8] a hybrid method has been developed to minimise the normalised Gibbs energy of the two liquid system.

Solving the LLE problem using Gibbs energy minimisation approaches is computationally expensive. In addition, as the number of components increases, the ability of the solver to find the global minimum potentially becomes compromised. Ideally we would like to derive an efficient solver that is unlimited with respect to the number of components in the condensed phase in order to probe the potential importance of LLE for atmospherically relevant systems. If the form of the equations defining mass transfer across multiple phase boundaries are the same then we can apply the same numerical solver to either process and develop a unified model framework of SOA formation. Revisiting the form of eqn (2), constraining the value of $\dfrac{f_i^\alpha}{f_i^\beta}$ enables the same method used by McFiggans *et al.*[14] to be applied to solving the problem of LLE. Moreover, the derivative function of eqn (2) is easily attained, ensuring efficiency and guaranteeing convergence. Constraining $\dfrac{f_i^\alpha}{f_i^\beta}$ is different from the approaches discussed previously. Here we want to force the algorithm to replicate 'expected' LLE behaviour based on initial predictions from the AIOMFAC model, or any other accurate activity coefficient model.

The concept is perhaps best illustrated using the following argument. Two liquids (A and B) will only form two liquid phases if the A–A and B–B interactions are much stronger than the A–B interactions. Two extremes exist for molecules A and B: i) molecules which only interact by London Dispersion forces (hydrocarbons) and ii) molecules that interact by dipole–dipole interactions and by charge transfer (water and ions). Hence water (and ions) will form a separate liquid phase when put into a mixture with a hydrocarbon. Organics of atmospheric interest are assumed to form a continuum between H-bonding/dipole interactions at one extreme and purely London Dispersion forces at the other, so will partition between the two phases (aqueous and hydrocarbon) to varying degrees. Since the troposphere is always moist and the ambient particulate will contain a substantial amount of liquid water at all reasonable ambient RH, water is assumed to be the dominant solvent in one of the phases. Binary solutions of each component are sequentially explored to discover the component least miscible with water using the AIOMFAC activity coefficient model. Once the component with lowest water affinity has been found, the values of $\dfrac{f_i^{\alpha}}{f_i^{\beta}}$ for each component are then calculated with the AIOMFAC model using fixed concentrations. In this way we follow the approach used to define octanol–water partitioning coefficients. Octanol–water partitioning coefficients are a measure of the equilibrium concentration between octanol and water, which indicates potential for partitioning between aqueous and non-aqueous mixtures. It is widely used in determining the fate of organic chemicals in the environment.[16] It can also be related to the ratio of activity coefficients of a given component in each phase.[17] When using this approach, octanol–water partitioning coefficients are evaluated at infinite dilution and assumed to be independent of concentration. Similarly, we assume the partitioning coefficient between the aqueous solution and the non-aqueous mixture to be independent of concentration and hence can be evaluated at a constant ratio of activity coefficients for each component. The partitioning coefficient we calculate using eqn (2) is between the aqueous solution and the non-aqueous mixture containing the component found to have had the highest activity coefficient in a binary aqueous solution. In summary, the approach we use quantitively calculates the grouping of compounds that want to be in the same liquid phase based on the predicted activity coefficient ratios using water and the compound with the highest activity coefficient in water. The model presented in this study is designed such that it can be used for complex mixtures with hundreds or thousands of components. In this case the important thing is to group the compounds into those favouring an aqueous environment and those preferring a hydrocarbon-like environment, for example. If the specific compound selected because it has the highest activity coefficient with water also has a very low abundance this is not a problem as there will be many other compounds of a very similar hydrophobic nature that do exceed their solubility limit in the aqueous solution and a second phase will be formed. While we use water and another organic compound to define those activity coefficient ratios, we are not forced into defining specific compositions of those two liquid phases, rather we force the algorithm to replicate expected LLE behaviour based on initial predictions from the AIOMFAC model, or any other accurate activity coefficient model. In the AIOMFAC model it is unnecessary to define separate reference states for solvent and solute and hence the activity coefficient of component i in a mixture with the component with least water affinity does not require the latter to be the solvent.

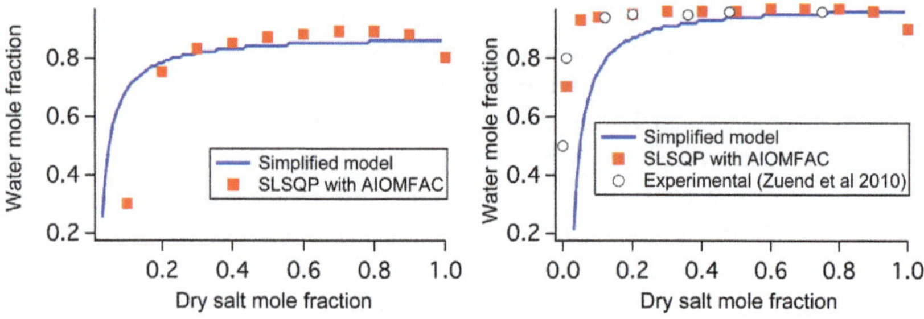

Fig. 1 Phase diagrams of glycerol–$(NH_4)_2SO_4$–H_2O (left) and *tert*-butanol–Na_2SO_4–H_2O (right). 'SLSQP' refers to the use of a global minimisation algorithm and the AIOMFAC activity coefficient model applied to eqn (4) as described in the main text. The blue lines are the results obtained from the simplified method described in this paper. Below each solid blue line, the simplified model presented in this work predicts the co-existence of two liquid phases. The experimental data in the right panel is taken from the compilation of Zuend *et al.*[10]

As the definition of chemical equilibrium was used in defining eqn (2), the value at the root ensures that this criteria is always satisfied. Fig. 1 illustrates the applicability of this simplified approach for 2 ternary systems following Zuend *et al.*[8] and comparison with the SLSQP method for solving eqn (4) and allowing activity coefficients to vary. For the glycerol–$(NH_4)_2SO_4$–H_2O system, the simplified model captures the trend very well, with increasing divergence at lower dry salt mole fractions. The degree to which solid precipitation would restrict the co-existence of two liquid phases at such low mole fractions of water is unclear and should be the focus of future work. According to Zuend *et al.*,[8] the electrolyte solution is saturated above the two liquid phase region across all dry salt mole fractions. For the *tert*-butanol–Na_2SO_4–H_2O system, again the simplified model replicates the trend. The 'SLSQP' method replicates measured data extremely well, discrepancies between that method and the simplified scheme increasing at lower dry salt mole fractions. It is expected that some discrepancies will exist between the simple approach described here and complex Gibbs energy approaches. However, these results give sufficient confidence in the ability of the approach to capture the behaviour of LLE within a specific system; the following section assesses impacts on predicted SOA mass by comparing with more complex predictions.

3 Coupled LLE–VLE model

The inclusion of LLE calculations within the standard VLE model simply requires inclusion of another iterative loop. In the following analysis we restrict the simulations to two liquid phases. It is possible however to account for any number of liquid phases using eqn (1) provided the phase rule is obeyed (consideration of further liquid phases simply requiring their definition and additional loops). The choice of components used to represent each liquid phase is defined by the ratio of activity coefficients for each possible binary pair. The potential importance of additional liquid phases will be explored in a future study. It is recognised that results from such a technique should be compared with a Gibbs energy minimisation approach. To the best of our knowledge, this is currently not possible with existing modelling techniques for complex atmospheric systems.

Table 1 Compounds taken from the study of Zuend et al.[10] The values for $\dfrac{f_i^\alpha}{f_i^\beta}$ calculated using the AIOMFAC activity coefficient model as described in the text. In this case the solvents were taken as water (phase β) and 1,2,10-decanetriol (phase α) based on the predictions of AIOMFAC. M is the molecular weight, P_{io} is the saturation vapour pressure for each compound and T_b (K) is the boiling point temperature

| Compound | Chemical formula | M (kg mol^{-1}) | |
T_b (K)	O : C ratio	$\dfrac{f_i^\alpha}{f_i^\beta}$	P_{io} (298 K) (Pa)
water	H_2O	1.80153×10^{-2}	3168.49
373	—	6.3043	
glycerol	$C_3H_5(OH)_3$	9.20940×10^{-2}	2.284×10^{-2}
588	1.0	2.14450	
1,6-hexanediol	$C_6H_{12}(OH)_2$	1.18172×10^{-1}	5.695×10^{-2}
525	0.33333	1.6933×10^{-2}	
1,2,5,8-octanetetrol	$C_8H_{14}(OH)_4$	1.78224×10^{-1}	6.725×10^{-5}
640	0.5	3.42535×10^1	
1,2,10-decanetriol	$C_{10}H_{19}(OH)_3$	1.90276×10^{-1}	1.826×10^{-4}
622	0.3	1.20061×10^{-4}	
ammonium sulphate	$(NH_4)_2SO_4$	1.32139×10^{-1}	(0.0)
—	—	3.45395×10^3	

3.1   Comparison with benchmark thermodynamic model

To test the applicability of the simplified model described by eqn (1)–(3), a set of coupled VLE–LLE partitioning calculations are compared with the results of Zuend et al.[10] for a system comprising ammonium sulphate and 4 organic compounds across a range of relative humidity (RH). The properties of the organic compounds are displayed in Table 1, taken from Zuend et al.[10] Also displayed are the values calculated for $\dfrac{f_i^\alpha}{f_i^\beta}$. The total amount of the organic compounds was set to 3.0×10^{-8} mol (30 nmol) and the amount of ammonium sulphate to 1.0×10^{-8} mol. The list of compounds studied and properties used are replicated in Table 1. In Fig. 2, the variability in predicted SOA mass is displayed with results using the Gibbs energy minimisation approach of Zuend et al.[10] and the method described here. Zuend et al.[10] found that at 20% RH, LLE results in a decrease in total SOA mass of 24.74% RH compared to an idealised one liquid phase simulation. On the graph, these are represented by the red and blue circles, respectively. In this study we predict a decrease in mass of −17.39%. At higher RH the results from both modelling approaches tend to converge. For example, at 60% RH, Zuend et al.[10] predict a decrease in mass of 12.19%, this study predicting a decrease of 12.7%. At higher RH the impact of LLE decreases as water becomes the dominant solvent.

3.2   Simulations of ambient aerosol

The primary and most considerable advantage of the current approach is that it is unrestricted in terms of the number of compounds that can be used within the calculations. It is possible to assess the potential importance of LLE in ambient

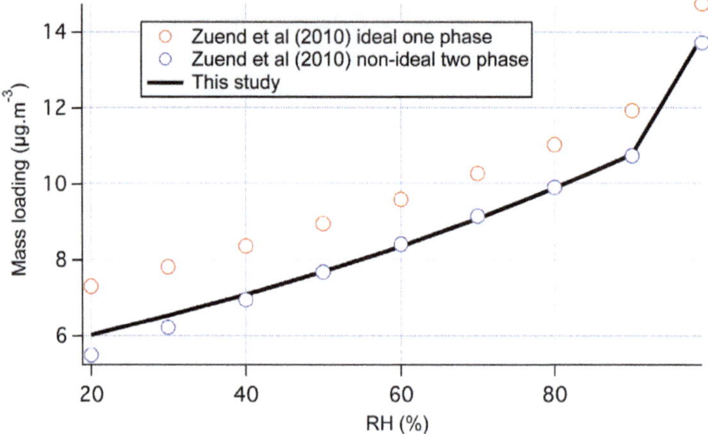

Fig. 2 Predicted mass loading at 298.15 K as a function of RH using the simplified model described in the text (black line) and the benchmark thermodynamic model of Zuend *et al.*[10] (blue circles).

aerosol using output from a near-explicit gas phase degradation model. Here we use the results described by Barley *et al.*[18] in which the Master Chemical Mechanism (MCM) was used to simulate the gas phase composition of air masses over multiple environments. In all, 2742 compounds are used to simulate a wide range of anthropogenic and biogenic scenarios, from UK industrial/urban (high anthropogenic and low biogenic inputs) through to rural background conditions (higher biogenic and low anthropogenic inputs). Emission scenarios were simulated by independently multiplying the Anthropogenic VOCs, biogenic VOCs (BVOCs) and NO$_x$ component of a base case emissions by factors of 0.01, 0.1, and 10. For these example simulations we use the 'biogenic' scenario, which was dominated by isoprene, α- and β-pinene emissions.

In these simulations we assume the same abundance of ammonium sulphate as used by Zuend *et al.*[10] but allow all of the 2742 compounds to re-equilibrate between the gaseous and condensed phases as a function of RH. The model also does not account for any solid precipitation, assuming the ammonium sulphate exists in a metastable state when below its deliquescence RH. As before, the AIOMFAC activity coefficient model was used to calculate the activity coefficient ratio $\dfrac{f_i^\alpha}{f_i^\beta}$ of each component in water and the reference organic solvent. As detailed earlier, the reference organic component used to calculate activity coefficients in the non-aqueous phase was chosen by cycling through the range of predicted activity coefficients in binary solutions in water and selecting the maximum value. The choice of reference component is responsive to the range of compounds used within the model simulations, and as such it is not possible to generalise a model compound for future studies. It should be noted here that the choice of component is solely on the basis of that which is least miscible in water as predicted by AIOMFAC and is independent of its abundance. Selection of the most dissimilar component in this way provides an appropriate starting point for the calculation. All other components are then allowed to partition to the separate phases containing these dissimilar components. It has been noted that the O : C ratio is a

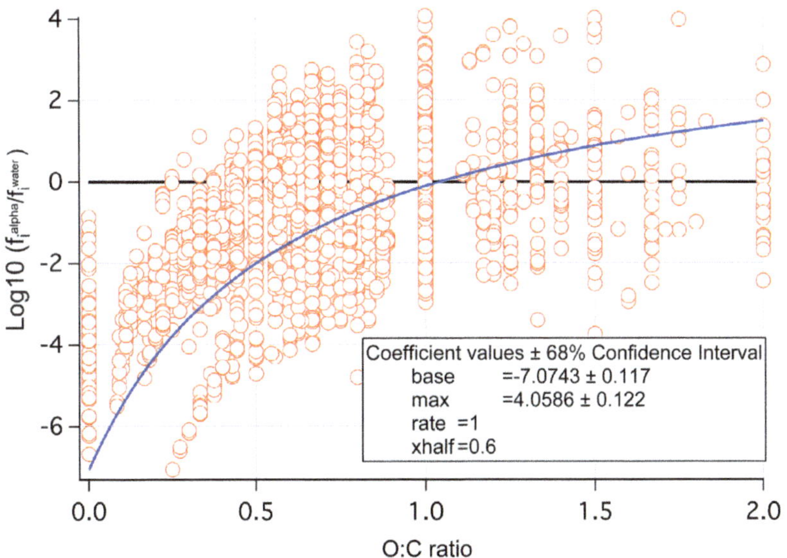

Fig. 3   Logarithm of $\frac{f_i^\alpha}{f_i^\beta}$ for all of the compounds used in the MCM *versus* O : C ratio where solvent β is taken as water.

good first-order proxy for hydrophilicity and hygroscopicity of organic compounds[4,19,20] and the expected degree of non-ideality due to organic–inorganic interactions in aqueous mixtures.[8] It is demonstrated here that it is not possible to generalise relationships between $\frac{f_i^\alpha}{f_i^\beta}$ and O : C ratio (Fig. 3). In this figure each value for $\frac{f_i^\alpha}{f_i^\beta}$ at 298.15 K is plotted as a function of O : C ratio and a Hill equation fit through the dataset (eqn (5)). This was taken as the best parametric fit from the IgorPro 6.2 data analysis software package and is used simply to represent the extracted trend in the data.

$$\log\left(\frac{f_i^\alpha}{f_i^\beta}\right) = \text{base} + (\text{max} - \text{base})\Big/\left\{1 + \left[\frac{'\text{xhalf}'}{\text{OCratio}}\right]^{\text{rate}}\right\} \tag{5}$$

where the values for base, max, xhalf and rate are given in the figure legend. Whilst there is a general trend of increasing $\frac{f_i^\alpha}{f_i^\beta}$ with O : C ratio, demonstrating more oxidised compounds will preferentially partition to the aqueous phase, there is a large spread of values. For example, at an O : C ratio of 1.0, $\frac{f_i^\alpha}{f_i^\beta}$ varies by seven orders of magnitude. Prescribing one value to compounds in this region could have significant consequences on the amount and composition of each liquid phase. In this case, each compound would be assumed to have the potential to partition roughly equally between water and the non-aqueous reference mixture as the value for $\frac{f_i^\alpha}{f_i^\beta}$ approaches unity. The magnitude of partitioning

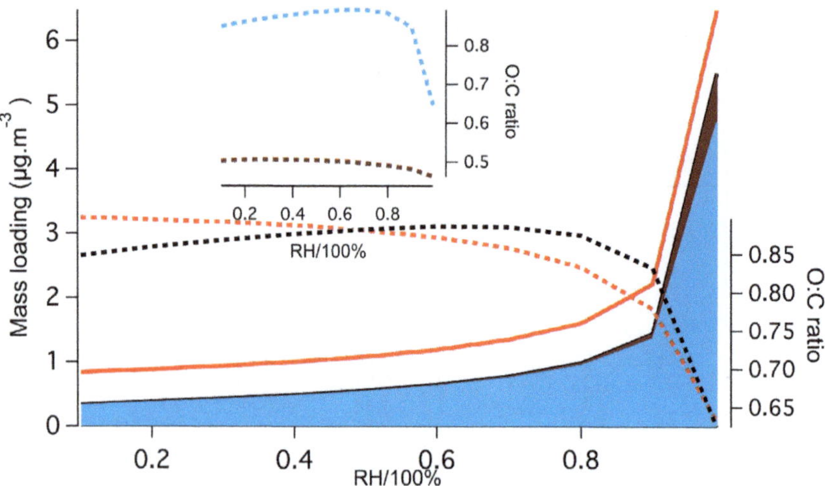

Fig. 4 Mass loadings as a function of RH by assuming an ideal one liquid phase system (solid red line), or assuming liquid–liquid phase separation (solid black line). The contributions from a 'high' O : C ratio phase (blue area) and 'low' O : C ratio phase (brown area) are displayed under the solid black line, the actual O : C ratio of each phase displayed in the subplot. The total O : C ratio of the idealised one liquid phase model (dashed red line) is compared with the two liquid phase model (dashed black line).

between separate liquid phases and the resulting impact on predicted SOA mass will depend on the relative concentrations of each and every compound. In the following discussion we use the prescribed abundances from the MCM to assess the potential impact of LLE on predicted SOA mass and O : C ratio.

Fig. 4 compares the predicted total SOA mass loadings for a 'biogenic' scenario at 298.15 K with and without the assumption of LLE, using the method of Nannoolal *et al.*[21,22] to predict the pure component vapour pressures and boiling points (hereafter referred to as the 'NN' method). The solid red lines show the variability in predicted SOA mass when assuming an ideal one liquid phase, the solid black line the variability in predicted mass when assuming LLE and accounting for predicted activity coefficients in each phase. The dashed lines display the predicted overall O : C ratio using the same modelling approaches. Under the solid black lines, the amount of each separate liquid phase is displayed using two colours. The O : C ratio of these phases are also displayed in the subplot as a function of RH. The blue area represents the 'high' O : C ratio phase and the brown area the 'low' O : C ratio phase. In both the one and two liquid phase simulations, mass increases as a function of RH. At the lowest RH, the two liquid phase simulation leads to a 59.8% decrease in mass which reduces to 15.2% at higher RH. Note the one phase simulation is an ideal model. As noted by Zuend *et al.*,[10] incorporation of non-ideality tends to lead to a decrease in predicted mass, whereas Barley *et al.*[18] indicated the effect is likely case dependent. In this case the mass always tends to decrease. At high RH, compounds with higher predicted saturation vapour pressures (P_{io}) are able to condense through a reduction in their equilibrium vapour pressure above the multicomponent solution. Zuend *et al.*[8] note that SOA mixtures with an average O : C > 0.7 may not exhibit phase separation in the presence of dissolved ammonium sulphate even at lower RH.[9,13]

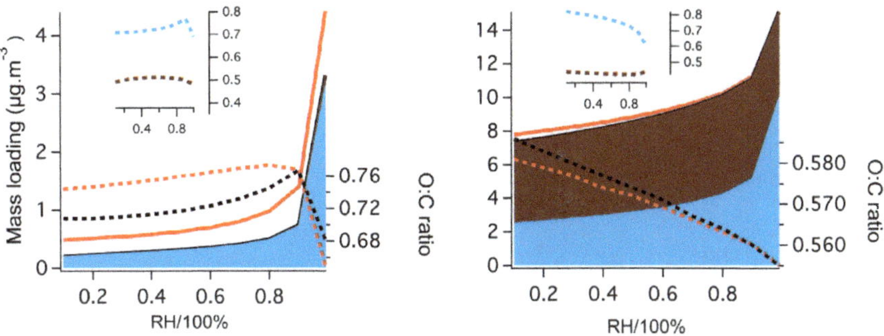

Fig. 5 Two methods for calculating saturation vapour pressures as described in the text (left panel: EVAPORATION (Compernolle *et al.*);[23] right panel: vapour pressure extrapolation of Nanoolal *et al.*[21] combined with boiling point predictions of Joback and Reid[24]). Mass loadings as a function of RH by assuming an ideal one liquid phase system (solid red line), or assuming liquid–liquid phase separation (solid black line). The contributions from a 'high' O : C ratio phase (blue area) and 'low' O : C ratio phase (brown area) are displayed under the solid black line, the actual O : C ratio of each phase displayed in the subplot. The total O : C ratio of the idealised one liquid phase model (dashed red line) is compared with the two liquid phase model (dashed black line).

However, the same authors note that for O : C < 0.5, assuming complete organic/ electrolyte phase separation over the entire RH range is a valid simplification. Using the simplified framework presented here, it is possible to confirm this through an analysis on methods used to predict P_{io}.

As demonstrated by McFiggans *et al.*,[14] whilst the mean condensed predicted O : C ratio from the MCM simulations can be quite high, the result is also dependent on the choice of method for predicting P_{io}. Saturation vapour pressures of organic components are currently poorly known, particularly for the least volatile compounds.[25] Uncertainty in this parameter is already known to introduce 4 orders of magnitude uncertainty in predicted mass.[14] McFiggans *et al.*[14] presented results displaying heterogeneity with regards to differences in calculated P_{io} as a function of O : C ratio and molecular weight. It is then reasonable to assume that choice of P_{io} will also affect the magnitude of liquid–liquid phase separation and therefore again influence the predicted mass and composition. This is demonstrated in Fig. 5 where predictions of mass with and without LLE are calculated using two additional two methods for calculating P_{io}: EVAPORA-TION (Compernolle *et al.*[23] – hereafter referred to as 'EVAP' method); vapour pressure extrapolation of Nanoolal *et al.*[21] combined with boiling point predictions of Joback and Reid[24] (hereafter referred to as the 'NJR' method). Note in this study the 'EVAP' method employed accounts for the diacid correction presented by Booth *et al.*[26]

Fig. 6 plots the absolute difference in predicted P_{io} between the vapour pressure method 'EVAP' and the combined vapour pressure/boiling point method 'NJR' as a function of O : C ratio. This figure shows that, whilst there is some heterogeneity covering up to 5 orders of magnitude, predictions from the 'NN' method systematically under-predict relative to the 'EVAP' method. It would be expected that use of the 'NJR' method would lead to higher amounts of predicted SOA. This is confirmed in Fig. 5 where the predicted mass varies from 0.478 to 7.76 µg m^{-3} (a difference of 1523%) between the 'EVAP' and 'NJR' method for the ideal case at 10% RH. At 90% RH these values increase to 1.39 and 11.2 µg m^{-3},

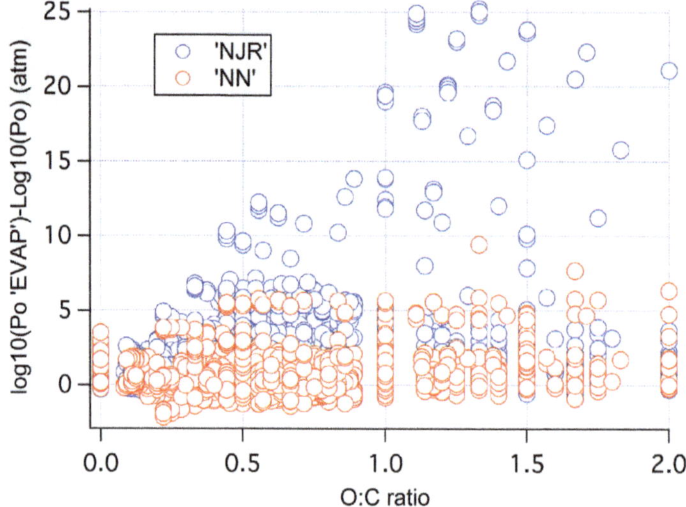

Fig. 6 Absolute differences in predicted pure component vapour pressures when comparing the EVAPORATION 'EVAP' method of Compernolle *et al.*[23] with vapour pressure extrapolation of Nanoolal *et al.*[21] combined with boiling point predictions of Joback and Reid[24] (NJR' method) and both vapour pressure extrapolation of Nanoolal *et al.*[21] combined with boiling point predictions of Nanoolal *et al.*[22] ('NN').

respectively. Assuming LLE in both cases at 10% RH leads to a decrease in mass to 0.201 and 7.34 μg m^{-3} (a difference compared with the ideal cases of 57% and 5.41%, respectively). In this case therefore, the uncertainty introduced through choice of vapour pressure method in the ideal case is much larger than that introduced through assumption of LLE. At 90% RH the same pattern emerges where absolute percentage difference in mass between the 'EVAP' and 'NJR' is +805%, the assumption of non-idealised LLE −58% and −35% for both methods, respectively. As demonstrated in Fig. 5, for both the ideal and non-ideal LLE method using the 'NN' technique the predicted O : C ratio of the aerosol is higher than that using the 'EVAP' method. Barley and McFiggans[25] reviewed available methods for predicting P_{io}. In that review the authors found that a popular method for predicting boiling points, that of Joback and Reid,[24] systematically leads to a large underprediction in P_{io}. This is demonstrated in Fig. 6 where the blue circles are predictions of the 'NJR' method for P_{io} for each MCM compound relative to the 'EVAP' technique. As with the 'NN' method, values of P_{io} are systematically lower than the 'EVAP' method, often by greater than 5 orders of magnitude. As expected, the overall mass predicted from the 'NJR' is much greater than the other two methods. At 10% RH the ideal mass increases to 7.76 μg m^{-3} as compared with 0.478 μg m^{-3} using the 'EVAP' technique, a huge 1523% difference. In addition, there is a clear splitting between a 'high' and 'low' O : C ratio liquid phase across all RH, the contribution from the 'low' O : C ratio liquid phase diminishing at higher RH, similar to the comparison with Zuend *et al.*[10] As mentioned previously, for O : C ratios approaching 0.5 these results suggest assumed complete phase separation over the entire RH range might be a valid approximation.

Table 2 Variability in total semi-volatile organic condensed material as a function of RH and ratio of HOA : inorganic, both totalling 2 µg m^{-3}

(Inorganic : HOA)	10%	20%	30%	40%	50%	60%	70%	80%	90%
SOA (1.99 : 0.01)	1.03	1.07	1.11	1.17	1.25	1.35	1.51	1.79	2.43
SOA (1 : 1)	1.82×10^{-1}	8.59×10^{-1}	9.01×10^{-1}	9.54×10^{-1}	1.02	1.12	1.26	1.50	2.06
SOA (0.01 : 1.99)	4.65×10^{-1}	4.96×10^{-1}	5.34×10^{-1}	5.82×10^{-1}	6.43×10^{-1}	7.28×10^{-1}	8.51×10^{-1}	1.05	1.51
Water (1.99 : 0.01)	9.47×10^{-2}	2.05×10^{-1}	3.40×10^{-1}	5.16×10^{-1}	7.59×10^{-1}	1.12	1.74	3.03	7.26
Water (1 : 1)	5.54×10^{-2}	1.21×10^{-1}	2.04×10^{-1}	3.12×10^{-1}	4.65×10^{-1}	6.98×10^{-1}	1.10	1.95	4.82
Water (0.01 : 1.99)	1.49×10^{-2}	3.44×10^{-2}	6.09×10^{-2}	9.84×10^{-2}	1.55×10^{-1}	2.46×10^{-1}	4.15×10^{-1}	7.98×10^{-1}	2.22

Under most atmospheric conditions, the ambient particle distribution might be expected to comprise both primary and secondary organic components. Indeed, strong circumstantial evidence for this has been provided by the ubiquitous identification of a hydrocarbon-like organic aerosol (HOA) fraction along with the oxygenated fractions in a compilation of Aerosol Mass Spectrometer (AMS) observations.[4] Whilst the internal mixing of HOA and oxygenated fractions has not been unambiguously demonstrated, assumption of their presence in the same particles provides a useful limiting case for atmospheric phase separation. A number of simulations have been conducted to investigate the equilibration of oxygenated organic components to particles containing using HOA and $(NH_4)_2SO_4$ as largely involatile water insoluble and water soluble components respectively. HOA was represented as triacontane,[27] a $C_{30}H_{62}$ alkane with molecular weight of 0.33866 kg mol^{-1}. The saturation vapour pressure was calculated using the same vapour pressure method for HOA and all other organic components. The mass ratio of HOA to $(NH_4)_2SO_4$ was varied from 1.99 : 0.01, 1 : 1 and

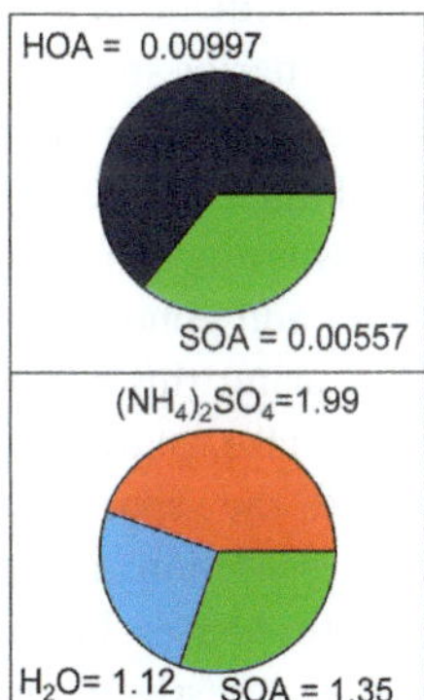

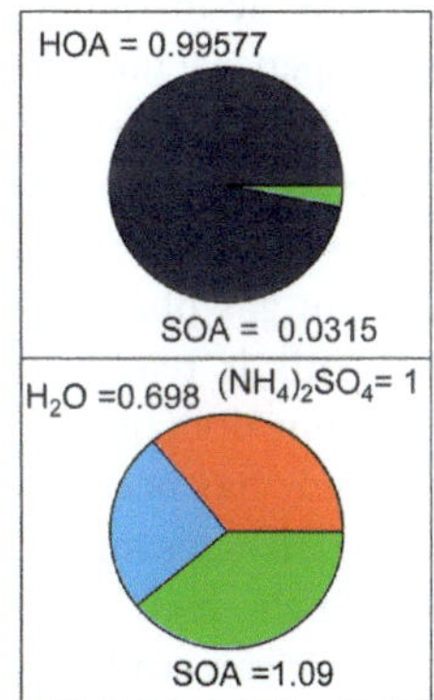

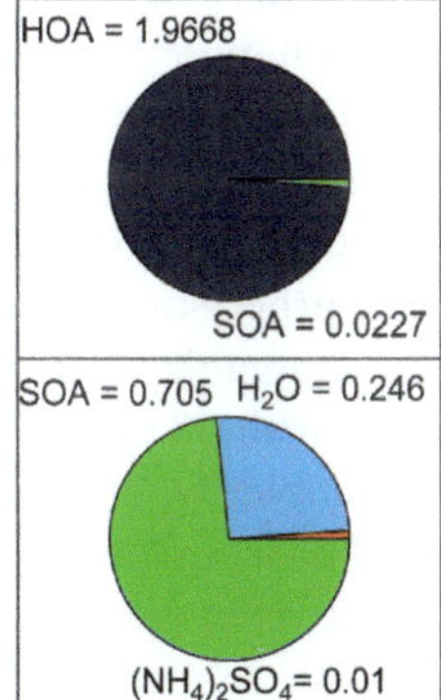

Fig. 7 Relative composition of two separate condensed liquid phases at 60% RH for three different relative amounts (µg m^{-3}) of HOA and ammonium sulphate $(NH_4)_2SO_4$ (left panels HOA = 0.01, $(NH_4)_2SO_4$ = 1.99; middle panels HOA = 1, $(NH_4)_2SO_4$ = 1; right panels HOA = 1.99, $(NH_4)_2SO_4$ = 0.01). The composition is displayed as contributions from HOA (black), SOA (green), $(NH_4)_2SO_4$ (red) and water (blue). Absolute amounts in each phase is displayed in each sub-panel.

0.01 : 1.99 to cover the range presented by Zhang *et al.*[3] Table 2 displays the total condensed non-HOA organic material as a function of RH and HOA : ammonium sulphate ratio. Fig.7 displays the relative composition of the predicted two liquid phases at 60% RH. For all simulations, it has been assumed that the inorganic salt exists in a metastable state and hence that phase separation is between the aqueous and organic phases, without solid precipitation.

The vast majority of the water is associated with the inorganic, providing the aqueous phase. The majority of the condensed non-HOA organic material is found to partition preferentially to this aqueous phase with a small amount partitioning to the predominantly HOA phase. As an increased fraction of HOA is prescribed, the total concentration of both water and non-HOA organic material decreases. At 10% RH and 90% RH there is a decrease of 54.85% and 37.86%, respectively. This is to be expected since i) water, as a polar solvent, has less electrolyte to solvate and ii) the majority of secondary organic material is heavily oxygenated and hence will preferentially partition to a polar solvent. As with all previous examples, the total concentration of condensed secondary organic material is reduced when LLE is considered. It may be expected that the phase separation of SOA species in the presence of an aqueous electrolyte solution and HOA is an extreme case and therefore that the partitioning of SVOC from VOC oxidation to the oxidation products of the so-called Intermediate Volatility Organic Compounds (IVOCs) will be somewhat less extreme. The effect of the presence of HOA on the partitioning of the IVOC oxidation products themselves will be dependent on their exact chemical nature and should be the focus of further study.

4   Discussion and conclusions

A simplified representation of liquid–liquid phase separation has been developed based on the derivation of partitioning coefficients between multiple liquid phases, constraining activity coefficient ratios for each compound in the mixture in separate solvents. Designed to represent the broad continuum of molecular interactions occurring within complex mixtures, comparisons with an existing benchmark gas-particle thermodynamic model shows the predictions compare very well with more complex and expensive methods. For scenarios studied here, inclusion of LLE decreases the predicted mass of condensed organic material by $\sim$10 to $\sim$50%, dependent on the concentration of semi-volatile components and ambient conditions. The generalisation of the validity of assuming complete LLE requires further studies on different gas phase precursors and conditions with different inorganic electrolytes.[8] This is readily possible using this simplified framework without the need to assume domain dependent phase splitting for any level of complexity. The current study has been restricted to the separation into two liquid phases. There is no theoretical reason for such a restriction and indeed there are circumstances under which three and even four phases have been observed.[28-33] The current framework can readily be used to investigate an arbitrary number of liquid phases and should be the focus of future work. However, we have shown that the uncertainty introduced by the inclusion or neglect of LLE is far lower than existing uncertainties in properties, such as pure component vapour pressures, where orders of magnitude differences in predicted mass are often encountered,[14,25,26] though the bias introduced when choosing a particular method for estimating the saturation vapour pressure will influence the magnitude of phase separation. Again, a sensitivity study across a more comprehensive range of property estimation techniques should be undertaken.

Importantly, in the real atmosphere, it should be recognised that there is an extremely broad range of sources of particles within a given airmass. Each air parcel may contain particles from such disparate sources as sea-spray generation, primary dust generation and a variety of anthropogenic (*e.g.* combustion) and biogenic (*e.g.* fungal spores) sources. It should therefore be recognised that multiple primary and secondary sources will contribute to particle composition throughout an airmass history and that mixing state is much more likely to be the primary driver for separation of components in the atmosphere. It is intuitively much more important than the thermodynamically driven 'phase separation' considered in our bulk analysis, though a consideration of the evolution of component mixing state through coagulation is beyond the scope of the current study.

Further additional uncertainties surround the influence of any solid precipitation that has been ignored in the present study. Such effects, potentially of importance at lower RH where LLE is most likely, could further influence the equilibrium partitioning of secondary components. Furthermore, the existence of metastable amorphous states, such as glasses, will be dependent on the particle composition. LLE could preferentially promote glass formation in one of the phases, thereby influencing the kinetics of phase partitioning as well as the final equilibrium state. Again, this is left for future work.

Topping and McFiggans[15] recently extended the one phase model described in this paper to account for diameter dependance and revisited the influence on re-equilibrating semi-volatile organics. It is possible to incorporate the Kelvin effect within the coupled VLE–LLE model, though it would be necessary to account for the surface energy between the separate liquid phases. Core–shell, partially engulfed structures and lensing have all been observed for laboratory generated particles, suggesting that such morphologies might also exist in tropospheric aerosols.[9] Such morphological influences may be included in future work by consideration of the surface energies at the phase boundaries for prescribed configurations.

Acknowledgements

Dr David Topping is funded by the National Centre for Atmospheric Science (NCAS). Dr Mark Barley was funded under NERC grant NE/H002588/1.

References

1 S. Twomey, *J. Atmos. Sci.*, 2009, **34**, 1149–1154.
2 B. A. Albrecht, *Science*, 1989, **245**, 1227–1230.
3 Q. Zhang, J. L. Jimenez, M. R. Canagaratna, J. D. Allan, H. Coe, I. Ulbrich, M. R. Alfarra, A. Takami, A. M. Middlebrook, Y. L. Sun, K. Dzepina, E. Dunlea, K. Docherty, P. F. De-Carlo, D. Salcedo, T. Onasch, J. T. Jayne, T. Miyoshi, A. Shimono, S. Hatakeyama, N. Takegawa, Y. Kondo, J. Schneider, F. Drewnick, S. Borrmann, S. Weimer, K. Demerjian, P. Williams, K. Bower, R. Bahreini, L. Cottrell, R. J. Griffin, J. Rautiainen, J. Y. Sun, Y. M. Zhang and D. R. Worsnop, *Geophys. Res. Lett.*, 2007, **34**, L13801.
4 J. L. Jimenez, M. R. Canagaratna, N. M. Donahue, A. S. H. Prevot, Q. Zhang, J. H. Kroll, P. F. DeCarlo, J. D. Allan, H. Coe, N. L. Ng, A. C. Aiken, K. S. Docherty, I. M. Ulbrich, A. P. Grieshop, A. L. Robinson, J. Duplissy, J. D. Smith, K. R. Wilson, V. A. Lanz,

C. Hueglin, Y. L. Sun, J. Tian, A. Laaksonen, T. Raatikainen, J. Rautiainen, P. Vaattovaara, M. Ehn, M. Kulmala, J. M. Tomlinson, D. R. Collins, M. J. Cubison, E. J. Dunlea, J. A. Huffman, T. B. Onasch, M. R. Alfarra, P. I. Williams, K. Bower, Y. Kondo, J. Schneider, F. Drewnick, S. Borrmann, S. Weimer, K. Demerjian, D. Salcedo, L. Cottrell, R. Griffin, A. Takami, T. Miyoshi, S. Hatakeyama, A. Shimono, J. Y. Sun, Y. M. Zhang, K. Dzepina, J. R. Kimmel, D. Sueper, J. T. Jayne, S. C. Herndon, A. M. Trimborn, L. R. Williams, E. C. Wood, A. M. Middlebrook, C. E. Kolb, U. Baltensperger and D. R. Worsnop, *Science*, 2009, **326**, 1525–1529.

5 D. O. Topping, G. B. McFiggans and H. Coe, *Atmos. Chem. Phys.*, 2005, **5**, 1205–1222.

6 A. S. Ansari and S. N. Pandis, *Atmos. Environ.*, 1999, **33**, 745–757.

7 A. S. Wexler and S. L. Clegg, *Journal of Geophysical Research*, 2009, **107**.

8 A. Zuend and J. H. Seinfeld, *Atmos. Chem. Phys.*, 2012, **12**, 3857–3882.

9 M. Song, C. Marcolli, U. K. Krieger, A. Zuend and T. Peter, *Atmos. Chem. Phys.*, 2012, **12**, 2691–2712.

10 A. Zuend, C. Marcolli, T. Peter and J. H. Seinfeld, *Atmos. Chem. Phys.*, 2010, **10**, 7795–7820.

11 Y. Ming and L. Russell, *AIChE J.*, 2002, **48**, 1331.

12 A. Zuend, C. Marcolli, B. P. Luo and T. Peter, *Atmos. Chem. Phys.*, 2008, **8**, 4559–4593.

13 A. K. Bertram, S. T. Martin, S. J. Hanna, M. L. Smith, A. Bodsworth, Q. Chen, M. Kuwata, A. Liu, Y. You and S. R. Zorn, *Atmos. Chem. Phys.*, 2011, **11**, 10995–11006.

14 G. McFiggans, D. O. Topping and M. H. Barley, *Atmos. Chem. Phys.*, 2010, **10**, 10255–10272.

15 D. O. Topping and G. McFiggans, *Atmos. Chem. Phys.*, 2012, **12**, 3253–3260.

16 E. Braekevelt, S. A. Tittlemier and G. T. Tomy, *Chemosphere*, 2003, **51**, 563–567.

17 G. Tse and S. I. Sandler, *J. Chem. Eng. Data*, 1994, **39**, 354–357.

18 M. H. Barley, D. Topping, D. Lowe, S. Utembe and G. McFiggans, *Atmos. Chem. Phys.*, 2011, **11**, 13145–13159.

19 N. M. Donahue, S. A. Epstein, S. N. Pandis and A. L. Robinson, *Atmos. Chem. Phys.*, 2011, **11**, 3303–3318.

20 J. Duplissy, P. F. DeCarlo, J. Dommen, M. R. Alfarra, A. Metzger, I. Barmpadimos, A. S. H. Prevot, E. Weingartner, T. Tritscher, M. Gysel, A. C. Aiken, J. L. Jimenez, M. R. Canagaratna, D. R. Worsnop, D. R. Collins, J. Tomlinson and U. Baltensperger, *Atmos. Chem. Phys.*, 2011, **11**, 1155–1165.

21 Y. Nannoolal, J. Rarey and D. Ramjugernath, *Fluid Phase Equilib.*, 2008, **269**, 117–133.

22 Y. Nannoolal, J. Rarey, D. Ramjugernath and W. Cordes, *Fluid Phase Equilib.*, 2004, **226**, 45–63.

23 S. Compernolle, K. Ceulemans and J.-F. Müller, *Atmos. Chem. Phys.*, 2011, **11**, 9431–9450.

24 K. G. Joback and R. C. Reid, *Chem. Eng. Commun.*, 1987, **57**, 233–243.

25 M. H. Barley and G. McFiggans, *Atmos. Chem. Phys.*, 2010, **10**, 749–767.

26 A. M. Booth, M. H. Barley, D. O. Topping, G. McFiggans, A. Garforth and C. J. Percival, *Atmos. Chem. Phys.*, 2010, **10**, 4879–4892.

27 H. Sakurai, H. J. Tobias, K. Park, D. Zarling, K. S. Docherty, D. B. Kittelson, P. H. McMurry and P. J. Ziemann, *Atmos. Environ.*, 2003, **37**, 1199–1210.

28 B. K. Das and R. B. Griffiths, *J. Chem. Phys.*, 1979, **70**, 5555.

29 J. L. Jr. and B. Widom, *Phys. A*, 1975, **81**, 190–213.

30 B. M. Knickerbocker, C. V. Pesheck, H. T. Davis and L. E. Scriven, *J. Phys. Chem.*, 1982, **86**, 393–400.

31 A. Martin, M. Klauck, A. Grenner, R. Meinhardt, D. Martin and J. Schmelzer, *J. Chem. Eng. Data*, 2011, **56**, 741–749.

32 A. Grenner, M. Klauck, R. Meinhardt, R. Schumann and J. Schmelzer, *J. Chem. Eng. Data*, 2006, **51**, 1009–1014.

33 M. Klauck, A. Grenner and J. Schmelzer, *J. Chem. Eng. Data*, 2006, **51**, 1043–1050.

Faraday Discussions

RSC Publishing

PAPER

Morphologies of mixed organic/inorganic/aqueous aerosol droplets

Mijung Song, Claudia Marcolli, Ulrich K. Krieger,[*] Daniel M. Lienhard and Thomas Peter

Received 28th March 2013, Accepted 29th April 2013

DOI: 10.1039/c3fd00049d

Despite major progress in the understanding of properties of tropospheric aerosol particles, it remains challenging to understand their physical state and morphology. To obtain more detailed knowledge of the phases, phase transitions and morphologies of internally mixed organic/inorganic aerosol particles, we evaluated liquid–liquid phase separation (LLPS), deliquescence relative humidity (DRH) and efflorescence relative humidity (ERH) of 33 organic/ammonium sulfate (AS)/H_2O systems from our own and literature data. The organic fraction consists of single compounds or mixtures with up to ten aliphatic and/or aromatic components with carboxylic acid, hydroxyl, carbonyl, ether, and ester functionalities, covering O : C ratios between 0.29 and 1.33. Thirteen out of these 33 systems did not show LLPS for any of the studied organic-to-inorganic mixing ratios, sixteen underwent LLPS showing core–shell morphology, and four showed both core–shell and partially engulfed configurations depending on the organic-to-inorganic ratio and RH. In all cases the organic fractions of the systems with partially engulfed configurations consisted of dicarboxylic acids. AS in mixed organic/AS/H_2O particles deliquesced between 70 and 84% RH. AS effloresced below 58% RH or remained in a one-liquid-phase state. AS in droplets with LLPS always showed efflorescence with ERH between 30 and 50% RH, providing clear evidence that the presence of LLPS facilitates AS efflorescence. Spreading coefficients of the organic-rich phase on the AS-rich phase for systems containing polyethylene glycol 400 (PEG-400) and a mixture of dicarboxylic acids are in agreement with the optically observed morphologies of droplets deposited on the hydrophobic substrate. Analysis of high resolution elastic Mie resonance spectra allowed the detection of LLPS for single levitated droplets consisting of PEG-400/AS/H_2O, whereas LLPS was difficult to detect in (2-methylglutaric acid + 3-methylglutaric acid + 2,2-dimethylsuccinic acid)/AS/H_2O. Measured Mie spectra of PEG-400/AS/H_2O at 93.5% and at 80.9% RH agreed with computed Mie spectra for a homogeneous and a core–shell configuration, respectively, confirming the results obtained from droplets deposited on a hydrophobic substrate. Based on the presented evidence, we therefore consider the core–shell morphology to be the prevalent configuration of liquid–liquid-phase-separated tropospheric organic/AS/H_2O particles.

Institute for Atmospheric and Climate Science, ETH Zurich, 8092, Zurich, Switzerland. E-mail: ulrich.krieger@env.ethz.ch; Fax: +41 44 6331058; Tel: +41 44 6334007

1   Introduction

Aerosol particles are an important atmospheric component, significantly affecting both climate and air quality.[1] The chemical composition, physical state, morphology and size influence their forcing and therefore should be known to accurately predict the impact of aerosols on climate.[2-4] Gas–particle partitioning of semivolatile compounds and atmospheric chemistry, specifically the heterogeneous reaction of N_2O_5, are linked with the physical state and morphology of the aerosol particles.[5-8]

Tropospheric aerosols comprise organic constituents, inorganic salts, and water. These components are mostly internally mixed.[9-11] The physical state of aerosol particles depends on chemical composition, RH and temperature.[12-14] Depending on these factors, atmospheric aerosol particles may be present as one liquid phase, two liquid phases, crystalline phases, in a glassy state and as mixture of these phases.[14] When aerosol particles are exposed to varying humidity in the atmosphere, they may undergo phase transitions such as liquid–liquid phase separation (LLPS), deliquescence, and efflorescence. Recent laboratory studies using optical microscopy showed that LLPS in internally mixed organic/ammonium sulfate (AS)/H_2O aerosol particles is expected to be a common feature in the atmosphere.[15,16] Since O : C reflects the polarity of organic components and their affinities to water and electrolytes, it can be an important parameter to determine the occurrence of LLPS in an aerosol.[15,16] Song *et al.*,[17] found for atmospheric relevant mixtures consisting of organics/AS/H_2O with up to 10 organic components that separation into two liquid phases always occurred for O : C < 0.56 and depended on the specific types of organic functional groups in the range of 0.56 < O : C < 0.80. There is no direct measurement technique available that is able to observe LLPS in ambient aerosols. Recently, ambient aerosols were collected on filters and the physical state was observed in the laboratory.[18,19] Aerosol samples collected near the city of Atlanta that is highly influenced by both anthropogenic and biogenic emissions and exhibited O : C of roughly less than 0.5 also showed LLPS.[19] Moreover, the presence of LLPS in Amazon aerosol samples, which contained a high share of aliphatic carboxylic acids and hydroxyl functional groups in the presence of inorganic microgranular material,[18] was in agreement with the O : C ranges of LLPS specified by Song *et al.*[17] The composition of the organic fraction and the mixing state of aerosol particles may influence deliquescence relative humidity (DRH) and efflorescence relative humidity (ERH) of inorganic salts during RH cycles.[15,16,20-24] No significant reduction of ERH and DRH was observed for moderately hydrophobic aerosol particles with O : C < 0.60 such as PEG-400/AS/H_2O, C7/AS/H_2O, and secondary organic material (SOM) (produced by dark ozonolysis of α-pinene)/AS/H_2O over the whole investigated dry (water-free) mass fractions of AS (mf_d(AS)) from 0–1.[16,23,25] Due to the high viscosity of the organic fraction in the organic-rich phase, the ERH was slightly decreased in systems consisting of C6/AS/H_2O and C7/AS/H_2O with mf_d(AS) < 0.25.[16] On the other hand, DRH and ERH of AS in hydrophilic organic/AS/H_2O particles (O : C > 0.70) were reduced significantly or even totally suppressed.[15]

LLPS in mixed aerosol particles may lead to different particle morphologies such as core–shell or partially engulfed configurations. Since there exist no *in situ* measurement techniques to monitor the ambient aerosols, determination of aerosol morphology is challenging. Using optical microscopy and Raman spectroscopy, a core–shell configuration consisting of an AS inner phase coated by an organic outer phase has been observed for $O:C < 0.70$ in phase separated organic/AS/H_2O particles deposited on a hydrophobically coated substrate.[15,25] Atlantic aerosol samples ($O:C \sim 0.5$) also adopted a core–shell structure when present in a two-liquid-phases state as observed by optical microscopy.[19] In contrast, recent laboratory studies using micrometer-sized levitated particles showed that for particles composed of aqueous salts and organics with low $O:C$ of ~ 0.1 a partially engulfed structure is the prevalent morphology.[12,13] Song *et al.*,[16] observed both morphologies in C6/AS/H_2O and C7/AS/H_2O particles on a substrate, but the partially engulfed structure was found only in a limited range of organic-to-inorganic mass ratio. Therefore, more laboratory experiments are necessary to obtain deeper insight into aerosol morphologies depending on their chemical composition and physical states.

The present study aims to explore how the deliquescence and efflorescence of AS in mixed organic/AS/H_2O particles is influenced by LLPS and to identify the corresponding morphologies with optical microscopy. To obtain conclusions as to whether the core–shell or the partially engulfed configuration is prevalent when there is no interaction with a substrate, we use an electrodynamic balance to investigate particles with compositions that led to core–shell or partially engulfed morphologies on a hydrophobic substrate. Furthermore, surface and interfacial tensions of model systems are obtained and evaluated.

2 Experiments

Phase transitions and morphological changes of internally mixed organic/AS/H_2O particles during humidity cycles were observed using an optical microscope equipped with a temperature- and humidity-controlled flow-cell. The detailed experimental procedure and method have been previously described by Song *et al.*[16,17] The solutions containing the organic components and AS were prepared in purified water (resistivity ≥ 18.0 MΩ cm). All substances were purchased from Sigma-Aldrich with purities $\geq 98\%$ and were used without further purification.

Micrometer-sized aqueous droplets (20–65 µm in diameter) were deposited on a hydrophobically coated substrate using a droplet generator,[26] and the substrate was mounted inside the flow-cell and equilibrated at $\sim 100\%$ RH for 15 min. The cell temperature was kept at 293.2 ± 0.1 K during the equilibration time and the humidity cycles. To observe the morphologies and phase transitions of the particles during humidity cycles, the water vapour mixing ratio of a constant total N_2/H_2O flow (180 sccm) through the cell was varied. After the equilibration at high RH, RH was continuously decreased down to 30% if (partial) efflorescence was observed or to 0% in the absence of efflorescence and subsequently, RH was continuously increased to $\sim 100\%$ at rates of both 0.1% and 0.4% RH min^{-1}. RH and temperature were determined by a G-TUCN.34 sensor (U.P.S.I., France) and the temperature was also measured directly under the substrate by a Pt100 temperature sensor (Minco, USA). RH calibration of the G-TUCN.34 sensor was done by observing the DRH of pure crystals of salts such as $(NH_4)_2SO_4$ (80.0%), NH_4NO_3 (65.5%) and NaI (38.0%) at 293 K. The uncertainty of the sensor was typically $\pm 1.5\%$ RH. To observe the morphological change of aerosol particles, a microscope equipped with a long working distance objective (Olympus BX-40, magnification 50, aperture 0.7) was used. We also recorded the optical images of the droplets during humidity cycles using a black and white CCIR video camera (25 frames s^{-1}).

Surface and interfacial tension measurements were carried out by IMETER (MessSysteme, Augsburg, Germany) using the ring method applying the DIN 53914, the ASTM D 1331 and the ISO 6889 norms for execution and evaluation. The ring method was corrected according to the German patent DE 4412405. The detection limit of the ring method for interfacial tension measurements is ~0.5 mN m^{-1} and depends among other factors on the flatness of the ring and the interface. Interfacial tensions were determined once in the phase separated systems. For surface tension measurements the two liquid phases were separated in a separating funnel. Samples for surface tension measurements were prepared and measured twice and the average of the two values was taken. Samples of PEG-400/AS/H$_2$O and C6 + C7/AS/H$_2$O were measured at 298 K and 293 K, respectively. Density measurements were carried out using a pycnometer (BRAND, Germany) at 293 K.

To investigate the potential impact of the substrate on particle morphology when using the microscope technique we performed additional experiments on single particles levitated in an electrodynamic balance (EDB). The setup has been described in detail previously.[27,28] While deliquescence and efflorescence are easily detected using temporal intensity fluctuations of scattered light[29] or spatial distortions in the two dimensional angular scattering pattern,[30] it proved impossible to detect LLPS using these techniques.[31] Therefore, we added to the setup a tunable diode laser with narrow line width (New Focus, model Velocity 6312), to measure high resolution Mie resonance spectra. The wavelength of the laser is monitored using a Michelson interferometer.[32] The levitated particle is illuminated from below; the beam is weakly focussed with 400 mm focal length lens. The detector (Si photodiode, New Focus femtowatt photoreceiver, model 2151) is placed at 90° relative to the incoming beam at the same height as the levitated particle; a lens integrates the scattered light of the particle over scattering angles ranging from 78° to 101°. The polarization of the incoming laser can be shifted using a motorized λ/2-wave plate so that the incident electric field is either parallel or perpendicular to the scattering plane (the plane defined by the forward direction of the laser beam and the scattering direction). Thus, either spectra of transversal electric polarization (TE) or transversal magnetic polarization (TM) are measured. Typically, we scan the laser from 765.0 nm to 781.5 nm using a scan speed of 0.04 nm s^{-1}, turn the λ/2-wave plate and measure TE and TM spectra subsequently. Refractive indices of bulk solutions, n_{D}, were measured with an Abbe type refractometer.

3　Results and discussion

3.1　Deliquescent and efflorescent behaviours depending on mixing states of aerosols

Table 1 gives literature data and results from this study for ERH, DRH, LLPS and the morphology of organic/AS/H$_2$O particles from optical microscopy and EDB measurements. It shows that AS in mixed organic/AS/H$_2$O particles deliquesced between 70 and 84% RH and effloresced below 58% RH or remained in a one-liquid-phase state when all mixtures are considered. AS in droplets with LLPS always effloresced between 30 and 50% RH and deliquesced between 70 and 84% RH.

Table 1 Optical microscope and EDB experiments collected from this study and literature. Abbreviations: COM, complex organic mixture; DCA, dicarboxylic acids; OAC, oxidized aromatic compounds; MFC, multifunctional compounds; SRH, onset RH of LLPS; Abs, no LLPS occurred; AS incs, AS inclusions in organic-rich phase; CS, core–shell configuration; PE, partially engulfed configuration; HA, homogeneous appearance; "?", not clear; "-", not measured or no information given; "0", particle did not crystallize. Literature[15–17,21,25,31,33–40] and "*", current study

| Mixture (O : C) | Organic components (wt% of component in organic mixture) | Studied mf_d(AS) | LLPS | | ERH | DRH |
			SRH	Morph.		
Optical microscopy						
COM1/AS/H$_2$O[17,*] (0.63)	2-methylglutaric acid (12.5); 3-methyladipic acid (12.5); diethylmalonic acid (12.5); malic acid (12.5); levoglucosan (12.5); pinonic acid (3.125); pinolic acid (3.125); 3-hydroxybenzoic acid (6.25); 3,5-dihydroxybenzoic acid (12.5); 1,2,7,8-octanetetrol (12.5)	0.14, 0.20, 0.26, 0.30, 0.33, 0.40, 0.50, 0.67, 0.80, 0.86, 0.91	Abs, Abs, Abs, 70.1, 74.2,[17] 81.0, 83.3, 84.5,[17] 83.1, 71.9,[17] Abs	HA, HA, HA, CS, CS, CS, CS, CS, CS, CS, HA	0, 21.0, 35.2, 46.6, 40.8, 43.9, 42.3, 43.6, 44.7, 44.9, 47.9	-, 72.3, 74.3, 75.3, 75.5, 76.2, 76.3, 77.0, 77.7, 78.0, 78.8
COM2/AS/H$_2$O[17,*] (0.68)	2-methylglutaric acid (12.5); 3-methyladipic acid (12.5); methylmalonic acid (12.5); malic acid (12.5); levoglucosan (12.5); pinonic acid (3.125); pinolic acid (3.125); 3-hydroxybenzoic acid (6.25); 3,5-dihydroxybenzoic acid (12.5); 1,2,7,8-octanetetrol (12.5)	0.14, 0.33, 0.67, 0.86	Abs, Abs,[17] 80.9,[17] 72.0[17]	HA, HA, CS, CS	0, 37.4, 42.7, 46.5	-, 74.7, 79.2, 78.1
COM3/AS/H$_2$O[17,*] (0.70)	2-methylglutaric acid (12.5); 3-methyladipic acid (12.5); malonic acid (12.5); malic acid (12.5); levoglucosan (12.5); pinonic acid (3.125); pinolic acid (3.125); 3-hydroxybenzoic acid (6.25); 3,5-dihydroxybenzoic acid (12.5); 1,2,7,8-octanetetrol (12.5)	0.33, 0.67, 0.86	Abs,[17] 79.4,[17] 73.3[17]	HA, CS, CS	33.6, 47.5, 44.4	74.1, 77.1, 77.7
COM4/AS/H$_2$O[17,*] (0.77)	Malonic acid (12.5); glutaric acid (12.5); methyl-malonic acid (12.5); malic acid (12.5); levoglucosan (12.5); pinonic acid (3.125); pinolic acid (3.125); 3-hydroxybenzoic acid (6.25); 3,5-dihydroxybenzoic acid (12.5); 1,2,7,8-octanetetrol (12.5)	0.33, 0.67, 0.86	Abs,[17] Abs,[17] Abs[17]	HA, HA, HA	27.0, 41.5, 47.9	71.9, 76.4, 77.5

Table 1 (*Contd.*)

Mixture (O : C)	Organic components (wt% of component in organic mixture)	Studied mf_d(AS)	LLPS			
			SRH	Morph.	ERH	DRH
OAC + DCA1/AS/ H$_2$O[17],* (0.68)	3-hydroxybenzoic acid (6.25); 4-hydroxybenzoic acid (6.25); 2,5-dihydroxybenzoic acid (12.5); 2,6-dihydroxybenzoic acid (6.25); 3,4-dihydroxybenzoic acid (6.25); 3,5-dihydroxybenzoic acid (37.5); malonic acid (12.5); malic acid (12.5)	0.67	Abs[17]	HA	49.4	77.0
OAC + DCA2/AS/ H$_2$O[17],* (0.61)	3-hydroxybenzoic acid (6.25); 4-hydroxybenzoic acid (6.25); 2,5-dihydroxybenzoic acid (18.75); 2,6-dihydroxybenzoic acid (6.25); 3,4-dihydroxybenzoic acid (6.25); 3,5-dihydroxybenzoic acid (43.75); malonic acid (6.25); malic acid (6.25)	0.86	Abs[17]	HA	45.0	78.4
OAC + DCA3/AS/ H$_2$O[17],* (0.56)	3-hydroxybenzoic acid (6.25); 4-hydroxybenzoic acid (6.25); 2,5-dihydroxybenzoic acid (18.75); 2,6-dihydroxybenzoic acid (6.25); 3,4-dihydroxybenzoic acid (6.25); 3,5-dihydroxybenzoic acid (43.75); 2-methylglutaric acid (6.25); diethylmalonic acid (6.25)	0.67, 0.86	Abs,[17] Abs[17]	HA, HA	45.1, 51.6	77.9, 78.8
Polyol + DCA1/AS/ H$_2$O[17],* (0.72)	Diethylmalonic acid (20); levoglucosan (20); glutaric acid (20); sorbitol (20); 1,2,7,8-octanetetrol (20)	0.33, 0.67, 0.86	70.7,[17] 81.4,[17] 73.6[17]	CS, CS, CS	42.1, 44.0, 42.8	74.5, 77.2, 78.1
Polyol + DCA2/AS/ H$_2$O[17],* (0.65)	2-methylglutaric acid (30); diethylmalonic acid (30); sorbitol (20); 1,2,7,8-octanetetrol (20)	0.33, 0.67, 0.86	85.2,[17] 87.0,[17] 78.0[17]	CS, CS, CS	41.9, 45.9, 43.3	76.0, 78.3, 78.7
OAC1/AS/H$_2$O[17],* (0.55)	2,5-dihydroxybenzoic acid (30); 3,5-dihydroxybenzoic acid (35); 2,4,5-trimethoxybenzoic acid (35)	0.33, 0.67, 0.86	92.6,[17] 95.3,[17] 87.0[17]	CS, CS, CS	43.7, 44.7, 46.4	78.2, 78.9, 79.1
OAC2/AS/H$_2$O[17],* (0.57)	2,5-dihydroxybenzoic acid (30); 3,5-dihydroxybenzoic acid (70)	0.67, 0.86	Abs,[17] Abs[17]	HA, HA	49.0, 46.8	77.8, 79.0

Table 1 (*Contd.*)

Mixture (O : C)	Organic components (wt% of component in organic mixture)	Studied mf_d(AS)	LLPS			ERH	DRH
			SRH	Morph.			
Polyol1/AS/H$_2$O[17,*] (0.71)	Sorbitol (50); 1,2,7,8-octanetetrol (50)	0.33, 0.67, 0.86	62.6,[17] 69.4,[17] 59.0[17]	CS, CS, CS		29.7, 45.2, 44.0	75.3, 76.8, 78.6
Polyol2/AS/H$_2$O[17,*] (0.76)	Sorbitol (60); 1,2,7,8-octanetetrol (40)	0.67, 0.86	Abs,[17] Abs[17]	HA, HA		41.1, 40.6	77.4, 78.5
MFC1/AS/H$_2$O[17,*] (0.67)	Maleic acid (20); dehydroacetic acid (20); kojic acid (20); 3,4-dihydroxy-2,2-dimethyl-4-oxo-2*H*-pyran-6-carboxylic acid (20); itaconic acid (20)	0.33, 0.67, 0.86	79.1,[17] 76.6,[17] 76.5[17]	CS, CS, CS		38.5, 43.9, 45.1	76.7, 78.2, 78.6
MFC2/AS/H$_2$O[17,*] (0.79)	Maleic acid (50); dehydroacetic acid (10); kojic acid (15); 3,4-dihydroxy-2,2-dimethyl-4-oxo-2*H*-pyran-6-carboxylic acid (10); itaconic acid (15)	0.33, 0.67, 0.86	Abs,[17] 59.0,[17] Abs[17]	HA, CS, HA		40.1, 49.5, 44.4	72.0, 77.3, 78.0
MFC3/AS/H$_2$O[17,*] (0.84)	2-oxoglutaric acid (1.00); maleic acid (30); dehydroacetic acid (5); kojic acid (15); 3,4-dihydroxy-2,2-dimethyl-4-oxo-2*H*-pyran-6-carboxylic acid (5); itaconic acid (15)	0.67, 0.86	Abs,[17] Abs[17]	HA, HA		46.7, 46.6	76.9, 78.3
DCA(C7)/AS/H$_2$O[16] (0.57)	3-methyladipic acid (33.3); 3,3-dimethylglutaric acid (33.3); diethylmalonic acid (33.3)	0.02–0.23, 0.34, 0.50–0.93	88.2–91.7, 89.6, 83.8–89.9	AS incs, PE, CS		33.5–41.7, 42.1, 40.3–48.2	70.1–79.8, 78.4, 78.4–81.3
DCA(C6)/AS/H$_2$O[16] (0.67)	2-methylglutaric acid (33.3); 3-methylglutaric acid (33.3); 2,2-dimethylsuccinic acid (33.3)	0.08–0.33, 0.50, 0.67, 0.76–0.89, 0.92	49.9–85.0, 72.9, 72.7, 72.3–72.8, Abs	AS incs, CS → PE, CS → PE, ?, HA		29.9–48.6, 43.7, 40.9, 39.7–45.8, 45.4	71.0–79.1, 78.1, 78.1, 78.2–80.4, 79.5
DCA(C5)/AS/H$_2$O[16] (0.80)	Glutaric acid (33.3); methylsuccinic acid (33.3); dimethylmalonic acid (33.3)	0.33, 0.50, 0.80	Abs, Abs, Abs	HA, HA, HA		32.4, 37.9, 40.8	75.1, 77.8, 77.9

Table 1 (*Contd.*)

Mixture (O : C)	Organic components (wt% of component in organic mixture)	Studied $mf_d(AS)$	LLPS			
			SRH	Morph.	ERH	DRH
DCA(C5 + C6 + C7)/AS/$H_2O^{17,*}$ (0.67)	Glutaric acid (11.1); methylsuccinic acid (11.1); dimethylmalonic acid (11.1); 2-methylglutaric acid (11.1); 3-methylglutaric acid (11.1); 2,2-dimethylsuccinic acid (11.1); 3-methyladipic acid (11.1); 3,3-dimethylglutaric acid (11.1); diethylmalonic acid (11.1)	0.11, 0.33, 0.50	Abs,[17] 70.3,[17] 74.0[17]	HA, AS incs, PE	18.7, 43.1, 47.8	71.8, 77.5, 78.2
DCA(C6 + C7)/AS/H_2O^* (0.61)	3-methyladipic acid (19.9); 3,3-dimethylglutaric acid (19.9); diethylmalonic acid (19.9); 2-methylglutaric acid (15.0); 3-methylglutaric acid (15.0); 2,2-dimethylsuccinic acid (10.0)	0.33, 0.50	83.3, 87.2	CS → PE, CS	44.7, 49.1	78.0, 79.0
Diethyl decanedioate/AS/H_2O^{15} (O : C = 0.29)	Diethyl decanedioate (100)	0.11–0.66, 0.80	97.5–100, Abs	CS, HA	34.1–36.0, 35.8	79.1–80.2, 80.4
Monomethyl octane-1,8 – dioate/AS/H_2O^{15} (O : C = 0.44)	Monomethyl octane- 1,8 – dioate (100)	0.18–0.68	~100	CS	33.2–35.2	78.4–79.6
1,2,6-trihydroxyhexane/AS/H_2O^{15} (O : C = 0.5)	1,2,6-trihydroxyhexane (100)	0.15–0.74	69.0–72.5	CS	32.5–36.7	79.7–82.8
α,4-dihydroxy-3-methoxybenzeneacetic acid/AS/H_2O^{15} (O : C = 0.56)	α,4-dihydroxy-3-methoxybenzeneacetic acid (100)	0.11, 0.24–0.66	Abs, 79.0–81.3	HA, CS	0, 37.8–40.3	-, 82.5–84.1
2,5-dihydroxybenzoic acid/AS/H_2O^{15} (O : C = 0.57)	2,5-dihydroxybenzoic acid (100)	0.33–0.42, 0.58, 0.58–0.80	Org crys, Abs, 61.5–64.2	HA, HA, CS	-, 32.9, 33.6–36.7	-, -, -

Table 1 (*Contd.*)

Mixture (O : C)	Organic components (wt% of component in organic mixture)	Studied mf_d(AS)	LLPS SRH	Morph.	ERH	DRH
2,2-dimethylbutane-dioic acid/AS/H$_2$O[15] (O : C = 0.67)	2,2-dimethylbutanedioic acid (100)	0.19–0.32, 0.37–0.42, 0.58–0.70	Org crys, Abs, 61.5–63.8	HA, HA, CS	-, -, 32.1–40.3	-, -, -
PEG-400/AS/H$_2$O[25] (~0.56)	Polyethylene glycol-400 (100)	0.11–0.33, 0.4–0.90	87–88, 87–89	As incs, CS	30–35, 30–37	77–80, 80
Adipic acid/AS/H$_2$O[33] (0.67)	Adipic acid (100)	0.10–0.97	—	—	35.3–40.5	79–81
Glutaric acid/AS/H$_2$O[34] (0.80)	Glutaric acid (100)	0.50–0.90	Abs[15]	HA[15]	29.5–34	79–79.5
Levoglucosan/AS/H$_2$O[35] (0.83)	Levoglucosan (100)	0.29–0.83	Abs[15]	HA[15]	10–30	72–80
Glycerol/AS/H$_2$O[35] (1.00)	Glycerol (100)	0.51–0.91	Abs[15]	HA[15]	5–30	72–80
Citric acid/AS/H$_2$O[36] (1.17)	Citric acid (100)	0.67–1.00	Abs[15]	HA[15]	24–35	—
Malonic acid/AS/H$_2$O[35] (1.33)	Malonic acid (100)	0.53–0.90	Abs[15]	HA[15]	0–32	72–80
EDB measurement						
Adipic acid/AS/H$_2$O[37] (0.67)	Adipic acid (100)	0.25, 0.67	—	—	35, 38	80, 79
Glutaric acid/AS/H$_2$O[38] (0.80)	Glutaric acid (100)	0.51, 0.87	—	—	40.8, 50.7	80.0, 80.9
Glutaric acid/AS/H$_2$O[39] (0.80)	Glutaric acid (100)	0.50	—	—	57.5	76.6
Glutaric acid/AS/H$_2$O[40] (0.80)	Glutaric acid (100)	0.50	—	—	33.5	79.0

Table 1 (*Contd.*)

Mixture (O : C)	Organic components (wt% of component in organic mixture)	Studied mf_d(AS)	LLPS		ERH	DRH
			SRH	Morph.		
Glutaric acid/AS/ H_2O^{37} (0.80)	Glutaric acid (100)	0.50	—	—	32	77.5
Succinic acid/AS/ H_2O^{40} (1.00)	Succinic acid (100)	0.48	—	—	30	80
Succinic acid/AS/ H_2O^{39} (1.00)	Succinic acid (100)	0.48	—	—	48.3	79
M5/AS/H_2O^{31} (1.08)	Malic acid (18); malonic acid (36); maleic acid (16); glutaric acid (22); methylsuccinic acid (8)	0.5	—	—	27	73.3
M5/AS/H_2O^{31} (1.08)	Malic acid (17); malonic acid (37); maleic acid (17.5); glutaric acid (20.5); methylsuccinic acid (8)	0.8	—	—	35	78.3
Malonic acid/AS/ H_2O^{21} (1.33)	Malonic acid (100)	0.41–0.71	—	—	0–30	—

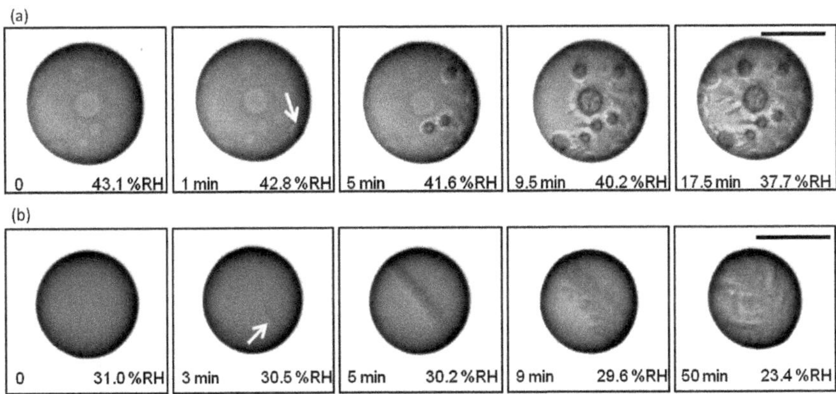

Fig. 1 Efflorescence process of (a) C7/AS/H$_2$O (O : C = 0.57), mf$_d$(AS) = 0.15, and (b) COM4/AS/H$_2$O (O : C = 0.77), mf$_d$(AS) = 0.33. Onset of crystallization is marked by an arrow. At time $t = 0$, particles are liquid. Size bar: 20 μm.

The process of crystallization in mixed organic/AS/H$_2$O particles is shown in Fig. 1a and b for optically monitored particles of C7/AS/H$_2$O (O : C = 0.57, mf$_d$(AS) = 0.15), and COM4/AS/H$_2$O (O : C = 0.77, mf$_d$(AS) = 0.33). In the particle of C7/AS/H$_2$O, which showed LLPS at 90% RH by nucleation and growth,[16] crystallization started from an aqueous AS inclusion at the rim of the particle at 42.8% RH (marked by an arrow, $t = 1$ min). Subsequently, AS needles slowly grew from the inclusion into the aqueous organic-rich phase, causing the crystallization of other aqueous AS inclusions when they came in contact with them. At the end of the experiment ($t = 17.5$ min), all AS inclusions had effloresced and the particle was covered with needles. On the other hand, the one-liquid-phase state particle of the COM4/AS/H$_2$O system crystallized by forming a needle from the rim of the particle at 30.5% RH ($t = 3$ min). After the needle reached the other side of the particle ($t = 5$ min), it started to branch into the organic-rich phase ($t = 9$ min). At the end of the experiment, the needles covered the particle ($t = 50$ min, RH = 23.4%). Due to the high viscosity, the growth of crystalline AS occurred on the timescale of minutes in organic-rich one-liquid-phase particles and in the organic-rich phase of mixed organic/AS/H$_2$O particles.[14] On the other hand, the low viscosity of the AS-rich phase allowed instantaneous crystallization (within milliseconds; see also Ciobanu *et al.*,[22]) of AS inclusions as soon as they were reached by a needle.

Fig. 2 presents the optical microscope measurements of DRH and ERH in terms of mf$_d$(AS) for organic/AS/H$_2$O particles from this work (circles) and literature (triangles) covering 0.29 < O : C < 1.33. To complete the picture, EDB measurements of DRH and ERH (crosses) of organic/AS/H$_2$O particles with 0.67 < O : C < 1.33 are also shown. To investigate the dependence of deliquescence and efflorescence of AS on the O : C ratio, we classify all available data of DRH and ERH into the O : C ranges of 0.20–0.60, 0.60–0.70, 0.70–0.80 and 0.80–1.40. Prediction curves for ERH and DRH of mixed organic/AS/H$_2$O particles parameterized by Bertram *et al.*,[15] are given in Fig. 2 (grey and black lines) according to this classification. Presence and absence of LLPS are represented as solid and open symbols, respectively (see also Table 1).

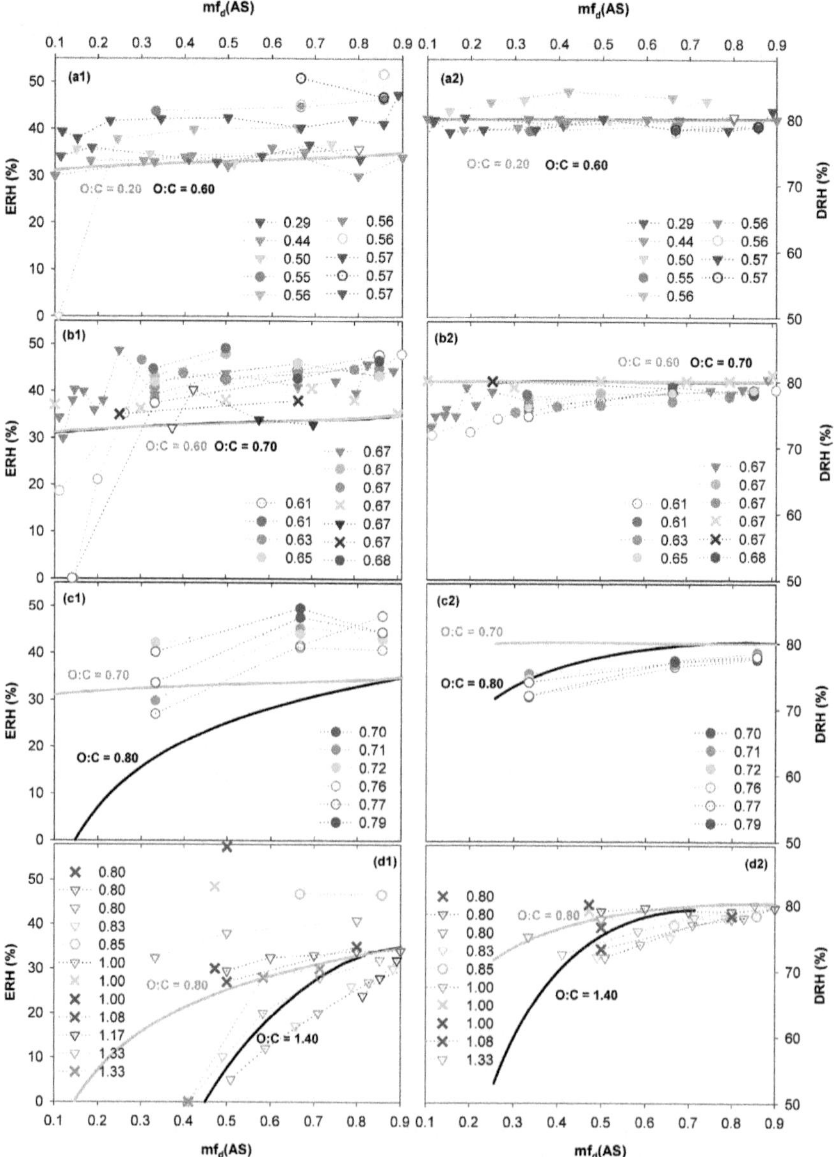

Fig. 2   ERH (panels a1–d1) and DRH (panels a2–d2) of organic/AS/H$_2$O particles as a function of AS dry mass fraction (mf$_d$(AS)) from this study and literature. The grey and black lines represent the prediction from Bertram *et al.*,[15] for AS ERH and DRH based on O : C ratios. Circles: systems measured within this study and Song *et al.*,[16] by optical microscopy. Triangles: systems measured by optical microscopy from literature. The solid and open symbols represent the presence or absence of LLPS, respectively. Crosses: systems mostly measured by EDB for which no LLPS information is provided. Panels a1 and a2: systems covering 0.2 < O : C < 0.6: O : C = 0.29 (red triangle), diethyl decanedioate; 0.44 (orange triangle), monomethyl octane-1,8-dioate; 0.50 (yellow triangle), 1,2,6-hexanetriol; 0.55 (olive circle), OAC1; 0.56 (green triangle), α,4-dihydroxy-3-methoxybenzeneacetic acid; 0.56 (dark green triangle), PEG-400; 0.56 (cyan circle), OAC + DCA3; 0.57 (blue triangle), C7; 0.57 (dark blue circle), OAC2; 0.57 (purple triangle), 2,5-dihydroxybenzoic acid. Panels b1 and b2: systems covering 0.60 < O : C < 0.70: 0.61 (red circle), OAC + DCA2; 0.61 (pink circle), C6 + C7; 0.63 (orange circle), COM1; 0.65 (yellow circle), Polyol + DCA2; 0.67 (olive triangle), C6; 0.67 (green circle), C5 + C6 + C7; 0.67 (dark green circle), MFC1; 0.67 (cyan cross), adipic acid; 0.67 (dark blue triangle), 2,2-dimethylbutanedioic acid; 0.67 (dark blue cross), adipic acid;

(continued)

0.68 (purple circle), COM2. Panels c1 and c2: systems covering 0.70 < O : C < 0.80: 0.70 (red circle), COM3; 0.71 (orange circle), Polyol1; 0.72 (yellow circle), Polyol + DCA1; 0.76 (green circle), Polyol2; 0.77 (blue circle), COM4; 0.79 (purple circle), MFC2. Panels d1 and d2: 0.80 < O : C < 1.40: 0.80 (red cross), glutaric acid; 0.80 (red triangle), glutaric acid; 0.80 (orange triangle), C5; 0.83 (yellow triangle), levoglucosan; 0.85 (green circle), MFC3; 1.00 (dark green triangle), glycerol; 1.00 (cyan cross), succinic acid; 1.00 (blue cross), succinic acid; 1.08 (purple cross), M5; 1.17 (black triangle), citric acid; 1.33 (grey triangle), malonic acid; 1.33 (grey cross), malonic acid.

In most mixtures covering O : C ratios from 0.20 to 0.60, LLPS occurred in the whole range of $0 < mf_d(AS) < 1$. AS effloresced in these phase-separated particles between 30.0–51.6% RH (Fig. 2, panel (a1)). The highest ERH was observed for the systems that did not show LLPS (cyan circle, OAC + DCA3/AS/H_2O, O : C = 0.56 and dark blue circle, OAC2/AS/H_2O, O : C = 0.57). Upon moistening, the effloresced particles deliquesced at 78.9 ± 2.6% RH (Fig. 2, panel (a2)). For 0.60 < O : C < 0.70, a clear dependence of ERH on the phase state of the particles can be seen (Fig. 2, panel (b1)). Systems consisting of Polyol + DCA2/AS/H_2O (yellow circle, O : C = 0.65), C6/AS/H_2O (olive triangle, O : C = 0.67), and MFC1/AS/H_2O (dark green circle, O : C = 0.67), which exhibited LLPS (solid symbols) in the entire range of $mf_d(AS)$, effloresced within values of 39.8 ± 4.8% RH. However, for low AS fractions, AS ERH decreased significantly when the particle was present in a one-liquid-phase state as can be seen for the systems COM1/AS/H_2O (orange circle, O : C = 0.63), C5 + C6 + C7/AS/H_2O (green circle, O : C = 0.67), and COM2/AS/H_2O (purple circle, O : C = 0.68). Such a decrease is also observed for AS deliquescence in particles with 0.60 < O : C < 0.70 as shown in panel (b2) of Fig. 2. For 0.70 < O : C < 0.80, almost constant ERH = 43.0 ± 0.96% and DRH = 76.6 ± 1.87% was observed for the system Polyol + DCA1/AS/H_2O (yellow circle, O : C = 0.72) which led to complete LLPS (Fig. 2, panels (c1) and (c2)). However, a significant decrease of ERH and DRH occurred for organic-rich compositions of COM3/AS/H_2O (red circle, O : C = 0.70), and MFC2/AS/H_2O (purple circle, O : C = 0.79) systems, which were present in a one-liquid-phase state. A reason for the low ERH (29.7%) and DRH (75.3%) in the Polyol1/AS/H_2O system (orange circle, O : C = 0.71) at $mf_d(AS) = 0.33$ could be that the system at this composition is close to the boundary of the miscibility gap. In this case, most AS is mixed in the organic-rich phase and the volume of the AS-rich phase is very small. Systems with 0.80 < O : C < 1.4 exhibited no LLPS and showed the largest decrease in ERH and even complete inhibition of AS efflorescence for low $mf_d(AS)$ (Fig. 2, panels (d1) and (d2)). AS ERH in glutaric acid/AS/H_2O particles shows a wide spread from 29.5 to 57.5% RH (Fig. 2, red cross and red triangles in panel (d1), see also Table 1) when results from different studies are compared. This spread is similar to the one observed for ERH of pure AS particles with diameters from 0.05 to 40 μm, which scattered between 31 and 48% RH when results from different studies and measurement techniques such as EDB, optical microscopy, HTDMA, and FTIR are compared.[21,22,41–46] Ciobanu *et al.*,[22] have shown that efflorescence of AS particles that are deposited on a hydrophobic substrate preferentially starts from the particle surface or the substrate/particle contact line rather than in the volume of the particle. Discrepancies between AS ERH reported by different studies might therefore result from different surface properties and interactions with the substrate also in the case of mixed organic/AS/H_2O particles. For pure AS particles, ERH values above 40% are suspected to occur due to heterogeneous impurities.[22] However, in the case of mixed particles, the high ERH values (>48%) of OAC + DCA3/AS/H$_2$O (cyan circle, O : C = 0.56), OAC2/AS/H$_2$O (dark blue circle, O : C = 0.57), and MFC2/AS/H$_2$O (purple circle, O : C = 0.79) shown in Fig. 2, panels (a1) and (c1), might be due to repulsive interactions between the ions of AS and the organic substances.

Overall, AS in mixed organic/AS/H$_2$O particles covering the O : C range of 0.29–1.33 deliquesced between 70 and 84% RH and effloresced below 58% RH or remained in a one-liquid-phase state. A more pronounced RH decrease was observed for AS efflorescence than for AS deliquescence. The decrease of ERH and DRH becomes more pronounced with decreasing AS mass fraction. The decrease in ERH may result from both thermodynamic and kinetic factors whereas the decrease in DRH is given by thermodynamics. Highly polar organic substances with high O : C may increase the solubility of AS in aqueous organic mixtures and thus decrease AS ERH and DRH.[20,31] Moreover, high viscosity in organic-rich particles may hinder the crystallization of AS.[14,16]

The parameterization of AS ERH as a function of organic-to-sulfate mass ratio and O : C by Bertram *et al.*,[15] is generally at the lower limit of the observed range of ERH in the mixed particles. It predicts almost no dependence of ERH on O : C for O : C < 0.7 and a drastic decrease for 0.7 < O : C < 0.8 and low organic-to-sulfate ratios. Such a dependence is in contrast to the experimental data in Fig. 2, which shows a rather gradual decrease of ERH with increasing O : C. Moreover, since this parameterization does not discriminate between liquid–liquid-phase-separated

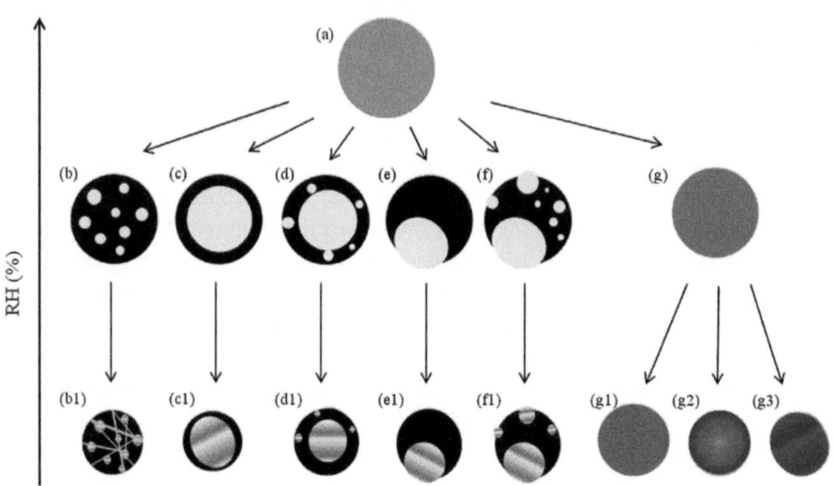

Fig. 3 Schematic diagram of possible morphologies of organic/AS/H$_2$O aerosol particles as a function of RH. Black: organic-rich phase, light grey: AS-rich phase, grey: mixed organic/AS phase. The following morphologies are expected: at high RH: (a) homogeneous droplet. Upon drying: (b) two liquid phases with aqueous AS inclusions in the organic-rich phase, (c) two liquid phases with a core–shell morphology, (d) two liquid phases with a core–shell morphology and additional AS inclusions, (e) two liquid phases with a partially engulfed configuration, (f) two liquid phases with a partially engulfed configuration and additional AS inclusions, and (g) a mixed liquid droplet. Further decrease in RH: (b1) crystalline AS phase with AS needles in the organic-rich liquid or glassy phase, (c1,d1,e1 and f1) crystalline AS phase surrounded by a liquid or a glassy organic-rich phase, (g1) mixed homogeneous droplet, (g2) mixed viscous droplet with concentration gradient, and (g3) crystallized particle with organic-rich phase in pores and veins.

and one-liquid-phase systems it does not capture the low AS ERH of organic-rich particles with O : C < 0.7 that do not show LLPS.

3.2 Investigation of aerosol morphologies

3.2.1 Possible morphologies of tropospheric aerosols. Fig. 3 illustrates the possible morphologies of tropospheric aerosol particles as a function of RH summarizing observations from literature[12,13,16,17,25] and this study. Grey spheres represent mixed one phase droplets. Black and light grey colours stand for an aqueous organic-rich phase and an aqueous AS-rich phase, respectively, and the light grey lines and stripes represent crystallized AS. High RH leads to a mixed organic/AS/H_2O particle consisting of one liquid phase (grey sphere, a). As water is released, organic/AS/H_2O particles may undergo LLPS (b, c, d, e and f) for organic compositions with low O : C or remain as a homogeneous one-liquid-phase particle for high O : C (g). For compositions with high fractions of organics, LLPS occurs by nucleation and growth,[47] resulting in a configuration of several small AS inclusions in an organic-rich phase (b). A further decrease in RH leads to crystallization of the AS inclusions and growth of AS needles (light grey lines) in the organic-rich phase (b1). The organic-rich phase is expected to remain in liquid state or turn glassy.[14,48] LLPS tends to occur by spinodal decomposition when organics and AS are present in similar fractions.[47] Spinodal decomposition is followed by coalescence of the initially formed inclusions and may result in a core–shell morphology (c), a core–shell morphology with AS inclusions (d), a partially engulfed configuration (e), or a partially engulfed configuration with AS inclusions (f).[14,16] Because surface forces and the Kelvin effect become more important with decreasing size of the droplets and the AS inclusions, accumulation mode particles with LLPS are expected to be either present in core–shell or partially engulfed morphologies without any additional AS inclusions. This is supported by Ciobanu et al.,[25] who found a decreasing number of AS satellite inclusions for decreasing diameters of PEG-400/AS/H_2O particles. For particles with high fractions of AS, Ciobanu et al.,[25] and Song et al.,[16] always observed a core–shell configuration (c) via 2nd phase formation from the surface of the particles. It depends on thermodynamic and kinetic factors whether AS effloresces at low RH. In droplets with LLPS, AS crystallization starts from the AS-rich phase and may proceed by the growth of needles into the organic-rich phase (b1, see also Fig. 1a). Effloresced droplets with core–shell and partially engulfed morphologies largely retain the morphology present before efflorescence (c1, d1, e1 and f1). Some satellite inclusions may disappear instead of crystallizing due to the diffusion of ammonium and sulfate ions to the crystallized regions of the particle (d → d1; f → f1). When homogeneous liquid particles effloresce (g → g3) the remaining liquid organic constituents may distribute in pores and veins and around the crystallized AS. Attractive interactions between organic functional groups and AS may suppress efflorescence in particles with a large organic fraction and high O : C (g1). AS crystallization may also be suppressed when particles become highly viscous or turn glassy at low RH (g1 and g2). Highly viscous particles may show concentration gradients (g2) instead of a homogeneous composition (g1).[14]

3.2.2 Identification of droplet morphologies. Table 1 summarizes the observed phases and phase changes of the organic/AS/H_2O systems investigated in this study together with the available literature data. Out of the 21 systems investigated in this study, eight systems did not show LLPS for all investigated organic-to-inorganic ratios, nine showed core–shell morphology and four showed both core–shell and partially engulfed configurations depending on the mass fraction of AS in the mixture and RH. Fig. 4 exemplifies the process leading to core–shell configuration in COM1/AS/H_2O for $mf_d(AS) = 0.40$, and the partially engulfed morphology in C6 + C7/AS/H_2O for $mf_d(AS) = 0.33$. Both systems underwent LLPS by spinodal decomposition. While the COM1/AS/H_2O particle showed LLPS at 82.8% RH forming a core–shell configuration and effloresced at 45.9% RH (Fig. 4a), the C6 + C7/AS/H_2O particle exhibited LLPS at 83.5% RH. Subsequently, the inner phase moved towards the edge of the particle between 78.0 and 71.3% RH within 30 min leading to a partially engulfed morphology and remained in this configuration down to 44.7% RH when AS effloresced (Fig. 4b).

Droplets that showed partially engulfed configurations all contained dicarboxylic acids as the organic fraction. C6/AS/H_2O mixtures with $mf_d(AS) = 0.50$ and 0.67 showed a transition from core–shell to partially engulfed morphology in the RH range from 70–66% and 63–50%, respectively. For C7/AS/H_2O with $mf_d(AS) = 0.34$ and C5 + C6 + C7/AS/H_2O with $mf_d(AS) = 0.5$ spinodal decomposition directly led to a partially engulfed configuration at 88% RH and 74% RH, respectively.

3.2.3 Calculation of equilibrium configurations. Since the optical microscope gives only the top view of the droplets on the substrate, it is difficult to determine whether the AS inclusion contacts the droplet/air or the droplet/substrate interface or whether it is floating within the droplet volume. Therefore, Song *et al.*,[16] performed confocal Raman scans through C6/AS/H_2O droplets revealing that the AS inclusion is at the bottom of the droplet. While the movement of the main AS inclusion from the middle to the edge of the particle is a clear indication of the transition from a core–shell to a partially engulfed configuration, it remains unclear whether an apparent core–shell morphology with the main AS inclusion in the middle of the droplet always fulfils the condition of complete wetting of the AS-rich phase by the organic-rich phase. If the top of the AS inclusion had direct contact to the gas phase, the top-view of the particle would be similar. Moreover, the hydrophobic glass slide might influence the droplet morphology. We therefore attempt to complement and confirm the optically

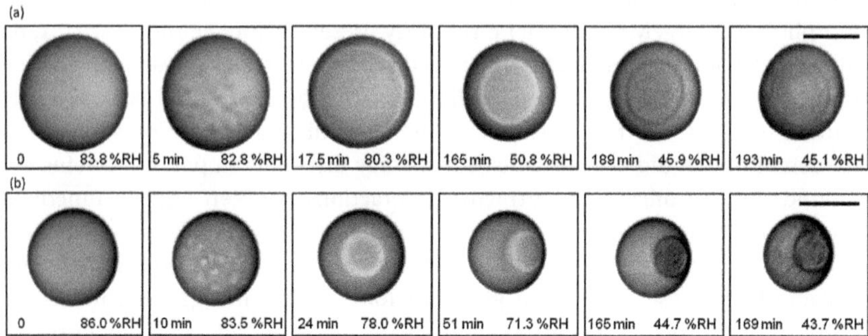

Fig. 4 Image sequence of LLPS and AS efflorescence leading to (a) core–shell morphology of COM1/AS/H_2O (O : C = 0.63, $mf_d(AS) = 0.40$) and (b) partially engulfed configuration of C6+C7/AS/H_2O (O : C = 0.61, $mf_d(AS) = 0.33$). Size bar: 20 μm.

observed morphologies by calculating the preferred morphology based on surface and interfacial tension measurements of the involved phases.

The spreading coefficient (S_i) gives an indication of the resulting equilibrium configuration of two immiscible liquid phases i and j within a gas phase k.[12,13,49] It can be calculated when the surface (σ_{jk}, σ_{ik}) and interfacial tensions (σ_{ij}) are known:

$$S_i = \sigma_{jk} - (\sigma_{ij} + \sigma_{ik}) \tag{1}$$

Since the surface tension of the AS-rich phase (σ_{AS}) is larger than the one of the organic-rich phase (σ_{org}), the aqueous AS-rich phase does not spread on the organic-rich phase. Therefore, the conceivable configurations are an organic-rich phase that fully (core–shell) or partially engulfs the AS-rich phase. The conditions for a core–shell configuration and a partially engulfed configuration are given by eqn (2) and (3), respectively:

$$\text{core–shell configuration: } S_{org} = \sigma_{AS} - (\sigma_{org/AS} + \sigma_{org}) > 0 \tag{2}$$

$$\text{Partially engulfed configuration: } S_{org} = \sigma_{AS} - (\sigma_{org/AS} + \sigma_{org}) < 0 \tag{3}$$

It should be noted that this use of spreading coefficient is only strictly valid for phases of infinite volume where the surface and interfacial area changes can be approximated as the same. This may render morphology predictions less reliable in cases of finite volumes when the spreading coefficient is close to zero. Surface and interfacial tensions of solutions can be measured by standard methods such as the ring method (see Section 2). Because these methods need bulk volumes, measurements are restricted to samples that are not supersaturated with respect to any component. Since solution saturation of AS is reached at water activity, $a_w = 0.8$, LLPS has to be present for $a_w > 0.8$. Systems that fulfil these conditions are PEG-400/AS/H$_2$O, which always showed a core–shell morphology and C6 + C7/AS/H$_2$O, which showed core–shell or partially engulfed configurations, depending on mf$_d$(AS) and RH (see Table 1). Interfacial and surface tensions were determined for the liquid–liquid-phase-separated systems prepared with mf$_d$(AS) = 0.5

Table 2 Water activities (a_w), densities (at 293 K), surface and interfacial tensions of PEG-400/AS/H$_2$O (at 298 K) and C6 + C7/AS/H$_2$O (at 293 K) of the upper organic-rich and lower AS-rich phases for mf$_d$(AS) = 0.5. Spreading coefficients are calculated according to eqn (2) and (3). Observed morphology is given for indicated a_w

Mixture	a_w	Density (g cm^{-3}) organic-rich	AS-rich	Surface tension (mN m^{-1}) organic-rich	AS-rich	Interfacial tension (mN m^{-1})	Spreading coefficient (mN m^{-1})	Observed morphology
PEG-400/	0.82	1.13	1.25	42.6 ± 1.58	51.2 ± 0.03	1.07	7.54	core–shell
AS/H$_2$O	0.86	1.12	1.21	42.1 ± 0.79	52.6 ± 0.42	0.46	9.97	
	0.89	1.12	1.19	42.7 ± 0.89	52.2 ± 1.07	0.08	9.43	
C6 + C7/	0.85 ±	1.19	1.21	40.7 ± 0.11	41.6 ± 0.01	below detection limit	0.4–0.8	core–shell
AS/H$_2$O	0.002							

and water contents leading to $a_w = 0.82$, 0,86, and 0.89 for PEG-400/AS/H$_2$O and $a_w = 0.85$ for C6 + C7/AS/H$_2$O. Table 2 presents the measured surface and interfacial tensions, together with the calculated spreading coefficients and the observed morphologies.

For the investigated a_w range of 0.82–0.89, the surface tensions of the organic-rich phase of PEG-400/AS/H$_2$O vary between 42.1–42.7 mN m^{-1} with no clear dependence on a_w. These values are slightly lower than the surface tension of pure PEG-400 (this study: 44.32 mN m^{-1}; Ciobanu *et al.*,[25]: 44.5 mN m^{-1}), probably due to a different arrangement of the PEG molecules on the surface of the solution.

We therefore assume that a decrease of a_w will not lead to a further decrease of the surface tension. Surface tension in saturated AS solutions is increased by 20% compared with the one of pure water (71.99 mN m^{-1}).[16] However, the surface tension of the AS-rich phase of PEG-400/AS/H$_2$O shows a strong decrease to values between 51.2–52.6 mN m^{-1} in the investigated a_w range of 0.82–0.89, indicating that small amounts of PEG-400 are present in the AS-rich phase and enriched on the surface. Interfacial tensions between the organic-rich and the AS-rich phase are low (0.08–1.07 mN m^{-1}) but show a clear tendency to larger values with decreasing a_w. Such an increase of interfacial tension with decreasing water content has also been observed by Wu *et al.*,[50] for PEG-1000/AS/H$_2$O, PEG-2000/AS/H$_2$O, and PEG-4000/AS/H$_2$O systems and can be rationalized by considering that the concentration of water as the common component is getting less and the two liquid phases become more dissimilar when a_w decreases. The spreading coefficient calculated from these interfacial and surface tension values is clearly positive (7.5–10 mN m^{-1}), indicating a core–shell morphology as the thermodynamically favoured configuration, which is in accordance with the optical appearance of the phase-separated droplets of this system on the hydrophobic substrate. If one assumes constant surface tensions of the two liquid phases and an increase of the interfacial tension of 1–2 mN m^{-1} for a decrease of a_w of 0.1, the spreading coefficient would be $\sim$0 mN m^{-1} when $a_w \approx$ ERH. A movement of the main AS inclusion to the edge of the particle has not been observed for this system, indicating that the core–shell morphology remains energetically favoured. However, it cannot be totally excluded that high viscosity of the organic-rich phase at low RH inhibits the movement of the AS inclusion within the droplet[25] and that a partially engulfed configuration is realized by incomplete wetting of the top of the AS inclusion by the organic-rich phase.

For the system C6 + C7/AS/H$_2$O the surface tension of the organic-rich phase (40.7 mN m^{-1}) at $a_w = 0.85$ is below the range estimated by Topping *et al.*,[51] for the surface tension of adipic acid (straight chain C6 dicarboxylic acid) in the supercooled liquid state (45–52 mN m^{-1}). Surprisingly, the surface tension of the AS-rich phase is also reduced to a similarly low value (41.6 mN m^{-1}). The interfacial tension of this system at $a_w = 0.85$ was too low to be measured. We therefore assume a value between 0.1–0.5 mN m^{-1}, which is just below the detection limit of the ring method (<0.5 mN m^{-1}). This assumption leads to a positive spreading coefficient of the investigated mixture, which is in accordance with the observed core–shell morphology of the droplet deposited on the hydrophobic substrate. If we assume an increase of the interfacial tension and constant surface tensions of the liquid phases with decreasing a_w – as it is the case for the PEG-400/AS/H$_2$O system – a transition from core–shell to partially engulfed morphology should occur when droplets of C6 + C7/AS/H$_2$O are exposed to decreasing RH. Such a transition is indeed observed between 78.0% RH and 71.3% RH in the C6 + C7/AS/H_2O mixture for $mf_d(AS) = 0.33$. We therefore assume that the interfacial tension reaches a critical value where the spreading coefficient of the organic-rich phase changes sign from a positive to a negative value indicating that a partially engulfed configuration becomes energetically favoured. Several systems with dicarboxylic acids as the organic fraction exhibited a partially engulfed configuration. We therefore consider the surface and interfacial tension values measured for C6 + C7/AS/H_2O as typical for these systems.

The comparison of the optically observed morphology of liquid–liquid-phase-separated droplets and calculated spreading coefficients shows that the movement of the main AS inclusion from the middle to the edge of a droplet deposited on a hydrophobic substrate is indicative for the transition from a core–shell to a partially engulfed configuration. The presence of the main AS inclusion in the middle of the droplet should therefore be a good indicator for a core–shell configuration as the energetically favoured one. This conclusion is certainly valid at intermediate and high RH when the viscosity of the organic-rich phase is low enough to allow movement of the AS inclusion within the droplet. In case of highly viscous particles at low RH, the energetically favoured configuration may not be realized or LLPS might even be totally impeded.[25] If an airborne particle with core–shell morphology is exposed to decreasing RH, it might therefore remain in a core–shell configuration because of kinetic restraints.

3.2.4   LLPS and morphology of single levitated aerosol particles. To investigate whether the contact to the substrate in optical microscope experiments influences the onset relative humidity of LLPS (SRH), the relative humidity range of the phase transition and – most important – the morphology of the phase separated aerosol, we performed measurements on single, levitated aerosol particles. We chose two systems for the comparison with bulk predictions (see 3.2.3) and the optical microscope experiments (see 3.2.2): PEG-400/AS/H_2O and C6/AS/H_2O both with $mf_d(AS) = 0.5$. Both optical microscopy and bulk measurements of interfacial tensions show that the morphology of the phase separated particle is core–shell for the PEG-400/AS/H_2O system over the entire RH range below SRH and above ERH, while a transition from core–shell morphology to partially engulfed morphology occurs below SRH for the C6/AS/H_2O system.

The experiments in the electrodynamic balance were performed at 291 K on at least three different particles with radii between 7 and 15 μm at 91% RH. The particles were exposed to cyclic changes in RH to pass through the LLPS transition but typically avoiding efflorescence.

For the PEG-400/AS/H_2O system thermodynamic data (a_w and volume fraction of the two phases) were determined previously.[31] In order to compute Mie resonance spectra for different morphologies, additional data for the densities and refractive indices of the involved aqueous solutions were required. Fig. 5 shows the results and gives the parameterizations in terms of a_w. We also used the data of Marcolli and Krieger[31] to parameterize the mass fraction of solute (mf_s) in terms of a_w as: $mf_s = \dfrac{(1 - a_W)^a}{1 + b \cdot a_W + c \cdot a_W^2}$; $1 > a_W > 0.8$ ($a = 0.818, b = -0.517, c = -0.0995$), and the volume ratio, $V_R = $ (volume of PEG-rich phase)/(volume AS-rich phase) as $V_R = 1.173 - 0.432 \cdot a_w$.

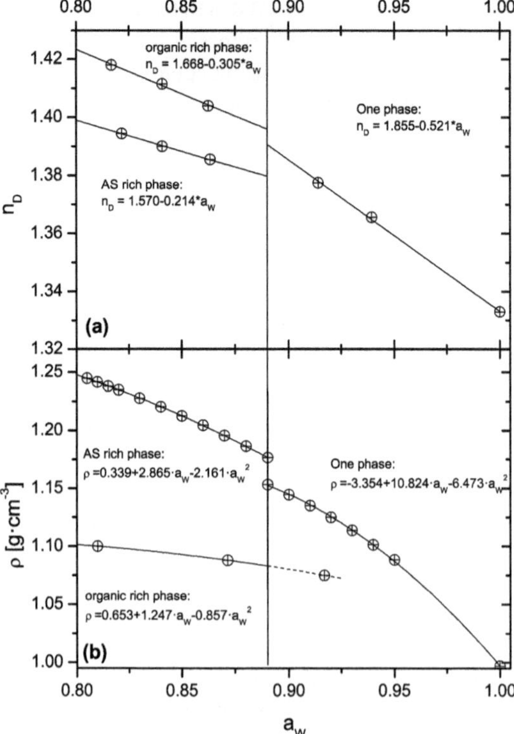

Fig. 5 Panel (a) shows refractive index data, n_D, measured in bulk samples for solute concentrations with one phase and for the two phases at concentrations lower than the one where LLPS occurs. Linear fits for the refractive indices in terms of a_w are provided in the figure. Panel (b) shows density data. We use density data of the binary PEG/H_2O solution and for the binary AS/H_2O solution to represent densities of the organic-rich and the AS-rich phases, respectively. We believe these are good approximations, since the AS-rich phase contains little PEG and the organic-rich phase little AS. Polynomial fits for density are provided as well.

With these data available we can compare measured Mie resonance spectra with computed ones. For the computations, we use the Mie scattering codes provided by Mätzler[52] for homogeneous spheres, Pena and Pal[53] for those of spherical symmetric core–shell morphology and that of Ngo *et al.*,[54] for core–shell particles with a nonconcentric core.

Fig. 6 shows such a comparison for a particle significantly above SRH, *i.e.* the morphology of the particle is a homogeneous liquid. Let us first note that tiny changes in RH (*e.g.* nominally 0.04%) at high humidities, with the associated large growth factors, lead to a significant shift in the resonance spectra; *cf.* the grey and black spectrum in panel (a1) of Fig. 6. This also means that small fluctuations of RH during a single scan will lead to a significant relative shift between the resonance peaks. For calculating Mie resonance spectra, a_w has to be known accurately but the uncertainty in our RH measurements is ±1.5%. In addition the parameterizations given above lead to uncertainties in the parameters needed to compute Mie resonance spectra. Hence a comparison with computed spectra should be based on the shape and occurrence of spectral features rather than absolute resonance positions. Accordingly we compare the

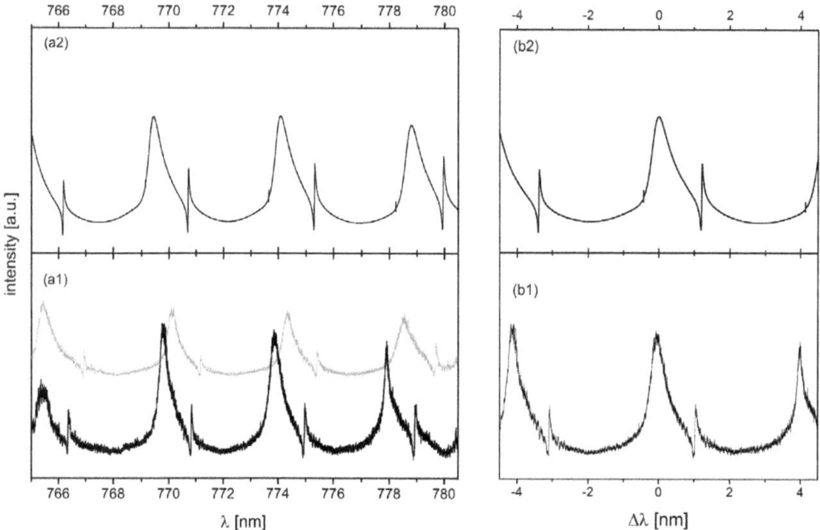

Fig. 6 Comparison of measured Mie resonance spectra (TE-polarization) with computed spectra at RH = 93.5%, T = 291 K. Panel (a1) shows two consecutively measured Mie resonance spectra, 7 min apart at nominally 93.58% RH (grey) and 93.54% RH (black). Panel (a2) shows the computed Mie spectrum for a_w = 0.9354 using a standard Mie code and the parameterizations as detailed above. Panels (b1) and (b2) show the same spectra but centered relative to the central peak of panel (a1) and (a2).

computed spectrum relative to the central peak (panel (b2) of Fig. 6) with the measured spectrum in panel (b1) of Fig. 6. The only free parameter in the computation is the size of the particle, which was fixed by evaluating a series of spectra at 90.5% RH. The satisfactory agreement between modelled and computed spectrum shows that our parameterizations based on bulk measurements can be successfully applied to levitated aerosol particles at least if the particle is homogeneous.

In Fig. 7 we show a series of spectra at different RH passing through the LLPS upon drying. We attribute the disappearance of the sharp resonance at about +0.5 nm relative to the central peak at 90.4% RH together with a general increase in noisy features in the spectrum to the onset of LLPS. This onset RH is slightly higher but agrees within the error of our RH sensor with the one determined from bulk measurements (SRH = 89.2%[31]) and the one determined using optical microscopy (SRH = 89.7%[25]). It is interesting to note that it takes almost 4 h slowly drying to about 87.1% RH until the noisy features reduce and a very small but sharp resonance appears at about +2.0 nm relative to the central peak (marked by an arrow in Fig. 7). Upon further drying the noise is further reduced, but the general spectral features are largely unchanged. When comparing these findings with the optical microscopic observations of Ciobanu et al.,[25] we speculate that this end of the transition marks the coalescence of satellite inclusions of the AS-rich phase with a central core AS-rich phase in the particle. This speculation is supported by the corresponding spectra upon subsequent humidification (Fig. 8).

We observe a clear reverse transition (note the disappearance of the very small, but sharp resonance about +2.5 nm relative to the central peak), but now the transition is much narrower in terms of the RH-range and time, namely between about 88.4% and 88.8% RH and about half an hour. The transition RH is slightly

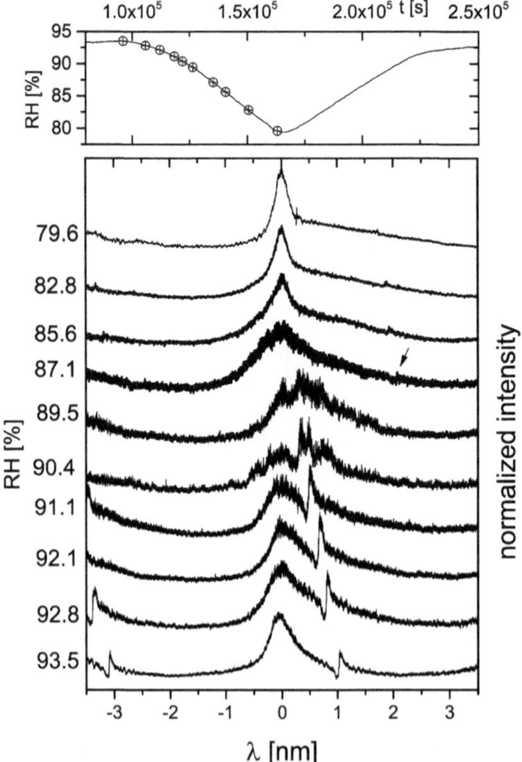

Fig. 7 Series of Mie resonance spectra (TE-polarization) upon slow drying from 93.5% RH to 79.6% RH. The scattered intensity is normalized to the maximum intensity in the spectra, the wavelength scale is relative to the peak center, for clarity spectra are shifted vertically to each other. The upper panel shows the change in RH with time. The symbols mark the times and humidities at which the spectra in the lower panel were recorded.

lower but within error the same as observed upon drying and in agreement with bulk and optical microscope experiments. We attribute the reduction in RH range for the transition from ΔRH $\sim$ 3.3% upon drying to ΔRH $\sim$ 0.4% upon humidification to the fact that coalescence of satellite inclusions occurs only when drying.

We compare the Mie resonance spectra (TE-polarization) at about 81% RH, significantly below SRH, to computed spectra for different morphologies in Fig. 9. First, note that at 81% RH a change of 0.2% in RH leads to roughly the same shift in the Mie Resonance spectra as does a shift of ΔRH = 0.04% at 93.5%. This is due to the significantly lower slope in the growth factor at lower humidity. Hence fluctuations in RH become less important at lower humidity. The uncertainty in the parameters needed for the Mie computations suggest that one should focus on the shape of the resonances rather than their absolute peak position. Therefore, the better agreement of peak position of the measured spectrum (black line) with the one computed for core–shell morphology may be accidental. If we compare the resonance shape of the broader resonance in panels (b1) and (b2) the agreement between measured and calculated spectrum is considerably better for the one with the core–shell morphology. We not only calculated Mie resonance

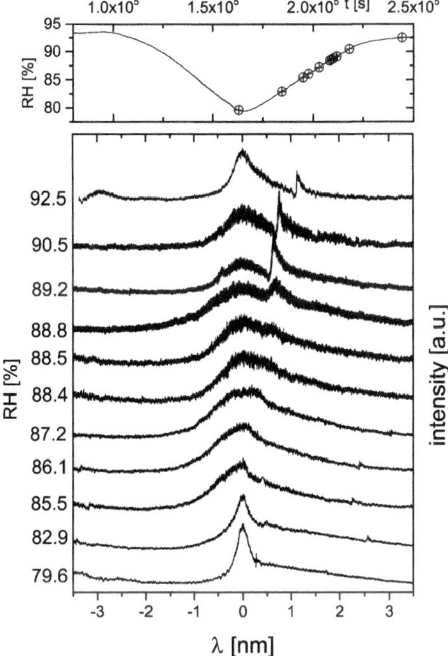

Fig. 8 Series of Mie resonance spectra (TE-polarization) upon humidification *cf.* Fig. 7.

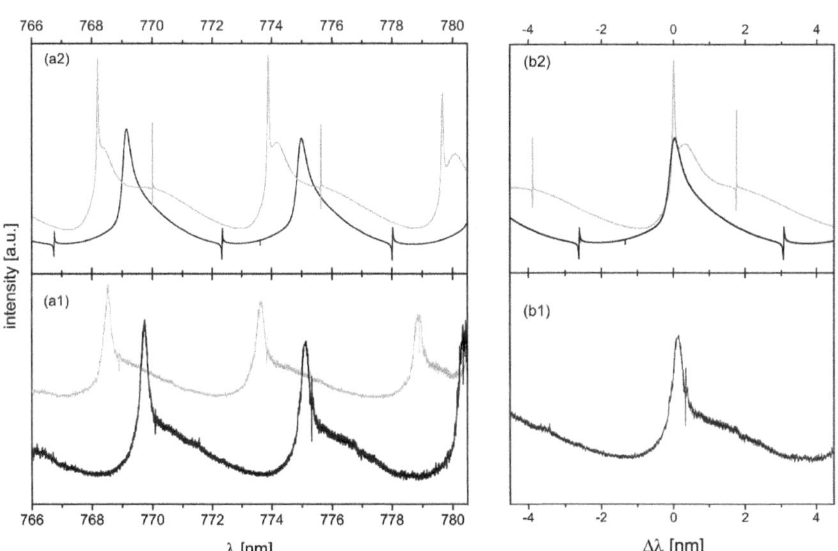

Fig. 9 Comparison of measured with computed spectra (TE-polarization) as in Fig. 6, but at RH = 80.9%, in the LLPS regime. Panel (a1) shows two subsequent spectra at RH = 80.9% (black line) and RH = 80.7% (grey line). The computed spectra in panels (a2) and (b2) are calculated for RH = 80.9% assuming a homogeneous droplet (grey line) and assuming a spherical symmetric core–shell morphology (black line, code of Pena and Pal[53]). Panels (b1) and (b2) show the same spectra but centered relative to the central peak of the spectra at 80.9% RH of panel (a1) and (a2).

spectra of spherical symmetric core–shell morphology, but also computed spectra with the core located off centre using the code of Ngo *et al.*[54] The shape of the broader resonances does not change when the core is no longer in the centre of the droplet, but the location of the sharpest resonances shift, especially if the core is located at the outer edge of the particle. Note that the sharp resonance close to the central peak of the measured spectrum is not present in both of the computed spectra. When comparing measured with computed spectra at much lower humidity (RH = 41.3%), close to the efflorescence humidity, we were not able to reproduce the measured spectra closely. While the spectra (see Fig. 10) exhibited consistent resonance features, including sharp resonances, all computed spectra look significantly different.

We varied size, volume ratio, refractive index and location of the core relative to the centre of the droplet (using the code of Ngo *et al.*[54]), but were not able to reproduce the measured spectra closely. At present we do not have any explanation for this disagreement, but suggest that further experiments with particles with varying total size may help to understand these differences. Nevertheless, we conclude that the particle exhibits core–shell morphology when phase separated in agreement with both predictions from bulk measurements of interfacial tensions and optical microscope images. We exclude partially engulfed morphology because spectra of particles with this morphology appear distinctly different (see Fig. 11 and discussion below).

To compare Mie resonance spectra of a system for which we expect core–shell morphology to one for which we expect a transition from core–shell to partially engulfed morphology, we also investigated the $C6/AS/H_2O$ system. A series of subsequent spectra upon drying are shown in Fig. 11. There is a continuous shift of the sharp resonance observed at initially (RH = 89.7%) + 6.3 nm (not shown) towards the central, broader peak upon drying. At 61.9% RH the sharp resonance

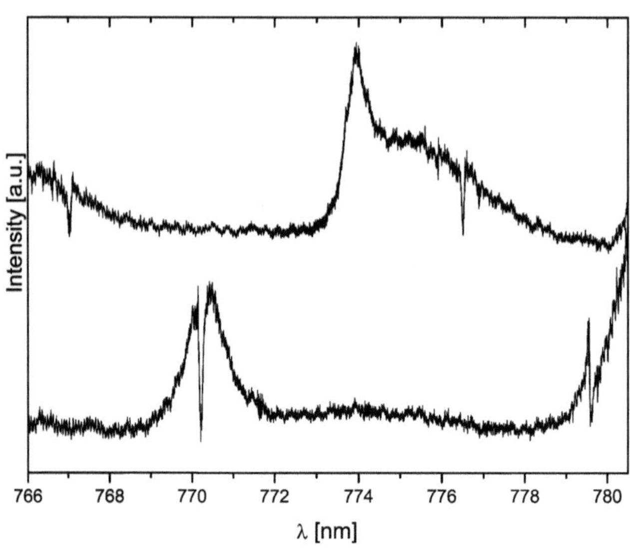

Fig. 10 Resonance spectra (lower trace TM-polarization, upper trace TE-polarization) of a second particle of the PEG-400/AS/H₂O system at low humidity (RH = 41.3%).

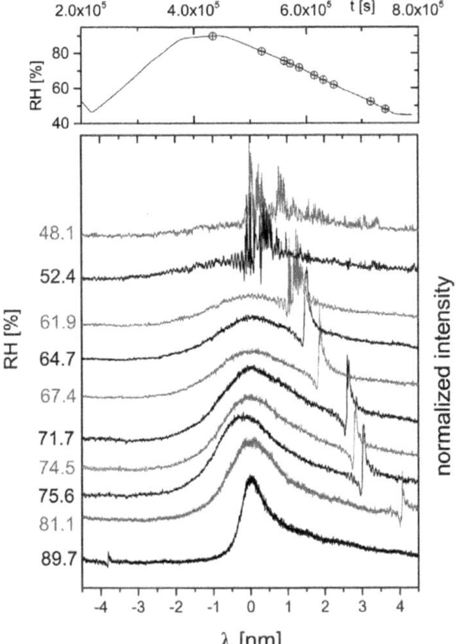

Fig. 11 Series of subsequent Mie resonance spectra (TE-polarization) recorded upon drying a particle of the C6/AS/H$_2$O system, analogous to Fig. 7 and 8.

becomes noisy and its spectral position scatters when recording the spectra in time. This noise increases further upon drying with only a broad noisy resonance being observed at RH around 50%. Subsequent drying cycles and repeated experiments with different particles always showed this pattern in the Mie resonance spectra with the described transition occurring between 65.6% RH and 61.9% RH. Upon humidifying the transition back to non-distorted spectra occurred at slightly higher RH around 66%.

According to our optical microscope experiments the onset RH of LLPS is 73.4%. No sudden changes in the Mie resonance spectra occurred around this humidity, probably indicating that the transition from homogeneous to core–shell morphology is fast and not accompanied over a broader humidity range with deviations of spherical symmetry. A comparison with computed spectra, which could reveal the transition, is difficult because the supportive data needed to perform such a calculation is lacking. The transition happens below the deliquescence humidity of AS, making bulk measurements of the needed parameters, like density, refractive index and volume ratio impossible.

The transition to partially engulfed morphology occurred at 67.2% RH in the optical microscope experiments, is slightly higher than the transition we observe with the levitated particle. However, the agreement is close enough to attribute the appearance of significant noise in the spectra of Fig. 11 to the transition from core–shell morphology to partially engulfed morphology. Also, the increase of the noise upon further drying supports this interpretation, since the deviation from spherical symmetry will become more pronounced with decreasing volume fraction of the AS-rich phase. Since the particle performs rotational Brownian motion when trapped in the EDB[55] and the position of the resonances for a non-spherical particle depends on scattering angle, the scattered intensity depends on wavelength and time, so that during slow scanning of the laser wavelength several resonances will be recorded closely spaced in time, which result then in a noisy spectrum in wavelength space. This is also consistent with the observations of Reid *et al.*,[13] that sharp whispering gallery mode resonances in stimulated Raman spectra are suppressed for particles with partially engulfed morphology. Let us note at the side, that in one particular experiment we observed a further transition back to core–shell morphology occurring at RH = 56%.

4   Conclusions

In this study, we analyzed and evaluated phases, phase transitions and morphologies of 33 organic/AS/H$_2$O systems investigated in this work and literature. The organic fractions of these systems consist of one to ten aliphatic and/or aromatic oxidized compounds with carboxylic acid, hydroxyl, carbonyl, ether, and ester functionalities and cover the O : C range from 0.29 to 1.33. Thirteen out of these 33 systems did not show LLPS for all studied organic-to-inorganic mixing ratios, sixteen showed core–shell morphology when present in a two-liquid-phases state and four showed both core–shell and partially engulfed configurations depending on the organic-to-inorganic ratio and RH. The organic fractions of the systems with partially engulfed configurations consisted in all cases of dicarboxylic acids. In some of these systems partially engulfed configuration was established immediately when LLPS occurred, other systems showed a transition from core–shell to partially engulfed configuration when RH was decreased. AS in mixed organic/AS/H$_2$O particles deliquesced between 70 and 84% RH and effloresced below 58% RH or remained in a one-liquid-phase state when all mixtures are considered. AS in droplets with LLPS always effloresced between 30 and 50% RH and deliquesced between 70 and 84% RH. The decrease of ERH and DRH became more pronounced with decreasing AS mass fraction and may be caused by an increased solubility of AS in the presence of highly polar organic substances or crystallization inhibition in highly viscous particles. These results clearly show that the presence or absence of LLPS influences AS efflorescence.

Whether aerosol particles consisting of two liquid phases show core–shell or partially engulfed morphologies depends on the balance of surface and interfacial tensions, which determine the spreading of the immiscible liquid phases on each other. The calculation of spreading coefficients for the systems PEG-400/AS/H$_2$O and C6 + C7/AS/H$_2$O confirmed that the movement of an AS inclusion from the middle to the edge of a droplet deposited on a hydrophobic substrate is indicative for the transition from a core–shell to a partially engulfed configuration. Interestingly, the surface tension of the AS-rich phase is strongly reduced compared with a pure aqueous AS solution and seems to depend more on the nature of the organic solutes than on the inorganic salt. The interfacial tensions between the organic-rich and the AS-rich phase are very low at the onset of LLPS and show a clear tendency to larger values with decreasing RH. A transition from core–shell to partially engulfed configuration with decreasing RH seems therefore to be triggered by this quantity. Analysis of high resolution elastic Mie resonance spectra allowed the detection of LLPS for single levitated droplets consisting of PEG-400/AS/H$_2$O. Mie spectra measured at 93.5% and at 80.9% RH agreed with computed

Mie spectra for a homogeneous and a core–shell configuration, respectively, confirming the results obtained from droplets deposited on a hydrophobic substrate. While the transition from a homogeneous particle to a core–shell configuration was not detected unambiguously in the C6/AS/H$_2$O system, the transition from core–shell to partially engulfed morphology could be clearly seen.

Based on the experimental evidence presented in this work, we consider the core–shell morphology as the prevalent configuration in organic/AS/H$_2$O particles in the atmosphere. This conclusion seems to be in conflict with Reid et al.,[13] who postulated a predominance of partially engulfed structures. However, this apparent disagreement might be resolved considering that the analysis by Reid et al.,[13] relies on systems with NaCl as the inorganic component and organic compounds that show LLPS also in the absence of an inorganic salt. Such hydrophobic substances are however not typical for the organic fraction in ambient aerosols. Further studies are needed to investigate how sensitively phase transitions and morphology of mixed organic/inorganic particles depend on the chemical composition of the inorganic fraction, temperature and particle size.

Acknowledgements

This work was supported by the Swiss National Foundation Project No. 200020-125151 and the ETH Research Grant ETH-0210-1.

References

1 J. H. Seinfeld and S. N. Pandis, *Atmos. Chem. Phys.: From Air Pollution to Climate Change*, Wiley & Sons, New York, 2006.

2 G. Hanel, *J. Atmos. Sci.*, 1976, **33**, 1120–1124.

3 S. T. Martin, *Chem. Rev.*, 2000, **100**, 3403–3453.

4 IPCC, *Climate Change 2007: The physical science basis. Contribution of working group I to the Fourth Assessment Report of the Intergovernmental Panel on Climate Change*, Cambridge, New York, 2007.

5 T. Anttila, A. Kiendler-Scharr, T. F. Mentel and R. Tillmann, *J. Atmos. Chem.*, 2007, **57**, 215–237.

6 L. M. Cosman, D. A. Knopf and A. K. Bertram, *J. Phys. Chem. A*, 2008, **112**, 2386–2396.

7 A. Zuend, C. Marcolli, T. Peter and J. H. Seinfeld, *Atmos. Chem. Phys.*, 2010, **10**, 7795–7820.

8 A. Zuend and J. H. Seinfeld, *Atmos. Chem. Phys.*, 2012, **12**, 3857–3882.

9 S. H. Lee, D. M. Murphy, D. S. Thomson and A. M. Middlebrook, *J. Geophys. Res.*, 2002, **107**(D1), 4003.

10 C. Marcolli, B. P. Luo, T. Peter and F. G. Wienhold, *Atmos. Chem. Phys.*, 2004, **4**, 2593–2599.

11 D. M. Murphy, D. J. Cziczo, K. D. Froyd, P. K. Hudson, B. M. Matthew, A. M. Middlebrook, R. E. Peltier, A. Sullivan, D. S. Thomson and R. J. Weber, *J. Geophys. Res.*, 2006, **111**, DOI: 10.1029/2006JD007340.

12 N. O. A. Kwamena, J. Buajarern and J. P. Reid, *J. Phys. Chem. A*, 2010, **114**, 5787–5795.

13 J. P. Reid, B. J. Dennis-Smither, N. O. A. Kwamena, R. E. H. Miles, K. L. Hanford and C. J. Homer, *Phys. Chem. Chem. Phys.*, 2011, **13**, 15559–15572.

14 U. K. Krieger, C. Marcolli and J. P. Reid, *Chem. Soc. Rev.*, 2012, **41**, 6631–6662.

15 A. K. Bertram, S. T. Martin, S. J. Hanna, M. L. Smith, A. Bodsworth, Q. Chen, M. Kuwata, A. Liu, Y. You and S. R. Zorn, *Atmos. Chem. Phys.*, 2011, **11**, 10995–11006.

16 M. Song, C. Marcolli, U. K. Krieger, A. Zuend and T. Peter, *Atmos. Chem. Phys.*, 2012, **12**, 2691–2712.

17 M. Song, C. Marcolli, U. K. Krieger, A. Zuend and T. Peter, *Geophys. Res. Lett.*, 2012, **39**, L19801.

18 C. Pöhlker, K. T. Wiedemann, B. Sinha, M. Shiraiwa, S. S. Gunthe, M. Smith, H. Su, P. Artaxo, Q. Chen, Y. F. Cheng, W. Elbert, M. K. Gilles, A. L. D. Kilcoyne, R. C. Moffet, M. Weigand, S. T. Martin, U. Poeschl and M. O. Andreae, *Science*, 2012, **337**, 1075–1078.

19 Y. You, L. Renbaum-Wolff, M. Carreras-Sospedra, S. J. Hanna, N. Hiranuma, S. Kamal, M. L. Smith, X. L. Zhang, R. J. Weber, J. E. Shilling, D. Dabdub, S. T. Martin and A. K. Bertram, *Proc. Natl. Acad. Sci. U. S. A.*, 2012, **109**, 13188–13193.
20 C. Marcolli, B. P. Luo and T. Peter, *J. Phys. Chem. A*, 2004, **108**, 2216–2224.
21 M. T. Parsons, J. L. Riffell and A. K. Bertram, *J. Phys. Chem. A*, 2006, **110**, 8108–8115.
22 V. G. Ciobanu, C. Marcolli, U. K. Krieger, A. Zuend and T. Peter, *J. Phys. Chem. A*, 2010, **114**, 9486–9495.
23 M. L. Smith, M. Kuwata and S. T. Martin, *Aerosol Sci. Technol.*, 2011, **45**, 244–261.
24 M. L. Smith, A. K. Bertram and S. T. Martin, *Atmos. Chem. Phys.*, 2012, **12**, 9613–9628.
25 V. G. Ciobanu, C. Marcolli, U. K. Krieger, U. Weers and T. Peter, *J. Phys. Chem. A*, 2009, **113**, 10966–10978.
26 D. A. Knopf, Diss. ETH Zurich, No. 15103, Zurich, Switzerland, 2003.
27 U. K. Krieger, C. A. Colberg, U. Weers, T. Koop and T. Peter, *Geophys. Res. Lett.*, 2000, **27**, 2097–2100.
28 C. A. Colberg, U. K. Krieger and T. Peter, *J. Phys. Chem. A*, 2004, **108**, 2700–2709.
29 U. K. Krieger and C. Braun, *J. Quant. Spectrosc. Radiat. Transfer*, 2001, **70**, 545–554.
30 C. Braun and U. K. Krieger, *Opt. Express*, 2001, **8**, 314–321.
31 C. Marcolli and U. K. Krieger, *J. Phys. Chem. A*, 2006, **110**, 1881–1893.
32 Y.-F. Te, Diss. University of Zurich, Zurich, Switzerland, 2011.
33 M. C. Yeung, A. K. Y. Lee and C. K. Chan, *Aerosol Sci. Technol.*, 2009, **43**, 387–399.
34 A. Pant, A. Fok, M. T. Parsons, J. Mak and A. K. Bertram, *Geophys. Res. Lett.*, 2004, **31**, L12111.
35 M. T. Parsons, J. Mak, S. R. Lipetz and A. K. Bertram, *J. Geophys. Res.*, 2004, **109**, D06212.
36 A. Bodsworth, B. Zobrist and A. K. Bertram, *Phys. Chem. Chem. Phys.*, 2010, **12**, 12259–12266.
37 A. A. Zardini, S. Sjogren, C. Marcolli, U. K. Krieger, M. Gysel, E. Weingartner, U. Baltensperger and T. Peter, *Atmos. Chem. Phys.*, 2008, **8**, 5589–5601.
38 M. N. Chan, A. K. Y. Lee and C. K. Chan, *Environ. Sci. Technol.*, 2006, **40**, 6983–6989.
39 M. Y. Choi and C. K. Chan, *Environ. Sci. Technol.*, 2002, **36**, 2422–2428.
40 T. Y. Ling and C. K. Chan, *J. Geophys. Res.*, 2008, **113**, D14205.
41 M. D. Cohen, R. C. Flagan and J. H. Seinfeld, *J. Phys. Chem.*, 1987, **91**, 4563–4574.
42 J. Xu, D. Imre, R. McGraw and I. Tang, *J. Phys. Chem. B*, 1998, **102**, 7462–7469.
43 J. H. Han and S. T. Martin, *J. Geophys. Res.*, 1999, **104**, 3543–3553.
44 T. B. Onasch, R. L. Siefert, S. D. Brooks, A. J. Prenni, B. Murray, M. A. Wilson and M. A. Tolbert, *J. Geophys. Res.*, 1999, **104**, 21317–21326.
45 C. L. Badger, I. George, P. T. Griffiths, C. F. Braban, R. A. Cox and J. P. D. Abbatt, *Atmos. Chem. Phys.*, 2006, **6**, 755–768.
46 A. Pant, M. T. Parsons and A. K. Bertram, *J. Phys. Chem. A*, 2006, **110**, 8701–8709.
47 P. Papon, J. Leblond and P. H. E. Meijer, *The Physics of Phase Transitions: Concepts and Applications*, Springer, New York, 1999.
48 B. Zobrist, C. Marcolli, D. A. Pedernera and T. Koop, *Atmos. Chem. Phys.*, 2008, **8**, 5221–5244.
49 S. Torza and S. G. Mason, *J. Colloid Interface Sci.*, 1970, **33**, 67–83.
50 Y. T. Wu, Z. Q. Zhu and L. H. Mei, *J. Chem. Eng. Data*, 1996, **41**, 1032–1035.
51 D. O. Topping, G. B. McFiggans, G. Kiss, Z. Varga, M. C. Facchini, S. Decesari and M. Mircea, *Atmos. Chem. Phys.*, 2007, **7**, 2371–2398.
52 C. Mätzler, *MATLAB Functions for Mie Scattering and Absorption*, 2002.
53 O. Pena and U. Pal, *Comput. Phys. Commun.*, 2009, **180**, 2348–2354.
54 D. Ngo, G. Videen and P. Chylek, *Comput. Phys. Commun.*, 1996, **99**, 94–112.
55 U. K. Krieger and A. A. Zardini, *Faraday Discuss.*, 2008, **137**, 377–388.

Faraday Discussions

RSCPublishing

DISCUSSIONS

General discussion

DOI: 10.1039/C3FD90033A

Professor McNeill opened the discussion of the paper by Professor Kroll: The average molecular weight inferred from AMS data may represent OA generated by more than one mechanism, for example, aqueous SOA formation pathways in addition to traditional condensational SOA formation. Do you have any ideas for how to expand your approach to include multiple targets?

Professor Kroll replied: In this study we focused on LV-OOA only, but this general approach could be applied to other aerosol (sub-)components as well. The AMS probably cannot distinguish aerosol formed by two different pathways if the elemental ratios of the products are roughly the same; however, if information was known about the products of one of these pathways (such as your example, aqueous SOA), then the two pathways could certainly be treated independently.

Dr Allan asked: Deriving a 'target' composition from a campaign average is perhaps inaccurate because this will, in most cases, contain a mixture of organics at different stages in their oxidation lifecycle. Because of this, the true end point will be more oxidised than these campaign averages would suggest. This being the case, what effect would this have on the areas covered by the potential pathways?

Professor Kroll answered: We used the LV-OOA PMF factor to estimate the average composition of "highly oxidized aerosol", but any factor is certainly a complex mixture containing a distribution of organic species that span a range of oxidation states. In order to at least begin examining how the most oxidized species within a given LV-OOA factor are formed, we treated the Mexico City LV-OOA (which to our knowledge is the most oxidized factor that has been reported) independently as an important case study (see Fig. 7 of our paper).

Professor Abbatt questioned: Is it a coincidence that the average LV-OOA formula has resemblance, in the carbon number, to a terpene given that many of the field sites are not biogenically impacted?

Professor Kroll answered: It's hard to say for sure, but given that the LV-OOA datasets are mostly from urban sites, it may well be a coincidence. Still, it would be interesting to apply this approach to a site that we know is biogenically impacted.

Dr Colussi commented: Poly-functional organic compounds in concentrated solutions, such as aerosol media, aggregate by supramolecular interactions. Therefore, their activities in mixtures cannot be calculated by assuming ideal behavior from vapour pressures of pure compounds – did you take this into account?

Professor Kroll replied: The effective vapor pressures were determined from thermodenuder measurements, rather than a calculation, so, to first order such interactions are inherent in the target $c*$ values used. But changes to $c*$ with each transformation were indeed determined using a structure–activity relationship based upon measurements of pure species (J. F. Pankow and W. E. Asher, *Atmos. Chem. Phys.*, 2008, **8**, 2773–2796.). This would lead to significant errors if the chemical environment of a molecule within the atmospheric aerosol is dramatically different from that of the molecule within a pure matrix. Whether this is the case for an oxidized molecule within an oxidized organic particle is unclear, but further study is certainly necessary. What would be particularly helpful is measurements of effective vapor pressures of low-volatility, highly oxidized organic species, in a range of different matrices.

Professor Nizkorodov said: I have a minor comment regarding your statement in the "Oligomerization reactions" section that "the target itself (of formula $C_{10.5}H_{13.4}O_{7.3}$) is not consistent with the oligomeric species measured in SOA, such as those from monoterpene oxidation (which typically have twenty or more carbon atoms and relatively low OSC values), isoprene photooxidation, or glyoxal isomerization". The average compositions obtained by the soft-ionisation high-resolution mass spectrometry methods in chamber studies, to the extent they can be trusted, are not too far from your estimated target. To cite an example from our own work, high-NO_x oxidation of isoprene results in $C_{12}H_{19}O_9N_{0.08}$ for the average formula of SOA compounds (T. B. Nguyen, A. Laskin, J. Laskin and S. A. Nizkorodov, *Phys. Chem. Chem. Phys.*, 2012, **14**, 9702–9714). The library of the average (and detailed) SOA composition from the soft-ionisation high-resolution mass spectrometry measurements will grow quickly in the coming years, and it should provide a useful alternative data set for your model. Soft-ionisation high-resolution mass spectrometry provides not just the average molecular formula but information about all (ionisable) compounds in SOA, which should be a better data set to test the oxidation mechanisms leading to SOA formation and photochemical aging.

Professor Kroll responded: I absolutely agree that speciated measurements of aerosol components are essential to understanding pathways of SOA formation, and it would be great if this retrosynthetic approach could be applied to such data. As for the isoprene SOA you mentioned, a big difference is the in H : C ratios: 1.6 for the laboratory SOA and 1.3 for the LV-OOA factors. In general, H : C ratios measured for isoprene oxidation (as well as glyoxal uptake) tend to be higher that what is seen in the field.

Professor McFiggans commented: A $c*$ taken from, or related to, field measured thermodenuder-derived values implicitly includes any non-ideal behaviour and other matrix effects as well as the pure component volatility.

Equations 1 to 5, relating c^*, carbon number and group abundances, appear to ignore any effects of the complex aerosol mixtures. Can you comment on the implications of the high variability in (and evolution of) the effective "activity coefficients" (and hence c^* values) that the components might be expected to experience within the lifetime of the aerosol and how this may impact on the transformations and trajectories across the parameter space within the model framework?

Professor Kroll responded: Right, the structure–activity relationship we used is derived from measurements of pure components and so the effects of transforms do not include any non-idealities. Thus we have assumed that a given molecule is present within a matrix that is chemically similar to itself, so that the activity coefficient is approximately 1. To examine the effects you are describing, the calculations would have to be run as a function of the chemical environment/matrix that a given target molecule is in – this would require explicit determination of activity coefficients. This approach has the additional complication that the chemical matrix would probably undergo chemical changes itself upon oxidation, in a way that is not well-defined. So right now it is difficult to speculate on what sort of effect these non-idealities will have, but it is certainly an important consideration.

Professor Coe opened the discussion of the paper by Professor Artaxo: The paper highlights the lack of observational evidence for new particle formation at ground based sites in Amazonia, except some infrequent events in the wet season. Given the predicted enhancements in sulphuric acid concentrations over Amazonia resulting from the Criegee radical generation of SO_3, shown in the Percival *et al.* paper, could you comment on the sources of particle formation across the region?

Professor Artaxo replied: The source of particle formation in Amazonia remains unknown. We could speculate formation of particles in the free atmosphere up downdraft transport to the canopy. We know that after precipitation, small (1-5 nm) clusters are produced, but we do not see the banana type of particle formation.

Professor Carslaw said: In connection with Paulo Artaxo's observations of very infrequent nucleation over the Amazon, this is something that global models also tend to predict, most likely because of a lack of sulphuric acid. In this connection, what is the origin of the well developed particle size distribution during the wet season when scavenging rates are very high and nucleation is rarely observed?

Professor Artaxo answered: The infrequent nucleation events are actually puzzling in Amazonia and they indeed could be caused by the very low sulfuric acid abundance, but further investigation is necessary. The almost constant accumulation mode with an average diameter of about 100 nm is a common feature observed at several sites. Very high precipitation rates (3000–4000 mm a year) very efficiently removes the particles and they quickly get back, possibly from the particle reservoir at the free troposphere. How new particles get there is still a mystery.

Professor Herrmann remarked: Can you comment on what organic speciation measurements are currently done and what is planned for the future?

Professor Artaxo answered: Right now, the organic aerosol measurements are limited to OC/EC Sunset Lab measurements as well as an Aerodyne ACSM instrument. It would be essential to do detailed organic aerosol speciation to separate the primary biogenic particles (PBAP) from the secondary organic aerosols (SOA). We are planning to do that in 2014–2015 as part of the GoAmazon experiment.

Professor Coe said: You observe low single scattering albedos at the Manaus site in the wet season. Could you provide more detail on the source of the particles which cause this? Why is this not observed at the Porto Velho site?

Professor Artaxo replied: The absorption observed in Manaus is due to fine mode SOA and PBAP. At this moment it is not possible to quantitatively apportion each type of aerosol to optical absorption. The observed relatively low SSA of 0.87 in the wet season in Amazonia is a feature that still needs to be explored in depth. In Porto Velho, a region with significant land use change, the increase in scattering aerosols and the much smaller impact of PBAP and forest SOA makes the SSA higher (at 0.92).

Professor Carslaw asked: You suggested that the wet season particle size distribution is possibly related to 'cloud processing', which I take to mean primarily aqueous phase sulphate production and growth of the Aitken particles into the accumulation mode. The amount of sulphate in the particles is very small. Could the Hoppel minimum in your observed size distribution also have another source, for example due to SOA chemistry? Could this generate such a size distribution?

Professor Artaxo responded: In Amazonia, certainly cloud processing is a key part of the aerosol properties, since vertical convection is very strong, and the hydrological cycle is very active. I have no clear idea if the observed Hoppel minimum in the size distribution could be due to SOA chemistry, because the organic chemistry involved is still not fully understood.

Professor McNeill opened the discussion of the paper by Ms Schindelka: In our model of aqueous aerosol chemistry, GAMMA (V. F. McNeill *et al.*, *Environ. Sci, Technol.*, 2012, **46**(15), 8075–8081), we predict very low formation of organo-sulfates by reaction of MACR and MVK with the sulfate radical, even if this reaction is assigned the maximum (diffusion-limit) rate constant. This is due to the very low Henry's Law constants of MACR and MVK and the low predicted concentrations of sulfate radicals in the model. Why do you think you observe significant organosulfate formation *via* this mechanism? Is it possible that sulfate radical formation is not well-characterized, leading to an underestimation of its formation in this and other models?

Ms Schindelka answered: The uptake coefficients of both carbonyl compounds are probably some of the most important sources of the uncertainty in this system. It has been shown by a field study about cloud and fog chemistry at the Schmücke hill (FEBUKO) that cloud water can contain up to two orders of magnitude higher concentrations of these compounds than the concentrations expected from the Henry's Law constants.[1] As there are not many laboratory measurements about the influence of the presence of organic compounds, ionic strength, acidity and reactivity on the partitioning coefficients, it is still not clear what drives the uptake of certain species beyond the expected particle or droplet phase concentrations estimated by the Henry's Law constants. Furthermore, as we discussed in the paper, this mechanism might be important at lower temperatures, *e.g.* in the free troposphere, as a partitioning of MACR and MVK to the particle phase is probably more pronounced. Therefore, the environmental conditions used in the model may also partly explain the discrepancy between the model results and the field/laboratory measurements.

We agree that the organosulfates produced by the sulfate radical induced oxidation add little mass to ambient organic aerosol and total sulfur concentration but it likely explains one of the mechanisms leading to small organosulfates, especially the C_4 organosulfates.

1 D. van Pinxteren, A. Plewka, D. Hofmann, K. Müller, H. Kramberger, B. Svrcina, K. Bachmann, W. Jaeschke, S. Mertes, J. L. Collett and H. Herrmann, *Atmos. Environ.*, 2005, **39**, 4305–4320.

Professor Herrmann responded: Please see the answer by J. Schindelka.

Professor Nizkorodov asked: Would you expect this mechanism to work for any water-soluble compound with a C=C bond, or is it limited to α,β-unsaturated carbonyls?

Ms Schindelka responded: Indeed, it should work for any compounds with a C=C double bond that partition into the aqueous particle phase. Buxton *et al.* have measured the rate of the reactions between sulfate radicals and monoterpenes as well as methacrolein. For all reactants the rate constants are in the same order of magnitude ($\sim 10^9$ L mol^{-1} s^{-1}) that is competitive with OH radical reactions. Of course, a carbonyl group adjacent to a C=C double bond can affect the reaction rate as the group is an electron withdrawing group. This leads to a hindrance of an electrophilic addition of the sulfate radical to the double bond. Nevertheless, a difference in the reaction rates between α-pinene and methacrolein is very small.

1 G. V. Buxton, G. A. Salmon and J. E. Williams, *J. Atmos. Chem.*, 2000, **36**, 111-134.

Dr Reid said: From the data in Figure 1 a, b, c and d, is it possible to determine relative branching between products in one set of conditions (or even between products in different sets of conditions), rather than just a relative intensity for one product as a function of the number of laser shots?

Ms Schindelka replied: Owing to the lack of authentic standard compounds we do not know the ionisation efficiencies of these products, hence branching ratios cannot be obtained from the data in the present study.

Dr Murphy commented: It is great to see structures identified for some previously unidentified peaks that we have seen in our laser ionization field data.

Ms Schindelka replied: The detection of these compounds in the field samples was a great motivation for us to study their possible formation mechanism, as non-radical mechanisms discussed in the literature do not explain the formation of these C_2–C_4 organosulfates well. We found it very interesting that signals for compounds with the same m/z values were found in higher altitudes[1] as environmental conditions at such altitudes (RH, T) could promote a partitioning of both semi-volatile carbonyls to the particle phase. We are curious if there is a dependency on the altitude and organosulfate.

1 K. D. Froyd, S. M. Murphy, D. M. Murphy, J. A. de Gouw, N. C. Eddingsaas and P. O. Wennberg, *Proc. Natl. Acad. Sci. U. S. A.*, 2010, **107**, 21360–21365.

Dr Hamilton asked: Is your ultimate aim to try and synthesise authentic standards to allow you to quantify your identified species?

Ms Schindelka responded: Yes, that is our ultimate goal. We are in contact with some groups and synthetic organic chemists, who are also interested in these compounds but the synthesis of some compounds will not be that easy. Another possibility would be the application of a universal detector for the clean samples from the lab experiments. Much work remains to quantify SOA marker compounds though.

Dr Alfarra commented: How would the hard UV used for initiating the radicals potentially affect the organic chemistry of the system being studied? In particular, I am questioning the possibility of photolysis of not only the parent hydrocarbon compounds but also their oxidation products.

Ms Schindelka responded: In the chamber experiments where MACR and MVK were present in the gas phase, photolysis was not a problem as both compounds absorb at higher wavelengths than 254 nm.[1] To measure a possible loss of both VOCs from photolysis or background reactions, we measured the mixing ratios of MACR and MVK with the PTR-MS but no significant decrease was observed. For bulk phase experiments a concentration ratio of carbonyl compound and peroxodisulfate was chosen so that peroxodisulfate was the major absorber and photolysis of the carbonyl compounds and the products was avoided. Indeed, these organosulfates cannot be detected by HPLC-UV at a wavelength above 200 nm.

1 T. Gierczak, J. B. Burkholder, R. K. Talukdar, A. Mellouki, S. B. Barone and A. R. Ravishankara, *J. Photochem. Photobiol., A*, 1997, **110**, 1–10

Dr Wenger remarked: Mercury lamps with a wavelength output at 254 nm were used in the chamber experiments. Since methylvinylketone and methacrolein both have significant absorption in this part of the ultraviolet region, direct photolysis of these OVOCs in the gas-phase may have occurred. Were the necessary tests performed, *e.g.* blank experiments with lights on but no aerosol present? If yes, was direct photolysis significant?

Ms Schindelka answered: The absorption maxima of both carbonyl compounds in the gas phase are in a higher wavelength region (~330 nm) than those in the aqueous phase where the absorption maxima are near to the 254 nm.[1] For the bulk phase experiments the concentration ratio of a carbonyl compound and peroxodisulfate was chosen so that most of the light is absorbed by the peroxodisulfate. Therefore, the extent of the photolysis of both carbonyl compounds should be negligible. In the chamber experiments the PTR-MS was used to follow a decrease in the mixing ratios of MACR and MVK in the gas phase from photolysis or background reactions but no significant decrease was observed.

1 T. Gierczak, J. B. Burkholder, R. K. Talukdar, A. Mellouki, S. B. Barone and A. R. Ravishankara, *J. Photochem. Photobiol., A*, 1997, **110**, 1–10

Dr Nozière addressed Ms Schindelka and Professor Herrmann: To avoid any risk of photolysis of the precursors and/or reaction products in the experiments, it should also be possible to generate the SO_4 radicals from the persulfate in the dark by heating up the solutions (30 °C).

Ms Schindelka replied: Despite the risk of the photolysis of the precursor compounds we decided to produce the sulfate radicals *via* irradiation. One of the reasons was the design of the chamber experiments. As the temperature has a significant influence on the partitioning of the VOCs, irradiation of seed particles containing peroxodisulfate was a more favourable method for the sulfate radical production. Furthermore, our chamber setup does not allow us to perform experiments at constant 30 °C. For the comparison of the chamber and bulk phase experiments, we performed the bulk phase experiments that produced sulfate radicals from the irradiation of a peroxodisulfate solution. An advantage of the laser induced radical formation is that we know the energy per laser pulse, unlike the thermal method.

Professor Herrmann responded: See answer of J. Schindelka

Professor Johnston opened the discussion of the paper by Professor Kroll: We often think about aged aerosol in the context of what we observe in a measurement. The real questions, though, are 1) what is truly there, 2) what does my measurement technique detect and 3) what does my measurement technique not detect? In that regard, the AMS identifies "LV-OOA" but not necessarily all organic matter in the particle. Which organic species doesn't AMS detect/identify well and how might the presence of this undetected matter affect our understanding of the chemical composition and properties of aged SOA?

Professor Kroll replied: Comparisons between AMS results and those of other instruments (SMPS, PILS, *etc.*) strongly suggest that there is not a significant fraction of the organic aerosol that remains unmeasured by the AMS, at least within the size range of the instrument. This makes sense given the underlying technique: vaporization at 600 °C (which is enough to flash-vaporize and/or chemically degrade most molecules) followed by ionization by electrons with energies of 70 eV (for which all organic molecules have a high cross section).

However, because these conditions are quite harsh, a good deal of chemical information about the molecules may be lost. For example, key classes of organic molecules, such as organosulfates, organic nitrates, and peroxides – all of which involve relatively weak bonds – may not be identified as such.

Dr Allan answered: What the AMS measures as 'LV-OOA' is really the thermal degradation products (chiefly CO_2 and H_2O) produced on the vaporiser surface from the complex multifunctional species that make up the bulk of heavily aged SOA. In theory, if there were organics with a high enough molecular weight such that they were involatile enough to be operationally considered 'refractory' but at the same time lacking the oxygenated functional groups (chiefly acids) that would permit the thermal degradation, they would not be detected by the AMS. So to answer the question, if a large fraction of SOA was to exist in such a form, it would necessitate a very major rethink because the SOA formation processes normally proposed involve functionalisation as a means to reducing volatility. But I do not consider this likely from a measurements perspective because volume closure exercises with ambient SMPS measurements tend to be quite favourable (at least in continental environments), so it is not likely that there is a significant involatile contributor to SOA mass that the AMS is failing to detect.

It should be stressed that the decomposition process does mean that any information on the molecular structure of the decomposing organics is lost completely, so the AMS does not 'identify' the species contributing to LV-OOA very well beyond saying that they are all very oxygenated.

A very well understood problem is that the AMS is limited to the range of particles covered by the aerodynamic lens. While this usually covers the accumulation mode (where the majority of SOA mass is thought to reside), anything happening to particles smaller than about 30 nm or bigger than a micron is missed completely.

Another possible issue (brought to my attention by Professor Volkamer during this meeting), as discussed by Kampf *et al.*,[1] is that smaller, more volatile molecules bound in the particle by ionic interactions (such as glyoxal) could be lost in the aerodynamic lens when the water present in the particles evaporates. In the context of the wider co-condensation phenomenon described by Topping *et al.*,[2] this could be an important issue to investigate. This being the case, this presents a challenge to the wider measurement community, as this issue is not confined to the AMS. Many other commonly used methods would also be blind to such species, including many of those often used as top-down constraints on particulate budgets such as SMPSs and gravimetric analyses.

1 C. J. Kampf *et al.*, *Environ. Sci. Technol.*, 2013, **47**(9), 4236–4244
2 D. Topping *et al.*, *Nat. Geosci.*, 2013, **6**, 443–446

Dr Allan asked: To clarify the point regarding organosulphates, it is not that the AMS cannot detect these, it is that it cannot identify them as organic, as they produce the same ions as inorganic sulphates. There is a similar problem with organonitrates, although it is not as bad because these produce some organonitrate fragments and ratios of NO and NO_2 ions that deviate from inorganic nitrate.

Regarding the point about inferring contributions to aerosol pH, most estimates of aerosol acidity based on AMS data are based on inorganic ions, principally nitrate, sulphate, ammonium and chloride. I am not aware of anyone quantitatively estimating the contribution of organic acids and bases to pH based on AMS data, but it should be pointed out that in most polluted environments, aerosol acidity is mainly dictated by inorganic ions anyway.

Professor Kiendler-Scharr commented: Elaborating on what James Allan said earlier, I would like to add that the method of looking into ion balances is inaccurate when one has large contributions of organic nitrate or organosulfate species. There the signal from the nitrate or sulfate group would tend to be added to inorganic species in standard AMS analysis. Also, in places like Cabauw, the Netherlands, with high ammonia concentrations, particulate organic acids can lead to ammonia uptake, which would then occur as "excess ammonia" in the AMS. Generally, the AMS community is certainly at a stage now where concepts beyond integral parameters such as O : C ratio are needed to advance our understanding of aerosol composition, especially of course the organic aerosol fraction.

Professor Herrmann commented: Isn't molecular speciation needed besides AMS measurements, especially for any retro-synthetic approach?

Professor Kroll responded: Molecular speciation is critical to unravel SOA chemistry. The point of this study (or any AMS study, I would argue) is not that ensemble measurements are sufficient to describe the chemistry of organic aerosol, or that speciated measurements are of limited utility. Rather, the AMS allows us to rapidly measure the oxidation state of the entire organic aerosol, and the extensive database of AMS field measurements makes for an excellent resource for understanding oxidation processes that govern aerosol formation and evolution in the atmosphere. That was our starting point for the study, which made for a good first application of this retrosynthetic approach. But the same approach could absolutely be used for individual molecular components of aerosol; this could provide substantial insight into the detailed oxidation pathways involved in aerosol chemistry. In fact this has been done already (B. K. Pun, C. Seigneur, D. Grosjean and P. Saxena, *J. Atmos. Chem.*, 2000, **35**, 199–223), but at a time when our understanding of organic aerosol chemistry and composition was not as developed as it is now. It is probably worth revisiting this approach based on more recent measurements.

Dr Hamilton asked: How do we attempt to bring both the bulk analysis type measurements of the AMS, such as LV-OOA, together with the detailed molecular characterization measurements such as high resolution mass spectrometry? The typical LV-OOA measurements from the AMS seem to be so different to what we can replicate at the molecular level. For instance, using FTICR-MS to look at aerosol collected in the tropical rainforest in Borneo, we obtain O : C values that are much lower than the AMS. Is this just a result of the different extraction and ionisation methods and we are fundamentally missing material in both techniques and so cannot hope to resolve this?

Professor Nizkorodov responded: There is currently no "do-it-all" method for the molecular-level analysis of environmental mixtures of the type of complexity found in atmospheric particles. The key advantage of the AMS instrument is the universality of the ionization method it relies on: every molecule will produce ionized molecular fragments in an electron impact. The price for this universality is extensive fragmentation leading to the loss of molecular-level information. Soft-ionization methods result in ionization of intact molecules but they are more selective to the molecular properties. For example, electrospray ionization (ESI) does not efficiently ionize saturated hydrocarbons, aromatic hydrocarbons, alcohols, simple peroxides, aliphatic carbonyls, and other types of molecules characterized by low proton or alkali metal affinities, but works reasonably well for carboxylic acids, amines, and multifunctional carbonyl compounds found in SOA. Chemical ionization mass spectrometry (CIMS) can detect a wider but still rather selective subset of compounds. Complementary analytical techniques are required for comprehensive molecular level analysis of SOA composition. I want to point out that for highly-oxidized, polar compounds found in biogenic SOA, the ESI-based high-resolution mass spectrometry methods and AMS provide reasonably good agreement for the O : C values (A. P. Bateman, J. Laskin, A. Laskin and S. A. Nizkorodov, *Environ. Sci. Technol.*, 2012, **46**, 8315–8324).

Professor Abbatt responded: Although CIMS analyzes a more selective set of aerosol compounds than AMS does, I bring your attention to a recent study from our group in which we have made simultaneous measurements of SOA properties (such as the O : C ratio) using both AMS and aerosol-CIMS: D. Aljawhary, A. K. Y. Lee, and J. P. D. Abbatt, *Atmos. Meas. Tech. Discuss.*, 2013, **6**, 6147–6186.

The agreement between the CIMS and AMS measurements (see Figure 8a of that paper) is really quite remarkable, when plotted in van Krevelen space. Closer examination (Figure 8b) shows that there are differences dependent on the reagent ion used, which is to be fully expected since CIMS does have selective sensitivity to different classes of compounds. Nevertheless, we believe that the appropriate use of reagent ions in CIMS can start to bridge the gap between the two classes of measurements referred to by Prof. Hamilton. In general, the difference in species sensitivity is much lower for complexing reagent ions, such as iodide than arises in the ESI spectra, where there can exist large differences in sensitivity between different classes of functionalized compounds.

Dr Reid commented: For AMS data such as that presented in Table 1 of your paper, can you give me a guide as to how to interpret the numerical values? For example, can you give me an indication of the typical errors and distributions in atomic ratios and oxidation state values, how the propagation of errors in AMS measurements is accounted for from the core observed ratios *etc.*? Is there an appropriate publication on this?

Professor Kroll replied: Errors associated with the determination of elemental ratios by the AMS arise largely from the complex fragmentation process that molecules may undergo upon ionization at 70 eV (as well as heating to 600 °C). In general the resulting ions will have a lower oxygen content than the parent molecule, due to the high likelihood of oxygenated species being lost as neutral molecules. To account for this effect, we apply a correction factor to the raw elemental ratios of the summed ions, based on measurements of a number of standards (see A. C. Aiken, *et al.*, *Environ. Sci. Technol.*, 2008, **42**, 4478–4485 and

A. C. Aiken *et al.*, *Anal. Chem.*, 2007, **79**, 8350–8358). This correction factor will be revised as the database of standards measured is expanded to include more highly oxidized species (M. R. Canagaratna *et al.*, manuscript in preparation) – this will have the effect of increasing reported O : C ratios somewhat, but not in a way that qualitatively changes results. The error associated with AMS-derived elemental ratios, based on measurements using standards, is ~30% for O : C and ~10% H : C; this is largely from the accuracy of the technique rather than precision of the measurement. However, errors are likely to be lower for ambient aerosol, given that it is made up of a complex mixture of a large number of individual species.

Dr Allan commented: Regarding the question concerning the evaluation of the breadth of chemical complexity: AMS, along with all the other 'bulk' techniques (*e.g.* HNMR, FTIR) provides no data on the breadth of complexity in isolation. On the other hand, speciated techniques such as chromatography and ESI can produce a large amount of data identifying a large number of chemical species, however it is difficult to be quantitative for all of these species. The challenge for the measurement community is to be able to quantitatively link the two data types, thereby delivering a comprehensive assessment of complexity.

Professor Lewis commented: Following the discussion of the relative responses of AMS and other instrumental techniques to particular organic aerosol components, this general question asks whether there is sufficient confidence in the quantitative nature of SOA techniques like AMS such that disparate data sets could be used in the future to determine long-term temporal trends.

Dr Allan replied: This is responded to later in the discussion. In addition, I would also add that the quantification of SOA as a subset of total OM is dependent on factorisation techniques (such as PMF and ME-2) in addition to the instrument itself quantifying OM correctly. This carries an inherent uncertainty because even if a standardised approach is adopted, there may be unforseen reasons for why the SOA : POA split is not accurately determined that are specific to a certain site or aerosol type. Parallels can be drawn to the thermal optical techniques, which are generally taken to be accurate for total particulate carbon, but the EC/OC split can sometimes be debatable.

Professor Donahue asked: If the AMS more or less measures all non-refractory organics quantitatively (on average), isn't the issue of comparing spectated analyses to the AMS just one of calibration (including accounting for compounds where a given method has a sensitivity of zero)?

Dr Hamilton replied: I would agree, but I guess the difficulty is knowing what your instrument sensitivity to those compounds is. It is clear that a water soluble extract will not include the least polar fraction, including larger hydrocarbons, but accounting for this material poses a real challenge if we want to calculate a O : C.

Dr Nozière reopened the discussion of the paper by Professor Donahue: To follow up on the discussion of the best approach for the analysis of organic aerosol material, AMS/global (VBS-type) *vs.* identifying individual classes of compounds: these approaches are not mutually exclusive as they address different questions. Thus it might not be necessary to try to compare or harmonize them (except to track potential artifacts). However the research efforts and visibility given to both types of approaches should be balanced.

Dr Taatjes commented: It seems to be suggested that determining the full chemical speciation of these aerosol particles is superfluous, and that for understanding the growth processes that are relevant in the atmosphere, only some more global parameterization should be pursued. I think that it is in fact necessary to investigate the chemical details. If the decrease in volatility that accompanies particle formation and growth really is a chemical process, as is stated, then a network of chemical reactions, unwieldy though it may be, represents actual physics that drives aerosol formation. This huge chemical kinetic system reflects the structure of the real process, even if its full complexity cannot be encompassed in working models. We encounter a similar situation in modeling combustion, the field in which I work. Here, knowledge of detailed chemistry remains incomplete, and turbulent reacting flow calculations cannot bear the computational burden of comprehensive chemistry models in any case. Nevertheless, the fact that the overall structure of the chemical mechanism *is* known allows reduced models to reflect the real structure of the more complex system more closely. Even incomplete knowledge of what is inside this "black box" of aerosol formation and growth will be crucial to determining whether a parameterization makes physical sense. In my view, predictive models must not neglect the real processes that are occurring.

Professor Rudich opened the discussion of the paper by Professor Artaxo: I have several questions: Are the absorbing aerosols secondary or primary biological particles? What is the source of absorption at 637 nm? Can it be chlorophyll? What are the indications that this is brown carbon (what is the angstrom exponent)?

Professor Artaxo responded: The absorbing aerosols are both secondary aerosols and also primary biological particles. At this moment it is not possible to apportion the absorption to each category. We will do in the next future detailed organic characterization to quantitatively discriminate the two sources.

The source of absorption at 637 nm can be a combination of SOA and PBAP. A fraction can be chlorophyll, since live organisms were detected as airborne particles.

This absorption is certainly not soot, since it was observed in the middle of the wet season, when heavy and daily rains do not allow biomass to burn. The nomenclature of brown carbon is not precise, so maybe you could call this absorption brown carbon. The angstrom exponent is larger than 1 for sure.

Professor McFiggans opened the discussion of the paper by Professor Kroll: Following the first 8 papers, and including the rest of those in this session, there appears to be a large number of processes and a vast amount of unknowns which we could spend an eternity resolving at an incredible level of detail. Is there anything we can borrow from other communities to identify which of the elements of the mechanisms that must be captured to describe the determinant processes in the atmospheric aerosol system and identify the minimum level of process description to capture the important details?

Dr Taatjes replied: In the combustion community there is a very similar situation, where a large web of chemical processes occur in conjunction with very complex fluid dynamics. It is not computationally feasible to describe the full detail of the chemical mechanism while also resolving the fluid dynamics fully (even if the chemical mechanism was known in its full detail!). There are many strategies that combustion modelers employ to simplify chemical mechanisms and reduce their dimensionality. It would surprise me if similar methods are not already being employed in tropospheric and aerosol chemistry to some degree.

Dr Reid asked: Can you more specifically define what you mean by "structure" in this context?

Dr Taatjes responded: The conversion of the volatile gas phase reactants to the eventual non-volatile aerosol constituents takes place by a network of chemical transformations. Molecules react and form products that are more oxidized and less volatile, and these products in turn react, *etc.* Were rate expressions known for every one of the chemical transformations, this would be a rigorous and comprehensive chemical kinetics model. But even with knowledge that is nowhere near that complete, we can say, for example, what types of molecules are connected to which other types by reasonable reactions. That set of relationships defines the "structure" of the conversion process.

Professor Donahue remarked: I agree that it would be useful to have a complete, enumerated list of all compounds in the organic aerosol, but only with calibrated abundance. The great value of this paper is that it addresses the average, or integrated behavior of the system, which the sum of constituents must match.

Professor Herrmann addressed Professor Nizkorodov and Dr Taatjes: I do agree to Craig Taatjes that our discussions resemble much of those in the combustion community. We need a molecular mechanistic picture of what is going on in aerosol chemistry even if certain effects regarding light absorption or health risks might only be caused by a few compounds, as Sergey Nizkorodov stated. Without such a molecular view, modelling and prediction beyond very broad descriptors will hardly be possible.

Professor Wahner said: This paper by Kelly Daumit, like others, depends in the modelling framework on the exsisting experimental parameters like C : H ratio and C : O ratio determined by the widely used AMS technique. As pointed out in the analysis of the model results: "Additionally, characterization of highly-oxidized OA using techniques other than AMS would be useful. The direct measurement of functional group distributions (*e.g.*, using FT-IR), can provide an alternative, and potentially more comprehensive, description of the relationships between chemical structure, volatility, and formation chemistry. Similarly, direct measurements of nC (*e.g.*, using UHR-ESI-MS) would improve the accuracy of the target chemical formula....". More generally this seems a central question in understanding the processes from gas-phase oxidation to organic aerosol formation and ageing: do we have sufficient different measurements to test the presented model approaches, also in terms of their predictive capabilities in preturbed conditions? And which parameters should be measured?

Dr Reid opened the discussion of the paper by Dr Wang: Only one measurement has been done under conditions of elevated (30%) RH. How important might it be to perform measurements on aerosols under humidified rather than dry conditions, particularly given the dependence of the mechanism on water/acid?

Dr Wang answered: The only experiment performed under conditions of elevated RH (30%) was an experiment using ozone as an oxidant without an OH scavenger, which is atmospherically relevant and allows the simulation of ambient photooxidation conditions much better than the experiments where OH radicals are generated by photolysis of H_2O_2. It thus appears that the photooxidation of isoprene is affected by humidity. As mentioned in our article, recent experiments by Berndt (T. Berndt *et al.*, *J. Atmos. Chem.*, 2012, **69**, 253) on the OH reaction of isoprene for low-NO_x conditions suggest a possible effect of water on the kinetics of the reaction mechanism. This new water-dependant gas-phase isoprene chemistry warrants to be explored in greater detail.

Dr Allan commented: There is much investigation of the formation of isoprene SOA under 'high NO_x' and 'low NO_x' conditions, but from an atmospheric measurement perspective, I am interested in the dividing line between the two regimes. Under what ambient conditions could one expect these processes to be important, in terms of a maximum NO_x concentration?

Dr Wang responded: The terminology "high NO_x" and "low NO_x" is rather confusing and there is, in our opinion, no strict division line between the two regimes. In our study, we have examined SOA formation from isoprene and C5–alkene diols in a smog chamber without added NO and denote these conditions as "low-NO_x" because there is always some background NO_x (about 30 ppt NO and 120 ppt NO_2). However, in a chamber, low NO_x levels may not necessarily lead to RO_2 + HO_2 chemistry: at high isoprene concentrations, RO_2 levels will be high, resulting in RO_2 + RO_2 reactions, as demonstrated in our study. One also has to consider that the use of H_2O_2 as an OH source can lead to high HO_2 levels, resulting in RO_2 + HO_2 reactions; hence, even at relatively high NO concentrations, this chemistry may predominate. Under ambient conditions, it is expected that very low (50–100 ppt) levels of NO are required for RO_2 + HO_2 reactions to dominate. However, ambient measurements of the C5–alkene triols, which are characteristic for a "low NO_x" regime and a product of RO_2 + HO_2 chemistry, during the 2002 LBA-SMOCC biomass burning experiment, show that they can be measured at high concentrations (on average, 54 ng m^{-3}) during the biomass burning (dry) episode that is characterized by high NO_x concentrations (*i.e.* 4.6 ppb) (M. Claeys *et al.*, *Atmos. Chem. Phys.*, 2010, **10**, 9319). For a discussion about "high NO_x" and "low NO_x" terminology, see the point of view article "Let's abandon the "high NO_x" and "low NO_x" terminology" (P. Wennberg, IGAC news,

Issue No 50, July 2013, p. 3). As argued in this article, the "low-NO$_x$"/"high-NO$_x$" terminology may lead to confusion; hence, the worthy suggestion is made to provide a description of the fate of peroxy radicals.

Dr Wenger asked: The chromatograms obtained from the ozonolysis experiments (rather than OH-initiated photooxidation) seem to be a closer match to the field samples. How can you explain this given that ozonolysis of isoprene is of very minor importance in the atmosphere?

Dr Wang responded: It needs to be pointed out here that the ozonolysis experiment was performed without an OH scavenger. It is therefore reasonable to assume that the isoprene SOA marker compounds that are detected in the chamber ozonolysis experiment are formed through the reaction with the OH radical, formed from ozone through impact of UV light (<310 nm) and in the presence of water vapor $[O_3 \rightarrow O(1D) + O_2; O(1D) + H_2O \rightarrow 2OH]$. We thus do not suggest that the isoprene SOA marker compounds that are detected in the ozonolysis experiment are due to ozonolysis reactions. Furthermore, it is indeed interesting to note that the isoprene SOA marker compound profile resembles that found for field samples, likely because $RO_2 + HO_2$ chemistry dominates, and not $RO_2 + RO_2$ chemistry.

Professor Kroll said: As a quick comment on the fate of RO_2 radicals: while it is true that very low (50–100 ppt) levels of NO are required in the atmosphere for RO_2 + HO_2 reactions to dominate, in chambers this cutoff might be substantially higher. At high levels of isoprene, RO_2 levels will be high, so $RO_2 + RO_2$ reactions may start being important. In addition, the use of H_2O_2 as an OH source can lead to high levels of HO_2, formed from the OH + H_2O_2 reaction. So the $RO_2 + HO_2$ reaction will also be fast. Under chamber conditions, such peroxy–peroxy reactions can dominate RO_2 chemistry even in the presence of 100s of ppt of NO. What really matters for this branching is the relative levels of peroxy radicals to NO, not their absolute levels. (At the same time, Paul Wennberg's group and others have shown that RO_2 unimolecular chemistry may be important as well. In that case, we need to run our chamber experiments at atmospherically relevant levels of NO, HO_2, and RO_2. This represents a major experimental challenge.)

Professor Heard responded: I completely agree with the statement above by Professor Kroll that the level of NO which still enables the reactions of $RO_2 + RO_2$ and $RO_2 + HO_2$ to dominate will be considerably higher than in the atmosphere owing to the considerably larger concentrations of HO_2 and RO_2 in this chamber study. Under atmospheric conditions of RO_2 and HO_2, the NO concentration needs to be very low (certainly <100 pptv, and perhaps even lower), otherwise RO_2 + NO and HO_2 + NO begins to become important. I agree that it is not the absolute NO levels which are important, rather the relative levels of peroxy radicals to NO. It is indeed very challenging to operate chambers or other laboratory chambers at these very low levels of NO$_x$.

Professor Heard asked: I am in the HO$_x$ field measurement community where the emphasis is more on the HO$_x$ production in the isoprene degradation process. Here the emphasis is on the other products that are formed, which are potential precursors for isoprene SOA. Absolute concentration measurements of HO_2 and RO_2 in this system would be very useful in order to gauge which of the mechanisms, A or B ($RO_2 + RO_2$ or $RO_2 + HO_2$, respectively) is the one that is operating here. I agree also that it is hard to be sure that the NO_x concentration is low. The limit of detection for the NO_x instrument is likely to be 50 ppt. However, the HO_2 and RO_2 concentrations are likely to be elevated compared with atmospheric levels, in which case the level of NO_x can be higher before $RO_2 + NO$ and $HO_2 + NO$ starts to dominate the loss of peroxy radicals. The measurement of water vapour concentrations is also important in order to further explore the possible effect of water on the reaction mechanism.

Dr Kahnt replied: Unfortunately, no analytical system was installed for HO_2 and RO_2 measurements in the chamber when the isoprene OH experiments were performed for this study. The comment raises an important issue that the concentrations of those radical species would enable us to assess whether the levels of NO_x and water vapour affect the fate of isoprene-derived peroxy-radicals. Higher levels of RO_2 and HO_2 radicals in the system could dominate the overall reactions under the experimental conditions used for this study (initial mixing ratios of isoprene, usage of H_2O_2 as OH radical source). However, based on the results it appears that higher levels of RO_2 resulted in $RO_2 + RO_2$ chemistry.

Professor Heard remarked: Could you please clarify what wavelengths are generated by the UV-Strahler lamp that is stated to generate UV-C light? The reason I ask is that peroxy radicals are known to absorb in the UV part of the spectrum, and more complex RO_2 species, for example those derived from the oxidation of isoprene, may have a longer wavelength absorption than for smaller RO_2 species. The longer absorption tail could result in some photolysis of RO_2 in your experiment which may impact on the final products that are formed.

Dr Kahnt answered: The UV-C lamps used in this study emit light at a wavelength of 254 nm and were used to photolyse the H_2O_2 to produce the OH radicals. We agree that RO_2 species can also absorb in the UV-light in the range of 200–300 nm and thus cannot exclude that some of the isoprene-derived RO_2 radicals undergo photolysis during the experiment. As the exact absorption cross section for RO_2 species is dependent on their individual structures, the overall impact of possible products that result from RO_2 photolysis is hard to predict. We are aware of a possible impact on the product distribution but we were unable to address this issue in the present study owing to large uncertainties and unknowns associated with this process. It will be interesting to evaluate whether a different type of UV lamp changes the final product distribution when other experimental conditions are kept the same. We believe this is something to be examined in a future study.

Professor Cappelletti commented: Hydrogen peroxide dissociates very fast to water on tubing, nozzle and walls and my experience is that it is not possible to introduce H_2O_2 in a chamber without adding water. How can this affect RH in your chamber and also the results of experiments in dry conditions?

Dr Wang answered: Yes, hydrogen peroxide dissociates quickly with water. In our experiments, the chamber was always flushed with dry air overnight before use. During experiments, 30% H_2O_2 was introduced into the chamber at a very slow flow rate which caused little RH effect. And the RH monitor remained unchanged throughout the experiments.

Professor Donahue opened the discussion of the paper by Dr Topping: Arguing with volatility is like arguing about a tautology. Volatility is simply the tendency of a compound to stay in the condensed phase. Activity coefficients are certainly interesting. Large ones are only a little interesting, because if there is any significant concentration of compounds with a large activity coefficient in the aerosol they will simply phase separate and make their own condensed phase. Very small ones, however, can certainly matter.

Professor Volkamer commented: Could you please elaborate on the role of activity coefficients. You mentioned that activity coefficients can be as large as 10^6. What are the smallest activity coefficients in your model? And for which molecules are they observed?

I note that we recently demonstrated activity coefficients of 1/500 for the salting-in of glyoxal in aerosols containing sulfate.[1] It would be interesting to see whether your model can rationalize these experimental findings. Also, could you please comment on what data is needed to evaluate your model.

1 C. J. Kampf, E. M. Waxman, J. G. Slowik, J. Dommen, L. Pfaffenberger, A. P. Praplan, A. S. H. Prevot, U. Baltensperger, T. Hoffmann, and R. Volkamer, *Environ. Sci. Technol.*, 2013, **47**(9), 4236–4244.

Dr Topping replied: In this application, activity coefficients are used to define the least miscible components and then a spectrum of partitioning coefficients between two separate liquid phases, based on activity coefficient ratios of each compound in these two reference solvents. Using the AIOMFAC model, we do see activity coefficient ratios as high as 10^6. For example, long chain hydrocarbons in water. For activity coefficients this high, the model should be able to allow said component to re-equilibrate with alternative liquid phase and the gas phase accordingly. For binary systems it is quite possible to demonstrate activity coefficients less than unity. In figure 9 of our 2011 paper in ACP (M. H. Barley, D. Topping, D. Lowe, S. Utembe and G. McFiggans, *Atmos. Chem. Phys.*, 2011, **11**, 13145–13159), we demonstrate that it is possible to have activity coefficients as low as 10^{-2} in non-reactive equilibrium calculations. It is difficult to prescribe specific compounds valid across all scenarios, as this will be dependent on the gas phase mechanism used. However, we show in that example that these compounds typically have a logC* of −4 to −2, with a relatively high O : C ratio (~1). In that study we assumed no interaction with an inorganic core, using the UNIFAC model to calculate activity coefficients of all organic components in the mixed aqueous solution. This could be important, as has been known for sometime. For example, Marcolli and Krieger (C. Marcolli and U. K. Krieger, *J. Phys. Chem. A*, 2006, **110**(5), 1881–93) state that NaCl is a salting-out agent for alcohols and organic acids, whereas ammonium nitrate and sulfate exhibited salting-in and salting-out tendencies depending on the nature and number of functional groups as well as on the concentration of the solution. For the salting-in of glyoxal in sulphate containing aerosols, you could use the AIOMFAC model to try and corroborate your findings (http://www.aiomfac.caltech.edu). However, you would have to assume a non-reactive system that might be important for glyoxal?

It is difficult to quantitatively determine the presence of LLE in systems for which the model in this study was designed (an unlimited number of compounds, in this paper we use ~3000). However, it is possible to assess the potential for LLE by comparing total SOA mass and detailed composition using models that account for, or neglect, LLE (*e.g.* chamber comparisons: A. Zuend and J. H. Seinfeld, *Atmos. Chem. Phys.*, 2012, **12**, 3857–3882). The challenge then is quantifying uncertainty in other parameters such as pure component vapour pressures, kinetic mass transfer limitations or condensed phase reactions to name a few. Focused laboratory studies on simple systems are useful in validating thermodynamic equilibrium models that calculate LLE (*e.g.* use of an electrodynamic balance or systems deposited on a hydrophobically coated substrate combined with optical microscopy and raman spectroscopy: M. Song, C. Marcolli, U. K. Krieger, A. Zuend and T. Peter, *Atmos. Chem. Phys.*, 2012, **12**, 2691–2712). These evaluations are then used to prescribe some confidence in modeling higher order systems. As we state in the paper, however, to our knowledge, there is no other model that can be used to treat systems with thousands of compounds. At present, in non-reactive equilibrium modeling frameworks, there are potentially larger uncertainties associated with pure component vapour pressures when predicting SOA mass and composition.

Professor Herrmann remarked: Can you clarify when liquid–liquid separation could be important ?

This might be the physical frames where reactions in the more organic liquid phase need to be further studied.

Dr Topping responded: Generally, results from this study agree with the more detailed thermodynamic investigations in that aerosols with an O : C ration >0.7 are unlikely to exhibit LLE. Defining how much of each liquid phase is present, along with the composition, can be performed using simple activity coefficient ratios as we have demonstrated here. With regard to condensed phase reactions, this is an interesting point. We have not considered any kinetic processes within our simulations. Whilst the impact on predicted total mass within an equilibrium model is small relative to other uncertainties, it could be that we need to consider the presence of the two immiscible phases for investigating the role of condensed phase reactions on, for example, gas-particle partitioning. It would also be interesting to investigate such impacts on predicted/measured viscosity of the two separate phases. It might be that the organic rich phase becomes highly viscous and thus impacts on total SOA mass loadings. The challenge would be how to model these potential effects with current frameworks and how to measure reaction rates/pathways in relevant mixtures.

Professor Herrmann addressed Professor McFiggans and Dr Topping: How do you judge if an effort should be made to better understand chemical conversions in the more organic particle phase, near sources or after liquid–liquid sepraration?

Dr Topping responded: It is difficult to judge this using existing model frameworks as, to the best of my knowledge, we have no capability to investigate this with any detail. We do know that condensed phase reactions are a potentially important process for improving predictions of SOA. Not only this, but composition dependent properties, including viscosity *etc.*, might also require a better understanding of such processes. With this in mind, there is no reason to neglect the impact of LLE on condensed phase reactions and vice-versa. The challenge would then be to incorporate any understanding we have on geometrical considerations of LLE, be it a core shell or partially engulfed model. Of course, one can keep adding processes to mechanistic models and find some new sensitivities. A fruitful course of action might be to use O : C ratio based LLE domains and prescribe a range of reaction pathways/rates to probe any potential sensitivity in a dynamic model. One could then use new statistical methods for highlighting the largest potential uncertainties which might inform laboratory studies.

Professor Herrmann asked: Could ensemble calculations be done?

Dr Topping replied: I assume this relates to the notion that mixing state is likely an important driver for phase separation? If so then our broad ensemble calculations, in that we assume all particles are represented by one bulk system, do not enable us to investigate this. For simulations within a binned size dependent dynamical model, then we would be able to do this, yes. Please let me know if this addresses the question.

Professor Abbatt said: Given that reactions may proceed differently in the aqueous phase and in the organic phase, such as radical oxidative processing, it does seem very important to be able to know when system phases separate.

Professor Herrmann responded: I do totally agree with Jon Abbatt's comment. The particle phase is setting the scene for those chemical reactions to occur either in more aqueous particles or particle parts or more organic particles or particle parts.

Professor Volkamer opened the discussion of the paper by Dr Schindelka: For aqueous aerosols that are organic–inorganic mixtures with water, could you please comment on what you expect is the relative importance of OH radical uptake from the gas-phase compared to *in situ* radical sources for OH radicals in the condensed phase?

We have conducted sensitivity studies on the relevance of OH radicals from the gas-phase to oxidize organics in the bulk, which show that OH is likely to react near the surface, and is strongly suppressed in the bulk.[1] There is arguably too little known about the *in situ* radical sources. Is organic matter to some degree sheltered from oxidation by hiding in the aerosol bulk?

1 E. M. Waxman, K. Dzepina, B. Ervens, J. Lee-Taylor, B. Aumont, J. L. Jimenez, S. Madronich, and R. Volkamer, *Geophys. Res. Lett.*, 2013, **40**(5), 978–982.

Professor Herrmann replied: Please see the available CAPRAM studies on this issue. Roughly, about a third of the aqueous phase OH comes from *in situ* sources and the rest from phase transfer.

However this will be different depending on the scenario and whether aqueous aerosol particles or cloud droplets are addressed.

For aqueous aerosol conditions, the Fenton reaction becomes important, provided that H_2O_2 and Fe are available and dominates the phase transfer, as known from CAPRAM.

Dr Reid opened the discussion of the paper by Dr Krieger: The images you present are 2D projections of a 3D particle. The hydrophobic phase at the edge of the deposited droplet wants to spread on the hydrophobic coated surface. To ensure that your inference of core–shell morphology is correct, have you attempted to calculate/estimate the Gibbs energies of configurations when the organic phase spreads and forms a doughnut rather than just staying as a lens on one side of the aqueous phase? *i.e.* to what extend does the presence of a hydrophobic substrate influence the morphology of the mixed phase particle. Indeed, when phase separation occurs and an aqueous phase forms within the organic volume, to what extent does the adhesion between the aqueous phase and the substrate inhibit the transition from an apparent core–shell geometry to a partially engulfed structure?

Dr Krieger answered: I completely agree that the interpretation of morphology from 2D projections is not unambiguous. However, we try to address this question in the paper by showing that for two typical systems our interpretation is consistent with (a) the spreading coefficients calculated from surface- and interfacial tensions measured in bulk solutions and (b) by showing that Mie-resonance spectra of levitated particles are consistent with the morphology observed by 2D microscopy. This holds for both a system exhibiting core–shell symmetry, as well as a system exhibiting a partially engulfed structure. At least for these systems the substrate did not influence the surface energy balance to alter morphology. Nor did adhesion inhibit the transition from core–shell morphology to partially engulfed morphology for the system shown in Fig. 4b.

Dr Reid remarked: There does appear to be a difference between the systems studied by Krieger *et al.* and Reid *et al.* with the former studying more highly oxgenated organic components and the latter studying less soluble/lower O : C compounds.

Dr Krieger responded: Yes, that is in my opinion the reason why we observe significantly more core–shell morphology compared to the studies of the Reid group. We believe that the higher O : C systems are more representative for the aged tropospheric SOA.

Dr Reid asked: The morphology of the mixed phase particle does depend on the relative volumes of the hydrophobic and hydrophilic phases. A greater volume fraction of the hydrophobic phase will lead to increased engulfing of the aqueous phase volume. For examples of this, please refer to: J. P. Reid, B. J. Dennis-Smither, N.-O. A. Kwamena, R. E. H. Miles, K. L. Hanford and C. J. Homer, *Phys. Chem. Chem. Phys.*, 2011, **13**, 15559–15572, N.-O. A. Kwamena, J. Buajarern and J. P. Reid, *J. Phys. Chem. A*, 2010, **114**, 5787–5795.

Dr Knopf said: It is stated that the core–shell morphology is the prevalent configuration of liquid–liquid-phase-separated tropospheric organic/AS/H$_2$O particles. These conclusions are drawn from experiments performed at room temperature. The average temperature in the troposphere is significantly colder thereby affecting particle viscosity, likely increasing viscosity. Does the core–shell morphology still represent the prevalent configuration for particles at lower temperatures?

Dr Krieger replied: Yes, this is an interesting point. We did not yet study temperature dependence. The first issue to consider is that a system showing liquid–liquid phase separation at room temperature may become miscible below a so called lower critical solution temperature (LCTS). For example, the higher molecular weight binary PEG/H$_2$O systems show such a LTCS. The second point is what you raise in your question: at sufficiently low temperatures spinodal decomposition may still occur, but diffusion may be so slow that the coalescence forming the two distinct liquid phases is kinetically impeded. At this time we do not have any measurements of this effect for atmospheric model systems.

Professor Freedman remarked: Your paper comments that liquid–liquid phase separation tends to occur by spinodal decomposition when the organic acid and the ammonium sulfate are present in approximately equal concentrations. Could you explain why phase separation occurs by spinodal decomposition rather than nucleation and growth in this case?

Dr Krieger replied: When the organic and the ammonium sulfate are present in approximately equal concentrations, we are close to the critical concentration (*i.e.* the concentration at which the onset relative humidity of liquid–liquid phase separation is the highest at constant temperature). At this point the spinodal line (the limiting line of metastability) merges with the coexistence (binodal) line. For nucleation and growth to occur your system needs to be in the metastable region of the phase diagram, *i.e.* between the binodal and spinodal line, which is not possible for concentrations exactly equal to the critical concentration and very unlikely close to this concentrations. For a detailed discussion see P. Papon, J. Leblond and P. Meijer, *The Physics of Phase Transitions*, Springer, Berlin Heidelberg, Germany, 2006, ch. 2, pp. 37.

Professor Abbatt commented: For the coating of a very small amount of immiscible organic onto an inorganic core, would you expect the same core–shell morphology to arise?

Dr Krieger responded: It depends on what a "very small amount" is. There will always be a finite solubility of the organic compounds in the aqueous phase, thus for very low amounts of organics I would not expect liquid–liquid-phase-separation at all. If the spreading coefficients predict core–shell morphology, but the volume of the organic phase is too small to spread over the inorganic core with a thickness exceeding, say 10 monolayers, a detailed calculation taking the respective volumes into account as described by Durant and Guillot (G. Durant and C. Guillot, *J. Colloid Polym. Sci.*, 1993, **271**, 607) or Kwamena *et al.* (N.-O. A. Kwamena, *J. Phys. Chem. A*, 2010, **114**, 5787) is advised.

Dr Reid remarked: Our conclusion that mixed hydrophobic/hydrophilic particles form partially engulfed structures comes from a survey of surface and interfacial tensions for a wide range of compounds and the dependence of the two on the inorganic salt concentration. (See J. P. Reid, B. J. Dennis-Smither, N.-O. A. Kwamena, R. E. H. Miles, K. L. Hanford and C. J. Homer, *Phys. Chem. Chem. Phys.*, 2011, **13**, 15559–15572.) The presence of the electrolyte leads to salting out of the organic component and increased suppression of the aqueous surface tension. Thus, the thermodynamic force for the hydrophobic component to spread over the aqueous surface to stabilise it is less significant. The determining factor is the difference between the change in the surface tension of the aqueous phase and the interfacial tension as the RH decreases. If the surface tension drops more quickly due to salting out, the system is driven towards a partially engulfed state. Measurements have not been made at electrolyte concentrations typical of atmospheric aerosol and so some extrapolations/estimations are required, typically using the Antonov rule for the interfacial tension. This predicts that the interfacial tension is just the difference between the organic and aqueous phase surface tensions. For the 65 compounds considered in the Reid *et al.* paper, the mean error in the interfacial tension estimate from Antonov's rule was found to be 1.3 mN m^{-1}. This was based on values reported by Demond and Lindner (A. H. Demond and A. S. Lindner, *Environ. Sci. Technol.*, 1993, **27**(12), 2318–2331). Thus, this leads to the conclusion that the actual change in the interfacial tension with decreasing RH will be, on average, marginally less than the change for the aqueous–air surface tension and partially engulfed structures will on average be preferred.

Dr Krieger answered: I agree with your argument. As already stated in the answer to the previous question, I am convinced that the main difference between our study and that of Reid *et al.* (Reid *et al.*, *Phys. Chem. Chem. Phys.*, 2011, **13**, 15559) is the selection of the compounds. The organic fraction in our study considers 33 components with O : C ratios ranging from 0.29 to 1.33. The Reid *et al.* study considers 65 organics with O : C ranging from 0 to 0.6. However, about 90% of their components have an O : C ratio smaller or equal 0.25 and only one component has an O : C ratio larger than 0.5. The difference in O : C between the two sets of organics is reflected in different spreading coefficients, yielding preferably partially engulfed morphologies for the set of organics considered by Reid *et al.* and preferably core–shell morphologies for the set considered by us. We claim that our set of organics better represents the O : C found in tropospheric aerosol.

Professor Nizkorodov addressed Dr Topping and Dr Krieger: As the papers discussed in this session state, there are no theoretical reasons why the number of chemically distinct phases in liquid–liquid equilibrium should be limited to two. Has anyone observed a separation of atmospherically relevant mixtures in more than two phases?

Dr Krieger replied: I agree that there is no fundamental reason for having more than two liquid phases in an atmospherically relevant complex mixture. However, I am not aware of any such observations.

Dr Topping replied: I think the problem in defining 'atmospherically relevant' mixtures here, based on laboratory evidence, becomes an issue of complexity. There is evidence from the literature, but for systems with limited number of compounds. For example:

1)   Acetonitrile–water–benzene–heptane   and   sulfalone–water–benzene–heptane both give 3 liquid phase regions with reasonable compositions (G M Hartwig *et al.*, *J. Am. Chem. Soc.*, 1955, **59**, 52–54);

2) Water–ethanol–benzene–ammonium sulphate can give 3 liquid phases (J. C. Lang and B. Widom, *Physica*, 1975, **81A**, 190–213.);

3) Water–ethylene glycol–C12-alcohol–C1–C2 nitroalkanes *etc.* reportedly gives 4 liquid phases. (B. K. Das and R. B. Griffiths, *J. Chem. Phys.*, 1979, **70**, 5555–5566). The data suggest that the 4 liquid phases exist over very narrow ranges of temperature (in this case between 37 and 42 degrees centigrade) and composition;

4) Water–alcohol–hydrocarbon–salt can give 3 liquid phases (B. M. Knicker-bocker *et al.*, *J. Phys. Chem.*, 1982, **86**, 393–400). Authors mapped out a compositional range for 2 and 3 liquid phase regions for 15 salts combined with 10 alcohols (all C1–C5 monohydric) and 6 hydrocarbons (C8, C10, C12, C14, C16). The phase region varies with salt but generally with C3 and C4 alcohols combined with high levels of salt and the higher hydrocarbons;

5) Water–phenol–hydrocarbon gives a 3 liquid phase region (A. Martin *et al.*, *J. Chem. Eng. Data*, 2011, **56**, 741–749);

6) Water–octane–aniline also gives a 3 liquid phase region (A. Grenner *et al.*, *J. Chem. Eng. Data*, 2006, **51**, 1009);

7) Water–octane or heptane–cyclohexylamine also gives a 3 liquid phase region (M. Klauck *et al.*, *J. Chem. Eng. Data*, 2006, **51**, 1043).

It appears evidence in the literature is quite sparse. For atmospheric systems it could be that it isn't important with regard to defining the number of liquid phases. For example, in a system such as water–hydrocarbon–isopropanol, the addition of isopropanol to the immiscible mixture of water and hydrocarbon reduces the activity coefficients of the water and hydrocarbon and narrows the miscibility gap. The isopropanol "bridges" the immiscibility of the alkane and water. Diagrams for the system water–isopropanol–ethyl acetate make a similar point:- see M. Rajendran *et al.*, *Fluid Phase Equilibria*, 1991, **70**, 65–106. In a complex mixture there is likely to be many such bridging compounds (across a wide range of polarity) that reduce miscibility gaps and hence reduce the number of liquid phases formed.

In conclusion, we can't dismiss the possibility of multiple (more than 2) liquid phases in aerosol samples but the one and two phase situations (ie. aqueous hydrocarbon immiscibility) will likely dominate most of the time for the reasons outlined above.

Professor Nizkorodov opened the discussion of the paper by Dr Topping: I was intrigued by the comment in the discussion section of your paper that differences between individual particles and the mixing state are more important than the liquid–liquid phase separation within a given particle. Can you or anyone else working in this area provide several examples where the liquid–liquid phase separation is going to critically affect important physical or chemical properties of an aerosol containing such phase-separated particles?

Mr Mason opened the discussion of the paper by Dr Krieger: Is the homogeneous nucleation/efflorescence relative humidity of an aqueous inorganic particle significantly dependent on the particle volume and/or temperature in addition to the relative humidity? Does homogeneous nucleation have a stochastic element to it; would we expect a range of efflorescence relative humidities for a set of particles of the same size/concentration *etc.*? If so are these effects significant?

Dr Krieger responded: The efflorescence relative humidity (ERH) for homogeneous nucleation is volume dependent according to classical nucleation theory, however the nucleation rate changes so steeply with relative humidity, that one can nevertheless assign a value for the ERH (see C. V. Ciobanu *et al.*, *J. Phys. Chem. A*, 2010, **114**, 9486).

Professor Abbatt said: One process that is very much dependent on whether there is phase separation in the aerosol particle is N_2O_5 hydrolysis. When the particle phases separate, the organic layer may have much lower reactivity than a mixed organic–water phase. Some field data can not be explained by these mixed systems, and might be explained by particles that have phase separated.

Dr Krieger added: I agree, this is probably the most prominent heterogeneous reaction to look for an effect of liquid–liquid-phase separation (see M. Folkers *et al.*, *Geophys. Res. Lett.*, 2003, **30**, 1644).

Professor Pandis opened the discussion of the paper by Dr Topping: Atmospheric and maybe laboratory measurements of the dependence of the concentration of organic aerosol on relative humidity can help constrain the importance of phase separation for the ambient atmosphere. Chemical transport models predict very different diurnal organic aerosol concentration profiles depending on the assumed magnitude of interactions [B. Pun, S. Y. Wu, S. Seigneur, J. H. Seinfeld, R. J. Griffin, and S. N. Pandis, *Environ. Sci. Technol.*, 2003, **37**, 3647–3661]. Does your model suggest a strong, medium, or weak dependence of the organic aerosol levels on ambient RH? How does this qualitative behavior compare with available measurements?

Dr Topping replied: I agree that laboratory measurements can offer insights into the potential role of LLE in dictating ambient loadings of SOA. With regard to dependency on RH, our results indicate a reduced organic mass loading dependency as a function of RH, both for the simplified laboratory system (indicated in figure 2) and using output from a detailed chemical mechanism to represent ambient scenarios (figures 4 and 5). This is particularly evident up to 90% RH. Note also the comparison with the simplified laboratory system is very good. The maximum decrease we found when compared with an ideal forced one phase system was ~50%. Zuend and Seinfeld (A. Zuend and J. H. Seinfeld *Atmos. Chem. Phys.*, 2012, **12**, 3857–3882) used a subset of compounds from alpha-pinene ozonolysis to show that inclusion of non-ideal LLE predictions substantially decreased the organic mass loading dependency on RH, particularly for lower SOA loading conditions. For certain conditions they found differences of ~200% with an almost zero dependency on RH. These separate studies on varying complexity indicate that more studies are required and it is difficult to prescribe a generic 'strong' or 'weak' dependency. It would be useful to assess the range of dependencies from different precursors in chamber environments. In particular, it would be fruitful to incorporate recommendations for approximating LLE in large-scale models to assess wider sensitivities over a range of emission scenarios. Aside from the model presented here, there are some quite easy domain based simplifications that could potentially be used. For example, at higher O : C ratios (>0.7) no LLE is often observed, with dependencies on the inorganic salt present.

The challenge then, from a mechanistic specific component modelling perspective, is quantifying uncertainty in other parameters such as pure component vapour pressures, kinetic mass transfer limitations or condensed phase reactions to name a few. We have not yet compared the ambient equilibrium simulations from this paper with available measurements, but the model presented here could be used. Nonetheless, as Zuend and Seinfeld (2012) demonstrate, the inclusion of LLE for modelling of specific systems can improve comparisons with total condensed mass.

Professor Donahue opened the discussion of the paper by Dr Krieger: From the perspective of the fate of organics as well as the total mass concentration of condensed-phase organics, the topic of phase separation in the condensed phase is vital, but we must never forget that this is coupled to chemistry in all phases. Most, if not all, semi-volatile organics are extremely short lived in the gas phase, at least with respect to oxidation by OH radicals.

Dr Reid opened the discussion of the paper by Dr Topping: There are large variations in values of vapour pressures of organic compounds estimated by different modelling techniques, as reported by Topping *et al.* previously and stated on page 11 of the Topping manuscript. To what extent is it important to reduce these model uncertainties and better constrain models to ensure better pure component estimates of vapour pressures?

Dr Topping replied: Specifically targeting LLE, we show in this paper that the choice of vapour pressure method can not only affect the total condensed mass but also the magnitude to which LLE is predicted to occur. This is simply due to heterogeneity with regard to correlations between predicted vapour pressures and molecular functionality. For example, using a method that is known to vastly under-predict VP leads to a 'large' increase in the abundance of two immiscible liquid phases. However, the uncertainty caused by predicted LLE is never outside of 50% for total condensed mass, much smaller than potential impacts by choice of the VP method. There is also an issue of complexity to consider here. For systems with a limited number of condensing compounds, the impact of uncertainties in predicted vapour pressures can be orders of magnitude. We, and others, have shown that increasing the complexity of the system can reduce the uncertainty to choose a predictive technique. There are still many questions to answer and it is difficult to prescribe a fixed uncertainty associated with vapour pressures at the present moment in time. For example, it is still unclear to what extent choice of laboratory technique affects measured vapour pressures. The body of work aimed at reducing these uncertainties will require combinations of mechanistic and semi-empirical modelling efforts, detailed and ensemble laboratory studies and large scale sensitivities.

Mr O'Meara opened the discussion of the paper by Dr Krieger: I would like to emphasise the need for more measurements of vapour pressures of chemical partitioning from gas to particulate matter, particularly semi-volatile compounds. These measurements are necessary for further development and testing of vapour pressure estimation methods.

Faraday Discussions

RSC Publishing

PAPER

Fluorescent lifetime imaging of atmospheric aerosols: a direct probe of aerosol viscosity

Neveen A. Hosny,[a] Clare Fitzgerald,[b] Changlun Tong,†[a] Markus Kalberer,[b] Marina K. Kuimova[a] and Francis D. Pope*[c]

Received 15th March 2013, Accepted 8th April 2013
DOI: 10.1039/c3fd00041a

The viscosity of atmospheric aerosol particles affects a number of key physical and chemical particle properties, such as composition and reactivity. However, determination of the microscopic viscosity of aerosol particles is a non-trivial task. We report a new method of imaging viscosity in a variety of model aerosol systems, based on a fluorescence lifetime determination of viscosity-sensitive fluorophores termed molecular rotors. We report the viscosity changes associated with the relative humidity dependent hygroscopicity of NaCl and sucrose aerosols, as well as reaction dependent changes in viscosity during ozonolysis of oleic acid aerosols. The Fluorescence Lifetime Imaging Microscopy (FLIM) of molecular rotors shows great promise in understanding important fundamental aerosol properties, which can be both time-dependent and spatially variable through the aerosol particle.

1   Introduction

Aerosol particles are central to many climate related processes, including: light scattering, light absorption, and cloud formation *via* nucleation.[1] They are also important in physiology due to their links with adverse human health effects, such as asthma and cancer.[2] Ambient aerosol composition varies widely both temporally and geographically and has often been found to be dominated by organic species.[3] The organic mass fraction in aerosol particles is a highly complex mixture of hundreds of compounds covering a wide range of physical–chemical properties, such as molecular weight, volatility and polarity. In general the composition of organic aerosols on a molecular level is only poorly known. Furthermore organic aerosol particles are susceptible to oxidation reactions, which can continuously change their chemical composition over their entire

[a]*Department of Chemistry, Imperial College London, London, SW7 2AZ, UK*

[b]*Department of Chemistry, University of Cambridge, Cambridge, CB2 1EW, UK*

[c]*School of Geography, Earth and Environmental Science, University of Birmingham, Edgbaston, B15 2TT, UK. E-mail: f.pope@bham.ac.uk; Tel: +0121 4149067*

† Present address: Institute of Environmental Science, Zhejiang University, Hangzhou 310058, P.R. China.

atmospheric lifetime. These ageing processes are also poorly understood but are known to significantly affect the cloud formation potential of aerosol particles and their light absorption properties.

Over the last few years much attention has been paid towards the need for a better understanding of atmospheric aerosol phase, viscosity and diffusion.[4-12] Diffusion is linked to viscosity *via* the Stokes–Einstein equation shown in E1, where D is the liquid diffusion coefficient, k is the Boltzmann constant, T is temperature, η is the dynamic viscosity that is measured in Pascal seconds (Pa.s), and r is the radius of a diffusing molecule which is assumed to be spherical.

$$D = kT/6\pi\eta r \tag{E1}$$

The phase of an aerosol can be classified by its viscosity and hence diffusion coefficient,[5] with liquids having viscosities less than 10^4 mPa.s (and diffusion coefficients greater than 10^{-10} cm^2 s^{-1}), semi-solids in the range 10^4–10^{14} mPa.s $(10^{-10}$–10^{-20} cm^2 s$^{-1})$, and solids greater than 10^{14} mPa.s ($<10^{-20}$ cm^2 s^{-1}). Some aerosols have been shown to have amorphous solid structures comprised of glassy states which possess very high viscosities.[7] It should be noted that the Stokes–Einstein equation has been shown to break down in the vicinity of the glass transition.[13,14]

The viscosity of an atmospheric aerosol can affect a number of key physico-chemical properties, such as water content and particle reactivity and hence composition.[15] The extent and time scales of aerosol processing can be strongly dependent on particle viscosity. For example, it has been shown, within the laboratory, that the high viscosity of organic aerosol systems under low relative humidity conditions can affect the mass transfer of water to and from the particle.[8,9] The ability of oxidants to diffuse through aerosol particles can determine aerosol reactivity, and the subsequent changes in aerosol composition.[10] Conversely changes in aerosol composition will likely change the aerosol viscosity. Modelling studies have shown that the bulk reactivity of short-lived chemical species is often limited by diffusion within aerosols.[6] Shiraiwa and Seinfeld have recently shown that the partitioning of secondary organic aerosol (SOA) can be severely affected by particle viscosity.[4] Aerosol viscosity determines the ability of chemical species to diffuse into the aerosol bulk, or be limited to surface reactions.

In spite of its importance, the measurement of aerosol viscosity presents great challenges, mainly because atmospheric particles and especially SOA particles are highly complex in their chemical composition. This complexity often leads to poor characterisation on a molecular level with high percentages of the organic mass unclassified. Thus it is not possible to mimic the composition of organic particles using synthetic mixtures for bulk phase analyses.

To the best of our knowledge, prior to this study there were no analytical techniques available that could directly quantify the viscosity of atmospheric aerosols. This significantly limits the mechanistic understanding of the aerosol processes that are affected by changes in particle viscosity. Standard mechanical methods (viscometers and rheometers) for measuring viscosity in bulk liquids require large sample volumes and are therefore not appropriate for aerosol measurements. The measurement of the bounce factor of aerosols, where the ability of aerosols to bounce from or stick to aerosol impactors is recorded, has been successful in determining the aerosol phase, but it cannot probe the aerosol viscosity directly.[11,12] A recent paper by Renbaum-Wolff *et al.* indirectly measured viscosity in the range of 10^{-1}–10^5 mPa.s using a novel method that measures the rate of circulation of small micron sized beads that are inserted into large supermicron (~50 micron) aerosols. The rate of circulation of the bead was linked to the aerosol viscosity.[16]

The problems associated with the determination of microscopic viscosity in aerosols are shared with that observed within biological systems. In the latter, several spectroscopic approaches were developed to estimate diffusion rates, such as single particle tracking,[17] fluorescence correlation spectroscopy (FCS)[18-20] and fluorescence recovery after photobleaching (FRAP).[21] However, all the above methods typically provide only single point measurements, which is less than ideal for monitoring viscosity in potentially highly heterogeneous systems.

A technique is required to directly investigate the viscosity of aerosol particles; ideally, one that can provide accurate measurements from liquid to semi-solid states, and is relevant to a large range of aerosol compositions that are of importance in atmospheric sciences. This paper describes the development of a methodology to quantitatively image viscosity in atmospheric aerosols by applying fluorescence lifetime imaging microscopy (FLIM) in conjunction with viscosity-sensitive fluorophores termed 'molecular rotors'.[22] This technique has previously been used to investigate the intra-cellular viscosity within individual organelles.[23-27] The fluorescence quantum yield and lifetime of molecular rotors show a remarkably strong dependence on the viscosity of their immediate environment, due to the competition between radiative decay and intramolecular rotation, which controls the rate of non-radiative processes. The different molecular rotor designs allow for measurements of different viscosity ranges, in both aqueous and lipophilic aerosol environments. Upon insertion into a system of interest, at a low concentration, the molecular rotor reports the viscosity of their microenvironment whilst including information on spatial variations and time-dependent changes.

In this paper we demonstrate the use of molecular rotors to measure the viscosity of particles composed of NaCl, sucrose, and oleic acid. Furthermore we demonstrate that the rotor technique can be used to observe the change in viscosity of oleic acid particles during ozonolysis.

Materials and methods

Calibration of viscosity probes

Molecular rotors 3,3'-diethylthiacarbocyanine iodide (Cy3, 36809, Sigma Aldrich), sulforhodamine B (SR, S1402, Sigma Aldrich) and *meso*-alkoxyphenyl-4,4'-difluoro-4-bora-3a,4adiaza-s-indacene (Bodipy-C_{10})[26] were used for microviscosity determination in aerosol droplets. SR and Cy3 are water-soluble dyes, which were used as viscosity probes in hydrophilic NaCl and sucrose aerosols. Conversely, Bodipy-C_{10} is highly hydrophobic and as such was considered suitable for viscosity determination in oleic acid aerosols. Dye stock solutions were prepared for SR, Cy3 and Bodipy-C_{10} at a concentration of 5 mM, each was respectively dissolved in HPLC water (VWR), dimethyl sulfoxide (DMSO) (Sigma-Aldrich) and chloroform (Sigma-Aldrich).

For calibration of viscosity response of each dye fluorescence lifetimes were measured either in quartz cuvettes (Fig. 1A) or in an eight well μ-Slide chamber (Ibidi) (Fig. 1B) *via* time correlated single photon counting (TCSPC). Cuvette measurements were performed using a Jobin Yvon IBH data station (5000F, HORBIA Scientific Ltd.) with a 467 nm 1 MHz pulsed NanoLED (N-467, HORBIA Scientific Ltd.) for excitation. Emission was captured at 580 ± 5 nm and a long pass filter at 560 nm (SR) and at 515 ± 5 nm and a long pass filter at 470 nm (Bodipy-C_{10}), until a peak count >3000 was reached; 1024 ADC and collection rate <2% was maintained. The eight well chamber was used for measuring the time resolved decays of Cy3 on the FLIM system as described below. The final concentrations for bulk TCSPC measurements were 2.4 μM, 10 μM and 10.6 μM for SR, Cy3 and Bodipy-C_{10}, respectively.

Viscosity calibration measurements were performed in a series of water/sucrose (Cy3) water/glycerol (SR) or methanol/glycerol (Bodipy-C_{10}) solutions of varying viscosities covering a range between 1 and 1000 mPa.s. Viscosities of the water/sucrose mixtures were measured on a Discovery Hybrid (DH) Rheometer (HR-3, TA Instruments) by performing a flow ramp test on each sample across a shear rate of 0.1–1000 s^{-1} with a 300 s duration; a 40 mm cone plate was used with the samples and the Newtonian nature of the fluids was confirmed by linear shear regions. The dynamic viscosity of water/glycerol and methanol/glycerol were measured using a Stabinger viscometer (SVM 3000, Anton Paar) with an accuracy and precision of ±0.35% and ±0.1%, respectively.

The solution viscosity was correlated to fluorescence lifetimes using the modified form of the Förster–Hoffmann equation,[22] as shown in E2.

$$\tau_f = \frac{z\eta^\alpha}{k_r} \tag{E2}$$

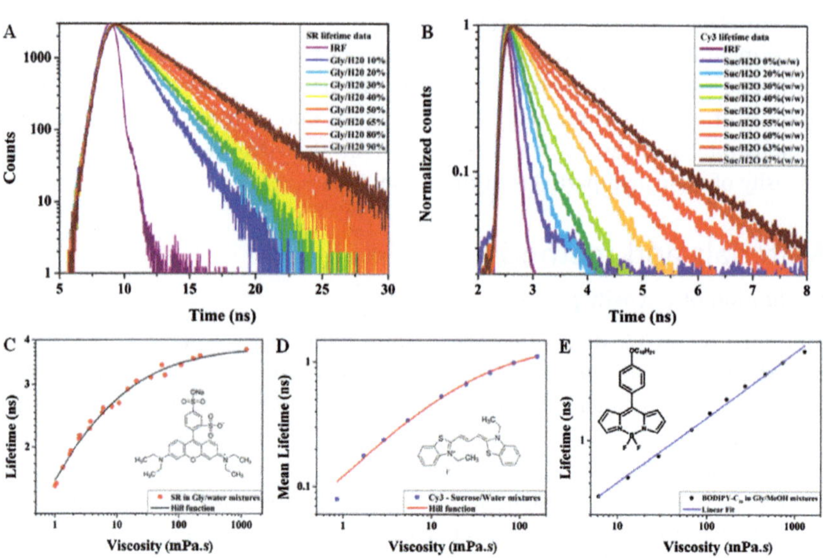

Fig. 1 Characterising viscosity response of molecular rotors. A) Lifetime decays of SR in glycerol/water mixtures between 0–90% glycerol. The decays can be fitted with a monoexponential function. B) Lifetime decays of Cy3 in sucrose/water mixtures between 0–67% w/w sucrose. The decays require biexponential fitting. Log–log calibration plot of lifetime *vs.* viscosity, fitted with a Hill function, inset chemical structure, C) SR (●) and D) Cy3 (●). E) Log–log calibration plot of BODIPY-C_{10} in glycerol/MeOH (●) with a linear fit, inset is the chemical structure of BODIPY-C_{10}.

Where τ_f is the fluorescence lifetime, k_r is the radiative rate constant, η is the dynamic viscosity, and z and α are constants. For calibration purposes it is useful to present this equation in the logarithmic form, as shown in E3.

$$\log \tau_f = \alpha \log \eta + \log\left(\frac{z}{k_r}\right) \tag{E3}$$

These calibration plots (Fig. 1C, D and E) allow the unequivocal conversion of experimentally measured lifetime (including those from microscopy FLIM experiments) to viscosity. According to E3 we observed a linear relationship between $\log \tau_f$ and $\log \eta$ in a large viscosity range 7.7–1140 mPa.s for Bodipy-C$_{10}$ and between 1–30 mPa.s for Cy3 and 1–10 mPa.s for SR. However, even though the data for Cy3 and SR deviate from linearity above 30 mPa.s and 10 mPa.s, respectively, it is possible to use interpolation (*e.g.* by applying a Hill function) to assign viscosities to these lifetime values.

Aerosol preparation

In order to prepare aqueous aerosols, SR and Cy3 stock solution were diluted with stock solutions of NaCl and sucrose in water to create final working concentrations of dye at 52 µM (SR) and 15.2 µM (Cy3), while NaCl and sucrose solutions were 0.4 M and 0.75 M, respectively; solutions were separately loaded into a plastic spray bottle. Oleic acid aerosols were generated using a paint sprayer (Model 250–2, Badger). The oleic acid/rotor solution was created by evaporating 5 µl of Bodipy-C$_{10}$ stock solution, which removes the chloroform but keeps the Bodipy-C10, and adding 1 ml of oleic acid to create the final dye concentration of 10.55 µM. Solutions were separately sprayed on to the siliconised coated cover-slips (22 × 22 mm, 0.2 mm thickness) (Hampton Research) either using the pump action (NaCl and sucrose) or N$_2$ gas supply (oleic acid) to deliver the aerosol. Droplets ranged in size between 5–200 µm, however only droplets less than 80 µm were investigated for internal viscosity. The contact angle of single droplets of either the aqueous solution or oleic acid on the siliconised coated glass coverslip was acquired using a stereoscope Olympus microscope mounted parallel to the cross-section of the droplets. It was determined that the wetting angle created by aqueous and oleic acid droplets was between 80–90° and 40–50°, respectively. The coverslip with the deposited droplets was mounted into a small custom-made chamber (dimensions: 50 × 25 × 20 mm), which allowed for the variation of the gas phase relative humidity (RH) and ozone concentration during the FLIM experiments. It also allowed for the observation of viscosity changes online. High vacuum grease (Dow Corning) was applied onto the chamber base for adhesion and to create a seal between the coverslip and the chamber (Fig. 2).

Industrial air was delivered to the chamber inlet *via* a humidifier at a pressure of 0.7 bar. The desired RH of the gas flow was generated by mixing a dry air flow with a water saturated air flow, which was produced by passing the air flow through a water bubbler. The air flows were controlled using two separate mass flow controllers (MKS), which could both be independently controlled between 0–500 sccm. The RH was measured using a digital humidity sensor (SHT-75, Sensirion) and controlled by varying the flow rate of the mass flow controller to attain a variable RH. Temperature was also monitored using the same sensor and ranged between 21–24 °C during the experiment. The accuracy of the RH

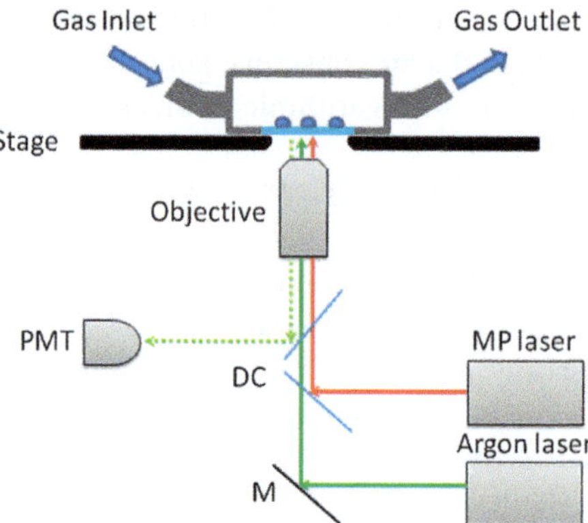

Fig. 2 A schematic diagram of the experimental layout. A) The sample inside the chamber is mounted on the microscope stage and has either a multiphoton (MP) or argon laser directed by either a mirror (M) or a dichroic (DC) through an inverted objective. The returning fluorescence is directed by a DC to the photomultiplier tube (PMT).

measurement is $\pm 1.8\%$ below 90% and rises linearly to $\pm 4\%$ in the range: 90–100%. The accuracy of the temperature measurement is ± 0.3 °C. After changing the RH at least 20 min equilibration time was allowed for the high RHs, and up to 1 h for the lower RHs. Equilibration times were based on the Shiraiwa *et al.* calculation that takes into account both particle size and particle viscosity.[5]

The viscosity change due to the oxidation of oleic acid aerosols by ozone was monitored by imaging of Bodipy-C_{10} lifetime. To produce ozone industrial air was passed through an ozoniser, consisting of a UV lamp (185 nm, Appleton Woods) at a flow rate of 300 sccm. The concentration of ozone (ppm) was measured using an ozone monitor (49i, Thermo Scientific) and a concentration of 1.5 ppm was attained. Ozone-enriched gas was directed through the inlet of the chamber off-line to expose the oleic acid droplets (Fig. 2), subsequent to the chamber exit ozone was captured by passing the gas through a 1.5 m charcoal denuder tube.

Droplet viscosity measurements

The chamber was mounted onto the microscope stage of an inverted scanning confocal microscope (SP5, Leica Microsystems Ltd.) and a ×63 (N.A. 1.2) HCX PL APO CS water immersion objective lens with correction collar (11506279, Leica Microsystems Ltd.) was used for imaging. For brightfield/intensity localisation imaging fluorescent dyes were excited at 561 nm (SR and Cy3) and 488 nm (Bodipy-C_{10}), emission was captured at 570–700 nm (SR and Cy3) and 500–660 nm (Bodipy-C_{10}).

A time correlated single-photon counting (TCSPC) module (SP830, Beckle&Hickl GmbH) was used for lifetime imaging (FLIM). A tuneable multi-photon laser (680–1080 nm, 80 MHz, 140 fs, Vision II, Coherent Inc.) was used as an excitation source in FLIM. Two-photon excitation of aerosols was performed at 905 nm for Cy3 and SR and at 800 nm for Bodipy-C_{10}. Laser power was maintained <400 mW prior to entering the microscope and image format was maintained at 128 × 128 pixels. The TCSPCs ADC were set to 256 and the collection time was maintained until maximum peak count >500 counts was reached. In the case of oleic acid aerosols it was determined that the lifetime of Bodipy-C_{10} was susceptible to laser exposure. The following low power settings were used to remove any effect caused by the laser irradiation: <100 mW, 64×64 image size, and collection times of 20–30 s. At these setting we ensured that the lifetime of Bodipy-C_{10} did not change in air over periods of less than 24 h of imaging.

Lifetime analysis

FLIM data was exported and analysed in TRI2 software (Version 2.7.6.1, Gray Institute for Radiation Oncology and Biology).[28] A mono-exponential model was fitted to all SR and BODIPY-C_{10} data, whilst a bi-exponential decay was applied to all Cy3 data, using the Levenberg–Marquardt algorithm. A mixture of images and single decays were collected. For all image analysis, pixels were binned to maintain at least a minimum peak count of 30–100 counts per pixel and only lifetimes with reduced chi-squares $0.8 > \chi_r^2 < 1.2$ were accepted in final images; thresholding was used to remove background noise. All single decays were exported to TRI2 Single TransImport software and exponential decays fitted using the same algorithm as above to extract the lifetime. A false rainbow colour scale was assigned to each fluorescence lifetime value in FLIM images: blue for a short lifetime and red for a long lifetime, to provide lifetime maps.

Mass spectroscopy

The ozonolysis products of oleic acid were detected by using ultra-high resolution mass spectrometry on material extracted from microscope slides at the end of FLIM experiments. Slides were washed with spectroscopic grade acetonitrile (Sigma-Aldrich) and tentative product identification was achieved using a LTQ Orbitrap Velos Mass spectrometer (Thermo Scientific) in direct infusion negative ion mode, with a Heated Electrospray Ionisation source (Thermo Scientific). The following settings were used: infusion flow rate 10 μl min^{-1}, vaporiser temperature 70 °C, ion transfer tube temperature 280 °C and auxiliary gas flow was off; the API stack, ion optics and Orbitrap mass analyser were left to the manufacturer's defaults settings. The sheath flow (SF) and spray voltage (V) settings were optimised for each use but common settings used were: SF = 17 au, V = 3.85 kV. The elemental composition of the peaks in the mass spectra was determined using Xcalibur (2.2 SP1.48, Thermo Scientific), and further evaluation processes as described in detail in Kourtchev *et al.*[29]

Results and discussion

NaCl aerosols

SR is a water soluble molecular rotor that was previously used for monitoring the mixing of solvents of different viscosity in microfluidic devices.[30] FLIM images of the SR rotor in NaCl aerosol particles were recorded in the RH range: 40–95%, as shown in Fig. 3. In total 68 particle images were analysed. The TRI2 software provides information on the intensity, amplitude (A), lifetime (Tau) and goodness-of-fit (Chisq) (Fig. 3A). In Fig. 3A the range of these parameters is represented as a colour hue beneath each image. The corresponding lifetime histogram is separately shown in Fig. 3B. The representative monoexponential fluorescence decay is shown in Fig. 3C. The monoexponential nature of the decay, as well as the

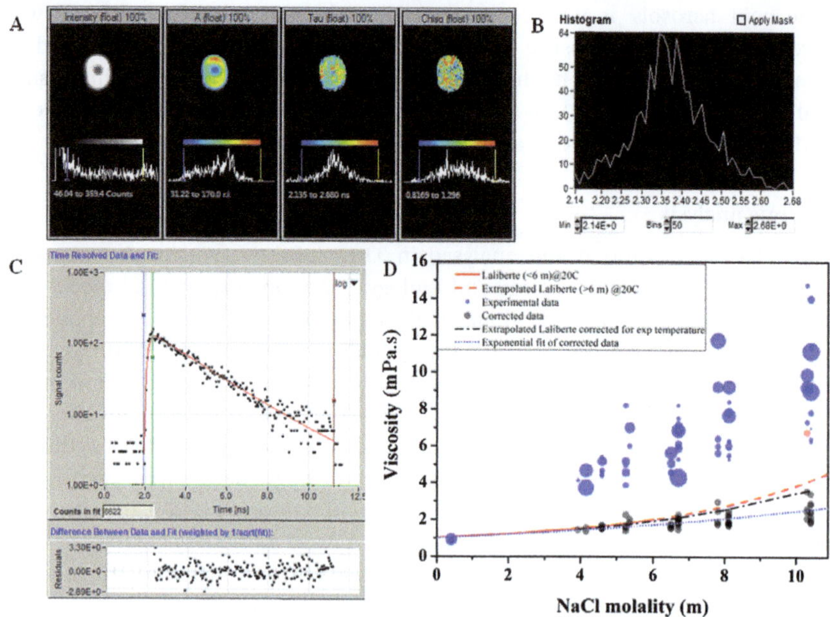

Fig. 3 Imaging of NaCl aerosol viscosity at an increasing relative humidity using FLIM of SR. A) FLIM data acquisition screen showing intensity, amplitude, lifetime and goodness of fit images for SR in NaCl droplet at 85 RH. B) The lifetime histogram extracted from A); C) A typical decay trace showing the monoexponential fit and residuals and D) the calculated viscosity of NaCl as a function of calculated molality at a measured RH; Size dependent experimental data ● where the size of circle indicates the relative size of particle measured, Corrected data ● (masked data point in red). Both the blue and grey data points are semi-transparent and hence darker colours represent regions of overlapping data points. Laliberté model <6 m at 20 °C (red line), extrapolated Laliberté model >6 m at 20 °C (dashed red line), extrapolated Laliberté model >6 m corrected for experimental temperature (dashed-dot black line) and exponential fit of corrected data (small dot blue line). The corrected data was obtained by rescaling the raw laboratory data to the Laliberté model data for molality values <6 m using a linear relationship ($y = 0.2x + 0.6252$, where y = Laliberté model viscosity, and x = laboratory measured viscosity). This relationship was then used to rescale the laboratory data over the entire molality range investigated.

low residuals of the fit (lower part of Fig. 3C) and the small range of Chisq values confirm that the SR rotor is in a homogeneous micro-environment.

For each particle image analysed, at a set RH, we took the mid-point value within the lifetime histogram as the average lifetime of the SR rotor. The lifetime of the SR rotor is correlated to viscosity using the calibration graph shown in Fig. 1C. We converted the measured RH value to NaCl particle molality using the Extended Aerosol Inorganics Model (E-AIM, http://www.aim.env.uea.ac.uk/aim/aim.php).[31] This allows for a plot of NaCl molality (for a given RH) against viscosity of the NaCl aerosol droplets to be produced, Fig. 3D.

The data points are compared to the modelled data of Laliberté, which is calibrated to bulk literature data.[32] The NaCl deliquescence point occurs at 75% and as such the bulk literature data only extend to a molality of 6. However, we extend the model to show predictions at higher molalities. We believe that our work provides the first measurement of viscosity below the NaCl deliquescence point. The lowest RH we managed to make a measurement at before efflorescence occurred was 51%. This value is slightly higher than that usually observed in the literature (40–50%),[33,34] and it is possible that the contact between the particle and coverslip is promoting crystallization more readily than would be the case for particles in the true aerosol phase.

As shown in Fig. 3D, the average measured viscosities varied between *ca.* 4–9 mPa.s in the RH range 86–55%. There was no correlation observed between particle size and particle viscosity. Whilst this trend in viscosity, at different RH, is similar to the Laliberté model, *i.e.* the viscosity increases with molality, we note that our method gives higher viscosity values, as compared to the model, with the exception of 0.4 M NaCl bulk droplet (starting value). It should be noted that the Laliberté model should be treated with caution at molalities greater than 6 because the model is not constrained by bulk measurements after this point. The calibration of the SR rotor was performed using mixtures of glycerol and water to achieve the range of viscosities needed. It is likely that organic glycerol solutions are a suboptimal calibration system for the inorganic salt system measured here. If the measured NaCl data is linearly rescaled to the subsaturated Laliberté model data (<6 m), then we achieve good agreement between the measured viscosity data and the Laliberté model predictions in the supersaturated regime (grey data points in Fig. 3D). We can apply a simple exponential to describe the data using this formula $y = \exp^{(0.09036x)}$, where y is viscosity and x is NaCl molality; one data point is masked in red to exclude it from the fitting.

Sucrose aerosols

Cy3 is a water soluble molecular rotor that was previously used for monitoring viscosity in cellular cytoplasm[35] and in other cellular organelles,[23,24] however, the viscosity-dependent lifetime of Cy3 was not rigorously characterised using TCSPC previously. We have established that in water/sucrose mixtures a biexponential decay model is required to fit the time resolved fluorescence decays of Cy3 and performed the calibration based on the mean fluorescence lifetime $(A_1\tau_1 + A_2\tau_2)$, where τ_1 and τ_2 are individual lifetimes from the fit and A_1 and A_2 are their amplitudes in %, Fig. 1D.

FLIM images of Cy3 in sucrose aerosol particles were recorded in the RH range: 75–95%, and are shown in Fig. 4. In total 44 particle images were analysed. We have extracted the average lifetime of biexponential decays for Cy3 from individual droplets, using the protocol similar to that described above for NaCl and the SR dye. The mean lifetime calculated can then be correlated to viscosity using the Cy3 calibration plot shown in Fig. 1D. In addition to the mean lifetime we have performed detailed analysis of the individual parameters of the fit (*i.e.* A_1 and τ_1, A_2 and τ_2) and ensured that all parameters correlate well to those obtained during the calibration. This gives us confidence that the lifetimes observed from sucrose droplets with changing RH are indeed due to changes in viscosity rather than the artefacts of binding and/or specific solvent interactions.

We record a marked increase in viscosity, from 10 mPa.s to 1000 mPa.s (Fig. 4A) upon RH decrease, observed from droplets ranging in diameter between 10–50 μm (Fig. 4B). It can be seen that our measurements are consistent with predictions of both the Laliberté and Swindells models.[32,36] Furthermore, our data suggests that the Laliberté model is superior to the Swindells model in the RH range investigated in this paper. Based on our direct measurements of viscosity in sucrose aerosol particles we can approximate the viscosity dependence on RH

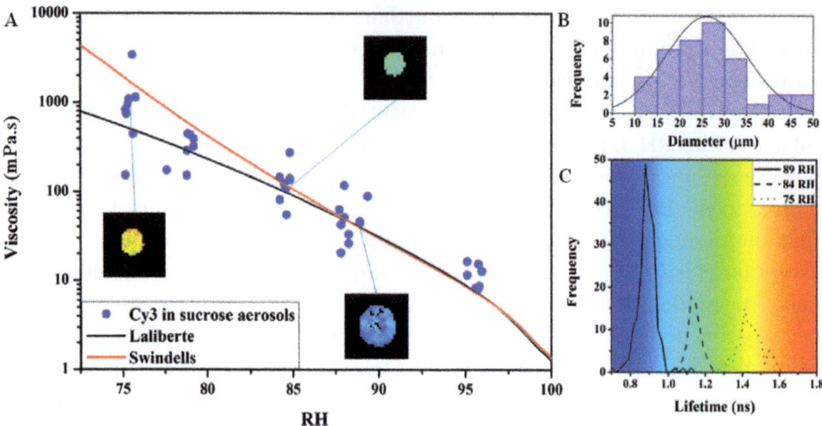

Fig. 4 Response of sucrose droplet viscosity to an increasing relative humidity measured *via* FLIM of Cy3. A) Droplet viscosity measured at a range of RH using lifetime/viscosity correlation for Cy3 in sucrose/water bulk solutions, Fig. 1D. Representative FLIM images are also shown as insets. B) Population distribution of aerosol droplets was normally distributed and was 26 (s.d. ±9) µm. C) Lifetime histograms of images at the three different humidities highlighted with the inserted images in A). The rainbow colour hue also corresponds to these images.

with a linear relationship: $\log y = 10.1568 - 0.0958x$, where y = viscosity (mPa.s), and x = RH (%), which is valid in the RH range from 75–95%.

Oleic acid

Next we set out to measure the changes of viscosity upon ozonolysis of oleic acid aerosol droplets. For this system we used the hydrophobic molecular rotor Bodipy-C_{10},[25,26] which was previously used for viscosity measurements in membranes of live cells. This dye should be capable of providing viscosity data in a wide range of viscosities between 10 and 1000 mPa.s, Fig. 1E.

Prior to the FLIM analysis we measured the temperature dependent viscosity of oleic acid using a strain controlled rheometer (ARES-LC, Rheometric Scientific) in the temperature range 14–46 °C controlled using a fluid bath. Our viscosity measurements give an excellent fit to the following function: $\eta = 10^{(A+B/T)}$, where the fitting constant values of $A = -2.3872$, and $B = 1141.3906$. These values are slightly below that recorded in Yaws, but are in good agreement with the measurements of Noureddini *et al.*[37] At 20 °C the viscosity of oleic acid is given by Yaws[41] and Noureddini *et al.* as 45.71 and 33.46 mPa.s, respectively.

The major products of oleic acid (C_{18}) ozonolysis are four compounds namely: nonanal (C_9 aldehyde), nonanoic acid (C_9 carboxylic acid), azelaic acid (C_9 dicarboxylic acid), and 9-oxononanoic acid (C_9 carboxylic acid and aldehyde functional group).[38–40] In addition to the C_9 products there are also higher molecular weight products that are attributed to the oligomerisation of the oleic acid precursor and C_9 products with reactive intermediates.[40] Nonanal has a high vapour pressure and partitions from the particle phase into the gas phase. At 20 °C the viscosity of pure nonanal and nonanoic acid are 1.47 and 9.84 mPa.s, respectively.[41] At room temperature azelaic acid is a crystalline solid and therefore there is no liquid viscosity data available, however the viscosity can be estimated from its higher temperature data to be 340 mPa.s.[41] We found no viscosity data in the literature for 9-oxononanoic acid but it is expected to have a viscosity between that of azelaic acid and nonanoic acid because of carboxylic acid and aldehyde moieties. Hence in an ideal mixture of liquids, at a set temperature, the relative viscosities of oleic acid and its main C_9 ozonolysis products is likely to have the following trend: $\eta_{\text{azelaic acid}} > \eta_{\text{9-oxononanoic acid}} > \eta_{\text{oleic acid}} > \eta_{\text{nonanoic acid}} > \eta_{\text{nonanal}}$. However, the relative position of $\eta_{\text{9-oxononanoic acid}}$ and $\eta_{\text{oleic acid}}$ is uncertain. The oligomer products are likely to have greater viscosities because of their longer chain lengths and similar functional groups. It should be noted that a multi-component mixture of several functional groups and carbon chain length is unlikely to form an ideal mixture.

FLIM images of oleic acid particles containing the Bodipy-C_{10} dye were recorded during ozonolysis (Fig. 5). In common with the NaCl and sucrose systems, the fluorescence intensity images do not allow a clear distinction between the viscosity of the droplets, but the lifetime images give very clear evidence of heterogeneity or changes in fluorescence lifetime and hence viscosity.

The unreacted oleic acid particles have measured viscosities of 80 mPa.s at 23 °C, which are independent of particle size as expected. Upon exposure to ozone the oleic acid droplets become oxidised and the FLIM measurement indicates that the viscosity of the droplets increase. The viscosity increase is dependent upon the particle size. The FLIM screen view showing the intensity, lifetime and the goodness of fit images is shown in Fig. 5. While intensity shows little variation, the lifetime histogram indicates that the distribution of lifetimes corresponds to a viscosity range of 191–317 mPa.s (1.93–2.43 ns; min–max). Furthermore the histogram below the lifetime image has three distinct populations. We infer that this is due to the fact that the bigger particle (blue) takes longer to react with ozone than the smaller (yellow and orange) particles. The validity of this result is confirmed by the fact that the goodness of fit (χ_r^2) for the image is uniform and indicates $0.8 > \chi_r^2 < 1.2$ for all droplets, suggesting no artefacts are present.

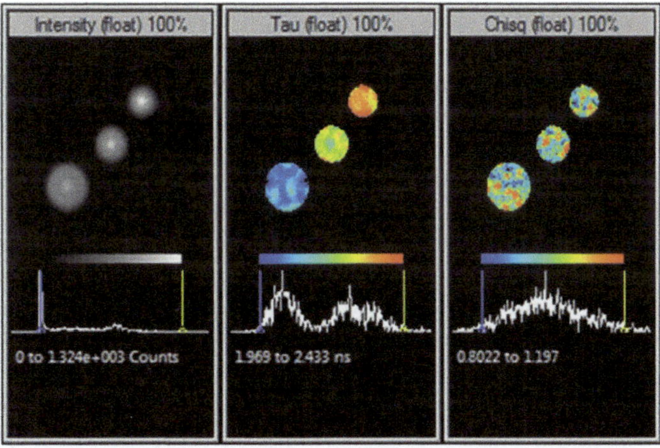

Fig. 5   FLIM of oleic acid aerosol droplets during ozonolysis. TRI2 output for the image collected at *ca.* 160 mins of exposure to ozone (1.5 ppm): Fluorescence intensity (grayscale), Fluorescence lifetime (Tau) and goodness of fit (χ_r^2). The histograms for each image are also shown.

The mass spectrum of the ozonolysed sample indicates that the particles contain azelaic acid, 9-oxononanoic acid, nonanoic acid, and dimer and trimer oligomers, as previously shown in Lee *et al.*[40] Nonanal was not observed due to its low ionisation efficiency and it is assumed that it volatilizes from the particles. The mass spectrum also included small amounts of oleic acid but it is unclear whether this is indicative of incomplete ozonolysis or merely gives the contribution from the oleic acid particles which were masked from the ozonolysis because of the chamber architecture. The highest signal in the mass spectrum was that of azelaic acid, but it should be noted that the signal intensities in mass spectra are dependent upon, but not directly proportional to, the concentration of the species measured. It is also dependent upon the ionization efficiency of the species. Hence the mass spectrum provides evidence, but is not conclusive, that the viscosity of the particle increases because oleic acid is oxidised to products that predominantly have higher viscosity than the oleic acid precursor.

Conclusions

We have developed a new technique that applies the precision in lifetime acquisition and spatial accuracy of a TCSPC FLIM system, with the environmental micro-sensitivity of fluorescent molecular rotors, to measure viscosity in atmospherically relevant aerosol particles. The technique works by exploiting the viscosity dependent fluorescence lifetime of the rotors. The molecular design of the rotors is important in determining their fluorescent properties, and hence the range of viscosities that can be probed. The design of the rotor is also responsible for its hydrophilicity and lipophilicity and therefore dictates which rotors are suitable for accessing different aerosol environments.

We have performed direct viscosity measurements in a variety of model aerosol systems: NaCl, sucrose and oleic acid, and recorded the changes in viscosity brought about by changes in two parameters that significantly affect particle phase in the atmosphere: relative humidity and oxidative reaction by ozone. For the first time we are able to demonstrate that changes in viscosity can be linked directly to changes in composition caused by atmospheric oxidation. This paper also reports the first determination of aerosol viscosity in a supersaturated NaCl system held at relative humidities below the NaCl deliquescence point. The study highlights the good spatial resolution of the FLIM technique, which should allow for the differentiation between surface and bulk phase aerosol reactivity, which is likely to be important for SOA particles.

The FLIM and rotor approach has been very successful in measuring relative viscosities. The measurement of absolute viscosities is more difficult and relies on good calibrations with suitable calibrants for the system being studied. At present the highest viscosity we have probed is ~1500 mPa.s, however higher viscosities measurements should be possible with both the Cy3 and BODIPY-C_{10} dyes. Furthermore, it is possible that new rotor designs will allow access to much higher viscosities, and in particular the semi-solid states of aerosol. At present most design effort has been guided towards use in biological systems and hence for lower viscosities. These first results now provide impetus to develop rotors suitable for higher viscosities.

The viscosity determination based on FLIM of molecular rotors shows great promise for the further study of the physico-chemistry of atmospheric aerosols.

We envisage that future applications of the technique will include the investigation of multiple phase aerosols. The judicious use of different rotors should allow measurement of the viscosity of both organic and inorganic phases within the same particle. The aerosol chamber will be extended to allow for the multi-parameter oxidation of particles, and the generation of the standard suite of atmospheric oxidants, including the hydroxyl radical (OH) and the nitrate radical (NO$_3$). In particular, viscosity measurements of SOA particles will be important and should help determine the mixing within this important class of particle. There are also good prospects of coupling the FLIM technique to single particle techniques, such as laser tweezers, electrodynamic balances (EDB), or acoustic trapping techniques, where the true aerosol phase can be probed and the influence (if any) of the cover slip can be removed.

Acknowledgements

MKK is thankful to the EPSRC for the Career Acceleration Fellowship (EP/E038980/1). We thank Dr Simon Butler for guidance on the use of the rheometer based in the Department of Chemical Engineering and Biotechnology (Cambridge) that was used for the bulk viscosity measurements of oleic acid. We would like to thank Dr Anita Jones and Dr Mathieu Bennet (Edinburgh University) for suggesting the use of SR as a molecular rotor. CF thanks NERC for her studentship NE/J500070/1.

References

1 P. Forster, V. Ramaswamy, P. Artaxo, T. Berntsen, R. Betts, D. W. Fahey, J. Haywood, J. Lean, D. C. Lowe, G. Myhre, J. Nganga, R. Prinn, G. Raga, M. Schulz and R. Van Dorland, Changes in Atmospheric Constituents and in Radiative Forcing, in *Climate Change 2007: The Physical Science Basis*, Cambridge University Press, 2007.

2 D. W. Dockery and C. A. Pope, *Annu. Rev. Public Health*, 1994, **15**, 107.

3 J. L. Jimenez, M. R. Canagaratna, N. M. Donahue, A. S. H. Prevot, Q. Zhang, J. H. Kroll, P. F. DeCarlo, J. D. Allan, H. Coe, N. L. Ng, A. C. Aiken, K. S. Docherty, I. M. Ulbrich, A. P. Grieshop, A. L. Robinson, J. Duplissy, J. D. Smith, K. R. Wilson, V. A. Lanz, C. Hueglin, Y. L. Sun, J. Tian, A. Laaksonen, T. Raatikainen, J. Rautiainen, P. Vaattovaara, M. Ehn, M. Kulmala, J. M. Tomlinson, D. R. Collins, M. J. E. Cubison, J. Dunlea, J. A. Huffman, T. B. Onasch, M. R. Alfarra, P. I. Williams, K. Bower, Y. Kondo, J. Schneider, F. Drewnick, S. Borrmann, S. Weimer, K. Demerjian, D. Salcedo, L. Cottrell, R. Griffin, A. Takami, T. Miyoshi, S. Hatakeyama, A. Shimono, J. Y. Sun, Y. M. Zhang, K. Dzepina, J. R. Kimmel, D. Sueper, J. T. Jayne, S. C. Herndon, A. M. Trimborn, L. R. Williams, E. C. Wood, A. M. Middlebrook, C. E. Kolb, U. Baltensperger and D. R. Worsnop, *Science*, 2009, **326**, 1525.

4 M. Shiraiwa and J. H. Seinfeld, *Geophys. Res. Lett.*, 2012, **39**, L24801.

5 M. Shiraiwa, M. Ammann, T. Koop and U. Pöschl, *Proc. Natl. Acad. Sci. U. S. A.*, 2011, **108**(27), 11003.

6 C. Pfrang, M. Shiraiwa and U. Pöschl, *Atmos. Chem. Phys.*, 2011, **11**, 7343.

7 T. Koop, J. Bookhold, M. Shiraiwa and U. Poschl, *Phys. Chem. Chem. Phys.*, 2011, **13**, 19238.

8 H. J. Tong, J. P. Reid, D. L. Bones, B. P. Luo and U. K. Krieger, *Atmos. Chem. Phys.*, 2011, **11**, 4739.

9 D. L. Bones, J. P. Reid, D. M. Lienhard and U. K. Krieger, *Proc. Natl. Acad. Sci. U. S. A.*, 2012, **109**(29), 11613.

10 F. D. Pope, P. J. Gallimore, S. J. Fuller, R. A. Cox and M. Kalberer, *Environ. Sci. Technol.*, 2010, **44**, 6656.

11 A. Virtanen, J. Joutsensaari, T. Koop, J. Kannosto, P. Yli-Pirila, J. Leskinen, J. M. Makela, J. K. Holopainen, U. Poschl, M. Kulmala, D. R. Worsnop and A. Laaksonen, *Nature*, 2010, **467**, 824.

12 E. Saukko, H. Kuuluvainen and A. Virtanen, *Atmos. Meas. Tech.*, 2012, **5**, 259.

13 D. Champion, M. Le Meste and D. Simatos, *Trends Food Sci. Technol.*, 2000, **11**, 41.

14 M. Rampp, C. Buttersack and H.-D. Lüdemann, *Carbohydr. Res.*, 2000, **328**, 561.

15 P. J. Gallimore, P. Achakulwisut, F. D. Pope, J. F. Davies, D. R. Spring and M. Kalberer, *Atmos. Chem. Phys.*, 2011, **11**, 12181.

16 L. Renbaum-Wolff, J. W. Grayson and A. K. Bertram, *Atmos. Chem. Phys.*, 2013, **13**, 791.

17 R. Iino and A. Kusumi, *J. Fluoresc.*, 2001, **11**, 187.

18 N. Thompson, *Top. Fluoresc. Spectrosc.*, 2002, 337.

19 J. Korlach, P. Schwille, W. W. Webb and G. W. Feigenson, *Proc. Natl. Acad. Sci. U. S. A.*, 1999, **96**, 8461.

20 N. L. Thompson, A. M. Lieto and N. W. Allen, *Curr. Opin. Struct. Biol.*, 2002, **12**, 634.

21 M. J. Dayel, E. Hom and A. Verkman, *Biophys. J.*, 1999, **76**, 2843.

22 M. K. Kuimova, *Phys. Chem. Chem. Phys.*, 2012, **14**(37), 12671.

23 E. Gatzogiannis, Z. Chen, L. Wei, R. Wombacher, Y.-T. Kao, G. Yefremov, V. W. Cornish and W. Min, *Chem. Commun.*, 2012, **48**, 8694.

24 X. Peng, Z. Yang, J. Wang, J. Fan, Y. He, F. Song, B. Wang, S. Sun, J. Qu and J. Qi, *J. Am. Chem. Soc.*, 2011, **133**, 6626.

25 M. K. Kuimova, G. Yahioglu, J. A. Levitt and K. Suhling, *J. Am. Chem. Soc.*, 2008, **130**, 6672.

26 J. A. Levitt, M. K. Kuimova, G. Yahioglu, P.-H. Chung, K. Suhling and D. Phillips, *J. Phys. Chem. C*, 2009, **113**, 11634.

27 M. K. Kuimova, S. W. Botchway, A. W. Parker, M. Balaz, H. A. Collins, H. L. Anderson, K. Suhling and P. R. Ogilby, *Nat. Chem.*, 2009, **1**, 69.

28 P. R. Barber, S. M. Ameer-Beg, J. Gilbey, L. M. Carlin, M. Keppler, T. C. Ng and B. Vojnovic, *J. R. Soc. Interface*, 2009, **6**, S93.

29 I. Kourtchev, S. Fuller, J. Aalto, T. Ruuskanen, M. W. McLeod, W. Maenhaut, R. Jones, M. Kulmala and M. Kalberer, *Environ. Sci. Technol.*, 2013, **47**(9), 4069.

30 M. A. Bennet, PhD Thesis, The University of Edinburgh, 2011.

31 S. L. Clegg, P. Brimblecombe and A. S. Wexler, *J. Phys. Chem. A*, 1998, **102**, 2155.

32 M. Laliberté, *J. Chem. Eng. Data*, 2007, **52**, 321.

33 I. Tang, H. Munkelwitz and N. Wang, *J. Colloid Interface Sci.*, 1986, **114**, 409.

34 F. D. Pope, B. J. Dennis-Smither, P. T. Griffiths, S. L. Clegg and R. A. Cox, *J. Phys. Chem. A*, 2010, **114**, 5335.

35 K. Luby-Phelps, S. Mujumdar, R. Mujumdar, L. Ernst, W. Galbraith and A. Waggoner, *Biophys. J.*, 1993, **65**, 236.

36 J. F. Swindells, C. F. Snyder, R. C. Hardy and P. E. Golden, *Supplement to National Bureau of Standards Circular*, 1958, **440**.

37 H. Noureddini, B. Teoh and L. Davis Clements, *J. Am. Oil Chem. Soc.*, 1992, **69**, 1189.

38 J. Zahardis and G. Petrucci, *Atmos. Chem. Phys.*, 2007, 7, 1237.

39 O. Vesna, M. Sax, M. Kalberer, A. Gaschen and M. Ammann, *Atmos. Environ.*, 2009, **43**, 3662.

40 J. W. L. Lee, V. Carrascon, P. J. Gallimore, S. J. Fuller, A. Bjorkegren, D. R. Spring, F. D. Pope and M. Kalberer, *Phys. Chem. Chem. Phys.*, 2012, **14**, 8023.

41 C. L. Yaws, *Handbook of Viscosity: Organic compounds C8 to C28*, Gulf Pub. Co., 1994.

Faraday Discussions

RSCPublishing

PAPER

Aqueous aerosol SOA formation: impact on aerosol physical properties†

Joseph L. Woo, Derek D. Kim, Allison N. Schwier, Ruizhi Li
and V. Faye McNeill*

Received 8th March 2013, Accepted 7th May 2013
DOI: 10.1039/c3fd00032j

Organic chemistry in aerosol water has recently been recognized as a potentially important source of secondary organic aerosol (SOA) material. This SOA material may be surface-active, therefore potentially affecting aerosol heterogeneous activity, ice nucleation, and CCN activity. Aqueous aerosol chemistry has also been shown to be a potential source of light-absorbing products ("brown carbon"). We present results on the formation of secondary organic aerosol material in aerosol water and the associated changes in aerosol physical properties from GAMMA (Gas–Aerosol Model for Mechanism Analysis), a photochemical box model with coupled gas and detailed aqueous aerosol chemistry. The detailed aerosol composition output from GAMMA was coupled with two recently developed modules for predicting a) aerosol surface tension and b) the UV-Vis absorption spectrum of the aerosol, based on our previous laboratory observations. The simulation results suggest that the formation of oligomers and organic acids in bulk aerosol water is unlikely to perturb aerosol surface tension significantly. Isoprene-derived organosulfates are formed in high concentrations in acidic aerosols under low-NO_x conditions, but more experimental data are needed before the potential impact of these species on aerosol surface tension may be evaluated. Adsorption of surfactants from the gas phase may further suppress aerosol surface tension. Light absorption by aqueous aerosol SOA material is driven by dark glyoxal chemistry and is highest under high-NO_x conditions, at high relative humidity, in the early morning hours. The wavelength dependence of the predicted absorption spectra is comparable to field observations and the predicted mass absorption efficiencies suggest that aqueous aerosol chemistry can be a significant source of aerosol brown carbon under urban conditions.

1 Introduction

Atmospheric aerosols impact Earth's climate *via* interactions with solar radiation, or by acting as nuclei for cloud droplet formation. Organic matter, a typical

---

Department of Chemical Engineering, Columbia University, New York, New York, USA. E-mail: vfm2103@columbia.edu; Fax: +1 212 854 3054; Tel: +1 212 854 2869

† Electronic supplementary information (ESI) available. See DOI: 10.1039/c3fd00032j

component of tropospheric particulate matter,[1] can affect the physical properties of aerosols which determine their influence on climate, such as optical properties[2-8] and surface tension.[6,7,9-11] One recently identified source of organic matter in tropospheric aerosols is aqueous aerosol secondary organic aerosol (aaSOA) formation. In this mechanism, water-soluble volatile organic compounds are absorbed into aerosol water, where they may react to form low-volatility secondary organic aerosol (SOA) material.[12-15] Aqueous aerosol SOA formation is likely to be significant in regions where aerosol sulfate concentrations, humidity, and volatile organic compound (VOC) concentrations tend to be high, such as the Southeastern United States.[16-18] Initial estimates of aaSOA in aerosols are as high as 0.9 $\mu g\ m^{-3}$ in rural environments and 0.12 $\mu g\ m^{-3}$ in urban environments.[14] Lin *et al.* observed that tracers of aqueous aerosol SOA formation from isoprene-derived epoxide precursors in Yorkville, GA comprised 7.9% of total organic aerosol mass.[18] Great uncertainty remains in the magnitude of the contribution of this pathway to total aerosol mass and its influence on aerosol physical properties.

Surface-active organic species partition to the gas–aerosol interface, where they may reduce the surface tension and act as a kinetic barrier to gas–aerosol mass transport.[9] Reduced surface tension at the moment of activation can reduce the energetic barrier to cloud droplet formation,[19] although the concomitant loss of solute molecules from the particle bulk due to partitioning to the surface may counteract the associated increase in cloud condensation nucleus (CCN) activity.[20] A surfactant film at the gas–aerosol interface can also reduce the rate of uptake of water and trace gases, impacting the rate of cloud droplet growth and aerosol heterogeneous chemistry.[9] Methylglyoxal, acetaldehyde, and formaldehyde, all species which have been suggested as aaSOA precursors, have been demonstrated to suppress surface tension in bulk solutions with compositions mimicking aqueous aerosols.[6,7,21,22] Small organic acids such as succinic acid and oxalic acid, typical aaSOA products, are also known to be surface-active.[9] Finally, aaSOA formation is thought to be an efficient pathway for the production of organosulfates, which may also be surface-active.[23]

The chemical composition of an aerosol particle can also influence its direct effects on climate. There is evidence from field studies of non-soot light-absorbing "brown carbon" in aerosols associated with biomass burning and pollution plumes.[24] The absorption of UV radiation by atmospheric brown carbon has the potential to perturb tropospheric photochemistry.[25] The formation mechanisms, prevalence, and physical properties of brown carbon in aerosols are poorly understood. Our recent laboratory studies showed that particle-phase chemical reactions between absorbed volatile organics and inorganic salts can lead to secondary products which absorb light in the UV and visible.[8,6,7,21] We characterized the mechanisms and kinetics of changes in the light absorption spectrum for bulk solutions containing glyoxal and/or methylglyoxal with $(NH_4)_2SO_4$. We found that, even in cross-reacting mixtures of glyoxal and methylglyoxal, light absorption by the reaction mixture could be modelled as a linear combination of the contribution of each reaction pathway in parallel.[7]

We recently developed a new photochemical box model, GAMMA (Gas–Aerosol Model for Mechanism Analysis),[14] that couples gas-phase chemistry with detailed aqueous phase chemistry, enabling speciated predictions of aqueous-phase SOA formation in atmospheric aerosols under ambient conditions. Besides allowing calculations of changes in aerosol mass due to aqueous SOA chemistry, these detailed speciated SOA calculations enable a bottom-up prediction of aerosol surface tension based on parameterizations for complex inorganic–organic mixtures. Surface-active species tracked by the model include several organic acids, dicarboxylic acids, aldehydes, methylglyoxal, and organosulfates. We have also developed a method to simulate the aaSOA brown carbon absorption spectrum based on predicted aerosol composition from GAMMA. Mechanisms for the dark formation of light-absorbing organics from methylglyoxal, glyoxal, and acetaldehyde which were been recently identified in laboratory studies are represented. Here we demonstrate the use of this module with GAMMA to evaluate the contribution of aqueous pathways of brown carbon formation to total observed BrC loadings.

2   Model description

GAMMA

GAMMA is a photochemical box that simulates coupled gas and aqueous-phase chemistry in order to predict aerosol composition under different atmospheric and laboratory test conditions. The model currently follows over 160 species participating in over 260 reactions. The species tracked by GAMMA were selected primarily due to high aqueous solubility (*i.e.*, high effective Henry's Law constant) and/or suspected contribution to SOA mass *via* aqueous processing. A complete, detailed description of the model, the parameters and initial conditions, and its application to modeling chemistry in deliquesced ammonium sulphate particles is available in McNeill *et al.* (2012).[14]

The general urban and remote conditions from CAPRAM 3.0 are used for initial gas-phase conditions and emissions and deposition rates.[26] Daytime conditions were simulated by scaling the photolysis rate constants semi-sinusoidally over 12 h. For night-time conditions, photolysis reactions and isoprene emissions were disabled. Only ammonium sulfate aerosols treated in the model; this aerosol type is a typical representative for deliquesced aerosols over continental regions. The size distribution from Whitby (1978) is used.[27] Aerosol liquid water content, aerosol sulphate/bisulphate concentrations, aerosol water fractions, and H^+ activity coefficients are calculated based on RH and aerosol pH using the Extended *AIM* Aerosol Thermodynamics Model, *E-AIM* (http://www.aim.env.uea.ac.uk/aim/aim.php).[28] For comparisons with the field data of Hecobian *et al.* (2010),[29] an aerosol sulphate mass of 4.0 ug m^{-3} was taken as representative for Yorkville, GA, USA[30] and 20 ug m^{-3} was used for S. DeKalb, GA, USA.[31]

Surface tension module

We predict particle surface tension (and thus the formation of organic surface films) based on the aerosol composition results from GAMMA following the approach outlined in Schwier *et al.* (2013).[22] In that study, we evaluated several approaches for their ability to model surface tension depression by reactive, complex mixtures of organic material in aqueous solutions containing 3.1 M ammonium sulphate. Other approaches for modelling surface tension in multi-component systems, including those based on thermodynamic approaches, are discussed in Schwier *et al.* (2013) and McNeill *et al.* (2013).[9,22] These approaches typically require additional information such as thermodynamic properties, adsorption isotherm data and surface pressures of surfactant species; such data is largely unavailable for many of the organic species tracked by GAMMA, making a semiempirical approach more practical.[32-36] The approach recommended by Schwier et al. (2013) for aqueous ammonium sulphate aerosols at RH ≤ 100% was a modified form of the Szyzskowski–Langmuir equation:[37]

$$\sigma = \sigma_{w}(T) - \sum_{i}\psi_i a_i T \ln(1 + b_i C) \tag{1}$$

Here, σ and σ_w are the surface tension of the solution with and without organics, respectively, T is T/K, C is the molality of the organic carbon (mol carbon (kg water)$^{-1}$), a_i and b_i are fit parameters of the ith organic compound (alone) in 3.1 M ammonium sulphate, and ψ_i is the molality fraction of compound i out of the total soluble carbon concentration in solution ($\psi_i = C_i C^{-1}$). Eqn (1) was first proposed by Henning et al. (2005),[38] but the implementation of the fit parameters obtained in concentrated salt solution was introduced by Schwier et al. (2013). We note that, since this approach uses surface-tension partitioning data measured for a bulk system (millimetre-sized droplets, in the case of pendant drop tensiometry[8,22]), and additional surface-bulk partitioning effects that may be important for submicron aerosol particles are not taken into account, we predict a lower bound of the surface tension of the aerosol.

Among the surface-active species for which parameterizations are available, those tracked by GAMMA include methylglyoxal, formaldehyde, oxalic acid, succinic acid, and acetaldehyde. Parameters a_i and b_i for those species are listed in Table 1. Other aaSOA species may be surface-active (see discussion below), but due to a lack of experimental data available for parameterization, they are not included in the model at this time.

Light absorption module

We previously observed and quantified changes in the absorption spectrum for reactive aqueous mixtures of glyoxal[8] and/or methyglyoxal[6,7] with ammonium sulphate. Nozière and coworkers made similar observations for acetaldehyde.[3] Changes in aerosol absorbance due to these complex reactive processes are tracked in GAMMA (see Supporting Information† for details). These reactions are not included in the mass balance for the precursor species since dark oligomerization is assumed to be accounted for via the use of the effective Henry's Law

Table 1　List of Szyzskowski–Langmuir constants for surface-active species traced by GAMMA, from Schwier et al. (2013)

Species Name	a_i	b_i
Methylglyoxal	0.0185	140
Formaldehyde	0.0119	50.23
Oxalic Acid	0.01297	1.0583
Succinic Acid	0.0349	0.74884
Acetaldehyde^a	0.02601	8.9282

^{a} Acetaldehyde treated as a proxy for all multicarbon (C > 2) aldehydes.

constants to describe uptake.[14] Other light-absorbing organic species tracked by GAMMA for which absorption data are available include oxalic acid and succinic acid,[39] as well as pyruvic acid.[40] Specific absorption data for each species are shown in the Supporting Information.†

Despite significant cross-reactions between methylglyoxal and glyoxal in aqueous ammonium sulphate solution, we showed that the increases in light absorption in a reactive mixture of the two could be modelled as if each precursor were reacting in parallel.[7] Therefore, in GAMMA we treat these processes as independent, and a composite UV-Visible absorption spectrum for the aerosol is calculated based on a linear combination of the contribution of each reactive pathway (in the case of glyoxal, methylglyoxal, and acetaldehyde) or individual organic species (for oxalic, succinic, and pyruvic acids) as follows:

$$A = \sum_i c_i \hat{A}_i \qquad (2)$$

where $\hat{A}_i$ is the specific absorbance of the ith species per molar concentration (L mol^{-1}) and c_i is the concentration of the ith species (mol L^{-1}).

We recently studied the aerosol-phase photochemical aging of light-absorbing material formed by methylglyoxal in aqueous ammonium sulphate solutions.[41] We observed that the lifetimes of these species are controlled by fast photolysis ($k^J = 0.0167\text{s}^{-1}$) which leads to volatilization. This process is represented in GAMMA as a first-order decrease in the absorption associated with the glyoxal, methylglyoxal, and acetaldehyde reactive pathways.

3   Results and discussion

36-hour simulations were performed, starting with sunrise at $t = 0$, for a variety of aerosol pH ($1 \leq \text{pH} \leq 4$) and RH (45%–80%) conditions, for urban (high-NO$_x$) and remote (low-NO$_x$) scenarios.

Aerosol surface tension

As we reported previously, aaSOA mass as predicted by GAMMA under simulated urban conditions is dominated by glyoxal uptake.[14] Glyoxal and its self-reaction products have not been found to depress surface tension in aqueous solutions.[8] Succinic acid was the largest contributor to surface tension depression, however total surface-active organic mass (*i.e.* the combined mass of the species listed in Table 1) did not exceed carbon concentrations of 2.2×10^{-3} kgC/kg water (<1% of total aaSOA mass). Predicted changes to net aerosol surface tension were negligible (<1% of the initial value).

Under low-NO$_x$ conditions, isoprene-derived epoxide (IEPOX)-derived tetrol and organosulfate species are predicted to comprise over 60% of the total formed SOA under all conditions tested (Fig. 1).[14] Succinic acid and the other species listed in Table 1 were again formed in quantities too small to significantly impact surface tension.

It is possible that other aaSOA species, the surface-bulk partitioning of which have not yet been characterized experimentally, may be surface active. Most individual species (*e.g.*, organic acids, photochemically-derived organosulfates) make a small mass contribution and therefore are unlikely to significantly impact the surface tension of the mixture. However, one class of unrepresented species

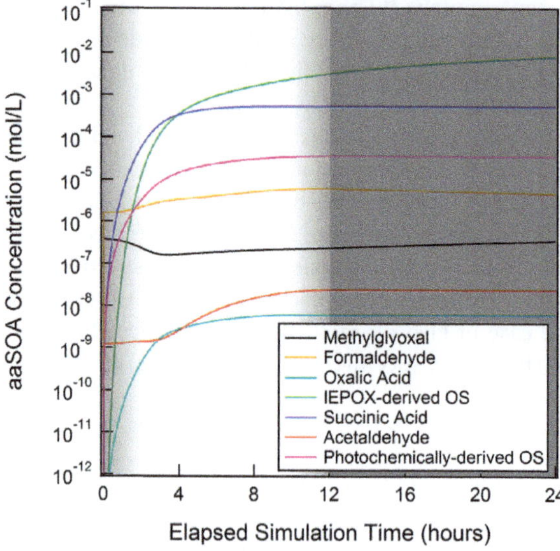

Fig. 1 Surface-active aqueous SOA concentrations *vs.* Time, pH 1 RH 65%, low-NO_x mode. Shaded regions indicate night-time conditions.

which makes a significant contribution to aaSOA mass, and is likely surface-active, is IEPOX-derived organosulfates. To our knowledge, the surface–bulk partitioning of IEPOX-derived organosulfates has not been characterized experimentally to date. In general, data for surface tension depression by organosulfates are sparse.[9] Noziére and coworkers studied surface tension depression by organosulfates generated by irradiating isoprene, methyvinylketone, methacrolein, or α-pinene in aqueous solutions containing sulphate salts.[23] For experiments initiated with 11 mM of isoprene, the minimum solution surface tension observed (after 200 min of reaction) was roughly 75% of the original value. Note that photochemically generated organosulfates of this type are formed in negligibly small quantities in our simulations (Fig. 1).[14] If we assume similar surface tension depression by IEPOX-derived organosulfate species, under conditions where their formation is predicted to be high, significant surface tension depression (∼18% at pH 1, 65% RH) may occur.

Aerosol light absorption

The computed composite absorption spectra were dominated by the reactive uptake of glyoxal under all conditions studied. Therefore the absorbance exhibited a diurnal cycle, increasing in the evening and night-time and decreasing to low levels in the early morning hours due to photolysis (Fig. 2). The baseline daytime absorption level was low, but increased with time due to the buildup of oxalic and pyruvic acids. While methylglyoxal oligomers are more strongly absorbing than those formed by glyoxal (Fig. S1†),[6,8,7] the relatively low effective Henry's Law constants for methylglyoxal and acetaldehyde made their contributions to the composite absorption negligible compared to glyoxal.

Fig. 3 shows the composite absorption spectra at maximum absorbance (24 h after the simulation was initiated) for low-NO_x and high-NO_x conditions. A peak is

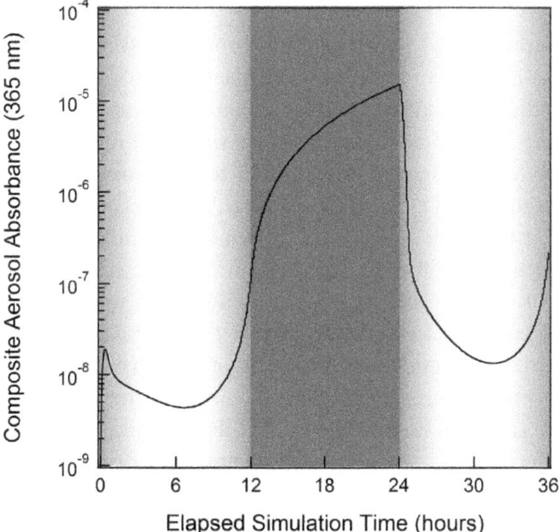

Fig. 2   Composite absorbance at 365nm *vs.* Time, 65% RH, pH 1, High NO_x mode. Shaded regions (12–24 h) indicate night-time conditions.

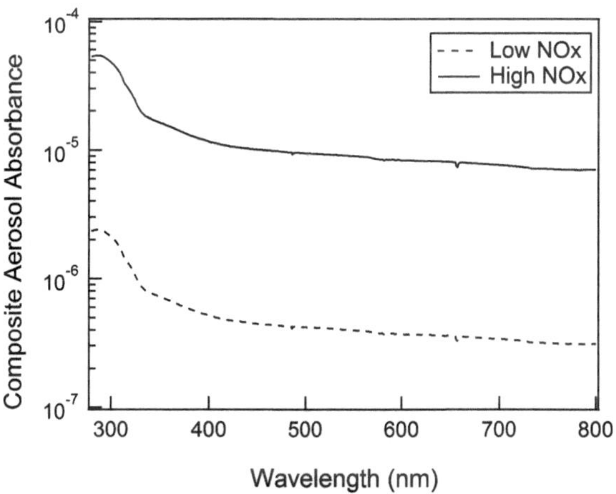

Fig. 3   Composite Aerosol Absorbance as predicted by GAMMA, pH 4, RH 65%.

apparent in the spectra at roughly 280 nm, consistent with light absorption by carbonyl-containing glyoxal oligomer species.[8] The maximum mass absorption efficiency (MAE) is predicted to be roughly 6.5 times higher under high-NO_x conditions ($0.13 \ \mathrm{m^2 gC^{-1}}$) than low-$NO_x$ conditions ($0.02 \ \mathrm{m^2 gC^{-1}}$) (see Supporting Information† for details of the MAE calculation). This is to be expected since, as previously mentioned, glyoxal uptake dominates SOA mass under high-NO_x conditions. Under low-NO_x conditions, IEPOX-derived species dominate total SOA. Their potential contribution to aerosol absorption is unknown, and therefore is not included in our calculations.

Model sensitivity to RH, temperature, and particle size

As mentioned previously, photolysis rate constants were modulated semi-sinu-soidally over 12-hour daytime periods and disabled during night-time conditions. Relative humidity, aerosol pH, aerosol diameter, and temperature were kept constant during simulations under both daytime and night-time conditions. An aerosol particle experiences fluctuations in temperature and relative humidity, and may undergo several cloud cycles in a day.[12] While GAMMA does not simulate cloud cycling at this time, the impact of relative humidity on aaSOA formation was quantified in McNeill *et al.* (2012). In order to assess the sensitivity of the model to a 10 K change in temperature, simulations were run at 288.15 K and 308.15 K, with constant RH. We observed a 13% decrease (at 288.15 K) or 23% increase (at 308.15 K) in generated SOA mass after 12 h of daytime simulation. We note that the temperature dependence of many of the rate and mass transfer parameters in GAMMA is not known. While the potential deviations due to temperature fluctuations are not insignificant, they do not change the conclusions of this study.

In order to estimate the effect of neglecting the increase in particle size associated with SOA formation on the mass transfer calculations, the predicted total SOA mass after 24 h was added to the total starting mass of the aerosol, and the starting and ending mass transfer coefficients were compared. Assuming pH 1, 45% RH conditions and an aaSOA density matching that of the initial aerosol (1480 kg m^{-3}), total aerosol diameter increased by 6%, resulting in a 7% decrease in the mass transfer coefficient.

4   Conclusions and atmospheric implications

We have performed detailed simulations of organic chemistry in the aerosol aqueous phase in order to predict changes in aerosol surface tension and light absorption due to the uptake and reaction of water-soluble volatile organic compounds.

We found that, under all conditions studied, the bulk concentrations of species known to depress surface tension (methylglyoxal, formaldehyde, acetal-dehyde, succinic acid, oxalic acid, and photochemically-generated organo-sulfates) are so small that their predicted effects on aerosol surface tension are negligible. Based on our recent aerosol chamber studies of the effects of gas-phase methylglyoxal and acetaldehyde on the CCN activity of aqueous ammo-nium sulphate particles, adsorption of these species at the gas–aerosol interface may be sufficient to depress aerosol surface tension and therefore affect aerosol CCN activity.[10] Relatively little is known about this phenomenon at this time; additional experiments are needed so that it may be parameterized for the model.

We predict significant brown carbon formation *via* aqueous aerosol-phase pathways, especially under urban conditions. Our predictions of a build-up of absorbing species throughout the night, followed by a decrease in absorption in the early morning hours due to photobleaching,[41] is consistent with field obser-vations of high brown carbon absorption in the early morning.[42] In Fig. 4, we compare the wavelength dependence of our predicted composite absorption spectra to the measurements of Hecobian *et al.* (2010), who performed UV-Vis spectrophotometry on aqueous filter extracts.[29] Note that the data from Hecobian

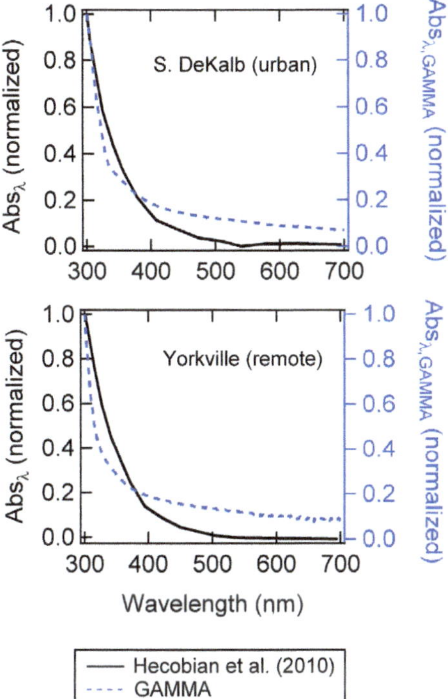

Fig. 4 Wavelength dependence of composite absorption spectra predicted by GAMMA (dashed line, right axis) compared to measurements of Hecobian *et al.* (2010) (solid line, left axis) for aqueous extracts of aerosol filter samples collected during periods with low biomass burning influence at representative urban and remote locations in Georgia, USA.

et al. (2010) were reported as absorption per volume of air sampled, and particle loading data for the sample period are not available, so normalized spectra are shown. The featureless spectra with absorption increasing with decreasing wavelength reported by that group is reproduced well by our model. They reported an MAE of 0.31 m^2 g C^{-1} at the S. Dekalb (urban) site, similar in magnitude to our predicted maximum MAE (0.12 m^2 g C^{-1}) under high-NO_x conditions. However, they reported a similar value of 0.29 m^2 g C^{-1} at the Yorkville (rural) site, whereas we predict significantly less absorbance under low-NO_x conditions. We note that our model does not account for primary sources of brown carbon such as biomass burning.[24] Furthermore, other reactive sources of brown carbon exist other than aqueous aerosol chemistry, such as reactions of condensationally formed monoterpene SOA with nitrogen-containing compounds such as ammonia, ammonium sulphate, or amines,[43,44] or the nitrate functionalization of aromatic species.[25,45,46] These processes are expected to occur in the aerosol organic phase rather than in the aqueous phase, and thus are not represented in GAMMA at this time. Additional light-absorbing species could be formed in the aerosol aqueous phase which are not accounted for by GAMMA. However, we note that any effects of unresolved species originating from reactions of carbonyl-containing VOCs were likely captured by our laboratory studies of these systems, upon which this work is based.[6–8,21,22] The surface tension and light absorption measurements described in those studies were performed for the same (or identically prepared) reactive solutions containing controlled initial concentrations of methylglyoxal, glyoxal, formaldehyde, and acetaldehyde. The final compositions of those solutions were not fully speciated, however the aggregate effects of the product populations on light absorption and surface tension depression was captured in the parameterizations used here for modeling.

Since IEPOX-derived tetrol and organosulfate species are predicted to contribute significantly to aqueous aerosol mass under low-NO_x conditions, experimental data are needed so that we can quantify their potential impacts on aerosol surface tension and light absorption.

References

1 J. L. Jimenez, M. R. Canagaratna, N. M. Donahue, A. S. H. Prevot, Q. Zhang, J. H. Kroll, P. F. DeCarlo, J. D. Allan, H. Coe, N. L. Ng, A. C. Aiken, K. S. Docherty, I. M. Ulbrich, A. P. Grieshop, A. L. Robinson, J. Duplissy, J. D. Smith, K. R. Wilson, V. A. Lanz, C. Hueglin, Y. L. Sun, J. Tian, A. Laaksonen, T. Raatikainen, J. Rautiainen, P. Vaattovaara, M. Ehn, M. Kulmala, J. M. Tomlinson, D. R. Collins, M. J. Cubison, E. J. Dunlea, J. A. Huffman, T. B. Onasch, M. R. Alfarra, P. I. Williams, K. Bower, Y. Kondo, J. Schneider, F. Drewnick, S. Borrmann, S. Weimer, K. Demerjian, D. Salcedo, L. Cottrell, R. Griffin, A. Takami, T. Miyoshi, S. Hatakeyama, A. Shimono, J. Y. Sun, Y. M. Zhang, K. Dzepina, J. R. Kimmel, D. Sueper, J. T. Jayne, S. C. Herndon, A. M. Trimborn, L. R. Williams, E. C. Wood, A. M. Middlebrook, C. E. Kolb, U. Baltensperger and D. R. Worsnop, *Science*, 2009, **326**, 1525–1529.
2 B. Nozière and W. Esteve, *Atmos. Environ.*, 2007, **41**, 1150–1163.
3 B. Nozière, P. Dziedzic and A. Córdova, *Phys. Chem. Chem. Phys.*, 2010, **12**, 3864–72.
4 B. Nozière, P. Dziedzic and A. Córdova, *Geophys. Res. Lett.*, 2007, **34**, L21812.
5 B. Nozière and A. Córdova, *J. Phys. Chem. A*, 2008, **112**, 2827–2837.
6 N. Sareen, A. N. Schwier, E. L. Shapiro, D. Mitroo and V. F. McNeill, *Atmos. Chem. Phys.*, 2010, **10**, 997–1016.
7 A. N. Schwier, N. Sareen, D. Mitroo, E. L. Shapiro and V. F. McNeill, *Environ. Sci. Technol.*, 2010, **44**, 6174–6182.
8 E. L. Shapiro, J. Szprengiel, N. Sareen, C. N. Jen, M. R. Giordano and V. F. McNeill, *Atmos. Chem. Phys.*, 2009, **9**, 2289–2300.
9 V. F. McNeill, N. Sareen and A. N. Schwier, *Top. Curr. Chem.*, 2013, DOI: 10.1007/128_2012_404.
10 N. Sareen, A. N. Schwier, T. L. Lathem, A. Nenes and V. F. McNeill, *Proc. Natl. Acad. Sci. U. S. A.*, 2013, **110**, 2723–2728.
11 A. Schwier, D. Mitroo and V. F. McNeill, *Atmos. Environ.*, 2012, **54**, 490–495.
12 B. Ervens, B. J. Turpin and R. J. Weber, *Atmos. Chem. Phys.*, 2011, **11**, 11069–11102.
13 B. Ervens and R. Volkamer, *Atmos. Chem. Phys.*, 2010, **10**, 8219–8244.
14 V. F. McNeill, J. L. Woo, D. D. Kim, A. N. Schwier, N. J. Wannell, A. J. Sumner and J. M. Barakat, *Environ. Sci. Technol.*, 2012, **46**, 8075–8081.
15 Y. Tan, A. G. Carlton, S. P. Seitzinger and B. J. Turpin, *Atmos. Environ.*, 2010, **44**, 5218–5226.
16 M. N. Chan, J. D. Surratt, M. Claeys, E. S. Edgerton, R. L. Tanner, S. L. Shaw, M. Zheng, E. M. Knipping, N. C. Eddingsaas, P. O. Wennberg and J. H. Seinfeld, *Environ. Sci. Technol.*, 2010, **44**, 4590–4596.
17 C. J. Hennigan, M. H. Bergin, A. G. Russell, A. Nenes and R. J. Weber, *Atmos. Chem. Phys.*, 2009, **9**, 3613–3628.
18 Y.-H. Lin, Z. Zhang, K. S. Docherty, H. Zhang, S. H. Budisulistiorini, C. L. Rubitschun, S. L. Shaw, E. M. Knipping, E. S. Edgerton, T. E. Kleindienst, A. Gold and J. D. Surratt, *Environ. Sci. Technol.*, 2012, **46**, 250–258.
19 H. Köhler, *Trans. Faraday Soc.*, 1936, **32**, 1152–1161.
20 N. L. Prisle, A. Asmi, D. Topping, A.-I. Partanen, S. Romakkaniemi, M. Dal Maso, M. Kulmala, A. Laaksonen, K. E. J. Lehtinen, G. McFiggins and H. Kokkola, *Geophys. Res. Lett.*, 2012, **39**, L05802.
21 Z. Li, A. N. Schwier, N. Sareen and V. F. McNeill, *Atmos. Chem. Phys.*, 2011, **11**, 11617–11629.
22 A. N. Schwier, G. A. Viglione, Z. Li and V. F. McNeill, *Atmos. Chem. Phys. Discuss.*, 2013, **13**, 549–580.

23 B. Nozière, S. Ekström, T. Alsberg and S. Holmström, *Geophys. Res. Lett.*, 2010, **37**, L05806.
24 M. O. Andreae and A. Gelencser, *Atmos. Chem. Phys.*, 2006, **6**, 3131–3148.
25 M. Z. Jacobson, *J. Geophys. Res.*, 1999, **104**, 3527–3542.
26 H. Herrmann, A. Tilgner, P. Barzaghi, Z. Majdik, S. Gligorovski, L. Poulain and A. Monod, *Atmos. Environ.*, 2005, **39**, 4351–4363.
27 K. Whitby, *Atmos. Environ.*, 1978, **12**, 135–159.
28 A. S. Wexler and S. L. Clegg, *J. Geophys. Res.*, 2002, **107**, DOI: 10.1029/2001JD000451, ACH 14-1 – ACH 14-14.
29 A. Hecobian, X. Zhang, M. Zheng, N. Frank, E. S. Edgerton and R. J. Weber, *Atmos. Chem. Phys.*, 2010, **10**, 5965–5977.
30 R. L. Tanner, K. J. Olszyna, E. S. Edgerton, E. Knipping and S. L. Shaw, *Atmos. Environ.*, 2009, **43**, 3440–3444.
31 J. L. Jimenez, J. T. Jayne, Q. Shi, C. E. Kolb, D. R. Worsnop, I. Yourshaw, J. H. Seinfeld, R. C. Flagan, X. Zhang, K. A. Smith, J. W. Morris and P. Davidovits, *J. Geophys. Res.*, 2003, **108**, DOI: 10.1029/2001JD001213.
32 Z. Li and B. C. Y. Lu, *Chem. Eng. Sci.*, 2001, **56**, 2879–2888.
33 V. B. Fainerman and R. Miller, *J. Phys. Chem. B*, 2001, **105**(46), 11432–11438.
34 V. B. Fainerman, R. Miller and E. V. Aksenenko, *Adv. Colloid Interface Sci.*, 2002, **96**(1–3), 339–359.
35 D. O. Topping, G. B. McFiggans, G. Kiss, Z. Varga, M. C. Facchini, S. Decesari and M. Mircea, *Atmos. Chem. Phys.*, 2007, **7**, 2371–2398.
36 A. M. Booth, D. O. Topping, G. McFiggans and C. J. Percival, *Phys. Chem. Chem. Phys.*, 2009, **11**(36), 8021–8028.
37 I. Langmuir, *Proc. Natl. Acad. Sci. U. S. A.*, 1917, **3**, 251–257.
38 S. Henning, T. Rosenørn, B. D'Anna, A. A. Gola, B. Svenningsson and M. Bilde, *Atmos. Chem. Phys.*, 2005, **5**, 575–582.
39 C. E. Lund Myhre and C. J. Nielsen, *Atmos. Chem. Phys.*, 2004, **4**, 1759–1769.
40 A. G. Rincón, M. I. Guzmán, M. R. Hoffmann and A. J. Colussi, *J. Phys. Chem. A*, 2009, **113**, 10512–10520.
41 N. Sareen, S. G. Moussa and V. F. McNeill, *J. Phys. Chem. A*, 2013, **117**(14), 2987–2996.
42 G. Paredes-Miranda, W. P. Arnott, J. L. Jimenez, A. C. Aiken, J. S. Gaffney and N. A. Marley, *Atmos. Chem. Phys.*, 2009, **9**, 3721–3730.
43 T. B. Nguyen, P. B. Lee, K. M. Updyke, D. L. Bones, J. Laskin, A. Laskin and S. A. Nizkorodov, *J. Geophys. Res.*, 2012, **117**, D01207.
44 D. L. Bones, D. K. Henricksen, S. A. Mang, M. Gonsior, A. P. Bateman, T. B. Nguyen, W. J. Cooper and S. A. Nizkorodov, *J. Geophys. Res.*, 2010, **115**, D05203.
45 J. N. Pitts, K. A. Van Cauwenberghe, D. Grosjean, J. P. Schmid, D. R. Fitz, W. L. Belser Jr., G. B. Knudson and P. M. Hynds, *Science*, 1978, **202**, 515–519.
46 X. Zhang, Y.-H. Lin, J. D. Surratt, P. Zotter, A. S. H. Prévôt and R. J. Weber, *Geophys. Res. Lett.*, 2011, **38**, L21810.

Faraday Discussions

RSCPublishing

PAPER

The effects of aircraft on climate and pollution. Part II: 20-year impacts of exhaust from all commercial aircraft worldwide treated individually at the subgrid scale

M. Z. Jacobson,[*a] J. T. Wilkerson,[a] A. D. Naiman[b] and S. K. Lele[c]

Received 8th March 2013, Accepted 16th April 2013
DOI: 10.1039/c3fd00034f

This study examines the 20-year impacts of emissions from all commercial aircraft flights worldwide on climate, cloudiness, and atmospheric composition. Aircraft emissions from each individual flight worldwide were modeled to evolve from the subgrid to grid scale with the global model described and evaluated in Part I of this study. Simulations with and without aircraft emissions were run for 20 years. Aircraft emissions were found to be responsible for ~6% of Arctic surface global warming to date, ~1.3% of total surface global warming, and ~4% of global upper tropospheric warming. Arctic warming due to aircraft slightly decreased Arctic sea ice area. Longer simulations should result in more warming due to the further increase in CO_2. Aircraft increased atmospheric stability below cruise altitude and decreased it above cruise altitude. The increase in stability decreased cumulus convection in favor of increased stratiform cloudiness. Aircraft increased total cloud fraction on average. Aircraft increased surface and upper tropospheric ozone by ~0.4% and ~2.5%, respectively and surface and upper-tropospheric peroxyacetyl nitrate (PAN) by ~0.1% and ~5%, respectively. Aircraft emissions increased tropospheric OH, decreasing column CO and CH_4 by ~1.7% and ~0.9%, respectively. Aircraft emissions increased human mortality worldwide by ~620 (−240 to 4770) deaths per year, with half due to ozone and the rest to particulate matter less than 2.5 micrometers in diameter ($PM_{2.5}$).

1   Introduction

This is a study to quantify the effects of all major components of aircraft exhaust and their resulting contrails, on global and regional climate, cloudiness, and atmospheric composition. The study also aims to examine the contribution of aircraft to Arctic warming. Previous studies of the effects of aircraft exhaust and

[a]Dept. of Civil and Environmental Engineering, Stanford University, Stanford, CA, USA. E-mail: Jacobson@stanford.edu

[b]Exponent Engineering and Scientific Consulting, CA, USA

[c]Dept. of Aeronautics and Astronautics, Stanford University, Stanford, CA, USA

the resulting contrails and gases on climate have been limited primarily to the global-scale radiative forcing due to the exhaust.[1-10] However, direct radiative forcing, which most studies have used to estimate climate effects, is not necessarily proportional to the climate response of aerosols, which is why recent studies have switched to using adjusted forcings and other metrics.[11] Because adjusted forcing itself is an approximate parameter since it requires sea-surface temperature to be held constant, and because it does provide information about air quality, simulating the full climate and air quality response of aircraft without approximation may be the cleanest method of estimating the climate and health effects of aircraft. This type of calculation is performed here.

Previously, Ponater *et al.*[12] and Rind *et al.*[13] examined the global climate response of aircraft water vapor or aircraft-enhanced cirrus. Jacobson *et al.*[14] examined the climate response of rerouting cross-polar flights around the Arctic Circle while treating each flight independently at the subgrid scale. Fromming *et al.*[15] examined the surface temperature response to changing the altitude of flight emissions. Previous studies have not examined the global climate response of major emissions from all commercial aircraft worldwide, particularly when the emissions from each aircraft are tracked individually.

This study uses a global model modified to treat the evolution of individual subgrid aircraft exhaust plumes to simulate the climate and pollution effects of contemporary emissions from all commercial flights worldwide. Short-term (20-year) effects are examined since this is the time scale of relevance to the Arctic sea ice, which is disappearing rapidly, and due to the significant computer time required to run the simulations here.

2   Model description

The model used for this study, GATOR-GCMOM, is described and evaluated with respect to aircraft-relevant parameters in Part I of this study, Jacobson *et al.*,[16] and in Whitt *et al.*[17] and with respect to climate response in Jacobson *et al.*[18] Briefly, it is a one-way nested global-through-local Gas, Aerosol, Transport, Radiation, General Circulation, Mesoscale, and Ocean Model that simulates climate, weather, and air pollution. It treats emissions; gas photochemistry; size- and composition-resolved aerosol and hydrometeor microphysics and chemistry; size- and composition-resolved aerosol-hydrometeor interactions; subgrid cumulus cloud thermodynamics; grid-scale stratiform thermodynamics; spectral radiative transfer for heating rates and photolysis; dynamical meteorology; 2-D ocean dynamics; 3-D ocean diffusion; 3-D ocean chemistry; ocean-atmosphere exchange; and ice, snow, land surface processes, and subgrid contrail formation and aerosol evolution from each individual flight worldwide.

With respect to aircraft, the model accounts for the microphysical evolution of contrails from all commercial aircraft worldwide at the subgrid scale (at their actual size).[16] Briefly, subgrid processes treated include (1) spreading and shearing of individual aircraft exhaust plumes,[19] (2) calculating a unique super-saturation for each plume based on water vapor and aerosol particles emitted during each flight added to existing water vapor and particles, (3) calculating time-dependent, discrete size-resolved ice deposition and condensation followed by freezing onto size-resolved aerosols, (4) calculating size-resolved coagulation to create and grow linear contrails at the subgrid scale, (5) solving for the discrete size-resolved evolution of grid-scale cirrus and other clouds from the remnants of subgrid linear contrails, and (6) determining radiative effects of subgrid linear contrails as a function of ice crystal size and composition and contrail plume shape and accounting for random overlap of contrails and clouds.

At the grid scale, three discrete (multiple size bin) aerosol size distributions and three discrete hydrometeor distributions, each with multiple components per bin, were treated.[14] The three aerosol distributions were an emitted fossil-fuel soot (EFFS) distribution, an emitted combined biofuel-soot and biomass-burning-soot (BFBB) distribution, and an ultimate internally-mixed (UIM) distribution. Each size distribution contained 14 size bins (2 nm–50 μm diameter). The three hydrometeor distributions included liquid, ice, and graupel, each with 30 size bins (0.5 μm to 8 mm in diameter). Each size bin of the EFFS distribution contained black carbon (BC), weakly-to-moderately-absorbing primary organic matter (POM), secondary organic matter (SOM), hydrated liquid water, $H_2SO_4(aq)$, HSO_4^-, SO_4^{2-}, NO_3^-, Cl^-, H^+, NH_4^+, $NH_4NO_3(s)$, and $(NH_4)_2SO_4(s)$. Each size bin of the BFBB distribution contained these same components plus tar balls, Na^+, K^+, Ca^{2+}, and Mg^{2+}. Each size bin of the UIM distribution contained the same components as the EFFS and BFBB distributions plus soil dust, pollen, spores, and bacteria. Each size bin of each hydrometeor distribution contained the same components as in all three aerosol distributions plus condensed liquid water or deposited ice. Gases, such as HNO_3, HCl, NH_3, H_2SO_4, and organics could condense onto or dissolve into grid-scale EFFS, BFBB, and UIM aerosol particles and dissolve within liquid hydrometeor particles or chemically react on ice and graupel hydrometeor particle surfaces.

Discrete size-resolved EFFS, BFBB, and UIM aerosol particles and liquid, ice, and graupel hydrometeor particles and their components could coagulate among each other as a function of size. Thus, grid-scale aerosol particles and their components were tracked within hydrometeor particles through cloud formation and precipitation.

BC, brown carbon (BrC), and soil dust absorption in clouds was treated with the dynamic effective medium approximation (DEMA), which accounted for aerosol particles distributed randomly throughout each hydrometeor particle in the model.[14] Inorganic gases, such as HNO_3, HCl, NH_3, and H_2SO_4, and organic gases could condense onto or dissolve into grid-scale EFFS, BFBB, and UIM aerosol particles. Liquid water could also hydrate to soluble species and ions could crystallize to solid within such particles. EFFS, BFBB, and UIM aerosol particles served as nuclei for new hydrometeor particles. In sum, the model treated indirect, semi-direct, and cloud absorption effects of aerosols on contrails and all other cloud types as a function of size, composition, and spectral wavelength.[14]

Flight-by-flight aircraft emissions used for this study were for 2006.[20] Size-resolved particle components from aircraft (BC, POM, $S(VI) = H_2SO_4(aq) + HSO_4^- + SO_4^{2-}$) and water vapor were emitted into subgrid plumes. Size- and composition-resolved contrail particles formed and grew onto the subgrid size-resolved aerosol particles by nucleation, water deposition, and coagulation. CO, CO_2, SO_2, NO_x, and speciated total hydrocarbons (THCs) from aircraft were emitted directly to the grid scale. When contrails in subgrid plumes dissipated due to plume expansion and dilution and ice sublimation, then water vapor and size- and composition-resolved aerosol cores containing BC, POM, and S(VI) from them were released from each plume to the grid scale in the EFFS aerosol distribution, where the particles affected clouds and radiation. The grid-scale EFFS distribution also contained emissions from other fossil-fuel sources.

3   Simulations

A pair of 20-year global (non-nested) simulations was first run. One simulation included all gas and particle aircraft emissions (baseline simulation), and the other (sensitivity simulation) did not. Six additional simulations, each with a random initial perturbation, were run to test the statistical significance of results through a two-sample t-test.

The horizontal grid scale resolution was 4°-SN × 5°-WE. The model included 68 vertical sigma-pressure layers from the ground to 0.219 hPa (≈ 60 km), with 15 layers from 0–1 km and 500 m resolution from 1–21 km. The center of the lowest model layer was 15 m above ground. Since all flights in the Enhanced Traffic Management System (ETMS) occur at 41 000 ft (12.5 km) or below, the 500 m resolution up to 21 km was sufficient.

All simulations were initialized with 1° × 1° reanalysis fields for 12 GMT of the start date of simulation.[21] Aircraft emissions for 2006 were repeated each year. Following initialization, the model was entirely prognostic (no data assimilation or nudging).

Although complete temperature responses, particularly of CO_2, require much more than 20 years, the purpose of this study is to examine the estimated impact of aircraft over the short term since that is relevant to the loss of Arctic sea ice, which may disappear over this time frame.

4   Results

Fig. 1–4 provide simulation-averaged difference (with minus without aircraft) information of the modeled aircraft effects on climate and atmospheric composition. A simulation-averaged value is one in which values of the parameter are summed over all time steps and the sum is divided by the number of time steps. In all cases, results are for grid-scale parameters or, in the case of contrail water, combined over all subgrid contrails.

4.1   Effects on atmospheric gas and aerosol composition

Aircraft emissions (at the subgrid scale) peak vertically between 10–12 km (265-194 hPa) above sea level.[20] This was the height range in which changes in modeled aerosol particle number, SO_2, NO_2, and contrail water peaked (Fig. 1a–d). Horizontally, such peaks occurred from 30–90 N.

Above 9 km (300 hPa), aircraft nearly doubled the concentration of BC in the EFFS aerosol size distribution (Fig. 1e) and increased BC by about 7% in the UIM distribution (Fig. 1f). BC emissions from aircraft represent a small percent (0.21%) of globally-emitted BC from all submicron fossil-fuel BC sources and even less (0.09%) from submicron fossil-fuel plus biofuel plus biomass-burning sources. Thus aircraft are a proportionately larger relative contributor to upper tropospheric and stratospheric BC than are other sources of BC. The result makes sense because most surface-emitted BC is removed with a lifetime of a few days[14] and the upper-troposphere/lower stratosphere (UTLS) is stable,[17] so surface BC

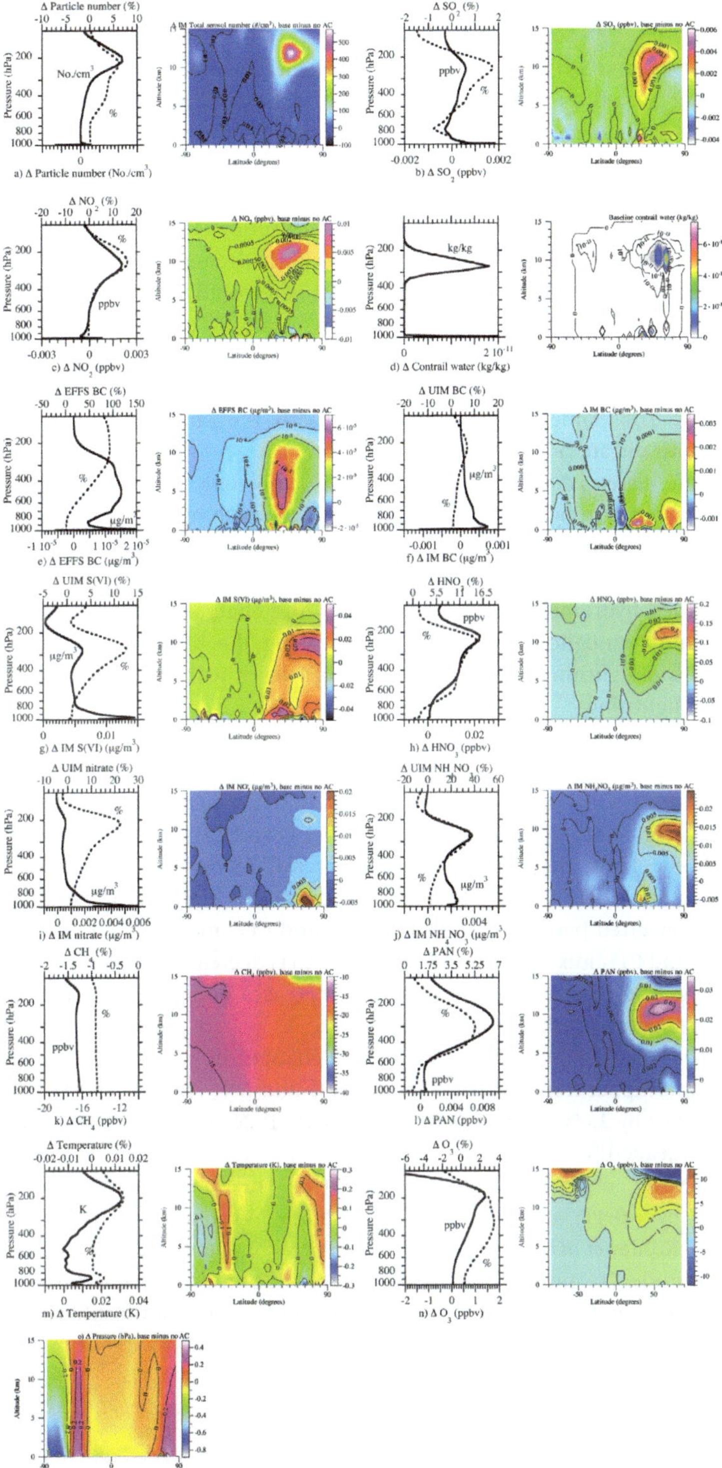

a) Δ Particle number (No./cm³)

b) Δ SO₂ (ppbv)

c) Δ NO₂ (ppbv)

d) Δ Contrail water (kg/kg)

e) Δ EFFS BC (μg/m³)

f) Δ IM BC (μg/m³)

g) Δ IM S(VI) (μg/m³)

h) Δ HNO₃ (ppbv)

i) Δ IM nitrate (μg/m³)

j) Δ IM NH₄NO₃ (μg/m³)

k) Δ CH₄ (ppbv)

l) Δ PAN (ppbv)

m) Δ Temperature (K)

n) Δ O₃ (ppbv)

o) Δ Pressure (hPa), base minus no AC

has difficulty rising to that height, whereas aircraft BC is emitted directly to the UTLS.

BC, POM, and S(VI) emitted into each aircraft exhaust plume either stayed as aerosol or served as nuclei for contrail formation. When the exhaust plume expanded sufficiently so that the contrail disappeared, the aerosol material was added to the grid-scale EFFS aerosol size distribution. Such aged and partly internally-mixed particles eventually coagulated with aerosol particles from other sources in the EFFS, BFBB, or UIM aerosol distributions to become UIM particles or larger EFFS particles, or they coagulated with hydrometeor particles or served as cloud deposition or condensation nuclei. EFFS coagulation with other aerosol particles resulted in an increase in internally-mixed BC and S(VI), particularly at cruise altitude from 30–90 N by about 7% and 12%, respectively (Fig. 1f and 1g). The increase in NO_x due to aircraft increased nitric acid available at cruise altitude (Fig. 1h). Some nitrate ion reacted with the ammonium ion to form ammonium nitrate (Fig. 1j) at cruise altitude.

Aircraft increased upper-tropospheric NO_x, not only by emitting anthropogenic NO_x but also by increasing lightning production of NO_x (Fig. 2a). Aircraft increased the number of liquid and ice cloud nuclei (*e.g.*, Fig. 1a), increasing the number of ice cloud particles in convective clouds, increasing the number of size-resolved hydrometeor particle collisions and charge-separating ice–ice, ice–liquid, and liquid–liquid bounceoffs, increasing the buildup of charges and probability of lightning. Jacobson and Streets[22] describe the equations describing modeled lightning.

The increases in aircraft-emitted H_2O (which reacts with $O(^1D)$ to form 2OH), NO (which reacts with HO_2 to form OH), HCHO, and CH_3CHO (which photolyze to OH) all increased tropospheric OH (Fig. 2b). Enhanced OH-oxidized CO (Fig. 2c) and CH_4 (Fig. 1k) to CO_2, decreasing column CO and CH_4 by ~1.7% and ~0.9%, respectively. Thus, despite the fact that aircraft increased CO emissions, CO mixing ratios decreased due to aircraft because the higher OH resulting from aircraft converted both emitted and background CO (and CH_4) to CO_2, reducing both CO and CH_4 mixing ratios. Most CO and CH_4 decreases were in the Northern Hemisphere in locations where most OH increases occurred. The reduction in CH_4 due to aircraft exhaust is well known.[10]

Higher NO_x (Fig. 1c) coupled with small changes in acetaldeheyde increased surface and upper-tropospheric PAN by 0.1% and 5%, respectively (Fig. 2t) and column PAN by 2.6%. Increases in tropospheric PAN peaked near where aircraft contrails peaked (Fig. 1h).

Higher NO_x, organics, and temperatures (Fig. 1m) due to aircraft caused O_3 to increase by ~0.4% at the surface and ~2.5% (~2.5 ppbv) at cruise-altitude (Fig. 1n). Köhler *et al.*[23] and Lee *et al.*[10] similarly concluded that aircraft increase cruise-altitude O_3. Immediately above cruise altitude, ozone decreased in the presence of aircraft because aircraft increased mid-to-lower stratospheric stability

Fig. 1 Globally- and simulation-averaged vertical profile differences and zonally- and simulation-averaged differences in several modeled parameters between the simulations with *versus* without aircraft. All parameters shown are grid-scale parameters, except that contrail water is summed over all subgrid contrails. For reference, altitudes are roughly correlated with the following pressures (5 km: 540.5 hPa; 10 km: 265 hPa; 15 km: 121.1 hPa).

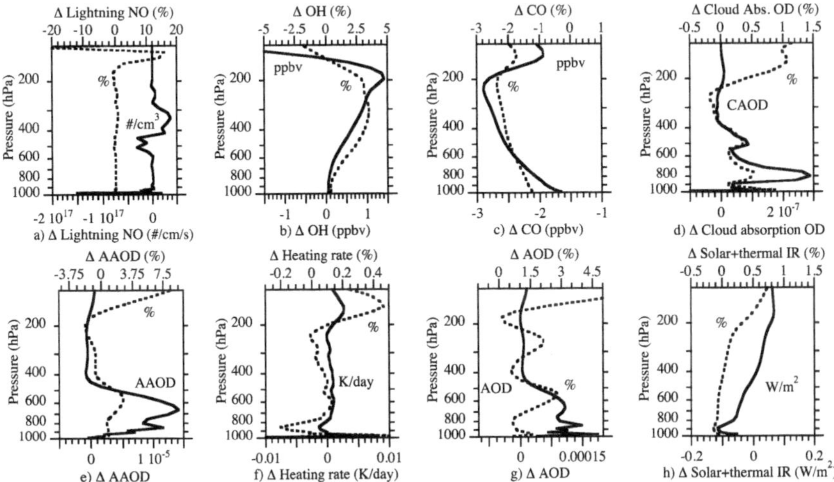

Fig. 2 Same as Fig. 1, but for globally- and simulation-averaged vertical profile differences for some additional parameters.

(not seen in Fig. 1m), reducing the downward transport of ozone from the ozone layer to the UTLS region.

Aircraft exhaust increased surface O_3 by 0.05 ppbv worldwide on average and by a peak of > 1 ppbv (Fig. 3a) by increasing emissions of ozone precursors. The estimated premature mortality rates due to enhanced 8-hr averaged ozone were calculated as ~310 (160 to 470) per year. Dry $PM_{2.5}$ and total $PM_{2.5}$ increased, in the global average, by 0.083 $\mu g\ m^{-3}$ and 0.54 $\mu g\ m^{-3}$ (Fig. 3d, e). Estimated premature mortalities, based on dry $PM_{2.5}$ are ~310 (−400 to 4300) per year. These mortality numbers were obtained during the model simulation by combining current mixing ratios or concentrations predicted by the model each time step in each model grid cell with population data for each cell and relative risks, all in a health effects equation, as described previously.[24]

4.2   Effects on contrails, clouds, and precipitation

Aircraft exhaust affected contrail, cumuliform, and stratiform cloud properties through both microphysical (indirect) and radiative effects of aerosols on clouds and through feedbacks of exhaust to temperatures, relative humidities, and winds. The Northern Hemisphere (NH) contrail cloud fraction change (+0.0030) due to aircraft exceeded that in the Southern Hemisphere (SH) (+0.000062) by a factor of 48 (Fig. 3f). The NH total cloud fraction increase (+0.12%) exceeded that in the SH (+0.018%) by a factor of 7 (Fig. 3g). Aircraft increased grid-scale in-cloud ice number concentration (Fig. 3h) due to the increased number of aerosol particles from aircraft that can serve as ice nuclei.

Cloud absorption optical depth (CAOD) was calculated by tracking BC, BrC, and soildust in each size bin of liquid, ice, and graupel hydrometeor particles and applying the iterative dynamic effective medium approximation (DEMA), assuming that nucleated and scavenged aerosol particles were randomly distributed in the hydrometeor particles.[14] Aircraft increase CAOD by up to 1% (Fig. 2d) and AAOD by up to 7.5% (Fig. 2e) in the global average above 11.8 km

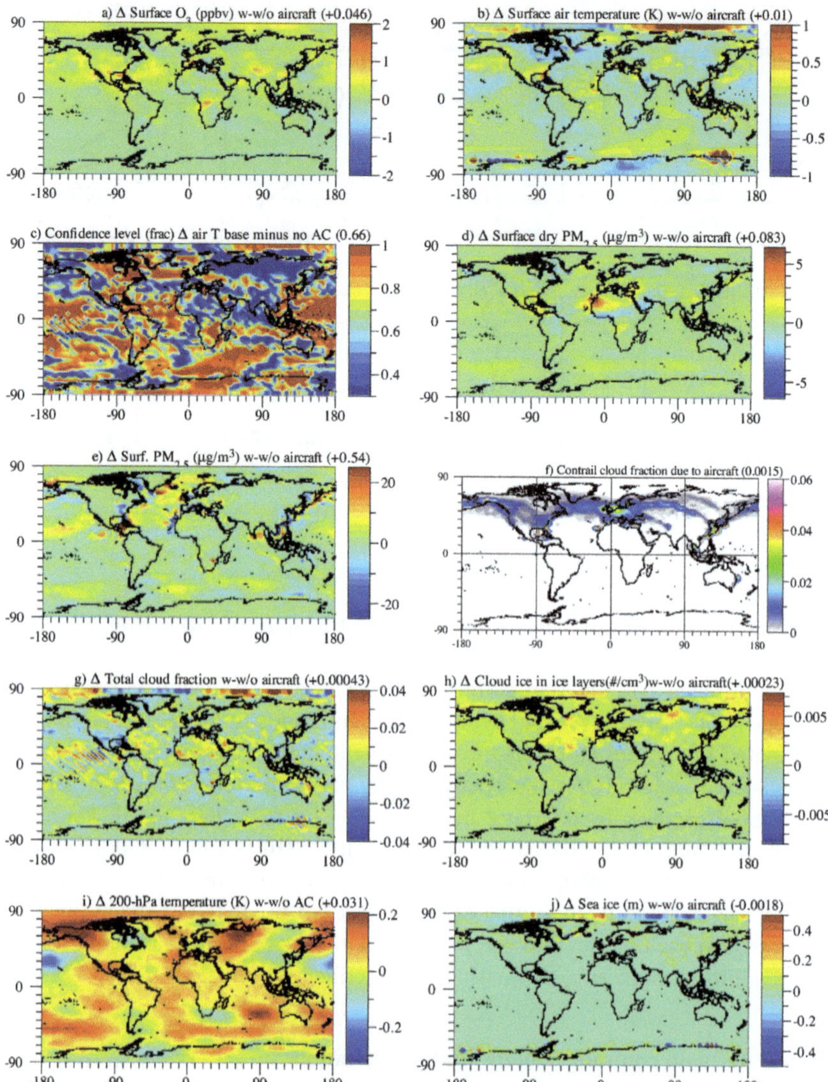

Fig. 3 Simulation-averaged spatial differences in several modeled parameters (surface or column-integrated) between when aircraft emissions of gases and particles were and were not included. Also shown, for one figure, is a corresponding plot of the confidence level of the results from a 2-sided t-test relative to six random perturbation simulations during the simulations.

(200 hPa). The enhanced absorption increased the heating rate in this altitude range by up to 0.5% (Fig. 2f) and the temperature by up to 0.015% in the global average (Fig. 1m). EFFS BC increased by up to 100% (Fig. 1e) and UIM BC, by up to 7% (Fig. 1f) in this region. The vertical spread of BC above cruise was helped by enhanced instability at and just above cruise altitude (Fig. 1m).

Increases in CAOD in a given location were due to both an increase in aircraft BC within clouds in that location and a local reduction in precipitation from clouds. A reduction in precipitation increases total cloud material, including BC, thus it increases CAOD. Aircraft decreased precipitation by 0.16% in the global

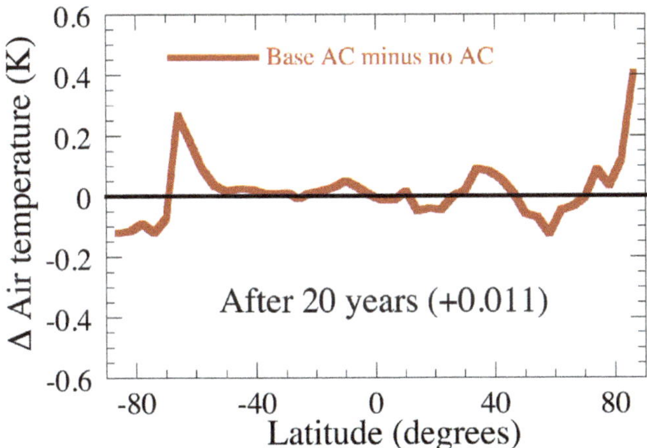

Fig. 4 Zonally- and simulation-averaged surface air-temperature difference between the baseline case with aircraft emissions and the simulation with no aircraft emissions.

average, 0.36% in the Northern Hemisphere, and 0.0038% in the Southern Hemisphere during the simulations.

Aircraft increased stratus (including cirrus-type) cloud fraction by 0.2%, decreased cumulus cloud fraction by 0.19%, and total cloud fraction by +0.065% in the global average. These numbers were +0.34%, −0.47%, and +0.12%, respectively, in the Northern Hemisphere. Total cloud fraction increases occurred significantly in the Arctic regions (Fig. 2f). The reduction in cumulus cloud fraction occurred because aircraft stabilized the troposphere below cruise altitude by increasing cruise-altitude temperatures more than surface temperatures (Fig. 1m). The reduction in cumulus also reduced precipitation, as previously discussed, and enhanced stratus since the water vapor in the air still needed to condense. Increases in cloud condensation nuclei (CCN) and ice deposition nuclei (IDN) due to an increase in particle number (Fig. 1a) also increased stratus.

The addition of aircraft exhaust slightly decreased the globally- and simulation-averaged in-cloud (for grid-scale clouds) geometric mean number diameter (integrated among all discrete size bins from 0.5 μm to 8-mm diameter) of liquid drops from 17.36 to 17.34 μm, of ice crystals from 42.9 to 41.0 μm, and of graupel particles from 54.5 to 54.2 μm. The reduction in cloud particle size is consistent with the increase in cloud optical depth due to aircraft (+0.51% globally and +0.87% in the Northern Hemisphere) (first indirect effect).

4.3 Effects on global temperatures, arctic temperatures, and sea ice

Aircraft exhaust increased globally-averaged near-surface air temperatures by ∼0.01 K and upper tropospheric air temperatures by ∼0.03 K, increasing tropospheric stability below cruise altitude and increasing it slightly above cruise altitude (Fig. 1m). The upper tropospheric warming represents ∼4% of the up to ∼0.8 K observed upper tropospheric global warming that has accumulated based on 1958–2007 radiosonde data.[25] The surface warming represents ∼1.3% of the 0.7–0.8 K surface global warming accumulated since 1850.[26]

The modest surface global warming due to aircraft was due to strong Arctic warming offset substantially by midlatitude cooling. Fig. 4 shows the zonal variation of the near-surface temperature change due to aircraft. It indicates that the strongest warming occurred north of 80 N. Fig. 3b shows that this warming exceeded 1 K over a third of the Arctic above 80 N. Although aircraft slightly cooled the surface at midlatitudes, they warmed the atmosphere at cruise altitude at midlatitudes (Fig. 1m).

Surface warming over the Arctic *versus* surface cooling at midlatitudes was generally due to a combination of (1) the stronger solar effects of aircraft-emitted BC over snow and sea ice than over dark ocean or land surfaces, (2) the fact that aircraft BC has a longer lifetime over the Arctic than at midlatitudes because BC emissions over the Arctic are mostly in the stratosphere,[17] (3) the greater infrared warming than solar cooling due to enhanced cloudiness from aircraft in the Arctic compared with midlatitudes since solar angles are much lower over the Arctic, and (4) the increase in surface pressure (causing air to descend and warm) over the Arctic relative to midlatitudes due to aircraft (Fig. 1o). The increase in Arctic surface pressure may have been due to enhanced poleward motion of air above cruise altitude due to the stronger elevated warming at midlatitudes than at the poles (Fig. 1m), resulting in a stronger pressure gradient between midlatitude and poles. At 11.8 km (200 hPa) and above, warming from 50–70 N exceeded that from 70–90 N. (Fig. 3i, 1m).

Although the global surface warming due to aircraft was small, it was statistically significant in the Arctic region where most warming occurred (Fig. 3c). About 26% of the global surface temperature change was statistically significant to a confidence level (CL) of 95% based on a 2-sided t-test. About 40% was significant to a CL of 85%.

IPCC [2007][9] estimates a 1960–2005 surface air temperature increase north of 65 N of ∼1.2 K. Based on computer-generated maps interpolating data,[27] the surface temperature increase north of 65 N from 1880–2008 relative to 1880–1920 was ∼2.5 K. In the model, aircraft increased surface air temperatures above the Arctic Circle by 0.1–0.4 K, with an average above the Arctic Circle of ∼0.15 K (6% of Arctic warming since 1880). Modeled Arctic ice area simultaneously decreased by 0.03% in the Northern Hemisphere. Sea ice thickness also decreased substantially in the Arctic Circle (Fig. 3j).

Previously, Jacobson *et al.* [2012][18] examined the effect of rerouting aircraft flying over the Arctic to around the Arctic Circle. The study concluded that rerouting would reduce BC and other emissions over the Arctic Circle by ∼83% but move the emissions to outside the Arctic Circle and slightly increase fuel use. Moving emissions out of the Circle puts them in regions of greater precipitation, causing a net reduction in particles, which are removed readily by precipitation. The study also found an Arctic cooling due to rerouting of up to 0.3 K in the zonal average, suggesting that higher temperatures over the Arctic were substantially due to local emissions over the Arctic Circle. The present study supports the previous finding that aircraft have a much stronger surface temperature impact over the Arctic region than in the global average. Arctic warming in the present study was up to 0.4 K in the zonal average (Fig. 4).

The results here suggest that, over the Arctic, the relative contribution of aircraft BC (with high BC emissions) to total column BC from all sources is ∼10%. Except for some local shipping emissions of BC, aircraft BC represent nearly all BC emissions directly over Arctic sea ice, and such emissions occur in the stratosphere at heights above most. As such, aircraft-emitted BC over the Arctic has a longer lifetime than does BC emitted from surface sources. BC emissions per unit ground surface area over the Arctic are about 17–33% of the global average aircraft emission density [Table 8 of Wilkerson *et al.*, 2010],[20] suggesting that the emission concentration of BC over the Arctic is significant.

BC emissions from aircraft affect air and surface temperatures in four major ways. First, contrails form on top of emitted aircraft particles, which contain BC coated by lubricating oil, unburned fuel oil, and sulfate. Some BC in the emission plume also coagulates with contrail particles, becoming an appendage or inclusion. When sunlight hits a contrail particle, the light is refracted into it and internally reflected, and some of this light is eventually absorbed by the BC inclusions, increasing cloud absorption. This absorption warms the ice crystal and hence the air around it, causing some melting and sublimation Jacobson [2012].[14] When a contrail particle sublimates, the resulting BC-containing aerosol particle is released to the air and can serve as a nucleus for cirrus or other types of cloud particles. Once incorporated within the cirrus particle, BC can warm the cirrus as well.

Second, aircraft-emitted aerosol particles containing BC that are not incorporated into contrails, but lie between them interstitially, warm the air by direct solar radiative heating (and some infrared heating). Scattering among contrail particles enhances this absorption significantly, as quantified in Jacobson [2012].[14]

Third, BC-containing particles in the air away from clouds reduce the single-scattering albedo of aerosol particles, warming the air due to solar absorption, heating the air, reducing the RH. This warming is enhanced for aged BC particles since such particles become more coated with volatile material as they age, and a coated BC particle warms the air much more than does an uncoated BC particle. This warming is also enhanced over the Arctic compared with midlatitudes due to the higher relative humidity over the Arctic (despite lower column water vapor), which increases the hydration of supercooled liquid water onto soluble material that coats BC, enhancing the coating more. Further, this warming is enhanced over snow, sea ice, and cloud surfaces compared with land, ocean, or cloud-free surfaces since strong reflection allows BC to absorb upward-reflected radiation along with downward radiation. This is one of the main reasons for the enhanced effects of BC over high-albedo surfaces. At night, all contrails and aerosol particles, including BC, increase the downward flux of thermal-IR, warming the surface. Since the Arctic is dark during much of the Northern Hemisphere winter, contrails and particles generally have only a warming effect during the winter.

Fourth, BC-containing particles ultimately deposit to snow and sea ice, primarily by precipitation. The combination of aircraft-emitted BC inclusions within snow and sea ice grains and all feedbacks of aircraft to climate reduced the surface albedo globally here by ~0.1%, warming and enhancing the melting/sublimation of snow and ice, potentially uncovering the lower-albedo ocean or land below.

Despite the fact that aircraft do not fly much over the Antarctic, Fig. 3b and 4 indicates that aircraft emissions globally affected temperatures at the edge of the Antarctic sea ice. Most warming was localized to one particular area (Fig. 3b). This warming was due primarily to the increase in subpolar surface pressure outside the Antarctic polar vortex (Fig. 1o). Temperature and pressure changes in the Northern Hemisphere resulted in enhanced south poleward movement of air. In the Southern Hemisphere, that air was blocked by the circumpolar jet stream wind system so descended outside the Antarctic. Over the Arctic, the polar vortex is weaker, so poleward air reaches the pole before descending. As the air descended outside the Antarctic, it compresses and warms, melting some snow and ice. In addition, aircraft CO_2 and BC permeated to the Antarctic, where it contributed to some warming there. Since sea ice at its borders is thin, small changes in temperature have a large impact on albedo, since melting uncovers the low albedo ocean below.

4.4   Effects on irradiance

The net (solar+thermal-IR) top-of-the-atmosphere (TOA, at 0.22 hPa or 59.43 km), tropopause (250 hPa or 10.4 km), and surface irradiance changes due to all feedbacks from aircraft were +54, +43, and -86 mW m^{-2}, respectively (Fig. 3h). These are climate response irradiance changes, not radiative forcings. Nevertheless, the 54 mW m^{-2} increase in irradiance at the TOA is not so far from the estimate of 55 mW m^{-2} from Lee et al. [2009][10] for the radiative forcing due to aircraft without considering cirrus enhancement and the 78 mW m^{-2} accounting for cirrus enhancement.

4.5   Uncertainties

Uncertainties in the model arise primarily from the coarse horizontal resolution of grid scale treatments, emission data (aircraft, other anthropogenic, and natural), the representation of clouds, the representation of aerosol-cloud interactions, and sub-grid turbulence in aircraft exhaust plumes. Quantifying the possible uncertainty in the mean results presented here due to the uncertainties of these treatments is not trivial. For example, the uncertainty due to resolution alone could not be determined because the computer time for the present simulation was over a year, so increasing the resolution by a factor of 2 in each of the x- and y- dimensions would result in simulations of over four years, and even this resolution is not sufficient for determining uncertainty. As such, it can only be said that the results presented are those based on the current setup and future work with higher resolution and even more detailed physical treatments is needed to determine the uncertainty of the present results.

5   Summary

This study examined the short term (20-year) effects of contemporary gas and particle exhaust from all commercial flights worldwide on climate and atmospheric composition. Results were obtained assuming constant emissions over this period. The model used, GATOR-GCMOM, treated the emissions and evolution of gas and particle exhaust from each commercial aircraft flight worldwide at the subgrid scale. Aircraft exhaust included CO, CO_2, H_2O, speciated hydrocarbons, NO_x, SO_2, BC, POM, and sulfate. The algorithms used and baseline model results were analyzed against data in Part I of this study, Jacobson et al. [2011].[16,18]

Results here indicate that aircraft emissions increased Arctic surface temperatures by $\sim$0.15 K, global averaged surface temperatures by -0.01 K, and globally-averaged upper tropospheric temperatures by $\sim$0.03 K, thereby increasing atmospheric stability below cruise altitude and decreasing it above cruise altitude. Strong Arctic surface warming due to aircraft was largely offset by midlatitude cooling. Cruise altitude temperatures increased substantially both at midlatitudes and over the poles. Overall, surface warming due to aircraft represents $\sim$6% of Arctic surface warming since 1880, $\sim$1.3% of accumulated net global surface warming since 1850, and $\sim$4% of observed upper-tropospheric warming from 1958–2007. Arctic warming due to aircraft slightly decreased sea ice area. Longer simulations should enhance warming due to the further reduction increases in CO_2.

Whereas aircraft emissions decreased cumulus convection by enhancing stability below cruise altitude, they increased stratiform and total cloud fraction on average. Aircraft emissions increased surface and upper tropospheric ozone by $\sim$0.4% and $\sim$2.5%, respectively and surface and upper-tropospheric PAN by $\sim$0.1% and $\sim$5%, respectively. Aircraft emissions increased tropospheric OH, decreasing column CO and CH_4 by $\sim$1.7% and $\sim$0.9%, respectively. Aircraft emissions increased mortality worldwide by $\sim$620 ($-$240 to 4770) deaths per year, with half due to ozone and the rest to $PM_{2.5}$.

These results suggest that aircraft, particularly their black carbon particle emissions, may have a significant impact on short-term climate, particularly over the Arctic, and chemical composition. The Arctic result is consistent with Jacobson *et al.* [2012], who examined the effect of rerouting aircraft around the Arctic Circle on climate. However, longer-term simulations are needed to account more fully for the effects of carbon dioxide from aircraft on climate and composition. Additional work is needed to improve aircraft emissions inventories, simulate better the contrail-cirrus transition, and reduce the uncertainty due to the coarse resolution of grid-scale processes and of physical treatments in the model.

Acknowledgements

This work was supported by the Partnership for AiR Transportation Noise & Emissions Reduction (PARTNER) and the Federal Aviation Administration (FAA). Any opinions, findings, and conclusions or recommendations expressed in this material are those of the authors and do not necessarily reflect the views of PARTNER or the FAA.

References

1 G. P. Brasseur, R. A. Cox, D. Hauglustaine, I. Isaksen, J. Lelieveld, D. H. Lister, R. Sausen, U. Schumann, A. Wahner and P. Wiesen, European Assessment of the atmospheric effects of aircraft emissions, *Atmos. Environ.*, 1998, **32**, 2329–2418.
2 M. Prather, R. Sausen, A. S. Grossman, J. M. Haywood, D. Rind, and B. H. Subbaraya, Potential climate change from aviation. in *"Aviation and Global Atmosphere", Intergovernmental Panel on Climate Changes*, ed. J. E. Penner, D. H. Lister, D. J. Griggs, D. J. Dokken and M. McFarland, Cambridge University Press, Cambridge, U.K, 1999.
3 P. Minnis, U. Schumann, D. R. Doelling, K. M. Gierens and D. W. Fahey, Global distribution of contrail radiative forcing, *Geophys. Res. Lett.*, 1999, **26**, 1853–1856.
4 G. Myhre and F. Stordal, On the tradeoff of the solar and thermal infrared radiative impact of contrails, *Geophys. Res. Lett.*, 2001, **28**, 3119–3122.
5 M. Ponater, S. Marquart and R. Sausen, Contrails in a comprehensive global climate model: Parameterization and radiative forcing results, *J. Geophys. Res.*, 2002, **107**, 4164, DOI: 10.1029/2001JD000429.

6 S. Marquart, M. Ponater, F. R. Mager and R. Sausen, Future development of contrail cover, optical depth, and radiative forcing: Impacts of increasing air traffic and climate change, *J. Clim.*, 2003, **16**, 2890–2904.

7 P. Minnis, J. K. Ayers, R. Palikonda and D. Phan, Contrails, cirrus trends, and climate, *J. Clim.*, 2004, **17**, 1671–1685.

8 R. Sausen, I. Isaksen, V. Grewe, D. Hauglustaine, D. Lee, G. Myhre, M. Köhler, G. Pitari, U. Schumann, F. Stordal and C. Zerefos, Aviation radiative forcing in 2000: An update on IPCC (1999), *Meteorol. Z.*, 2005, **14**, 555–561.

9 Intergovernmental Panel on Climate Change (IPCC), Fourth Assessment Report, *The Physical Science Basis*, Cambridge University Press, 2007.

10 D. S. Lee, G. Pitari, V. Grewe, K. Gierens, J. E. Penner, A. Petzold, M. J. Prather, U. Schumann, A. Bais, T. Bernsten, D. Iachetti, L. L. Lim and R. Sausen, Transport impacts on atmosphere and climate: Aviation, *Atmos. Environ.*, 2010, **44**, 4678–4734.

11 T. C. Bond, *et al.*, Bounding the role of black carbon in the climate system: A scientific assessment, *J. Geophys. Res.*, 2013, **118**, 5380–5552.

12 M. Ponater, S. Brinkop, R. Sausen and U. Schumann, Simulating the global atmospheric response to aircraft water vapour emissions and contrails. A first approach using a GCM, *Ann. Geophys.*, 1996, **14**, 941–960.

13 D. Rind, P. Lonergan and K. Shah, Modeled impact of cirrus cloud increases along aircraft flight paths, *J. Geophys. Res.*, 2000, **105**, 19927–19940.

14 M. Z. Jacobson, Investigating cloud absorption effects: Global absorption properties of black carbon, tar balls, and soil dust in clouds and aerosols, *J. Geophys. Res.*, 2012, **117**, D06205, DOI: 10.1029/2011JD017218.

15 C. Fromming, M. Ponater, K. Dahlmann, V. Grewe, D. S. Lee and R. Sausen, Aviation-induced radiative forcing and surface temperature change in dependency of the emission altitude, *J. Geophys. Res.*, 2012, **117**, D19104, DOI: 10.1029/2012JD018204.

16 M. Z. Jacobson, J. T. Wilkerson, A. D. Naiman and S. K. Lele, The effects of aircraft on climate and pollution. Part I: Numerical methods for treating the subgrid evolution of discrete size- and composition-resolved contrails from all commercial flights worldwide, *J. Comput. Phys.*, 2011, **230**, 5115–5132, DOI: 10.1016/j.jcp.2011.03.031.

17 D. B. Whitt, J. T. Wilkerson, M. Z. Jacobson, A. D. Naiman and S. K. Lele, Vertical mixing of commercial aviation emissions from cruise altitude to the surface, *J. Geophys. Res.*, 2011, **116**, D14109, DOI: 10.1029/2010JD015532.

18 M. Z. Jacobson, J. T. Wilkerson, S. Balasubramanian, W. W. Cooper, Jr. and N. Mohleji, The effects of rerouting aircraft around the Arctic Circle on Arctic and global climate, *Clim. Change*, 2012, **115**, 709–724, DOI: 10.1007/s10584-012-0462-0.

19 A. D. Naiman, S. K. Lele, J. T. Wilkerson and M. Z. Jacobson, Parameterization of subgrid plume dilution for use in large-scale atmospheric simulations, *Atmos. Chem. Phys.*, 2010, **10**, 2551–2560.

20 J. T. Wilkerson, M. Z. Jacobson, A. Malwitz, S. Balasubramanian, R. Wayson, G. Fleming, A. D. Naiman and S. K. Lele, Analysis of emission data from global commercial aviation: 2004 and 2006, *Atmos. Chem. Phys.*, 2010, **10**, 6391–6408.

21 GFS (Global Forecast System), $1° \times 1°$ reanalysis fields, http://nomads.ncdc.noaa.gov/data/gfs-avn-hi/, 2012, Accessed January 28, 2013.

22 M. Z. Jacobson and D. G. Streets, The influence of future anthropogenic emissions on climate, natural emissions, and air quality, *J. Geophys. Res.*, 2009, **114**, D08118, DOI: 10.1029/2008JD011476.

23 M. O. Köhler, G. Radel, O. Dessens, K. P. Shine, H. L. Rogers, O. Wild and J. A. Pile, Impact of perturbations to nitrogen oxide emissions from global aviation, *J. Geophys. Res.*, 2008, **113**, D11, DOI: 10.1029/2007JD009140.

24 M. Z. Jacobson, The enhancement of local air pollution by urban CO_2 domes, *Environ. Sci. Technol.*, 2010, **44**, 2497–2502, DOI: 10.1021/es903018m.

25 W. J. Randel, *et al.*, An update of observed stratospheric temperature trends, *J. Geophys. Res.*, 2009, **114**, D02107, DOI: 10.1029/2008JD010421.

26 P. Brohan, J. J. Kennedy, I. Harris, S. F. B. Tett and P. D. Jones, Uncertainty estimates in regional and global observed temperature changes: A new data set from 1850, *J. Geophys. Res.*, 2006, **111**, D12106, DOI: 10.1029/2005JD006548.

27 GISS (Goddard Institute for Space Studies), GISS Surface Temperature Analysis, http://data.giss.nasa.gov/gistemp/maps/, 2009, accessed Jan. 19, 2009.

Faraday Discussions

RSC Publishing

PAPER

Contact freezing efficiency of mineral dust aerosols studied in an electrodynamic balance: quantitative size and temperature dependence for illite particles

Nadine Hoffmann,[a] Denis Duft,[a] Alexei Kiselev[a] and Thomas Leisner[*ab]

Received 8th March 2013, Accepted 19th March 2013

DOI: 10.1039/c3fd00033h

Contact freezing has long been discussed as a candidate for cloud ice formation at temperatures warmer than about −25 °C, but until now the molecular mechanism underlying this process has remained obscure and little quantitative information about the size and temperature dependent contact freezing properties of the various aerosol species is available. In this contribution, we present the first quantitative measurements of the freezing probability of a supercooled droplet upon a single contact with a size selected illite mineral particle. It is found that this probability is a strong function of temperature and aerosol particle size. For the particles investigated and on the minute time scale of the experiment, contact freezing indeed dominates immersion freezing for all temperatures.

Introduction

At temperatures warmer than about −36 °C, atmospheric ice formation requires the presence of heterogeneous freezing nuclei.[1] The abundance of such nuclei thus governs precipitation initiation *via* the Wegener–Bergeron–Findeisen process in most tropospheric clouds and thereby has an important influence on their lifetime and optical properties. The temperature, size and material dependent ice nucleation propensity of insoluble aerosol particles is still poorly understood and a subject of ongoing research. Generally, four modes of heterogeneous ice nucleation are distinguished; these are deposition nucleation, condensation freezing, immersion freezing and contact freezing.[2] Deposition freezing is the direct formation of ice on the surface of a dry particle from gas phase water vapor, while in the case of condensation freezing, a transient liquid

---

[a]*Karlsruhe Institute of Technology, Institute for Meteorology and Climate Research, P.O. Box 3640, Karlsruhe, Germany. E-mail: Nadine.Hoffmann@kit.edu; Denis.Duft@kit.edu; Alexei.Kiselev@kit.edu; Thomas.Leisner@kit.edu*

[b]*Universität Heidelberg, Institut für Umweltphysik, Im Neuenheimer Feld 229, 69120 Heidelberg, Germany. E-mail: Thomas.Leisner@iup.uni-heidelberg.de*

phase is assumed. Accordingly, deposition freezing can occur at relative humidity above ice saturation, while condensation freezing needs water saturation. Immersion freezing describes the formation of ice on the surface of an insoluble particle immersed in a supercooled liquid droplet, while contact freezing describes the formation of ice in the wetting process occurring during the contact between a liquid drop and a dry particle entering from the outside. For contact freezing to be relevant, the free energy of the formation of the critical ice germ has to be reduced during the wetting as compared to the fully immersed particle. Since the first descriptions of contact freezing,[3] there have been several experimental studies[4–10] which confirmed that for a given particle type and size, ice forms at warmer temperatures in contact mode than in immersion mode. In those experiments, this behavior could not be formulated quantitatively, as the concentration and nature of contact nuclei, as well as the rates of contact, were not determined sufficiently well.[11,12] Therefore, in numerical models contact freezing was included only cursorily at best and its importance remained unclear. Only recently, new experimental approaches have revived the interest in contact nucleation and have provided more quantitative data.[11,13,14]

In this contribution, we report nucleus size- and temperature-dependent contact freezing probabilities which quantitatively describe the probability of droplet freezing upon a single contact between an illite mineral dust particle and a supercooled droplet.

Experimental

Our newly developed apparatus for contact freezing measurements was described recently,[15] so here only a brief recollection of the most important features is given. Fig. 1 shows a scheme of the experimental setup. For the sake of clarity, the components that have been added as compared to our standard EDB setup[15] are highlighted. It combines a cooled electrodynamic balance (EDB) for the storage of individual supercooled droplets under well-defined conditions with a source of dry aerosol particles. The aerosol under investigation is generated either by a fluidized bed generator or by the dispersion of a suspension in an atomizer with subsequent dryer. Particle size selection is achieved by a multistage impactor followed by a

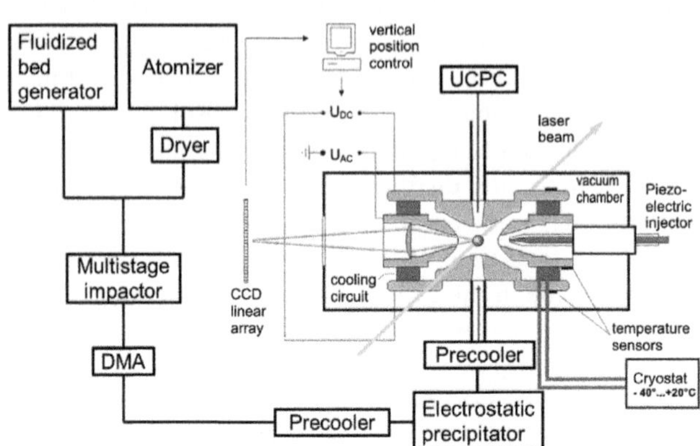

Fig. 1 Scheme of the experimental setup (UCPC: Ultrafine Condensational Particle Counter, DMA: Differential Mobility Analyzer).

differential mobility analyzer (DMA) in order to minimize the presence of multiply charged particles. The aerosol flow is precooled to the EDB temperature and passed vertically through the EDB cell. The frost point temperature of the aerosol stream is always kept well below the EDB temperature ($\Delta T > 30$ K) in order to avoid deposition nucleation of ice on the aerosol particles. For calibration and comparison purposes, an electrostatic precipitator mounted close to the EDB allows removal of all aerosol particles on demand without changing the flow conditions.

Individual droplets of about 100μm in diameter from high purity water are created by a piezo-driven droplet dispenser and are injected into the EDB, which is continuously held at the low temperature of interest. At the EDB center, the droplets are illuminated by an He–Ne laser and imaged by a microscope objective onto a position sensitive detector (PSD). Using the PSD signal, the AC frequency and the DC voltages on the EDB are adjusted under computer control to allow for stable trapping of the levitated droplets at the EDB center for several minutes. A second optical system projects the scattered light onto a 2D CCD camera. Hereby, the scattering phase function in an angular range from 35° to 55° is recorded and used to determine the size and the mass of the droplet using the known density and index of refraction of water. The droplet charge is then inferred from the DC potential applied to the balance. Droplet freezing is easily detected either by the breakdown of the Mie light scattering pattern or even more rapidly by a strong intensity increase in the PSD signal due to light scattering from the internal interfaces of the polycrystalline particle at the moment of freezing.

Experiments are performed by applying a fixed aerosol type and size distribution at a fixed temperature and measuring the time interval between droplet injection and freezing for a large number of individual supercooled water droplets. The aerosol concentration and flow is adjusted such that freezing of the droplets occurs on a minute timescale which is several orders of magnitude longer than the timescale of droplet thermalization. Frozen droplets are ejected from the EDB and fresh liquid droplets are injected under computer control. In a typical experiment, several hundred individual droplets are frozen under identical conditions in order to acquire a sufficient amount of data for statistical analysis (see discussion below).

In order to determine the fraction of collisions that lead to droplet freezing, we analyze the observed ensemble of droplets with respect to the fraction of liquid particles N_{liq}/N_0 as a function of their residence time in the EDB (τ). This freezing rate can be described as a combination of two stochastic processes, the collision process described by a collision rate and the contact freezing probability per collision. If the collision rate is known sufficiently well, the freezing probability per collision can be obtained by fitting the experimental data. Without immersion freezing of the scavenged aerosol particles, $N_{liq}/N_0(\tau)$ should drop exponentially with τ. This is illustrated in Fig. 2, where measured and fitted $N_{liq}/N_0(\tau)$ curves are displayed using a logarithmic ordinate. From the slope of the linear fitting curve, the contact freezing probability is obtained by division by the collision rate. By a Monte Carlo model of the statistics of droplet freezing we have confirmed that the relative error (1 sigma) of the thus-determined freezing rate is given by $N^{-1/2}$ where N is the number of droplets in the ensemble investigated. This error is propagated on a 1 sigma level into the error bars of the result plots given below.

In order to determine the collision rate between droplets and aerosol particles, we exploit the known electrical charge on the levitated water droplets, which acts

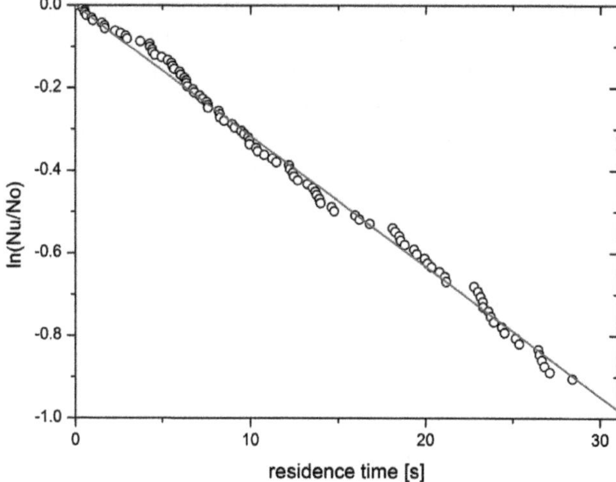

Fig. 2 Typical $N_{liq}/N_0(\tau)$ curve (see text) obtained with illite particles of 320 nm diameter at a temperature of 243.1 K and linear fit to the data.

to attract the aerosol particles due to the Coulomb force and to charge-induced dipole interaction with the polarizable aerosol particles.[16,17] Without these forces, the aerosol particle–droplet collision rates under our experimental conditions would be negligible. Based on fluid dynamics calculations for the particle free flow field in the exact geometry of the balance, we perform numerical simulations of the motion of singly charged aerosol particles under the influence of the above mentioned forces in order to obtain a collision cross section for every given experimental condition, *i.e.* droplet charge, aerosol size and type, temperature and aerosol volume flow. Especially for strongly aspherical particles like kaolinite,[18] these calculations are hampered by the fact that the aerosol polarizability and the friction force on the aerosol particles depend on particle shape and orientation and are therefore not well known. In order to verify our fluid dynamical modeling and to determine the aerosol particle shape factor assumed in the calculations, we eject liquid droplets from the balance after a given exposure to the aerosol flow and collect them on a clean silicon surface. After evaporation of the droplet, the residual aerosol particles can be detected and counted with the help of an electron-microscope. A typical microscopic image of droplet residuals is given in Fig. 3.

The thereby verified scavenging cross section is used to calculate the aerosol–droplet collision rate as the product of aerosol number concentration, scavenging cross section and flow velocity, as obtained from the numerical fluid dynamics simulation. Details of the numerical modeling of the collection efficiency and its experimental verification can be found in a recently submitted manuscript.[18]

A complete model of the freezing process has to take into account that every contact not leading to freezing creates a new immersion freezing nucleus inside the droplet with its own heterogeneous immersion freezing rate J_{imm}. This process can easily be included in the numerical analysis of the data and leads to a deviation of the modeled $\ln N_{liq}/N_0(\tau)$ curve from the linear form. Conversely, the absence of such a deviation indicates that contact freezing dominates immersion

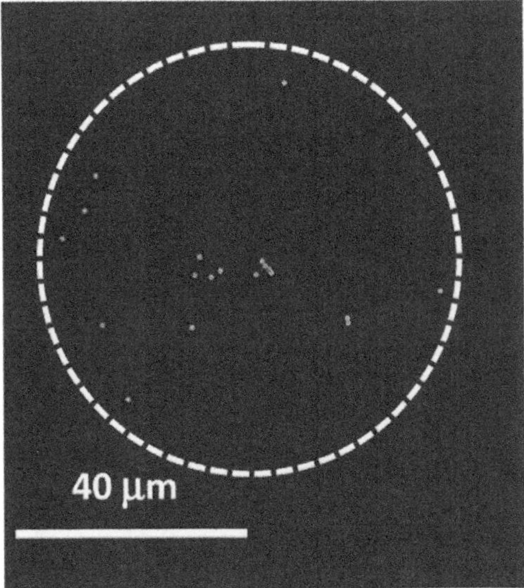

Fig. 3 Electron micrograph of a collection of hematite particles found as residuals from one microdroplet exposed to hematite aerosol flow in the EDB for 40 s. The circle represents the footprint of the droplet on a substrate.

freezing on the timescale of the experiment. The transition from contact freezing to immersion freezing will be discussed in more detail in a forthcoming publication.

Results

So far, this apparatus has been used to investigate the size dependent contact freezing properties of polystyrene latex (PSL) spheres, hematite particles, kaolinite particles and illite particles. In this contribution we focus on illite particles, which constitute a major fraction of the atmospheric mineral dust aerosol and are well known for their good ice nucleation potential.[19–21]

For the experiments reported here, commercial Arginotec® NX illite powder from "B + M Nottenkämper" was used. According to the manufacturer, it contains 86% illite, 10% kaolinite, 4% calcite, as well as quartz and feldspar in trace amounts. The main component (illite) is a dioctahedral 2 : 1 layered mineral[22] of the mica group, characteristic of mineral dusts transported from arid areas. Illite particles are agglomerates composed of nano-crystals with a size of 20 nm on average, as confirmed by electron microscopy (*cf.* Fig. 4).

We investigated four size classes characterized by a mobility diameter of 150 nm, 320 nm, 550 nm and 750 nm, in the following referred to as small, medium, large and very large particles, respectively. Each size class is characterized by a roughly Gaussian size distribution with the width being about 20% of the nominal diameter. For each size class, the temperature was varied from the onset of homogeneous freezing, which occurs at $-36\,^{\circ}\mathrm{C}$ in our experiment, up to a temperature where hardly any heterogeneous freezing could be detected. Within that range, several hundred freezing events have been recorded and analyzed for every temperature point. For all sizes and at all temperatures investigated, is was

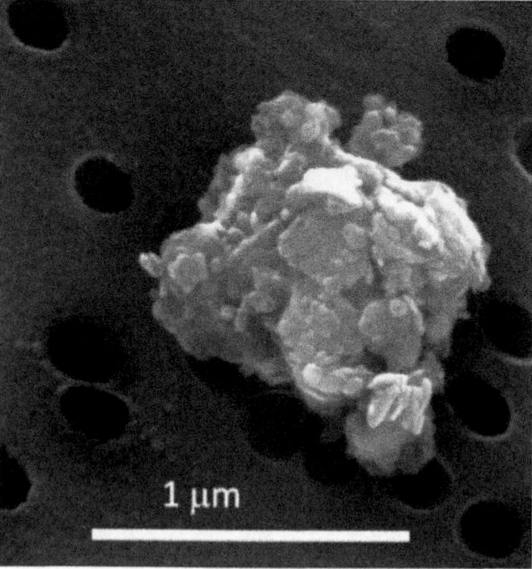

Fig. 4   Scanning electron micrograph of a typical "very large" (see text) illite particle.

found that immersion freezing did not contribute significantly within the experimental time frame of 30 s per droplet so that contact freezing rates could be obtained by fitting linear functions to the observed $\ln N_{\mathrm{liq}}/N_0(\tau)$ curves, as detailed above. Fig. 5 shows a summary of the results obtained for illite.

The results for the four size classes are well separated. For every size class, the contact freezing probability is a roughly exponential function of the temperature, with the slope decreasing with increasing particle size. We observe a strong size dependence; the small illite particles reach a 10% contact freezing probability only at about 8 K colder temperatures than the very large particles.

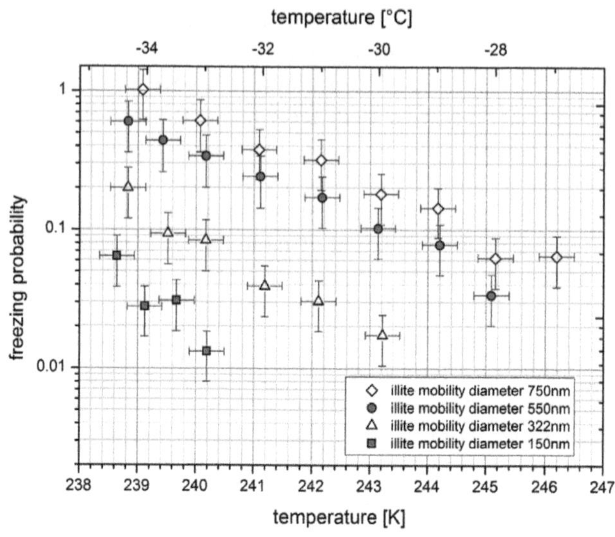

Fig. 5   Temperature and size dependence of the contact freezing probability for illite aerosol particles.

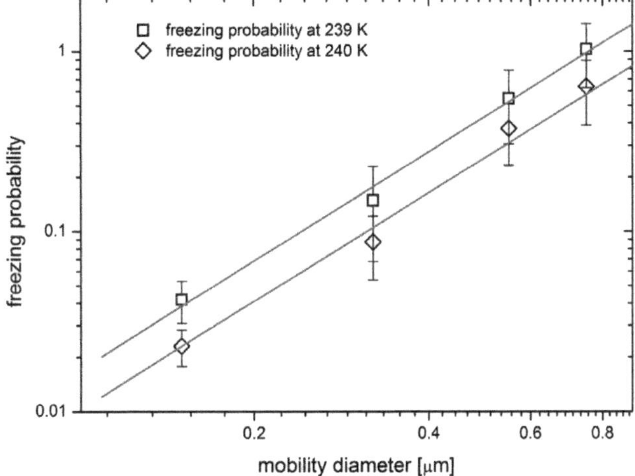

Fig. 6 Size dependence of the freezing probability at two fixed temperatures. Solid grey lines show a quadratic fit to the experimental data.

In Fig. 6, the freezing probability is plotted as a function of particle diameter at two fixed temperatures. It is nicely fitted by a quadratic curve, indicating that the freezing probability is proportional to the surface area of the particle.

Discussion and outlook

Here we report the first quantitative size and temperature resolved measurements of the contact freezing probability of illite particles as found in the atmosphere. For every size class this quantity exhibits a strong temperature dependence, the slope of which increases with decreasing size. At a given temperature, the contact freezing probability is proportional to the surface area. These observations are in general accordance with the notion of ice-active sites (IN) which are distributed over the surface at a certain density, which in itself is a function of temperature. This approach has been successfully applied to parameterize immersion freezing on mineral dusts[23] and is here shown to be applicable to contact freezing as well. For larger particles, the larger spread in IN-activity might explain the less steep temperature dependence.

The surface area dependence of the contact freezing probability suggests that the concept of a first *point* contact between the droplet and the particle triggering the freezing of supercooled water is not valid. It rather appears that the whole surface and probably the most ice-active site is involved in the contact freezing process.

Assuming that the ice-active sites that are active in immersion and the contact freezing modes are the same, the following two questions arise: (a) Why are these sites more active in the contact freezing mode? (b) Is this enhancement independent of the IN nature or are there any ice nuclei that do not exhibit enhanced ice nucleating properties in contact mode?

Currently, experiments to clarify these questions are ongoing in our laboratory using longer observation times, modulated aerosol flow conditions and different contact ice nuclei.

Acknowledgements

The authors acknowledge Daniel Rzesanke for setting up and characterizing an initial version of the experiment. This work was funded by the German Science foundation under contract LE8343/2-3 and FOR 1525 INUIT.

References

1 H. R. Pruppacher and J. D. Klett, *Microphysics of clouds and precipitation*, Kluwer Academic Publishers, Dordrecht; Boston, 1997.
2 G. Vali, *Bull. Am. Meteorol. Soc.*, 1985, **66**, 1426–1427.
3 W. Rau, *Wetter und Klima: Monatsschrift für angewandte Meteorologie; Organ des Deutschen Meteorologischen Dienstes im Französischen Besatzungsgebiet*, 1949, vol. 2, pp. 81–92.
4 N. Fukuta, *J. Atmos. Sci.*, 1975, **32**, 1597–1603.
5 W. A. Cooper, *J. Atmos. Sci.*, 1974, **31**, 1832–1837.
6 N. R. Gokhale and J. Goold, *J. Appl. Meteorol.*, 1968, 7, 870–874.
7 A. P. Fornea, S. D. Brooks, J. B. Dooley and A. Saha, *J. Geophys. Res.–Atmos.*, 2009, 114.
8 R. A. Shaw, A. J. Durant and Y. Mi, *J. Phys. Chem. B*, 2005, **109**, 9865–9868.
9 K. Diehl, S. Matthias-Maser, R. Jaenicke and S. K. Mitra, *Atmos. Res.*, 2002, 61, 125–133.
10 R. L. Pitter and Pruppach.Hr, *Q. J. R. Meteorol. Soc.*, 1973, **99**, 540–550.
11 L. Ladino, O. Stetzer, F. Luond, A. Welti and U. Lohmann, *J. Geophys. Res.–Atmos.*, 2011, 116.
12 E. A. Svensson, C. Delval, P. von Hessberg, M. S. Johnson and J. B. C. Pettersson, *Atmos. Chem. Phys.*, 2009, 9, 4295–4300.
13 C. Gurganus, A. B. Kostinski and R. A. Shaw, *J. Phys. Chem. Lett.*, 2011, 2, 1449–1454.
14 K. W. Bunker, S. China, C. Mazzoleni, A. Kostinski and W. Cantrell, *Atmos. Chem. Phys. Discuss.*, 2012, 12, 20291–20309.
15 D. Rzesanke, J. Nadolny, D. Duft, R. Muller, A. Kiselev and T. Leisner, *Phys. Chem. Chem. Phys.*, 2012, **14**, 9359–9363.
16 H. F. Kraemer and H. F. Johnstone, *Ind. Eng. Chem.*, 1955, **47**, 2426–2434.
17 E. Bichoutskaia, A. L. Boatwright, A. Khachatourian and A. J. Stace, *J. Chem. Phys.*, 2010, 133.
18 N. Hoffmann, A. Kiselev, D. Rzesanke, D. Duft and T. Leisner, *Atmos. Measurement Tech. Discuss.*, 2013, 6, 3407–3437.
19 P. J. DeMott, D. J. Cziczo, A. J. Prenni, D. M. Murphy, S. M. Kreidenweis, D. S. Thomson, R. Borys and D. C. Rogers, *Proc. Natl. Acad. Sci. U. S. A.*, 2003, **100**, 14655–14660.
20 M. S. Richardson, P. J. DeMott, S. M. Kreidenweis, D. J. Cziczo, E. J. Dunlea, J. L. Jimenez, D. S. Thomson, L. L. Ashbaugh, R. D. Borys, D. L. Westphal, G. S. Casuccio and T. L. Lersch, *J. Geophys. Res.–Atmos.*, 2007, 112.
21 P. J. DeMott, K. Sassen, M. R. Poellot, D. Baumgardner, D. C. Rogers, S. D. Brooks, A. J. Prenni and S. M. Kreidenweis, *Geophys. Res. Lett.*, 2003, 30.
22 A. Meunier, *Clays*, Springer, 2005, ISBN 978-3-540-21667-4.
23 M. Niemand, O. Möhler, B. Vogel, H. Vogel, C. Hoose, P. Connolly, H. Klein, H. Bingemer, P. DeMott, J. Skrotzki and T. Leisner, *J. Atmos. Sci.*, 2012, **69**, 3077–3092.

Faraday Discussions

RSC Publishing

PAPER

Kinetic limitations in gas-particle reactions arising from slow diffusion in secondary organic aerosol

Shouming Zhou,[*a] Manabu Shiraiwa,[bc] Robert D. McWhinney,[a] Ulrich Pöschl[c] and Jonathan P. D. Abbatt[a]

Received 8th March 2013, Accepted 16th May 2013

DOI: 10.1039/c3fd00030c

The potential for aerosol physical properties, such as phase, morphology and viscosity/ diffusivity, to affect particle reactivity remains highly uncertain. We report here a study of the effect of bulk diffusivity of polycyclic aromatic hydrocarbons (PAHs) in secondary organic aerosol (SOA) on the kinetics of the heterogeneous reaction of particle-borne benzo[a]pyrene (BaP) with ozone. The experiments were performed by coating BaP-ammonium sulfate particles with multilayers of SOA formed from ozonolysis of α-pinene, and by subsequently investigating the kinetics of BaP loss *via* reaction with excess ozone using an aerosol flow tube coupled to an Aerodyne Aerosol Mass Spectrometer (AMS). All reactions exhibit pseudo-first order kinetics and are empirically well described by a Langmuir–Hinshelwood (L-H) mechanism. The results show that under dry conditions (RH < 5%) diffusion through the SOA coating can lead to significant mass transfer constraints on the kinetics, with behavior between that previously observed by our group for solid and liquid organic coats. The reactivity of BaP was enhanced at $\sim$50% relative humidity (RH) suggesting that water uptake lowers the viscosity of the SOA, hence lifting the mass transfer constraint to some degree. The kinetics for $\sim$70% RH were similar to results obtained without SOA coats, indicating that the SOA had sufficiently low viscosity and was sufficiently liquid-like that reactants could rapidly diffuse through the coat. A kinetic multi-layer model for aerosol surface and bulk chemistry was applied to simulate the kinetics, yielding estimates for the diffusion coefficients (in cm^2 s^{-1}) for BaP in α-pinene SOA of 2×10^{-14}, 8×10^{-14} and >1×10^{-12} for dry (RH < 5%), 50% RH and 70% RH conditions, respectively. These results clearly indicate that slow diffusion of reactants through SOA coats under specific conditions can provide shielding from gas-phase oxidants, enabling the long-range atmospheric transport of toxic trace species, such as PAHs and persistent organic pollutants.

[a]Department of Chemistry, University of Toronto, Ontario, ON M5S 3H6, Canada. E-mail: szhou@chem. utoronto.ca; Fax: +1 416 946 7359; Tel: +1 416 946 7359

[b]Division of Chemistry and Chemical Engineering, California Institute of Technology, Pasadena, CA91125, USA

[c]Multiphase Chemistry Department, Max Plank Institute for Chemistry, Mainz, D-55128, Germany

1 Introduction

The question we address in this paper is whether coatings of secondary organic aerosol (SOA) can provide a kinetic barrier to the rates of aerosol chemistry. In particular, while it is known that SOA frequently contributes the majority of the sub-micron aerosol mass,[1] the physical properties of atmospheric SOA and their impacts remain poorly understood.

Based on an equilibrium gas/particle partitioning theory,[2,3] SOA has generally been treated as a well-mixed liquid in atmospheric models[3–5] in which the gas-phase oxidation products are assumed to quickly adopt gas-particle equilibrium. However, a number of recent studies have shown that SOA behaves as a highly viscous, amorphous solid or semi-solid. In this case the equilibration timescale of SOA partitioning can be longer and the assumption of equilibrium partitioning may be in question.[6]

Virtanen et al.[7] investigated the bounce behavior of biogenic SOA formed from ozone reaction with plant-emitted VOCs on smooth and hard surfaces and demonstrated that the organic-rich atmospheric aerosol can exist in the amorphous solid state. Later work of Vaden et al.[8] and Abramson et al.[9] studied evaporation of α-pinene SOA at room temperature and found evaporation kinetics were orders of magnitude slower than those expected from well-mixed liquid droplets. As well, Cappa and Wilson[10] observed that the volatility of organic aerosol was significantly lower than that of liquid aerosols by studying the thermal desorption properties of α-pinene SOA. Finally, Perraud et al.[11] reported that the partitioning of organic nitrates formed from the reaction of α-pinene with NO_3 radicals to α-pinene ozonolysis SOA can only be explained by a non-equilibrium, kinetically limited condensation mechanism rather than the equilibrium partitioning commonly assumed to occur between gas- and liquid-particles.

If SOA exists in an amorphous, semi-solid state, then its properties are expected to be affected by environmental conditions. Saukko et al.[12] showed that α-pinene SOA remains solid or semi-solid at RH < 50% and a transition to more liquid-like behavior was seen when RH is higher than 50%. Recent work of Kuwata and Martin[13] reported that gas-phase ammonia uptake by α-pinene SOA under dry conditions (RH < 5%) was consistent with adsorption on highly viscous semi-solid particles, whereas the adsorption/absorption of ammonia on SOA particles was significantly increased at RHs close to saturation, suggesting that the SOA had become liquid.

As evidence accumulates showing that SOA possesses solid or semi-solid properties,[14] it is important to understand how the fundamental physical properties of SOA-containing particles affect heterogeneous reactivity. In particular, are all reactants available for surfaces reactions, implying rapid diffusion through the particle? Or, does semi-solid SOA provide shielding from gas-phase oxidants? Through the analysis of the kinetics of multi-component heterogeneous reactions, we should be able to derive estimates for the diffusivity of reactants through SOA, noting these are largely lacking for this hard-to-measure quantity. In particular, to the best of our knowledge, only one value of a bulk diffusion coefficient of 2.5×10^{-17} cm^2 s^{-1} for pyrene diffusion in α-pinene SOA has been estimated to date.[9]

To this end, we present a study on the effect of α-pinene SOA viscosity/diffusivity on the kinetics of gas-phase ozone reacting heterogeneously with particle-borne benzo[*a*]pyrene (BaP) that was initially buried under the SOA coat. The effects of water-uptake induced changes in SOA on the heterogeneous reactivity were also investigated. As a model heterogeneous reaction system, the kinetics of particle-borne BaP reacting with gas-phase ozone have been studied previously.[15–17] In particular, our group has demonstrated that this reaction can be effectively shut down when a solid organic coating is deposited on top of adsorbed-BaP, whereas liquid coatings impose no mass transfer limitations on the kinetics.[17] Kinetic models for aerosol surface and bulk chemistry (K2-SURF[18] and KM-SUB[19]) are applied to simulate the BaP degradation kinetics and to estimate the diffusivity of BaP in α-pinene SOA at room temperature and interpret the effect of changing RH on the kinetics.

2 Experimental

2.1 Particle generation, particle coating, and aerosol kinetics

A schematic representation of the experimental set-up is shown in Fig. 1. The particles were generated by nebulizing a ~1 mM ammonium sulfate (AS) (solid, ≥99.5%, Fluka) aqueous solution using an atomizer (TSI model 3076). A fraction of the poly-disperse AS particles (~300 sccm) was coated with sub-monolayer benzo[*a*]pyrene (BaP) (solid, ≥98%, Sigma-Aldrich) in a manner analogous to our previous work, by passing through a heated tube containing BaP.[17] The BaP-coated AS particles (AS-BaP) were then mixed with gas-phase α-pinene in a mixing flow tube (Fig. 1). The gas-phase α-pinene was introduced into the tube by passing 0.5–1 sccm nitrogen over the headspace of α-pinene liquid (≥99%, liquid) which had been placed in a bubbler. The bubbler was cooled to −15 °C and the α-pinene mixing ratio varied between 250–500 ppb, as measured by a

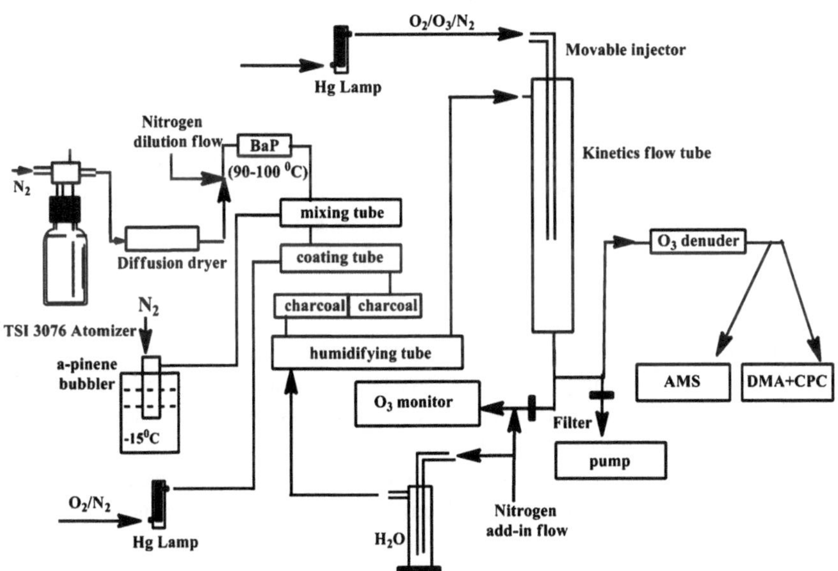

Fig. 1 A schematic representation of the experimental set-up.

proton-transfer-reaction mass spectrometer. The AS-BaP particles and the gas-phase α-pinene were then sent to a second coating tube, where ozone was added and the AS-BaP particles were coated with SOA from the ozonolysis reaction with α-pinene. Ozone (~100 sccm, 2–3 ppm) was generated by ultraviolet irradiation of a mixed flow of N_2/O_2 in a Pyrex glass chamber with a mercury pen-ray lamp (UVP Inc.). The SOA coated AS-BaP particles (AS-BaP-SOA) then passed through two charcoal denuders to remove the gas-phase reactants and products before entering a humidifying tube. Using a scanning mobility particle sizer (SMPS) consisting of a differential mobility analyzer (DMA, model 3080, TSI Inc.) and a condensation particle counter (CPC 3025, TSI Inc.), no SOA formation was observed with only ozone and α-pinene present in the coating flow tube, confirming that the SOA measured in this work all arises from the condensation of low volatility products from ozonolysis of α-pinene onto AS-BaP particles, with negligible new particle formation.

A 1000 sccm dry nitrogen flow was added into a horizontally oriented humidifying tube making a total particle flow of ~1.9 slpm in the kinetics flow tube (RH < 5%). A ~50% RH in the kinetics flow tube was achieved by passing the dry nitrogen flow through a bubbler containing deionized water, whereas ~70% RH was achieved by also adding ~50 ml deionized water into the horizontally oriented humidifying tube. The residence time for the particles and water vapor in the humidifying tube was ~60s. The dry or humidified AS-BaP-SOA particles were then introduced into a vertically oriented kinetics flow tube (Fig. 1), where a flow (~100 sccm) containing ozone was added through a movable stainless steel injector.

The kinetics flow tube (6 cm i.d. and 96 cm in length) was operated at room temperature and atmospheric pressure (296 ± 3 K and 1 atm) under laminar flow conditions. The reaction time (up to 66 s) between BaP and ozone was varied by setting the injector to different positions along the flow tube. The total flow exiting the kinetics flow tube was divided into three: 1) ~100 sccm was introduced into a UV photometric O_3 analyzer (Thermo Model 49i) with a dilution flow of 1.3 slpm dry nitrogen; 2) ~400 sccm passing through an ozone denuder, which can remove more than 90% of ozone, was sampled by the SMPS and an Aerodyne Aerosol Mass Spectrometer (C-ToF AMS) to measure particle mobility size distributions and chemical composition, respectively; and 3) the remaining aerosol flow was removed by a diaphragm pump (Fig. 1). The ozone was in large excess to BaP in the kinetics flow tube and its mixing ratio was controlled by varying the ratio of O_2 to N_2 passing the pen-ray lamp.

2.2 Aerosol composition measurement

The aerosol chemical composition was measured by Aerodyne AMS. The instrument operation and sample analysis have been detailed in our previous work.[17] Briefly, the C-ToF AMS was operated in either mass spectrum (MS) mode or particle time-of-flight (PTOF) mode, with the former producing the mass spectra of non-refractory components of submicron particles and the latter providing size-resolved mass spectra. The molecular ion of m/z 252 was used to detect particle-borne BaP and was normalized by the sulfate mass loading in the kinetics studies. Test experiments showed that the SOA contributed between 5–20% to the total m/z 252 signal intensities of the AS-BaP-SOA particles. This contribution is found to be proportional to the total organic mass loadings and was unaffected by the presence of ozone in the flow tube. Therefore, the contribution of the SOA to the m/z 252 signal is easily accounted for. In addition, as has been pointed out in our previous work,[17] a small amount of organics, arising either from the deionized milli-Q water or laboratory contamination, have always been observed by the AMS in the AS particles. However, the background organics can only account for a maximum of 10% and 2% of the total organic mass loadings after 'thin' and 'thick' SOA coatings, respectively, were applied. The amount of BaP coated onto the AS particles was very small, representing well less than a monolayer coverage, assuming uniform coverage.

3   Results and discussion

3.1   Particle characterization

Fig. 2 provides an example of the particle-size-resolved measurements from the AMS before (a) and after (b) the SOA coating is applied. The size distributions of the three materials, *i.e.* organic, sulfate and m/z 252, are all shifted to larger sizes after the AS-BaP particles are coated with SOA, with the mass loadings for organic and sulfate increased while that for m/z 252 decreased (note the different scales for organic mass loadings in Fig. 1 before (a) and after (b) SOA is coated). The increase in sulfate mass loading is likely due to the enhanced collection efficiency of the AMS with SOA coated AS-BaP particles, implying that the SOA formed from ozonolysis of α-pinene is not as rigid as solid AS-BaP particles. The decrease in m/z 252 mass loading is because of a small degree of reaction of surface-bound BaP with ozone in the SOA coating tube. In addition, we note that the size distributions for these three materials are qualitatively similar before SOA is coated. However, after the SOA coating is applied, sulfate and m/z 252 remain with generally similar size distributions, whereas the larger particles are more organic rich than the smaller particles.

The SOA coating thickness is estimated using a method detailed in our previous work (in supplementary information of Zhou *et al.*[17]) to be 5–10 nm and 20–80 nm in thickness for 'thin' and 'thick' coatings, respectively. In particular, we use the AMS data for these estimates, assuming uniform spherical coats. The estimated absolute thickness of the coated organic is almost independent of particle size for the 'thin' coating experiments; for the 'thick' coats, however, the

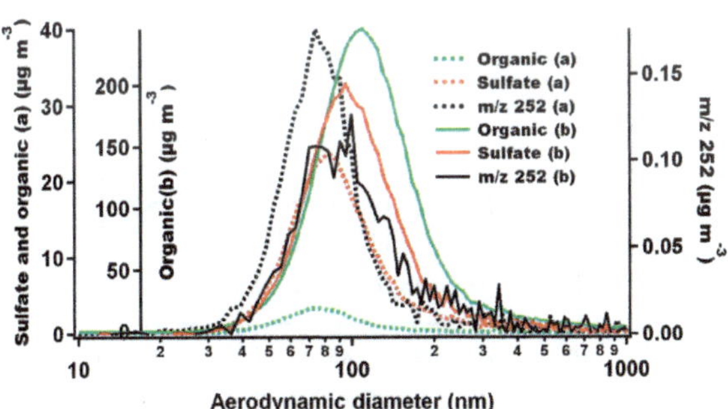

Fig. 2   Size-resolved mass distributions for chemical species obtained from the PTOF mode of the AMS for the AS-BaP particles before (a) and after (b) SOA coating is applied.

coating thickness increased to some degree with particle size, consistent with the observation above that the larger particles are more organic rich.

3.2   Kinetics of reaction of the SOA coated BaP particles with ozone

Fig. 3 gives an example of the BaP concentrations plotted against reaction time for thin SOA coating experiments. The uncertainty in each data point was taken as the standard error of the measurement. Control experiments were conducted in a manner analogous to the kinetic measurements without ozone present. The small changes in the BaP concentrations as a function of reaction time have been accounted for in the data analysis.

The good linearity of the kinetic plots in Fig. 3 demonstrates that the reaction of particle-borne BaP with ozone exhibits pseudo-first order kinetics in terms of BaP loss. The linear least-squares fit of the data in Fig. 3 yields the first-order rate constant (k_1) at a specific ozone concentration. All the k_1 obtained in this work are summarized in Table 1. The uncertainties in k_1 are given as a combination of the least-squares 1-σ uncertainties in the kinetic measurements and control experiments. The k_1 in Table 1 are plotted as a function of the corresponding ozone concentrations and presented in Fig. 4 and 5 for dry (<5% RH) and high RH experiments, respectively.

All the plots in Fig. 4 and 5 (except for the thick solid organic coating in Fig. 4 taken from our previous work[17]) show saturation of the kinetics at high ozone concentrations. This is consistent with a Langmuir–Hinshelwood (L-H) mechanism, indicating a surface reaction between particle-borne BaP and ozone (or a reactive intermediate).

Using the following L-H equation, two parameters, *i.e.* k_{max} and K_{O_3}, can be obtained by fitting the data given in Fig. 4 and 5:

$$k_1 = \frac{k_{max}K_{O_3}[O_3]}{1 + K_{O_3}[O_3]}$$

where k_{max} is the maximum first-order rate constant, K_{O_3} is the gas-to-particle partition coefficient of ozone, and $[O_3]$ is the gas-phase ozone concentration in the kinetics flow tube.

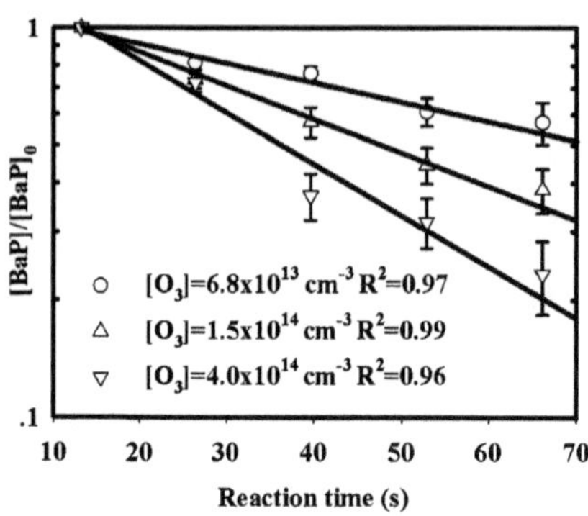

Fig. 3   Examples of the kinetic plots for the reaction of gas-phase ozone with BaP-AS particles with thin SOA coats. Symbols represent different ozone concentrations in the kinetics flow tube.

Table 1 The pseudo-first-order rate constants (k_1) for the heterogeneous reactions of gas-phase ozone (in concentration units of 10^{14} molecules cm^{-3}) with BaP with different thickness of SOA coatings under dry and humid conditions

Thin SOA dry		Thick SOA dry		Thick SOA 50% RH		Thick SOA 70% RH	
[O_3]	k_1 (s^{-1})	[O_3]	k_1 (s^{-1})	[O_3]	k_1 (s^{-1})	[O_3]	k_1 (s^{-1})
0.37	0.008 ± 0.002	1.94	0.007 ± 0.003	0.74	0.010 ± 0.004	0.77	0.013 ± 0.003
0.68	0.010 ± 0.003	3.18	0.009 ± 0.003	1.24	0.016 ± 0.005	1.05	0.014 ± 0.002
1.52	0.018 ± 0.003	4.60	0.013 ± 0.003	2.26	0.015 ± 0.003	1.38	0.023 ± 0.003
2.95	0.024 ± 0.004	4.98	0.012 ± 0.004	2.57	0.016 ± 0.002	2.13	0.019 ± 0.004
3.22	0.028 ± 0.005	5.91	0.014 ± 0.004	3.20	0.017 ± 0.003	2.55	0.021 ± 0.003
3.98	0.028 ± 0.004			4.09	0.017 ± 0.002	2.55	0.022 ± 0.003
				4.60	0.017 ± 0.002	4.34	0.028 ± 0.005
				4.65	0.022 ± 0.004	5.44	0.034 ± 0.004
				4.79	0.022 ± 0.004	6.15	0.034 ± 0.005
				5.09	0.023 ± 0.003	6.83	0.028 ± 0.003
				5.98	0.021 ± 0.003		
				6.03	0.019 ± 0.004		

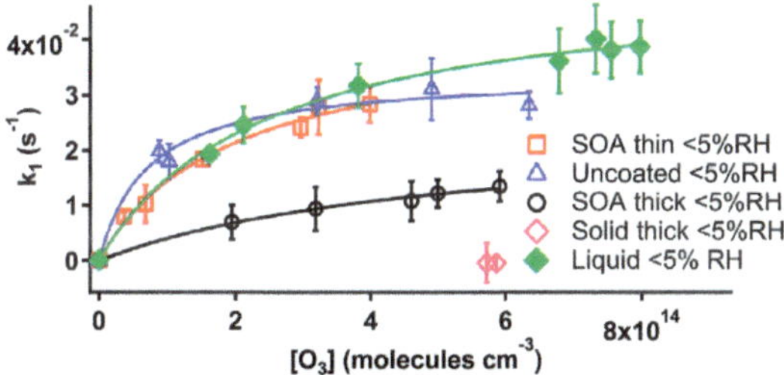

Fig. 4 Pseudo-first-order rate constants k_1 as a function of gas-phase ozone concentrations for the reaction of particle-borne BaP and ozone with different organic coatings under dry conditions. Note that the "Solid thick" and "Liquid" coating data come from Zhou *et al.*,[17] where a thick coating of a solid alkane was deposited onto BaP-AS particles.

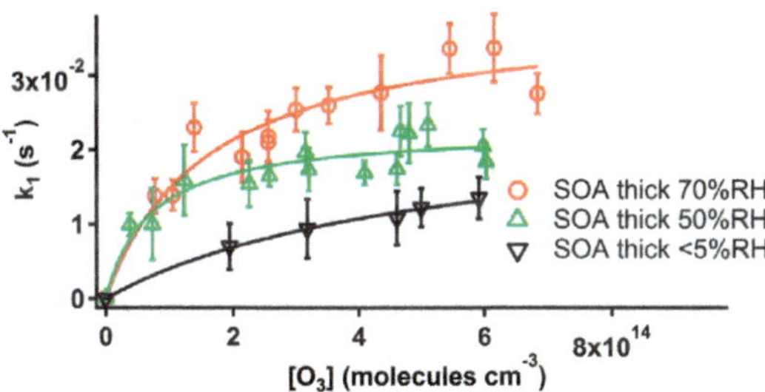

Fig. 5 Pseudo-first-order rate constants k_1 as a function of gas-phase ozone concentrations for the reaction of particle-borne BaP and ozone with thick SOA coatings under different RH.

Table 2 summarizes the best-fit k_{max} and K_{O_3} values obtained in this work with associated standard deviations and those from previous studies on the heterogeneous reaction of ozone with BaP or anthracene. The k_{max} values obtained in

Table 2 Comparison of results from the present work and previous studies of the heterogeneous reaction of particle-borne BaP or anthracene with gas-phase ozone (TTTS = 1,1,5,5-tetraphenylte-tramethylthisiloxane, BES = bis(2-ethylhexyl)sebacate, PSO = phenylsiloxane oil)

PAH	Aerosol substrate	K_{O_3} (10^{-15} cm^3)	k_{max} (s^{-1})	References
BaP	Soot	280 ± 20	0.015 ± 0.001	15
BaP	Azelaic acid	1.2 ± 0.4	0.048 ± 0.008	16
Anthracene	TTTS	100 ± 40	0.010 ± 0.003	20
Anthracene	Azelaic acid	2.2 ± 0.9	0.057 ± 0.009	20
BaP	AS	14 ± 4	0.034 ± 0.002	17
BaP	AS-BES	4.1 ± 0.1	0.051 ± 0.001	17
BaP	AS-PSO	13 ± 4	0.047 ± 0.004	17
BaP	AS-thin SOA dry	5.1 ± 0.9	0.042 ± 0.004	This work
BaP	AS-thick SOA dry	2.3 ± 0.4	0.022 ± 0.003	This work
BaP	AS-SOA 50% RH	14 ± 4.4	0.023 ± 0.004	This work
BaP	AS-SOA 70% RH	6.2 ± 1.1	0.039 ± 0.004	This work

this work for the thin SOA dry and thick SOA ~70% RH of 0.042 s^{-1} and 0.039 s^{-1} (Table 2), respectively, are in good agreement with those for uncoated and liquid-organic coated BaP particles, as well as most of the literature data (Table 2).

3.3   Effect of SOA phase on the kinetics

From Fig. 4 it is clear that under dry conditions (RH < 5%) a thin SOA coating provides little-to-no mass transfer limitation on the kinetics of particle-borne BaP with ozone compared to the uncoated AS-BaP particles, whereas the reactivity of BaP towards ozone is substantially reduced by a thick SOA coating. In our previous work,[17] where we investigated the effects of organic coatings of different physical states, we found that the reactivity of BaP towards ozone was unaffected by liquid oil coatings regardless of whether the coating was thin or thick (Fig. 4). By contrast, the reactivity of BaP was reduced significantly when a thin solid alkane coating was applied and the reaction was effectively shut off with a thick solid coating (Fig. 4).[17]

These observations were explained by the different diffusivity of the reactants through the organic coats. For liquid organic coatings, the diffusion of the BaP through the coated materials was estimated to be fast, on a timescale of nano- to microseconds.[17] As a result, the kinetics were not limited by mass transfer. For the solid organic, the reactant diffusivity was likely to be orders of magnitude lower so kinetics were controlled by solid phase diffusion.[17]

We interpret the present results in a similar manner. The negligible effect of the thin SOA coating may arise because diffusion is sufficiently fast to replace BaP being lost at the surface of the particle through an interfacial process. Note that we cannot rule out the possibility that the BaP may not be fully buried under the SOA material, leaving the BaP directly exposed to gas-phase ozone. However, in a similar coating procedure of SOA onto ammonium bisulfate particles, we observed that equally thin coats provided a dramatic reduction in the reactivity of N_2O_5.[21]

Compared to the kinetic plots for the uncoated and liquid-organic-coated (Fig. 4) particles,[17] the reduction in BaP reactivity by a thick SOA coating under dry conditions (<5% RH) is likely a result of the mass transfer constraints that the

SOA coats place on the kinetics, indicating that the α-pinene SOA is not a low viscosity liquid under these conditions. This is consistent with previous studies suggesting that SOA can be solid or semi-solid.[7,10,14] Moreover, the difference in the kinetic plots for the thick SOA and solid organic coatings (Fig. 4), with the latter entirely shutting off the reaction, implies that the SOA under dry conditions is less viscous than a rigid solid, allowing the reactants to diffuse through the organic coats to some extent over the experimental timescale of the order of minutes. This is consistent with previous work showing that the SOA particles possess some degree of fluidity.[8,13,22,23]

The effects of organic coatings on heterogeneous reactions have been investigated by a number of research groups. For example, studies on the reactive uptake of N_2O_5 on different surfaces, such as AS particles,[24] aqueous H_2SO_4 solution/particles[25–28] and sea salt aerosols,[29–32] and the uptake of ozone on deliquesced potassium iodide (KI) particles[33] and aqueous pyrene solutions,[34] have shown a decrease in reactivity in the presence of a monolayer surfactant coating. A decrease of the uptake of N_2O_5 on ammonium bisulfate particles coated with SOA has also been reported.[21,35] We note, however, that this is a reaction with a substantial bulk-phase component, so the results are interpreted by the effects of the organic coating on the ability of N_2O_5 to access water in the bulk of the particle. By contrast, several studies have shown an enhancement of PAH uptake at air–water interface,[34] gas-phase HCl uptake on aqueous H_2SO_4 solutions[36–38] as well as O_3 reactivity on aqueous pyrene solutions[39] in the presence of soluble organic surfactants. These are likely related to the changes in mass accommodation coefficients. Thus, there is a complexity that can arise in the effects of coatings on heterogeneous reactions beyond that which is explored in this paper, likely related to the phase states of the organic coats and their specific interactions with the reactants.

3.4   Effect of humidity-induced viscosity/diffusivity changes

It can be seen from Fig. 5 that the heterogeneous reactivity of BaP towards ozone is enhanced when RH increases. The higher reactivity is likely due to water uptake by SOA resulting in a lower viscosity of the SOA materials and hence increased diffusivity of the reactants through the coats. The similarities in the kinetic plots for ~70% RH and uncoated BaP, as well as those for liquid organic coats (Fig. 4, 5 and Table 2), suggest a transition of the SOA from semi-solid to lower viscosity, more-liquid-like particles between 50 to 70% RH. We note that the RH in the flow tube was kept lower than the AS deliquescence RH (DRH) of 80%[40] to maintain a uniform aerosol substrate; it has been reported that AS mixed with α-pinene ozonolysis SOA has a DRH which is only 4% lower than pure AS.[41] This lends confidence that the AS in the particles remains in the solid state at ~70% RH.

These results are in accord with previous studies. As mentioned above, humidity-induced modification of biogenic SOA particles has been investigated by Saukko et al.[12] by studying particle bounce, where it was concluded that α-pinene photooxidation SOA remains solid or semi-solid at RH < 50%. A phase change from solid to liquid-like substances was suggested in the RH range of 50–64% RH.[12] A similar observation was made in previous work on the multiphase oxidation kinetics of protein substrates with ozone,[42] where the increased ozone uptake with RH was explained by hygroscopic water uptake of amorphous organics leading to an increase of the diffusivity of the reactants through the organic matrix.

While a number of laboratory and modeling studies have reported that the multi-component organic/inorganic particles may undergo liquid–liquid phase separation at high RH conditions,[43–47] it is impossible for us to determine if this is the case for the high RH experiments in the present work. Nevertheless, the kinetic results indicate that the enhanced RH lowers the viscosity of the SOA, lifting the mass transfer constraint on the heterogeneous reactivity of BaP with ozone. Also, we note that the results that have been observed in this work are specific to the form of SOA studied. Subsequent studies of other SOA types are required to determine the extent to which the results are generalizable.

3.5 Kinetic model for degradation kinetics of BaP with ozone

3.5.1 Degradation kinetics of uncoated BaP.

A kinetic double-layer model for aerosol surface chemistry (K2-SURF)[18] is applied to analyze the experimental data (Fig. 6a). The observed nonlinear dependence of k_1 on $[O_3]_g$ can be well reproduced under the assumption of a simple L-H mechanism formalism, in which an adsorbed O_3 molecule reacts with BaP in a surface reaction,[15,48] as shown in Fig. 6c. The O_3 surface-residence time or desorption lifetime (τ_{d,O_3}) inferred from kinetic data based on simple L-H mechanims is more than milliseconds.

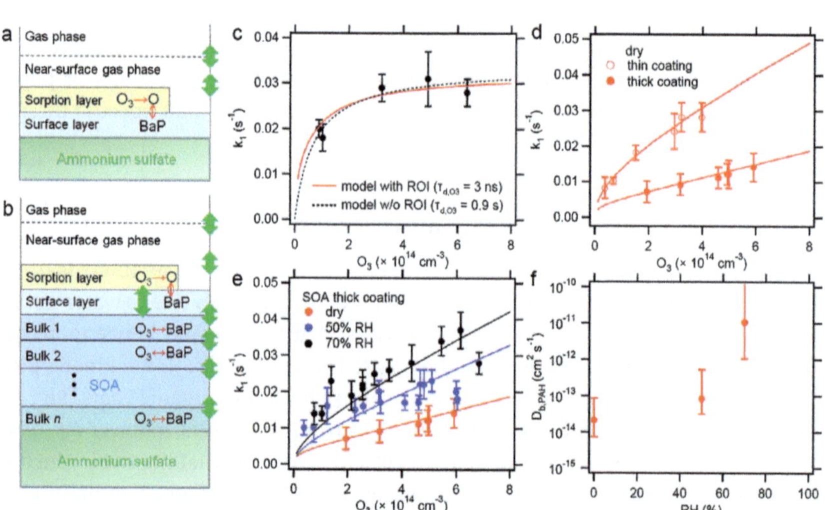

Fig. 6 Kinetic modeling of ozonolysis of BaP on ammonium sulfate aerosol particles. (a) Schematics of kinetic double-layer model (K2-SURF)[18] applied for ozonolysis of uncoated BaP (experimental data taken from our previous work[17]). (b) Schematics of kinetic multi-layer model (KM-SUB)[19] applied for ozonolysis of BaP coated by α-pinene SOA. It consists of a sorption layer, a surface layer, and a number of n bulk layers. Red arrows show chemical reactions and green arrows show mass transport. (c) Pseudo first-order decay rate coefficient (k_1) for uncoated BaP as a function of gas-phase O_3 concentration. The red line is from a model with ROI formation using a desorption lifetime for O_3 (τ_{d,O_3}) of 3 ns. The black dotted line is modeled using a simple L-H formalism without ROI formation with τ_{d,O_3} of 0.9 s. (d) k_1 for BaP coated by α-pinene SOA thinly and thickly under dry conditions. The red lines are modeled by KM-SUB. (e) k_1 for BaP with thick α-pinene SOA coats at different RH (dry, ~50% and ~70% RH). The lines are modeled by KM-SUB. (f) Bulk diffusion coefficient (cm^2 s^{-1}) of BaP in α-pinene SOA as a function of RH derived by KM-SUB.

However, according to quantum mechanical calculations based on density functional theory (DFT), τ_{d,O_3} should be only on the order of nanoseconds,[49] more than six orders of magnitude less. This implies that the actual reaction mechanism is a multi-step L-H mechanism involving the decomposition of surface ozone and formation of long-lived reactive oxygen intermediates (ROI). The dissociated products are molecular oxygen such as a chemisorbed oxygen atom bound to the delocalized π-electrons of an aromatic surface.[49–51] The chemical equations of the multi-step L-H mechanism can be described as follows:

$$O_3 \text{ (g)} \leftrightarrow O_3 \text{ (s)} \tag{R1}$$

$$O_3 \text{ (s)} \leftrightarrow O \text{ (s)} + O_2 \text{ (s)} \tag{R2}$$

$$O \text{ (s)} + PAH \text{ (ss)} \rightarrow O\text{–}PAH \text{ (ss)} \tag{R3}$$

where g, s and ss denote the gas phase, sorption and surface layer, respectively. The weakly bound physisorbed O_3 molecule can be desorbed thermally to the gas phase with a desorption lifetime of nanoseconds, or it can overcome an activation barrier ($E_{a,pc}$), undergo dissociation and enter into a state of stronger binding to the surface. The second activation barrier ($E_{a,ox}$) is the reaction between O atoms (or the appropriate form for the ROI) and BaP to form stable oxidation products. With this formulation K2-SURF can reproduce the kinetics observations very well with parameters listed in Table 3 (red line in Fig. 6c). The estimated activation energies are very similar to those for the ozone reaction with BaP on soot[15,51] and consistent with that estimated from DFT calculations.[49,50]

3.5.2 Degradation kinetics of PAH coated by SOA. For modeling the degradation kinetics of BaP coated by SOA the kinetic multi-layer model for aerosol surface and bulk chemistry (KM-SUB)[19] is applied. Fig. 6b shows the schematics of KM-SUB, which consists of a sorption layer, a quasi-static surface layer, and a number of bulk layers. The thickness of each bulk layer is set to be 0.5 nm, corresponding to the molecular diameter of O_3.[18] KM-SUB treats the following processes explicitly: gas-phase diffusion and reversible adsorption of O_3, bulk

Table 3 Kinetic parameters used in the model simulation assuming a multi-step Langmuir–Hinshelwood mechanism for the ozonolysis of BaP on ammonium sulfate[a]

Parameter	Description	Value
$\alpha_{s,0,O_3}$	surface accommodation coefficient of O_3	1
τ_{d,O_3} (s)	desorption lifetime of O_3	3.2×10^{-9}
D_{b,O_3} (cm^2 s^{-1})	bulk diffusivity of O_3 in SOA	10^{-10}
$D_{b,PAH}$ (cm^2 s^{-1})	bulk diffusivity of PAH in SOA	*
$E_{a,pc}$ (kJ mol^{-1})	activation energy from physisorbed O_3 to ROI	42
$E_{a,ox}$ (kJ mol^{-1})	activation energy from ROI to oxidized PAH	80
k_{BR} (cm^3 s^{-1})	second-order bulk reaction rate coefficient between PAH and O_3	5×10^{-18}
$K_{sol,cc}$ (mol cm^{-3} atm^{-1})	Henry's law coefficient of O_3	6×10^{-4}

[a] * The values are shown in Fig. 6f depending on RH.

diffusion of O_3 and PAH in SOA matrix, surface reaction including formation of ROI and subsequent reaction with PAH, which diffuses through the SOA coating to the surface, and a bulk reaction between O_3 and PAH. The bulk reaction between O_3 and SOA is assumed to be negligible, as the C=C should have been consumed by the gas-phase ozonolysis reaction, forming oxygenated products that then adsorb/condense to form the SOA.[52,53] Before particles are exposed to O_3, they are humidified for ~60 s during which diffusion of BaP commences. To make the most direct comparison between the measurements and the models, the ozone exposure is turned on after a delay period of 60 s. This time matches the residence time in the humidifying tube in the experimental system. Kinetics simulations without this delay time yield essentially the same results.

The kinetic parameters required for simulations are listed in Table 3. $\alpha_{s,0,O_3}$, τ_{d,O_3} and the activation energies are kept the same as in the uncoated PAH case. The Henry's law coefficient of ozone ($K_{sol,cc}$) in α-pinene SOA matrix is not known. $K_{sol,cc}$ in organics generally lies in the range of 10^{-4}–10^{-3} mol cm^{-3} atm^{-1}.[48,54,55] Thus, $K_{sol,cc}$ are varied in this range and other parameters are also varied systematically and iteratively to fit to the experimental data. The sensitivity studies revealed that the bulk diffusivity of PAH in SOA ($D_{b,PAH}$) is the most sensitive parameter and other parameters are not as critical. D_{b,O_3} is assumed to be 10^{-10} cm^2 s^{-1} which is a typical value of small oxidants in amorphous solids, as percolation theory would predict only a slight increase of D_{b,O_3} up to ~80% RH.[42] k_{BR} is estimated to have a low value of 5×10^{-18} cm^3 s^{-1}, which is in line with previous studies reporting that the bulk reaction between O_3 and PAH proceeds slower than surface reaction whose reaction rate coefficient ranges from 10^{-17}–10^{-16} cm^3 s^{-1}.[20,56–58] It is important to note that if a larger value of k_{BR} is used, the modeled k_1 do not show saturation behavior with increasing ozone concentrations but rather exhibit a linear increase trend.

Fig. 6d shows experimentally observed k_1 values for BaP when coated by α-pinene SOA under dry conditions. The SOA coating thicknesses are taken as 8 nm and 40 nm for thin and thick coatings, respectively. KM-SUB successfully fits to both data sets by a single parameter set listed in Table 3. $D_{b,PAH}$ is estimated to be 2×10^{-14} cm^2 s^{-1} under dry conditions. k_1 for BaP coated thickly at ~50% and ~70% RH were also well fit with $D_{b,PAH}$ of 8×10^{-14} cm^2 s^{-1} and 10^{-11} cm^2 s^{-1}, respectively (Fig. 6e). Note that at ~70% RH KM-SUB is not sensitive enough to constrain $D_{b,PAH}$ well but gives more accurately a lower limit of 10^{-12} cm^2 s^{-1}, as k_1 is suppressed only slightly by SOA coating. The obtained $D_{b,PAH}$ are summarized in Fig. 6f, showing the clear increasing trend as RH increases. As discussed previously, this trend corresponds to humidity-induced changes in viscosity/diffusivity of α-pinene SOA.[14,40,59] Note that Abramson et al.[9] estimated the bulk diffusivity of pyrene in α-pinene SOA to be 2.5×10^{-17} cm^2 s^{-1} under dry conditions (<5% RH) based on measurements of evaporation kinetics, which is three orders of magnitude lower than our estimate.

Fig. 7 shows the radial distribution and temporal evolution of the bulk concentration of (a) O_3 and (b) BaP with thick SOA coatings under dry conditions with a gas phase O_3 concentration of 2.5×10^{14} cm^{-3}. The left axis indicates the radial distance from the ammonium sulfate core. Due to low diffusivity and reactive consumption both O_3 and BaP exhibit a steep concentration gradient; the O_3 concentration is higher near the surface and the BaP concentration is higher in the inner bulk. The profile of the degradation rate of BaP (cm^{-3} s^{-1}) is shown in

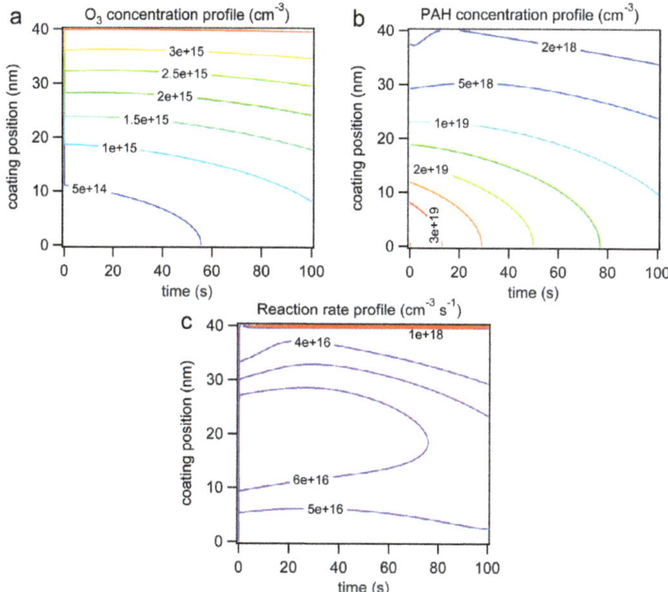

Fig. 7 Radial profile of bulk concentration (cm^{-3}) of (a) ozone and (b) BaP and (c) BaP degradation rate (cm^{-3} s^{-1}) simulated by KM-SUB with dry conditions and a gas phase ozone concentration of 2.5×10^{14} cm^{-3}. The left axis is radial position in SOA coating (0 nm = ammonium sulfate; 40 nm = particle surface).

Fig. 7c. The highest reaction rate is observed at the surface ($>10^{18}$ cm^{-3} s^{-1}), clearly showing that reaction proceeds faster at the surface than in the bulk. The slight radial gradient of reaction rate in the bulk reflects the concentration gradient of O$_3$ and PAH.

We note that these modeling estimates of the diffusion coefficient are dependent upon the assumed mechanism whereby BaP diffuses to the surface to react with ozone at the interface. Other mechanisms cannot be fully ruled out but we note that if we assume a faster bulk phase reaction, whereby ozone diffusivity could end up being rate limiting, the model results yielded a linear dependence of the BaP loss rate constant on ozone concentration. As well, there will be uncertainty arising in our inferred values from the thickness of the SOA coats, which we assume to be uniform.

4   Conclusions and atmospheric implications

This paper illustrates that mass transfer limitations can arise in atmospheric heterogeneous reactions involving SOA. In particular, the effects of viscosity/diffusivity of α-pinene ozonolysis SOA on the heterogeneous reactivity of particle-borne BaP towards gas-phase ozone were investigated. The results show that under dry conditions the SOA coating can lead to mass transfer constraints on the kinetics, with behavior between that of solids and liquids, *i.e.* a semi-solid. The BaP kinetics were enhanced with RH indicating that the water uptake by SOA organics lowers the viscosity and lifts the mass transfer constraint. By application of a kinetic multi-layer model for aerosol surface and bulk chemistry(KM-SUB), the diffusivity (in $cm^2\ s^{-1}$) of BaP in SOA was estimated to be 2×10^{-14}, 8×10^{-14} and $>1 \times 10^{-12}$ for dry (RH < 5%), ~50% RH and ~70% RH, respectively.

It is expected that PAHs, such as BaP, will become coated with SOA after their formation from combustion sources. Both modeling studies and other experimental studies suggest that PAHs buried under multi-layer organic films could be protected from heterogeneous oxidation.[17,60–62] It is quite likely that this provides a mechanism for long range transport of such species to occur in the atmosphere to remote environments such as the Arctic.[63,64] The diffusion coefficients obtained in this work, even at low RH, still imply quite rapid diffusion within the particle, *i.e.* longer than the experimental timescale but shorter than an atmospheric residence time of a week or so. However, this estimate is for one type of SOA only. As well it is not for oxidatively aged SOA, and it has been reported that the viscosity of aged α-pinene SOA is higher than fresh SOA by a factor of three.[9] As well, it is expected that the diffusivity will be markedly lower at lower temperatures, *e.g.* after a particle is lofted out of the boundary layer. Mapping out these dependencies may allow for a more accurate modeling of the lifetime of such species in the atmosphere.

Acknowledgements

Financial support of this work by NSERC (Canada) is gratefully acknowledged.

References

1 M. Hallquist, J. C. Wenger, U. Baltensperger, Y. Rudich, D. Simpson, M. Claeys, J. Dommen, N. M. Donahue, C. George, A. H. Goldstein, J. F. Hamilton, H. Herrmann, T. Hoffmann, Y. Iinuma, M. Jang, M. Jenkin, J. L. Jimenez, A. Kiendler-Scharr, W. Maenhaut, G. McFiggans, Th. F. Mentel, A. Monod, A. S. H. Prévôt, J. H. Seinfeld, J. D. Surratt, R. Szmigielski and J. Wildt, *Atmos. Chem. Phys.*, 2009, **9**, 5155–5236.
2 J. F. Pankow and T. F. Bidleman, *Atmos. Environ., Part A*, 1991, **25**, 2241–2249.
3 J. F. Pankow, *Atmos. Environ.*, 1994, **28**, 189–193.
4 J. R. Odum, T. Hoffmann, F. Bowman, D. Collins, R. C. Flagan and J. H. Seinfeld, *Environ. Sci. Technol.*, 1996, **30**, 2580–2585.
5 N. M. Donahue, A. L. Robinson, C. O. Stanier and S. N. Pandis, *Environ. Sci. Technol.*, 2006, **40**, 2635–2643.
6 M. Shiraiwa and J. H. Seinfeld, *Geophys. Res. Lett.*, 2012, **39**, L24801.
7 A. Virtanen, J. Joutsensaari, T. Koop, J. Kannosto, P. Yli-Pirila, J. Leskinen, J. M. Makela, J. K. Holopainen, U. Pöschl, M. Kulmala, D. R. Worsnop and A. Laaksonen, *Nature*, 2010, **467**, 824–827.
8 T. D. Vaden, D. Imre, J. Beranek, M. Shrivastava and A. Zelenyuk, *Proc. Natl. Acad. Sci. U. S. A.*, 2011, **108**, 2190–2195.
9 E. Abramson, D. Imre, J. Beránek, J. Wilson and A. Zelenyuk, *Phys. Chem. Chem. Phys.*, 2013, **15**, 2983–2991.
10 C. D. Cappa and K. R. Wilson, *Atmos. Chem. Phys.*, 2011, **11**, 1895–1911.
11 V. Perraud, E. A. Bruns, M. J. Ezell, S. N. Johnson, Y. Yu, M. L. Alexander, A. Zelenyuk, D. Imre, W. L. Chang, D. Dabdub, J. F. Pankow and B. J. Finlayson-Pitts, *Proc. Natl. Acad. Sci. U. S. A.*, 2012, **109**, 2836–2841.
12 E. Saukko, A. T. Lambe, P. Massoli, T. Koop, J. P. Wright, D. R. Croasdale, D. A. Pedernera, T. B. Onasch, A. Laaksonen, P. Davidovits, D. R. Worsnop and A. Virtanen, *Atmos. Chem. Phys.*, 2012, **12**, 7517–7529.
13 M. Kuwata and S. T. Martin, *Proc. Natl. Acad. Sci. U. S. A.*, 2012, **109**, 17354–17359.
14 T. Koop, J. Bookhold, M. Shiraiwa and U. Pöschl, *Phys. Chem. Chem. Phys.*, 2011, **13**, 19238–19255.
15 U. Pöschl, T. Letzel, C. Schauer and R. Niessner, *J. Phys. Chem. A*, 2001, **105**, 4029–4041.
16 N.-O. A. Kwamena, J. A. Thornton and J. P. D. Abbatt, *J. Phys. Chem. A*, 2004, **108**, 11626–11634.
17 S. Zhou, A. Lee, R. D. McWhinney and J. P. D. Abbatt, *J. Phys. Chem. A*, 2012, **116**, 7050–7056.

18 M. Shiraiwa, R. M. Garland and U. Pöschl, *Atmos. Chem. Phys.*, 2009, **9**, 9571–9586.
19 M. Shiraiwa, C. Pfrang and U. Pöschl, *Atmos. Chem. Phys.*, 2010, **10**, 3673–3691.
20 N.-O. A. Kwamena, M. G. Staikova, D. J. Donaldson, I. J. George and J. P. D. Abbatt, *J. Phys. Chem. A*, 2007, **111**, 11050–11058.
21 E. N. Escorcia, S. J. Sjostedt and J. P. D. Abbatt, *J. Phys. Chem. A*, 2010, **114**, 13113–13121.
22 A. Zelenyuk, J. Yang, C. Song, R. A. Zaveri and D. Imre, *Environ. Sci. Technol.*, 2008, **42**, 8033–8038.
23 T. D. Vaden, C. Song, R. A. Zaveri, D. Imre and A. Zelenyuk, *Proc. Natl. Acad. Sci. U. S. A.*, 2010, **107**, 6658–6663.
24 C. L. Badger, P. T. Griffiths, I. George, J. P. D. Abbatt and R. A. Cox, *J. Phys. Chem. A*, 2006, **110**, 6986–6994.
25 S. C. Park, D. K. Burden and G. M. Nathanson, *J. Phys. Chem. A*, 2007, **111**, 2921–2929.
26 D. A. Knopf, L. M. Cosman, P. Mousavi, S. Mokamati and A. K. Bertram, *J. Phys. Chem. A*, 2007, **111**, 11021–11032.
27 L. M. Cosman and A. K. Bertram, *J. Phys. Chem. A*, 2008, **112**, 4625–4635.
28 L. M. Cosman, D. A. Knopf and A. K. Bertram, *J. Phys. Chem. A*, 2008, **112**, 2386–2396.
29 D. J. Stewart, P. T. Griffiths and R. A. Cox, *Atmos. Chem. Phys.*, 2004, **4**, 1381–1388.
30 J. A. Thornton and J. P. D. Abbatt, *J. Phys. Chem. A*, 2005, **109**, 10004–10012.
31 V. F. McNeill, J. Patterson, G. M. Wolfe and J. A. Thornton, *Atmos. Chem. Phys.*, 2006, **6**, 1635–1644.
32 V. F. McNeill, G. M. Wolfe and J. A. Thornton, *J. Phys. Chem. A*, 2007, **111**, 1073–1083.
33 A. Rouviere and M. Ammann, *Atmos. Chem. Phys.*, 2010, **10**, 11489–11500.
34 B. T. Mmereki and D. J. Donaldson, *Phys. Chem. Chem. Phys.*, 2002, **4**, 4186–4191.
35 M. Folkers, T. F. Mentel and A. Wahner, *Geophys. Res. Lett.*, 2003, **30**, 1644, DOI: 10.1029/2003GL017168.
36 S. V. Glass, S.-C. Park and G. M. Nathanson, *J. Phys. Chem. A*, 2006, **110**, 7593–7601.
37 D. K. Burden, A. M. Johnson and G. M. Nathanson, *J. Phys. Chem. A*, 2009, **113**, 14131–14140.
38 S. C. Park, D. K. Burden and G. M. Nathanson, *Acc. Chem. Res.*, 2009, **42**, 379–387.
39 S. N. Henderson and D. J. Donaldson, *J. Phys. Chem. A*, 2012, **116**, 423–429.
40 S. T. Martin, *Chem. Rev.*, 2000, **100**, 3403–3453.
41 M. L. Smith, M. Kuwata and S. T. Martin, *Aerosol Sci. Technol.*, 2011, **45**, 244–261.
42 M. Shiraiwa, M. Ammann, T. Koop and U. Pöschl, *Proc. Natl. Acad. Sci. U. S. A.*, 2011, **108**, 11003–11008.
43 C. Marcolli and U. K. Krieger, *J. Phys. Chem. A*, 2006, **110**, 1881–1893.
44 V. G. Ciobanu, C. Marcolli, U. K. Krieger, U. Weers and T. Peter, *J. Phys. Chem. A*, 2009, **113**, 10966–10978.
45 A. K. Bertram, S. T. Martin, S. J. Hanna, M. L. Smith, A. Bodsworth, Q. Chen, M. Kuwata, A. Liu, Y. You and S. R. Zorn, *Atmos. Chem. Phys.*, 2012, **11**, 10995–11006.
46 M. Song, C. Macolli, U. K. Krieger, A. Zuend and T. Peter, *Atmos. Chem. Phys.*, 2012, **12**, 2691–2712.
47 A. Zuend and J. H. Seinfeld, *Atmos. Chem. Phys.*, 2012, **12**, 3857–3882.
48 M. Ammann, U. Pöschl and Y. Rudich, *Phys. Chem. Chem. Phys.*, 2003, **5**, 351–356.
49 A. Maranzana, G. Serra, A. Giordana, G. Tonachini, G. Barco and M. Causa, *J. Phys. Chem. A*, 2005, **109**, 10929–10939.
50 A. Giordana, A. Maranzana, G. Ghigo, M. Causa and G. Tonachini, *J. Phys. Chem. A*, 2008, **112**, 973–982.
51 M. Shiraiwa, Y. Sosedova, A. Rouviere, H. Yang, Y. Zhang, J. P. D. Abbatt, M. Ammann and U. Pöschl, *Nat. Chem.*, 2011, **3**, 291–295.
52 M. Kanakidou, J. H. Seinfeld, S. N. Pandis, I. Barnes, F. J. Dentener, M. C. Facchini, R. Van Dingenen, B. Ervens, A. Nenes, C. J. Nielsen, E. Swietlicki, J. P. Putaud, Y. Balkanski, S. Fuzzi, J. Horth, G. K. Moortgat, R. Winterhalter, C. E. L. Myhre, K. Tsigaridis, E. Vignati, E. G. Stephanou and J. Wilson, *Atmos. Chem. Phys.*, 2005, **5**, 1053–1123.
53 J. H. Kroll and J. H. Seinfeld, *Atmos. Environ.*, 2008, **42**, 3593–3624.
54 G. D. Smith, E. Woods, C. L. DeForest, T. Baer and R. E. Miller, *J. Phys. Chem. A*, 2002, **106**, 8085–5095.
55 M. D. King, A. R. Rennie, K. C. Thompson, F. N. Fisher, C. C. Dong, R. K. Thomas, C. Pfrang and A. V. Hughes, *Phys. Chem. Chem. Phys.*, 2009, **11**, 7699–7707.
56 T. F. Kahan, N. O. A. Kwamena and D. J. Donaldson, *Atmos. Environ.*, 2006, **40**, 3448–3459.
57 C. Liu, P. Zhang, B. Yang, Y. Wang and J. Shu, *Environ. Sci. Technol.*, 2012, **46**, 7575–7580.
58 J. McCabe and J. P. D. Abbatt, *J. Phys. Chem. C*, 2009, **113**, 2120–2127.
59 E. Mikhailov, S. Vlasenko, S. T. Martin, T. Koop and U. Pöschl, *Atmos. Chem. Phys.*, 2009, **9**, 9491–9522.
60 V. S. Semeena, J. Feichter and G. Lammel, *Atmos. Chem. Phys.*, 2006, **6**, 1231–1248.

61 G. Lammel, A. M. Sehili, T. C. Bond, J. Feichter and H. Grassl, *Chemosphere*, 2009, **76**, 98–106.
62 A. Zelenyuk, D. Imre, J. Beránek, E. Abramson, A. Wilson and M. Shrivastava, *Environ. Sci. Technol.*, 2012, **46**, 12459–12466.
63 R. W. MacDonald, L. A. Barrie, T. F. Bidleman, M. L. Diamond, D. J. Gregor, R. G. Semkin, W. M. J. Strachan, Y. F. Li, F. Wania, M. Alaee, L. B. Alexeeva, S. M. Backus, R. Bailey, J. M. Bewers, C. Gobeil, C. J. Halsall, T. Harner, J. T. Hoff, L. M. M. Jantunen, W. L. Lockhart, D. Mackay, D. C. G. Muir, J. Pudykiewicz, K. J. Reimer, J. N. Smith, G. A. Stern, W. H. Schroeder, R. Wagemann and M. B. Yunker, *Sci. Total Environ.*, 2000, **254**, 93–234.
64 R. Lohmann and G. Lammel, *Environ. Sci. Technol.*, 2004, **38**, 3793–3803.

Faraday Discussions

RSC Publishing

PAPER

Tropospheric aerosol as a reactive intermediate†

Agustín J. Colussi,[*a] Shinichi Enami,[*bcd] Akihiro Yabushita,[e] Michael R. Hoffmann,[a] Wei-Guang Liu,[f] Himanshu Mishra[af] and William A. Goddard, III[f]

Received 14th March 2013, Accepted 3rd April 2013

DOI: 10.1039/c3fd00040k

In tropospheric chemistry, secondary organic aerosol (SOA) is deemed an end product. Here, on the basis of new evidence, we make the case that SOA is a key reactive intermediate. We present laboratory results on the catalysis by carboxylate anions of the disproportionation of NO_2 'on water': $2NO_2 + H_2O = HONO + NO_3^- + H^+$ (R1), and supporting quantum chemical calculations, which we apply to reinterpret recent reports on (i) HONO daytime source strengths *vis-à-vis* SOA anion loadings and (ii) the weak seasonal and latitudinal dependences of NO_x decay kinetics over several megacities. HONO daytime generation *via* R1 should track sunlight because it is generally catalyzed by the anions produced during the photochemical oxidation of pervasive gaseous pollutants. Furthermore, by proceeding on the everpresent substrate of aquated airborne particulates, R1 can eventually overtake the photolysis of NO_2: $NO_2 + h\nu = NO + O(^3P)$ (R2), at large zenith angles. Thus, since R1 leads directly to ·OH-radical generation *via* HONO photolysis: $HONO + h\nu = NO + ·OH$, whereas the path initiated by R2 is more circuitous and actually controlled by the slower photolysis of O_3: $O_3 + h\nu (+H_2O) = O_2 + 2·OH$, the competition between R1 and R2 provides a mechanistic switch that buffers ·OH concentrations and NO_2 decay (*via* R1 and/or $NO_2 + ·OH = HNO_3$) from actinic flux variations.

---

[a]Ronald and Maxine Linde Center for Global Environmental Science, California Institute of Technology, Pasadena, California 91125, USA. E-mail: ajcoluss@caltech.edu

[b]The Hakubi Center for Advanced Research, Kyoto University, Kyoto 606-8302, Japan. E-mail: enami.shinichi. 3r@kyoto-u.ac.jp

[c]Research Institute for Sustainable Humanosphere, Kyoto University, Uji 611-0011, Japan

[d]PRESTO, Japan Science and Technology Agency, Kawaguchi 332-0012, Japan

[e]Department of Molecular Engineering, Kyoto University, Kyoto 615-8510, Japan

[f]Materials and Process Simulation Center, California Institute of Technology, Pasadena, California 91125, USA

† Electronic supplementary information (ESI) available: Further experimental details, additional data and information. See DOI: 10.1039/c3fd00040k

1   Introduction

More than half of the world population inhabits rapidly expanding conurbations with increasing *per capita* energy needs.[1] Assessing the impact of these combined trends on air pollution, human health and quality of life in megacities is therefore a pressing issue.[2–4] This goal is being approached *via* three-dimensional chemical transport models that include pollutant emissions, transport, removal, gas-phase chemistry and aerosol physics.[5–10] Since 'the key to understanding tropospheric chemistry begins with the ·OH radical'[7] a mandatory test for any model is how well it reproduces ·OH radical concentrations under diverse conditions.[11] The chemistry of nitrogen oxides (NO_x) plays an essential role in the production of ·OH radicals, which can be triggered by the photolysis of NO_2, reaction R2, followed by reactions R3–R5:[7,12,13]

$$NO_2 + h\nu \ (\lambda < 422 \text{ nm}) = NO + O(^3P) \tag{R2}$$

$$O_2 + O(^3P) = O_3 \tag{R3}$$

$$O_3 + h\nu \ (\lambda < 340 \text{ nm}) = O_2 + O(^1D) \tag{R4}$$

$$O(^1D) + H_2O = 2\,·OH \tag{R5}$$

The augmented strength of nitrogen oxides (NO_x) emissions in megacities (from internal combustion engines used in transportation) poses a critical test to current models.[1,2,14,15]

Reliable air quality forecast models should have the ability to simulate the composition of the troposphere in most locations throughout the year.[3,16] This is a major challenge, because some pollutants, such as particulate matter,[17] are always present in significant amounts,[18] whereas others, such as O_3, peak in summer. Disappointingly, current models generally fare poorly in predicting how ·OH concentrations respond to actinic flux and NO_x levels,[11] and systematically underpredict HONO concentrations.[11,16,19] The strong direct correlation between tropospheric ·OH concentrations and solar ultraviolet radiation: $[OH] \propto J(O^1D)^\beta$,[20] detected at a pristine site yearlong,[21] and episodically at a polluted one,[22] is not well reproduced by standard chemical models.[22–24] Even more troubling is that, contrary to expectations, noon-time ·OH concentrations (in Birmingham) during winter were found to be only a factor of 2 smaller than in summer, despite a factor of 15 reduction in $J(O^1D)$ over the same period.[25] Given the key role of ·OH, this represents a major deficiency in our understanding of tropospheric chemistry. It has been suggested that the diminished solar photolysis of O_3 in winter could be supplemented by that of HONO,[16,26–30] reaction R6,

$$HONO + h\nu \ (\lambda < 396 \text{ nm}) = NO + ·OH \tag{R6}$$

which is driven at longer wavelengths than R4 and, hence, will still proceed at significant rates at low zenith angles. Since R6 is fast (HONO half-life ~10 min) this proposal in effect calls for an unspecified process that causes HONO daytime production to be competitive with NO_2 photolysis, R2.[31–34] The missing process is likely heterogeneous,[19,35–39] must obviously track sunlight and be able to reduce $N(IV)O_2$ into $HON(III)O$ under the prevailing oxidizing conditions.[40] Reaction R1:

$$2NO_2 + H_2O = HONO + NO_3^- + H^+ \tag{R1}$$

may meet these requirements but, since it entails the concomitant termination of NO_2 as NO_3^-, its likelihood rests on whether is consistent with the kinetics of NO_2 decay in the field. Herein, we address this outstanding issue in tropospheric chemistry *via* a *meta*-analysis of recent field measurements on HONO daytime sources and NO_2 decay in urban plumes from the perspective of our new results on the catalysis by organic anions of NO_2 uptake on water.

The objective of this paper is to present an explicit chemical mechanism for R1, and show that its operation on wet SOA accounts for a suite of apparently unrelated, hitherto unexplained observations. The paper is organized as follows: we (1) report new laboratory results obtained by a novel technique developed in our laboratory[41–44] on the catalysis by dicarboxylic acid anions of the hydrolytic disproportionation of $NO_2(g)$ on aqueous surfaces (reaction R1)[45–47] as well as supporting quantum mechanical calculations and (2) show that such anions are normally produced in urban SOA from the photochemical oxidation of volatile organic compounds (VOCs).[48] We then review and summarize pertinent information on aerosol optical depth time series data over megacities,[49,50] and solar photolysis frequencies $J(NO_2)$, $J(O^1D)$ and $J(HONO)$ as functions of season and latitude.[20] On this factual basis, we provide a novel interpretation of recently reported data on the quantification of the unknown HONO daytime source,[26] and seasonally-averaged daytime NO_2 decay lifetimes over several megacities.[51]

2 Experimental results

In our experiments we expose continuously flowing aqueous microjets to a steady, orthogonal stream of $NO_2(g)$ (at ppmv levels in a N_2 gas carrier) for a few tens of μs under ambient (1 atm total pressure, at 293 K) conditions, while detecting the NO_3^- produced in reactive gas $NO_2(g)$–liquid collisions *via* online electrospray ionization mass spectrometry (ESI-MS). A full description of the instrument and its operation in similar experiments can be found in previous publications from our laboratory.[41–43,47] Specific details are provided as Supplementary Information (ESI),† which includes a diagram of the reaction zone (Fig. S1†). The decisive advantages of online ESI-MS over other techniques are that it (i) operates *in situ*, *i.e.*, avoids sample manipulation, (2) is fast: gas–liquid reaction times are a few tens of μs,[52] and products are detected within ~1 ms, (3) is mass-selective, thereby providing unequivocal identification of ionic reactants and products, (4) has high sensitivity, with routine detection limits down to 0.1 μM, allowing experiments approach realistic conditions, (5) minimizes surface contamination, because reactive gas–liquid events take place on fast-flowing, continuously refreshed liquid microjets. We have previously demonstrated that our experiments are surface specific by showing that anion signal intensities in the mass spectra of equimolar salt solutions, rather than being identical follow a normal Hofmeister series,[53,54] and detect products of gas–liquid reactions that could only be formed at the air–water interface.[41]

The basic information provided by these experiments is how $m/z = 62$ NO_3^- signal intensities, I_{62}, vary as functions of the composition of the aqueous microjets and the partial pressure of $NO_2(g)$ in the impinging streams. The initial composition of the aqueous microjets ranged from deionized water to sub-mM solutions of malonic acid (MA, HOOC–CH_2–COOH), glutaric acid (GTR, HOOC–$(CH_2)_4$–COOH) or glutamic acid (GTM, HOOC–$(CH_2)_3$–$CH_2(NH_2)$–COOH) as

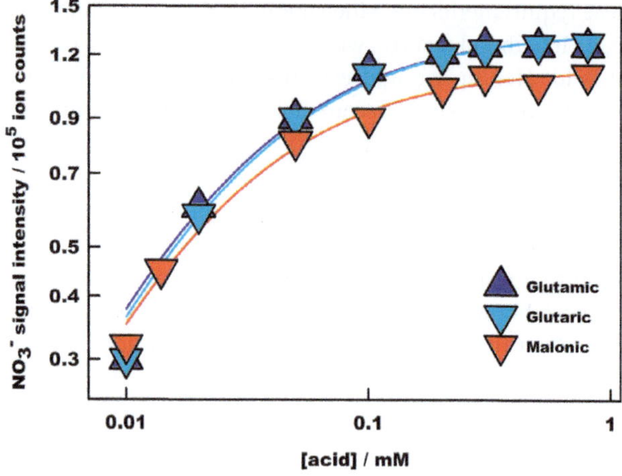

Fig. 1 ESI-MS NO_3^- ($m/z = 62$) signal intensities from aqueous microjets containing glutamic, malonic or glutaric acids at variable concentrations, adjusted to pH 7 and exposed to 2 ppmv [NO_2(g)] for ~10 µs. All experiments under 1 atm N_2(g) at 293 K.

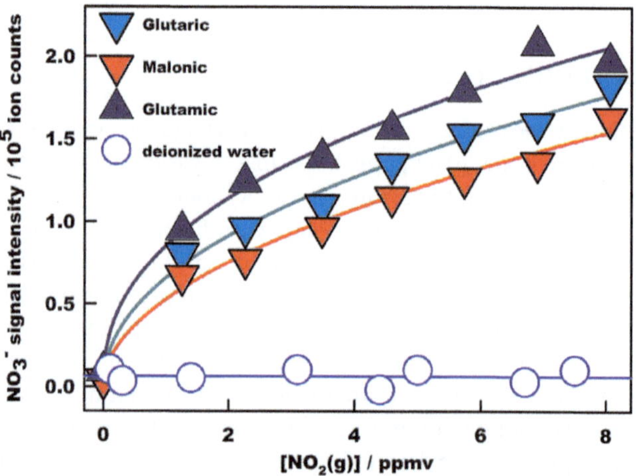

Fig. 2 ESI-MS NO_3^- ($m/z = 62$) signal intensities from aqueous microjets containing 1 mM glutamic, malonic or glutaric acids at pH 7 exposed to variable concentrations of NO_2(g) for ~10 µs. All experiments under 1 atm N_2(g) at 293 K.

representatives of the dicarboxylic acids found in SOA.[48,55] We monitored NO_3^- formation in experiments comprising (1) aqueous microjets of varying dicarboxylic acid concentrations under 2 ppmv NO_2(g) (Fig. 1) (1 ppmv $= 2.5 \times 10^{13}$ molecules cm^{-3} at 1 atm, 293 K. See supplementary Fig. S2†), (2) 1 mM carboxylic acid microjets exposed to variable NO_2(g) concentrations (Fig. 2).

Confirming previous experiments in our laboratory[45–47] and others,[56–58] we found that NO_3^- signals remained at or below noise level on deionized water microjets up to the highest NO_2(g) concentrations (NO_2(g) < 8 ppmv $= 2 \times 10^{14}$ molecules cm^{-3}, Fig. 2). This observation is consistent with the generations of <0.1 µM NO_3^- (<6×10^{13} ions cm^{-3}) in the interfacial layers of the microjets under such conditions (See Fig. S2, ESI†). The surface density, $S(NO_3^-)$ [molecules cm^{-2}], of the NO_3^- produced in liquid layers of thickness δ during $NO_2(g)$ collisions on aqueous microjets, is given by the kinetic theory of gases, eqn (E1):[59]

$$S(NO_3^-) = \delta[NO_3^-] = (1/4)\,\gamma_w\,c\,\tau\,[NO_2(g)]/2 \qquad (E1)$$

where $\delta \sim 5 \times 10^{-8}$ cm is the average thickness of interfacial layers, γ_w is the uptake coefficient of $NO_2(g)$ on water, $c = 3.72 \times 10^4$ cm s^{-1} is the mean speed of NO_2 molecules at 300 K, and $\tau > 10$ μs is a lower bound to $NO_2(g)$/microjets contact time.[52] Thus, we estimate that $\gamma_w < 2 \times 10^{-7}$, in accordance with previous results on the negligible uptake of $NO_2(g)$ on pure water obtained by independent techniques.[57,58] The exceedingly low probability of R1 on pure water is supported by MD calculations showing that the hydrophobic free radical NO_2 has strong propensity for the surface of neutral clusters $NO_2(H_2O)_n$.[60]

In striking contrast (but in accordance with the cloud-chamber Berkeley experiments on NaCl-seeded droplets)[61] we found that the uptake of $NO_2(g)$ increases dramatically in the presence of anions in the sub-millimolar range. Anions are known to be selectively enriched at the air–water interface according to their size.[54,62,63] Under present conditions, dicarboxylic acids ($pK_1 = 2.8$ (MA), 4.3 (GTR), 2.1 (GTM)) are largely present as monoanions. The curves in Fig. 1 correspond to Langmuir adsorption functionals: $I_{62} = I_{62,MAX} \times [acid]/(K_{1/2} + [acid])$, with $K_{1/2} \sim (20 \pm 5)$ μM in all cases. The curves in Fig. 2 correspond to a $I_{62} \propto [NO_2]^\alpha$ functional with $\alpha \sim 0.5$ in all cases. An apparent kinetic order of $\alpha \sim 0.5$ is consistent with a bimolecular reaction R1 proceeding *via* a multistep mechanism whose rate determining step involves one NO_2 molecule (see below). Since at $[NO_2(g)] = 1$ ppmv, $[N_2O_4]/[NO_2] = K_{DIMERIZATION}\,[NO_2] = 2.5 \times 10^{-19}$ cm^3 molecule^{-1} $[NO_2] < 1 \times 10^{-5}$, the participation of N_2O_4 in these events is

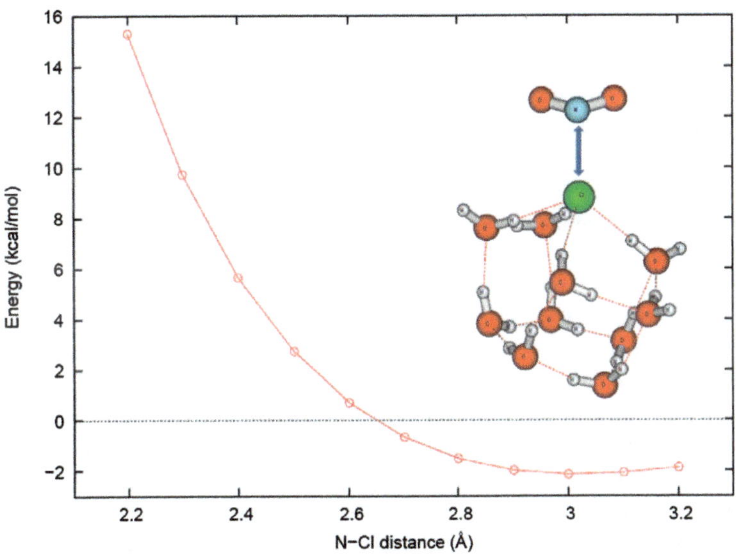

Fig. 3 Potential energy diagram for the interaction between NO_2 and a $[(H_2O)_{10}–Cl]^-$ cluster. The energy between NO_2 and $[(H_2O)_{10}–Cl]^-$ cluster is defined as the energy to separate NO_2 from the optimized clusters to infinity without geometry relaxation. See main text for details on the calculations.

insignificant. From the ratio of I_{62} signal intensities in the presence and absence of carboxylate anions: $I_{62}(\text{anions})/I_{62}(\text{water}) \sim \gamma/\gamma_w$, we estimate that reactive NO_2 uptake coefficients increase with anion concentrations from $\gamma_w \sim 10^{-7}$ up to $\gamma \sim 10^{-4}$–10^{-3} values in the μM to mM range.[46,47] The range of our reported γ values is a conservative estimate based on the reproducibility of a series of experiments in which we alternated the use of water and electrolytes as liquids.

3 Computational results

To gain deeper molecular insights into the interactions underlying our experimental observations, we performed quantum mechanical calculations on NO_2–$[(H_2O)_{10}–Cl]^-$ and NO_2–$[(H_2O)_{10}]$ clusters. From the rather similar effects of different anions, X^-, on γ we observed in our previous experiments,[46,47] we inferred the operation of a non-specific NO_2–X^- interaction. This contention is confirmed by our calculations. The energy of the clusters was optimized at the MP2/6-31G level, and their single point energies were refined by CCSD(T)/6-31G augmented with d-orbitals on N and O of NO_2 and Cl. The interaction energy E_{int} between NO_2 and $[(H_2O)_{10}–Cl]^-$ or $[(H_2O)_{10}]$ clusters is defined as the energy to separate NO_2 from the optimized clusters to infinity without geometry relaxation. MP2 and CCSD(T) calculations were carried out with GAMESS[64] and NWChem,[65,66] respectively. Ten water molecules were considered the minimum number required to solvate Cl^- on the water surface. We found that $E_{int} = -0.5$ kcal mol^{-1} for NO_2–$[(H_2O)_{10}]$ and $E_{int} = -2.2$ kcal mol^{-1} NO_2–$[(H_2O)_{10}–Cl]^-$. The minimum of the potential energy surface (PES) along the N–Cl distance occurs at a 3.0 Å separation (Fig. 3). The small value of $E_{int} = -2.2$ kcal mol^{-1} is consistent with a non-covalent interaction between N and Cl^-. The significant difference between the gas-phase electron affinities, EA, of Cl (EA = 3.61 eV) and NO_2 (EA = 2.27 eV): $\Delta EA = 31$ kcal mol^{-1}, which would be barely compensated by the creation of an incipient single Cl–NO_2 bond (33 kcal mol^{-1} in nitryl chloride),[12] prevents the development of covalent bonding *via* partial charge transfer from Cl^- to the semi-occupied molecular orbital of NO_2. We also found that in the optimized NO_2–$[(H_2O)_{10}–Cl]^-$ cluster the negative charge largely remains localized on Cl^- (the Mulliken charge of Cl^- is -0.80 at the MP2 level). In fact, the electrostatic interaction between Cl^- and NO_2 in a point-like charge model, $E_{int} = 2.1$ kcal mol^{-1}, closely matches the magnitude of interaction energy calculated from CCSD(T). We infer that the binding between NO_2 and a $[(H_2O)_{10}–Cl]^-$ cluster is fully accounted for by electrostatic, charge–dipole interactions. The fact that the interaction is purely electrostatic implies that anions in general will catalyze reaction R3, as we found experimentally. This interaction is long-ranged and prolongs the lifetime of NO_2 on the surface, thereby enhancing the probability of reaction with a second NO_2 molecule. A previous theoretical study[67] revealed that two NO_2 molecules held together by a Cl^- dimerize to *cis*-$ONONO_2$ *via* a modest 1.9 kcal mol^{-1} enthalpy barrier; *cis*-$ONONO_2$ can then proceed to *trans*-$ONONO_2$ (*via* a 2.4 kcal mol^{-1} enthalpy barrier) or react with water to form HONO, H^+ and NO_3^-. Njegic *et al.*[68] found that the activation energy for *cis*-ONO-NO_2 reacting with H_2O is only 8.7 kcal mol^{-1} (at the CCSD(T)/cc-pVTZ level). However, Zhu *et al.* found the hydrolysis barrier for *cis*-ONO-NO_2 is 7.1 kcal mol^{-1} and for *trans*-ONO-NO_2 is 2.3 kcal mol^{-1} at the CCSD(T)/6-311++G(3df,2p) level.[69] With such low enthalpy barriers, the overall reaction rate is determined by the free energy barrier, *i.e.*, by the concentration of NO_2 in the gas-phase. Thus, we conclude that the interaction between NO_2 and a $[(H_2O)_{10}-Cl]^-$ cluster is purely electrostatic and, hence, confirm that all anions (X^-) should catalyze reaction R1, in accord with our experimental observations.[45–47] Our calculations are also consistent with recent molecular dynamics calculations.[70]

The preceding observations are accounted for by the following mechanism:

$$X^-_{\text{interface}} + NO_2(g) \xrightarrow{\text{slow}} [X-NO_2{}^{\cdot-}]\text{interface} \tag{R7}$$

$$[X-NO_2{}^{\cdot-}]_{\text{interface}} + NO_2(g) + H_2O \xrightarrow{\text{fast}} X^- + HONO + NO_3{}^- + H^+ \tag{R8}$$

In which $NO_2(g)$ disproportionates in two steps, reactions R7 and R8, through the intermediacy of persistent, interfacial $X-NO_2{}^{\cdot-}$ adducts. According to this mechanism, the overall rate of R1 is controlled by the slow uptake of NO_2 *via* R7. The probability that a second $NO_2(g)$ molecule reacts with interfacial $X-NO_2{}^{\cdot-}$ adducts, reaction R8, is significantly higher than that of reaction R7. The catalytic role of anions in R7 has been clearly demonstrated in similar experiments in this setup, in which the enhanced production of $NO_3{}^-$ in the presence of various X^- was not accompanied by X^- losses except in the case of $X^- = I^-$, which was partially oxidized to $I_2{}^{\cdot-}$.[47] Thus, the enhancement of $NO_2(g)$ uptake by anions is simply a case of general base catalysis.

4　Tropospheric aerosols: optical depths and anion loadings

It is a remarkable fact that the global distribution of tropospheric aerosols, as characterized by satellite sightings, shows weak seasonal and interannual variations.[49,50] For example, a 67-month time series of daytime column aerosol optical depths (AOD) over land: AOD (532 nm) = 0.145 ± 0.0228 (all-sky) and 0.179 ± 0.0257 (cloud-free) merely display ~15% variations overall.[50] This finding is a direct indication that aerosol has multiple, uncorrelated sources, and that the photochemistry involved is complex, indeed more complex than currently envisioned. Thus, although the composition of SOA varies daily with sunlight, as expected from components of photochemical origin,[48] seasonal trends tend to be rather flat (within a factor of 2) over most megacities,[49] such as Los Angeles,[71] Istanbul,[72] Paris,[73] Moscow,[74] Seoul,[75] Beijing, Chongqing, Shanghai, Chengdu and Guangzhou,[18,76] Pearl River Region (China),[77] Delhi,[78] and Tokyo.[48] For example, the gravimetric mass concentration of ultrafine particles (*i.e.*, the major contributors to number density and specific surface) in Los Angeles, changes from ~6 µg m^{-3} in spring to ~10 µg m^{-3} in fall and winter.[71] The ionic mass over Istanbul steadily increased, *i.e.*, without showing major seasonal oscillations, from 10 µg m^{-3} in November 07 to 20 µg m^{-3} in June 09.[72] Perhaps not unexpectedly, Moscow is less typical: atmospheric aerosol number densities show more pronounced seasonal variability, peaking in winter at values 3–4 times higher than at midsummer.[74]

We have shown in the laboratory that all anions, organic and inorganic,[45–47] catalyze reaction R1. Organic acids are produced as ubiquitous components of

SOA *via in situ* oxidation of the water-soluble compounds generated in the photo-oxidation of volatile organic compounds VOCs.[7,18,72,78–80] Mono and dicarboxylic acids in urban atmospheres systematically increase during the morning peaking at noon, as reported by Kawamura *et al.*,[48] and are highly correlated with non-sea salt sulfate (the most abundant anion throughout) by sharing a common photochemical origin.[79,81–83] Summing up, extensive ground-based and remote data show that mass concentrations and anion makeup of tropospheric aerosols display daily cycles that average out into seasonal variations within a factor of $\sim$2.

5　Seasonal and latitudinal variations of $J(NO_2)$, $J(O^1D)$ and $J(HONO)$ photolysis frequencies

The frequencies at which NO_2, O_3 and HONO are photo-dissociated by sunlight *via* reactions R2, R4 and R6, respectively, are strong functions of zenith angle φ. Thus, $J(NO_2) = 1.7 \times 10^{-2}\ s^{-1} = 4.5\ J(HONO) = 222\ J(O^1D)$ at $\varphi = 0°$, and $J(NO_2) = 5 \times 10^{-3}\ s^{-1} = 4.7\ J(HONO) = 1280\ J(O^1D)$ at $\varphi = 70°$ (Fig. S3†).[20] Since $J(O^1D)$ is the lowest frequency under all conditions, it might be inferred that photo-stationary $\dot{O}H$ concentrations will be always determined by ozone photodissociation.[11] The linear $[\dot{O}H] \propto J(O^1D)$ correlation found at a remote site over a 5-year period seems to confirm this contention,[21] but it should not be expected to hold in general.[25] In fact, there is overwhelming evidence that tropospheric $[\dot{O}H]$ is well buffered against $J(O^1D)$ variations in most locations. The failure of current tropospheric chemistry models to account for such crucial phenomenon has prompted various suggestions about alternative $\dot{O}H$ sources.[39] The generalized perception is that such sources are linked to yet unidentified NO_x chemistry for no better reason that the same models systematically underpredict HONO field measurements.

6　Observational constraints on the unknown HONO daytime source

Three decades after the first spectroscopic detection of HONO in the atmosphere,[84] and much work since,[27,38,85–92] the mechanism of HONO daytime formation remains a puzzle.[93] However, recent work has provided strong constraints on the unknown HONO daytime source and its relation to NO_2 chemistry.[26] Key findings were that the rates of HONO production during daytime, R_{HONO}, at a given site (1) are an order of magnitude faster than extrapolated nighttime rates of the hydrolytic disproportionation of NO_2 on wet surfaces,[27] (2) systematically increase with solar flux, peaking at noon but, however, (3) can be accounted neither by the reaction of excited NO_2^* with $H_2O(g)$,[28] reaction R9 nor by the reduction of NO_2 photosensitized by irradiated soot, $[C–H]_{red}^*$, reaction R10:[46,94]

$$NO_2^* + H_2O = HONO + \dot{O}H \tag{R9}$$

$$NO_2 + [C–H]_{red}^* = HONO + [C]_{ox} \tag{R10}$$

Measured R_{HONO} at noon correspond to the conversion of 10% to 43% NO_2 into HONO per hour.[26] Since a 50% $NO_2 \rightarrow$ HONO conversion is the stoichiometric limit imposed by R1, the implication is that the 'unknown HONO daytime source' must also be a significant daytime sink (as NO_3^-) for NO_2, and should therefore impact NO_2 decay lifetimes. At present, it is not obvious how the heterogeneous 'dark' reaction R1 could be significantly enhanced by sunlight,[26] or whether such enhancements are compatible with the kinetics of NO_2 removal,[95] which is generally deemed to occur exclusively in the gas-phase *via* reaction R11:[96]

$$NO_2 + {}^\bullet OH + M = HNO_3 + M \qquad\qquad (R11)$$

7 Satellite data on NO_2 decay lifetimes in urban plumes

The conversion of the free radical NO_2 into HONO in urban smog amounts to the reduction of N(IV) into N(III) under oxidizing atmospheric conditions, an event made possible by the fact that endoergic NO_2 can function both as oxidant and reductant in reaction R1. A substantial production of HONO at daytime therefore requires a heterogeneous process with R1 stoichiometry. However, the fast daytime conversion of NO_2 into HONO *via* R1 is concomitant fast termination of NO_2 as NO_3^-. Such outcome is at odds with the assumption that NO_2 is removed exclusively by gas-phase $^\bullet$OH *via* R11. Could such assumption be incorrect? Could the diurnal cycle of HONO production and the uniform long-term removal of NO_2 be reconciled by the operation of reaction R1?

Beirle *et al.* recently reported NO_2 decay lifetimes and emission strengths in urban plumes deduced from direct satellite observations.[51] By monitoring various sites ranging from Singapore at 1.3° N to Moscow at 55.8° N to throughout the year, they derived NO_2 emission strengths from direct NO_2 column and lifetime measurements. Up to now, NO_2 emission inventories have been based on e-fold NO_2 lifetimes inferred from eqn (E2):

$$\tau_{NO_2+OH,model} = (k_{NO_2+OH} \times [OH]_{model})^{-1} \qquad\qquad (E2)$$

(*i.e.*, by assuming that NO_2 is removed *via* R11) by using $^\bullet$OH concentrations estimated by atmospheric chemistry models: $[OH]_{model}$. Fig. 4 shows how NO_2 decay lifetimes should have increased from mid-summer to mid-winter from Singapore to Madrid if they were inversely proportional to $[^\bullet OH]$ (eqn (E1))[20] *and* $[OH] \propto J(O^1D)$. However, defying all expectations, seasonal averages of measured NO_2 decay lifetimes, $\langle \tau_{NO_2,measured} \rangle$, were found to fall within a (4 ± 1) h band over all cities below 40° N without significant seasonal trends. Even the outlier $\langle \tau_{NO_2,measured} \rangle$ values over Madrid (40.4° N) and Moscow (55.8° N) were only a factor of 2 longer in winter than in summer. Clearly, the calculated $\tau_{NO_2+OH,model}$ increases vastly exceed the observational dispersion of $\langle \tau_{NO_2,measured} \rangle$ in all cases (Fig. 4) except, of course, over seasonless Singapore. The implications are that (i) if NO_2 were to decay exclusively *via* R11, actual $^\bullet$OH concentrations should not change appreciably with solar irradiance, at variance with expectations based on putative direct $[OH] \propto J(O^1D)^\beta$

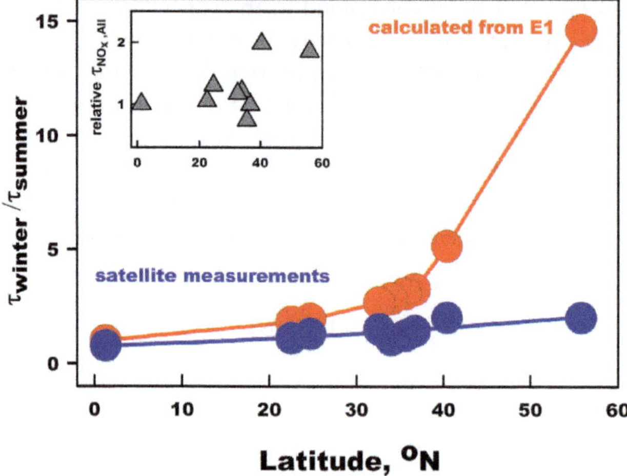

Fig. 4 Red circles and line: Ratio of e-fold NO_2 decay lifetimes $\tau_{winter}/\tau_{summer} = \langle\tau_{NO_2+OH}\rangle_{midwinter}/\langle\tau_{NO_2+OH}\rangle_{midsummer}$ calculated from eqn (E1) (see main text) by assuming that ·OH concentrations are proportional to $J(O^1D)$ (from ref. 20) the frequency of $O(^1D)$ atom production in the solar photolysis of O_3, at zero elevation, surface albedo 0.4, under an ozone column of 300 Dobson units, at noon, in mid-winter and mid-summer over: (1) Singapore 1.3° N, (2) Pearl River Delta 22.5° N, (3) Riyadh 24.6° N, (4) Isfahan 32.6° N, (5) Los Angeles 34° N, (6) Tokyo 35.6° N, (7) Four Corners 36.7° N, (8) Madrid 40.4° N, and (9) Moscow 55.8° N. Herein, mid-summer is Aug. 1st, mid-winter is Feb. 1st. Blue circles and line: Ratio of seasonal mean NO_2 decay lifetimes derived from satellite sightings in plumes downwind from the same locations (ref. 51). Inset: $\langle\tau_{NO_x,All}\rangle$ relative to $\langle\tau_{NO_x,All}\rangle$ at Singapore.

correlations,[21] or (ii) NO_2 also decays by a competitive pathway that simultaneously buffers ·OH concentrations.

8   How R1 buffers tropospheric OH-radicals

Rates of the reactive dissolution of NO_2 in aerosol particles wet with aqueous electrolytes can be calculated from the kinetic theory of gases as: $r(R1) = k_d [NO_2]$, with $k_d = 0.25\ \gamma\ c\ (S/V)$, where (S/V) is the combined surface of wet particles per volume of air.[46,59] Typical (S/V) values vary from $<2 \times 10^{-5}$ cm^{-1} in clear weather to $>5 \times 10^{-5}$ cm^{-1} during foggy events. It should be realized, however, that (S/V) and total electrolyte concentrations of typical cloud and fog droplets and, hence, γ will display wide variability over daily and weekly periods.[97] Thus, by adopting a representative average value for the uptake coefficient of $NO_2(g)$ on tropospheric aerosols: $\gamma \sim 3 \times 10^{-4}$ (see above), we estimate $k_d \sim 6 \times 10^{-5}$ s^{-1}, which corresponds to $\sim$20% $NO_2(g)$ hourly conversions in clear-day conditions, $i.e.$, in the range of those observed in the field.[26] By assuming that all HONO produced in reaction R1 is rapidly photolyzed via R6, the rate of ·OH production at the terminus of this route, $r_{+OH}(R1)$, will therefore be given by eqn (E3):

$$r_{+OH}(R1) = r(R1) = 9.3 \times 10^3\ \gamma\ (S/V)\ [NO_2] \tag{E3}$$

In contrast, the production of ·OH initiated by $O(^3P)$ from R2 via R3–R5, $r_{+OH}(R2)$, has higher losses.[7] As a result, the rate of ·OH production from this pathway is significantly smaller than that of the initiation step: $r(R2) = J(NO_2)$

[NO$_2$]. By assuming the photostationary state relation: [O$_3$] $= J$(NO$_2$) [NO$_2$]/(k_{NO+O3} [NO]),[7] and taking into account that only a fraction θ of O(^{1}D) from reaction R4 forms $^{\bullet}$OH (θ is an increasing function of H$_2$O concentration),[7] we obtain eqn (E4):

$$r_{+OH}(R2) = \theta J(O^1D)\,[O_3] \sim \theta J(O^1D) \times J(NO_2)\,[NO_2]/(k_{NO+O3}[NO]) \qquad (E4)$$

Since, typically, J(NO$_2$)/(k_{NO+O3} [NO]) ~ 1,[7] we get

$$r_{+OH}(R1)/r_{+OH}(R2) \sim 9.3 \times 10^3\, \gamma(S/V)\,(\theta J(O^1D))^{-1} \qquad (E5)$$

For representative $\theta = 0.2$, J(O^{1}D) $\sim 4 \times 10^{-5}$ s^{-1} values (Fig. S4†), the switching condition $r_{+OH}(R1)/r_{+OH}(R2) = 1$, is reached when $\gamma(S/V) \sim 1 \times 10^{-9}$ cm^{-1}, which should be met on realistic wet aerosols (see above). Thus, at small zenith angles (low latitudes, summertime) $^{\bullet}$OH production is controlled by $r_{+OH}(R2)$, but, when the actinic flux declines, $^{\bullet}$OH production switches to (and never falls below) $r_{+OH}(R1)$, which depends on $\langle\gamma(S/V)\rangle$ values rather than solar flux, eqn (E3) (Fig. S4†). We believe that the preceding considerations lie at the core of the mechanism by which $^{\bullet}$OH concentrations are buffered in polluted air.

Summing up, we report laboratory results showing that carboxylate anions, such as those found in real atmospheric aerosol particles, dramatically enhance the conversion of NO$_2$(g) to HONO(g) on the surface of water, and quantum mechanical calculations confirming that most anions will trap NO$_2$ *via* electrostatics on the surface of water clusters. We also point out that recent satellite observations of NO$_2$ evolution over several megacities exclude a dominant role for the removal of NO$_2$ by gas-phase OH-radicals. On this evidence, we propose that the anion-catalyzed heterogeneous disproportionation of NO$_2$ (2NO$_2$ + H$_2$O $=$ H$^+$ + NO$_3^-$ + HONO) on pervasive SOA, whose carboxylate anion content tracks photochemical activity, may both account for the time dependence of the HONO daytime source day and the dominant sink of NO$_2$ over the course of the year. Our insights provide a causal link that bridges episodic and long-term, apparently unrelated and hitherto unexplained, observations on NO$_2$/HONO in the lower troposphere.

Acknowledgements

SE is grateful to the Japan Science and Technology Agency (JST) PRESTO program and Steel Foundation for Environmental Protection Technology. AY is grateful to grant-in-aid from JSPS (Grant 23651014). This work was supported, in part, by NSF Grant AC1238977.

References

1 A. Sarzynski, *Urban Studies*, 2012, **49**, 3121–3138.
2 A. M. Fiore, V. Naik, D. V. Spracklen, A. Steiner, N. Unger, *et al.*, *Chem. Soc. Rev.*, 2012, **41**, 6663–6683.
3 M. J. Cooper, R. V. Martin, A. van Donkelaar, L. Lamsal, M. Brauer and J. R. Brook, *Environ. Sci. Technol.*, 2012, **46**, 8523–8524.
4 B. R. Gurjar, A. Jain, A. Sharma, A. Agarwal, P. Gupta, A. S. Nagpure and J. Lelieveld, *Atmos. Environ.*, 2010, **44**, 4606–4613.
5 A. J. Haagensmit, *Ind. Eng. Chem.*, 1952, **44**, 1342–1346.
6 A. J. Haagensmit, *Proc. Natl. Acad. Sci. U. S. A.*, 1970, **67**, 887–897.

7 J. H. Seinfeld and S. N. Pandis, *Atmospheric chemistry and physics: from air pollution to climate change*, 2nd edn, Wiley, Hoboken, N.J., 2006.

8 T. M. Butler and M. G. Lawrence, *Environ. Chem.*, 2009, **6**, 219–225.

9 J. B. Cohen and R. G. Prinn, *Atmos. Chem. Phys.*, 2011, **11**, 7629–7656.

10 K. M. Wagstrom, S. N. Pandis, G. Yarwood, G. M. Wilson and R. E. Morris, *Atmos. Environ.*, 2008, **42**, 5650–5659.

11 D. Stone, L. K. Whalley and D. E. Heard, *Chem. Soc. Rev.*, 2012, **41**, 6348–6404.

12 NASA-JPL Chemical Kinetics and Photochemical Data for Use in Atmospheric Studies, Evaluation Number 14, 2003, http://rpw.chem.ox.ac.uk/JPL_02-25_01_intro_rev00.pdf.

13 Y. Matsumi and M. Kawasaki, *Chem. Rev.*, 2003, **103**, 4767–4781.

14 M. Kulmala, A. Asmi, H. K. Lappalainen, U. Baltensperger, J. L. Brenguier, *et al.*, *Atmos. Chem. Phys.*, 2011, **11**, 13061–13143.

15 M. Labriet, N. Caldes and L. Izquierdo, *International Journal of Global Warming*, 2009, **1**, 144–159.

16 C. X. Cai, C. Hogrefe, P. Katsafados, G. Kallos, M. Beauharnois, *et al.*, *Atmos. Environ.*, 2008, **42**, 8585–8599.

17 P. Alpert, O. Shvainshtein and P. Kishcha, *American Journal of Climate Change*, 2012, **1**, 117–131.

18 M. Lin, J. Tao, C. Y. Chan, J. J. Cao, Z. S. Zhang, L. H. Zhu and R. J. Zhang, *Aerosol Air Qual Res*, 2012, **12**, 1049.

19 J. L. An, Y. Li, Y. Chen, J. Li, Y. Qu and Y. J. Tang, *Adv. Atmos. Sci.*, 2013, **30**, 57–66.

20 Tropospheric Ultraviolet and Visible (TUV) Radiation Model, http://cprm.acd.ucar.edu/Models/TUV/Interactive_TUV/.

21 F. Rohrer and H. Berresheim, *Nature*, 2006, **442**, 184–187.

22 K. D. Lu, F. Rohrer, F. Holland, H. Fuchs, B. Bohn, *et al.*, *Atmos. Chem. Phys.*, 2012, **12**, 1541–1569.

23 S. A. Montzka, M. Krol, E. Dlugokencky, B. Hall, P. Jockel and J. Lelieveld, *Science*, 2011, **331**, 67–69.

24 W. P. L. Carter and J. H. Seinfeld, *Atmos. Environ.*, 2012, **50**, 255–266.

25 D. E. Heard, L. J. Carpenter, D. J. Creasey, J. R. Hopkins, J. D. Lee, *et al.*, *Geophys. Res. Lett.*, 2004, **31**, DOI: 10.1029/2004gl020544.

26 M. Sorgel, E. Regelin, H. Bozem, J. M. Diesch, F. Drewnick, *et al.*, *Atmos. Chem. Phys.*, 2011, **11**, 10433–10447.

27 J. Kleffmann, *ChemPhysChem*, 2007, **8**, 1137–1144.

28 S. P. Li, J. Matthews and A. Sinha, *Science*, 2008, **319**, 1657–1660.

29 A. R. Reisinger, *Atmos. Environ.*, 2000, **34**, 3865–3874.

30 R. Zhang, G. Sarwar, J. C. H. Fung, A. K. H. Lau and Y. H. Zhang, *Adv. Meteorol.*, 2012, DOI: 10.1155/2012/140932.

31 Y. Sosedova, A. Rouviere, T. Bartels-Rausch and M. Ammann, *Photochem. Photobiol. Sci.*, 2011, **10**, 1680–1690.

32 M. Qin, P. H. Xie, H. Su, J. W. Gu, F. M. Peng, *et al.*, *Atmos. Environ.*, 2009, **43**, 5731–5742.

33 C. George, R. S. Strekowski, J. Kleffmann, K. Stemmler and M. Ammann, *Faraday Discuss.*, 2005, **130**, 195–210.

34 C. E. Kolb, R. A. Cox, J. P. D. Abbatt, M. Ammann, E. J. Davis, *et al.*, *Atmos. Chem. Phys.*, 2010, **10**, 10561–10605.

35 J. P. D. Abbatt, A. K. Y. Lee and J. A. Thornton, *Chem. Soc. Rev.*, 2012, **41**, 6555–6581.

36 R. Van Dingenen, J. Van Aardenne, F. Dentener, E. Marmer, E. Vignati, P. Russ, L. Szabo and F. Raes, *Geochim Cosmochim Ac*, 2009, **73**, A1369–A1369.

37 D. J. Jacob, *Atmos. Environ.*, 2000, **34**, 2131–2159.

38 G. Lammel and J. N. Cape, *Chem. Soc. Rev.*, 1996, **25**, 361–369.

39 G. Li, W. Lei, M. Zavala, R. Volkamer, S. Dusanter, P. Stevens and L. T. Molina, *Atmos. Chem. Phys.*, 2010, **10**, 6551–6567.

40 K. Stemmler, M. Ammann, C. Donders, J. Kleffmann and C. George, *Nature*, 2006, **440**, 195–198.

41 S. Enami, M. R. Hoffmann and A. J. Colussi, *Proc. Natl. Acad. Sci. U. S. A.*, 2008, **105**, 7365–7369.

42 H. Mishra, S. Enami, R. J. Nielsen, L. A. Stewart, M. R. Hoffmann, W. A. Goddard and A. J. Colussi, *Proc. Natl. Acad. Sci. U. S. A.*, 2012, **109**, 18679–18683.

43 H. Mishra, S. Enami, R. J. Nielsen, M. R. Hoffmann, W. A. Goddard and A. J. Colussi, *Proc. Natl. Acad. Sci. U. S. A.*, 2012, **109**, 10228–10232.

44 S. Enami, M. R. Hoffmann and A. J. Colussi, *J. Phys. Chem. Lett.*, 2012, **3**, 3102–3108.

45 S. Enami, M. R. Hoffmann and A. J. Colussi, *J. Phys. Chem. B*, 2009, **113**, 7977–7981.

46 T. Kinugawa, S. Enami, A. Yabushita, M. Kawasaki, M. R. Hoffmann and A. J. Colussi, *Phys. Chem. Chem. Phys.*, 2011, **13**, 5144–5149.

47 A. Yabushita, S. Enami, Y. Sakamoto, M. Kawasaki, M. R. Hoffmann and A. J. Colussi, *J. Phys. Chem. A*, 2009, **113**, 4844–4848.

48 K. Kawamura and O. Yasui, *Atmos. Environ.*, 2005, **39**, 1945–1960.

49 D. M. Giles, AERONET Aerosol Optical Depth, Goddard Space Center, http://aeronet.gsfc.nasa.gov/new_web/index.html.

50 D. M. Winker, J. L. Tackett, B. J. Getzewich, Z. Liu, M. A. Vaughan and R. R. Rogers, *Atmos. Chem. Phys.*, 2013, **13**, 3345–3361.

51 S. Beirle, K. F. Boersma, U. Platt, M. G. Lawrence and T. Wagner, *Science*, 2011, **333**, 1737–1739.

52 S. Enami, M. R. Hoffmann and A. J. Colussi, *J. Phys. Chem. Lett.*, 2010, **1**, 1599–1604.

53 S. Enami, C. D. Vecitis, J. Cheng, M. R. Hoffmann and A. J. Colussi, *Chem. Phys. Lett.*, 2008, **455**, 316–320.

54 J. Cheng, C. D. Vecitis, M. R. Hoffmann and A. J. Colussi, *J. Phys. Chem. B*, 2006, **110**, 25598–25602.

55 J. S. Jung, B. Tsatsral, Y. J. Kim and K. Kawamura, *J. Geophys. Res.*, 2010, **115**, DOI: 10.1029/2010jd014339.

56 Y. N. Lee and S. E. Schwartz, *J. Geophys. Res. Atmos.*, 1981, **86**, 1971–1983.

57 S. E. Schwartz and Y. N. Lee, *Atmos. Environ.*, 1995, **29**, 2557–2559.

58 J. Kleffmann, K. H. Becker and P. Wiesen, *Atmos. Environ.*, 1998, **32**, 2721–2729.

59 P. Davidovits, C. E. Kolb, L. R. Williams, J. T. Jayne and D. R. Worsnop, *Chem. Rev.*, 2006, **106**, 1323–1354.

60 D. S. Bezrukov and Y. V. Novakovskaya, *Struct. Chem.*, 2004, **15**, 77–81.

61 A. Bambauer, B. Brantner, M. Paige and T. Novakov, *Atmos. Environ.*, 1994, **28**, 3225–3232.

62 J. Cheng, M. R. Hoffmann and A. J. Colussi, *J. Phys. Chem. B*, 2008, **112**, 7157–7161.

63 J. Cheng, E. Psillakis, M. R. Hoffmann and A. J. Colussi, *J. Phys. Chem. A*, 2009, **113**, 8152–8156.

64 M. W. Schmidt, K. K. Baldridge, J. A. Boatz, S. T. Elbert, M. S. Gordon, *et al.*, *J. Comput. Chem.*, 1993, **14**, 1347–1363.

65 M. Valiev, E. J. Bylaska, N. Govind, K. Kowalski, T. P. Straatsma, *et al.*, *Comput. Phys. Commun.*, 2010, **181**, 1477–1489.

66 S. Hirata, *J. Phys. Chem. A*, 2003, **107**, 9887–9897.

67 W.-G. Liu and W. A. Goddard, *J. Am. Chem. Soc.*, 2012, **134**, 12970–12978.

68 B. Njegic, J. D. Raff, B. J. Finlayson-Pitts, M. S. Gordon and R. B. Gerber, *J. Phys. Chem. A*, 2010, **114**, 4609–4618.

69 R. S. Zhu, K. Y. Lai and M. C. Lin, *J. Phys. Chem. A*, 2012, **116**, 4466–4472.

70 G. Murdachaew, M. E. Varner, L. F. Phillips, B. J. Finlayson-Pitts and R. B. Gerber, *Phys. Chem. Chem. Phys.*, 2013, **15**, 204–212.

71 N. Daher, S. Hasheminassab, M. M. Shafter, J. J. Schauer and C. Sioutas, *Environm. Sci. Processes Impacts*, 2013, **15**, 283–295.

72 C. Theodosi, U. Im, A. Bougiatioti, P. Zarmpas, O. Yenigun and N. Mihalopoulos, *Sci. Total Environ.*, 2010, **408**, 2482–2491.

73 M. Bressi, J. Sciare, V. Ghersi, N. Bonnaire, J. B. Nicolas, *et al.*, *Atmos Chem Phys*, 2013, **13**, 29391–29442.

74 N. O. Plaude, E. A. Stulov, I. P. Parshutkina, E. V. Sosnikova, N. A. Monakhova and V. V. Yakhno, *Russ. Meteorol. Hydrol.*, 2012, **37**, 21–27.

75 W. Kim, H. Lee, J. Kim, U. Jeong and J. Kweon, *Atmos. Environ.*, 2012, **56**, 101–108.

76 T. G. Theofanous, V. V. Mitkin, C. L. Ng, C. H. Chang, X. Deng and S. Sushchikh, *Phys. Fluids*, 2012, **24**, 022104.

77 K. F. Ho, S. S. H. Ho, S. C. Lee, K. Kawamura, S. C. Zou, J. J. Cao and H. M. Xu, *Atmos. Chem. Phys.*, 2011, **11**, 2197–2208.

78 L. Zhang, D. J. Jacob, E. M. Knipping, N. Kumar, J. W. Munger, C. C. Carouge, A. van Donkelaar, Y. X. Wang and D. Chen, *Atmos. Chem. Phys.*, 2012, **12**, 4539–4554.

79 J. Z. Yu, X. F. Huang, J. H. Xu and M. Hu, *Environ. Sci. Technol.*, 2005, **39**, 128–133.

80 N. Takegawa, T. Miyakawa, Y. Kondo, J. L. Jimenez, Q. Zhang, D. R. Worsnop and M. Fukuda, *J. Geophys. Res.*, 2006, **111**, D11206.

81 S. Myriokefalitakis, K. Tsigaridis, N. Mihalopoulos, J. Sciare, A. Nenes, K. Kawamura, A. Segers and M. Kanakidou, *Atmos. Chem. Phys.*, 2011, **11**, 5761–5782.

82 Y. Kanaya, R. Q. Cao, H. Akimoto, M. Fukuda, Y. Komazaki, *et al.*, *J. Geophys. Res.*, 2008, **112**, DOI: 10.1029/2007JD008671.

83 Y. Kanaya, M. Fukuda, H. Akimoto, N. Takegawa, Y. Komazaki, Y. Yokouchi, M. Koike and Y. Kondo, *J. Geophys. Res.*, 2008, **113**, DOI: 10.1029/2007JD008670.

84 D. Perner and U. Platt, *Geophys. Res. Lett.*, 1979, **6**, 917–920.

85 B. Aumont, F. Chervier and S. Laval, *Atmos. Environ.*, 2003, **37**, 487–498.

86 J. Stutz, B. Alicke, R. Ackermann, A. Geyer, S. H. Wang, A. B. White, E. J. Williams, C. W. Spicer and J. D. Fast, *J. Geophys. Res.*, 2004, **109**, DOI: 10.1029/2003JD004135.

87 R. M. Harrison and G. M. Collins, *J. Atmos. Chem.*, 1998, **30**, 397–406.

88 Y. Yu, B. Galle, A. Panday, E. Hodson, R. Prinn and S. Wang, *Atmos. Chem. Phys.*, 2009, **9**, 6401–6415.

89 K. Acker, D. Beysens and D. Moller, *Atmos. Res.*, 2008, **87**, 200–212.

90 J. Notholt, J. Hjorth and F. Raes, *Atmos. Environ.*, 1992, **26A**, 211–217.

91 X. Li, T. Brauers, R. Haseler, B. Bohn, H. Fuchs, *et al.*, *Atmos. Chem. Phys.*, 2012, **12**, 1497–1513.

92 P. Wojtal, J. D. Halla and R. McLaren, *Atmos. Chem. Phys.*, 2011, **11**, 3243–3261.

93 B. J. Finlayson-Pitts, *Phys. Chem. Chem. Phys.*, 2009, **11**, 7760–7779.

94 M. E. Monge, B. D'Anna, L. Mazri, A. Giroir-Fendler, M. Ammann, D. J. Donaldson and C. George, *Proc. Natl. Acad. Sci. U. S. A.*, 2010, **107**, 6605–6609.

95 B. J. Finlayson-Pitts, L. M. Wingen, A. L. Sumner, D. Syomin and K. A. Ramazan, *Phys. Chem. Chem. Phys.*, 2003, **5**, 223–242.

96 B. J. Finlayson-Pitts and J. N. Pitts, *Chemistry of the upper and lower atmosphere*, Academic Press, San Diego, CA, 2000.

97 A. Marinoni, P. Laj, K. Sellegri and G. Mailhot, *Atmos. Chem. Phys.*, 2004, **4**, 715–728.

Faraday Discussions

RSC Publishing

General discussion

DOI: 10.1039/C3FD90034G

Professor McNeill opened the discussion of the paper by Dr Pope: Are your fluorescent probe molecules vulnerable to O_3 oxidation during your experiments?

Dr Pope responded: The SR and Bodipy-C10 dyes are resistant to ozonolysis. The Bodipy-C10 dye was chosen for the oleic acid ozonolysis work because of this resistance and for its ability to dissolve in lipophillic systems. The double bonds on the Cy3 dye would undergo ozonolysis, and as such this dye would not be suitable for ozonolysis experiments.

Professor Heard commented: I liked this technique very much as a way to infer viscosity.

I have two related questions. (1) Why is one set of curves (Fig. 1A) best fit by a monoexponential, whereas the other set (Fig. 1B) is best fit by a biexponential function? Is there some physical reason for this associated with the type of molecular interactions that are important in each case, or that there is more than one excited state from which fluorscence is occurring? (2) For Fig. 1C, 1D and 1E, why are some correlations curved, whereas others linear?

Dr Pope replied: The biexponential excited state decay of Cy3 is a special feature of the photophysics of this dye, which possibly indicates the presence of an equilibrium between the bright (emissive) and dark (non-emissive) excited states. We do not see any evidence for the emission from the second (dark) excited state, as verified by studying the wavelength dependence of the time resolved decays. The deviations from monoexponential behaviour are minor, yet using a monoexponential fitting algorithm can result in major errors when fitting aerosol data at high viscosities. We note that in the recent literature [references 23 and 24 in the main paper], the decays of Cy3 were assumed to be monoexponential, however we found this approximation inaccurate and used biexponential analysis in our present study. Molecular rotors SR and Bodipy display monoexponential decays at all viscosities studied, which we interpret as a non-equilibrium decay from the bright state to the lower lying dark state.

Response to (2) - The precise nature of the non-linear appearance of the calibration graph is not known, although it is most certainly the result of complicated viscosity-dependent photophysics of the rotors, where rotation effectively stops when a certain viscosity is reached (so no further photophysical effect is seen). These limiting viscosities are different for all 3 compounds used as molecular rotors in this study. The Bodipy rotor used for oleic acid aerosol measurements has the highest active viscosity range and does not show saturation until approximately 4000 cP [Y. Wu *et al.*, *Phys. Chem. Chem. Phys.*, 2013, **15**(36), 14986–14993]. We would like to emphasise that the Forster Hoffmann equation that predicts the linear log–log plot was previously shown only to work in a limited range of viscosity, for selected dyes.

Professor Heard asked: What is the depth from within the aerosol which the fluorescence can escape from, and which is detected? Presumably this will depend in part on the concentration of the dye used? The viscosity might be expected to vary as a function of depth within the aerosol? Also, is there any local heating of the aerosol following absorption of the laser pulse, which could lead to a lowering of the viscosity?

Dr Pope replied: We have not performed a rigorous investigation into the variation of fluorescence intensity with depth of measurement. However, since the technique requires information about the fluorescence lifetime rather than fluorescence intensity, the technique will continue to work as long as sufficient signal is retrieved with which to perform the lifetime analysis. The largest particles analyzed were 80 microns in diameter and possessed a wetting angle of approximately 90° which results in a maximum depth of 40 microns. No obvious changes in fluorescence intensity were observed at this depth when compared to shallower measurements. This is also consistent with the fact that we use intense two-photon excitation (which is not significantly absorbed by the dye or any other aerosol components) and so should allow a very significant depth penetration.

It is not possible for us to measure localized heating within the aerosol but the wavelength of the excitation laser for FLIM should only generate significant absorption within the molecular rotors rather than the bulk aerosol material. Since the molecular rotors are present at very low concentrations, we expect the local heating to be negligible and hence the change in viscosity to also be negligible. Furthermore, we observe no change in aerosol viscosity over time unless we change some environmental parameters (*e.g.* RH or introduce ozone), hence we are confident that the laser is not heating the aerosol sample.

Ms Steimer said: You mention in the article that you had some problems with the calibration for sulforhodamine B. Do you understand what exactly influences your signal apart from viscosity? How would you select your calibration system to avoid these problems? This will be particularly important if you manage to extend the measurement range to the higher viscosities, which are interesting in regard to limited transport, since using bulk measurements of the respective system for calibration (such as is done with sucrose in this study) are unlikely to succeed. Investigations of material which is of unknown composition or limited availability (such as SOA) would also heavily rely on having a reliable calibration system.

Dr Pope replied: As noted in the paper we believe the calibration system of glycerol water mixtures is not the most suitable calibrant for the ionic solutions of NaCl. We believe this is due to the great polarity difference between NaCl and glycerol that perturbs the ability of SR to rotate. With regard to organic aerosol mixtures (including SOA), it is much easier to find calibrants with similar polarities to the systems tested. We are at present building a database of calibrant systems which are suitable for different dye and aerosol classes. We are confident that we will have reliable calibration systems for interesting systems such as SOA.

Dr Reid asked: Viscosity models such as that reported by Laliberté and Swindells[1,2] are usually based on mass fraction (or mole fraction) scale. The measurements presented here are reported in terms of water activity. What treatment has been used to go from water activity to mass/mole fraction to estimate the viscosity of the solution?

The Laliberté model is fine tuned in fitting parameters for ternary mixtures with inorganic salts, whereas the Swindells model is reported for binary solutions. In addition, these measurements can often only be reported up to the solubility limit of the solute (corresponding to a water activity of 0.89 for sucrose). Thus, how accurate are the water activity predictions of viscosity from the Laliberté and Swindells parameterisations when coupled with the uncertainty in the water activity/mass fraction of solute? How do these treatments compare with the more frequently used parameterisations of Chenlo and Quintas[3–6] (notably the Quintas measurements extend down to a water activity of 0.54)?

1 M. Laliberté, *J. Chem. Eng. Data*, 2007, **52**, 321
2 J. F. Swindells, C. F. Snyder, R. C. Hardy and P. E. Golden, *Supplement to National Bureau of Standards Circular*, 1958, 440
3 M. Quintas, T. R. S. Brandao, C. L. M. Silva and R. L. Cunha, *J. Food Eng.*, 2006, **77**(4), 844–852.
4 M. Quintas, T. R. S. Brandao, C. L. M. Silva and R. L. Cunha, *J. Phys. Chem. B*, 2007, **111**(12), 3192–3196.
5 F. Chenlo, R. Moreira, G. Pereira and A. Ampudia, *J. Food Eng.*, 2002, **54**, 347–352.
6 F. Chenlo, R. Moreira, G. Pereira and P. Bello, *Int. J. Food Prop.*, 2006, **9**(2), 149–156.

Dr Pope replied: We used the Extended Aerosol Inorganics Model (E-AIM) to convert from water activity to mole fraction for NaCl. We used the parameterization of Zobrist *et al.*[1] to convert from water activity to mole fraction for sucrose. This should have been noted in the paper and we apologise to Zobrist *et al.* for this oversight. The full reference is given below.[1]

For this work we did not compare our results to the papers mentioned. We restricted ourselves to parameterizations of Swindells and Laliberté.[2,3] Subsequent investigation of the Quintas *et al.* paper shows that within experimental error our results are in reasonable agreement with their data set.[4] It should be noted that for atmospheric aerosols we typically do not care about mass fractions of individual components, we want to know the bulk particle properties. As such it is usually much more useful to compare particle viscosity to relative humidity, rather than mass fraction.

1 B. Zobrist *et al.*, *Phys. Chem. Chem. Phys.*, 2011, **13**(8), 3514–3526.
2 M. Laliberté, *J. Chem. Eng. Data*, 2007, **52**, 321
3 J. F. Swindells, C. F. Snyder, R. C. Hardy and P. E. Golden, *Supplement to National Bureau of Standards Circular*, 1958, 440
4 M. Quintas, T. R. S. Brandao, C. L. M. Silva and R. L. Cunha, *J. Food Eng.*, 2006, **77**(4), 844–852.

Professor Lewis said: Could you expand on any potential chemical interaction between the dyes used in this study and the behaviour of the composition?

Would the dye behave the same (in terms of viscosity response) in a complex mixture of organics as it does in a model solution?

Dr Pope responded: For our dyes, we ensure that the viscosity is the biggest factor determining the fluorescence lifetime, compared to polarity, pH and ionic strength, for example. Thus, even in complex aerosol mixtures we remain confident that the dye principally senses the viscosity of its environment. This will of course not be the case if any specific interactions (hydrogen bonding, binding) occur. Therefore we perform the necessary tests to ensure this does not happen, by choosing calibration systems that have a similar composition to the aerosol.

Professor Rudich asked: I wonder, what is the source for the broad spread in the viscosity results presented? Is it inherent in the method or is it the variability of the aerosol?

Dr Pope answered: Our paper represents the first results from a new technique. At present it is unknown what the biggest contributor to the experimental scatter is. The spread of lifetimes is not inherent to the method, *i.e.* the lifetime distribution will be narrow for a sample of a homogeneous solvent. We have to conclude that the spread of lifetimes (=viscosities) is due to the heterogeneity in aerosol particles. Better temperature stabilization of the environmental cell is likely to reduce the experimental scatter since temperature has such a large effect on viscosity. It is possible that the greater scatter within the sucrose data is due to the formation of microdomains within the particle and our future work will investigate this possibility.

Professor Donahue opened the discussion of the paper by Professor McNeill: The Gamma model in Fig. 4 of your paper shows quite a flat wavelength dependence to the red of 500 nm, whereas the field data go more or less to zero. Does this suggest that those analyzing the field data may be misattributing some absorption as black carbon that may really be brown carbon?

Professor McNeill responded: Hecobian *et al.*[1] reported absorption of WSOC, so it is unlikely that black carbon caused any interference. That being said, in that study, absorption for all wavelengths was referenced to absorption at 700 nm in order to remove baseline drift, perhaps artificially driving absorption at longer wavelengths to zero.

1 A. Hecobian, X. Zhang, M. Zheng, N. Frank, E. S. Edgerton and R. J. Weber, *Atmos. Chem. Phys.*, 2010, **10**, 5965–5977.

Professor Lewis said: Rapid photobleaching of the aerosol is shown in Fig. 2 in the paper during sunrise. What does the model show as being the most important species and bond types that are broken in this process?

Professor McNeill replied: The kinetics of photobleaching were taken from our recent experimental study (N. Sareen *et al.*, *J. Phys. Chem. A*, 2013, **117**(14), 2987–2996). Based on that study we believe that photolysis of carbonyl groups in the oligomer species formed by glyoxal, methylglyoxal, and acetaldehyde in the particle phase is the primary driver for photobleaching.

Professor Abbatt asked: What drives the very fast formation of light absorbing species during the night, given that laboratory studies show somewhat slower chemistry?

Professor McNeill responded: We use the kinetics and rate laws for brown carbon formation from glyoxal and methylglyoxal derived from our experimental studies, which were performed in bulk solution (A. N. Schwier *et al.*, *Environ. Sci. Technol.*, 2010, **44**(16) 6174–6182). For acetaldehyde, we use the rate data reported by Nozière *et al.* (B. Nozière *et al.*, *Phys. Chem. Chem. Phys.*, 2010, **12**, 3864–3872), also for bulk solutions. The much higher concentration of ammonium ion in the aerosols as compared to the bulk solutions leads to the fast rate in the model.

Dr Colussi commented: We found that colour development is due to supramolecular rather than covalent interactions among chromophores. How did you calculate the absorbance?

Professor McNeill answered: In the studies I have just mentioned, the change in net light absorption for the complex reaction mixture was tracked. If supramolecular effects were significant in these systems, their contribution to the observed light absorption would have been accounted for by the experimental approach.

Professor Herrmann commented: How complete is your kinetic data set? Were all the rate constants readily available or did they need to be estimated?

Professor McNeill answered: As mentioned above, we use the kinetics and rate laws derived from our experimental studies for the net increase in absorption (brown carbon formation) from reactions of glyoxal and methylglyoxal in ammonium sulfate solutions. For acetaldehyde, we use the rate data reported by Nozière *et al.* (B. Nozière *et al.*, *Phys. Chem. Chem. Phys.*, 2010, **12**, 3864–3872).

Dr Colussi asked: What is the chemistry involved in 'browning'?

Professor McNeill responded: In the glyoxal/ammonium sulfate system, the absorption spectrum of the light-absorbing products shows a peak at around ~280 nm with a long tail, consistent with carbonyl-containing oligomers. Nitrogen-containing products such as imidazoles may also contribute to the absorption. In the case of acetaldehyde and methylglyoxal, aldol condensation to form unsaturated oligomers also contributes to browning.

Dr Nozière commented: In the absorbance that you measure and predict for your reactions mixtures, did you have any contribution of the imidazoles (near 270–300 nm)?

Professor McNeill answered: We use optical properties data from our experimental work (*e.g.* A. N. Schwier *et al.*, *Environ. Sci. Technol.*, 2010, **44**(16)

6174–6182). It is possible that imidizoles contribute to the net absorption we observed in the glyoxal–ammonium sulfate system, although we generally attribute the absorption near 270–300 nm to carbonyl groups.

Dr Reid commented: What are the complications of including the effects of gas phase absorption on surface tension in your model? Does this require you to account for the detailed chemistry, including pH, *etc.*?

Professor McNeill responded: We account for the effects of *absorption* of water soluble gas-phase organic species on aerosol surface tension in this study, using our model GAMMA which does include detailed pH-dependent chemistry in the aqueous phase. We use a model of surface tension depression in aqueous mixtures containing ammonium sulfate and many organic components based on measurements made with bulk solutions (A. N. Schwier, G. A. Viglione, Z. Li and V. F. McNeill, *Atmos. Chem. Phys. Discuss.*, 2013, **13**, 549–580). While there are refinements to this approach that could be made, we believe Dr Reid may be asking about the effects of *adsorption* of gas-phase surfactants at the gas-aerosol interface. Based on our recent aerosol chamber studies (N. Sareen, A. N. Schwier, T. L. Lathem, A. Nenes and V. F. McNeill, *Proc. Natl. Acad. Sci. U. S. A.*, 2013, **110**, 2723–2728) we believe that accounting for these effects will be necessary in order to fully represent aerosol surface tension depression in this system. A theoretical approach based on the work of Donaldson (D. J. Donaldson, *J. Phys. Chem. A*, 1999, **103**, 62–70) and Wexler and Dutcher (A. S. Wexler and C. S. Dutcher, *J. Phys.*

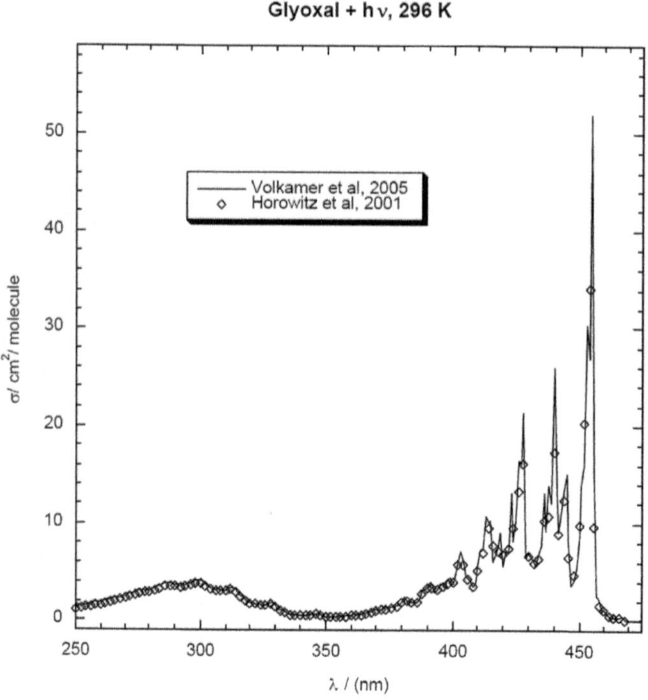

Fig. 1 UV–Visible spectrum of Glyoxal

Chem. Lett., 2013, **4**(10), 1723–1726) is planned. Very little experimental data exist in order to provide physical parameters and challenge such a model at this time, so further laboratory work is also planned.

Dr Cox commented: My comment concerns the relationship of gas phase glyoxal to the oligomers responsible for the composite aerosol absorbance, Fig. 3 of your paper. The glyoxal spectrum in the gas phase has a strong structured absorption in the UV–Vis region (see Fig. 1 of this discussion) which is absent in the composite spectrum. Presumably this reflects in part the loss of low-lying electronic states during oligomerisation of $(CHO)_2$?

Professor McNeill answered: Yes, glyoxal spontaneously hydrates in aqueous solution to form the *gem*-diol, and forms acetal dimers, resulting in a very different absorption spectrum from the gas-phase species.

Professor Pandis asked: The model simulations suggest that brown carbon formed in the aqueous phase during the night rapidly becomes white after sunrise. Does this mean that the contribution of this specific source/type of brown carbon on direct radiative forcing is very small?

Professor McNeill responded: Yes, any effect would be confined to the first hour or so after sunrise.

Professor Carslaw opened the discussion of the paper by Professor Jacobson: How do you treat particle formation in the free troposphere? In our model we see a substantial impact of new aerosol surface area on nucleation, which can cause substantial impacts on local particle concentrations.

Professor Jacobson answered: We treat ternary nucleation above a low ammonia threshold and binary below that. Nucleation is solved simultaneously with growth to allow competition between the two, as described in Section 4 of M. Z. Jacobson, *J. Geophys. Res.*, 2002, **107**, 4366.

Professor Carslaw asked: What nucleation mechanism do you include in the upper free troposphere and lower stratosphere, and could the rate of particle formation be perturbed by the aged aircraft particles due to increases in condensation sink for nucleating vapours?

Professor Jacobson replied: Ternary nucleation is from Merikanto *et al.* (J. Merikanto *et al.*, *J. Geophys. Res.*, 2007, **112**, D15207) Below a low NH_3 threshold, binary nucleation is calculated by Vehkamäki *et al.* (H. Vehkamäki *et al.*, *J. Geophys. Res.*, 2002, **107**, 4622). Yes, contrail particles compete with non-aircraft grid-scale nucleation for available vapor once contrail particles have been released from their subgrid plumes to the grid scale.

Professor Carslaw questioned: Could the response of nucleation to the additional condensation sink affect nucleation, and hence affect the total number of particles that you calculate due to aircraft emissions?

Professor Jacobson responded: Yes, definitely, once particles and gases have been released from the subgrid aircraft plume to the grid scale, where grid scale nucleation is calculated.

Professor Abbatt remarked: Does your model have any sensitivity to the efficiency of ice nucleation on black carbon?

Professor Jacobson answered: Yes, for background conditions, I use an ice nucleation parameterization that is a function of temperature. However, in the case of aircraft exhaust plumes, the initial supersaturation is so high that virtually all particles should nucleate, and coagulation will combine any interstitial particles with nucleated particles so fast to remove remaining ones; however, as the ambient partial pressure of water drops during this process, the smallest particles will sublimate/evaporate as they are subsaturated due to the Kelvin effect.

Professor Rudich asked: Is it possible to verify these predictions by remote sensing?

Professor Jacobson answered: Some predictions from aircraft can be compared with satellite data. Most notably, contrail cloud fraction data are now available by satellite. For other parameters, it is difficult to distinguish aircraft-caused changes from background pollution or natural sources, except in the case of vertical profiles of particulate matter number, black carbon, and some gases.

Dr Reid remarked: The distribution of flights in the northern and southern hemispheres is very different. To what extent can this allow you to look at/isolate the influence of aircraft emissions and contrails?

Professor Jacobson answered: From the modeling point of view, the difference in emission patterns becomes evident in the difference in cloud, gas, aerosol, and temperature effects. From a measurement point of view, however, it is difficult if not impossible to determine impacts of aircraft because the measurements show data only with aircraft present, not absent, and the differences due to aircraft are generally small relative to the variation in the measurements. However, some peaks in data at cruise altitude can be seen and potentially ascribed to aircraft. Contrail cloud fraction is one. Peaks in particles and gases occasionally can be seen at cruise altitude as well. Peaks generally won't occur in the SH.

Professor Lewis asked: Could you give some indication of the likely scale of aviation impacts, relative to the past twenty years presented in the paper, given the likely reductions in emissions from new aircraft balanced with increases in overall air traffic?

Professor Jacobson answered: This is something we arc looking at now, so I can't answer for sure, but my guess is, as you say, there will be a decline in emissions per aircraft-km but an increase in the number of km flown. Sulfur emissions may decline most significantly, but NO_x and CO_2 may increase. Conversion to a cryogenic hydrogen fleet, which is technically feasible but faces

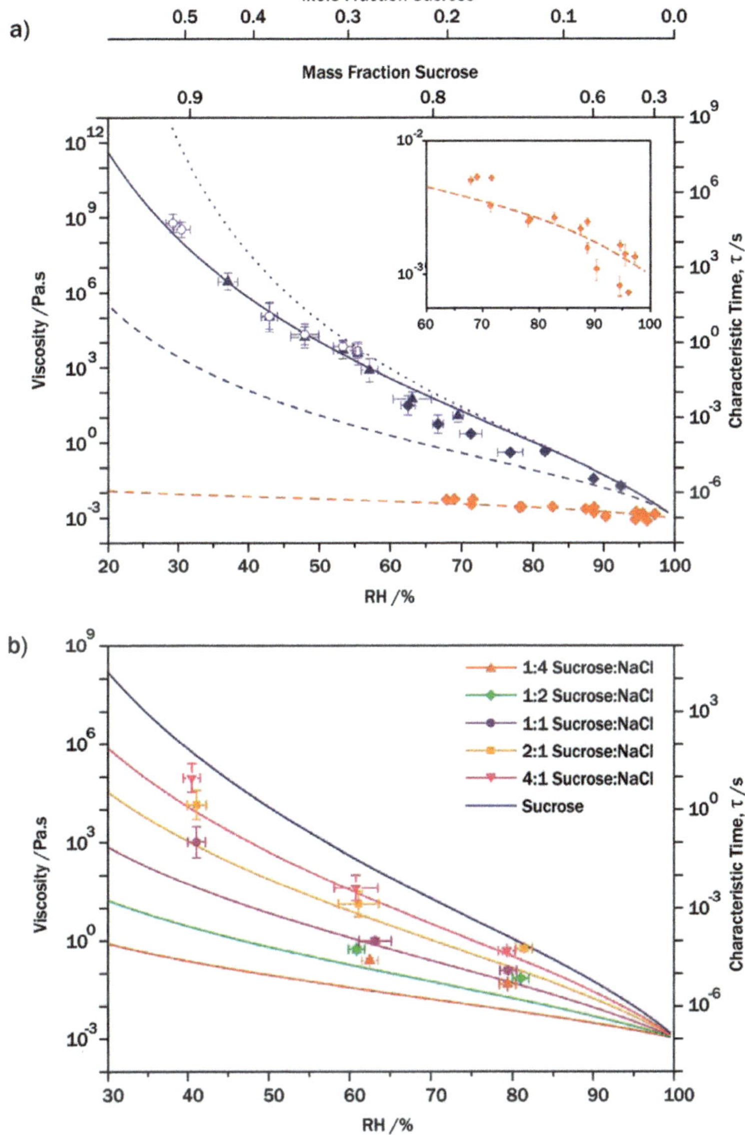

Fig. 2 Viscosity measurements of aerosols. (a) Measurements for aqueous sucrose (blue) and sodium chloride (red) droplets (light scattering, diamonds; brightfield imaging, triangles; Raman, circles) compared with model predictions for binary solutions using different thermodynamic treatments for water activity (see Chemical Science paper for details). Measurements for sucrose have been grouped in 5% RH intervals with up to a maximum of 11 droplet pairs for any one point (total of 85 droplet pairs). Error bars correspond to one standard deviation in both the measured RH and calculated viscosities. (b) Measurements of viscosity for aqueous sucrose–sodium chloride mixtures, compared with model predictions.

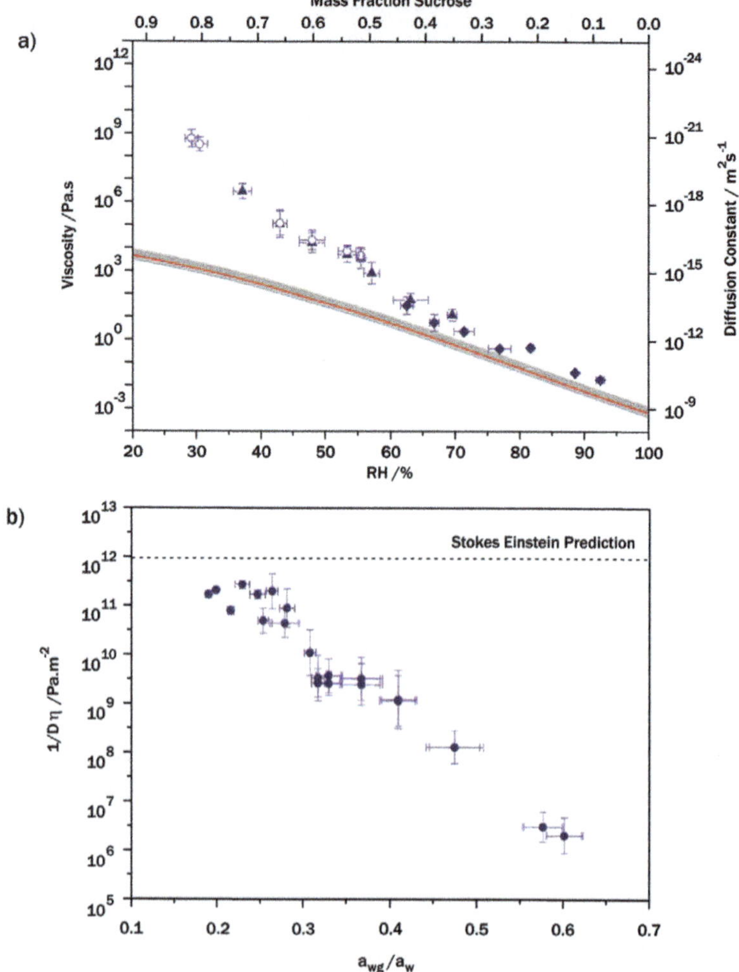

Fig. 3 Decoupling of bulk viscosity and the diffusion constant of water. (a) Measurements of viscosity for aqueous sucrose droplets (light scattering, diamonds; brightfield imaging, triangles; CERS, circles) compared with viscosities calculated using the parameterisation of Zobrist et al.[1] (line and shaded envelope) for the compositional water activity of sucrose solutions and the water activity dependent diffusion constant of water in viscous sucrose aerosol. The predictions assume adherence to the Stokes–Einstein equation with the shaded envelope corresponding to an uncertainty in the molecular diameter of water of 0.2 ± 0.1 nm. (b) Stokes–Einstein plot for sucrose. The Stokes–Einstein equation relationship predicts that $1/D\eta$ will have a constant value at constant temperature. $1/D\eta$ is estimated using the parameterisation of Zobrist et al. for the diffusion constant and the viscosity data from this study.

obstacles, would reduce emissions the most since it would eliminate sulfur, CO_2, CO, VOC, BC and OC emissions, allowing only primarily NO_x, and H_2O emissions.

Dr Allan remarked: The enhancement of the absorption efficiency of the interstitial BC particles by externally mixed ice is interesting, but is it not the case that within a discrete aircraft plume, they will become separated due to the difference in fall speeds?

Professor Jacobson answered: Yes, interstitial BC will fall more slowly than the ice crystals will themselves, so over time there will be a separation due to fall speed differences. Some of the vertical motion, though, is due to turbulent mixing, which will mix the two roughly similarly.

Professor Pandis said: How important is the simulation of the individual plumes (compared to the approaches that have been used in the past by global models) for the estimated effects of aircraft on climate and pollution?

Professor Jacobson answered: They are important for simulating contrail cloud fraction and particular; not so much the gas chemistry although there do seem to be differences in grid scale *versus* plume scale treatment of gas chemistry, particularly early on. These differences decrease over time but the results (grid-scale *versus* plume scale) don't converge due to the nonlinear behavior of chemistry.

Dr Reid opened the discussion of the paper by Dr Pope: We have developed an approach for measuring directly the viscosity of aerosols at specified relative humidities. See R. M. Power, S. H. Simpson, J. P. Reid and A. J. Hudson, *Chem. Sci.*, 2013, **4**(6), 2597–2604.

Two particles are captured in two optical tweezer traps and manipulated, initiating coalescence. The timescale for relaxation in the shape of the particle can then be used to estimate the bulk viscosity of the particles. Light scattering measurements on microsecond timescales can allow us to probe the under-damped, low viscosity range extending down to viscosities of dilute aqueous solutions of 1 mPa s. Brightfield imaging of the coalescence and relaxation in shape from the millisecond to many hours can also be used to probe more viscous particles. Finally, the disappearance/reappearance of whispering gallery modes in the Raman signature from the trapped particles can provide access to extremely highly viscous particles and very slow relaxation times. In combination, the viscosities of particles spanning the range 1 mPa s to 10^9 Pa s can be measured, 12 orders of magnitude. We have used this approach to measure the viscosity of sucrose particles spanning behaviour from dilute solutions through to viscosities approaching the glass transition (see Fig. 2a for sucrose solutions and Fig. 2b for mixtures with sodium chloride). The Stokes–Einstein equation is often used to relate the diffusion constant, D, for molecules or radius a in the bulk to the viscosity, η, at a temperature T. Specifically, $D=k_BT/(6na\eta)$ However, the Stokes–Einstein equation is strictly valid only for the diffusion of a large (spherical) molecule moving in a continuum. This is demonstrated in Fig. 3a, which compares diffusion constants of water estimated from direct viscosity measurements using optical tweezers to the diffusion constants estimated from the parameterisation of Zobrist *et al.*[1] derived from water transport kinetic measurements. This comparison clearly shows that viscosity measurements and diffusion constants measured separately are not self consistent (as indicated in Fig. 3b which shows that the inverse of the product of the diffusion constant and viscosity is not a constant). Indeed, the water diffusion constants are larger than would be estimated from using the Stokes–Einstein equation and values of the viscosity. These data suggest that the Stokes–Einstein equation can be used to estimate the viscosity (and *vice versa*) within an order of magnitude only up to values of 10 Pa s, equivalent to a diffusion constant greater than 10^{12} m^2 s^{-1}. In summary, the Stokes–Einstein equation should be used with great caution when considering the relationships between viscosities and diffusion constants and molecular diffusion may be much faster within aerosol particles than the designation as a semi-solid or glass would suggest.

1 B. Zobrist, V. Soonsin, *et al.*, *Phys. Chem. Chem. Phys.*, 2011, **13**(8), 3514–3526

Dr Murray commented: I'd like to reinforce Prof. Reid's comments on the Stokes–Einstein relationship between viscosity and diffusion coefficients. My group (in particular Hannah Price) have made measurements of diffusion coefficients of water in ultra-viscous and glassy aqueous solutions. We use a technique which involves using a Raman microscope to quantify the diffusion of D_2O into ultra-viscous and glassy aqueous solutions. Our results are consistent with Prof. Reid's, *i.e.* viscosity in combination with the Stokes–Einstein equation is not a good predictor of diffusion coefficient close to the glass transition.

Professor Lewis asked: Is it possible to conduct viscosity type determination using more representative organic matrices, *e.g.* a spread of polarities in humic material, rather than model compounds such as sucrose? Would you anticipate this to have an impact on the observed deviations from Stokes–Einstein behaviour?

Dr Reid answered: Yes, this is entirely possible and we intend to study a wider range of systems. We have started with sucrose only because this system is well characterised and provides a good benchmark.

Dr Murray commented: The range of viscosity over which measurements can be made with the current set up are limited. Do you think it will be possible to extend the range of viscosity measurements into the ultra-viscous and glassy range in the future?

Dr Pope answered: We believe that we can synthetically tune the sensitivity of the molecular rotors to viscosity. For example, by lowering the barrier for rotation we should be able to obtain molecular rotors sensitive to higher viscosity range. At this stage we would not like to speculate how high in viscosity we can push the technique.

Dr Krieger remarked: How do you feel should we proceed to estimate the diffusion constant of *e.g.* ozone into SOA when the application of the Stokes–Einstein relationship proves to be very problematic at high viscosities.

Dr Reid replied: We need first and foremost to develop techniques that give us spatially resolved signatures of diffusion so that we can measure diffusion constants directly for different reactive and unreactive species. We can then see how these relate to predictions from the Stokes–Einstein equation.

Professor Donahue addressed Dr Pope and Dr Marcolli: Does the breakdown of Stokes–Einstein for small solute molecules suggest that it is also problematic when considering the transport of larger organic molecules in organic particles consisting of a complex mixture, or is Stokes–Einstein likely adequate for those species?

Dr Marcolli responded: Stokes–Einstein is adequate when the transport of molecules of similar sizes is considered. When viscosity is high, the molecules cannot pass each other. Small molecules, on the other hand, can move fast through empty spaces of the matrix that is made up by the larger molecules. This is explained well in the perspective by Koop *et al.* (T. Koop *et al.*, *Phys. Chem. Chem. Phys.*, 2011, **13**, 19238–19255).

Dr Pope responded: The Stokes–Einstein relationship has been shown to work well over the liquid viscosity range. As such we expect the methodology, presented in this work, to work well with complex mixtures of organic particles in this viscosity range. At higher viscosities care will have to be taken in using the Stokes–Einstein relationship. It should be noted that corrections to the Stokes–Einstein relationship will likely allow it to be used in viscosity regions where it has been shown to break down.

Professor Nizkorodov asked: In this study, SOA material was prepared, collected, dissolved, mixed with a solution of probe molecules, and re-aerosolised. Is there a good way of introducing an involatile organic tracer molecule directly into growing SOA particles in a smog chamber in such a way that the tracer is uniformly dispersed inside the particles? In other words, can one avoid the dissolution–re-aerosolisation step?

Dr Pope answered: We are at present investigating methods to avoid the dissolution–re-aerosolisation step but cannot comment further here.

Dr Allan opened the discussion of the paper by Paulo Artaxo: In response to the question regarding the use of the AMS for going beyond process studies to assessing spatiotemporal trends: As with any measurement, the usefulness in this regard is dependent on ensuring comparability between measurements and this is becoming more important with the proliferation of the ACSM, which is more suited to routine measurements than the AMS. As the measurement has matured, I would say that the quality assurance protocols have reached the stage where we can be confident of the total mass measurements, but more work needs to be done to ensure repeatability of the markers within the organic mass spectra (note that elemental ratios derived from HR analysis appear to be more robust than using individual marker peaks). I would say that Aerodyne recognise this issue and efforts are being made to address this. These quality assurance procedures should also be traceable and documented such that they are accessible to third parties and data users.

Dr Marcolli commented: Modeling tools have become available during the last few years that allow predicting the phase state of aerosol particles consisting of organic compounds and inorganic salts. To account for the non-ideality in liquid phases and to detect liquid–liquid phase separation, a thermodynamic activity coefficient model like AIOMFAC can be used together with an algorithm to minimize the Gibbs free energy of the aerosol particle. For the calculation of gas-particle partitioning, a vapor pressure estimation model is needed in addition. Such calculations rely on the assumption of thermodynamic equilibrium. If there are mass transfer limitations in the condensed phase due to high viscosity, the timescales needed to reach thermodynamic equilibrium have to be also taken into account. The prediction of liquid–liquid phase separation for an aerosol particle of a given chemical composition can be considered as quite reliable. Much larger uncertainties are connected with the estimation of pure compound vapor pressures. Zuend and Seinfeld (A. Zuend and J. H. Seinfeld, *Atmos. Chem. Phys.*, 2012, **12**, 3857–3882) have published an example, how modeling tools can be combined to predict gas-particle partitioning and the phase state of the condensed phase for a mixture of reaction products from alpha-pinene ozonolysis and ammonium sulfate.

Dr Reid opened the discussion of the paper by Francis Pope: In many cases, based on the extended comment made, it is more important to determine molecular diffusion constants for water, reactive species and (semi-)volatile organic compounds rather than bulk particle viscosities.

Dr Pope responded: As noted in the paper, and resulting discussion, viscosity can be related to diffusion *via* the Stokes–Einstein relationship as long as the viscosity range is not near the glass transition. As such a spatially resolved measurement of viscosity is a perfect method for retrieving spatially resolved diffusion constants as long as there is linearity between viscosity and diffusivity. The FLIM based measurement presented in our paper provides a very effective method of following viscosity, and hence diffusivity, both spatially and temporally within aerosol particles.

Professor Carslaw opened the discussion of the paper by Professor Jacobson: How important is this discussion about viscosity for ice formation in the upper troposphere? Is this a major uncertainty?

Professor McFiggans commented: Is it possible/is there an intention to factor in measurements of intermediate or semi-volatile compounds from aircraft to investigate the influence of SOA as IN based on chamber findings of the formation of glassy SOA particles, or is there an insensitivity of contrail formation to aerosol number?

Professor Jacobson answered: Aircraft emit so many particles that secondary particle formation is probably not so important as primary particle emissions.

Dr Murray added: This is a comment on a comment about black carbon emissions influencing contrails. Contrails are probably not particularly sensitive to heterogeneous ice nuclei which might be emitted from the engine. The engine exhaust becomes so supersaturated as it cools that aerosol particles swell and freeze both heterogeneously and homogeneously. The emission of particulates from jet engines is potentially much more important for altering the population of ice nuclei in the background upper troposphere and therefore may influence subsequent cirrus cloud formation.

Professor Jacobson replied: I agree with this assessment.

Dr Knopf asked: Recent studies by our group have shown that organic particles possess the potential to initiate ice formation under tropospheric conditions. Furthermore, these experiments indicate that the viscosity of the organic material may play a crucial role in governing the ice nucleation mode, *e.g.* water uptake and subsequent immersion freezing *versus* deposition ice nucleation.

Dr Murray opened the discussion of the paper by Dr Leisner:† This is a comment focussed on the statement at the end of the abstract that contact mode nucleation is more efficient. Yes, in these specific experiments contact nucleation triggers freezing at a higher temperature than immersed particles. But, does this mean contact nucleation is more efficient? This depends on how you define 'efficiency'. The way it is currently defined is experiment specific and does not mean that contact nucleation will be more efficient in the atmosphere. The importance of contact nucleation in the atmosphere depends on collision frequency and probability of collision, resulting in freezing. Since the collision frequency is highly dependent on the saturation (*i.e.* if the droplet is growing, collisions are rare) as well as other variables, it is non-trivial to make statements about which mode is more important. I fear that the statement made in the abstract is not sufficiently well caveated and there is a danger that this will be incorrectly interpreted to mean that contact nucleation is more important than immersion freezing in the atmosphere. My opinion is that we do not know how important contact mode nucleation is and these experiments are an important part of solving this problem.

Dr Saathoff answered: I agree with the remark. Even if the sentence prior to the statement in question clarifies that we look at specific single particle properties in a well controlled laboratory environment, a word of caveat about the relevance in the atmosphere might have been appropriate.

Professor Donahue commented: Given that you observe a surface-area dependence and deduce a substantial penetration of the contact nucleus into a supercooled droplet, is this "contact" nucleation mechanism really much different from immersion freezing, other than the presence of a three-phase junction?

Dr Saathoff replied: Yes, contact nucleation is different from immersion freezing in the sense that we observe a much higher ice nucleation probability for contact nucleation than the same nucleus would exhibit if immersed in the liquid for the time of the wetting process. Both contact freezing probability and immersion freezing rate can be analyzed in our experiment for the very same nuclei.

Dr Knopf asked: The observed size dependency of contact freezing is surprising. Did you compare measured contact freezing rates with immersion

---

† Dr Leisner's paper was presented by Dr Saathoff, Karlsruhe Institute of Technology, Karlsruhe, Germany.

freezing rates published previously? Referring to an idea formulated previously in the literature, could it be that ice forms first on the approaching mineral dust particle before impacting the supercooled droplet? Could this process explain the observed size dependency?

Dr Saathoff responded: Since the focus of this paper is on contact freezing, we did not include a discussion or comparison with immersion freezing data. In principle, the deposition activation could lead to a surface area dependence. However, while detailed calculations are still to be completed, we did some estimates showing that deposition freezing in the vicinity of the levitated droplet under our experimental conditions should not play a significant role. We have modelled the relative humidity field around a levitated droplet of 100 μm in diameter in an air flow of 0.5 m s^{-1}. The region with supersaturation relevant for deposition freezing is in the order of 1/10 of the droplet diameter which corresponds to less than 0.3 ms residence time for an aerosol particle in the supersaturated region, depending on the trajectory approaching the droplet surface. Even with a relatively high deposition freezing rate of 10^7 cm^{-2} s^{-1} and a typical particle surface area of 6×10^{-9} cm^2, this time is too short to lead to a significant frozen fraction. Calculations for a variety of different experimental conditions and particles sizes are currently being done but will most likely not change the conclusion from this estimate. Furthermore, we plan to do independent measurements of the deposition freezing rates with the same particles.

Dr Marcolli commented: The abstract of this paper states that "for the particles investigated and on the minute time scale of the experiment, contact freezing indeed dominates immersion freezing for all temperatures." However, no experimental evidence is presented in the paper that would support this statement. If the authors want to adhere to this statement they should add a figure to the manuscript that shows the experimental data it relies on.

Dr Saathoff replied: We adhere to the statement that "for the particles investigated and on the minute time scale of the experiment, contact freezing indeed dominates immersion freezing for all temperatures." It follows directly from the fact that we are not able to detect any quadratic term in the temporal dependence of the logarithm of the fraction of frozen droplets as a function of residence time at a given supercooling. We feel that it is not necessary to reproduce all these curves in the manuscript, but please see the ESI of the paper.

Professor Carslaw asked: Could you explain more clearly the evidence for contact *vs.* immersion freezing? How would you expect the form of Fig. 2 to change in an experiment in which immersion and contact nucleation were acting together? If the nucleation rate is proportional to the surface area of the ice nucleus, doesn't this mean that the particles become immersed in the droplets and then initiate freezing?

Dr Saathoff responded: An appreciable contribution of contact freezing would lead to a negative quadratic component in the ln $N_{\mathrm{liq}}/N_0\ (\tau)$ curve given in Fig. 2. This is due to the fact that the effective area of immersion nucleus surface in the droplet is increasing linearly with time. The detailed derivation of this can be found in ref. 18 of the manuscript (N. Hoffmann *et al.*, *Atmos. Measurement Tech. Discuss.*, 2013, **6**, 3407-3437).

Dr Allan opened the discussion of the paper by Professor Abbatt: Regarding the structure, is it possible that if the coating becomes deliquesced, that a hydro-phobic inclusion could be ejected to the surface of the droplet rather than remaining in the centre?

Professor Abbatt replied: It is unlikely that the SOA coating will become fully deliquesced. Instead, it is expected that a small amount of water will dissolve into it. Nevertheless, I think it is indeed quite likely that the PAH will indeed diffuse through the SOA coating given the nature of the observed kinetics. In particular, the Langmuir–Hinshelwood-like kinetics imply a surface reaction, probably involving the PAH and ozone at the surface of the coated particle.

Professor Pandis asked: Can the cycling of ambient particles through cycles of relative humidity change the corresponding kinetic limitations observed? For example, does the observed kinetic limitation for a particle at low relative humidity persist after the particle has been exposed to high (say >70%) relative humidity?

Professor Abbatt responded: It would be very interesting to cycle these particles through different clues of RH to see if there is a hysteresis in these kinetics effects. Unfortunately, due to experimental constraints we were only able to expose the particles to elevated RH for about one minute before the kinetics experiments and we were unable to cycle the RH.

Professor Pandis questioned: Did the particles contain water during these experiments?

Professor Abbatt answered: On purpose, we kept the relative humidity below the value that would lead to deliquescence of a mixed ammonium sulfate–organic particle, *i.e.* we did not go above 70% RH. There will certainly be some degree of water diffusing through the SOA leading, we believe, to its decreased viscosity, but hygroscopic growth experiments have shown these amounts to be quite small for SOA.

Dr Reid asked: Is there a significance to the diffusion constant of ozone remaining 10^{-10} cm^2 s^{-1}? Does this suggest that the reaction is solely a surface reaction?

Professor Abbatt responded: This value is not constrained by the model and is highly uncertain. It is expected to be larger than that for PAHs because ozone is a much smaller molecule. Note that the higher assumed diffusion constant for ozone implies that the model ozone diffuses throughout the particle. However, the bulk reaction between dissolved PAH and O_3 appears to not occur in our experiment because we observe saturation effects with increasing ozone concentrations. Accordingly, the bulk phase rate constant was chosen to be sufficiently small to not significantly affect the kinetics. The end result is that the reaction (in the model) occurs mostly at the surface, with a reactive intermediate resulting from ozone.

Dr Reid questioned: For the kinetic fits presented in Fig. 6, how unique are the fit parameters retrieved? Are there no other plausible fit solutions?

Professor Abbatt replied: The set of the kinetic parameters is not absolutely unique and it is possible to assume a different mechanism and arise at a different set of parameters (*e.g.* a heterogeneous reaction could potentially occur between PAHs adsorbed to ammonium sulfate with dissolved ozone). Under the mechanistic assumptions stated in the paper, the sensitivity studies revealed that the bulk diffusivity of PAH in SOA (Db,PAH) is the most sensitive parameter and other parameters are not as critical. Thus, Db,PAH could be estimated (with relatively large uncertainties) by constraining other parameters in the reasonable range that is consistent with previous studies.

Professor Nizkorodov commented: In your 2011 paper by Shiraiwa *et al.* (Ref. 51 in the text) you argued that ozone is first transformed into a relatively long-lived surface-bound reactive oxygen intermediate (ROI), which then reacts with other surface adsorbed species. In the case of soot, one possible structure for the ROI was atomic oxygen bound to the pi-electrons of the surface. Can you speculate on the molecular nature of the ROI that could be formed on the surface of the secondary organic aerosol material (which does not contain a lot of aromatic compounds)?

Professor Abbatt replied: This is a good question. There is a remarkable consistency in the kinetics of ozone loss across a very wide array of substances, ranging from pure PAHs, to soot, to mineral dust, to unsaturated self-assembled monolayers, to halide solutions. In particular, each of these systems demonstrates saturation effects in the kinetics at high ozone gas phase concentrations. The first order rate constants for substrate modification in this limit are all within one or two orders of magnitude of each other, as pointed out in McCabe and Abbatt.[1] This implies a common rate determining step in the heterogeneous reactions of ozone involving the transformation of ozone to a more reactive form. Recent work by Markus Ammann's group[2] has shown clear evidence for electron transfer from electron rich substrates to ozone, leading to the weakening of O–O bonds in ozone. Whether or not the ROI formed from this chemistry is the same in each reaction system is not known (although it seems unlikely), but it is possible that the pathway for breaking the O_3 molecule apart may be similar in these reactions.

1 J. McCabe and J. P. D. Abbatt, *J. Phys. Chem. C*, 2009, **113**, 2120–2127.
2 M. Lampimaeki, V. Zelenay, A. Krepelova, Z. Liu, R. Chang, H. Bluhm and M. Ammann, *Chem. Phys. Chem.*, 2013, **14**, 2419–2425.

Professor Lewis commented: PAH have rather complex life-cycles and move between air, water and soil. Is there any evidence that current estimates of PAH burden are incorrect and that their environmental lifetimes are longer than we had previously thought, as would be suggested by the paper?

Professor Abbatt replied: My feelings are that the atmospheric processes involving the larger PAHs, such as their lifetimes when adsorbed to particles, are sufficiently uncertain that we cannot infer inaccuracies about PAH sources/burden as yet. For example, if the general implications from this paper – *i.e.* that PAHs are shielded from atmospheric oxidants when coated – are valid, then this provides a ready explanation for their ability to undergo long range transport. On the other hand, if the fast oxidation lifetimes observed for surface-adsorbed species are found to be unaffected by coatings, then it does appear as though our understanding of PAH chemistry is incomplete. One aspect of this science that has confused me is the evidence that PAHs reversibly adsorb and desorb from particles according to diurnal temperature cycles. How that can occur with very short (minutes to hours) lifetimes with respect to heterogeneous oxidation for adsorbed species is unclear to me.

Professor Heard opened the discussion of the paper by Dr Colussi: Some of the statements in the introduction are rather sweeping.

For example:

(1) Page 2 line 25 of the paper states that models generally fare poorly in predicting how OH concentrations respond to actinic flux and NO_x levels. Actually, there are many examples of good agreement between field measurements of OH and model calculations, for a range of light levels and NO_x. (2) Also on page 2, line 30, it states that even more troubling is that, contrary to expectations, noon time OH in Birmingham in winter is only a factor of 2 less than in summer, despite a factor of 15 reduction in J(O1D). In fact, this is not surprising, and is explained in ref. 25 of the paper,[1] that a dominant source of OH in this environment is the reaction of ozone with alkenes, which would not be expected to have a significant change between summer and winter, and is one reason to account for the relatively small change in OH levels between summer and winter. In Section 8, the expression (E3) gives the production rate of OH from HONO photolysis. However, OH also reacts with NO to give HONO, and so it is the net production of OH from HONO photolysis taking into account loss of OH back to HONO by reaction of OH with NO, that is important. Also, the reaction of O_3 with alkenes does not seem to be included in the steady–state analysis as a source of OH, and will influence how important HONO is as a source of OH, and when the switchover takes place and the ratio given in (E5).

1 D. E. Heard, L. J. Carpenter, D. J. Creasey, J. R. Hopkins, J. D. Lee, et al., *Geophys. Res. Lett.*, 2004, **31**, DOI: 10.1029/2004gl020544.

Dr Colussi responded: Ref. 25 (Heard *et al.*[1]) arrived at the conclusion that the dominant source of OH radicals (in Birmingham) was the reaction of ozone with alkenes on the basis of the chemistry outlined in Paulson and Orlando[2], Carslaw *et al.*[3] and Faloona *et al.*.[4] Since the issue of OH yields in the ozonation of olefins is the subject of active ongoing debate in connection with the unexpectedly high levels of OH radicals detected over forests, Heard *et al.*'s early conclusion must be held in suspense at present.[2] I also wish to point out that the high levels of olefins detected in Birmingham may not be expected to be found in general. More to the point, the high reactivity of olefins precludes high levels of olefins in the urban plumes monitored from satellites.[5] Therefore, I stand by my statement that current atmospheric chemistry models fare poorly in predicting how OH concentrations respond to actinic flux and NO_x levels! Regarding the net rate of OH production from the photolysis of HONO, it is reasonable to assume that the high reactivity of OH radicals towards all species will minimize their recombination with NO. For the reasons outlined above, the lack of olefins in urban plumes (*i.e.*, far from their sources) implies that the production of OH radicals from the ozonation of olefins is unimportant in such environments.

1 D. E. Heard, L. J. Carpenter, D. J. Creasey, J. R. Hopkins, J. D. Lee, et al., *Geophys. Res. Lett.*, 2004, **31**, DOI: 10.1029/2004gl020544.
2 S. E. Paulson and J. J. Orlando, *Geophys. Res. Lete.*, 1996, **23**, 3727–3730
3 N. Carslaw *et al.*, *Atmos. Environ.*, 2001, **35**, 4725–4737
4 I. Faloona *et al.*, *J. Geophys. Res.*, 2001, **106**, 24315–24333
5 S. Bierle *et al.*, *Science*, 2011, **333**, 1737–1739

(515:[515]515) **Dr Reid** asked: What are the accepted reasons for the low degree of seasonable variability in OH or is it dependent on site/location *etc.*?

Professor Heard answered: It is very dependent upon the site/location. For example, when the production of OH is controlled largely by the photolysis of ozone, followed by the reaction of O(1D) atoms with water vapour, then the seasonal change is expected to be quite large – for example in mid/high latitudes in clean, marine environments. In polluted environments, where there are higher levels of NO_x and VOCs, the dominant sources of OH can be quite different, for example the reaction of O_3 with alkenes, for which there is less of a diurnal or seasonal change in the rate of OH production. In tropical environments, where there is less of a seasonal change of water vapour or sunlight, the production rate of OH may not vary as much with season. Also, sinks of OH can play a seasonal role. For example, isoprene levels in temperate continental regions have a strong seasonal signature, being much higher in summer, and hence the rate of removal of OH by reaction with isoprene will have a strong seasonal signature – and this can be balanced out by the change in photolysis rate of species (acting as sources of OH), or seasonal changes in NO_x, which leads to a smaller overall seasonal variability of OH.

Dr Cox commented: Whilst this paper presents some interesting ideas relating to NO_2 conversion in urban atmospheres, I have two main criticisms.

1. The title does not reflect the content of the paper. My understanding of the term "reactive intermediate" is a constituent that is rapidly produced and removed by chemical reaction in a mechanism, which may act to promote/catalyse overall chemical change, *e.g.* an atom, free radical or other oxidizing agent. It stretches pausibility how this applies to the tropospheric aqueous aerosol in this context.
2. The analysis of mechanisms for NO_2 conversion to HONO in the atmosphere does not take adequate account of the various works demonstrating the photosensitized heterogeneous reactions involving organic species (Prof C George and coworkers, U. Lyon) and metal oxides (e.g.TiO_2[1]).

1 R. J. Gustafsson, A. Orloff, P. T. Griffiths, R. A. Cox and R. M. Lambert, *Chem. Commun.*, 2006, **37**, 3936–3938.

Dr Colussi replied: It must be emphasized that our proposal does *not* exclude other HONO sources. The inescapable fact that alternative mechanisms of HONO production must face is that, in contrast with our proposal, they may not apply everywhere at all times. The goal is to account for *all* independent observations, rather than episodic events in particular locations at specific times.

Mr Held added: Could you speculate on the relative contribution of aerosol surfaces *vs.* ground surfaces such as soil or plant surfaces to heterogeneous production of HONO? For many ground-based observations, the aerosol surface is minor compared to ground surfaces. From our own observations, we typically find HONO profiles indicating HONO production close to the ground. These HONO profiles are not consistent with a HONO volume source as expected from heterogeneous production on aerosols.

Dr Colussi replied: Your observations, and interpretation, of HONO production at ground level are, of course, not incompatible with the more general NO_2 to $HONO + NO_3^-$ conversion process responsible for satellite sightings.

Dr Murphy opened the discussion of the paper by Professor Abbatt: Do we have a good systematic idea of which heterogeneous reactions are most sensitive to surface coatings and why?

Professor Abbatt responded: There are very few reactions of atmospheric importance that have been shown to definitely be affected by the presence of surface coatings. From the perspective of global oxidizing capacity, the most important heterogeneous reactions are N_2O_5 hydrolysis, HO_2 reactive uptake to aerosol, HNO_3 uptake to mineral dust, and some of the reactions in the halogen family that either give rise to active halogens or else recycle them, *e.g.* $HOBr + Cl^-$ or Br^-. Of these, the N_2O_5 reaction has been shown by a number of workers to be sensitive to organic coatings through non-reactive mechanisms, including work from our own research group (Thornton *et al.*;[1,2] Escorcia *et al.*[3]) and those of Joel Thornton (McNeill *et al.*[4]), Thomas Mentel (Folkers *et al.*;[5] Anttila *et al.*[6]), Allan Bertram (Cosman *et al.*;[7] Cosmann and Bertram[8]), Gil Nathanson (Park *et al.*[9]) and others. It is still unclear the degree to which these coatings affect uptake coefficients of N_2O_5 measured/inferred on ambient aerosol (Bertram *et al.*[10]). In the case of the other reaction systems, the experiments with uncoated particles are still sufficiently uncertain that experiments with coated particles have largely yet to be attempted. Where N_2O_5 uptake is affected by slow mass transfer, other processes that can be affected by organic coatings include the reactive uptake of oxidants such as OH and O_3. The atmospheric importance here is the potential for particle oxidation and not for effects on the atmospheric oxidizing capacity. An example is from our own group where we shut down the oxidation of bromide by OH when we added a soluble surfactant to the aqueous halide solution (Frinak and Abbatt[11]). Other effects arise in the oxidation of adsorbed PAHs by ozone, where organic coatings have been demonstrated to have measurable effects (Henderson and Donaldson;[12] Zhou *et al.*[12]).

1 J. A. Thornton and J. P. D.Abbatt, *J. Phys. Chem. A*, 2005, **109**, 10004–10012.
2 J. A. Thornton, C. F. Braban and J. P. D. Abbatt, *Phys. Chem. Chem. Phys.*, 2003, **5**, 4593–4603.
3 E. N. Escorcia, S. J. Sjostedt and J. P. D. Abbatt, *J. Phys. Chem. A*, 2010, **114**, 13113–13121.
4 V. F. McNeill, J. Patterson, G. M. Wolf and J. A. Thornton, *Atmos. Chem. Phys.*, 2006, **6**, 1635–1644.
5 M. Folkers, T. F. Mentel and A. Wahner, *Geophys. Res. Lett.*, 2003, **30**, 1644.
6 T. Anttila, A. Kiendler-Scharr, R. Tillmann and T. F. Mentel, *J. Phys. Chem. A*, 2006, **110**, 10435–10443.
7 L. M. Cosman, D. A. Knopf and A. K. Bertram, *J. Phys. Chem. A*, 2008, **112**, 2386–2396.

8 L. M. Cosman and A. K. Bertram, *J. Phys. Chem. A*, 2008, **112**, 4625–4635.
9 S. C. Park, D. K. Burden and G. M. Nathanson, *J. Phys. Chem. A*, 2007, **111**, 2921–2929.
10 T. H. Bertram, J. A. Thornton, T. P. Riedel, A. M. Middlebrook, R. Bahreini, T. S. Bates,
   P. K. Quinn and D. J. Coffman, *Geophys. Res. Lett.*, 2009, **36**, L19803
11 E. K. Frinak and J. P. D. Abbatt, *J. Phys. Chem. A*, 2006, **110**, 10456–10464.
12 E. A. Henderson and D. J. Donaldson, *J. Phys. Chem. A*, 2012, **116**, 423–429.
13 S. Zhou, A. K. Y. Lee, R. D. McWhinney and J. P. D. Abbatt, *J. Phys. Chem. A* 2012, **116**,
   7050–7056.

Dr Cox commented: This paper demonstrates beautifully how slow diffusion through a coating of SOA can inhibit heterogeneous reaction in the bulk of an organic particle. For most inorganic compounds undergoing ionic processes in the aqueous phase diffusion in the bulk is important in controlling uptake. In earlier work on N_2O_5 hydrolysis it was discovered that the uptake coefficient of N_2O_5 was retarded significantly in the presence of SOA (Folkers *et al.*[1]), or on humic acid particles (Badger *et al.*[2]). This has been attributed to a reduction in the accommodation coefficient α. Can slow diffusion in a kinetic multilayer model account for these phenomena?

1 M. Folkers, T. F. Mentel and A. Wahner, *Geophys. Res. Lett.*, 2003, **30**, 1644.
2 C. L. Badger, P. T. Griffiths, I. George, J. P. D. Abbatt, and R. A. Cox, *J. Phys. Chem. A*, 2006,
   110, 6986.

Professor Abbatt replied: This is a very good question. The work of Anttila *et al.*,[1] follow-up work from the original studies of the Julich group, analyzed the suppressed kinetics of N_2O_5 hydrolysis on coated sulfate particles *via* a model including slow diffusion (and limited solubility) in an SOA coat. The previous work with mono-layer thick films referred to in our response to Dan Murphy's question shows that mass accommodation coefficient effects are entirely possible when organics are deposited onto aqueous surfaces, but slow diffusivity/low solubility may also affect the kinetics when thicker films are present.

1 T. Anttila, A. Kiendler-Scharr, R. Tillmann and T. F. Mentel, *J. Phys. Chem. A*, 2006, **110**,
   10435–10443.

Professor Villenave added: The heterogeneous chemistry of PAHs has been extensively studied the last 30 years and it is difficult to hear from the audience and somehow confirmed by Jon Abbatt that there is little information explaining the presence of PAHs far away from their sources (for instance in polar regions). Among others (cited by Prof. Abbatt), we have in Bordeaux published maybe 20 papers in the last 10 years about the reactivity of PAHs in many different conditions, studying the effect of the nature of the particles (carbonaceous or mineral, model or natural, different specific areas and porosity, temperature, humidity), of the type of oxidant (light, ozone, OH, NO_x), the identification of oxidation products, *etc.* I am sure that even if most of our work is fundamental, has been performed in the darkness, in conditions sometimes not relevant for atmospheric chemistry (in terms of concentration, *etc.*), a part of our work merits citation, as for instance i) it confirms that heterogeneous reactions may be much faster than gaseous ones (but for OH in the case of PAHs, with both experimental and theoretical approaches) ii) it proposes chemical pathways for the formation of some oxidation products with branching ratio determination associated to elementary reaction steps, iii) it shows the presence of large parts of parent PAHs not oxidized but depending on the nature of the particle and the way the PAHs are present on the particles (chemi- or physisorbed) (the famous "plateau", first studied during his PhD by Doug Lane) iv) the difference of the various isomers that may be formed during emissions, homogeneous or heterogeneous processes (allowing to separate such processes in the atmosphere and to propose/confirm some tracers) and iv) field measurements of oxy- and nitro-PAHs in different environments.

Professor Abbatt replied: Yes, this is a good point. It is entirely appropriate to highlight the important work from this group. The papers that Eric refers to (Esteve *et al.*;[1,2] Perraudin, *et al.*;[3,4] Albinet, *et al.*;[5] Oubal, *et al.*;[6] Ringuet, *et al.*[7]) reported a series of lab studies on the kinetics and mechanisms of atmospheric oxidation of PAHs as well as some nitrated and oxygenated PAHs (NPAHs and OPAHs) with gas-phase oxidants, *e.g.* OH, O3, NO3, NO2 and NO at atmospheric relevant conditions. The research demonstrated the importance of the heterogeneous reactions of PAHs in the atmosphere and revealed the possible mechanisms for the formation of the OPAHs and NPAHs, known to be associated with human health. The work from this group has definitely extended our knowledge on the heterogeneous reaction of PAHs and contributed to our understanding of atmospheric degradation of PAHs.

1 W. Esteve, H. Budzinski and E. Villenave, *Atmos. Environ.*, 2004, **38**, 6063–6072.
2 W. Esteve, H. Budzinski and E. Villenave, *Atmos. Environ.*, 2006, **40**, 201–211.
3 E. Perraudin, H. Budzinski and E. Villenave, *J. Atmos. Chem.*, 2009, **56**, 57–82.
4 E. Perraudin, H. Budzinski and E. Villenave, *Atmos. Environ.*, 2007, **41**, 6005–6017.
5 A. Albinet, E. Leoz-Garziandia, H. Budzinski and E. Villenave, *Atmos. Environ.*, 2008, **42**, 54–64; A. Albinet, E. Leoz-Garziandia, H. Budzinski and E. Villenave, *Atmos. Environ.*, 2008 **42**, 43-54.
6 M. Oubal, G. Hantal, S. Picaud, P. N. M. Hoang, D. Liotard, M. T. Rayez, J. C. Rayez and E. Villevace, *Comp. Theo. Chem.*, 2011, **965**, 259–267.
7 J. Ringuet, A. Albinet, E. Leoz-Garziandia, H. Budzinski and E. Villenave, *Atmos. Environ.*, 2012, **61**, 15–22.

Dr Wenger commented: The observed effect of relative humidity is very interesting. Did you determine the liquid water content of the particles used in the experiments?

Professor Abbatt replied: We did not measure the water content of the particles but prior hygroscopic growth experiments conducted with SOA have demonstrated that only very small amounts of water are taken up, up to the value of 70% RH.

Dr Wenger asked: The paper by Zhou *et al.* provides further evidence that relative humidity can influence the nature of organic aerosols. How does the relative humidity and hence the liquid water content of aerosols affect the reactive uptake of semi-volatile organics and the subsequent growth of secondary organic aerosol?

Professor Donahue responded: We have performed experiments in which we attempted to separate any chemical effect of humidity from any thermodynamic or dynamical effect of water activity in particles (for example in N. L. Prisle *et al.*,

Geophys. Res. Lett., 2010, **37**, L01802). The bottom line is that we do not see a strong effect.

Professor McFiggans commented: Even after extensive efforts to investigate SOA formation and properties in the presence of water in the moist atmosphere as illustrated by a number of the papers in the first three sessions, I remain very confused about role of water in SOA formation and SOA–water interactions once formed. There is evidence that timescales for complete water uptake are not realised within instruments and ambient particles may experience such limitations to equilibration. There remain discrepancies in experimentally-determined hygroscopicities depending on the instruments/techniques employed and RH experienced. Predictions and experiments are in apparent conflict. Some modelling approaches choose to ignore water and others make it a central focus. There appear to be a lot of unresolved questions (from the role in nucleation, participation as a major solvent in phase separation, determination of the morphology thereafter, enhancement/reduction of SOA mass and associated change in composition through to co-evaporation/co-condensation with changing RH or saturation ratio), even before consideration of aqueous phase reactions and its numerous potential implications. It appears timely to try to reconcile, or at least summarise, the evidence relating to the roles of water in multiphase systems in the moist atmosphere.

Professor Artaxo responded: I think that there is no question that water plays a critical role in SOA formation or aerosol chemistry, specially for sites like Amazonia, when RH at night time is always at 95–100%, and during daytime is never below 70% in the wet season. There are still serious issues in aerosol analysis at high RH, as well as modeling such complex reactions. Over the surface of natural particles in Amazonia, a film of water should always be present. When we try to measure the chemical (ACSM) and optical properties, certainly we are changing surface properties that are reflected in the bulk particle characteristics.

Professor Herrmann added: How does RH and hence water change reactivity on organic surfaces on a molecular level?

Professor Abbatt replied: This is an important question, one for which we do not have a general understanding, with the effects being dependent on the phase transitions and changes to particle properties that will occur. For example, if the material does take up water substantially, then the possibility of hydrolysis reactions increases, *e.g.* of N_2O_5. That being said, increasing the RH over solid azelaic acid particles to 85% lead to no measurable increase in the uptake of N_2O_5 in experiments conducted in our lab (Thornton *et al.*, 2003). Similarly, we found that high RH did not affect the reactive uptake of O3 by oleic acid films (Thornberry and Abbatt[1]).

T. Thornberry and J. P. D. Abbatt, *Phys. Chem. Chem. Phys.*, 2004, **6**, 84–93; J. A. Thornton, C. F. Braban and J. P. D. Abbatt, *Phys. Chem. Chem. Phys.*, 2003, **5**, 4593–4603.

Professor Artaxo added: Yes, it is very clear that the difficult to deal with aqueous phase chemistry in small aerosol particles makes it impossible to model

SOA formation in Amazonia. Another complication is that in Amazonia, 80–95% of aerosols mass is organic, a very large fraction, and only part of this organic material comes from oxidation of isoprene. Thousands of other biogenic compounds also contribute to SOA, and with 100% relative humidity most of the time, heterogeneous liquid–gas reactions take place with hundreds of organic compounds. A very difficult picture for SOA modelers.

Faraday Discussions

RSC Publishing

PAPER

Online and offline mass spectrometric study of the impact of oxidation and ageing on glyoxal chemistry and uptake onto ammonium sulfate aerosols†

Jacqueline F. Hamilton,[*a] M. Teresa Baeza-Romero,[b] Emanuela Finessi,[a] Andrew R. Rickard,[c] Robert M. Healy,[d] Salvatore Peppe,[e] Thomas J. Adams,[f] Mark J. S. Daniels,[f] Stephen M. Ball,[f] Iain C. A. Goodall,[f] Paul S. Monks,[f] Esther Borrás[g] and Amalia Muñoz[g]

Received 4th April 2013, Accepted 13th May 2013

DOI: 10.1039/c3fd00051f

Recent laboratory and modelling studies have shown that reactive uptake of low molecular weight α-dicarbonyls such as glyoxal (GLY) by aerosols is a potentially significant source of secondary organic aerosol (SOA). However, previous studies disagree in the magnitude of the uptake of GLY, the mechanism involved and the physicochemical factors affecting particle formation. In this study, the chemistry of GLY with ammonium sulfate (AS) in both bulk laboratory solutions and in aerosol particles is investigated. For the first time, Aerosol Time of Flight Mass Spectrometry (ATOFMS), a single particle technique, is used together with offline (ESI-MS and LC-MS^2) mass spectrometric techniques to investigate the change in composition of bulk solutions of GLY and AS resulting from aqueous photooxidation by OH and from ageing of the solutions in the dark. The mass spectral ions obtained in these laboratory studies were used as tracers of GLY uptake and chemistry in AS seed particles in a series of experiments carried out under dark and natural irradiated conditions at the outdoor European Photo-reactor (EUPHORE). Glyoxal oligomers formed were not detected by the ATOFMS, perhaps due to inefficient absorption at the laser wavelength. However, the presence of organic nitrogen compounds, formed by reaction of GLY with ammonia was confirmed, resulting in an increase in the absorption efficiency of the aerosol, and this increased the number of particles successfully ionised by the ATOFMS.

[a]Department of Chemistry, University of York, York, YO10 5DD, UK. E-mail: jacqui.hamilton@york.ac.uk; Tel: +44 1904 324076

[b]Escuela Ingeniería Industrial Toledo, Universidad Castilla la Mancha, Toledo 45071, Spain

[c]National Centre for Atmospheric Science (NCAS), University of York, York, YO10 5DD, UK

[d]Southern Ontario Centre for Atmospheric Aerosol Research, University of Toronto, Canada

[e]School of Earth and Environment, University of Leeds, Leeds, LS2 9JT, UK

[f]Department of Chemistry, University of Leicester, Leicester, LE1 7RH, UK

[g]EUPHORE Laboratories, Instituto Universitario CEAM-UMH, 46980, Valencia, Spain

† Electronic supplementary information (ESI) available. See DOI: 10.1039/c3fd00051f

A number of light absorbing organic nitrogen species, including 1*H*-imidazole, 1*H*-imidazole-2-carboxaldehyde, 2,2'-bis-imidazole and a glyoxal substituted 2,2'-bisimidazole, previously identified in aqueous laboratory solutions, were also identified in chamber aerosol and formed on atmospherically relevant timescales. An additional compound, predicted to be 1,2,5-oxadiazole, had an enhanced formation rate when the chamber was open and is predicted to be formed *via* a light activated pathway involving radical oxidation of ammonia to hydroxylamine, followed by subsequent reaction with glyoxal to form an intermediate glyoxime.

Introduction

Atmospheric aerosols are ubiquitous throughout the Earth's atmosphere owing to a wide variety of primary and secondary anthropogenic and biogenic sources. However, despite much scientific effort over the last decade, the true impact of aerosols on climate, air quality and human health has yet to be elucidated owing to large uncertainties and lack of fundamental knowledge on their sources, chemical composition (and hence physical properties) and formation mechanisms.[1] Recent laboratory and modelling studies have shown that, despite their high volatility, reactive uptake of low molecular weight α-dicarbonyls such as glyoxal (GLY; CH(O)CHO) by aerosols is a potentially significant source of secondary organic aerosol (SOA).[2-7] Global sources of GLY[4,8] are estimated between 45 and 56 Tg y^{-1}, representing a significant reservoir of atmospheric carbon. GLY is highly soluble in water[9] and its multiphase chemical processing could potentially change the optical properties, climate effects and heterogeneous chemistry of the seed aerosol. For example, Volkamer and co-workers measured gaseous GLY directly in the urban atmosphere of Mexico City[10] and subsequently demonstrated that a SOA source from GLY is needed to balance GLY sources/sinks, contributing $\geq$15% to SOA observed.[11,12] Fu *et al.*,[3,4] using a global chemical transport model, showed that cloud/aqueous aerosol processing of α-dicarbonyls could represent a significant global source of SOA (3 and 8 Tg y^{-1} for GLY and methylglyoxal, MGLY, respectively) and demonstrated their potentially important contribution (17–27%) to ambient concentrations of water soluble organic carbon aerosol measured from the surface and aircraft over the eastern US.

Much work has been carried out over the past decade on the reactive uptake of α–dicarbonyls onto/into atmospheric aerosol, leading to particle phase chemistry that is highly complex.[5,13,14] Volkamer *et al.*,[15] have investigated SOA formation from the reaction of OH with acetylene and looked at the effect of different seed aerosol (organic, inorganic and mixed) on SOA growth under both photochemical and dark conditions. The main product from this reaction is GLY with >99% of the SOA formed attributed to it. SOA growth was found to be enhanced by several orders of magnitude under photochemical conditions and SOA yields increase linearly with aerosol liquid water content but no correlation was seen with the size of the organic fraction of the seed aerosol.

Galloway *et al.*,[16] have also performed chamber studies of glyoxal uptake onto aqueous ammonium sulfate (AS) seed aerosol, in the presence and absence of light, in order to investigate the processes governing reactive uptake. Aerosol growth, dominated by reversible glyoxal monomer and oligomer formation, was observed under dark conditions. Using high resolution aerosol mass spectrometry (HR-AMS) they also observed the irreversible formation of light absorbing organic nitrogen (ON) compounds, with 1*H*-imidazole-2-carboxaldehyde identified for the first time as an AS aerosol condensed phase C–N product. Formation of imidazoles from glyoxal in the presence of ammonia in bulk solution phase chemistry has been known for over a century[17,18] No organosulphate compounds were observed in the Galloway *et al.*,[16] study in the dark but under irradiated conditions, photochemically active uptake was observed, with various acids (glyoxalic, glycolic and formic) and glycolic acid sulphate ($C_2H_3SO_6{}^-$) detected by liquid chromatography tandem mass spectrometry (LC-MS^2), indicating photo-oxidative radical chemistry is occurring. However, overall aerosol growth was reduced in the presence of light, compared to the dark, in contrast to the results of Volkamer *et al.*,[15] although no OH source was present in the Galloway *et al.*,[16] study.

In a subsequent study, Galloway *et al.*,[19] (2011) performed a similar set of photochemical chamber studies to Volkamer *et al.*,[15] and Galloway *et al.*,[16] (2009), this time in the presence of hydroxyl radicals. Aerosol chemical composition and growth in the presence of light was shown to be independent of light and OH. Galloway *et al.*,[19] (2011) suggest that in their experiments, they may have inadvertently coated the AS with residual chamber organics, creating a barrier to photochemical glyoxal uptake, but which does not affect dark uptake. This observation suggests that (fast) photochemical and (slower) dark types of uptake may, therefore, be governed by surface and bulk processes, respectively. The source of GLY (direct injection *vs. in situ* formation from an acetylene precursor) could also possibly explain the difference between the Galloway *et al.*,[16] (2009) and Volkamer *et al.*,[15] results. Nakao *et al.*,[20] also show significant fast, reversible uptake of GLY onto deliquesced AS seed under dark conditions.

As already mentioned, prior to the identification of imidazoles in glyoxal/AS seed aerosol, light absorbing C–N bond containing compounds have previously been identified and characterised in aged glyoxal/AS bulk solution studies.[21–25] Noziere *et al.*,[22] used LC-HRMS and UV absorption techniques to investigate product distributions from reactions of glyoxal in aqueous solutions containing various ammonium salts, identifying two possible irreversible reaction pathways; (a) a Brønsted acid (proton donor) pathway leading to the formation of C–O bond containing products and/or oligomerisation; (b) an iminium pathway leading to the formation of C–N containing compounds and/or oligomerisation. This finding was backed up by Shapiro *et al.*,[21] who also suggested a mechanism involving $NH_4{}^+$ ions that could form high MW, light absorbing compounds possibly containing C–N bonds.

Yu *et al.*,[24] identified the formation of imidazole, imidazole-2-carboxaldehyde, formic acid (co-product of imidazole), and *N*-glyoxal substituted imidazole using a combination of 1 and 2D NMR techniques (^{1}H and ^{13}C). UV-vis and NMR studies on solutions containing AS indicated that the light absorbing compounds must contain nitrogen. Kampf *et al.*,[25] also confirmed the identity of 2,2′-bisimidazole, as well as imidazole and imidazole-2-carboxaldehyde in GLY/AS aqueous mixtures using HPLC-ESI-MS/MS. Formamides and formic acid ester imidazole derivatives are also tentatively identified as possible organic nitrogen products. Trainic *et al.*,[23] used AMS and CRD spectroscopy to investigate the chemical and optical properties of hydrated AS aerosol exposed to gaseous GLY, observing an increase in optical activity of the aerosols as they grow. The main products of the heterogeneous GLY reaction were glyoxal oligomers, with a minor contribution from imidazoles, with the ratio of C–N containing compounds (including imidazoles) increasing with increasing RH.

Previous mass spectrometric studies of GLY uptake onto AS have predominantly used AMS, where the mass spectra produced are based on the average of many particles. Alternatively, Aerosol Time-of-flight Mass Spectrometry (ATOFMS) has been employed in this study. ATOFMS has been previously used in chamber experiments to detect oxidation products and oligomers in SOA from aromatics.[26–29] However, for chamber experiment applications, ATOFMS has predominantly been used to monitor organic particles formed by self-nucleation of oxidised vapours, rather than SOA grown on pre-existing inorganic seed particles. This task is challenging due to the low sensitivity of the instrument to certain inorganic seed particles (AS in this case[30]) and organic matrices[26] that absorb only very weakly at the ATOFMS laser's wavelength. However, an important advantage of ATOFMS over instruments like AMS is that it records both positive and negative MS and potentially provides information at the single particle level. Negative MS provides heteroatom composition, for example the presence of S-, O- and N-containing organic compounds.[31] Single particle analysis also offers the advantage of separating any background particles in the chamber from the AS-GLY particles of interest.

The general aim of this study is to investigate the chemistry of glyoxal with aqueous ammonium sulfate in both bulk laboratory solutions and in chamber aerosol particles. Online (ATOFMS) and offline (ESI-MS and LC-MS^2) mass spectrometric techniques were used to investigate the change in composition as a result of aqueous oxidation by light-induced OH formation from H_2O_2 and from ageing of the solutions. The mass spectral ions obtained were used as tracers of GLY uptake and chemistry in AS seed particles in a series of experiments carried out in the outdoor European Photo-reactor (EUPHORE) in order to assess dark and photochemical (under natural solar irradiative conditions) uptake of GLY produced homogeneously *in situ* from the oxidation of acetylene.

Experimental

Laboratory solutions and materials

Glyoxal (40% wt. solution), 1*H*-imidazole (>99.5%), 1*H*-imidazole-2-carbaldehyde (97%), 2,2′-bis(4,5-dimethylimidazole) (technical grade), hydrogen peroxide (>30% wt. solution, traceSELECT®), formic acid (>95%), oxalic acid (>99%), glycolic acid (99%, reagentplus®) and glyoxylic acid (50% wt. solution) were obtained from Sigma Aldrich. Ammonium sulfate (>99%), Acetonitrile (LC-MS Optima grade®) and water (LC-MS Optima grade®) were obtained from Fisher Scientific. Optima grade water and acetonitrile were used for laboratory prepared solutions, filter samples extractions, dilutions and as mobile phases in the HPLC analyses. The glyoxal purity was not checked prior to its use. However, it was tested with ^{1}H NMR and there were no peaks that would indicate degradation of the glyoxal to acids or other species.

Laboratory solutions at molar concentrations were prepared in batches of 10 mL in pre-fired transparent glass vials. Solutions at millimolar concentrations were prepared in 100 mL round-bottom transparent glass flasks. For the experiments in the dark, the vials (or flasks) were wrapped in aluminum foil for protection from exposure to light. GLY and ammonium sulfate AS solutions were prepared containing initial concentrations ranging from 1 mM to 1 M at various GLY/AS ratios. Solutions at millimolar or molar concentrations were used as proxy for cloud/fog waters or aerosols respectively. In general, solutions at millimolar concentrations showed a mildly acidic or neutral pH (4–7) while solutions at molar concentrations achieved pH < 4, closer to that of ambient aerosols. Aqueous photo-oxidation experiments were conducted for 1 mM solutions of GLY and AS only to more closely resemble cloud waters conditions. OH radicals were formed *in situ* by photolysis of 15 mM of hydrogen peroxide (H_2O_2) exposed to UV light. All solutions were kept at room temperature, *i.e.* at 20 $\pm$ 3 °C. Discrete samples were taken from the laboratory solutions at different intervals of time ranging from a few hours to several weeks and used for analyses. No catalase was added to the discrete samples used for the analyses to destroy any remaining H_2O_2. For solutions at molar concentrations, 100 fold diluted samples were typically used for the offline analyses. 10 ml of solutions were nebulized into the ATOFMS inlet directly with no additional drying steps to ensure they were consistent with the chamber aerosol.

Chamber experiments

All chamber experiments reported in this paper were performed in the European Photo-Reactor (EUPHORE), located in Valencia, Spain during the period of the 2nd to the 31st of July to 2012. EUPHORE is a large scale atmospheric simulation chamber, which is used for studying the mechanisms of photochemical atmospheric processes. Extensive details of the chamber and its instrumentation are given elsewhere.[32,33] Briefly, it consists of two 200 m^3 volume simulation chambers. The hemispherical shaped chambers are formed from fluorine–ethene–propene (FEP) Teflon foil (127 μm thickness) fitted with housings that exclude ambient light. Each chamber is fitted with large horizontal and vertical fans to ensure rapid mixing. Relevant analytical instrument employed include a relative humidity sensor (Walz Hygrometer TS-2), O_3 monitor (monitor Labs 9810), and a SMPS (TSI Incorporated: DMA, model 3081, and a condensation particle counter (CPC), model 3775 operated at a 5 min scan rate and measured size distributions in the 14–737 nm diameter range. Sheath and aerosol sampling flows were 3 L min^{-1} and 0.30 L min^{-1}, respectively. Chamber dilution was monitored using FTIR measurements of SF_6. Relative humidity was maintained throughout the experiment by 45 $\pm$ 10% using an automatic feedback system and MilliQ water. A new cleaning regime was implemented to decrease background contamination and full details can be found in the Supplementary Information.†

Aqueous solutions of 60 mM AS were nebulized (using a six-Jet Atomizer (TSI. Model 9306)), neutralized (GRIMM Am source 241) and introduced into the chamber, prior to the start of each experiment. Seed particles were poly-disperse with a median diameter of around 100–150 nm. In order to minimize aerosol loss to the walls and thus extend the length of experiments, the mixing fans were used only during the injection of reactants prior to the experiments. Therefore data obtained from the SMPS during injection and mixing periods were not included in the analysis.

The uptake of GLY onto AS aerosol was investigated using a series of different types of experiments as outlined in Table 1. Experiment type 1 involved the injection of gas phase GLY into the chamber with AS aerosols in the dark and the

Table 1 EUPHORE experiment types and conditions, mixing ratios of precursors and initial aerosol mass loadings

Date	Type of Experiment	Precursors initial mixing ratio (ppb if not indicated)	$[Gly]_{max}{}^{a}$/ppb	Initial aerosol mass/μg m^{-3}	RH/%	Natural Light	OH Radicals
02/07/2012	Background aerosols	—	—	156	50–40	No	No
03/07/2012	Type 1: AS/Gly	—	160	190	40–62	No	No
05/07/2012	Type 3: AS/C$_2$H$_2$/T2B/O$_3$	$[C_2H_2]_0 = 20$ ppm; $[T2B]_0 = 1000$; $[O_3] = 800$	175	181	40–62	No	Yes
16/07/2012	Type 2: AS/C$_2$H$_2$/HONO/light	$[C_2H_2]_0 = 10$ ppm; $[HONO]_0 = 143, 176$	362	280	40–64	Yes	Yes
23/07/2012	Type 2: AS/C$_2$H$_2$/HONO/light	$[C_2H_2]_0 = 5$ ppm; $[HONO]_0 = 151, 115$	167	279	50–70	Yes	Yes
30/07/2012	Type 2: AS/C$_2$H$_2$/HONO/light	$[C_2H_2]_0 = 5$ ppm; $[HONO]_0 = 130$	164	270	50–54	Yes	Yes

^{a} According to BBCEAS measurements.

resulting growth observed over a period of ~18 h. Experiment type 2 was a photochemistry experiment, where GLY was formed *in situ* by the oxidation of acetylene (ACET) (Abello-Linde, 99.5%) with OH radicals formed from HONO photolysis after the chamber cover was opened. Once the HONO was depleted, the chamber was closed and additional HONO added. The chamber was reopened for a further period (~1 h), before being closed again and the evolution of the aerosol was monitored over the next 16 h in the dark. Experiment type 3 was a dark OH chemistry experiment, where gaseous GLY was again formed *in situ* by the oxidation of acetylene (Abello-Linde, 99,5%) by OH radicals formed from the ozonolysis of *trans*-2-butene (T2B) (Abello-Linde, ≥99%) and again the evolution of the chamber and aerosol composition was monitored for a further 16 h. T2B reacts relatively quickly with ozone (k_{O3} (298 K) = 1.90 × 10^{-16} cm^3 molecule^{-1} s^{-1}) with an OH yield of 0.64.[34] This reaction also has a reported GLY yield of 10–18%.[34] At the end of each experiment the chamber was flushed with clean air to investigate the reversibility of the aerosol growth using the ATOFMS. The advantage of the EUPHORE experiments over previous studies is that the photochemical reaction takes place using natural sunlight rather than artificial light produced by arc and black lamps, which is often more intense in the UV region. Additionally, due to the large size of this chamber, longer experiments can be carried out before deposition to the wall and floor depletes the aerosol, and larger volumes of air can be removed for filter sampling, allowing collection of filters at different times within an experiment. In this work, typically 3–5 filter samples were collected for each experiment. Filter samples were collected at 10, 20 or 40 l min^{-1} onto pre-fired 47 mm quartz filters (Whatman QMA), which were kept frozen until analysis. Sampling times ranged from 1 up 15 h depending on the stage of the experiment (lower sampling times associated with initial stages and longer ones with later stages). Filter samples were wrapped in aluminum foil, sealed in plastic bags and stored in the freezer immediately after collection and kept frozen until analysis.

Pure GLY monomer was prepared from the solid trimer-dihydrate (95%) using the methods described in ref. 35 with minor modification. Cold fingers containing pure unpolymerized GLY were temporarily kept at liquid nitrogen temperature prior to experimental use. Pure GLY was introduced into the EUPHORE chamber by passing a small flow of nitrogen through a cold-trap whilst allowing the trap to warm gently. A small impurity of MGLY was seen by the BBCEAS in the experiment on 03/07/12 when this method was used.

Broadband cavity enhanced absorption spectroscopy

Gas-phase concentrations of GLY (and MGLY) were measured using broadband cavity enhanced absorption spectroscopy (BBCEAS).[36–38] Gas was sampled from the EUPHORE chamber into the BBCEAS instrument through a PFA line (1.2 m length, 6.35 mm outside diameter, 2 litres/min flow rate) that passed through a bulkhead fitting in the chamber floor and protruded 40 cm above the chamber floor in order to sample well-mixed gas uncompromised by wall effects. GLY, MGLY, NO$_2$, water vapour and oxygen's O$_2$–O$_2$ collision complex all possess distinctive structured absorption features within the 430–488nm spectral bandwidth of the instrument. The gas-phase concentrations of these species were obtained by applying differential optical absorption spectroscopy (DOAS) to fit reference absorption cross sections to the absorption spectra recorded by BBCEAS. The remaining unstructured continuum contribution to the absorption spectra, which is primarily due to aerosol extinction, was fitted with a polynomial function. Wavelength-resolved measurements of the aerosol absorption co-efficient (at 10 nm intervals; 440–480 nm) were obtained from the continuum absorption measurements, after making any minor corrections necessary to account for other weakly structured, gas-phase absorption contributions (such as from ozone).

Baseline measurements and Allan variance analyses show the BBCEAS instrument has 1σ detection limits of 25 pptv for GLY, 40 pptv for NO_2 and 350 pptv for MGLY[39] against a zero absorption background for an integration time of 10 s. However for the present experiments, light scattering by aerosol particles (whether from the addition of seed aerosol or from organic aerosol generated during the oxidation of VOCs inside the chamber) reduced the effective path length of the intra-cavity absorption measurement, and so reduced the sensitivity of the BBCEAS instrument. Consequently a longer integration time of 1 min was used. Typical detection limits were 55 pptv for GLY, 60 pptv for NO_2 and 800 pptv for MGLY (1σ in 1 min) in the presence of aerosol with an absorption coefficient of 1×10^{-6} cm^{-1} ($=100$ Mm^{-1}). The overall accuracy of the BBCEAS measurements is estimated to be 7% for GLY, 7% for NO_2 and 10% for MGLY. A full discussion of the BBCEAS instrument and its performance can be found in Daniels *et al.*[39]

Proton transfer reaction mass spectrometry

A PTR-TOF-MS instrument (Series I, Kore, UK) was used to detect the gaseous organic compounds. The PTR-TOF-MS technique is based on chemical ionization by the hydronium ion (H_3O^+) of trace VOC present in the atmospheric samples by proton transfer reactions, which produces a protonated molecular ion (VOC-H$^+$) for each VOC molecule. More details on this instrumental technique are discussed elsewhere.[40]

Aerosol time-of-flight mass spectrometer

An Aerosol Time-of-flight Mass Spectrometer (ATOFMS; TSI 3800) was used to analyze particles on-line and has been described previously in detail.[41,42] It operates in a continuous sampling mode, allowing for real-time analysis of particles with high time resolution. An aerodynamic lens system (TSI- Model 3800–100) was used and the desorption/ionization laser (266 nm) was operated with a power output of $\sim$0.9–0.95 mJ pulse^{-1}. The ATOFMS instrument sampled directly from the EUPHORE chamber through a 0.25$''$ *o.d.* stainless steel sampling line. ATOFMS sampling typically began before introduction of the precursor to analyze typical background aerosol in the chamber, and continuously sampled aerosol for the duration of each experiment. Typical hit rates (HR, defined as number of particles for which a mass spectra is obtained *vs.* number of sized particles) during these experiments were 30–40% for chamber background particles and $\sim$1% at the beginning of an experiment when fresh AS seeds particles were introduced and increasing as the aerosol composition evolved during the experiments (especially in experiments with light).

The nebulized laboratory solution exhibited a highly homogeneous composition. Thus, the individual particle mass spectra were averaged without any pre-sorting of the particles. Between 300 and 1000 particles were used to obtain an average MS for each solution. Threshold values of 15, 15 and 0.001% for peak height, area and relative area, respectively were used when importing mass spectral data. In all experiments described here, peak detection was carried out up to (1000 Da). Both the laboratory and the chamber datasets were imported into Enchilada for analysis.[43] The K-means algorithm was then used to remove any background particles from the chamber datasets prior to further analysis.[44] This step was necessary because although background particles were present at low number concentrations, they are ionized more efficiently than AS-GLY particles. Average mass spectra were generated every 30 min for the chamber experiments to monitor the temporal evolution of specific ions.

Offline analysis

Aerosol collected onto the filters during chamber experiments were extracted into 4 ml of high purity water (LC-MS Optima grade®) by sonication for 60 min. The extracts were filtered by using PTFE membranes (0.45 μm pore size), concentrated by vacuum evaporation and re-dissolved with 0.4 ml acetonitrile/water (50/50, v/v). Lab solutions and concentrated filter extracts were analyzed both directly with electrospray ionization and with prior separation using HPLC.[45,46] Mass spectrometry analysis was performed using an HCT-Plus ion trap mass spectrometer with electrospray ionisation (Bruker Daltonics GmbH). Electrospray ionisation (ESI) was carried out at 350 °C with a nebuliser pressure of 65 psi and a nitrogen drying gas flow of 11 l min^{-1}. Reversed phase liquid chromatography (1100 series, Agilent) equipped with a Zorbax Eclipse XDB-C18 column, 150 mm × 4.6 mm I.D., 5 μm pore size (Agilent). Samples (60 μl) were injected then separated by step gradient elution with solvents A (0.1% v/v formic acid water) and B (acetonitrile) at a flow rate of 0.6 mL min^{-1} (B at 3% for 2 min then increased to 60% at 20 min. B kept at 60% for 2 min and decreased to 3% by 30 min). Two separations were carried out for each sample, one in the positive and one in negative MS ionization mode, scanning from m/z 50–600 and with automatic selection of the three most abundant ions for MS^2 by collision induced dissociation (CID). Direct injections of samples in the ESI source were carried out using a syringe pump at 5 μl min^{-1}.

Samples were also analysed at high mass resolution using a Bruker APEX 9.4 T Fourier Transform Ion Cyclotron Resonance Mass Spectrometer (FTICRMS). Samples were sprayed at a flow rate of 2 μL min^{-1}, into an Apollo II electrospray interface with ion funnelling technology. Spectra were acquired in both positive and negative ion mode over the scan range m/z 50–600 using the following MS parameters: 1.1 L min^{-1} (nebulising gas flow), 4 L min^{-1} (drying gas flow), 200 °C (drying temperature), 0.05–0.5 s (collision cell accumulation), and data acquisition size of 2 Mb (yielding a target resolution of 130 000 at m/z 400). Data were analysed using DataAnalysis 4.0 software (Bruker Daltonics, Bremen, Germany). The instrument was externally calibrated using protonated (positive ion mode) or deprotonated (negative ion mode) arginine clusters and internally calibrated using a series of polyethylene glycol contamination peaks.

Results

Measurements of lab solutions using online and offline mass spectrometry

A series of lab solution were investigated using ESI-MS, LC-MS^2 and ATOFMS. ESI-MS was used to investigate the overall molecular composition of the GLY/AS solutions using soft ionisation, although it is recognized that this may cause artifact formation. The molecular structures were investigated using LC-MS[2], to isolate the species formed and reduce artifact formation. Finally, the solutions were also nebulised and introduced into the ATOFMS. Each complementary technique has a different ionization mechanism and will favour the formation of different ions. The aim of these tests was to identify tracer ions for GLY chemistry in the chamber experiments.

Glyoxal aqueous solution. The ESI-MS of GLY (shown in Supplementary Fig. 1†) in water showed oligomer ions similar to those seen previously,[14,47,48] with m/z 117, 161, 175 and 219 representing tracers of GLY oligomers as shown in Table 2. However, the ATOFMS did not ionize aqueous GLY particles efficiently, presumably due to poor absorption efficiency at 266 nm and thus an average mass spectrum could not be generated for this solution.

Glyoxal in aqueous ammonium sulfate solution. The positive ionization mass spectra obtained using ESI-MS and ATOFMS for the GLY + AS (1M) solution aged for one week are shown in Fig. 1a and b. The ESI-MS spectrum in Fig. 1a contains ions related to both GLY oligomers (m/z 117, 161, 175) and N-containing species from aqueous GLY reactions with ammonia. A number of previously identified ON species were observed,[16,24,25] including m/z 69 (1H-Imidazole, IM), m/z 97 (1H-Imidazole-2-carboxaldehyde, IC), m/z 115 (hydrated 1H-Imidazole-2-carbox-aldehyde, HIC), m/z 135 (2,2′-bisimidazole, BI), m/z 145 (hydrated glyoxal substituted 1H-imidazole, HGI), m/z 173 (GLY substituted hydrated 1H-Imidazole-2-carboxaldehyde GHIC), m/z 193 (GLY substituted 2,2′-bisimidazole GBI) and m/z 203 (hydrated GLY dimer substituted imidazole, HGGI). Compound structures and abbreviations are given in Supplementary Table 1.† Identification of these compounds have been supported through comparison with the retention time of standards (IM), comparison to MS[2] fragments in Kampf *et al.*,[25] and accurate mass measurements using FTICR-MS. A number of previously unidentified ions are also observed including m/z 127, 157, 221, 261, 279. FTICR-MS was used to investigate the molecular formulae of these unknowns and the predicted molecular formulae and errors (ppm) are shown in Table 2. The FTICR-MS was internally re-calibrated using IM, HIC, BI and GBI to improve the mass accuracy, hence in Table 2 no error values are given for these species. The ions at m/z 221, 261 and 279 represent formulae that correspond to HGGI substituted with H_2O, GLY and H_2O + GLY respectively. The ion at m/z 127 has a predicted molecular formula equivalent to GLY substituted IM. The ion at m/z 157 is unidentified at present.

The AS + GLY laboratory solution nebulized particles were ionized more efficiently by the ATOFMS than the GLY only aqueous solution. The average positive ionization mass spectrum is shown in Fig. 1b ($N = 455$). Signals for the ammonium (m/z 18 $[NH_4]^+$) and organic aerosol (m/z 27 $[C_2H_3]^+$, 39 $[C_3H_3]^+$) are clearly visible. There are also peaks normally associated with aromatic fragment ions (m/z 50 $[C_4H_2]^+$, 51 $[C_4H_3]^+$, 77 $[C_6H_5]^+$), although the source of these is unclear. The most striking feature of the mass spectrum is the appearance of an ion at m/z 69, which is assigned here as protonated imidazole (IM) $[C_3H_4N_2 + H]^+$ based on the associated ESI measurements. IM has been observed previously in chamber studies of dark GLY uptake using online High Resolution Aerosol Mass Spectrometry (HR-AMS) at m/z 68, which is the molecular ion using electron impact ionization.[16] $[IM + H]^+$ has a pK_a of 7.0, therefore in AS solutions at 1M (pH ∼ 3) it is likely to be mostly protonated. In addition, its aromatic structure will increase

Table 2 Details of laboratory solutions and prominent ions observed by ATOFMS, ESI-MS and FTICR-MS. Ions also observed in the filter samples from the experiment on 23/07/12 are indicated in italics

Lab solution	ATOFMS positive ions (m/z)	Proposed ion formula	ESI-MS major positive ions (m/z of $[M + X]^+$)	Proposed ion formula	FTICR-MS error (ppm)	Identity	Reference
1M GLY in water	None detected		117	$[C_2H_6O_4Na]^+$	—[b]	GLY oligomer	
Aged 1 week			143	$[C_4H_8O_4Na]^+$	0.10	GLY oligomer	
			161	$[C_4H_{10}O_5Na]^+$	0.76	GLY oligomer	
			175	$[C_4H_8O_6Na]^+$	0.48	GLY oligomer	
			189	$[C_5H_{10}O_6Na]^+$	0.49	GLY oligomer	
			203	$[C_6H_{12}O_6Na]^+$	0.64	GLY oligomer	
			219	$[C_6H_{12}O_7Na]^+$	0.50	GLY oligomer	
			233	$[C_6H_{10}O_8Na]^+$	0.47	GLY oligomer	
1M GLY + 1M AS	*18*	$[NH_4]^+$					
Aged 1 week	*27*	$[C_2H_3]^+$					
	39	$[C_3H_3]^+$					
	50	$[C_4H_2]^+$					
	51	$[C_4H_3]^+$					
	69	$[C_3H_5N_2]^+$	69	$[C_3H_5N_2]^+$	—[b]	1*H*-Imidazole (IM)	17,24
	77	$[C_6H_5]^+$					
			97	$[C_4H_5N_2O]^+$	—[b]	1*H*-Imidazole-2-carboxaldehyde (IC)	24
	115	$[C_4H_7N_2O_2]^+$	115	$[C_4H_7N_2O_2]^+$	[c]	Hydrated 1*H*-Imidazole-2-carboxaldehyde (HIC)	25
			127	$[C_5H_7N_2O_2]^+$	−0.52	—[b]	
	132	—[b]					
	134	—[b]					
	135	$[C_6H_7N_4]^+$	135	$[C_6H_7N_4]^+$	0.03	2,2′-bisimidazole (BI)	25
			145	$[C_5H_9N_2O_3]^+$	[c]	Hydrated glyoxal substituted IM (HGI)	24
			157	$[C_6H_9N_2O_3]^+$	0.01	—[b]	

Table 2 (*Contd.*)

Lab solution	ATOFMS positive ions (*m/z*)	Proposed ion formula	ESI-MS major positive ions (*m/z* of [M + X]$^+$)	Proposed ion formula	FTICR-MS error (ppm)	Identity	Reference
			173	$[C_6H_9N_2O_4]^+$	c	*N*-Glyoxal substituted, hydrated 1*H*-Imidazole-2-carboxaldehyde (GHIC)	25
			193	$[C_8H_9N_4O_2]^+$	c	Glyoxal substituted 2,2′-bisimidazole (GBI)	25
			203	$[C_7H_{11}N_2O_5]^+$	0.11	Hydrated GLY dimer substituted imidazole (HGGI)	25
	219	—[b]					
			221	$[C_7H_{13}N_2O_6]^+$	0.56	HGGI + H$_2$O	
			261	$[C_9H_{13}N_2O_7]^+$	0.6	HGGI + GLY	
			279	$[C_9H_{15}N_2O_8]^+$	0.35	HGGI + H$_2$O + GLY	
1M GLY + 1M AS Aged 6 months	69	$[C_3H_5N_2]^+$	69	$[C_3H_5N_2]^+$	1.88	1*H*-Imidazole (IM) - very small peak	17,24
			97	$[C_4H_5N_2O]^+$	−0.22	1*H*-Imidazole-2-carboxaldehyde (IC)	24
	107	—[b]					
			115	$[C_4H_7N_2O_2]^+$	0.02	Hydrated 1*H*-Imidazole-2-carboxaldehyde (HIC)	25
	121	—[b]	120	$[C_6H_6N_3]^+$	−0.03	—[b]	
	135	$[C_6H_7N_4]^+$	135	$[C_6H_7N_4]^+$	0.06	2,2′-bisimidazole (BI)	25
	149		149	$[C_7H_9N_4]^+$	−0.03	—[b]	
	163	—[b]					
			170	$[C_4H_{12}NO_6]^+$	0.09		
	177	—[b]	177	$[C_8H_9N_4O]^+$	−0.07		
	191	$[C_6H_{11}O_5N_2]^+$				Hydrated glyoxal hydrated 1*H*-imidazole-2-carboxaldehyde (HGHIC)	25
	193	$[C_8H_9O_2N_4]^+$	193	$[C_8H_9O_2N_4]^+$	0.07	Glyoxal substituted 2,2′-bisimidazole (GBI)	25
			203	$[C_7H_{11}N_2O_5]^+$	0.08	Hydrated GLY dimer substituted imidazole (HGGI)	25

Table 2 (*Contd.*)

Lab solution	ATOFMS positive ions (*m/z*)	Proposed ion formula	ESI-MS major positive ions (*m/z* of [M + X]$^+$)	Proposed ion formula	FTICR-MS error (ppm)	Identity	Reference
			214	$[C_8H_{12}N_3O_4]^+$ and $[C_6H_{16}NO_7]^+$	0.2	—[b]	
	220	—[b]					
			228	$[C_6H_{14}NO_8]^+$	0.55	—[b]	
			246	$[C_6H_{16}NO_9]^+$	0.68	—[b]	
1 mM GLY + 1 mM AS + 15 mM H$_2$O$_2$ UV 1 week	39	$[C_3H_3]^+$	Not analysed				
	43	$[CH_3CO]^+$					
	57	$[C_3H_5O]^+$					
	71	$[C_3H_3O_2]^+$ *or* $[C_2H_3N_2O]^+$					
	77	$[C_6H_5]^+$					
	91	$[C_7H_7]^+$ *or* $[C_2H_3O_4]^+$					
	115	$[C_4H_7N_2O_2]^+$					
	135	$[C_6H_7N_4]^+ = BI^a$					
	149	—[b]					

[a] tentative. [b] not identified. [c] used for calibration.

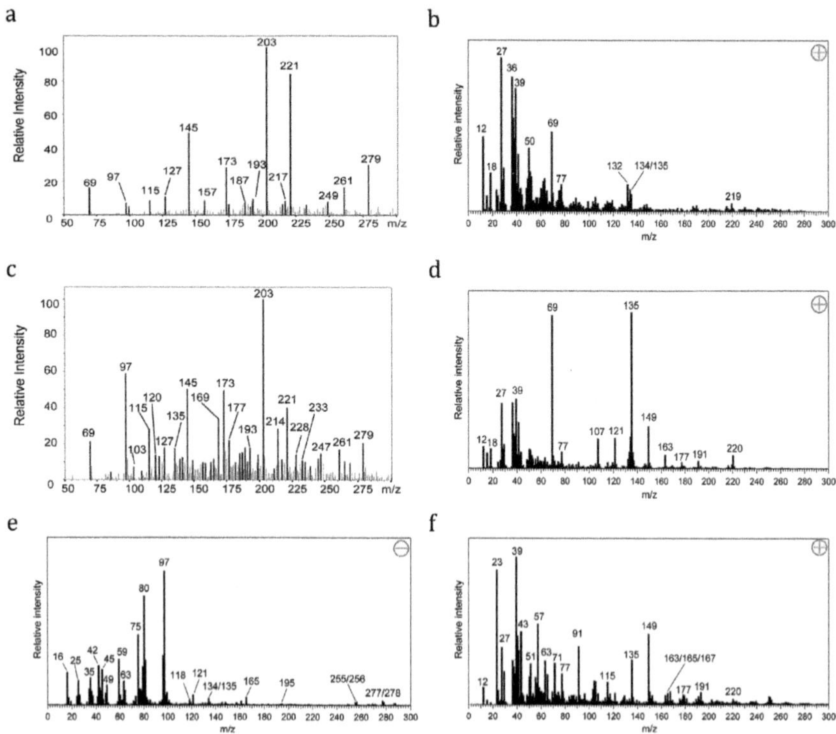

Fig. 1 Mass spectra of GLY + AS lab solutions. Upper: 1 M GLY + 1 M AS aged in the dark for 1 week (a) ESI-MS and (b) ATOFMS. Middle: 1 M GLY + 1 M AS aged in the dark for 6 months (c) ESI-MS and (d) ATOFMS. Lower: 1 mM GLY + 1 mM AS + 15 mM H_2O_2 exposed to light for 1 week. (e) ATOFMS negative ionization and (f) ATOFMS positive ionisation.

the stability of IM relative to oxidized oligomers and this may explain why it does not fragment during ionization. The peak at m/z 135 is assigned as 2,2′-bisimidazole (BI) as seen using ESI. The UV absorption properties of these species (between 200–300 nm)[22] may contribute to the enhanced absorption of the ATOFMS laser ($\lambda = 266$ nm) even at low concentration. The peak at m/z 115 is tentatively assigned as the protonated and hydrated form of 1H-imidazole-2-carbaldehyde (HIC) $[C_4H_6N_2O_2 + H]^+$, which was also observed in the ESI-MS analysis, although it has a very minor contribution to the average spectrum. Unlike the ESI analysis, there is no signal for IC in this mass spectrum at m/z 97, and it is possible that IC decomposes to IM during ionization, although it has been observed by AMS in a previous study.[16] There are a number of peaks in the ATOFMS that remain unidentified including m/z 132, 134, 219, which are not seen in the ESI or LC analysis.

Effect of ageing in GLY + AS solutions. A GLY + AS (1M) solution was left to age in the dark for 6 months and reanalysed using ESI-MS and ATOFMS to investigate the change in composition. The solution had turned a dark brown colour as expected.[17] The major peak in the ESI-MS spectrum, as shown in Fig. 1c, is the m/z 203 $[HGGI + H]^+$ and the mono-imidazole ring species, m/z 69 $[IM + H]^+$ and 97 $[IC + H]^+$, are also dominant. Minor peaks are seen for m/z 115 $[HIC + H]^+$ and the glyoxal substituted imidazoles including m/z 145 $[HGI + H]^+$ and 173 $[GHIC + H]^+$.

The GLY oligomer ions, including m/z 117, 161, 175 and 219, are now absent from the spectrum. A large increase is seen in the imidazole dimers, including 2,2-bisimidazole at m/z 135 and glyoxal substituted BI (GBI) $[C_8H_8O_2N_4 + H]^+$ at m/z 193. An intense ion is observed at m/z 120, which has an odd molar mass and therefore must contain an odd number of N atoms. An ion at m/z 120.05563 was observed in the FTICR-MS analysis and is assigned as $[C_6H_5N_3 + H]^+$, suggesting an aromatic compound. A series of even mass ions were observed (Supplementary Fig. 2†) at m/z 170 $[C_4H_{11}NO_6 + H]^+$, 214 $[C_6H_{16}NO_7 + H]^+$, 228 $[C_6H_{14}NO_8 + H]^+$ and 246 $[C_6H_{16}NO_9 + H]^+$ that contain one nitrogen and between 6 and 9 oxygen atoms. They have a very high H:C ratio suggesting they are either NH_4^+ adducts or contain a formamide group as suggested by Kampf *et al.*[25] The FTICR-MS of the aged solution is very complex and contains ions with up to 5 N atoms, suggesting extensive aqueous chemistry is occurring over longer timescales. Further work is needed to investigate the functionality and formation mechanisms.

The aged solution was nebulised to investigate if the ATOFMS could detect the increase in the light absorbing organics from GLY + AS reactions. The ATOFMS average positive ion mass spectrum for this solution is shown in Fig. 1d ($N = 527$) and shows similar OC fragment ions as the solution aged for 1 week. The increase in the N-containing species is pronounced, with the m/z 69 $[IM + H]^+$ and m/z 135 $[BI + H]^+$ peaks dominating the mass spectrum. A series of new peaks, separated by 14 Da (CH_2, N or 0.5CO), were observed at m/z 107, 121, 149, 163, 177, 191 and 193. The last two are tentatively identified as the protonated ions of the hydrated glyoxal substituted HIC, HGHIC $[C_6H_{10}O_5N_2 + H]^+ = m/z$ 191 and GBI $[C_8H_8O_2N_4 + H]^+ = m/z$ 193 as seen in the ESI-MS analysis and Kampf *et al.*[25] There is also a stronger signal at m/z 77 $[C_6H_5]^+$, which may indicate that this is related to a GLY product. It is clear that both the ESI-MS and ATOFMS mass spectra are dominated by ON species, but the differences in ionization mechanism favour the production of different ions.

Negative ions were also observed in a small fraction of the ATOFMS mass spectra collected ($\sim$9%). The dominant ions were m/z −97 $[HSO_4]^-$ and −195 $[HSO_4^-.H_2SO_4]$ and are related to sulfate ions. The m/z −195 ion is typically observed in acidic particles and reflects the acidic pH of the solution after many months of ageing.[49]

Effect of OH oxidation. In addition to the dark experiments, a GLY + AS (1 mM) solution was subjected to oxidation using H_2O_2 (15mM) and UV light for 1 week. The main oxidation products of GLY are glyoxylic acid and oxalic acid,[48] both of which have high limits of detection on the ESI-MS system used. Thus the low concentrations and extended period of ageing lead to a complex ESI-MS spectrum that had no significant marker ions and will not be discussed further.

The positive ATOFMS mass spectrum, shown in Fig. 1f shows some similarities with the non-oxidatively aged solutions, with abundant OC fragment ions at m/z 27, 29 and glyoxal related peaks at m/z 115 (HIC), 121, 135 (BI), 149, 163, 191 and 193 (GBI). Although there is a large peak at m/z 135 $[BI + H]^+$, the peak at m/z 69 corresponding to $[IM + H]^+$ has reduced dramatically compared to the dark reaction solutions. Above m/z 100, the major ion is now m/z 149 (as seen in Carlton *et al.*,[48]). There is also a large increase in the relative intensity of m/z 43 $[CH_3CO]^+$ and m/z 57 $[C_3H_5O]^+$, which is due to an increase in oxidised OC. Other fragment ions that have increased in relative intensity include m/z 63, 65, 71, 77 and 91. The m/z 91 ion was tentatively assigned to an aromatic fragment in previous solutions.

However, the major oxidation product of GLY is oxalic acid (90 Da) and its protonated form is also a possible assignment for this ion. However, there is no evidence for the expected corresponding negative ion at m/z −89 and so this remains an uncertainty at present. There are also additional ions at higher MW present up to around m/z 260. The increase at m/z 23 is attributed to trace Na^+ contamination from the H_2O_2 solution.[31]

ATOFMS negative ion mass spectra were obtained for ∼5% of the particles ionised. The average negative mass spectrum in Fig. 1e shows peaks related to organic acids at m/z −45 $[CHOO]^-$, m/z −59 [acetic acid–H]$^-$ and m/z 75 [glycolic acid-H]$^-$. Sulfate ions are observed at m/z −97 $[HSO_4]^-$, m/z 81 $[HSO_3]^-$, m/z 80 $[SO_3]^-$ and m/z 64 $[SO_2]^-$. OC fragments ions are seen at m/z 25 $[C_2H]^-$ and m/z 42 $[C_2H_2O]^-$. There are also a small number of ions at higher m/z (−118, −121, −134, −135, −165 and −277). A corresponding peak for the ion at m/z −165 was also seen in the LC-MS^2 analysis of this solution. The m/z −165 parent ion fragmented to m/z −121 in the LC-MS^2 analysis, with a loss of 44 Da, indicating a carboxylic acid group. This may correspond to the same compound that produces m/z +167 in the ATOFMS mass spectra.

A summary of the laboratory GLY and GLY + AS product tracer ions, along with the details of the solutions, is shown in Table 2.

Chamber experiments

Impact of OH chemistry on GLY uptake: dark _versus_ light. The chamber background experiment (02/07/12) showed no growth in aerosol mass measured by the SMPS (corrected for dilution and wall losses) of the aerosol over the course of 8 h. In the type 1 experiment (03/07/12), where gas phase GLY was introduced into the dark chamber in the presence of AS seed particles, the aerosol mass increased steadily for 16 h until flushing of the chamber began. This dark non-oxidative uptake is consistent with previous studies. The impact of oxidation by OH radicals on GLY uptake on aqueous AS seed was investigated by the _in situ_ formation of GLY from acetylene oxidation (ACET + OH) under both irradiated and dark conditions. In the irradiated experiments (experiments Type 2), OH was formed by HONO photoloysis when the chamber was opened, which may then react with both ACET and any subsequent GLY formed. In this set up, GLY photolysis also occurs,[50] therefore the impact of light and OH cannot be disentangled. In the dark experiment (experiments Type 3), OH radicals are formed from the ozonolysis of T2B, which has an OH yield of 0.64. The majority of oxidation products from the T2B ozonolysis (except for GLY) are predicted to be too volatile to have significant uptake to the particles, although this cannot be confirmed here. Thus the T2B ozonolysis experiments facilitate the investigation of the role of OH chemistry without the impact of light-induced effects. The formation of GLY (green squares) is shown in Fig. 2 for the experiments on the 30/07/12 (light + HONO + ACET) and 05/07/12 (dark + T2B + O_3 + ACET). The rapid formation of GLY is seen in both experiments, where the peak [GLY] is similar (164–175 ppbv).

Fig. 2 also shows the SMPS mass concentrations (black points; corrected for dilution and wall losses) and the aerosol extinction measured by the BBCEAS (red, mean continuum 440 nm (average aerosol extinction between 439 nm and 441 nm) and blue, mean continuum 480 nm (average aerosol extinction between

479 nm and 481 nm); not corrected for dilution or wall losses). At the start of the irradiated OH experiment (30/07/12), the AS seed was added and the chamber opened without any reactants to investigate the chamber background (09:55–11:06). A steady decrease in particle surface and mass were seen, probably as a result of an increase in chamber temperature leading to evaporation of some of the aerosol water (although RH was kept constant ± 5%). It is clear that no growth from the chamber background organics was observed. The chamber was then closed and the reactants (ACET + HONO) were added. Both the aerosol extinction

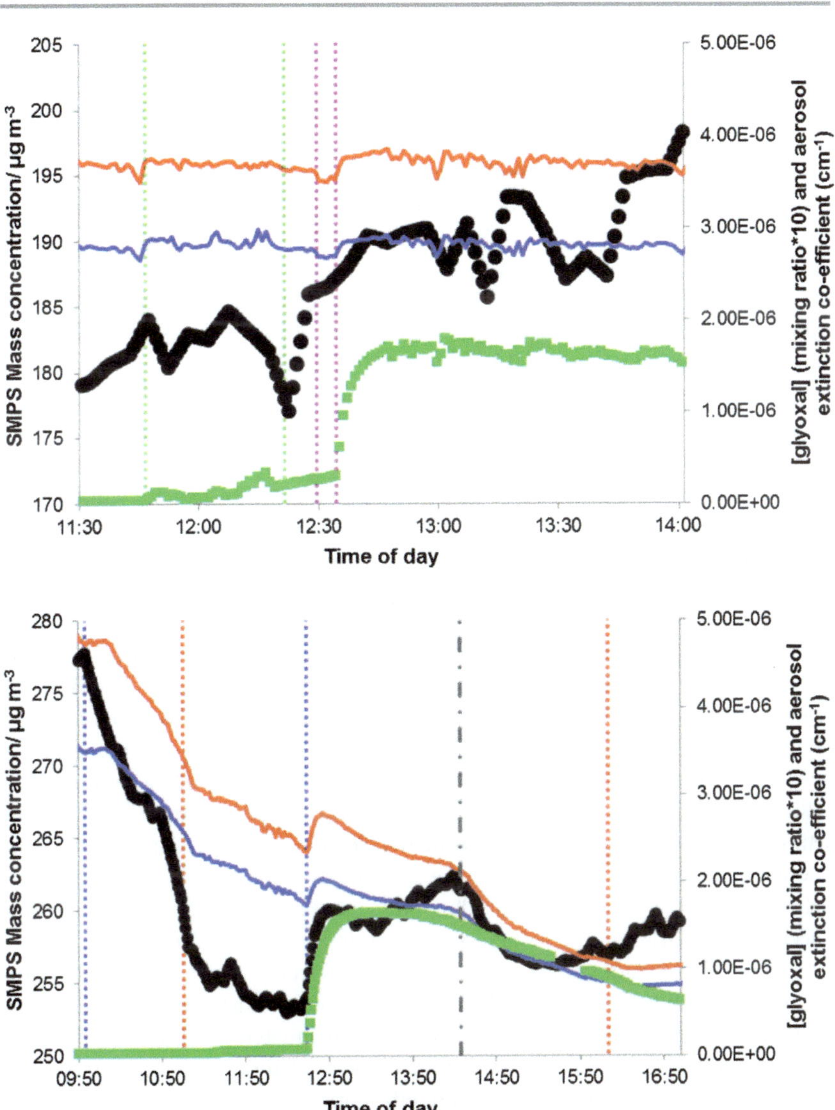

Fig. 2 GLY mixing ratio (multiplied by 10; green), aerosol extinction coefficient (red: mean continuum 440 nm and blue: mean continuum 480 nm) and SMPS mass concentrations (black; corrected for dilution and wall loses) for the experiments on the 30/07/12 (lower) and 05/07/12 (upper). Vertical lines indicate the chamber condition; Blue = chamber opened, Red = chamber closed, Grey dashed = chamber partially closed due to high winds, Green = O_3 addition, Purple = T2B addition.

and mass showed a rapid increase on opening the chamber at 12:34, indicating increased uptake of the GLY formed, followed by a steady or slight decrease until the chamber is finally closed. Once the chamber roof is finally closed at 16:10, the corrected aerosol mass indicates slow growth in the dark as a result of non-oxidative glyoxal uptake. It is clear that there is a relatively short lived period of rapid growth on opening the chamber (4–5 min), after which there is no significant fast growth of the aerosol mass under irradiated conditions.

In contrast, the experiment with OH chemistry in the dark (05/07/12) does not lead to a significant rapid increase in aerosol extinction measured by BBCEAS or in aerosol mass measured by the SMPS, although a very small increase is seen in the aerosol extinction data. Slow non-oxidative GLY uptake can be seen over the course of the experiment, which is consistent with all dark ageing experiments in this study. Gas phase oxidation of GLY is likely to result in products that have a higher volatility and lower Henry's law constants, with the exception of oxalic acid, organosulfates and any oligomers of oxidation products.[16] These results suggest that OH chemistry alone does not lead to significant enhancement of GLY uptake, but that the presence of light is the determining factor for the short term rapid growth as seen by Volkamer *et al.*[15]

The double opening of the chamber in the experiment on the 30/07/12 is similar to the Galloway *et al.*, study.[19] No aerosol growth was observed when the chamber was opened in the absence of GLY with only AS present, unlike in the Galloway study. This discrepancy between the two studies may be a result of lower background organic levels in the chamber in this study, due to the extensive cleaning regime, which are not sufficient to cause a coating on the AS.

Unfortunately, no oxidation products from GLY could be measured by the PTR-MS in the dark as a result of a large acetone signal (~200 ppbv), which is used to stabilize pure ACET, and high levels of T2B oxidation products formed at masses similar to GLY oxidation products.

Photochemistry followed by dark ageing

Online analysis. The effect of photochemistry and dark uptake on the aerosol composition was investigated by forming glyoxal *in situ* in the chamber, followed by ageing overnight. The experiment on the 23/07/2012 will be discussed here as it had the most complete data coverage and a good ATOFMS hit rate (typically 2–37%). However, the other type 2 experiments on the 16/07/12 and 30/07/12 (Fig. 2) show similar trends. The SMPS data was corrected for wall losses and dilution as outlined in the experimental section and the data is shown in Supplementary Fig. 3† for the entire experiment. After the periods of photochemistry, there was significant slow dark uptake of the glyoxal formed from acetylene oxidation to the AS aerosol as seen in previous studies.

The ATOFMS data required extensive analysis and clustering due to the inhomogeneity in the chamber particles. 54 000 particle spectra were obtained over the experiment on 23 and 24/07/2012. Of these approximately 2000 were classified as background potassium-rich combustion particles. Although present at very low concentrations, these are ionized more efficiently than GLY + AS and so have an important contribution to the global averaged MS. Thus, these were removed by clustering the data, to allow exclusive analysis of the AS particles. Four generations of K-means analysis ($K = 5$) were required to produce a 'pure' homogeneous GLY + AS cluster. The averaged mass spectra for the GLY + AS cluster (containing 52 000 particles) during the chamber open period (23/07/12 14:49–15:55) and after the dark ageing period (24/07/12 05:00–07:30) are shown in Fig. 3. Negative ion mass spectra were only produced for a very small fraction of particles and are therefore not included. However, the relative intensities of certain negative ions did increase during the experiment and are discussed below.

The spectra observed contain similar ions to those in both the GLY + AS (aged) and GLY + AS + H_2O_2 laboratory solutions, with the m/z 18, 27, 43 and 69 clearly present. The temporal trends of the ions discussed below are seen in all illuminated experiments (16/07/2012 and 30/07/2012) and show a consistent pattern.

It is clear that IM, or a species that fragments to the IM ion, has formed during the experiment as evidenced by the peak at m/z 69. Rather than look at the evolution of the organic : sulfate ratio, as done in previous AMS studies, here the change in relative ion intensity can be plotted as shown in Fig. 4. An essentially linear increase in the m/z 69 = $[IM + H]^+$ signal is observed throughout the experiment as shown in the upper panel in Fig. 4. The ion at m/z 87 shows a highly similar temporal profile to m/z 69. A plot of ion intensity of m/z 69 against m/z 87 showed a very strong correlation with an R^2 = 0.98, although there is some

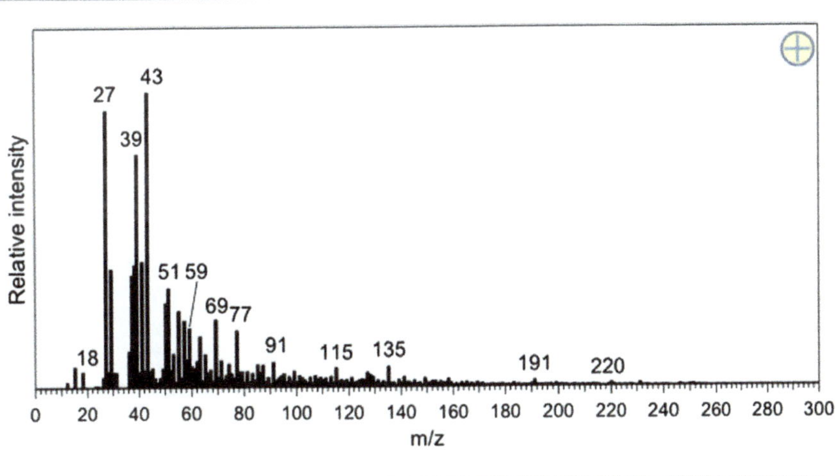

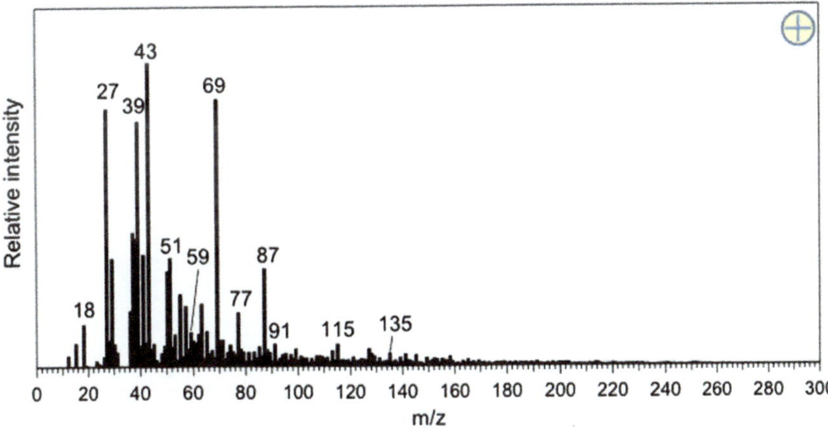

Fig. 3 Average mass spectra obtained from ATOFMS during ACET + HONO + light experiment (23-24/07/12). Upper: Chamber open period (23/07/12 14:49–15:55). Lower: Following dark ageing prior to flushing (24/07/12 05:00–07:30).

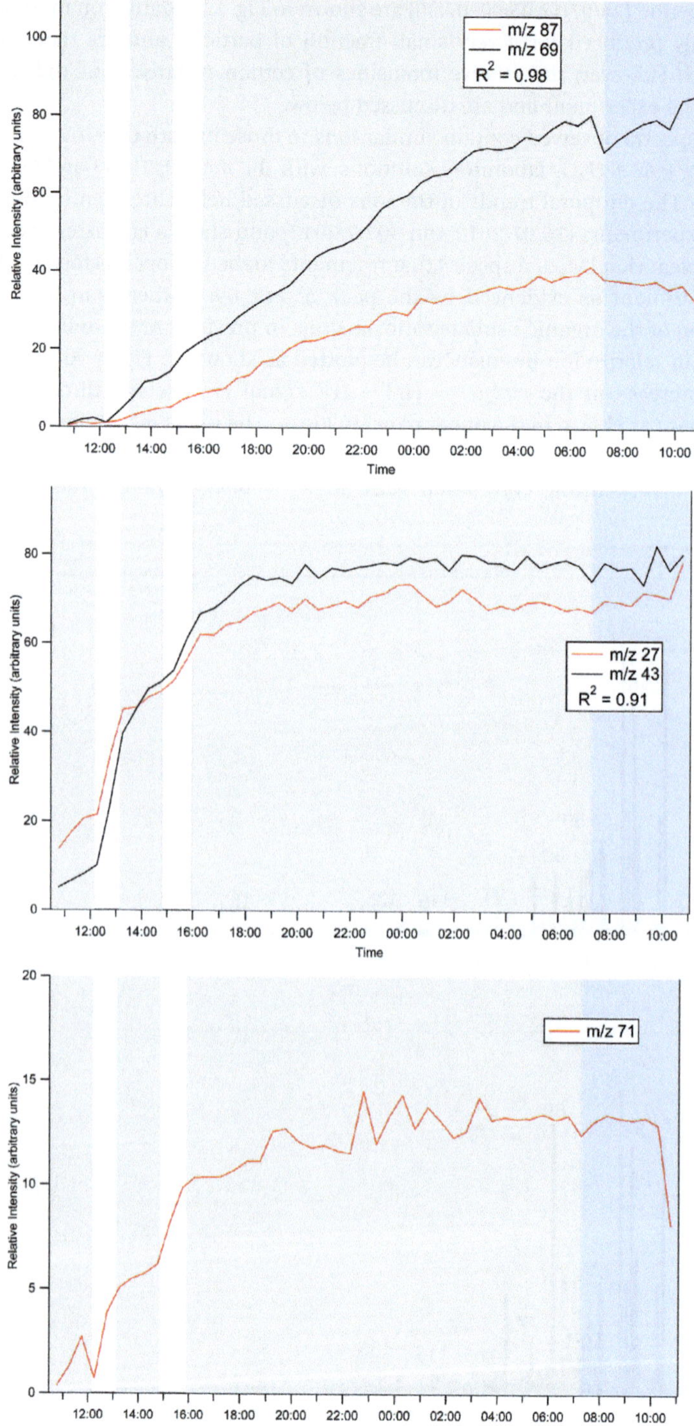

Fig. 4 Temporal evolution of selected ATOFMS ions during the experiment on 23-24/07/2012. Shaded grey areas = chamber closed. White shaded areas = chamber open. Blue shaded areas = flushing.

divergence on flushing. A possible assignment for m/z 87 is a water adduct of the imidazole $[IM + H_2O + H]^+$, although it is unclear why this is not seen in the lab solutions. The IM ion does not decrease significantly in relative intensity during dilution/ flushing (after 07:15, blue shaded area in Fig. 4), as seen in Galloway *et al.*,[16] highlighting the irreversible nature of this reaction product compared to the GLY oligomers. There is also no evidence of gas phase imidazole in the PTR-MS spectra (which would be expected to ionize well and be stable) confirming its very low volatility. It is possible that the IM is present in the form of an imidazolium salt with sulfate or other negative ions, decreasing its volatility further.

Other ions in the lab solution related to IM, including m/z 115 (HIC) and 135 (BI), do not correlate with m/z 69 and do not increase in intensity during the experiment. The HIC and BI can be seen in the LC-MS analysis of the chamber filter samples however, and so are present at low levels. A very similar temporal profile to m/z 69 is seen for the ion at m/z −45, assigned to formic acid ($R^2 = 0.83$) (shown in Supplementary Fig. 2†). Formic acid has been proposed as a co-product of IM formation[24] and this data provides further evidence for this mechanism. The aromatic ions seen in the lab solution m/z 51 and 77 also increase continually throughout the experiment, indicating they may be from unknown GLY + AS reaction products.

A number of ions have also been identified that show different temporal profiles to the IM. Ion fragments related to oxidized organic aerosol, including m/z 27 and 43, increase when the chamber is opened to sunlight as shown in the middle panel of Fig. 4. These ions show a steady increase until about 1 h after the chamber is closed and then there is very little change in their intensity. It is interesting that even when the chamber is diluted there is little change in their relative intensity, suggesting they are fragments of relatively involatile material.

The difference in formation mechanisms of the m/z 43, which is enhanced in the photochemical period, and the m/z 69, which is associated with dark uptake can be seen in Fig. 5. During the two periods of photochemistry and the

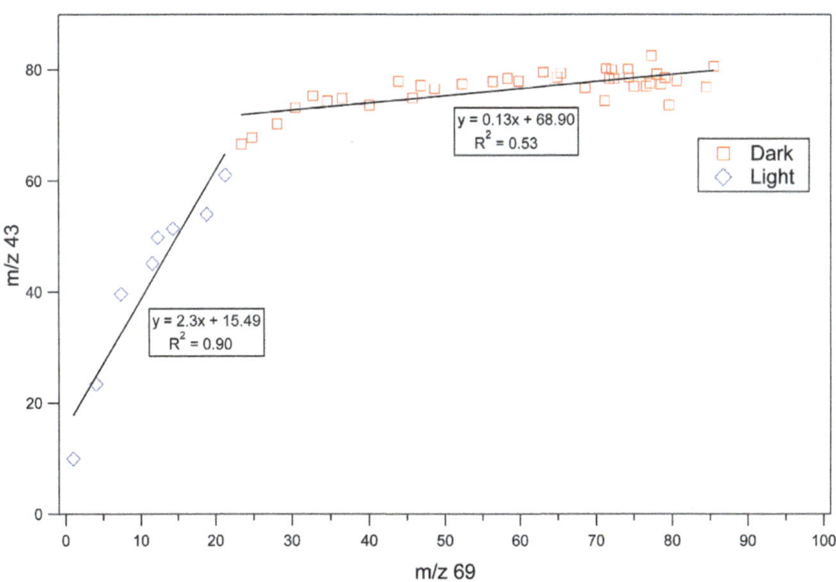

Fig. 5 Correlation of relative ion intensity of m/z 69 *versus* 43.

intermediate dark period, the m/z 43 shows a strong correlation ($R^2 = 0.9$) with the m/z 69 ion. However, under prolonged dark conditions there is less correlation between these ions and a drop in the production of m/z 43, indicating that different mechanisms control their formation/uptake.

The ion at m/z 71 also shows a rapid increase during the periods when the chamber is open and then very little growth in the dark as shown in Fig. 4 (lower). This ion was also enhanced in the laboratory solution exposed to light and OH. Possible assignments of this ion include $[C_3H_3O_2]^+$, $[C_3H_5NO]^+$ or $[C_2H_3N_2O]^+$. Here we proposed that this ion is protonated 1,2,5-oxadiazole $[C_2H_2N_2O + H]^+$ formed from the dehydration of glyoxime.[51] A possible formation mechanism is shown in Fig. 6. The mechanism requires hydroxylamine and a possible formation route is *via* a photo-enhanced radical oxidation of aqueous ammonia or ammonium in solution.[52] This can then react with the GLY *via* nucleophilic attack, followed by protonation and dehydration, to form glyoxime. Dehydration of the glyoxime would result in the formation of 1,2,5-oxadiazole. Olofson and Michelman,[51] formed 1,2,5-oxadiazole by glyoxime dehydration under mildly acidic conditions at high temperatures (did not proceed at room temperature). Therefore, it is unclear if this forms as a result of the ATOFMS ionization or by some other mechanism in the particle. It could not be seen in the ESI-MS spectra and further work is needed to confirm the formation of this compound. However, this species may represent a previously unobserved, light activated, route to form CN containing species from GLY + AS, that may be involved in further aqueous phase chemistry.

The average mass spectra in Fig. 3 also show an ion at m/z 59. This ion increases on opening the chamber and then drops to a rather steady level throughout the rest of the experiment. Possible assignments for this peak include GLY $[C_2H_2O_2 + H]^+$ and acetone $[CH_3C(O)CH_3 + H]^+$, and this ion was not seen in the lab solutions. The lab solutions all contain GLY oligomers primarily and no free GLY. This ion could represent GLY being taken up onto the particle, prior to hydration and reaction. The relative intensity of this ion increases with flushing, as shown in Supplementary Fig. 4,† and may represent intermediate GLY formation as the reversible GLY oligomers are volatilized back into the gas phase.

The average ATOFMS mass spectra for both the type 1 (03/05/12) and type 3 (05/07/12) experiments showed organic ion fragments at m/z 27, 43 and 59. ON peaks at m/z 69 and 71 were also observed but at much lower relative intensities than those seen in type 2 experiments. A much higher fraction, but equivalent number, of background particles was seen in these experiments, with between 30–50% of particles removed by clustering. It appears that the photochemical/photolysis stage of the type 2 experiments enhance aerosol absorption at 266 nm, resulting in higher ionization efficiency.

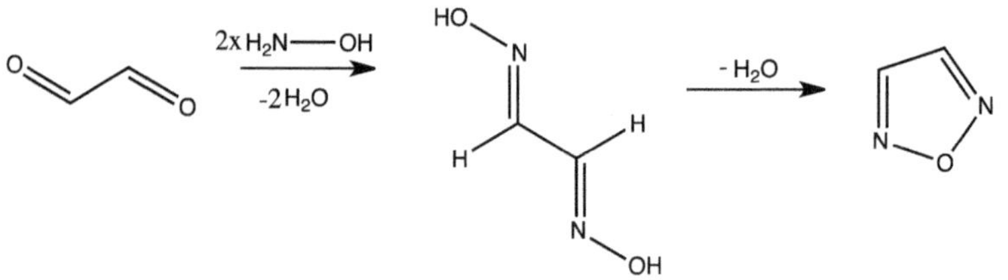

Fig. 6 Formation of glyoxime and 1,2,5-oxadiazole from GLY + hydroxylamine.

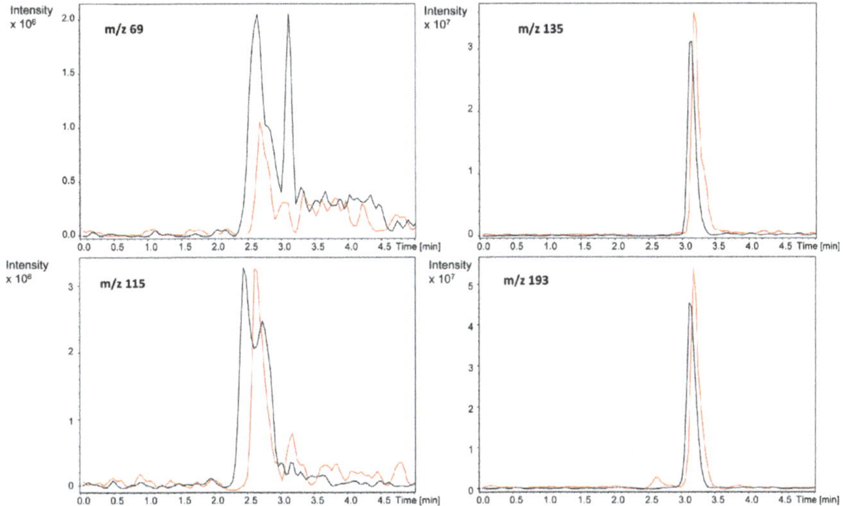

Fig. 7 LC-MS chromatograms of ON species found in the filters collected during 23/07/12 experiment. Red; F2 collected between 18:00 and 01:00 during dark ageing period. Black: 1M GLY + 1M AS lab solution, aged 1 week. Slight differences in retention time are a result of different pH and concentrations of extracts.

Offline Analysis. Filter samples were collected for offline analysis and comparison to laboratory generated aqueous samples of GLY + AS. In total 5 filters were collected during the experiment on 23/07/2102, to allow some degree of time resolved behavior to be observed while maximising the amount of aerosol collected. Only the first 2 filters (F1 and F2) will be discussed here, with F1 collected between 12:24 and 15:53 (chamber open) and F2 collected between 18:00 and 01:00 (chamber closed). Filters 3–5 were collected later in the experiment and there was insufficient aerosol mass left in the chamber for offline analysis.

The mass spectrum obtained using FTICR-MS for filter 2 from the 23/07/12 experiment is shown in Supplementary Fig. 2† lower, with the mass spectrum of the aged GLY + AS solution shown in the upper section. The chamber spectrum looks very different and much more complex than the lab solutions, perhaps as a consequence of working so close to the detection limit of the FTICR-MS, but a number of nitrogen containing species are present (italic ion formulae in Table 2). The FTICR-MS observed ions with matching molecular formulae of HIC, BI, HGI, GHIC, GBI and $m/z\ 157 = [C_6H_8N_2O_3 + H]^+$. For further confirmation, LC-MS^2 was carried out and the chromatographic peaks obtained for 4 species are shown in Fig. 7 for Filter F2 and for the GLY + AS aged laboratory solution. The retention times and MS^2 fragmentations obtained matched the laboratory solutions and previous literature. The ON species are seen in both F1 and F2, indicating they are formed relatively quickly and confirms the ATOFMS assignment of $m/z\ 69$ as $[IM + H]^+$, which may be IM itself or a stable fragment ion from other ON species. Unfortunately, quantification was not possible due to lack of standards. The concentrations of GLY and aerosol used in the chamber were considerably higher than the ambient atmosphere, never the less, we can conclude that the uptake of GLY onto AS clearly produces a range of ON species from interactions with aqueous ammonia, changing the absorption properties of the aerosol on atmospherically relevant timescales. More work is needed to investigate the complex composition of organic aerosol produced during the uptake of GLY onto inorganic seed particles.

Discussion and conclusions

It is clear that the interaction of glyoxal with aqueous ammonium sulfate particles is extremely complex with many competing factors. The use of laboratory bulk solutions allowed mass spectral marker ions of GLY + AS chemistry to be determined, which were used to investigate the uptake of GLY onto AS seed particles in chamber simulations. The ATOFMS has rarely been used to study the uptake of organics/SOA formation onto pre-existing inorganic seed particles, due to the poor ionization efficiency at the laser wavelength. However, the formation of nitrogen containing organic compounds that absorb in the UV region, such as the imidazole functionality species, increases the number of particles observed by the ATOFMS, allowing it to be a powerful instrument for observing the reactive uptake of GLY in chamber experiments. The m/z 69 ion may be a good tracer of GLY uptake onto ambient particles. A number of organic aerosol clusters produced from ambient ATOFMS data contain an ion at m/z 69, which is usually not identified.[53,54] In addition, more complex nitrogen containing organics, such as imidazole oligomers, previously seen in laboratory bulk solutions, have been observed for the first time using offline analysis of GLY reactive uptake onto AS aerosols in chamber experiments. The use of a large volume chamber allowed high volumes of aerosol to be collected, overcoming the sensitivity issues usually associated with LC-MS and ESI-MS analysis. If these ON species, such as HIC, BI, HGI, GHIC, GBI, are observed in ambient aerosol they may be potential source markers for GLY SOA.

In the chamber experiments carried out here, the role of OH and light could not be fully rationalized. The extensive cleaning regime limited the number of experiments that could be carried out. The experiments confirm the results of Volkamer *et al.*,[15] where there is a period of fast photochemical uptake on exposure to light and OH, which proved to be relatively short lived (~15 min) in this study. The experiment set up was very similar, with GLY formed *in situ* by ACET oxidation. However, this is the first time that the GLY + AS system has been studied in the presence of natural sunlight. The temporal trends of ions and increased hit rate on the ATOFMS were only observed during experiments that included an illuminated period. OH oxidation chemistry in the dark (using alkene ozonolysis to form OH) did not show rapid GLY uptake onto the AS particles and the ATOFMS produced mass spectra for a significantly lower number of particles. The reasons for the discrepancies seen in the literature are not resolved here, although the results suggest that light rather than OH may be the key factor for enhanced GLY uptake.

The exposure to light seems to activate the AS particles and make them more visible to the ATOFMS. The ion at m/z 71, tentatively identified as 1,2,5-oxadiazole, is clearly enhanced when the chamber is open and indicates that some nitrogen containing species are formed *via* a light activated pathway. Nguyen *et al.*,[55] recently reported the formation of heterocyclic species with two nitrogen atoms from the photolysis of isoprene SOA extracted into water, and identified them as substituted N-oxides of 1,2,5-oxadiazole (furoxans). Further work is needed to investigate the formation mechanisms, for example the ion at m/z 71 may be the 1,2,5-oxadiazole or a stable fragment ion of more complex compounds containing the oxadiazole functionality, as well as the absorption properties of these complex heterocyclic species.

Acknowledgements

The authors gratefully acknowledge the UK Natural Environment Research Council for funding of the PhoSOA project (NE/H021108/1) and TJA, MJSD & ICAG acknowledge NERC PhD studentships. ARR and SMB acknowledges financial and logistical support from the NERC National Centre for Atmospheric Science – Composition. FTICR-MS facilities were made available by the University of York Centre of Excellence in Mass Spectrometry, which was created thanks to a major capital investment through Science City York, supported by Yorkshire Forward with funds from the Northern Way Initiative. The Instituto Universitario CEAM-UMH is partly supported by Generalitat Valenciana, Fundación Bancaja, and the projects GRACCIE (Consolider-Ingenio 2010) and FEEDBACKS (Prometeo – Generalitat Valenciana). EUPHORE instrumentation is partly funded by the Spanish Ministry of Science and Innovation, through INNPLANTA project: PCT-440000-2010-003. We also thank EUPHORE staff for their work in the experiments.

References

1 M. Hallquist, J. C. Wenger, U. Baltensperger, Y. Rudich, D. Simpson, M. Claeys, J. Dommen, N. M. Donahue, C. George, A. H. Goldstein, J. F. Hamilton, H. Herrmann, T. Hoffmann, Y. Iinuma, M. Jang, M. E. Jenkin, J. L. Jimenez, A. Kiendler-Scharr, W. Maenhaut, G. McFiggans, T. F. Mentel, A. Monod, A. S. H. Prevot, J. H. Seinfeld, J. D. Surratt, R. Szmigielski and J. Wildt, *Atmos. Chem. Phys.*, 2009, **9**, 5155–5236.
2 M. S. Jang and R. M. Kamens, *Environ. Sci. Technol.*, 2001, **35**, 4758–4766.
3 T.-M. Fu, D. J. Jacob and C. L. Heald, *Atmos. Environ.*, 2009, **43**, 1814–1822.
4 T.-M. Fu, D. J. Jacob, F. Wittrock, J. P. Burrows, M. Vrekoussis and D. K. Henze, *Journal of Geophysical Research-Atmospheres*, 2008, **113**, DOI: 10.1029/2007JD009505.
5 J. Liggio, S. M. Li and R. McLaren, *Journal of Geophysical Research-Atmospheres*, 2005, **110**, DOI: 10.1029/2004JD005113.
6 C. L. Heald, D. J. Jacob, R. J. Park, L. M. Russell, B. J. Huebert, J. H. Seinfeld, H. Liao and R. J. Weber, *Geophys. Res. Lett.*, 2005, **32**, DOI: 10.1029/2005GL023831.
7 J. H. Kroll, N. L. Ng, S. M. Murphy, V. Varutbangkul, R. C. Flagan and J. H. Seinfeld, *Journal of Geophysical Research-Atmospheres*, 2005, **110**, DOI: 10.1029/2005JD006004.
8 S. Myriokefalitakis, M. Vrekoussis, K. Tsigaridis, F. Wittrock, A. Richter, C. Bruehl, R. Volkamer, J. P. Burrows and M. Kanakidou, *Atmos. Chem. Phys.*, 2008, **8**, 4965–4981.
9 H. S. S. Ip, X. H. H. Huang and J. Z. Yu, *Geophys. Res. Lett.*, 2009, **36**, DOI: 10.1029/2008GL036212.
10 R. Volkamer, J. L. Jimenez, F. San Martini, K. Dzepina, Q. Zhang, D. Salcedo, L. T. Molina, D. R. Worsnop and M. J. Molina, *Geophys. Res. Lett.*, 2006, **33**, DOI: 10.1029/2006GL026899.
11 K. Dzepina, R. M. Volkamer, S. Madronich, P. Tulet, I. M. Ulbrich, Q. Zhang, C. D. Cappa, P. J. Ziemann and J. L. Jimenez, *Atmos. Chem. Phys.*, 2009, **9**, 5681–5709.
12 E. Waxman, B. Ervens and R. Volkamer, *Abstracts of Papers of the American Chemical Society*, 2011, **242**.
13 J. P. D. Abbatt, A. K. Y. Lee and J. A. Thornton, *Chem. Soc. Rev.*, 2012, **41**, 6555–6581.
14 J. Liggio, S. M. Li and R. McLaren, *Environ. Sci. Technol.*, 2005, **39**, 1532–1541.
15 R. Volkamer, P. J. Ziemann and M. J. Molina, *Atmos. Chem. Phys.*, 2009, **9**, 1907–1928.
16 M. M. Galloway, P. S. Chhabra, A. W. H. Chan, J. D. Surratt, R. C. Flagan, J. H. Seinfeld and F. N. Keutsch, *Atmos. Chem. Phys.*, 2009, **9**, 3331–3345.
17 H. Debus, *Philos. Trans. R. Soc. London*, 1858, **148**, 205–209.
18 D. O. De Haan, A. L. Corrigan, K. W. Smith, D. R. Stroik, J. J. Turley, F. E. Lee, M. A. Tolbert, J. L. Jimenez, K. E. Cordova and G. R. Ferrell, *Environ. Sci. Technol.*, 2009, **43**, 2818–2824.

19 M. M. Galloway, C. L. Loza, P. S. Chhabra, A. W. H. Chan, L. D. Yee, J. H. Seinfeld and F. N. Keutsch, *Geophys. Res. Lett.*, 2011, **38**, DOI: 10.1029/2011GL048514.

20 S. Nakao, Y. Liu, P. Tang, C. L. Chen, J. Zhang and D. R. Cocker, III, *Atmos. Chem. Phys.*, 2012, **12**, 3927–3937.

21 E. L. Shapiro, J. Szprengiel, N. Sareen, C. N. Jen, M. R. Giordano and V. F. McNeill, *Atmos. Chem. Phys.*, 2009, **9**, 2289–2300.

22 B. Noziere, P. Dziedzic and A. Cordova, *J. Phys. Chem. A*, 2009, **113**, 231–237.

23 M. Trainic, A. A. Riziq, A. Lavi, J. M. Flores and Y. Rudich, *Atmos. Chem. Phys.*, 2011, **11**, 9697–9707.

24 G. Yu, A. R. Bayer, M. M. Galloway, K. J. Korshavn, C. G. Fry and F. N. Keutsch, *Environ. Sci. Technol.*, 2011, **45**, 6336–6342.

25 C. J. Kampf, R. Jakob and T. Hoffmann, *Atmos. Chem. Phys.*, 2012, **12**, 6323–6333.

26 D. S. Gross, M. E. Galli, M. Kalberer, A. S. H. Prevot, J. Dommen, M. R. Alfarra, J. Duplissy, K. Gaeggeler, A. Gascho, A. Metzger and U. Baltensperger, *Anal. Chem.*, 2006, **78**, 2130–2137.

27 R. M. Healy, Y. Chen, I. Kourtchev, M. Kalberer, D. O'Shea and J. C. Wenger, *Environ. Sci. Technol.*, 2012, **46**, 11813–11820.

28 M. Huang, W. Zhang, L. Hao, Z. Wang, W. Zhao, X. Gu, X. Guo, X. Liu, B. Long and L. Fang, *J. Atmos. Chem.*, 2007, **58**, 237–252.

29 M.-Q. Huang, W.-J. Zhang, L.-Q. Hao, Z.-Y. Wang, W.-W. Zhao, X.-J. Gu and L. Fang, *Journal of the Chinese Chemical Society*, 2008, **55**, 456–463.

30 R. J. Wenzel, D. Y. Liu, E. S. Edgerton and K. A. Prather, *Journal of Geophysical Research-Atmospheres*, 2003, **108**, DOI: 10.1029/2001JD001563.

31 P. J. Silva and K. A. Prather, *Anal. Chem.*, 2000, **72**, 3553–3562.

32 I. Barnes, *The European Photoreactor EUPHORE* Wuppertal, 2001.

33 K. H. Becker, *The European Photoreactor EUPHORE*, Wuppertal, 1996.

34 J. G. Calvert, *The mechanisms of atmospheric oxidation of the alkenes*, Oxford University Press, 2000.

35 K. J. Feierabend, J. E. Flad, S. S. Brown and J. B. Burkholder, *J. Phys. Chem. A*, 2009, **113**, 7784–7794.

36 J. M. Langridge, S. M. Ball, A. J. L. Shillings and R. L. Jones, *Rev. Sci. Instrum.*, 2008, **79**.

37 R. A. Washenfelder, A. O. Langford, H. Fuchs and S. S. Brown, *Atmos. Chem. Phys.*, 2008, **8**, 7779–7793.

38 S. M. Ball, J. M. Langridge and R. L. Jones, *Chem. Phys. Lett.*, 2004, **398**, 68–74.

39 S. R. T. N. M. J. S. Daniels, J. Patel, E. Paterson, M. Rodenas and S. M. Ball, *in preparation for Atmospheric Measurement Techniques*, 2013.

40 R. S. Blake, P. S. Monks and A. M. Ellis, *Chem. Rev.*, 2009, **109**, 861–896.

41 E. Gard, J. E. Mayer, B. D. Morrical, T. Dienes, D. P. Fergenson and K. A. Prather, *Anal. Chem.*, 1997, **69**, 4083–4091.

42 K. A. Prather, T. Nordmeyer and K. Salt, *Anal. Chem.*, 1994, **66**, 1403–1407.

43 D. S. Gross, R. Atlas, J. Rzeszotarski, E. Turetsky, J. Christensen, S. Benzaid, J. Olson, T. Smith, L. Steinberg, J. Sulman, A. Ritz, B. Anderson, C. Nelson, D. R. Musicant, L. Chen, D. C. Snyder and J. J. Schauer, *Environ. Modell. Software*, 2010, **25**, 760–769.

44 R. M. Healy, S. Hellebust, I. Kourtchev, A. Allanic, I. P. O'Connor, J. M. Bell, D. A. Healy, J. R. Sodeau and J. C. Wenger, *Atmos. Chem. Phys.*, 2010, **10**, 9593–9613.

45 J. F. Hamilton, M. R. Alfarra, K. P. Wyche, M. W. Ward, A. C. Lewis, G. B. McFiggans, N. Good, P. S. Monks, T. Carr, I. R. White and R. M. Purvis, *Atmos. Chem. Phys.*, 2011, **11**, 5917–5929.

46 J. F. Hamilton, A. C. Lewis, T. J. Carey and J. C. Wenger, *Anal. Chem.*, 2008, **80**, 474–480.

47 W. P. Hastings, C. A. Koehler, E. L. Bailey and D. O. De Haan, *Environ. Sci. Technol.*, 2005, **39**, 8728–8735.

48 A. G. Carlton, B. J. Turpin, K. E. Altieri, S. Seitzinger, A. Reff, H.-J. Lim and B. Ervens, *Atmos. Environ.*, 2007, **41**, 7588–7602.

49 X. Yao, P. J. G. Rehbein, C. J. Lee, G. J. Evans, J. Corbin and C.-H. Jeong, *Atmos. Environ.*, 2011, **45**, 6251–6256.

50 R. J. Salter, M. A. Blitz, D. E. Heard, T. Kovacs, M. J. Pilling, A. R. Rickard and P. W. Seakins, *Phys. Chem. Chem. Phys.*, 2013, **15**, 4984–4994.

51 R. A. Olofson and J. S. Michelman, *J. Org. Chem.*, 1965, **30**, 1854–1859.

52 L. Huang, L. Li, W. Dong, Y. Liu and H. Hou, *Environ. Sci. Technol.*, 2008, **42**, 8070–8075.

53 M. Dall'Osto and R. M. Harrison, *Atmos. Chem. Phys.*, 2012, **12**, 4127–4142.

54 C. Giorio, A. Tapparo, M. Dall'Osto, R. M. Harrison, D. C. S. Beddows, C. Di Marco and E. Nemitz, *Atmos. Environ.*, 2012, **61**, 316–326.

55 T. B. Nguyen, A. Laskin, J. Laskin and S. A. Nizkorodov, *Phys. Chem. Chem. Phys.*, 2012, **14**, 9702–9714.

Faraday Discussions

RSCPublishing

PAPER

Brown carbon formation from ketoaldehydes of biogenic monoterpenes†

Tran B. Nguyen,‡[*a] Alexander Laskin,[b] Julia Laskin[c] and Sergey A. Nizkorodov[a]

Received 10th March 2013, Accepted 10th April 2013

DOI: 10.1039/c3fd00036b

Sources and chemical composition of brown carbon are poorly understood, and even less is known about the mechanisms of its atmospheric transformations. This work presents molecular-level investigations of the reactive compound ketolimononaldehyde (KLA, $C_9H_{14}O_3$), a second-generation ozonolysis product of limonene ($C_{10}H_{16}$), as a potent brown carbon precursor in secondary organic aerosol (SOA) through its reactions with reduced nitrogen compounds, such as ammonium ion (NH_4^+), ammonia, and amino acids. The reactions of synthesized and purified KLA with NH_4^+ and glycine resulted in the formation of chromophores nearly identical in spectral properties and formation rates to those found in similarly-aged limonene/O_3 SOA. Similar chemical reaction processes of limononaldehyde (LA, $C_{10}H_{16}O_2$) and pinonaldehyde (PA, $C_{10}H_{16}O_2$), the first-generation ozonolysis products of limonene and α-pinene, respectively, were also studied, but the resulting products did not exhibit the light absorption properties of brown carbon, suggesting that the unique molecular structure of KLA produces visible-light-absorbing compounds. The KLA/NH_4^+ and KLA/GLY reactions produce water-soluble, hydrolysis-resilient chromophores with high mass absorption coefficients (MAC = 2000–4000 cm^2 g^{-1}) at $\lambda \sim 500$ nm, precisely at the maximum of the solar emission spectrum. Liquid chromatography was used to isolate the light-absorbing fraction, and UV-Vis, FTIR, NMR and high-resolution mass spectrometry (HR-MS) techniques were

[a]Department of Chemistry, University of California, Irvine, California, USA 92697. E-mail: tbn@calech.edu; Tel: +1-626-395-8650

[b]Environmental Molecular Sciences Laboratory, Pacific Northwest National Laboratory, Richland, Washington, USA 99352

[c]Physical Sciences Division, Pacific Northwest National Laboratory, Richland, Washington, USA 99352

† Electronic supplementary information (ESI) available: Table S1 – NMR data for identification of PA, LA, and KLA; Fig. S1 – proton NMR spectra of LA, KLA, and PA confirming their purity; Fig. S2 – LC-UVIS separation of the C_{500} fraction; Fig. S3 – control experiment with evaporation of KLA alone with reduced nitrogen compounds added; Fig. S4 – comparison of the rates of aging of KLA and limonene SOA by gaseous ammonia; Fig. S5 – proton NMR spectra of the C_{500} fraction in D_2O and in H_2O; Fig. S6 – absorption spectra recoded by the PDA detector at different elution times; Fig. S7 – graphical representation of the compounds detected by HR-MS; Scheme S1 – reactions of KLA leading to low DBE compounds and supporting references. See DOI: 10.1039/c3fd00036b

‡ Currently at: Division of Geological and Planetary Sciences, California Institute of Technology, Pasadena, CA 91125.

used to investigate the structures and chemical properties of the light-absorbing compounds. The KLA browning reaction generates a diverse mixture of light-absorbing compounds, with the majority of the observable products containing 1–4 units of KLA and 0–2 nitrogen atoms. Based on the HR-MS product distribution, conjugated aldol condensates, secondary imines (Schiff bases), and N-heterocycles like pyrroles may contribute in varying degree to the light-absorbing properties of the KLA brown carbon. The results of this study demonstrate the high degree of selectivity of organic compound structures on the light-absorbing properties of SOA.

Introduction

Atmospheric aerosols scatter and absorb solar radiation; however, the magnitude and sign of their direct climate forcing remain uncertain.[1] Most climate models assume primary and secondary organic aerosols (POA and SOA) have only scattering components, resulting in negative climate forcing. However, the presence of black carbon and brown carbon[2] compounds in organic aerosols may significantly alter their climate forcing. While it is well-accepted that the dominant sources of black carbon are primary combustion emissions, the sources of brown carbon are not well-understood. Brown carbon can be directly emitted from combustion sources[2–4] as part of POA or created by reactions in the atmosphere, such as in the formation or aging of SOA. This *secondary brown carbon* can be formed from multiple precursors and a variety of reaction pathways, *e.g.*, the reactions of 1,2-dicarbonyls with ammonium (NH_4^+) salts,[5–10] the aqueous photooxidation of hydroxyacids,[11–13] the acid-catalysed condensation of volatile aldehydes,[14–21] and the nitration of aromatic compounds.[22–24]

Brown carbon formation has also been observed in the reaction of NH_4^+(aq), amino acids (aq), and NH_3(g) with SOA derived from the ozonolysis of limonene ($C_{10}H_{16}$).[25,26] Interestingly, SOA components generated from the ozonolysis of α-pinene, an isomeric monoterpene, do not form brown carbon when reacted with the same nitrogen species.[26,27] Both limonene and α-pinene are abundant volatile organic compounds, so it is important to understand their potential to modify aerosol optical properties. For example, limonene and α-pinene contribute 3–18% and 23–53%, respectively, to the total monoterpene emissions from forests in the United States.[28] Additionally, limonene is one of the five most abundant volatile organic compounds in indoor air.[29] α-Pinene and limonene are structural isomers; however, α-pinene has one fewer oxidisable double bond (Scheme 1). The differences in the aging characteristics of limonene and α-pinene ozonolysis SOA suggest that rather specific chemistry, involving second-generation products, may be important in the brown carbon formation reaction.

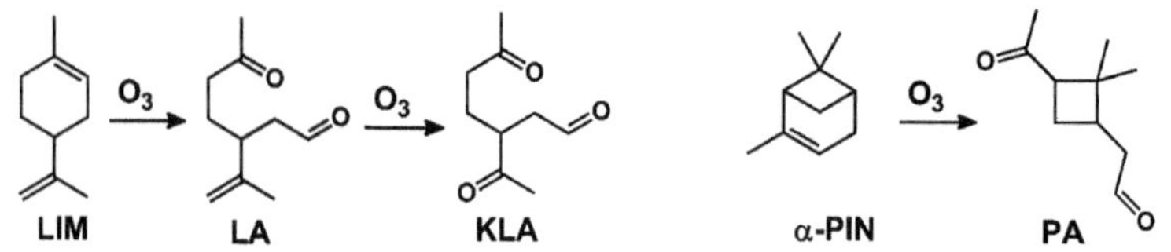

Scheme 1　Products of limonene (LIM) ozonolysis: limononaldehyde (LA) and ketolimononaldehyde (KLA), and α-pinene (α-PIN) ozonolysis: pinonaldehyde (PA).

Because the SOA mixture from the limonene/O_3 SOA does not contain significant amounts of 1,2-dicarbonyls,[30,31] its NH_4^+-mediated browning mechanism may be different from that previously-proposed for glyoxal and methylglyoxal, which produce oligomeric imidazole-based compounds in reactions with NH_4^+ and amino acids.[5–7,9,32,33] As such, the active brown-carbon precursors in the limonene/O_3 SOA mixture have yet to be identified. The study of Bones *et al.* (2010) and related studies[5,8,10,26,27,34] offer important insights on the nature of the chromophore structure and the NH_4^+-mediated reaction. These studies suggest that the chromophores likely contain a $>$C=N– (imine) motif in their structures. Furthermore, it has been demonstrated that chromophores account for less than ~1% of the total aerosol mass, and have high molar extinction coefficients (possibly as high as ~10^5 M^{-1} cm^{-1}), which makes it possible for them to dominate the optical properties of the aerosol despite their low relative abundance. The NH_4^+-mediated browning of SOA proceeds through aqueous reactions under slightly acidic (pH ~ 4) to neutral conditions, and is dramatically accelerated by water evaporation from the NH_4^+ + SOA solution.[35]

Ketoaldehydes limononaldehyde (LA), ketolimononaldehyde (KLA) and pinonaldehyde (PA) are the most abundant products of ozonolysis of limonene and α-pinene. For example, PA product yields greater than 50% have been reported in ozonolysis of α-pinene,[36] while LA and KLA dominate both the gas- and particle-phase products of ozonolysis of limonene.[30,31,37] Note, this work uses the naming convention for the monoterpene oxidation products introduced by Larsen *et al.* (1998);[38] the systematic nomenclature is reported in the Electronic Supplementary Information (ESI)† section. The first-generation ketoaldehydes are semivolatile, *i.e.*, they exist mainly in the gas phase but can also be found in the aerosol phase, while the second-generation products are less volatile and are predominantly found in the aerosol phase.[30]

We hypothesized that because aldehydes are prone to nucleophilic addition (including hydration, oxidation, imine formation, *etc.*), unlike the much more stable alcohols and acids that are also present in the SOA mixture, ketoaldehydes are excellent candidates for brown carbon precursors.[39,40] To test this hypothesis, we synthesized PA, LA and KLA and aged them in the presence of NH_4^+(aq), aqueous glycine (GLY, HOOC–CH_2–NH_2), and NH_3(g) to reproduce the browning chemistry (or lack of thereof) observed in limonene/O_3 and α-pinene/O_3 SOA.

Experimental

Synthesis and purification of ketoaldehydes

LA was synthesized by the oxidative ring opening of the 1,2–epoxide of limonene, according to the procedure reported by Binder *et al.* (2008).[41] 1 mmol of limonene oxide (Aldrich, 99% purity) was continuously stirred with 2 molar equivalents of sodium metaperiodate ($NaIO_4$, Alfa Aesar, 99.8% purity) at room temperature in 40 mL of a solvent mixture of 2 : 1 tetrahydrofuran (Aldrich, ACS grade) and water (Aldrich, HPLC grade). The reaction progress was monitored with thin layer chromatography (TLC). The concentrated crude was separated on a silica gel column (60 Å pores, 32–63 μm particles, Sorbent Tech.) with 3 : 1 hexanes/ethyl acetate mobile phase, which recovered ~80 mg of pure material after concentration under vacuum.

KLA and PA were synthesized by low-temperature ozonolysis of the corresponding pure monoterpenes using a modified procedure of Griesbaum *et al.*

(1997).[42] Specifically, 1 mmol of limonene (Acros, 97% purity) or α-pinene (Acros, 99% purity) was dissolved in ∼10 mL of dichloromethane (Aldrich, ACS grade) and ozonized to completion by bubbling 1 standard litre per minute (SLM) flow of O_3/O_2 from a commercial ozone generator through the solution with continuous stirring at −78 °C. After all the double bonds had reacted, ozone started to accumulate in the solution (evidenced by a blue tint) and the flow was immediately stopped. 1.2 mole equivalent (per C=C bond) of triphenylphosphine (Alfa Aesar, 99% purity) was added and left to stir at −78 °C for 1 h, followed by stirring at room temperature for 24 h. The KLA crude was separated with a semi-prep reverse-phase liquid chromatography column (Luna C_{18}, 10 × 250 mm, 100 Å pores, 5 μm particles, Phenomenex, Inc.) coupled to a photo-diode array (PDA) UV-VIS detector (LC-UVVIS) with 2 mL min^{-1} of acetonitrile (ACN, Aldrich HPLC grade) mobile phase, affording ∼35 mg of the pure material after concentration under vacuum. The PA crude was separated with semi-prep normal-phase LC-UVVIS (10 × 250 mm Luna Si, 100 Å pores, 5 μm particles, Phenomenex, Inc.) with 2 mL min^{-1} hexanes/ethyl acetate (70 : 30) mobile phase flow, affording ∼70 mg of the pure material after concentration under vacuum. The appearance of the distinctive n → π* band (∼280 nm) of KLA and PA in the real-time PDA spectrum was used as a prompt for the collection of carbonyl fractions, prior to purity analysis.

The purified ketoaldehyde products were characterized with ^{1}H nuclear magnetic resonance (NMR) spectroscopy with a 500 MHz instrument (Bruker, DRX500). The NMR spectra taken in $CDCl_3$ are shown in Fig. S1 (a–c) and the chemical shifts are reported in Table S1 of the ESI section.† The spectra are identical to those previously reported in the literature for PA,[43] LA,[44] and KLA.[42,45] Minor residual solvent impurities from ACN and HOD[46] are noted in the spectra, where applicable. Based on ^{1}H NMR, the purity of all synthesized compounds was conservatively estimated to be greater than 95%.

Aging experiments

The synthesized and purified ketoaldehydes were dissolved in water (18 MΩ cm, Barnstead Nanopure) with 60 °C heating and a few seconds of sonication. The aging results were negligibly affected by dissolution method. Solutions had the concentration range of 10^{-4}–10^{-3} M and were used within 2 h. Aging experiments were performed at room temperature. Small (μL) volumes of stock (20 g L^{-1}) ammonium sulphate (AS, $(NH_4)_2SO_4$) and glycine (GLY) solutions were added to 2–3 mL of the ketoaldehyde solutions at desired molar ratios ([N] : [KLA] = 0–3, where [N] refers to $[NH_4^+]$ or [GLY]), to avoid significantly diluting the ketoaldehyde solutions. We refer to AS-induced reactions as "NH_4^+-mediated" because replacing the anions in control experiments with $(NH_4)_2SO_4$, NH_4NO_3 or NH_4Cl was found to have small-to-negligible effects on the browning reactions (primarily through modification of the solution pH). The aqueous solutions of ketoaldehydes containing NH_4^+ or GLY were then evaporated to dryness using a rotary evaporator (Buchi R-210) at a bath temperature of $T = 45$ °C. The dry residue was redissolved in water to the original dilution. Unless stated otherwise, "aging" in this manuscript refers to the evaporative aging, which was demonstrated to yield the maximal chromophore formation over the course of a few minutes, as opposed to solution-phase aging or aging with NH_3 vapour, which require several days to produce the same concentration of chromophores at $\lambda = 500$ nm.[27,35] Due to their distinctive absorption maxima, the chromophore(s) for the NH_4^+ and GLY-aged samples are referred to as "C_{500}" and "C_{520}", respectively. Control experiments without nitrogen-containing additives were performed for PA, LA and KLA and showed no detectable changes in their UV-Vis absorbance spectra.

To quantify the degree of visible light absorption, UV-Vis absorbance spectra were acquired using a dual-beam spectrometer (Shimadzu UV-2450) with purified water as the reference. The data were recorded in 1 cm quartz cuvettes in the wavelength range of 300–700 nm. Some of the samples required dilution to stay within the linearity limit of the spectrometer and data were corrected for dilution. The degree of browning is reported as the effective mass absorption coefficient (MAC, units of $cm^2 \, g^{-1}$), normalized to the concentration of the organic material in SOA. MAC and ΔMAC are calculated from the base-10 absorbance A_{10} of the ketoaldehyde solution with mass concentration C_{mass} ($g \, cm^{-3}$) measured over pathlength b (cm):[47]

$$\text{MAC}\,(\lambda) = \frac{A_{10}\,(\lambda)\ \times\ \ln 10}{b\ \times\ C_{mass}} \tag{1}$$

$$\Delta\text{MAC}\,(\lambda) = (\text{MAC})_{aged} - (\text{MAC})_{control} \tag{2}$$

MAC values ($cm^2 \, g^{-1}$) correspond to the mass concentration of the starting organic material in the solution and not those of the nitrogen-containing additives that do not absorb radiation above $\lambda = 300$ nm.

Characterization of the products

Aged samples were separated isocratically with reversed-phased LC-UVVIS at 2 mL min^{-1} in (1 : 1) ACN : H_2O in order to isolate the coloured fraction. The separation was monitored in real time using a UV-VIS PDA. The C_{500} fraction containing eluted compounds that absorbed at $\lambda \sim 500$ nm was collected. The C_{500} fractions obtained from repeated injections were then combined and concentrated under vacuum. Fig. S2 in the ESI section† provides an example of one such separation. The dried residues of the separated chromophores were redissolved in solvents suitable for further analysis.

Vibrational spectra were recorded on a single-beam Fourier-transform infrared (FTIR) spectrometer (Mattson Galaxy 5000) in the spectral range of 1000–4000 cm^{-1} with a resolution of 4 cm^{-1}. For FTIR measurements, a film of the KLA was deposited onto the surface of 3 mm thick zinc-selenide (ZnSe) optical windows (Cradley Crystals Corp.). Previous experiments showed that limonene/O_3 SOA films aged with humid NH_3 vapour produced the C_{500} chromophore similar to that observed with evaporative or aqueous aging with NH_4^+.[25,27] A window with deposited KLA film was placed into a small polyethylene dish and suspended on top of a 0.5 M ammonium nitrate solution (NH_4NO_3, Fluka ACS grade) in a sealed glass container with 40 mL headspace filled with $\sim$100 ppb of NH_3 and saturated with water vapour (predicted RH $\sim$ 99%, measured RH $\sim$ 85–90%).[48] The window was removed at time intervals of 0, 12, 36, and 84 h and dried under vacuum in a desiccator for several hours to eliminate condensed water prior to FTIR analyses.

Proton (^{1}H) and carbon (^{13}C) NMR experiments were recorded on a Bruker DRX500 (500 MHz) instrument equipped with a liquid-nitrogen cooled probe. The separated C_{500} fraction was dissolved in a (v/v) 90% H_2O (Aldrich, HPLC) /10%

D_2O (Cambridge Isotope Lab, >99.8% D) solvent mixture, and analysed with pulse-field gradient (PFG) water suppression in order to view exchangeable protons. Water-based solvent systems were used because C_{500} was found to be poorly soluble in nonpolar organic solvents (an important clue about the high polarity of the chromophores).

The separated C_{500} fraction was also dissolved in 90% H_2O/10% ACN (v/v) and analysed with high-resolution mass spectrometry (HR-MS). Data acquisition was performed with positive ion mode direct infusion electrospray ionization (ESI) using a linear-ion-trap (LTQ)-Orbitrap ($R = 100\ 000$) mass spectrometer; data analyses were performed as described in our previous studies.[49,50] In direct infusion ESI/HR-MS experiments samples were injected through pulled fused silica capillary tip (50 μm i.d.) at flow rates of 0.5–2 mL min^{-1}, and the spraying voltage of 4.0 kV. All analyte compounds were observed as sodiated $(M + Na)^+$ or protonated $(M + H)^+$ species. The mass range for acquisition was m/z 100–2000 and the ESI spray voltage was 4 kV. Signals from ^{13}C were removed and molecular formulas are reported for the corresponding neutral compounds (M) by correcting for the mass of a proton or sodium atom.

We also carried out several LC-UVVIS-ESI/HRMS experiments in which the solution containing coloured products was separated on a reverse-phase column (Luna C_{18}, 5 × 150 mm, 100 Å pores, 5 μm particles, Phenomenex, Inc.), with the eluate simultaneously analysed with a photodiode array detector (Thermo Finnigan) and by HR-MS. The samples were prepared by: (a) evaporating aqueous solutions of KLA with AS, GLY, or H_2O (control), and dissolving the residue in ACN; (b) exposing a KLA film to humid NH_3 vapour for 2 days and subsequently extracting it into ACN. The absorbance of the PDA was referenced to pure methanol before each experiment. The gradient elution used a 200 μL min^{-1} flow of H_2O/ACN mixture as the eluent: 0–3 min hold at 95% H_2O, 3–43 min linear gradient to 5% H_2O, 43–53 min hold at this level, 53–56 min linear gradient back to 95% H_2O, and 56–60 min hold in anticipation of the next injection. The standard IonMAX™ ESI source was used for LC-UVVIS-ESI/HRMS experiments making it possible to send the 200 μL min^{-1} flow from the LC directly into the ion source. The settings were: 3 units of sheath gas flow, 0 units of auxiliary gas flow, 10 units of sweep gas flow, 5 kV spraying potential. The instrument was calibrated using a standard mixture of caffeine, MRFA, and Ultramark 1621 (calibration mix MSCAL 5, Sigma-Aldrich, Inc.).

Results and discussion

Ketoaldehydes as brown carbon precursors (UV-Vis)

Fig. 1 shows the quantitative wavelength-dependent ΔMAC values (eqn (2)) for NH_4^+-evaporative aging experiments at 1 : 1 molar ratios of NH_4^+ to KLA, LA and PA. Of the three ketoaldehydes, only KLA produced a significant increase in MAC upon aging. The same trends were observed for GLY-mediated aging, where KLA was once again the only ketoaldehyde that produced a coloured product mixture. Absorption spectra obtained following evaporation of a pure water solution of KLA (Fig. S3 in the ESI section†) demonstrated that KLA does not undergo browning without GLY or NH_4^+ additives. Because all three compounds (KLA, LA and PA) are 1,6-ketoaldehydes we conclude that this particular combination of functional groups is not conducive to formation of brown carbon products. KLA is

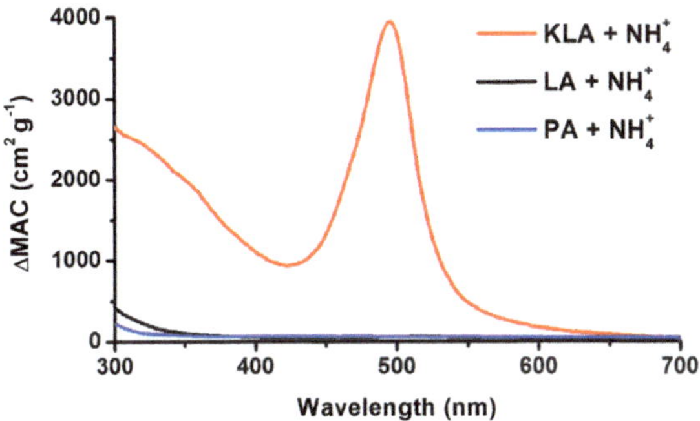

Fig. 1 Comparison of the net increase in the mass absorption coefficients ($\Delta MAC = MAC_{sample} - MAC_{control}$) between the NH_4^+-aged KLA, LA and PA.

the only compound that has an additional ketone group, providing 1,5-diketone and 1,4-ketoaldehyde moieties. The importance of these structural motifs will be illustrated in the mechanistic schemes below. Both LA and PA must be reactive towards ammonia and amino acids, but because neither LA nor PA can form brown carbon products, the remainder of the manuscript will focus on reactions of KLA.

Fig. 2 compares the absorption spectra of KLA aged by (a) NH_4^+ and (b) GLY with the spectra obtained for the aqueous extract of limonene/O_3 SOA aged in a similar manner. Compared to SOA, the shapes of the absorption spectra are almost identical. The absorption maximum for the KLA brown carbon is slightly shifted to 495 nm (compared to 500 nm in SOA), likely due to a reduction in complexity of the aged KLA mixture. Another distinctive property of brown carbon formation from limonene/O_3 SOA is the production of a 430 nm band in the initial

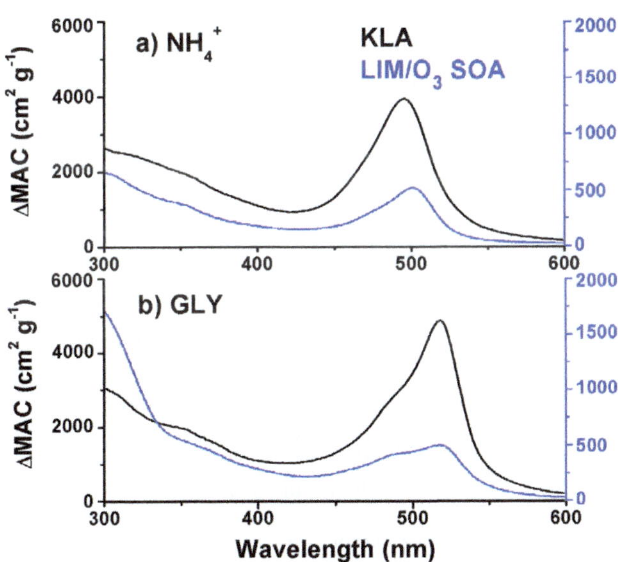

Fig. 2 Comparison of absorption spectra for brown carbon from aged KLA (black traces and left axes) and LIM/O_3 SOA (blue traces and right axes) by (a) NH_4^+ and (b) glycine.

stages of aqueous aging, which is obscured by the 500 nm band at longer time-scales and upon evaporation.[26,35] Even this temporal evolution of chromophores in limonene/O_3 SOA can be reproduced by KLA (Fig. S4 in the ESI section†), offering strong evidence of similar chemistry. In the GLY-mediated aging, the absorption maxima appear at 520 nm and 518 nm for SOA and KLA, respectively. The smaller band of GLY-aged SOA at ~490 nm is also observed in the spectra of aged KLA. The observation that a single compound can reproduce the aging behaviour of the entire SOA extract suggests that the structural elements of KLA and/or its aging products are essential for brown carbon formation in the limonene/O_3 SOA organic mixture.

The aged KLA has a maximum ΔMAC value of 2000–5000 cm^2 g^{-1} at peak maximum, depending on aging method. Inherent uncertainties in the evaporation method (such as dependence of the degree of browning on the evaporation temperature)[35] lead to relatively large uncertainties in the absolute ΔMAC values. However, the relative ΔMAC values can be determined with greater accuracy by keeping the evaporation conditions identical for a series of different samples. For example, the GLY-aged KLA has systematically higher ΔMAC than NH$_4^+$-aged KLA. The ΔMAC values for KLA alone are higher than those for the SOA (maximum ~ 500–800 cm^2 g^{-1}), which is expected because SOA contains a significantly higher fraction of non-absorbing organic mass that is included in the calculation. Based on ΔMAC ratios between the NH$_4^+$-aged KLA and SOA, and the assumption that KLA-like compounds are the only C_{500} precursors in limonene/O_3 SOA, we expect KLA (and its derivatives) to be initially present in the SOA at roughly 10–40% mass. Although a KLA yield in limonene ozonolysis has not been reported, this value is qualitatively in agreement with the expectation that KLA is the most abundant single component of limonene/O_3 SOA.[30]

Effective stoichiometry of reaction (UV-Vis)

We investigated the dependence of the evaporative KLA + NH$_4^+$ and KLA + GLY reactions on the molar concentration of KLA and added nitrogen compounds. Fig. 3 shows that for a fixed initial concentration of KLA, an increase in the

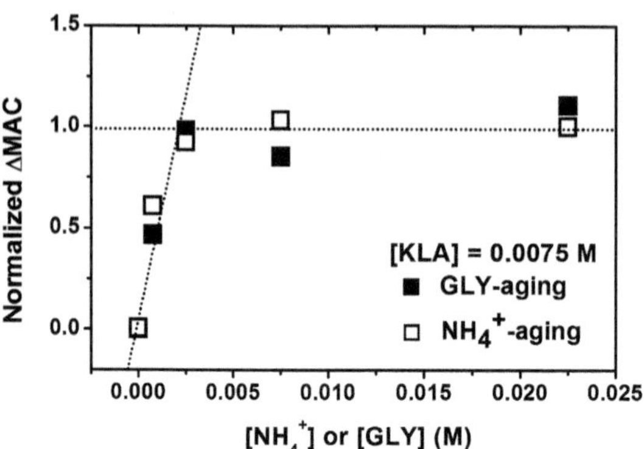

Fig. 3 Dependence of chromophore formation on the molar concentration of reduced nitrogen compound added. The MAC values for NH$_4^+$ and GLY aging are normalized by the maximum to be on the same scale for clarity. The effective reaction stoichiometry of [N]/[KLA] ~ 1/3 can be determined from the location of the break in the concentration dependence.

amount of added nitrogen compounds enhances ΔMAC until a plateau is reached. We previously showed that the absorbance by the chromophores obeys Beer's law,[27] therefore, the ΔMAC values of chromophoric species should be directly proportional to the overall concentration of chromophoric products. If we assume that reaction $a(NH_4^+$ or GLY$) + b$KLA $\rightarrow$ P goes to completion upon evaporation (P = chromophoric products), the effective stoichiometric ratio a/b can be determined from the concentration of [N]$_{plateau}$ (N = GLY or NH_4^+) at which the break in the dependence occurs. For both NH_4^+ and GLY the plateau is reached at $\sim$2.5 mM when the initial KLA concentration is 7.5 mM indicating the reaction stoichiometric ratio is $(a/b) = $ [N]$_{plateau}$/[KLA]$_0 \sim$ 1/3. Experiments in which the initial [N] was fixed and [KLA] varied led to a similar conclusion: the relative initial slopes in the MAC $vs.$ [KLA] and MAC $vs.$ [N] plots also gave a/b $\sim$ 1/3. The reactions of KLA with NH_4^+ and GLY have similar concentration dependence, presumably because NH_4^+ and amines initially react with aldehydes by the same mechanism. With the assumption that reduced nitrogen reacts with only the aldehyde group, a 1/1 stoichiometry would be expected for the reaction. The observed stoichiometry of N/KLA $\sim$ 1/3 indicates that the reactions leading to the chromophores (P = C_{500} or C_{520}) may involve as many as three KLA units. As we discuss below, a number of compounds observed in HR-MS experiments with KLA + NH_4^+ have chemical formulas of $C_{27}H_xNO_y$, which can be formed by combining one ammonia molecule and three molecules of KLA ($C_9H_{14}O_3$). However, a number of other products with different N/KLA combinations are also observed, therefore, the empirically determined 1/3 stoichiometry should be regarded as an average over a large number of light-absorbing products.

Importance of the aldehyde to imines conversion in browning (FTIR and NMR)

Because the absorption spectra and reaction of the NH_4^+- and GLY-mediated chromophores are qualitatively similar, and because of the larger environmental relevance of NH_4^+, we focused on structure analysis of only the NH_4^+-based chromophore (C_{500}). Direct application of FTIR to the evaporated residues or redissolved KLA + NH_4^+ mixtures was complicated by interference from solvent and inorganic additives. As an alternative, we investigated aging of a pure KLA film with humid NH_3 vapour that was monitored with FTIR at four time intervals (Fig. 4). Although aging of a film with humid NH_3 vapour produces similar chromophores to those formed in evaporative aging,[25] it is much slower and may be limited by the surface area of the film. However, this method of aging eliminates interference from the inorganic additives, for example, the interfering bands such as the strong NH bending modes of NH_4^+ (expected at $\sim$1400 cm^{-1})[51] are absent in the spectra.

Based on the expected chemistry,[25,27] the characteristic bands associated with the vibrations of the reactive carbonyl groups in KLA should be augmented or replaced by the new bands corresponding to C–N (amines) and C=N (imines) and other aging products. The pure KLA (Fig. 4, black trace) film is characterized by aliphatic CH_3/CH_2 stretches (2800–3000 cm^{-1}) and bends (1300–1500 cm^{-1}) that are relatively broad, suggesting a disordered film. A broad peak at $\sim$3430 cm^{-1} indicate a hydrogen-bonded OH stretching band, presumably of an intramolecular aldol of KLA, diol (hydrated KLA), or carboxylic acid (oxidized KLA). The relatively small width and symmetric shape of the C=O stretch at 1710 cm^{-1}

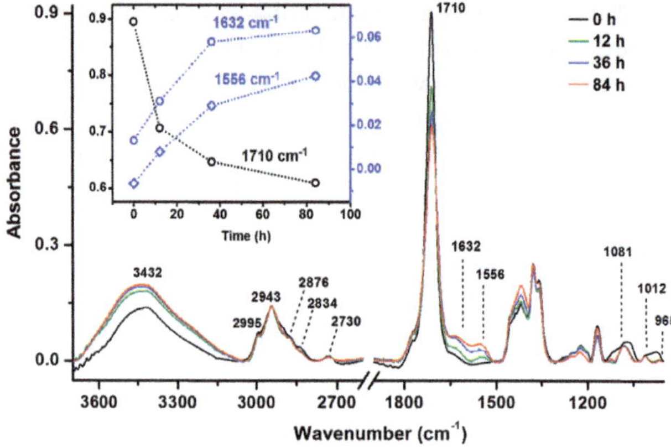

Fig. 4 Infrared spectra of the KLA film aged with humid NH_3 vapours at four aging times. Inset shows time profiles of select bands (see text).

rules out high acid contribution to the film at the early stages of reaction (^{1}H NMR spectra indicate that the pure KLA does not have COOH contamination). The broad peak around 3430 cm^{-1} overlaps the hydrogen-bonded N–H stretches, of pyrroles for example,[52,53] or of absorbed water, which limits its analytic usefulness. The aldehyde functionality is confirmed by two weak bands unique to aldehydes: the C(O)–H stretch at 2834 cm^{-1} and the first overtone of the in-plane C(O)–H bend at 2730 cm^{-1}, which gains intensity because of the Fermi resonance with the stretch.[54,55] The two C(O)–H bands and the carbonyl C=O band at 1710 cm^{-1} are reduced in intensity relative to the skeletal C–H vibrations as the reaction progresses confirming that the carbonyls groups in KLA react with NH_3 and water vapour. After 84 h of exposure of the KLA film to the humid NH_3 vapour, the reaction is not complete, *i.e.*, most of the original vibrational bands of KLA are still present. Because the dried KLA is viscous, substantial diffusion limitation likely exists in the film; therefore, the reaction may be limited to the surface and eventually ceases, even when unreacted KLA is still present.

As the film experiences aging by $NH_3(g)$, several new bands appear at 1012, 1082, 1070, 1556 and 1632 cm^{-1}. In addition, the bands between 1300 and 1500 cm^{-1} slightly increase in intensity. The new peaks with the highest diagnostic potential are those at 1632 and 1556 cm^{-1}, which are consistent with the presence of nitrogen-containing bonds in the products. The growth of these peaks and the decay of the carbonyl peak are shown in the insert in Fig. 4. The alternative assignment of the 1632 cm^{-1} band to a conjugated C=C stretch, that would also appear in the 1610–1640 cm^{-1} spectral region, can be ruled out because of the lack of the corresponding vinyl CH stretches in the spectrum (>3000 cm^{-1}, frequency increasing with the level of conjugation). In particular, the 1556 cm^{-1} band may correspond to an N–H bending mode and the 1632 cm^{-1} band may correspond to the C=N stretching mode of substituted imines (or Schiff bases)[56,57] or N-heterocycles like pyrroles.[58] Bones *et al.* (2010) also suggested that a protonated Schiff bases may play a role in the visible light absorption,[26] as they can be highly absorbing when only moderately conjugated.[59] Given the nature of reaction, the presence of pyrroles may be equally likely (discussed in more detail below). Amides, which have so called amide-I and amide-II bands in this frequency range,[55] are not expected to form from carboxylic acids under the aging conditions employed in our experiments (and there is no evidence for them in the NMR experiments described next).

The C_{500} fraction was analysed by NMR spectroscopy to supplement FTIR observations. We note that NMR does not necessarily probe the same compounds as FTIR because: (a) NMR was performed on the C_{500} fraction of the KLA + NH_4^+ evaporated sample, whereas FTIR was performed on the entire KLA + $NH_3(g)$ mixture with a large signal contribution from the intact KLA in the film throughout the reaction and (b) the NMR samples are dissolved in $CDCl_3$ or H_2O/D_2O and the FTIR samples have no solvent apart from the residual absorbed water. Furthermore, the same molecules may exist in different chemical forms, such as open, aldol, hydrated, *etc.*, depending on the type of the sample, presence or absence of water, and measurement method. We have taken these factors in consideration in our data analysis.

Fig. 5a shows the KLA ^1H NMR spectrum in $CDCl_3$, which can be compared with the ^1H spectrum of C_{500} fraction taken in H_2O, with water suppression, shown in Fig. 5b. Based on the shape of the peaks in the alkyl region, C_{500} fraction contains one major component and several minor ones. The major C_{500} products

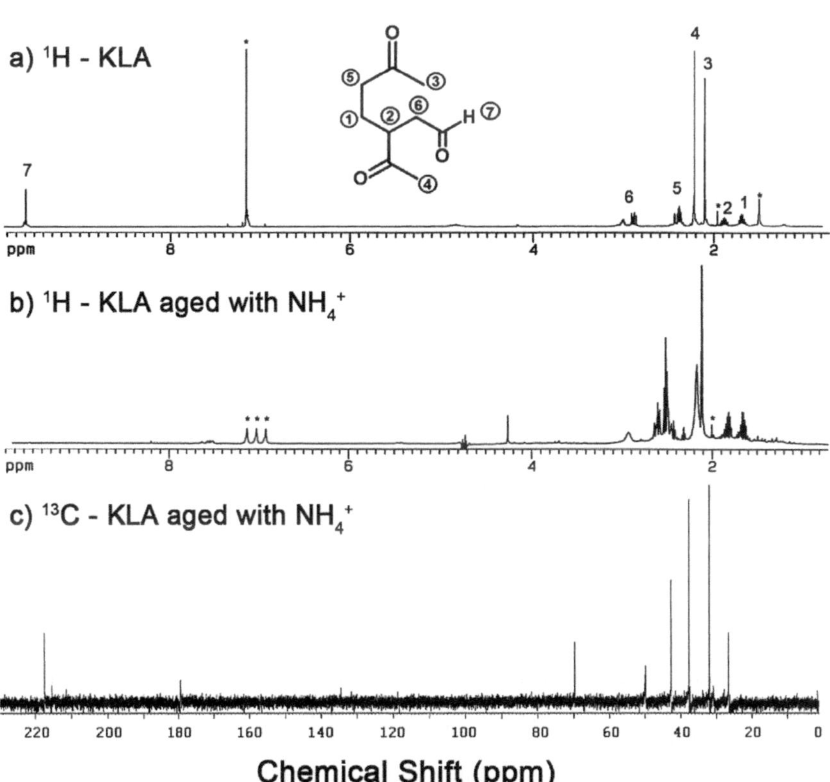

Fig. 5 Proton (^1H) NMR spectra of (a) KLA taken in $CDCl_3$ and (b) the separated coloured fraction of KLA aged with NH_4^+ taken in H_2O/D_2O (9 : 1) with PFG water suppression. The carbon (^{13}C) spectrum of the aged/separated KLA fraction taken in H_2O/D_2O (9 : 1) is shown in panel (c). Residual solvent or inorganic resonances are shown with asterisks, from right to left: (a) HOD, ACN, $CDCl_3$; (b) ACN, NH_4^+; (c) none.

retained several resonances characteristic to KLA including the ketone CH_3 proton groups (although one of them is broadened) and protons from the main carbon skeleton such as those labelled "1" and "2" in Fig. 5a. The aldehyde proton at 9.7 ppm (labelled "7" in Fig. 5a and Fig. S1† for pure KLA) is no longer present, in good agreement with the reduction of the C(O)–H vibrations observed in FTIR. Aldehydes are known to hydrate in water but if hydration was the only mechanism operating in the solution, we would still observe the aldehyde proton based on the expected hydration equilibrium ratio for aldehydes of this size. The removal of the aldehyde group most likely results from an intramolecular aldol formation *via* a ketone enol adding to the aldehyde group, which can be catalysed by NH_4^+,[60] or from the nucleophilic addition of reduced nitrogen compounds. These types of reactions are slow in solution but are promoted by evaporation because water is a product. Several different enols and intramolecular aldols can be formed from KLA, but based on product yields from similar 1,6-ketoaldehydes, *e.g.*, 6-oxo-heptanal,[61] the aldol with a 5-member ring should be preferentially formed from nucleophilic attack at the aldehyde site.

Several unique major and minor resonances for C_{500} are observed in the alkyl (1–3 ppm) region, which cannot be assigned conclusively. The singlet at 4.3 ppm is most consistent with an alcohol proton. This singlet does not show up in the C_{500} spectrum taken in D_2O (Fig. S5 in the ESI section†) because alcohol protons are exchangeable with the solvent (OH $\rightarrow$ OD) and become silent. The observation of the alcohol proton is also consistent with a cyclic aldol as a major product. The trio of peaks at around 7 ppm belongs to the NH protons of the ammonium ion.[33] Although the C_{500} fraction has been separated, NH_4^+ may be present in the sample due to complexation with water-soluble organics or decomposition of nitrogen-containing compounds. These NH protons also exchange with the solvent and consequently are also absent in the D_2O spectrum (Fig. S5†).

The ^{13}C proton-decoupled spectrum of C_{500} taken in H_2O, shown in Fig. 5c, is more straightforward to interpret. Each resonance belongs to a carbon atom in a unique electronic environment and unlike ^1H NMR, ketones may be straightforwardly identified. The first useful observation is that the number of the observed peaks is relatively small, of the order of 10. This implies that the major C_{500} fraction's component is a molecule that is comparable in size to KLA which has 9 chemically distinguishable carbon atoms. The 25–45 ppm region corresponds to various types of alkyl carbons and transitions from 1°–3° aliphatic carbons in the direction of higher chemical shift. The 27 ppm peak can be assigned to a cyclic CH_2 group. The most downfield shifts at 218 ppm correspond to cyclic ketones (less shielded than an aliphatic ketone). Another small peak at 216 ppm may also be assigned to a ketone from a minor species. Similarly to the ^1H spectrum, no aldehyde carbons, ~200 ppm, can be observed. The resonance at 70 ppm is most likely due to carbon atom attached to hydroxyl functional group. The 50 ppm resonance may be assigned to a 1°–2° carbon atom attached to nitrogen by a single bond (C–NH), although in some cases 2°–3° alkyl carbons can show up in this region.

The ^{13}C NMR data are consistent with the suggestion by ^1H NMR that a cyclic aldol with a ketone functional group (but no aldehydes) is the primary component in the aqueous solution of the C_{500} fraction, which co-exists in solution with a few minor organics that may or may not contain nitrogen. The dominant structures implicated by NMR are not the species that absorb visible light because all possible monomeric KLA aldols are expected to be colourless. Although we used LC-UVVIS to specifically isolate the light-absorbing compounds, the aldol compound(s) could coincidentally co-elute with them or be produced from them by hydrolysis of C_{500} in water.

Based on this and previous works,[35] the chromophore is likely a *minor species* whose signals are smaller, and possibly below the detection limit, compared to the more-dominant FTIR and NMR signals. Very weak ^{13}C resonances in the 120–140 ppm region of the ^{13}C spectrum provide useful indications that conjugated and/or aromatic carbons ($-C=C-$) exist in the C_{500} mixture. Another weak resonance at approximately 180 ppm may be assigned to either a COOH or $C=N$ carbon; however, $C=N$ is more consistent with the FTIR observations. We suspect these weak resonances are most indicative of the brown carbon chromophores, because it may take substantial conjugation and nitrogen substitution to promote light absorbance in the visible region. The fact that no obvious vinyl CH stretches (>3000 cm^{-1}) were discernible in the aged FTIR spectra is also consistent with small relative fraction of the chromophore amongst the products of aging. In summary, these observations suggest that the major functional groups in the C_{500} fraction compounds are hydroxyls and carbonyls, with minor contributions from amines, imines (or N-heterocycles that cannot be distinguished from secondary imines by NMR), and carboxyls.

Charged nature of the coloured compounds (HR-MS)

In order to provide molecular level information about the chromophores, the C_{500} fraction was analysed by direct infusion high-resolution ESI-MS. The coloured residues formed in the evaporation of KLA + NH_4^+ and KLA + GLY mixtures, as well as the products formed by browning of a KLA film with humid NH_3 vapour, were additionally analysed with LC-UVVIS-ESI/HRMS. The analysis was also performed for the products of reactions of limonene/O_3 SOA and NH_4^+ for comparison. Representative chromatograms corresponding to the integrated 400–600 nm absorbance are shown in Fig. 6. The SOA + NH_4^+ (Fig. 6a) and KLA + NH_4^+(Fig. 6b) chromatograms are qualitatively similar but the former contains far more peaks than the latter. In fact, the NH_4^+-aged KLA chromatogram has only two distinct features: two overlapping sharp peaks eluting at 2–3 min and a single broad peak eluting at 13–36 min. The early-eluting fraction corresponds to the C_{500} fraction isolated on the LC separation stage and discussed extensively in the previous sections. The 500 nm absorption band is distinctively observed in the absorption spectra for both early- and late-eluting coloured fractions (Fig. S6 in the ESI section†). GLY-aged and NH_3-aged KLA resulted in qualitatively similar chromatograms (not shown in Fig. 6).

The early peaks in the chromatograms elute close to the column's dead time together with the charged inorganic mixture constituents, such as HSO_4^-. We therefore suspect that a significant fraction of compounds eluting early either carry a permanent charge or are zwitterions. Indeed, the observed molecular formulas of the compounds eluting early (Table 1) correspond to fairly large organic molecules (mostly C_9–C_{27}), which would be expected to elute considerably later than at 2–3 min if they remained neutral during their passage through the column. This is consistent with our observation that the LC-separated C_{500} fraction, which also eluted close to the column's dead time (Fig. S2 in the ESI

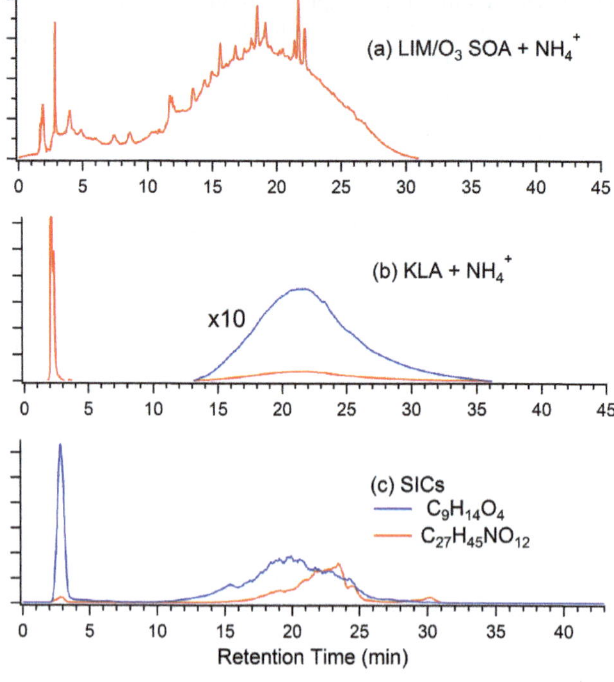

Fig. 6 Sample LC-UV/VIS-ESI/HRMS (integrated 400–600 nm absorbance) chromatograms of the NH$_4^+$-aged (a) aqueous SOA mixture and (b) aqueous KLA extract. The blue trace in panel (b) corresponds to a tenfold magnification of the late-eluting signal. Panel (c) shows selected ion chromatograms (SICs) for protonated C$_9$H$_{14}$O$_4$ and C$_{27}$H$_{45}$NO$_{12}$. SICs for most other ions displayed a similarly poor level of chromatographic resolution.

section†), was not soluble in non-polar solvents but dissolved readily in water and ACN. It is likely that KLA-derived brown carbon contains ionic or zwitterionic compounds for which the reversed-phase chromatography used in this work is not sufficient.

The lack of resolution in the late eluting broad peak for the KLA + NH$_4^+$, KLA + GLY, and KLA + NH$_3$ samples was a surprise. We note the separation column works quite well for many mixture components in the SOA + NH$_4^+$ sample, as evidenced by multiple well resolved peaks appearing throughout the chromatogram (Fig. 6a). One possible explanation for the lack of resolution is a very large number of products with comparable elution times. However, this assumption is not consistent with selected ion chromatograms (SICs), which display a similarly low level of chromatographic resolution. For example, Fig. 6c shows a SIC for protonated C$_{27}$H$_{45}$NO$_{12}$, which elutes as a broad peak between 15 and 30 min. This behaviour is characteristic of most other ions we have examined. Therefore, the peak broadening must arise from a chemical reason, perhaps from keto–enol tautomerization, multiprotic acid–base, and hydration–dehydration equilibria occurring during the separation.

The most abundant molecular formulas reproducibly observed in the LC-ESI/HRMS and the direct infusion ESI/HRMS are listed in Table 1. Most of the observed compounds have carbon skeletons traceable to KLA, *e.g.*, monomers (C$_9$), dimers (C$_{18}$), trimers (C$_{27}$), and tetramers (C$_{36}$). However, in source

Table 1 The most abundant neutral molecular formulas observed in the positive ion mode direct infusion ESI/HRMS measurements on the C_{500} fraction in mixed water/ACN (50 : 50 v) solvent and in the LC-UVVIS-ESI/HRMS measurements on the KLA + NH_3(g) sample in H_2O/ACN eluent at mixing ratios of 95 : 5 v (2–3 min elution time) and 75 : 25–39 : 61 v (14–30 min elution time). We note that differences in mixing ratios of solvent as well as difference in ionization sources may be responsible for the distribution of the MS peak intensities of the observed species. The charge carrier in the LC-ESI/HRMS measurements was H^+ in most cases. The last three columns contain relative intensities, scaled to 100%, for the compounds eluting between 14–30 min and 2–3 min in the LC-ESI/HRMS experiments, and compounds detected by direct infusion ESI-MS, respectively

# C	# H	# O	# N	DBE	LC-ESI/HRMS eluted at 14–30 min	LC-ESI/HRMS eluted at 2–3 min	ESI/HRMS direct infusion
7	8	1	0	4	9.1	1.6	
7	8	2	0	4	8.9	2.7	
7	10	2	0	3	57.0	5.5	
7	10	3	0	3	23.2	1.6	
8	10	1	0	4	5.8	2.4	
8	10	2	0	4	4.2	1.1	
8	10	3	0	4	1.7	3.2	
8	12	2	0	3	13.2	2.3	
9	10	2	0	5	51.8	51.8	14.5
9	10	3	0	5	4.5	4.1	
9	11	2	1	5		0.4	1.7
9	12	2	0	4	100.0	100.0	2.7
9	12	3	0	4	71.0	49.9	18.5
9	12	4	0	4	5.8	3.1	1.3
9	13	1	1	4		0.9	1.3
9	13	2	1	4			1.1
9	14	3	0	3	4.0	2.8	4.7
9	14	4	0	3	26.4	11.0	100.0
9	14	5	0	3	0.8	0.9	7.6
9	16	5	0	2			1.4
9	17	4	1	2			2.4
10	13	1	1	5	2.1	54.7	
10	13	2	1	5	0.3	11.0	
11	15	1	1	5		10.6	
11	15	2	1	5	0.6	19.3	
11	16	1	2	5		4.9	
18	20	3	0	9	0.3		
18	23	3	1	8	0.3	3.8	0.1
18	24	6	0	7	3.4		
18	26	7	0	6	9.3	1.2	
18	26	8	0	6	5.3	0.4	
18	28	8	0	5	2.0		0.6
18	28	9	0	5	1.0		0.7
18	31	8	1	4	25.8		
18	31	9	1	4	17.2	1.4	
18	33	10	1	3	17.0	1.0	
18	33	11	1	3	4.4	0.5	
20	24	2	2	10	0.4	10.7	
20	24	3	2	10	0.4	7.6	
20	26	4	2	9	0.2	2.8	3.1
21	24	2	2	11		4.1	
21	26	2	2	10	0.9	28.9	
21	26	3	2	10	0.6	21.9	
21	27	2	3	10	0.2	7.5	

Table 1   (*Contd.*)

# C	# H	# O	# N	DBE	LC-ESI/HRMS eluted at 14–30 min	LC-ESI/HRMS eluted at 2–3 min	ESI/HRMS direct infusion
21	29	2	3	9			4.0
27	43	10	1	7	6.3		
27	43	11	1	7	3.5		
27	45	11	1	6	5.2		
27	45	12	1	6	14.4		
27	45	13	1	6	16.6	0.5	
27	45	14	1	6	6.0		
27	47	13	1	5	4.9	0.4	
27	47	14	1	5	5.8	0.6	
36	59	16	1	8	3.3		
36	59	17	1	8	3.4		

fragmentation, multistep chemistry (*e.g.*, oligomerization combined with decomposition reactions like decarboxylation), or direct contamination introduced by the synthesis and evaporation may be responsible for the appearance of molecular formulas with carbon numbers that are not multiples of 9, such as C_7, C_8, C_{10}, C_{11}, *etc.* Several dominant species that appear at essentially all retention times, such as protonated $C_9H_{12}O_2$, likely correspond to common fragments of larger compounds. For example, the protonated $C_9H_{14}O_4$, a compound that cannot be a chromophore itself in view of its small size and DBE, co-elutes with both early-eluting and late-eluting absorbers (Fig. 6c). It is also the most dominant compound detected in the direct infusion ESI/HRMS experiments.

The majority of the observed products contain 0–2 N-atoms in their molecular formulas and double bond equivalencies (DBE = the total number of double bonds and rings) in the range of 3–10. A significant fraction of the observed compounds appear to be "built" from three KLA molecules (C_{27} skeleton) and one N-atom, in agreement with the 3 : 1 stoichiometry suggested by the optical absorption measurements (Fig. 3). However, a number of 1 : 1, 2 : 1, and other KLA : N combinations are also observed. Fig. S7 in the ESI section† provides a pictorial representation of the observed compounds as a graph of the (O + N)/C ratio *vs.* the number of C-atoms in the molecules. Reactions that form imines from carbonyls are expected to conserve the (O + N)/C ratio, while the condensation reaction leading to oligomerization and loss of water are expected to lower it relative to the starting (O + N)/C ratio of 0.33 in KLA. While most compounds do follow this trend a number of other compounds have (O + N)/C ratios in excess of 0.33, implying that some products are oxidized (increased O) and/or are present as the β-hydroxy secondary amines (*e.g.*, from hemiaminal addition to the aldehyde prior to dehydration, see Scheme 1 or Scheme S1†). Indeed the major ion detected by ESI-MS is $C_9H_{14}O_4$ = KLA oxidized to ketolimononic acid (existing in its intramolecular aldol form according to the NMR observations).

Some compounds with high DBE (= 10–11), and 2 nitrogen atoms cannot be traced directly to the KLA structure. One may speculate that the higher nitrogen content, higher DBE and unusual carbon number of compounds like $C_{21}H_{24}N_2O_2$ (DBE = 11) are due to reactions favouring the elimination of carbon to yield an aromatic product. For example, the hetero retro Diels–Alder reactions of cyclic imines,[62–64] a reaction that may occur at room temperature and is enhanced by protic solvents. The result may be the oligomeric N-heterocyclic compounds that are suggested to play a role in the light-absorbing properties of other types of secondary brown carbon.[10,33,34,65] However, due to the uncertainty in their formation pathways starting from KLA, a representative structure for these 2N compounds cannot be determined based on the results of this study.

Possible production pathways of light-absorbing species

Most of the compounds observed by NMR and HR-MS are not expected to absorb visible radiation due to their low DBE and small molecular size. The proposed formation pathways of monomers and oligomers corresponding to many of the molecular formulas detected in HR-MS, based on a few well-known reactions, are shown in Scheme S1 of the ESI section.† Full discussion of compounds that don't absorb visible light is outside the scope of this work. We focus our attention on the molecular formulas with high DBE, which may correspond to chromophoric compounds. Although these molecules produce a much lower signal in HR-MS relative to the most abundant ions, they are likely to dominate the visible absorbance by the mixture. This makes their identification comparable to finding a "needle in a haystack".

Another complication is that multiple isomeric structures may correspond to a given molecular formula inferred from HR-MS. However, we can narrow down the spectrum of probable candidates based on the available analytical evidence (from UV-VIS, FTIR, NMR, and LC) and provide examples of reactions that may generate plausible structures. Our observations support the following picture: (1) fast disappearance of the aldehyde group to form aldols and compounds with $>$C=N– groups (imines or N-heterocycles); (2) a relatively large product pool with an abundance of simple monomers that don't absorb visible light and a smaller mass fraction of brown carbon chromophores; and (3) compounds containing 0–2 N-atoms found in the mixture with high degrees of unsaturation that may contribute to the light-absorption properties of C_{500}. Scheme 2 shows examples of the proposed formation routes to three 0N, 1N and 2N compounds that have the

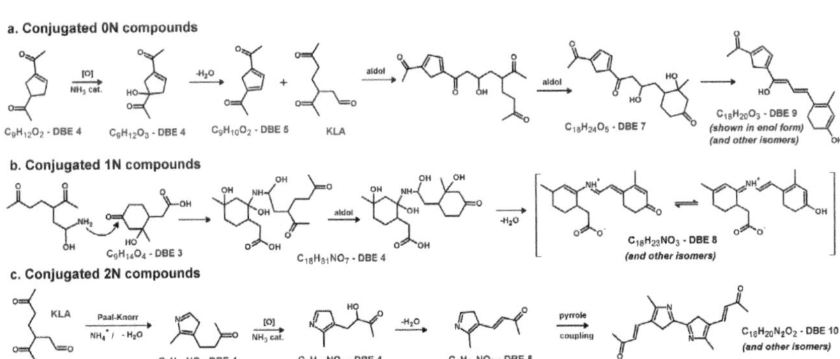

Scheme 2 Proposed formation of plausible highly-conjugated species in C_{500} that may contribute to the visible-light-absorbing properties of C_{500}. Likely, many isomers and higher-order oligomers exist. Molecular formulas shown in the scheme, with their proposed corresponding structures, have been detected with HR-MS (Table 1).

highest DBE in the mixture (Table 1). We are not suggesting that these are the actual structures of the chromophores and the precise mechanism leading to their formation – most likely, a number of alternate pathways exist that lead to similar compounds or their structural isomers. Furthermore, although the products shown in Scheme 2 have the highest DBE observed in this work, it does not imply that those explicit compounds contain enough conjugation and the correct functional group distribution to absorb visible light. Electronic structure calculations would be useful in answering this question in the future.

Scheme 2a shows the formation of a conjugated compound with molecular formula $C_{18}H_{20}O_3$ (DBE = 9), starting with the dominant aldol condensate from KLA. The ketone group can undergo NH_3- or amino acid-catalysed α-hydroxylation,[66] or in the case of α,β-unsaturated ketones, γ-hydroxylation,[67] both yielding a cyclic diene diketone with molecular formula $C_9H_{12}O_2$ (DBE = 5) that is detected in C_{500} with significant signal (Table 1). In the α-hydroxylation of a ketone in the presence of ammonia, the ketone is first transformed into an enolate or enamine, and subsequent oxidation of the C=C bond in water affords the α-hydroxylated product that can dehydrate further to form a conjugated product. The diketone may react with KLA by intermolecular enol addition to the KLA aldehyde. Subsequent aldol condensations affords $C_{18}H_{20}O_3$ (DBE = 9), a compound that is highly-conjugated if most of its ketones are present as enols, expected to be the dominant form in situations where a large π- system results from the tautomerization.[61] The proposed structure is not expected to be permanently charged, in good agreement with its absence in the early-eluting chromatographic peak. Its ability to tautomerize may lead to its detection as part of the broad peak at later retention times (Table 1). The product of Scheme 2a has 7 linearly-conjugated double bonds. Comparatively, the highly-coloured carotenoids (a large class of mainly $C_xH_yO_z$ chromophores) that absorb around 500 nm have 10–13 conjugated double bonds. It is possible that a conjugated KLA trimer can be formed in a similar manner as shown in Scheme 2a, which would further extend the π-system for 0N chromophores, but its signal lies below the detection limit of our instrument.

Scheme 2b shows the formation of a conjugated compound with molecular formula $C_{18}H_{23}NO_3$ (DBE = 8), one of the observed products of KLA aging (Table 1). This compound was also one of the major N-containing species detected by the DESI/HR-MS analysis of limonene SOA exposed to NH_3 vapour.[25] The proposed scheme starts with the formation of a hemiaminal intermediate of KLA that can nucleophilically add to a carbonyl,[68] such as the aldol acid version of $C_9H_{14}O_4$. The product is a α-hydroxy secondary amine, with subsequent dehydration yielding a Schiff base ($R_1R_2C = NR_3$, $R_3 \neq H$). The end product in Scheme 2b has 5 linearly-conjugated double bonds. Because the proposed $C_{18}H_{23}NO_3$ Schiff base is an organic compound containing both an imine group and a carboxylic acid group, strong intramolecular H-bonding (*via* a 7-member ring) may lead to zwitterion formation by protonation of the imine by the acid.[69] We note there are other reaction pathways leading to a $C_{18}H_{23}NO_3$ molecule stemming from the observed compounds in aqueous media, such as organic acid-catalysed Mannich-type reaction between enols and imines[70,71] or transimination of an imine and amine.[72-74] However, the proposed reaction mechanism satisfies the following criteria: it requires only mild organic acid catalysis and it occurs in aqueous solution. Because absorbance of Schiff bases can shift significantly to the red due to protonation at the nitrogen site, the protonation of the imine is expected to greatly enhance the visible light absorption of this molecule. For example, the non-protonated form of the 11-*cis*-retinal + *n*-butylamine Schiff base (6 conjugated double bonds) absorbs at $\sim$350 nm in methanol, but its protonated form absorbs at $\sim$450 nm.[75,76] The suggestion of a protonated Schiff base is supported by these molecules' short retention time in a C_{18} column and the similarity in the shape of absorbance spectra and high extinction coefficients ($>10^4$ M^{-1} cm^{-1}) of C_{500} with that of the *cis*-11 retinal Schiff base.[59] Therefore, even though the product of Scheme 2b contains less conjugation than the product of Scheme 2a, the presence of a zwitterionic secondary imine makes it more likely to contribute to the visible light absorption.

Scheme 2c shows the formation of a conjugated compound with molecular formula $C_{18}H_{20}N_2O_2$ (DBE = 10), starting with the Paal–Knorr pyrrole formation from the 1,4-dicarbonyl moiety of KLA with either NH_4^+ or amino acid.[77,78] Presumably, a hemiaminal is formed at the keto site from the nitrogen addition, which adds to aldehyde and subsequent dehydration forms the pyrrole. Pyrroles couple easily, especially in water at neutral pH.[79] Dimerization produces a 2N compound joined at the α-carbon. Multiple stages of the reaction may occur for the pyrrole compound, producing higher-order oligomers; however, pyrroles from KLA do not have a second unsubstituted α-carbon site to produce polymers. Furthermore, pyrrole polymers, although highly-absorbing, are also insoluble or poorly-soluble in polar solvents,[80] in stark contrast to C_{500} where the brown colour disappears upon dissolution of the aged KLA in nonpolar solvents. Pyrroles may stabilize a positive charge reasonably well due to their basic N, which may be the reason $C_{18}H_{20}N_2O_2$ elutes only at the early retention time. The product of Scheme 2c has 8 linearly-conjugated double bonds; however, similar structures have not been studied in the literature so a direct comparison of optical characteristics is not possible at this time. It's worth noting that the polypyrroles that are insoluble in water have absorption spectra remarkably similar to C_{500}, and the degree of conjugation can be very high.[81,82] Bipyrrole ($C_8H_8N_2$) itself does not absorb at wavelengths greater than 300 nm;[83] however, it is not clear how additional π-conjugation will change its absorption characteristics.

Conclusions and implications

The reaction of KLA was demonstrated to form brown carbon chromophores that are nearly identical in properties to those found in the limonene/O_3 SOA upon aging with NH_4^+ and GLY. In contrast, comparable NH_4^+- and GLY-mediated reactions with LA or PA did not result in brown carbon formation. The analytical insights obtained from spectrometric methods suggest the contribution of conjugated compounds, *e.g.*, aldol condensates, secondary imines (Schiff bases), or N-heterocycles like pyrroles, to the optical properties of the chromophores. They form by different mechanisms than the imidazole-based oligomers previously linked to secondary brown carbon in the atmospheric chemistry literature.[7–10,33,34] As the chemistry that produces brown carbon is diverse, the structures of the chromophoric compounds may be similarly varied in nature.

This work emphasizes the importance of molecular structure in chemistry affecting important physical properties of aerosols, namely their optical absorption coefficient. It would not be possible to predict the occurrence of the reactions described in this manuscript from average properties of aerosols alone, such as the O/C ratio, which is frequently used in correlating physical properties to the average composition. Even information about presence or absence of specific functional groups in the aerosol may not be sufficient. For example, KLA, LA, and PA are all ketoaldehydes but only one of them browns in reactions with ammonium ions. Therefore, it is not enough to characterize the average aerosol composition and the functional group content – the detailed molecular structures of aerosol constituents must be characterized in order to predict its aging chemistry and optical properties.

Acknowledgements

The authors would also like to thank the following people for helpful discussions: Dr John DeLorbe, Dr Phil Dennison, Dr Hyun Ji (Julie) Lee, Dr Jiri Misek, Dr Theresa McIntire, Dr John Greaves, Prof. Richard Chamberlain, and Prof. James Nowick. The UCI group acknowledges support by the NSF grants CHE-0909227 and AGS-1227579. TBN thanks UCI for the Chancellors Club fellowship. The PNNL group acknowledges support from the Chemical Sciences Division (JL), Office of Basic Energy Sciences of the U.S. DOE, and the W.R. Wiley Environmental Molecular Sciences Laboratory (EMSL) – a national scientific user facility located at PNNL, and sponsored by the Office of Biological and Environmental Research of the U.S. PNNL is operated for US DOE by Battelle Memorial Institute under Contract No. DE-AC06-76RL0 1830.

References

1 S. Solomon, D. Qin, M. Manning, Z. Chen, M. Marquis, K. B. Averyt, M. Tignor and H. L. Miller, *Climate Change 2007: The Physical Science Basis. Contribution of Working Group I to the Fourth Assessment Report of the Intergovernmental Panel on Climate Change.*, IPCC, 2007.

2 M. O. Andreae and A. Gelencser, *Atmos. Chem. Phys.*, 2006, **6**, 3131–3148.

3 T. Bond and R. Bergstrom, *Aerosol Sci. Technol.*, 2006, **40**, 27–67.

4 D. T. L. Alexander, P. A. Crozier and J. R. Anderson, *Science*, 2008, **321**, 833–836.

5 M. M. Galloway, P. S. Chhabra, A. W. H. Chan, J. D. Surratt, R. C. Flagan, J. H. Seinfeld and F. N. Keutsch, *Atmos. Chem. Phys.*, 2009, **9**, 3331–3345.

6 B. Noziere, P. Dziedzic and A. Cordova, *J. Phys. Chem. A*, 2009, **113**, 231–237.

7 E. L. Shapiro, J. Szprengiel, N. Sareen, C. N. Jen, M. R. Giordano and V. F. McNeill, *Atmos. Chem. Phys.*, 2009, **9**, 2289–2300.

8 N. Sareen, A. N. Schwier, E. L. Shapiro, D. Mitroo and V. F. McNeill, *Atmos. Chem. Phys.*, 2010, **10**, 997–1016.

9 M. Trainic, A. Abo Riziq, A. Lavi, J. M. Flores and Y. Rudich, *Atmos. Chem. Phys.*, 2011, **11**, 9697–9707.

10 D. O. De Haan, M. A. Tolbert and J. L. Jimenez, *Geophys. Res. Lett.*, 2009, **36**, L11819, DOI: 10.1029/12009gl037441.

11 A. Gelencser, A. Hoffer, G. Kiss, E. Tombacz, R. Kurdi and L. Bencze, *J. Atmos. Chem.*, 2003, **45**, 25–33.

12 A. Hoffer, G. Kiss, M. Blazso and A. Gelencser, *Geophys. Res. Lett.*, 2004, **31**, L06115, DOI: 10.1029/2003GL018962.

13 J. L. Chang and J. E. Thompson, *Atmos. Environ.*, 2010, **44**, 541–551.

14 H. E. Krizner, D. O. De Haan and J. Kua, *J. Phys. Chem. A*, 2009, **113**, 6994–7001.

15 B. Noziere and W. Esteve, *Atmos. Environ.*, 2007, **41**, 1150–1163.

16 M. T. Casale, A. R. Richman, M. J. Elrod, R. M. Garland, M. R. Beaver and M. A. Tolbert, *Atmos. Environ.*, 2007, **41**, 6212–6224.

17 B. Noziere, D. Voisin, C. A. Longfellow, H. Friedli, B. E. Henry and D. R. Hanson, *J. Phys. Chem. A*, 2006, **110**, 2387–2395.

18 J. Zhao, N. P. Levitt and R. Zhang, *Geophys. Res. Lett.*, 2005, **32**, L09802, DOI: 10.1029/2004GL022200.

19 B. Noziere and W. Esteve, *Geophys. Res. Lett.*, 2005, **32**, L03812, DOI: 10.1029/2004GL021942.

20 W. Esteve and B. Noziere, *J. Phys. Chem. A*, 2005, **109**, 10920–10928.

21 R. M. Garland, M. J. Elrod, K. Kincaid, M. R. Beaver, J. L. Jimenez and M. A. Tolbert, *Atmos. Environ.*, 2006, **40**, 6863–6878.

22 M. Z. Jacobson, *J. Geophys. Res.*, 1999, **104**, 3527–3542.

23 J. N. Pitts, Jr., K. A. Van Cauwenberghe, D. Grosjean, J. P. Schmid, D. R. Fitz, W. L. Belser, Jr., G. B. Knudson and P. M. Hynds, *Science*, 1978, **202**(4367), 515–519.

24 N. O. A. Kwamena and J. P. D. Abbatt, *Atmos. Environ.*, 2008, **42**, 8309–8314.

25 J. Laskin, A. Laskin, P. J. Roach, G. W. Slysz, G. A. Anderson, S. A. Nizkorodov, D. L. Bones and L. Q. Nguyen, *Anal. Chem.*, 2010, **82**, 2048–2058.

26 D. L. Bones, D. K. Henricksen, S. A. Mang, M. Gonsior, A. P. Bateman, T. B. Nguyen, W. J. Cooper and S. A. Nizkorodov, *J. Geophys. Res.*, 2010, **115**, D05203, DOI: 10.1029/2009jd012864.

27 K. M. Updyke, T. B. Nguyen, sa. S. A. and sa. Nizkorodov, *Atmos. Environ.*, 2012, **63**, 22–31.

28 C. Geron, R. Rasmussen, R. R. Arnts and A. Guenther, *Atmos. Environ.*, 2000, **34**, 1761–1781.

29 S. K. Brown, M. R. Sim, M. J. Abramson and C. N. Gray, *Indoor Air*, 1994, **4**, 123–134.

30 S. Leungsakul, M. Jaoui and R. M. Kamens, *Environ. Sci. Technol.*, 2005, **39**, 9583–9594.

31 H. Hakola, J. Arey, S. M. Aschmann and R. Atkinson, *J. Atmos. Chem.*, 1994, **18**, 75–102.

32 H. Debus, *Justus Liebigs Ann. Chem.*, 1858, **107**, 199–208.

33 G. Yu, A. R. Bayer, M. M. Galloway, K. J. Korshavn, C. G. Fry and F. N. Keutsch, *Environ. Sci. Technol.*, 2011, **45**, 6336–6342.

34 D. O. De Haan, L. N. Hawkins, J. A. Kononenko, J. J. Turley, A. L. Corrigan, M. A. Tolbert and J. L. Jimenez, *Environ. Sci. Technol.*, 2011, **45**, 984–991.

35 T. B. Nguyen, P. B. Lee, K. M. Updyke, D. L. Bones, J. Laskin, A. Laskin and S. A. Nizkorodov, *J. Geophys. Res.*, 2012, **117**, D01207, DOI: 10.1029/2011JD016944.

36 T. Berndt, O. Boge and F. Stratmann, *Atmos. Environ.*, 2003, **37**, 3933–3945.

37 M. L. Walser, Y. Desyaterik, J. Laskin, A. Laskin and S. A. Nizkorodov, *Phys. Chem. Chem. Phys.*, 2008, **10**, 1009–1022.

38 B. R. Larsen, M. Lahaniati, A. Calogirou and D. Kotzias, *Chemosphere*, 1998, **37**, 1207–1220.

39 D. Johnson and G. Marston, *Chem. Soc. Rev.*, 2008, **37**, 699–716.

40 Y. Ma, A. T. Russell and G. Marston, *Phys. Chem. Chem. Phys.*, 2008, **10**, 4294–4312.

41 C. M. Binder, D. D. Dixon, E. Almaraz, M. A. Tius and B. Singaram, *Tetrahedron Lett.*, 2008, **49**, 2764–2767.

42 K. Griesbaum, Y. Dong and K. J. McCullough, *J. Org. Chem.*, 1997, **62**, 6129–6136.

43 C. W. Dicus, D. Willenbring and M. H. Nantz, *J. Labelled Compd. Radiopharm.*, 2005, **48**, 223–229.

44 S.-T. Liu, K. V. Reddy and R.-Y. Lai, *Tetrahedron*, 2007, **63**, 1821–1825.

45 D. Yang and C. Zhang, *J. Org. Chem.*, 2001, **66**, 4814–4818.

46 H. E. Gottlieb, V. Kotlyar and A. Nudelman, *J. Org. Chem.*, 1997, **62**, 7512–7515.

47 Y. Chen and T. C. Bond, *Atmos. Chem. Phys.*, 2010, **10**, 1773–1787.

48 S. L. Clegg, P. Brimblecombe and A. S. Wexler, *J. Phys. Chem. A*, 1998, **102**, 2137–2154.

49 T. B. Nguyen, A. P. Bateman, D. L. Bones, S. A. Nizkorodov, J. Laskin and A. Laskin, *Atmos. Environ.*, 2010, **44**, 1032–1042.

50 T. B. Nguyen, J. Laskin, A. Laskin and S. A. Nizkorodov, *Environ. Sci. Technol.*, 2011, **45**, 6908–6918.

51 L. F. H. Bovey, *J. Opt. Soc. Am.*, 1951, **41**, 836–848.

52 Y. Futami, Y. Ozaki, Y. Hamada, M. J. Wojcik and Y. Ozaki, *Chem. Phys. Lett.*, 2009, **482**, 320–324.

53 I. V. Rubtsov, K. Kumar and R. M. Hochstrasser, *Chem. Phys. Lett.*, 2005, **402**, 439–443.

54 E. L. Saier, L. R. Cousins and M. R. Basila, *J. Phys. Chem.*, 1962, **66**, 232–235.

55 G. Socrates, *Infrared and Raman Characteristic Group Frequencies*, John Wiley and Sons, Ltd., Chichester, 3rd edn, 2001.

56 R. D. Bicca Alencastro, S. Badilescu, L. S. Lussier, C. Sandorfy, H. L. Thanh and D. Vocelle, *Int. J. Quantum Chem.*, 1990, **38**, 173–179.

57 K. J. Rothschild, P. Roepe, J. Lugtenburg and J. A. Pardoen, *Biochemistry*, 1984, **23**, 6103–6109.

58 H. Kato, O. Nishikawa, T. Matsui, S. Honma and H. Kokado, *J. Phys. Chem.*, 1991, **95**, 6014–6016.

59 E. Zhukovsky and D. Oprian, *Science*, 1989, **246**, 928–930.

60 P. Dziedzic, A. Bartoszewicz and A. Córdova, *Tetrahedron Lett.*, 2009, **50**, 7242–7245.

61 T. W. G. Solomons and C. B. Fryhle, *Organic Chemistry*, John Wiley & Sons, Inc., Hoboken, NJ, 8th edn, 2004.

62 J. W. Wijnen and J. B. Engberts, *Liebigs Ann./Recl.*, 1997, 1085–1088.

63 P. A. Grieco, D. T. Parker, M. Cornwell and R. Ruckle, *J. Am. Chem. Soc.*, 1987, **109**, 5859–5861.

64 J. W. Wijnen and J. B. F. N. Engberts, *J. Org. Chem.*, 1997, **62**, 2039–2044.

65 A. Laskin, J. S. Smith and J. Laskin, *Environ. Sci. Technol.*, 2009, **43**, 3764–3771.

66 B. Plietker, *Tetrahedron: Asymmetry*, 2005, **16**, 3453–3459.

67 R. Howe and F. J. McQuillin, *J. Chem. Soc.*, 1958, 1513–1518.

68 V. Amarnath, D. C. Anthony, K. Amarnath, W. M. Valentine, L. A. Wetterau and D. G. Graham, *J. Org. Chem.*, 1991, **56**, 6924–6931.

69 L. Carlton, B. Staskun and T. van Es, *Magn. Reson. Chem.*, 2006, **44**, 510–514.

70 C. Mukhopadhyay, A. Datta and R. J. Butcher, *Tetrahedron Lett.*, 2009, **50**, 4246–4250.

71 Q.-X. Guo, H. Liu, C. Guo, S.-W. Luo, Y. Gu and L.-Z. Gong, *J. Am. Chem. Soc.*, 2007, **129**, 3790–3791.

72 J. L. Hogg, D. A. Jencks and W. P. Jencks, *J. Am. Chem. Soc.*, 1977, **99**, 4772–4778.

73 P. S. Tobias and R. G. Kallen, *J. Am. Chem. Soc.*, 1975, **97**, 6530–6539.

74 C.-H. Jun, C. W. Moon and D.-Y. Lee, *Chem.–Eur. J.*, 2002, **8**, 2422–2428.

75 R. S. Becker and K. Freedman, *J. Am. Chem. Soc.*, 1985, **107**, 1477–1485.

76 H. Akita, S. P. Tanis, M. Adams, V. Balogh-Nair and K. Nakanishi, *J. Am. Chem. Soc.*, 1980, **102**, 6370–6372.

77 C. Paal, *Ber. Dtsch. Chem. Ges.*, 1884, **17**, 2756–2767.

78 L. Knorr, *Ber. Dtsch. Chem. Ges.*, 1884, **17**, 2863–2870.

79 C. P. Schwartz, J. S. Uejio, A. M. Duffin, A. H. England, D. Prendergast and R. J. Saykally, *J. Chem. Phys.*, 2009, **131**, 114509–114508.

80 J. Y. Lee, D. Y. Kim and C. Y. Kim, *Synth. Met.*, 1995, **74**, 103–106.

81 W. Wang, D. Yu and F. Tian, *Synth. Met.*, 2008, **158**, 717–721.

82 K. Kijewska, P. Głowala, K. Wiktorska, M. Pisarek, J. Stolarski, D. Kępińska, M. Gniadek and M. Mazur, *Polymer*, 2012, **53**, 5320–5329.

83 D. Birnbaum and B. E. Kohler, *J. Chem. Phys.*, 1991, **95**, 4783–4789.

Faraday Discussions

RSC Publishing

PAPER

The magnitude and sources of uncertainty in global aerosol

Kenneth S. Carslaw,[*] Lindsay A. Lee, Carly L. Reddington, Graham W. Mann and Kirsty J. Pringle

Received 20th March 2013, Accepted 10th May 2013

DOI: 10.1039/c3fd00043e

Aerosol radiative forcing over the industrial period has remained the largest forcing uncertainty through all IPCC assessments since 1996. Despite the importance of this uncertainty for our understanding of past and future climate change, very little attention is given to the problem of uncertainty reduction in its own right, mainly because most uncertainty analysis approaches are not appropriate to computationally expensive global models. Here we show how a comprehensive understanding of global aerosol model parametric uncertainty can be obtained by using emulators. The approach enables a Monte Carlo sampling of the model uncertainty space based on a manageable number of simulations. This allows full probability density functions of model outputs to be generated from which the uncertainty and its causes can be diagnosed using variance decomposition. We apply this approach to global concentrations of particles larger than 3 and 50 nm diameter (N_3 and N_{50}) to produce a ranked list of twenty-eight processes and emissions that control the uncertainty. The results show that the uncertainty in N_{50} is much more strongly affected by emissions and processes that control the availability of gas phase H_2SO_4 than by uncertainties in the nucleation rate itself, which cause generally less than 10% uncertainty in N_{50} in July. Secondary organic aerosol production is assumed to be very uncertain (5–360 Tg a^{-1} for biogenic emissions) but the effect on global N_3 uncertainty is <3% except in a few hotspots, and generally <2% for N_{50}. A complete understanding of the model uncertainty combined with global observations can be used to determine plausible and implausible parts of parameter space as well as to identify model structural weaknesses. In this direction, a preliminary comparison of the model ensemble with observations at Hyytiala, Finland, suggests that an organic-mediated boundary layer nucleation mechanism would help to optimise the behaviour of the model.

Introduction

Almost every proposal or publication related to aerosol processes and climate includes some kind of motivational statement that the work is being done to

University of Leeds, School of Earth and Environment, Leeds, UK. Tel: +44 113 3431597. E-mail: k.s.carslaw@leeds.ac.uk

reduce the uncertainty in aerosol effects on climate. Certainly, incomplete knowledge introduces uncertainty, so making more measurements of complex processes should ultimately help to reduce it. However, as a research community we pay surprisingly little attention to uncertainty as a scientific problem in its own right; mostly it is implied that more knowledge and complexity of processes in our models will, by themselves, lead to reduced uncertainty. In many cases, while the aim of a piece of research may ostensibly be to reduce uncertainty, the outcome is often just a different (and usually more complex) description of some process, or a different model prediction. These multiple interpretations, model simulations and predictions are then periodically assessed, such as by the IPCC, to come up with some measure of uncertainty. This approach to atmospheric aerosol science has led to huge gains in our understanding of processes, but has resulted in very little reduction in the uncertainty in model predictions of aerosol effects on climate. Certainly we would have difficulty tracing back any change in uncertainty to particular developments in processes.

Uncertainty is not at the heart of our current approach to aerosol research. Our knowledge of "key processes" motivates research, but the research process is not carried through to demonstrating success in terms of reduced or better characterised uncertainty. Because we have an incomplete (or biased) knowledge of how processes contribute to the uncertainty, we may also inadvertently devote effort to problems that have little bearing on the uncertainty. With the current explosion of aerosol measurements and process studies, we face a significant challenge to convert our improved understanding of processes into useful and robust models; that is, models with predictive skill, but which are not so expensive to run that we can only run specific case studies. At present, the representation of processes in global models lags significantly behind what we know from laboratory, field and detailed model studies. But what are the truly key processes that will make global models "better"?

The ultimate aim of process-based research feeding into global models should be to enable models to make robust decisions. A robust decision requires a full characterisation of model uncertainty so that the model can provide useful information about the real world in spite of uncertainties in the parameters or underlying processes Robustness is related to the *fidelity* of a model: the degree to which a simulation reproduces the behaviour of the real world – the model realism. A model would not be robust if the calculated aerosol radiative forcing could be either positive or negative depending on the secondary organic aerosol model used. But just because a model compares well with some observations does not mean it is robust. It is possible to be right for the wrong reasons, especially if only a small part of the plausible parameter space has been tested. And several model parameter settings within plausible ranges may match observations within acceptable limits but give very different results in future projections (called *equifinality* of a model[1]). If we haven't explored the model behaviour across the whole of its uncertainty space, then we cannot rule out these possibilities. The consequence is that our future projections will not be robust and we are likely to underestimate the true uncertainty. To know if a model is robust we need to be able to characterise the uncertainties and understand how they affect our predictions.

Another role of process-based research should be to help define the development path of models. Out of the many processes we are currently working on, which ones should be included in global models five or ten years from now? How do we prioritise their inclusion in the models? More complex aerosol processes need to be justified against equally important developments in model resolution, cloud physics *etc*, so we will always have to be selective. The complexity of various processes in different global models varies quite substantially, but this may be related more to the research interests of the groups involved than to the effect of the complexity on model robustness. Again, our criterion for inclusion of a more complex process should be that it improves model robustness, which means understanding the uncertainties and how they change when a new process is added to the model.

How can we characterise and understand the effect of uncertainties in models, and therefore show they are robust? The *modus operandi* in aerosol modelling is to do one-at-a-time parameter perturbations to obtain some measure of model sensitivity. But this approach, while computationally cheap, has many limitations: it only probes a tiny part of the parameter space;[2] it ignores interactions between parameters; it cannot generate any useful information about prediction uncertainty (*i.e.*, a proper plus/minus error bar); it doesn't provide any information that would help to decide if the model is right for the wrong reasons; and it doesn't tell us about equifinal models, which may lead us to be overconfident in our model predictions.

In this paper we describe an approach that avoids these problems by using model emulators to generate model output across the entire space of the uncertain parameters, even for a fairly computationally expensive global model. In essence the approach is straightforward: model outputs (*e.g.*, for particle concentration) are generated from a number of simulations in different parts of the model uncertainty space (*i.e.*, different parameter settings) and are used to train a Bayesian emulator. The emulator is used as a way of filling the gaps in parameter space so that the outputs can be generated for any parameter setting we choose (within the limits of the training runs). By using the emulator in place of the full model (after having tested the emulator against the model) we can then do a full Monte Carlo sampling across the whole parameter space, which opens up possibilities like variance analysis and variance decomposition to understand how different parameters contribute to the overall uncertainty.

Here we present an uncertainty analysis of global particle concentrations using a global 3-D aerosol microphysics model. There are three main aims of this paper. First, we aim to get to the point where we can generate what would be recognised as a "proper error bar" for various aerosol quantities at any global location. Second, we aim to pick apart the error bar and determine which uncertain processes contribute to it in different locations and times of year. Third, we take the first steps towards model calibration in which the model behaviour across the full parameter space is confronted with observations in order to identify model structural weaknesses and to determine regions of plausible and implausible parameter space. With such information we can then begin to learn about the key processes – defined to be those that have the greatest bearing on the uncertainty in our predictions.

Methodology

The GLOMAP global aerosol model

The GLObal Model of Aerosol Processes (GLOMAP-mode)[3] is a 3-D global aerosol microphysics model that simulates the size distribution and chemical composition of aerosol particles on a global 3-D grid with horizontal resolution of $2.8° \times 2.8°$ and 31 vertical levels between the surface and 10 hPa. The aerosol and chemical species are transported by meteorological fields read in from the European Centre for Medium-Range Weather Forecasts ERA-40 reanalyses.

The global model is fairly sophisticated compared to the majority of models that have been used in IPCC assessments; it simulates the full aerosol number size distribution, internally and externally mixed particles and all the important microphysical processes, while the majority of climate assessment models lack such microphysics. But the model is also used in the HadGEM climate model, so it includes some necessary simplifications so the model can be run for hundreds of years. The parameter perturbations that we have performed here have been designed to explore whether the magnitude of the parameters associated with these simplified processes affect the model uncertainty, which would suggest a benefit of increasing the complexity. However, we need to keep in mind that oversimplification of some processes may lead to misleading conclusions about the importance of that process. We discuss this possibility in the context of a few specific parameters below.

The aerosol size distribution is defined by 7 log-normal modes: one nucleation mode and soluble and insoluble modes covering the Aitken, accumulation and coarse size ranges. The aerosol chemical components are sulphate, sea salt, black carbon (BC), particulate organic matter (POM) and dust. The POM enters the aerosol particles from primary emissions and from gas phase oxidation of precursors. Secondary organic aerosol (SOA) is produced from the first stage oxidation products of biogenic monoterpenes and anthropogenic volatile organic carbon (VOC) compounds, and is assumed to have zero vapour pressure. It is combined with the POM component after condensation on the particles.

The microphysical model includes the main processes that determine the particle size distribution: new particle formation, coagulation, gas-to-particle transfer, cloud processing (aqueous oxidation of SO_2 to sulphate), and dry and wet deposition. Nucleation occurs by two separate processes. In the boundary layer we assume a rate $j = A[H_2SO_4]$ (ref. 4) and in the whole atmosphere (but mostly important in the higher free troposphere) binary homogeneous nucleation of H_2SO_4 and H_2O can occur.[5] Nuclei in the boundary layer can grow by condensation of H_2SO_4 from 2 nm and SOA from 3 nm diameter.

Wet deposition of particles occurs by two processes. In-cloud nucleation scavenging in which activated particles form cloud drops and are removed in precipitation and below-cloud impaction scavenging by falling raindrops. ECMWF meteorological fields are used to diagnose large-scale frontal precipitation and sub-grid convective precipitation is assumed to occur in 30% of the affected grid box area. Low-level stratiform clouds are read in separately from International Satellite Cloud Climatology Project (ISCCP) D2 data. In these clouds we assume that aerosol particles are activated and subsequently undergo 'cloud processing' in which sulphate mass is added to activated aerosol due to aqueous phase oxidation of sulphur dioxide.

The model has previously been evaluated against observations[3] and compared against simulations of another version of the model that treats the aerosol size distribution using a sectional approach.[6] The GLOMAP models have been widely used and evaluated against global measurements of particle number concentrations,[7,8] CCN,[9,10] chemical components[11,12] and cloud drops.[13]

Table 1 Parameters and their ranges used in the model simulations

Param	Parameter name	Description	Uncertainty range	Effect[a]
P1	BL_NUC	Boundary layer nucleation rate (j) based on $j = A$ [H$_2$SO$_4$]. Rate coeff A was perturbed.	3.2×10^{-7}–2×10^{-4} s^{-1}	Absolute
P2	FT_NUC	Free troposphere nucleation rate based on binary homogeneous nucleation mechanism.	0.01–10	Scaled
P3	AGEING	Ageing 'rate' from insoluble to soluble. Monolayers of soluble material required to be water-soluble and potential CCN.	0.3–5 monolayer	Absolute
P4	ACT_DIAM	Cloud drop activation dry diameter.	50–100 nm	Absolute
P5	SO2O3_CLEAN	pH of cloud drops in clean regimes. Controls SO$_2$ + O$_3$ → SO$_4$, which affects important sulphate production mechanism.	pH 4–6.5	Absolute
P6	SO2O3_POLL	pH of cloud drops (SO$_2$ + O$_3$) in polluted conditions. As P5.	pH 3.5–5	Absolute
P7	NUC_SCAV_DIAM	Nucleation scavenging diameter offset dry diameter. The size above which activated aerosols are scavenged by in-cloud nucleation scavenging.	0–50 nm	Absolute
P8	NUC_SCAV_ICE	Nucleation scavenging fraction (accumulation mode) in mixed and ice clouds ($T < -15$C).	0–1	Scaled
P9	DRYDEP_AER AIT	Dry deposition velocity of Aitken mode aerosol.	0.5–2	Scaled
P10	DRYDEP_AER ACC	Dry deposition velocity of accumulation mode aerosol.	0.1–10	Scaled
P11	ACC_WIDTH	Modal width (accumulation soluble/insoluble).	1.2–1.8	Absolute
P12	AIT_WIDTH	Modal width (Aitken soluble/insoluble).	1.2–1.8	Absolute
P13	NUCAIT_WIDTH	Mode separation diameter (nucleation/Aitken). Defines the size limits of these modes.	9–18 nm	Absolute
P14	AITACC_WIDTH	Mode separation diameter (Aitken/accumulation). As P14.	(0.9–2.0) × ACT_DIAM	Scaled[a]

Table 1 (*Contd.*)

Param	Parameter name	Description	Uncertainty range	Effect[a]
P15	FF_EMS	BCOC mass emission rate (fossil fuel). Global uniform scaling.	0.5–2	Scaled
P16	BB_EMS	BCOC mass emission rate (biomass burning and wildfires). Global uniform scaling.	0.25–4	Scaled
P17	BF_EMS	BCOC mass emission rate (biofuel). Global uniform scaling.	0.25–4	Scaled
P18	FF_DIAM	BCOC emitted mode diameter (fossil fuel).	30–80 nm	Absolute
P19	BB_DIAM	BCOC emitted mode diameter (biomass burning and wildfires).	50–200 nm	Absolute
P20	BF_DIAM	BCOC emitted mode diameter (biofuel).	50–200 nm	Absolute
P21	PRIM_SO4_FRAC	Mass fraction of SO_2 converted to new sulphate particles in sub-grid plumes.	0–1%	Scaled
P22	PRIM_SO4_DIAM	Mode diameter of new sub-grid sulphate particles.	20–100 nm	Absolute
P23	SS_ACC	Sea spray mass flux (coarse/accumulation).	0.2–5	Scaled
P24	ANTH_SO2	SO_2 emission flux (anthropogenic).	0.6–1.5	Scaled
P25	VOLC_SO2	SO_2 emission flux (volcanic).	0.5–2	Scaled
P26	DMS_FLUX	DMS emission flux.	0.5–2	Scaled
P27	BIO_SOA	Biogenic monoterpene production of SOA.	5–360 Tg a^{-1} [b]	Absolute
P28	ANTH_SOA	Anthropogenic VOC production of SOA.	2–112 Tg a^{-1} [b]	Absolute

[a] For the scaled parameters the magnitude of the parameter was multiplied by the factor in column 4. For absolute adjustments, the parameter was set at an absolute value within the range given in column 4. [b] The biogenic SOA production parameter (P27) conflates the uncertainty in the emissions of the precursor gases (biogenic VOCs) and the uncertainty in the yield of secondary organic aerosol (SOA) material following oxidation reactions. SOA is produced in the model through oxidation of transported alpha-pinene by OH, NO_3 and O_3. There are also uncertainties in the volatility of different compounds that we do not account for here. The range of emissions used here has been shown to span the range of global *in situ* measurements of OA.[11] Anthropogenic SOA production (P28) is treated in a similar way to biogenic SOA. We used the same approach as in Spracklen *et al.*[11] by scaling gridded CO emissions over a range known to span the range of observed global OA.

Parameter perturbations

Twenty-eight parameters describing aerosol and precursor gas emissions, aerosol processes, and definitions of the aerosol log-normal modes were perturbed across a range defined by expert elicitation (Table 1). Full details about how each parameter is represented in the model are provided elsewhere.[14] Although 28 parameters is more than has been included in any previous uncertainty analysis, the choice does not cover every model process that might affect aerosol, so our error estimates are likely to be an underestimate. For example, we do not perturb the host model physics (convection scheme, rainfall distribution, boundary layer mixing, *etc.*). It is interesting to speculate whether a larger fraction of aerosol uncertainty could be attributable to such non-aerosol processes than to the details of the aerosol microphysics and chemistry that most interest us. It is also important to note that the bounds on the parameters ultimately determine their ranking. We believe these bounds, based on expert elicitation, are a reasonable first estimate, although there is clearly scope for future improvements as knowledge improves.

Model emulation

Full details about the steps involved in model emulation and statistical analysis have been presented in earlier papers.[14–16] Here we outline the main steps.

Gaussian process emulation[15,17] was used to estimate model predictions at untried points throughout the space of the uncertain model parameters. An emulator can be built for any model output (particle concentration, PM, CCN, *etc*) over any spatial domain (single grid boxes or averaged over a larger regions) and for any time period (a single output step or averaged over a longer period). Here we build emulators for each grid box on one model level in the boundary layer based on monthly mean aerosol.

To train the emulators, an ensemble of model runs was performed for 168 parameter combinations sampled from a maximin Latin hypercube, which generates a distribution of points (parameter settings) through the 28-dimensional parameter space. The number of runs required to build a good emulator is a matter of trial and error; here we use 6 runs per parameter, which was sufficient to give an emulator with an error lower than the parametric uncertainties we are trying to compute.

The model was run for each of the parameter settings through the year 2008 following a spin up period of 6 months. The emulators were validated using 84 additional model runs to ensure that the emulator uncertainty around its mean is low compared to the parametric uncertainty.

Model variance and sensitivity analysis

The uncertainty, or variance, of each model output (here N_3 and N_{50}) was calculated by sampling from the emulator using the extended-FAST method.[18] We sampled 5000 points from the emulator to obtain a probability distribution of each output in each grid square of the model. Although 168 model runs might seem sufficient for performing such a variance analysis, it is actually a very sparse sampling of the model in 28 dimensions, so the emulator is needed to enable a much denser Monte Carlo sampling.

Variance-based sensitivity analysis[19] is used to decompose the uncertainty in the model predictions into the uncertainty attributable to the model parameters. Two measures of sensitivity were calculated: the *main effect index* (or the one-at-a-time effect) measures by how much the variance will be reduced if the parameter can be learnt precisely, and the *total effect index* measures both the individual effect and the interaction effect of each parameter with all others. The two sensitivity measures are compared to assess the sensitivity of the model output to interactions. Note that a non-linear response of the model output to a parameter across the specified range is accounted for in the main effect variance.

Results

We show results for the year 2008 meteorology, with emissions from the AERO-COM1 inventory[20] for the year 2000.

Global patterns and causes of aerosol uncertainty

Fig. 1 shows the global distribution of N_3 and N_{50} concentrations for January and July, together with the uncertainty (one standard deviation, σ) and coefficient of variation ($\sigma_{N_3}/\overline{N_3}$ and $\sigma_{N_{50}}/\overline{N_{50}}$). The $1 - \sigma$ uncertainty varies between about 20% and 100% and is fairly similar for N_3 and N_{50}. The N_3 uncertainty map has more hotspots over remote regions and is generally higher over high latitude southern hemisphere regions. Below we show how the model predictions and uncertainties compare with measurements, but in general the uncertainty range spans the typical model-observation biases that we highlighted in a previous analysis of 36 global sites[8].

To understand the causes of the uncertainty we decompose the variance into the contributions from each parameter in each grid box. This is shown in Fig. 2 for the ten most uncertain parameters, again in terms of the absolute uncertainty in N_3 and N_{50} (σ caused by each uncertainty individually) and the relative uncertainty ($\sigma_{N_3,i}/\overline{N_3}$ and $\sigma_{N_{50},i}/\overline{N_{50}}$) where i refers to each parameter. We do not attempt here to explain every feature of these plots, but will draw attention to a few of the more interesting ones.

The high latitude southern hemisphere (winter) uncertainty can be attributed to a combination of DMS emissions (DMS_FLUX), sea spray flux (SS_ACC), and the oxidation of SO_2 in cloud water (parameter SO2_O3_CLEAN), but principally

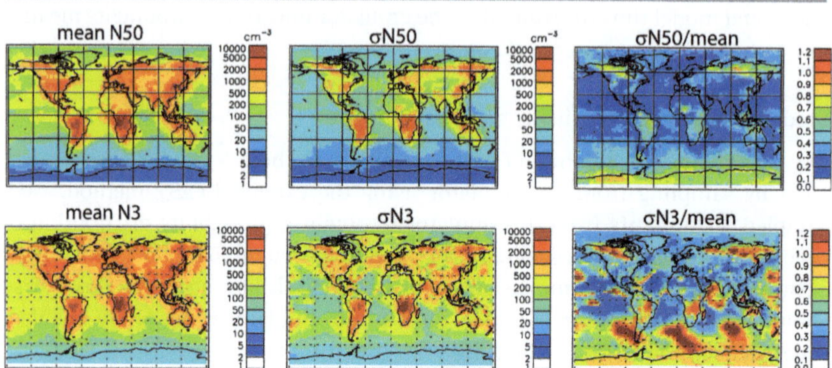

Fig. 1 Monthly mean concentrations and uncertainties for particles larger than 3 nm diameter (N_3) and 50 nm (N_{50}) for July 2008. Left column, particle concentrations; middle column, $1 - \sigma$ uncertainty in concentration; right column, coefficient of variation (σ/mean). Concentrations are given on the 915 hPa model level (approximately 850 m asl) to be relevant to cloud base. The fields were generated from an emulator built for each grid box.

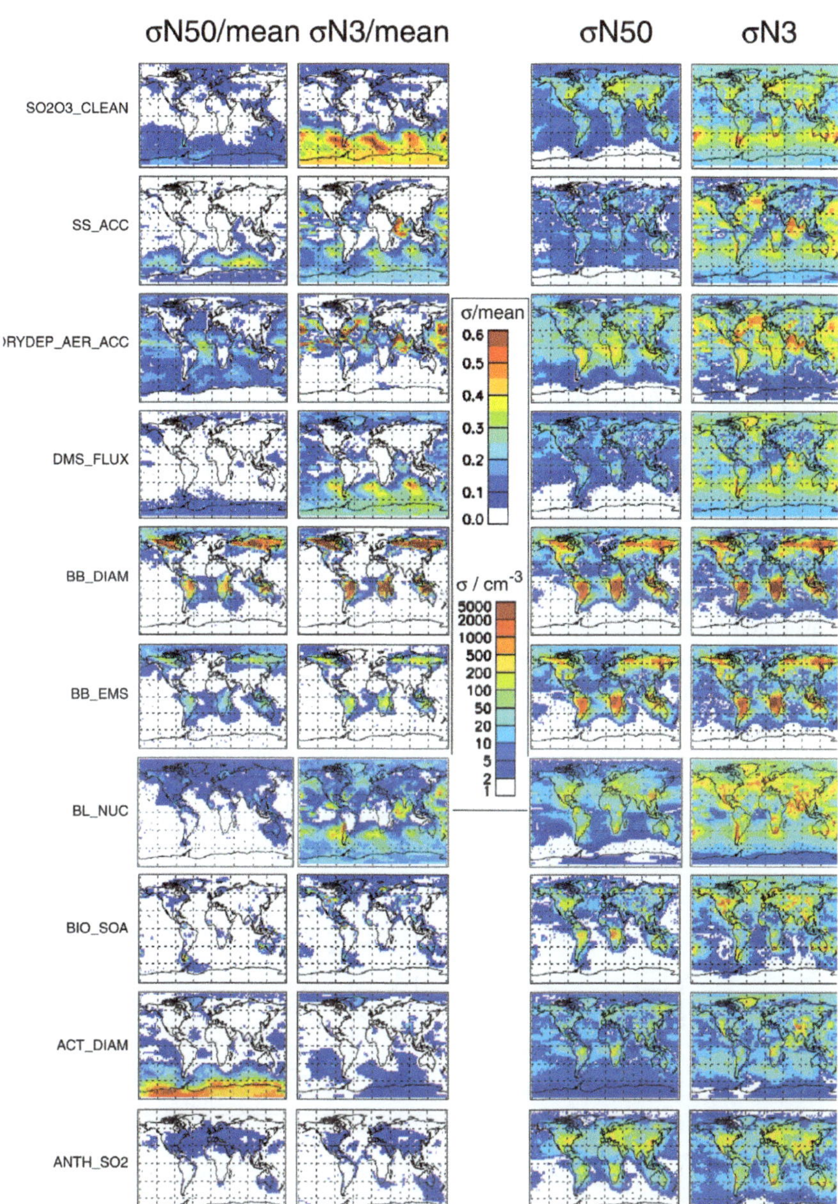

Fig. 2 The contribution of the ten parameters to the uncertainty in N_3 and N_{50}. The magnitude of uncertainty is shown as the standard deviation (σ, right columns) and the relative uncertainty, or coefficient of variation (σ/mean, left columns). The uncertainties refer to monthly mean concentrations on the 915 hPa model level (approximately 850 m asl) to be relevant to cloud base.

the latter parameter, which causes a relative uncertainty of 30% in N_{50} and 90% in N_3. Although sulphate production in clouds produces aerosol mass, it also suppresses the formation of new particles by depleting gas phase SO_2 and H_2SO_4 and by enhancing the condensation sink. Since nucleation accounts for between 20 and 80% of N_{50} globally,[21] this coupling is an important process. In all regions where this parameter is important, a faster SO_2 aqueous oxidation rate causes a decrease in N_3 and N_{50} (shown in Fig. 3 for one particular site). Thus, parameterisations of cloud drop number *versus* sulphate mass in climate models[22] may not capture the correct dependency if particles larger than 50 nm dry diameter are the relevant CCN. This suppression of N_{50} by aqueous sulphate production was highlighted by Woodhouse *et al.*[23] as the main reason why the sensitivity of CCN to DMS emissions (the CLAW hypothesis) could be weak, since the relation between DMS emission and CCN production is not direct. Fig. 1 also shows that sea spray affects N_3 and N_{50} uncertainty with different spatial patterns. For N_3, increased sea spray causes a decrease in N_3 because of suppression of nucleating and condensing H_2SO_4 (so the spatial pattern matches that of the DMS effect on N_3), but for N_{50} the effect is opposite because sea spray particles are a direct source of N_{50}-sized particles. Thus, remote marine CCN-sized particles are regulated by coupled processes, which may make predictions of future CCN and cloud drop concentrations in remote regions complex. The emulator could now be used to explore all such plausible interactions without rerunning the model.

The impact of very large biogenic SOA uncertainty (5–360 Tg a^{-1} OA) on global N_3 and N_{50} is quite small. There are a few hotspots over Scandinavia, N America, South America and Australia where it is the dominant uncertainty, but globally averaged it is ranked 8th and $\sigma_{N_3}/\overline{N_3}$ is less than 4% and $\sigma_{N_{50}}/\overline{N_{50}}$ is less than about 2%. The effects of biogenic SOA are likely to be open to discussion because of the relatively simple way that we treat it *versus* what is known from experiments: we do not account for the effects that organic compounds have on nucleation rate[24] and we assume that the first stage oxidation products of monoterpenes have zero volatility, so they condense on all particles at the diffusive flux rate. In GLOMAP, an increase in biogenic SOA causes a decrease in N_{50}

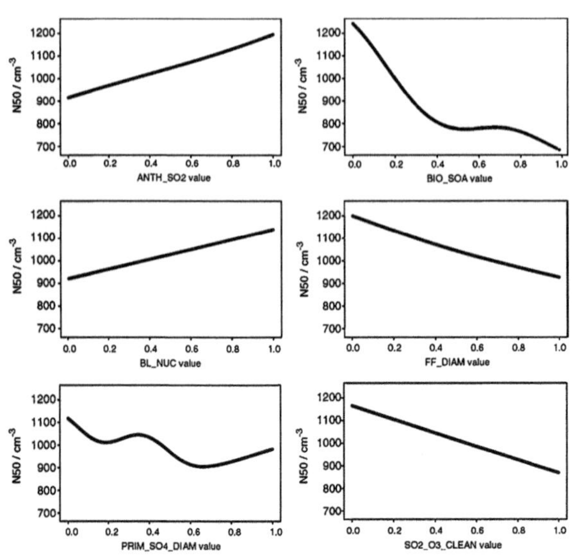

Fig. 3 Response of N_{50} to changes in several parameters across the full range of the parameter. Results are shown for one site, Hyytiala in Finland. The horizontal axis is scaled between the maximum and minimum values in Table 1.

(see Fig. 3 for Hyytiala). This is a result of increasing condensation sink, which suppresses nucleation. Whether biogenic SOA enhances or suppresses N_{50} depends on the balance of nucleation and sink rates and on the availability of small primary particles to grow to 50 nm diameter, all of which are quite uncertain. Globally, we find that SOA generally suppresses N_3 but can enhance or suppress N_{50} depending on location and season (note that the low global mean value of $\sigma_{N_{50}}/\overline{N_{50}}$ is not the result of cancellation of regional effects because the absolute magnitude of σ is used in the average).

Fig. 4 shows how the size distribution responds to SOA in an environment where nucleation is important and where it is completely unimportant. In the latter case (over the rainy Amazon in January), additional SOA reduces CCN because it continually grows particles up to size where they can be scavenged by rain. Globally, however, the net effect of biogenic SOA uncertainty on N_{50} is small. In environments where nucleation is a dominant CCN production process (mainly over land areas) particle growth from 3 to 50 nm diameter is quite rapid irrespective of the level of SOA production and there is some compensation between the effect of SOA on the condensation sink and growth rate. In many other environments dominated by primary emissions there is insufficient supply of sub-50 nm diameter particles to grow to 50 nm, so growth by uptake of SOA is not a critical process. There may be higher sensitivity to SOA for larger particles like N_{100}.

How might these sensitivities to SOA change if SOA formation were described differently in the model, and could our results be misleading? Our global SOA production range (5–360 Tg a^{-1} OA) is very large and probably covers the full range of any conceivable SOA model, at least in terms of mass (the continental PM standard deviation often exceeds 5 μg m^{-3} and is between 0.1 and 0.5 μg m^{-3} over most northern hemisphere marine regions). Another consideration is the neglect of SOA volatility. If we assumed that some SOA is semi-volatile[25] and therefore condenses only on larger particles then this would likely increase the nucleation sink effect of the SOA (because it shifts the size distribution to larger particles), but it is likely that the resulting sink rates are already contained within our very large SOA uncertainty range.

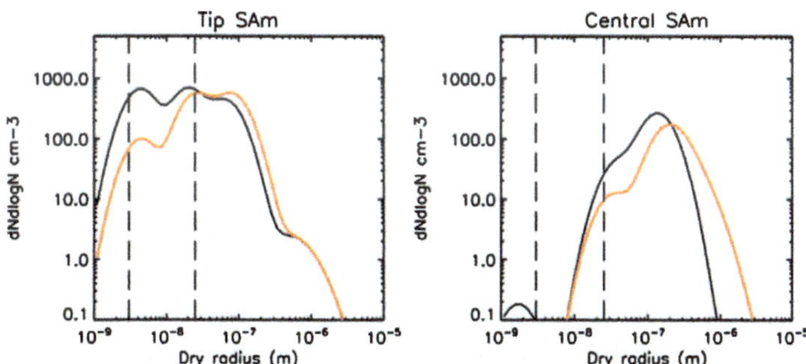

Fig. 4 The response of the particle size distribution in two locations to a one-at-a-time perturbation to the SOA production rate between the median and the upper limit in Table 1. Black line, median SOA; red line, high SOA. The left plot (southern tip of Chile) shows a case where nucleation is an important process and the right plot (Amazon) shows a case where nucleation is not important.

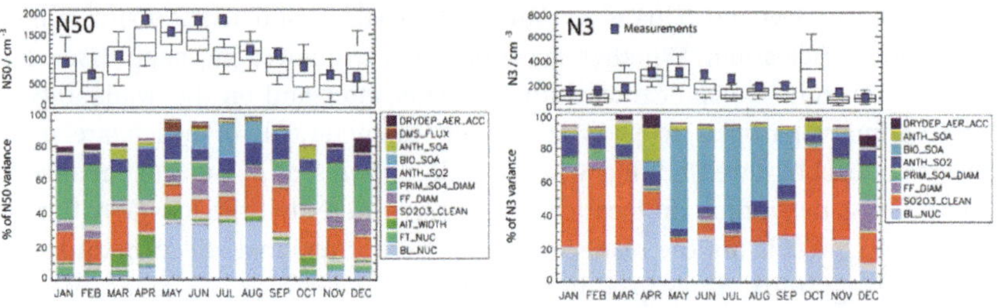

Fig. 5 Particle concentrations and causes of uncertainty at Hyytiala, Finland for the year 2008. The top row shows the modelled and observed monthly mean N_3 and N_{50} concentrations. The blue dot shows the observations and the model range is shown as the interquartile range (box) and 9th/91st confidence interval (whisker). The bottom row is a stacked bar chart showing the contribution of each process to the total variance. Contributions to variance less than 5% are shown grey and the white space at the top indicates the fraction of variance due to parameter interactions.

The importance of boundary layer nucleation for N_{50} uncertainty is small in this model. In July, the uncertainty in N_{50} is less than 10%. Note that this contribution to uncertainty is not the same as the net contribution of nucleation to CCN in previous studies where the process has often been switched on or off,[21,26] which is not realistic. In terms of model development, this result is interesting because it suggests that we do not need to know the nucleation rate with any precision, a point also made by Tunved *et al.* for Scandinavia.[27] Thus, nucleation is a key process in terms of global CCN formation[21,28] (and therefore cannot be removed from the model), but it is not a key process in terms of uncertainty reduction. A large fraction of the N_{50} uncertainty can be attributed to the indirect effects of other processes on nucleation (acting through condensation/coagulation sinks and growth rates), such as biogenic SOA, sea spray and in-cloud sulphate production discussed above. It is important to establish how organic compounds enhance nucleation rates.[24] If it is a strong effect, then uncertainties in N_{50} formation will combine with those associated with coagulation losses and growth rates, which could increase the importance of SOA for global N_{50} uncertainty. By comparing the ensemble of model runs with observations (see below) we suggest there may be evidence for the need for organic-enhanced nucleation rates.

Uncertainties at individual locations

Fig. 5 shows the seasonal variation in N_3 and N_{50} at Hyytiala together with the observations, model uncertainty (the box and whisker), and the fraction of variance explained by the different parameters (the stacked bar chart). An alternative way of visualising how each parameter contributes to the total uncertainty (for July) is shown in Fig. 6 in terms of the modelled interquartile range (the box) and 9/91% confidence intervals (the whisker). The ranges were calculated from a sample of 5000 points across the parameter space using the emulator. Thus, to obtain the same statistical information would require a prohibitively large number of model simulations compared to the 168 that were actually used to train the emulator. Fig. 5 and 6 provide direct information about how the uncertainty in N_3 and N_{50} would be reduced at Hyytiala if a given parameter could be determined precisely. For example, the fraction of variance in N_{50} due to

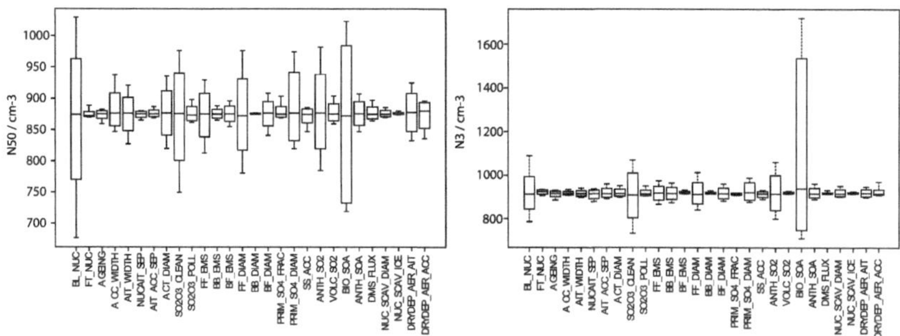

Fig. 6   The uncertainty contributions to N_3 and N_{50} for all parameters at Hyytiala for July 2008. The box shows the interquartile range and the whisker is the 9th/91st confidence interval.

biogenic SOA in July is about 30%, so if this parameter were known precisely the variance would fall by 30%.

These statistical analyses show which parameters contribute to the uncertainty in N_3 and N_{50}, but how do the associated processes cause the uncertainty? Fig. 3 showed the dependence of N_{50} on six of the most important parameters: Anthropogenic SO_2, biogenic SOA, boundary layer nucleation, the diameter of fossil fuel particles, the diameter of sub-grid "primary" sulphate particles, and the oxidation of SO_2 in cloud water. There is a clear link between N_{50} and SO_2 emissions and the boundary layer nucleation rate. The response to boundary layer nucleation is approximately in line with previous studies at Hyytiala – *i.e.*, about a 10% change in N_{50} across the full range of rates we have used.[21,28] Increases in anthropogenic SO_2 emissions also lead to higher N_{50} because of more nucleation and growth. Increasing the diameter of fossil fuel particles reduces N_{50} because, for a fixed emission mass flux, larger particles must be fewer in number. The oxidation of SO_2 to sulphate in cloud water (parameter SO2_O3_CLEAN) suppresses N_{50}. As discussed above, although aqueous phase SO_2 oxidation produces aerosol sulphate mass, it also leads to a feedback on nucleation rates by increasing the condensation sink without producing any new N_{50}. This effect will manifest itself as a reduction in the number of nucleation event days at the site. Finally, the effect of SOA is to decrease N_3 and N_{50}. This effect makes about a 75% contribution to N_3 variance at Hyytiala (Fig. 5). The effect of SOA on nucleation mode particles has also been simulated in other model studies.[27]

Fig. 5 shows that the modelled N_3 and N_{50} are generally in good agreement with the observations at Hyytiala. The most obvious discrepancy is in the summer, when both modelled N_3 and N_{50} are lower than observed by almost a factor of two. We now explore what can be learned by comparing individual model simulations with the observations, which is a first step towards calibration of the model.

Calibration using observations and the multi-parameter model space

The statistical information generated by sampling model behaviour across the full uncertainty space of the parameters opens up the possibility of doing model calibration. Models are almost always 'tuned' to obtain the best agreement with observations, but this is normally done just on one or two parameters at a time. Data assimilation is usually applied to the state variables of a model rather than the processes. In the Global Aerosol Synthesis and Science Project (gassp.org.uk) we aim to use the model statistical information together with extensive *in situ* aerosol measurements to address the following questions:

1 From within the very large sampled parameter space, can we identify implausible parameter combinations and even a best parameter combination?

2 Is the model right for the right reasons? For example, is the space of plausible parameters the same when the model is optimised against particle concentration measurements, CCN or component mass concentrations?

3 Can we identify structural model weaknesses in cases where none of the model simulations overlap with the measurements?

4 In which environments or locations should measurement campaigns be directed to have the maximum effect on model uncertainty in, say, radiative forcing?

Fig. 7 shows an example of how the plausible parameter space can be identified. We show the annual cycle of modelled and observed N_3 and N_{50} at Hyytiala for the 168 model runs *versus* observations. The bottom two panels of Fig. 7 show some of the model runs which agree best with the N_{50} observations (in terms of lowest root mean squared error (RMSE) and highest seasonal correlation r^2). Clearly, the best agreement with N_{50} is not always consistent with the best agreement with N_3.

Run 10 agrees best with the N_{50} measurements ($r^2 = 0.82$, RMSE = 55 cm^{-3}). However, this model run is biased high compared to the N_3 measurements ($r^2 = 0.45$, RMSE = 950 cm^{-3} and up to 3000 cm^{-3} in spring). The reason is that this run assumes large diameters of emitted sub-grid sulphate particles and fossil fuel combustion particles (resulting in low N_{50}) but has a very high boundary layer nucleation rate (1.5×10^{-4} s^{-1}), which compensates, but also causes N_3 to be too high. Sulphate production by O_3 oxidation in clouds is also very low in this run

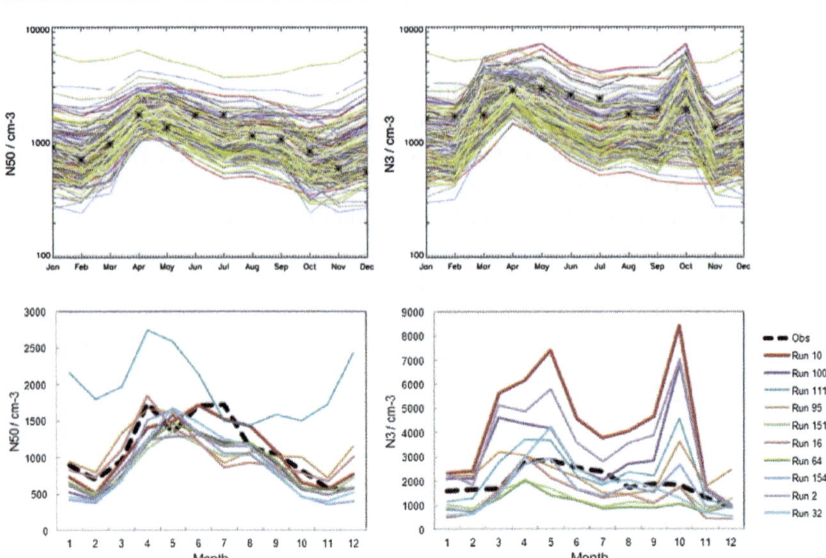

Fig. 7 Comparison of the ensemble of 168 model runs with measurements at Hyytiala, Finland for the year 2008. Top panels, all model runs; bottom panels, a selection of model runs discussed in the main text.

(pH = 4.3), which results in a high fraction of SO_2 being oxidised in the gas phase to H_2SO_4 vapour (again, enhancing nucleation rates). This is a good example of how a model can be right for the wrong reasons.

Run 32 is the best run when compared to both N_3 and N_{50}. This again has high nucleation rate (1.0×10^{-4} s^{-1}) and moderate sizes of primary particles (as in run 10), but the rate of in-cloud sulphate production is high (a cloud pH of 6.4), so gas phase SO_2, H_2SO_4 vapour, and hence nucleation rates, are suppressed.

Another example is run 16. This run captures the winter and spring N_{50} reasonably well but the summer N_{50} is lower than observed, consistent with low N_3, which suggests very little nucleation despite a high assumed boundary layer nucleation rate of 2.0×10^{-4} s^{-1}. This behaviour can be explained by the very high biogenic SOA production (344 Tg a^{-1}), which occurs mostly in the summer and suppresses nucleation as described above. In general, runs with high biogenic SOA (> about 50 Tg a^{-1} OA) tend to produce too low N_3 and N_{50} in summer.

There are also several runs that have similar skill in terms of simulating N_3 and N_{50} but have very different parameter settings (some with high primary emissions which are compensated by low nucleation rates). These are *equifinal* models. It is likely that simulations of future aerosol using equifinal models will result in different predictions of N_3 and N_{50} – *i.e.*, they are likely diverge because each model depends differently on primary emissions and nucleation.

We can take this comparison with observations only so far because 168 runs in 28 dimensions is very sparse and it is not combinations of model runs we need to compare with measurements, but regions of parameter space, and these can be identified only by using the emulator. But a pattern is emerging just by comparing individual runs against two metrics at one site. The majority of the 168 model runs have lower N_3 and N_{50} in summer than measured, despite exploring the full parameter space of SO_2, biogenic SOA, primary emissions, *etc.* This may already suggest that the model is structurally incomplete. A possible structural deficiency is the neglect of SOA in the nucleation mechanism, as suggested by laboratory measurements.[24] This would preferentially increase particle concentrations in the summer (during peak biogenic emissions) and improve the simulation of the seasonal variation. It would allow the high summer N_3 and N_{50} to be captured without excessive nucleation rates in spring and autumn, which are clearly inconsistent with the N_3 measurements.

Comparison with measurements at other sites, such as Barrow in Fig. 8, suggest that there are model structural weaknesses that may be unrelated to any of the 28 aerosol process and emission parameters we have perturbed, or that some parameters have not been perturbed enough. The high particle concentrations at Barrow are possibly related to incorrect representation of low drizzling clouds, which account for much of the scavenging in the Arctic summer.[29]

Clearly, multiple independent measurements at multiple sites are needed to attempt a model calibration, in order to avoid error compensation. In the above example for Hyytiala, an additional measurement of the sulphate PM (or, better, the size distribution of sulphate from an AMS instrument) might help to constrain the in-cloud sulphate production rate. Clearly, measurements of SOA (ideally separated into primary and secondary) would constrain the SOA production. It is important to note that different locations in the atmosphere provide different constraints on the parameter space. Central European ground sites, for example, will help to constrain fossil fuel particle emissions, which may not be possible at Hyytiala. Free

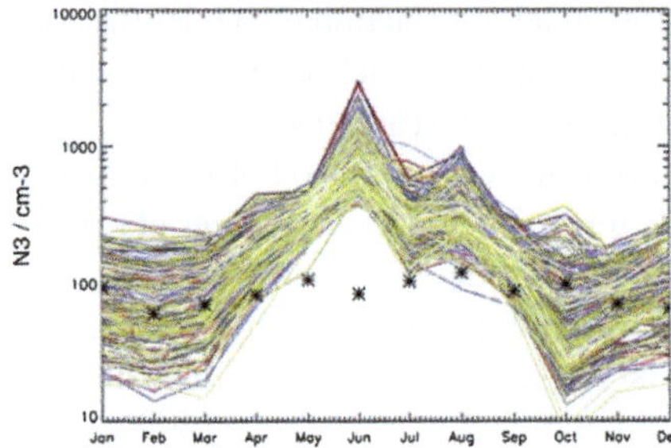

Fig. 8 Comparison of the ensemble of 168 model runs with measurements at Barrow, Alaska for the year 2008.

tropospheric sites (from ground stations and aircraft) will help to constrain free tropospheric nucleation rates. Finally, we note that the uncertainty in each output (N_3, N_{50} *etc*) over most parts of the world is affected only by around five parameters (Fig. 5 and ref. 14), so we are not searching the entire 28-dimensional space for the plausible models at a given time and location.

Conclusions

In this paper we have argued that model uncertainty is a significantly under-developed area of aerosol research. This situation is at odds with the importance of uncertainty in aerosol radiative forcing, which limits our ability to attribute past climate change and the predict climate change in the near future. The standard approach in developing, improving and evaluating complex global aerosol models is to perform one-at-a-time parameter perturbations experiments. However, this approach samples only a very small part of the model's uncertainty space and provides limited information to quantify and understand model uncertainty.

We have used Bayesian emulators to generate model outputs across the complete uncertainty space of 28 parameters related to aerosol emissions and processes in a global 3-D aerosol microphysics model. Because the emulator is much faster to run than the original model, it enables a Monte Carlo sampling of the uncertainty space, which then allows the magnitude and causes of model uncertainty to be quantified using variance analysis. This approach leads to several results of value to the development of global models:

First, we have generated an "error bar" due to the parametric uncertainty in the model. This is quite distinct from a range of model results caused by perturbations to single parameters because the error bar takes into account all parameters simultaneously, including their interactions. This additional information now allows us to compare the model with observations and to express the model fidelity in terms of the likelihood that it matches the observations.

Second, the model uncertainty can be attributed to particular parameters (here, describing processes and emissions) by using variance decomposition of the Monte Carlo data. It is now possible to quantify by how much the uncertainty would fall if the magnitude of a particular parameter could be better constrained. This information enables us to prioritise research that will maximise the reduction in model uncertainty. In terms of N_3 and N_{50} particle concentrations, the key parameters are associated with primary emissions and processes that affect the concentration of nucleating H_2SO_4 vapour, such as the in-cloud oxidation of SO_2. Comparison of the model ensemble with observations at a single site (Hyytiala, Finland) hints at the need for an organic-mediated nucleation mechanism in order to explain the seasonal cycle of particle concentrations. Having explored the full space of 28 parameters, there is a general failure to capture the peak in summertime concentrations without compromising the model–observation agreement at other times of year. This is an example of how a complete exploration of the model parameter space could enable structural uncertainties to be identified – *i.e.*, missing or poorly parameterised processes. Here, we have used just the ensemble of 168 model runs to reach this conclusion, which only very sparsely fill the 28-dimensional uncertainty space. In future, more sophisticated approaches will be developed that exploit the emulator.

Third, this approach makes it possible to tackle the problem of model *equifinality*, meaning that observations can be explained with equal skill using different combinations of parameters. This possibility is sometimes cited as a major limitation of such multi-parameter approaches. However, equifinality is telling us something very important about our understanding of the model. In particular, it implies that observations have not provided sufficient constraint on the model. If a set of observations can be explained by several settings of the model parameters, then the actual model uncertainty is likely to be larger than we thought. For example, as we discussed, if primary emissions and nucleation rates can compensate, and cannot be separated by comparing with observations, then future projections using these models could diverge depending on how nucleation rates and primary emissions evolve. Identification of such model behaviour will point us towards valuable observations and it will enable us to make a more realistic assessment of the true model uncertainty.

An important open question is how these results will differ between models and whether it is useful to go to such lengths to quantify the causes of uncertainty in one model. The issue of multi-model *versus* multi-parameter ensembles is an active area of research in the climate modelling community. Both approaches have their value, although it has been shown that the multi-parameter prediction range can span the multi-model range if the experiment is well designed.[30] Our priority is to combine the aerosol process and emission perturbations with perturbations to the host physics model (boundary layer schemes, cloud processes, updraught velocities, radiation, *etc*), which are likely to be an important additional cause of model-observation discrepancy.

Acknowledgements

We are grateful to M. Kulmala and researchers at the Hyytiala station for providing particle concentration data. This research was funded by the Natural Environment Research Council AEROS project (Aerosol Model Robustness and Sensitivity Analysis) and GASSP (Global Aerosol Synthesis and Science Project).

References

1 K. Beven and J. Freer, *J. Hydrol.*, 2001, **249**, pp. 11–29.
2 A. Saltelli and P. Annonia, *Environ. Modell. Software*, 2010, **25**, 1508–1517.
3 G. W. Mann, K. S. Carslaw, D. V. Spracklen, D. A. Ridley, P. T. Manktelow, M. P. Chipperfield, S. J. Pickering and C. E. Johnson, *Geosci. Model Dev.*, 2010, **3**, 519–551.

4 M. Kulmala, K. E. J. Lehtinen and A. Laaksonen, *Atmos. Chem. Phys.*, 2006, **6**, 787–793.
5 H. Vehkamaki, M. Kulmala, I. Napari, K. E. J. Lehtinen, C. Timmreck, M. Noppel and A. Laaksonen, *J. Geophys. Res.*, 2002, **107**, 4622.
6 G. W. Mann, *et al.*, *Atmos. Chem. Phys.*, 2012, **12**, 4449–4476.
7 C. L. Reddington, *et al.*, *Atmos. Chem. Phys.*, 2011, **11**, 12007–12036.
8 D. V. Spracklen, *et al.*, *Atmos. Chem. Phys.*, 2010, **10**, 4775–4793.
9 D. V. Spracklen, *et al.*, *Atmos. Chem. Phys.*, 2011, **11**, 9067–9087.
10 H. Korhonen, *et al.*, *J. Geophys. Res.*, 2008, **113**, D15204.
11 D. V. Spracklen, *et al.*, *Atmos. Chem. Phys.*, 2011, **11**, 12109–12136.
12 C. L. Reddington, *et al.*, *Atmos. Chem. Phys. Discuss.*, 2012, **12**, 26503–26560.
13 K. J. Pringle, *et al.*, *Atmos. Chem. Phys.*, 2012, **12**, 11647–11663.
14 L. A. Lee, K. J. Pringle, C. L. Reddington, G. W. Mann, P. Stier, D. V. Spracklen, J. R. Pierce and K. S. Carslaw, *Atmos. Chem. Phys. Discuss.*, 2013, **13**, 6295–6378.
15 L. A. Lee, K. S. Carslaw, K. J. Pringle, G. W. Mann and D. V. Spracklen, *Atmos. Chem. Phys.*, 2011, **11**, 12253–12273.
16 L. A. Lee, K. S. Carslaw, K. J. Pringle and G. W. Mann, *Atmos. Chem. Phys.*, 2012, **12**, 9739–9751.
17 A. O'Hagan, *Reliab. Eng. Syst. Saf.*, 2006, **91**, 1290–1300.
18 A. Saltelli, S. Tarantola and K. P.-S. Chan, *Technometrics*, 1999, **41**, 39–56.
19 A. Saltelli, K. Chan and M. E. Scott, *Sensitivity Analysis*, New York, Wiley, 2000.
20 F. Dentener, *et al.*, *Atmos. Chem. Phys.*, 2006, **6**, 4321–4344.
21 J. Merikanto, D. V. Spracklen, G. W. Mann, S. J. Pickering and K. S. Carslaw, *Atmos. Chem. Phys.*, 2009, **9**, 8601–8616.
22 A. Jones, *et al.*, *J. Geophys. Res.*, 2001, **106**, 20293–20310.
23 M. T. Woodhouse, G. W. Mann, K. S. Carslaw and O. Boucher, *Atmos. Chem. Phys.*, 2013, **13**, 2723–2733.
24 A. Metzger, B. Verheggen, J. Dommen, J. Duplissy, A. S. H. Prevot, E. Weingartner, I. Riipinen, M. Kulmala, D. V. Spracklen, K. S. Carslaw and U. Baltensperger, *Proc. Natl. Acad. Sci. U. S. A.*, 2010, **107**, 6646–6651.
25 T. E. Lane, N. M. Donahue and S. N. Pandis, *Atmos. Environ.*, 2008, **42**, 7439–7451.
26 R. Makkonen, *et al.*, *Atmos. Chem. Phys.*, 2009, **9**, 1747–1766.
27 P. Tunved, D. G. Partridge and H. Korhonen, *Atmos. Chem. Phys.*, 2010, **10**, 10161–10185.
28 V.-M. Kerminen, *et al.*, *Atmos. Chem. Phys.*, 2012, **12**, 12037–12059.
29 J. Browse, K. S. Carslaw, S. R. Arnold, K. Pringle and O. Boucher, *Atmos. Chem. Phys.*, 2012, **12**, 6775–6798.
30 M. Collins, B. B. B. Booth, B. Bhaskaran, G. R. Harris, J. M. Murphy, D. M. H. Sexton and M. J. Webb, *Clim. Dyn.*, 2011, **36**, 1737–1766.

Faraday Discussions

RSC Publishing

A water activity based model of heterogeneous ice nucleation kinetics for freezing of water and aqueous solution droplets†

Daniel A. Knopf* and Peter A. Alpert

Received 8th March 2013, Accepted 23rd April 2013

DOI: 10.1039/c3fd00035d

Immersion freezing of water and aqueous solutions by particles acting as ice nuclei (IN) is a common process of heterogeneous ice nucleation which occurs in many environments, especially in the atmosphere where it results in the glaciation of clouds. Here we experimentally show, using a variety of IN types suspended in various aqueous solutions, that immersion freezing temperatures and kinetics can be described solely by temperature, T, and solution water activity, a_w, which is the ratio of the vapour pressure of the solution and the saturation water vapour pressure under the same conditions and, in equilibrium, equivalent to relative humidity (RH). This allows the freezing point and corresponding heterogeneous ice nucleation rate coefficient, J_{het}, to be uniquely expressed by T and a_w, a result we term the a_w based immersion freezing model (ABIFM). This method is independent of the nature of the solute and accounts for several varying parameters, including cooling rate and IN surface area, while providing a holistic description of immersion freezing and allowing prediction of freezing temperatures, J_{het}, frozen fractions, ice particle production rates and numbers. Our findings are based on experimental freezing data collected for various IN surface areas, A, and cooling rates, r, of droplets variously containing marine biogenic material, two soil humic acids, four mineral dusts, and one organic monolayer acting as IN. For all investigated IN types we demonstrate that droplet freezing temperatures increase as A increases. Similarly, droplet freezing temperatures increase as the cooling rate decreases. The $\log_{10}(J_{het})$ values for the various IN types derived exclusively by T and a_w, provide a complete description of the heterogeneous ice nucleation kinetics. Thus, the ABIFM can be applied over the entire range of T, RH, total particulate surface area, and cloud activation timescales typical of atmospheric conditions. Lastly, we demonstrate

---

Institute for Terrestrial and Planetary Atmospheres/School of Marine and Atmospheric Sciences, Stony Brook University, Stony Brook, NY 11794-5000, USA. E-mail: Daniel.Knopf@Stonybrook.edu; Fax: +1 631 632 6251; Tel: +1 631 632 3092

† Electronic supplementary information (ESI) available: Supplementary Fig. S1–S10. See DOI: 10.1039/c3fd00035d

that ABIFM can be used to derive frozen fractions of droplets and ice particle production for atmospheric models of cirrus and mixed phase cloud conditions.

1　Introduction

Clouds containing ice particles cover a significant portion of the Earth's surface.[1,2] Their importance for establishing radiative equilibrium between Earth and the Sun and hydrological fluxes are well recognized, however, not well understood.[3–12] This lack of understanding is in part due to our insufficient knowledge of the role of atmospheric aerosol particles in the underlying cloud microphysical processes leading to ice crystal formation.[3,4,13] It is well established, that ice can nucleate homogeneously from aqueous droplets or heterogeneously from various physically and chemically distinct particles at different atmospheric conditions of temperature (T) and relative humidity (RH).[13–15] In addition, heterogeneous ice nucleation can occur through various nucleation modes, further complicating the prediction of their effects on ice cloud formation. Immersion freezing can be initiated when an ice nucleus is immersed in a supercooled aqueous droplet; deposition ice nucleation takes place on the ice nucleus which is in a supersaturated water vapour environment; condensation freezing can occur as a two step process at supercooled conditions in which water condenses prior to immersion freezing; and contact freezing can happen due to the physical contact of an ice nucleus with a supercooled aqueous droplet.[15,16] Insoluble particles that nucleate ice heterogeneously can be responsible for the formation of cirrus[17–22] as well as mixed-phase clouds.[8,10,23–25] Previous studies suggest that aqueous-phase dependent heterogeneous ice formation, *i.e.* immersion and contact freezing, can play important roles in ice formation for cirrus[21,26–28] and mixed-phase clouds, in which supercooled aqueous droplets and ice crystals co-exist.[10,29–34]

In this study we introduce a new model of immersion freezing parameterized by T and water activity, a_w, which is drawn from the water activity based homogeneous ice nucleation theory.[35] The application of our model allows prediction of freezing temperatures and rate coefficients, frozen fractions, and ice particle production rates thereby accounting for changes in ice nuclei (IN) surface areas as well as the time that the droplets remain supersaturated with respect to ice, S_{ice}. To facilitate the introduction of the underlying concepts of this new model for immersion freezing we first describe the concepts of homogeneous ice nucleation and follow by adding the effects of IN on the ice nucleation kinetics.

1.1　Homogeneous ice nucleation

The process of homogeneous ice nucleation has been shown to be stochastic in nature and to scale with the volume of the aqueous solution and time period that the solution is at $S_{ice} > 1$.[15,36–38] For the remainder of the manuscript this time period will be referred to as nucleation time. Classical nucleation theory (CNT) is commonly used to describe the freezing rate in terms of a rate coefficient, J_{hom}, in units cm^{-3} s^{-1}, which is dependent on T and S_{ice}.[38] Koop *et al.*[35] introduced the parameter water activity, a_w, to describe homogeneous ice nucleation and to predict freezing temperatures and J_{hom} for pure water and aqueous solutions droplets without knowledge of solute composition. For an aqueous droplet in equilibrium with the surrounding water partial pressure, its a_w will be equal to the

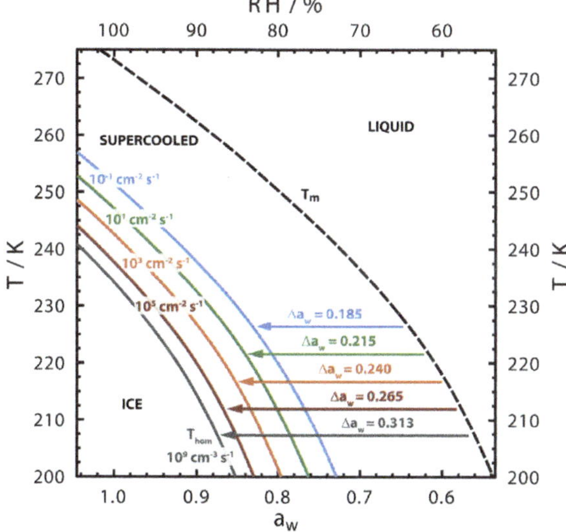

Fig. 1 Graphical representation of the conceptual framework underlying water activity based homogeneous ice nucleation[35] and water activity based immersion freezing. This figure shows the dependency of homogeneous and heterogeneous freezing temperatures and homogeneous and heterogeneous ice nucleation rate coefficients (in units $cm^{-3}\ s^{-1}$ and $cm^{-2}\ s^{-1}$, respectively) on the water activity criterion, Δa_w, implicitly accounting for changes in nucleation time and available ice nuclei surface area. Exemplary numbers for the ice nucleation rate coefficients are provided for respective immersion freezing curves.

RH.[35] The authors have shown that when homogeneous freezing temperatures for different types of aqueous solutions and various concentrations are expressed as a function of a_w, all freezing data fall along one freezing curve which can be constructed from the thermodynamically defined ice melting curve.[35] A positive horizontal shift in the ice melting curve by Δa_w toward values of higher a_w completely represents the homogeneous freezing curve.[35,39] Since the ice melting curve is fixed, Δa_w provides pairs of T and a_w (or RH) for any freezing point. This concept is demonstrated by the example in Fig. 1. Furthermore, each constructed freezing point curve representing a specific Δa_w value describes a single J_{hom} value. A decrease in Δa_w translates into increased freezing temperatures and a decreased J_{hom} value with J_{hom} converging to zero as T approaches the ice melting point. Hence, Δa_w, which provides a complete description of the nucleation process with freezing temperatures and kinetics being a function of a_w, can be applied to all nucleation times and aqueous-phase volumes. The a_w based description of homogeneous ice nucleation kinetics has been found to be generally valid experimentally[35,37,40–44] and when used in cloud parcel and climate models.[45–49] However, a significant drawback of the water activity based homogeneous ice nucleation theory is that for many aqueous solutions, a_w is determined at or above the ice melting point and obtained by extrapolation at supercooled temperatures using thermodynamic models.[50–55] If no data or model are available it is assumed that a_w does not vary significantly with decreasing temperature, a sufficient assumption for many inorganic and inorganic/organic aqueous solutions.[35,37,39,40,42,43,56,57] In general, using a_w greatly simplifies the prediction of homogeneous freezing temperatures and kinetics for aqueous solution droplets.

1.2 Heterogeneous ice nucleation

In laboratory experiments, numerous inorganic and organic categories of particles serve as surrogates for atmospheric IN and have been investigated for their efficiency to nucleate ice. An extensive overview of the literature is provided by previous summaries and review articles.[13,14,58–61] Mineral dust and clay particles, for example, are known to act as immersion IN by elevating the freezing temperature above the homogeneous freezing limit.[15,39,57,60–70] Terrestrial soil dust particles contain humic like substances (HULIS) which can act as immersion IN[71–73] and marine biogenic and terrestrial biological particles can also initiate the ice phase heterogeneously.[39,41,43,58,74,75] Previous studies have shown that organic surfactants, which form monolayer coatings on aqueous droplets, can trigger ice nucleation at much warmer temperatures than for homogeneous freezing.[14,57,76–78] Finally, field-collected and laboratory generated organic particles also show a strong propensity for heterogeneous ice nucleation.[59,72,79–84]

While there exists extensive literature on IN active particles in pure water droplets, only a few studies have investigated immersion freezing for $a_w < 1$ involving aqueous solutions with various solute concentrations in which freezing occurs at subsaturated conditions with respect to water. For example, it has been shown that mineral dust can nucleate ice heterogeneously in aqueous solutions of $(NH_4)_2SO_4$,[39,63,66] H_2SO_4,[64–66] NaCl, malonic acid, polyethylene glycol, and organic-inorganic mixtures.[39,66] Solid phase oxalic acid acts as IN in aqueous solutions for a variety of organic and inorganic solutes.[56] Bacterial ice nucleators from *Pseudomonas syringae* strains have been shown to efficiently freeze water in aqueous succinic acid, glucose, and $(NH_4)_2SO_4$ solutions.[39] Biogenic particles derived from insects in various aqueous solutions enhance freezing temperatures.[39] Highly ordered organic monolayers on the surface of aqueous solution droplets can nucleate ice.[14,57,66,78] In micrometer sized aqueous NaCl solution droplets, marine phytoplankton are observed to act as ice nuclei.[41,43,75] Humic acid soil particles immersed in aqueous solutions of $(NH_4)_2SO_4$[73] can induce ice nucleation.

As mentioned above, most of the previous immersion freezing studies have used pure water droplets as the surrounding medium of the IN, and therefore have not considered solute effects and the role of a_w, but have focused instead on how to parameterise the data from freezing experiments. A review of these numerous mathematical parameterisations is beyond the scope of this paper and the reader is referred to previous literature.[67–69,73,85–88] However, summarising, these analyses have fit functions based on CNT and deterministic (*i.e.* a time dependence of the nucleation process is ignored) descriptions with different assumptions regarding the ice nucleating sites to reproduce the frozen fractions of water droplets containing IN for the specific experimental conditions. Recently, our group extended immersion freezing descriptions for the first time to include organic IN immersed in aqueous $(NH_4)_2SO_4$ solution droplets for a range of a_w.[73] We found that previous immersion freezing fit-based parameterisations can neither be readily modified to include solute effects or that they represent a continuum of a_w.[73] Furthermore, they do not allow the extension of laboratory derived nucleation parameters to varying atmospheric conditions.[88] Given that a continuum of RH and thus droplet a_w exists under atmospheric conditions, a more appropriate inclusive and flexible approach for this descriptive process would be to apply Δa_w as a determinant for immersion freezing temperatures and nucleation rates. Additionally, the application of Δa_w would greatly simplify inclusion of immersion freezing in atmospheric cloud system resolving models. Kärcher and Lohmann[89] first attempted to describe immersion freezing as a function of a_w in a cloud resolving model by parameterisation of the water activity based homogeneous ice nucleation theory.[35,89] Due to a lack of a physical model and sufficient experimental data J_{hom} values as a function of a_w were shifted with respect to a_w to yield heterogeneous ice nucleation rate coefficients, J_{het}, assuming a nucleation rate of $1\ \text{s}^{-1}$ at $S_{ice} = 1.2$. In other words, heterogeneous freezing rates are tied to homogeneous freezing rates, resulting in similar dependencies on water activity and exactly the same temperature dependency despite their different underlying nucleation mechanisms. Thus, homogeneous and heterogeneous ice nucleation are not independent of each other in the parameterisation by Kärcher and Lohmann.[89] A physical model of heterogeneous ice nucleation should be decoupled from homogeneous ice nucleation as outlined below.

1.3   Water activity based immersion freezing model

Previous laboratory studies have shown that immersion freezing temperatures of specific IN follow a freezing curve generated by a horizontal shift in water activity, Δa_w, of the ice melting curve independent of surrounding aqueous solutions[39,41,43,56,57,63,64,66,73,75] as conceptualised in Fig. 1. Considering the different IN types and vast array of different solutes employed in these experiments, the parameter Δa_w has performed exceptionally well to predict the respective freezing temperatures. However, prediction of heterogeneous ice nucleation rate coefficients, J_{het}, in units $\text{cm}^{-2}\ \text{s}^{-1}$, using Δa_w to determine the frozen droplet fractions and ice particle numbers under atmospheric conditions, remains elusive. As of now, there has only been a single immersion freezing study that investigated ice nucleation kinetics as a function of Δa_w.[57] In that study, we found that median freezing temperatures of 1-nonadecanol coated aqueous solution droplets can be described by Δa_w. A decrease in the 1-nonadecanol surface area decreases the freezing temperatures and increases J_{het},[78] which in turn shifts the immersion freezing curve to lower temperatures and increases Δa_w,[57] as shown in Fig. 1. Each of these individual immersion freezing curves represents a constant J_{het} value, a result corroborated by immersion freezing studies of different IN.[41,57,64,73,75] Parameterizing immersion freezing data of only pure water droplets coated by 1-nonadecanol, *i.e.* yielding $J_{het}(\Delta a_w)$ evaluated at $a_w = 1.0$, successfully reproduced J_{het} for aqueous NaCl solution droplets, $a_w < 1.0$, coated by 1-nonadecanol.[57] This is a remarkable finding given that the analysis was done without considering any solute effects and is based only on the premises that Δa_w is the determinant of the ice nucleation process, corroborating the fact that ABIFM can be applied to immersion freezing of aqueous solutions droplets.

ABIFM predicts that decreasing Δa_w values translate into increasing freezing temperatures and decreasing J_{het} values, as shown in Fig. 1. It should be noted that J_{het} is a coefficient and independent of IN surface areas or nucleation time, and dependent only on T and a_w. Imagine two independent sets of droplets containing the same IN cooled at the same rate, the first set having a greater total IN surface area than the second. Assume that median freezing temperatures for the first set of droplets at any aqueous solution concentration would fall *e.g.* on the green line in Fig. 1, and so $J_{het} = 10^1\ \text{cm}^{-2}\ \text{s}^{-1}$. According to ABIFM, under the same conditions of T and a_w, the second set of droplets with smaller IN surface areas would have the exact same J_{het} as the first set. However, a smaller IN surface area would yield a smaller droplet freezing probability (or nucleation rate in units s^{-1}) and the droplets, which would remain liquid, cool below the green freezing curve, and access higher J_{het} values as indicated in Fig. 1. The second set of droplets then freezes at conditions represented by the orange freezing curve where $J_{het} = 10^3$ cm^{-2} s^{-1}. The same reasoning can be applied to two sets of IN containing droplets exposed to different cooling rates or nucleation time periods. If the argument is correct, then the ABIFM could be used to predict freezing temperatures and nucleation kinetics for all possible changes in the parameters controlling the nucleation process. It therefore would represent a holistic model of immersion freezing.

To assess the general validity of the ABIFM, we will challenge it to address the following questions in this manuscript: Which particle types acting as immersion IN follow the ABIFM? How do changes in IN surface area and cooling rate affect immersion freezing temperatures and nucleation kinetics, and are those effects captured by the ABIFM? Are experimentally derived nucleation kinetics characterized by J_{het} accurately reproduced as a function of solely Δa_w? Can the ABIFM be applied to predict atmospheric ice particle numbers at relevant atmospheric conditions of T and S_{ice}? We ultimately show that the answers to these questions are positive and that Δa_w can serve as the determinant for a complete description of immersion freezing from water and aqueous solution droplets, allowing prediction of freezing temperatures and ice nucleation rates, frozen fractions, and ice particle production rates.

To address the above questions we apply the ABIFM for analysis of laboratory obtained immersion freezing data. This analysis includes a thorough re-examination of previous immersion freezing data from our group including the freezing from aqueous $(NH_4)_2SO_4$ droplets containing particles of Pahokee Peat and Leonardite serving as surrogates of HULIS, aqueous NaCl droplets containing two distinct species of phytoplankton *Nannochloris atomus* and *Thalassiosira pseudonana*, aqueous NaCl droplets coated by 1-nonadecanol, and new freezing data from aqueous $(NH_4)_2SO_4$ droplets containing Illite dust. In addition, new ice nucleation data for these IN are presented showing the dependence of freezing temperatures and kinetics on changes in IN surface area and cooling rate in both pure water and aqueous solution droplets. Finally, we reproduce data from previous laboratory studies investigating IN surface area dependent immersion freezing using only pure water droplets. The suitability and application of the ABIFM for inclusion in cloud system resolving, air parcel, and climate models for prediction of ice particle production and numbers will be discussed.

2   Experimental

2.1   Droplet preparation

Suspensions of pure water or aqueous solution droplets containing *N. atomus*, *T. pseudonana*, Pahokee Peat, Leonardite, and Illite particles were prepared for either micrometer or millimeter sized droplets. Growth of *N. atomus* and *T. pseudonana* and solution preparation is described fully in previous studies[41,43,75] and only briefly presented here. Axenic unialgal cultures are grown for one week in a 14 h light : 10 h dark cycle. Cells are washed and resuspended in either pure water or an aqueous solution with a known amount of NaCl. Cell numbers are targeted at 10^6, 10^7, and 10^8 mL^{-1} through dilution by adding solution and concentrating through centrifugation. A calibrated microliter pipette is used to deposit about 30 millimeter sized pure water or aqueous solution droplets, 2 μL in volume containing cells onto a hydrophobically coated glass plate. The cell concentration in 2 μL droplets is determined by measuring the cell concentration in solution employing a hemocytometer having an accuracy of ±5% and scaling down by volume. A droplet generation technique uses a piezo-electric single drop on demand dispenser[41,43,73,75] to create 30–60 monodisperse droplets with diameters of 40–70 μm typically having 3 cells each on a hydrophobic coated glass plate.[41,43,75] Surface area per droplet is determined by observation of cell numbers per droplet by optical microscopy and known surface area per cell.[41,43,75]

Pahokee Peat, Leonardite, and Illite particles in pure water or aqueous $(NH_4)_2SO_4$ solutions used for droplet sample preparation follows Rigg *et al.*[73] Using a mortar and pestle, Pahokee Peat, Leonardite, and Illite are crushed, suspended in pure water, sonicated for 30 min and then roughly agitated. Larger particles are allowed to settle to the bottom of the container. The aqueous supernatant is carefully removed and is filtered through a 3 μm nuclepore filter. Aliquots of 0.04 mL are then pipetted into aluminum weigh boats, dried to remove all water, and weighed to determine the particle weight per volume of water. Aluminum boats are dried for 2 h at 60 °C and 24 h at 130 °C, and weighed with an ATI CAHN C-33 microbalance to an accuracy of ±0.001 mg before and after the addition of the aliquot. Particle surface areas in solution are determined from the measured weight per volume and the specific surface area using the Brunauer–Emmett–Teller (BET)[90] gas adsorption technique as described in Rigg *et al.*[73] for Pahokee Peat and Leonardite and in Dogan *et al.*[91] for Illite. Surface area per volume is varied by diluting the solution with water. Pure water solutions or solutions with a known amount of $(NH_4)_2SO_4$ added are used to make 2 μL and micrometer sized droplets. The surface area per droplet is determined by scaling the known surface area per volume in solution accounting for the added $(NH_4)_2SO_4$ when applicable.[73] All droplet sample preparation steps are done in a clean bench to reduce the possibility of contamination from airborne particles.

2.2   Droplet water activity conditioning of droplets

The a_w of micrometer-sized aqueous solution droplets is preset using an aerosol conditioning cell (ACC) that allows droplets to equilibrate with controlled RH as detailed previously.[41,42,57,73,75] Droplets are exposed to a humidified flow of N_2 gas with known dew point temperature, T_{dew}. The uncertainty of T_{dew} is <±0.15 K. The droplet temperature, T_{drop}, is controlled within ±0.2 K. From this, $\mathrm{RH} = [p_{H_2O}(T_{dew})]/[p^0_{H_2O}(T_{drop})]$ can be derived,[92] resulting in an uncertainty in a_w of ±0.01. RH calibration is achieved by determination of the ice melting point and deliquescence relative humidities of various inorganic salts.[41,42,57,65,73,75,93] Droplet samples are allowed to come into equilibrium with RH and thus, $a_w = \mathrm{RH}$ is set. The sample is sealed against the environment using a second hydrophobically coated glass slide cover and a tin foil spacer coated with high vacuum grease. When using 2 μL droplets, the a_w is given by the composition of the aqueous solution and by using thermodynamic models to determine equilibrium RH.[50–55] The change in the water amount of these large droplets due to evaporation or condensation of water vapor from the ambient air during the short moments of sample preparation is negligible compared to the condensed phase water amount.[50,94]

2.3 Ice nucleation experiments

Two ice nucleation experimental setups, both consisting of a cryo-cooling stage coupled to an optical microscope and equipped with digital imaging capabilities are used, one for micrometer sized droplets[41–44,73,75,93] and the other for 2 μL droplets.[57] They allow controlled cooling and heating of droplets in the temperature range of about 150–350 K. The first uses $100\times$ magnification to view the 1.5 mm sample area for micrometer droplets.[42] Ice nucleation is investigated for a range of cooling rates from 1–15 K min^{-1} until all droplets freeze. Heating rates of 0.5 K min^{-1} are used to determine the melting temperatures. The second setup employs a macro lens to view the entire 25 mm sample area for 2 μL droplets and uses cooling rates from 1–5 K min^{-1} for freezing and heating rates of 0.1 K min^{-1} for melting experiments. Temperature calibration is conducted by measuring known melting points of ice and various organic species covering the investigated freezing temperatures[42,57,93] resulting in a temperature uncertainty of $\leq\pm0.1$ K. During the freezing and melting cycles, images are recorded at 0.2 K intervals that contain experimental time and temperature. At least two independently prepared droplet samples are applied for each investigated a_w, IN surface area, and cooling rate and each droplet sample is used for a maximum of three cooling and melting cycles. After completion of the experiment droplet freezing and melting temperatures are collected and analysed.

2.4 Chemicals

N_2 (99.999%) was purchased from Praxair. $(NH_4)_2SO_4$ (99.95%) was purchased from Alfa Aesar. NaCl (99.95%) was purchased from EMD Chemical Inc. Humic Acid Reference Pahokee Peat (R103H-2) and Humic Acid Standard Leonardite (1S104H-5) were purchased from the International Humic Substances Society (IHSS). Illite IMt-1 Cambrianshale was purchased from Clay Mineral Society. Millipore water (resistivity $\geq$ 18.2 MΩ cm) was used for preparation of aqueous solutions.

3 Results and discussion

Median freezing temperatures, $\tilde{T}$, for six different IN representing four different particle-type classes; inorganic, organic, marine biogenic, and organic surfactant in water and aqueous solution droplets are shown in Fig. 2 as a function of a_w. A total of about 18 000 individually analysed droplet freezing events from our group are presented. Previous immersion freezing data by Zobrist *et al.*[66,78] and Broadley *et al.*[95] are additionally included. $\tilde{T}$ for the diatom *Thalassiosira pseudonana*[41,43] and the microalga *Nannochloris atomus*[75] in water and aqueous NaCl droplets are shown in panel (A) and (B), respectively. Freezing data of Pahokee Peat and Leonardite immersed in pure water and aqueous $(NH_4)_2SO_4$ solution droplets[73] are shown in panel (C) and (D), respectively. Panel (E) shows immersion freezing temperatures of Illite mineral dust in water and aqueous $(NH_4)_2SO_4$ solution droplets as a function of a_w along with previous data.[95] Lastly, panel (F) shows $\tilde{T}$

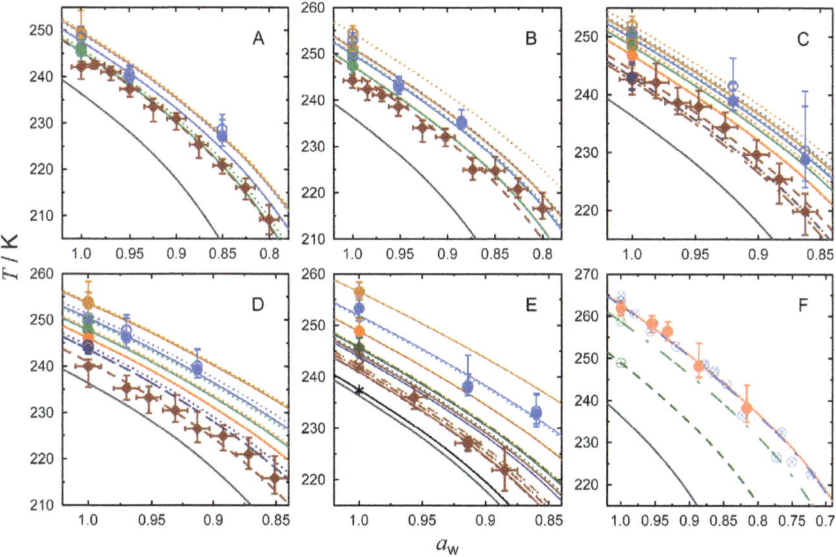

Fig. 2 Median heterogeneous ice nucleation temperatures are shown as a function of a_w for water and aqueous solution droplets containing (A) *Thalassiosira pseudonana*,[41,43] (B) *Nannochloris atomus*,[75] (C) Pahokee Peat,[73] (D) Leonardite,[73] and (E) Illite particles (this study and Broadley et al.[95]), and (F) coated by 1-nonadecanol.[57,78] The temperature error gives the 10th and 90th percentiles of all freezing events and the error in a_w for aqueous solutions droplets is ±0.01. Lighter to darker colors represent droplets containing greater and smaller IN surface area, where light brown, light blue, light red, light green, orange, dark green, dark brown, dark blue, dark red, and black represent IN surface area of about 1 cm^2, 1×10^{-1} cm^2, 5×10^{-2} cm^2, 1×10^{-2} cm^2, 1×10^{-3} cm^2, 1×10^{-4} cm^2, 5×10^{-5} cm^2, 1×10^{-5} cm^2, 1×10^{-6} cm^2, 1×10^{-7} cm^2 per droplet, respectively. Symbol/line combinations represent varying cooling rates employed where open circles/dotted, filled circles/solid, filled diamonds/dash, and open diamonds/dot-dash-dotted freezing points obtained by cooling with 1 K min^{-1}, 5 K min^{-1}, 9.5 K min^{-1}, 14.2 K min^{-1}, respectively. Data from Broadley et al.[95] and Zobrist et al.[78] are included in (E) and (F) and have differently labeled symbol/line combinations where cooling rates of 1 K min^{-1}, 3 K min^{-1}, 5 K min^{-1}, 7.5 K min^{-1}, and 10 K min^{-1} correspond to asterisks/dotted, crossed-circles/dash-dash-dotted, stars/solid, crosses/dash-dotted, and plus-circles/dash, respectively. The homogeneous freezing of pure water droplets 5–10 μm in diameter is shown as the solid gray line.[35,39]

for monolayer coatings of 1-nonadecanol on pure and aqueous NaCl solution droplets.[57,66,78] Presented in Fig. 2 are 10 different IN surface areas, where each colored line represents an investigated IN surface area from 1×10^{-7}–1.0 cm^2 per droplet. Various cooling rates in the range of 1.0–15.0 K min^{-1} are shown as different symbols. Freezing curves representing each set of $\tilde{T}$ having the same IN surface area per droplet and cooling rate are given as a function of a_w. For all data, droplet melting temperatures (not shown) are in agreement with the ice melting point curve.[39,41,43,57,66,73,75,78,95] It is clear from Fig. 2 that all particle types act as efficient IN in pure water and aqueous solution droplets. Marine biogenic particles, terrestrial organic HULIS particles, and the mineral dust Illite can nucleate ice as high as 15–20 K above the homogeneous freezing limit. Freezing temperatures of 1-nonadecanol monolayers can be observed at temperatures 10 K cooler than the ice melting point.

The highest observed immersion freezing temperatures shown in Fig. 2 are due in part to the large IN surface area per droplet, A (~1 cm^2 per droplet). In both pure water and aqueous solution droplets, decreasing A results in decreasing $\tilde{T}$ for all investigated IN types, as shown in Fig. 2 and detailed in the ESI† Fig. S1–S5. Comparisons of the dependence of $\tilde{T}$ on A were made only between those data sets having the same cooling rate and a_w. We derive the difference in median freezing temperature, $\Delta\tilde{T}$, between adjacent values of $\tilde{T}$ given in Fig. S1–S5 (ESI†) per one order of magnitude change in IN surface area. It is important to note that $\Delta\tilde{T}$ may change as a function of T. Freezing temperature dependence on A for Pahokee Peat is the smallest on average having $\Delta\tilde{T} \simeq 1.2$ K per order of magnitude change in IN surface area. For *N. atomus*, *T. pseudonana*, Leonardite, and Illite, $\Delta\tilde{T}$ is approximately equal to 1.7 K, 2.0 K, 2.2 K and 2.9 K, respectively. For all particle types it is clear that $\tilde{T}$ and A are exponentially related. Previous laboratory studies investigating ice nucleation of mineral dust for example, have observed a similar trend of increasing immersion freezing temperatures with increasing IN surface area[70] even for particles as small as hundreds of nanometers corresponding to IN surface areas of $\sim 10^{-9}$ cm^2 present in individual droplets.[64,67–69] This observed behavior is consistent with predictions by CNT.[15]

Fig. 2 illustrates the appropriateness of the ABIFM for describing immersion freezing of water and aqueous solution droplets for varying A. For any fixed value of a_w, Δa_w increases with decreasing temperature, as shown in Fig. 1. As freezing temperatures decrease with decreasing A, Δa_w must increase. Observe in Fig. 2D for the case of Leonardite particles acting as IN, that the freezing curve which best describes $\tilde{T}$ for the filled, light blue circles with $A = 10^{-1}$ cm^2, is described by $\Delta a_w = 0.20$. For droplets containing smaller IN surface area, as is shown for the dark red circles at $A = 7.7 \times 10^{-7}$ cm^2, $\Delta a_w = 0.2703$. When comparing our observations of immersion freezing of Illite particles with previous studies[95] (Fig. 2E), it also holds that as A decreases, Δa_w increases: $A = 10^{-4}$ cm^2 and $A = 10^{-7}$ cm^2 which correspond to $\Delta a_w = 0.23$ and $\Delta a_w = 0.29$, respectively. Values of $\tilde{T}$ for $a_w \leq 1.0$ and various A, are well described by Δa_w for all particle types, in particular when assuming the same error in Δa_w as for the a_w based homogeneous ice nucleation description of ± 0.025.[35,37] Overall, the ABIFM can capture not only the relationship between immersion freezing temperatures and IN surface area, but also the solute effect on freezing point depression explicit through Δa_w.

Our results indicate a cooling rate, r, dependence on $\tilde{T}$ as indicated in Fig. 2 and in more detail in Fig. S6–S10 (ESI†). Comparisons of the dependence of $\tilde{T}$ on r were made only between those data sets having the same A and a_w. In general, applying slower r results in warmer $\tilde{T}$. *N. atomus* and Leonardite particles both exhibit $\Delta\tilde{T} \simeq 1.4$ K higher freezing temperatures when decreasing r by an order of magnitude. For particles of Illite, Pahokee Peat, and *T. pseudonana* particles, $\Delta\tilde{T}$ is approximately 0.5 K, 1.8 K, and 3.0 K, respectively, per one order of magnitude change in r. These results support a time dependent heterogeneous ice nucleation mechanism in the ongoing debate regarding the influence of nucleation time on immersion freezing.[60,61] The argument concerns whether increasing the nucleation time of a collection of droplets by slowing the cooling rate results in ice nucleation at warmer temperatures. Hoose and Möhler[60] summarize numerous laboratory experiments on immersion freezing of the same particle type with varying cooling rates and found no clear consensus on a cooling rate dependence due to the large scatter in data. This may be due in part to temperature uncertainties[38] and uncertainties in applied IN surface areas. However, the most recent ice nucleation studies clearly demonstrate a time dependence of immersion freezing,[95,96] in agreement with our findings.

Fig. 2 demonstrates the appropriateness of the ABIFM for describing immersion freezing of water and aqueous solution droplets for varying r. As freezing temperatures increase with decreasing r for any value of a_w, Δa_w must increase. For example, in the case of Leonardite particles, the filled, light blue circles in Fig. 2D are $\tilde{T}$ measured for $r = 5$ K min^{-1} as a function of a_w. Note that the solid light blue freezing curve having $\Delta a_w = 0.20$ best represents these data. For the same IN surface area, droplets cooled at $r = 1$ K min^{-1} shown by the light blue open circles are characterized by the dotted light blue freezing curve having $\Delta a_w = 0.19$. The difference between Δa_w derived for 5 K min^{-1} and 1 K min^{-1} is minimal due to the relatively small change in r. In order to better establish a cooling rate dependence, r would need to be varied by orders of magnitude, similar to the case when investigating an IN surface area dependence. Despite this limitation, variation in cooling rates shows the consistent trend that $\tilde{T}$ increases with decreasing r, which extends to different IN types, IN surface areas, and droplet compositions investigated here (see Fig. S6–S10, ESI†) and application of different experimental methods.[69,95,96] This suggests that the trend should also be observed for other particle types.

In summary, for the first time we can account for IN surface area, cooling rate, and solute effect dependencies on immersion freezing temperatures explicitly through the use of Δa_w. We now address our first and second questions for which our data demonstrates a time-dependent and IN surface area dependent heterogeneous ice nucleation mechanism for immersion freezing which can be entirely described by the ABIFM for the various IN types investigated here and for other IN types that can be described by Δa_w.[39,41,43,56,57,63,64,66,73,75]

Previous studies have confirmed that freezing curves characterized by Δa_w represent a single J_{het} value.[41,43,57,64,73,75] In other words, J_{het} is the same at median freezing temperatures for different a_w. However, it should be emphasized that this relationship is not only true at $\tilde{T}$ where the cumulative freezing probability of droplets is equal to 0.5, but for every freezing probability from 0.0 to 1.0.[73] A freezing probability of < and >0.5 yields smaller and larger Δa_w values, respectively. Thus, we can express each individual freezing point in terms of Δa_w, thereby implicitly accounting for each individual droplet's freezing temperature, a_w, A, and r, using

$$\Delta a_w^{\,j} = a_w^{\,j} - a_w^{ice}(T^j),\tag{1}$$

where $a_w^{ice}(T^j)$ is the value of water activity along the ice melting curve evaluated at the jth freezing temperature, T^j, and $a_w^{\,j}$ is the water activity value corresponding to T^j. Thus, $\Delta a_w^{\,j}$ is the difference between the jth droplet's water activity and the ice melting curve. We now compare J_{het} values for each individually recorded freezing event with calculated $\Delta a_w^{\,j}$.

Fig. 3 shows $\log_{10}(J_{het})$ as a function of Δa_w for all freezing events and particle types discussed above. For each particle type, J_{het}, increases exponentially for increasing values of Δa_w. Eqn (1) shows that Δa_w depends on temperature and previous immersion freezing experiments have observed that J_{het} exponentially increases as temperature decreases[41,64,67,73,75,78] in agreement with CNT. Therefore, it is not surprising that greater values of Δa_w, indicative of greater supercooled

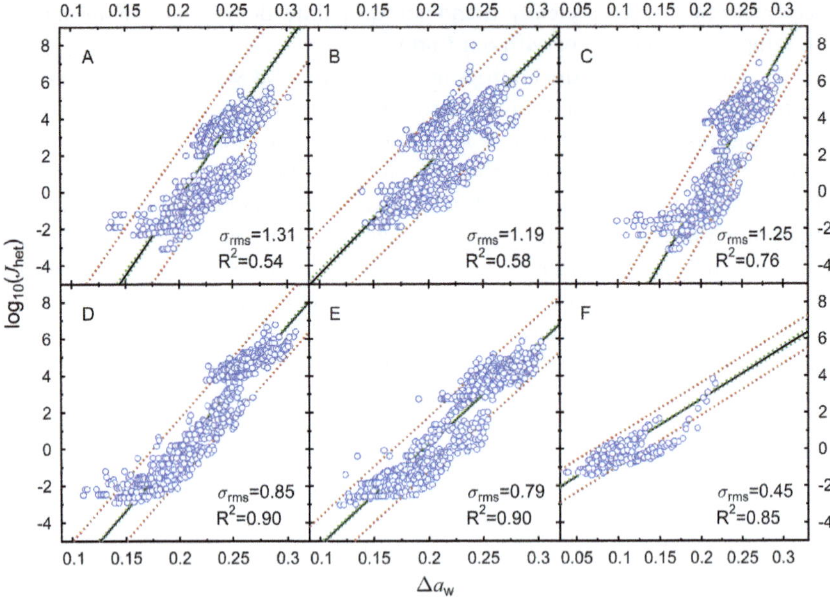

Fig. 3 The decadal log of the heterogeneous ice nucleation rate coefficients, $\log_{10}(J_{het})$, are shown as a function of Δa_w for individually analysed freezing events, initiated by the different IN types investigated in this study and previous work.[41,43,57,66,73,75,78,95] $\mathrm{Log}_{10}(J_{het})$ are shown for (A) *Nannochloris atomus*, (B) *Thalassiosira pseudonana*, (C) Pahokee Peat, (D) Leonardite, (E) Illite, and (F) 1-nonadecanol. The solid black line is a linear fit where dashed green and red lines represent confidence intervals and prediction bands at 95% level. The root mean square error, σ_{rms}, and the adjusted coefficient of determination, R^2, are given in each panel.

temperatures and increased S_{ice}, yield an exponential increase in J_{het}. At lower temperatures and higher S_{ice}, freezing is more likely and thus, will proceed at a faster rate corresponding to greater values of J_{het}. For all investigated IN types we find a consistent relationship between J_{het} and Δa_w, irrespective of applied IN surface areas, cooling rates, and solute concentrations, representing a corollary of ABIFM. The range of J_{het} values for a single Δa_w value spans about 2–4 orders of magnitude. However, this may not be the case on average, as data point density at some locations is extremely high and outliers may misguide the eye. In attempts to parameterise and estimate the prediction error for J_{het} as a function of Δa_w, we apply a linear fit to $\log_{10}(J_{het})$ for all IN types as shown in Fig. 3. In addition to the root mean square error, σ_{rms}, the adjusted coefficient of determination, R^2, confidence intervals and predictions bands at 95% are plotted in Fig. 3. Fitting parameters, their standard error and upper and lower confidence levels are given in Table 1. As can be seen in Fig. 3 the linear fit captures the behavior of $\log_{10}(J_{het})$ *versus* Δa_w for all investigated IN. For particles of Pahokee Peat and Leonardite, Illite and 1-nonadecanol, R^2 values are exceptionally high around 0.80 implying that the fit can account for the majority of the variance in the data. Additionally, σ_{rms} is on the order of 1.0. This indicates, that even so values of J_{het} are spanning 10 orders of magnitude, an average error of only 1 order of magnitude for reproducing J_{het} can be expected. For the two marine phytoplankton, R^2 values are modest around 0.55. However, σ_{rms} values for $\log_{10}(J_{het})$ are only slightly larger

Table 1 Parameters for derivation of the heterogeneous ice nucleation rate coefficient as a function of the water activity criterion, Δa_w, according to $\log_{10}(J_{het}) = c + m\Delta a_w$ are given. σ, LCL, and UCL represent standard deviation, and lower and upper confidence levels at 95%, respectively, for the fit parameters

Ice nucleus	c	σ_c	LCL_c	UCL_c	m	σ_m	LCL_m	UCL_m
N. atomus	−17.12381	0.27733	−17.66751	−16.58010	83.95603	1.17979	81.64302	86.26904
T. pseudonana	−10.43450	0.17629	−10.78012	−10.08889	59.66992	0.79104	58.11905	61.22079
Pahokee Peat	−15.77884	0.18337	−16.13839	−15.41929	78.30951	0.79777	76.74529	79.87374
Leonardite	−13.40148	0.09931	−13.59621	−13.20675	66.90259	0.42911	66.06118	67.74400
Illite	−10.66873	0.07544	−10.81666	−10.5208	54.48075	0.33316	53.8275	55.13400
1-Nonadecanol	−2.92414	0.04322	−3.00894	−2.83934	28.13797	0.35129	27.44871	28.82723
Kaolinite	−10.54758	0.38782	−11.31377	−9.78140	54.58834	1.39323	51.83590	57.34078
Al_2O_3	1.60671	0.58271	0.34784	2.86558	14.96639	2.14742	10.32717	19.6056
Fe_2O_3	1.42411	0.64593	0.07672	2.77149	17.62106	2.50034	12.40544	22.83668
Fungal Spores	0.97931	1.29892	−3.15444	5.11306	15.47856	4.7061	0.50162	30.45549

compared to the other IN types and fall around 1.25 orders of magnitude, thus, implying a similar result, that J_{het} can be well descried by Δa_w over many orders of magnitude change in J_{het}. In summary, we find that ice nucleation kinetics are well captured as a function of Δa_w for all investigated IN particle types.

These results allow us to address our third question of whether J_{het} is well represented as a function of Δa_w for all investigated IN types. From the calculated prediction bands at a 95% confidence level as shown Fig. 3, we assign a conservative uncertainty of ± 2 orders of magnitude for the calculation of J_{het} using the presented fitted parameterisation. This is a much smaller uncertainty than usually found, e.g. for the case of homogeneous ice nucleation[35,37] and implies that for immersion freezing the underlying heterogeneous ice nucleation kinetics are well defined by the ABIFM.

We have shown that immersion freezing follows the ABIFM and the physically based parameterisation of J_{het} as a function of Δa_w is found to be accurate for vastly different particle types in various aqueous solutions. This strongly suggests that the ice nucleation kinetics of other particles, including mineral dusts, combustion type, organic, and other various biological particles not included in the above analysis, can nevertheless be represented by ABIFM as well. As an example, we apply the ABIFM to capture immersion freezing by Kaolinite, aluminum and iron oxide particles, and fungal spores investigated using different techniques.[64,70,96,97]

In the studies of Murray et al.,[96] J_{het} for Kaolinite is given as a function of temperature at only $a_w = 1.0$ while Pinti et al.[70] investigated the freezing of water droplets containing Kaolinite with various IN surface areas at constant cooling rates. Freezing temperatures are reported for which 1% of the droplets freeze. This equates to a frozen fraction, $f(T) = 0.01$ and from this, $J_{het}(T)$ can be derived using the following equation

$$f(T) = 1 - \exp[-J_{het}(T)At_{nuc}], \tag{2}$$

where t_{nuc} is the nucleation time. In order to derive J_{het} from this information, it is necessary to know t_{nuc}, or have some knowledge of the sequence of nucleation events in order to integrate as a function of T from the melting point. Since the latter is unavailable we choose a reasonable nucleation time of 1 s. Therefore, upon rearranging eqn (2), we obtain

$$J_{het}(T) = \frac{-\ln[1 - f(T)]}{At_{nuc}}. \tag{3}$$

Fig. 3A shows that J_{het} from Murray et al.[96] and Pinti et al.[70] are in agreement with each other. We then follow the same procedure of fitting $\log_{10}(J_{het})$ versus Δa_w and find that within an uncertainty of $\leq \pm 2$ orders of magnitude, the ABIFM can account for immersion freezing by Kaolinite spanning orders of magnitude in IN surface areas and employing different cooling rates. Corresponding fit parameters are given in Table 1.

For metal oxide particles, Al_2O_3 and Fe_2O_3, J_{het} determined by Archuleta et al.[64] for various a_w are applied. Using eqn (1), we determine Δa_w and fit the data as described above. Also in this case the ABIFM accurately reproduces the kinetic freezing data as shown in Fig. 4B and C and corresponding fit parameters are given in Table 1.

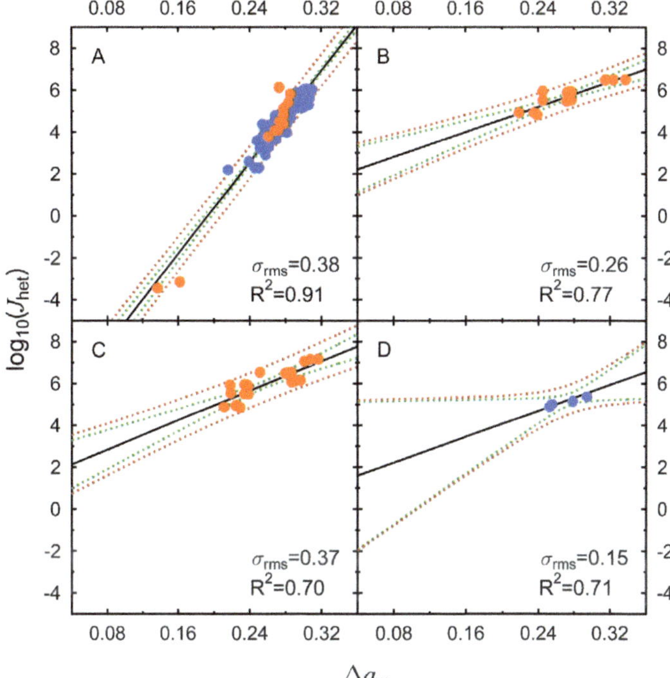

Fig. 4 The decadal log of heterogeneous ice nucleation rate coefficients, J_{het}, are shown as a function of Δa_w for (A) Kaolinite,[70,96] (B) aluminum oxide particles,[64] (C) iron oxide particles,[64] and (D) fungal spores.[97] The blue and red circles in (A) are taken from Murray et al.[96] and Pinti et al.,[70] respectively. The solid black line is a linear fit where dashed green and red lines represent confidence intervals and prediction bands, respectively, at a 95% level. The root mean square error, σ_{rms}, and the adjusted coefficient of determination, R^2, are given in each panel.

Finally, we use immersion freezing data from experiments by Iannone et al.[97] with fungal spores which reported median freezing temperatures, $f(\tilde{T}) = 0.5$, and corresponding fungal spore surface area in pure water droplets. Using eqn (3), $t_{nuc} = 1$ s, and calculating Δa_w from eqn (1), we derive J_{het} as a function of Δa_w and corresponding parameterisation as shown in Fig. 4D. Fit parameters are given in Table 1.

In the case of Al_2O_3 and Fe_2O_3 (Fig. 4B and C) the parameterisation shows increasing uncertainty as Δa_w decreases. However, even at $\Delta a_w = 0.08$ the uncertainty in J_{het} is $\leq \pm 1.5$ orders of magnitude. In the case of fungal spores (Fig. 4D) acting as IN there is considerable uncertainty as indicated by the calculated confidence and prediction bands for lower Δa_w. Adding more experimental freezing data obtained at specific IN surface areas and cooling rates to fill the voids with respect to Δa_w will benefit the fit quality. However, it can be expected that the presented parameterisation will not change significantly upon adding new data for the following reasons. When applying immersion freezing data of aqueous solutions with various a_w to describe the freezing point depression due to a solute, a relatively constant J_{het} and Δa_w value is observed by many studies.[41,43,57,73,75] In other words, one immersion freezing curve represents a single J_{het} and Δa_w value for varying a_w. Also, changes in A and t_{nuc} do not affect

a_w, T, or J_{het}. The thermodynamic state of the aqueous solution is not governed by changes in A and t_{nuc} and thus experiments employing different A and t_{nuc} will not derive freezing data that will affect J_{het} and Δa_w. It should be emphasised again that the kinetic process of nucleation is expressed by the thermodynamic parameters a_w and T.[35] Lastly, we have shown that the exponential relationship between J_{het} and Δa_w holds within an uncertainty of ± 2 orders of magnitude over a wide range in Δa_w. Therefore, under the assumption that the actual freezing data is sufficiently accurate to establish the relation between J_{het} and Δa_w, additional immersion freezing data will likely accumulate close to the parameterisations given in Fig. 4B, C, and D. These predictive capabilities are similar to those of the water activity based homogeneous ice nucleation theory that can, *e.g.*, be applied to uncover changes in a_w with temperature when freezing data do not follow the expected homogeneous freezing curve.[40,42,44,98]

Laboratory observations of immersion freezing can be described by the ABIFM only if J_{het} can be derived and Δa_w can be calculated from T and a_w using eqn (1). Any attempt to compare the appropriateness of the ABIFM without knowledge of either J_{het} or Δa_w is difficult. Previous studies typically report $f(T)$ and some of those give additional information allowing derivation of J_{het} and thus evaluation of ABIFM. We point out that previous studies which report $f(T)$ that can not be analyzed to yield J_{het}, can be compared with those that we have found to follow the ABIFM as presented in Fig. 4. For example, observations of $f(T)$ for immersion freezing by Illite dust in pure water by Niemand *et al.*[99] are in agreement with data from Broadley *et al.*[95] when comparing a similar IN surface area per droplet. This implies that freezing data from Niemand *et al.*[99] should also be well described by the ABIFM using the relationship given in Fig. 3E. It is important to note that Niemand *et al.*[99] also found a similar behavior of $f(T)$ for laboratory dusts as for natural dusts collected from the Saharan desert, Asia, Canary Island, and Israel. Thus, we would expect that the ABIFM of Illite dust can be applied to these mineral dust species for the whole atmospherically relevant ranges of T and a_w, and applicable for any IN surface areas and nucleation times.

We have shown that Δa_w can serve as the determinant for description of immersion freezing of aqueous solution droplets containing IN. The qualitative reasons why Δa_w can act as a determinant for immersion freezing is based on the fact that when a solute is added to pure water, a_w decreases since water molecules are bound in hydration shells of ions or by hydrogen bonds with other *e.g.* organic molecules.[40,44] Therefore water self-molecular interactions and thus molecular density fluctuations to form ice clusters are reduced. This in turn decreases the rate of formation of a critical ice embryo necessary to overcome the activation energy required for ice nucleation. This also means that the amount of time for ice nucleation to proceed at constant T will take longer for lower a_w. Density fluctuations and thus, formation of critical ice clusters, are considered in relation to particle surface area. A greater IN surface area allows a larger number of critical ice clusters to form by enhancing the probability for ice nucleation to occur. The result of a larger IN surface area is a decrease in nucleation time. Therefore, the effects of water-ion or hydrogen bonding molecular interactions[35] on bulk water can qualitatively explain the dependency of immersion freezing kinetics on a_w accounting for changes of IN surface area and nucleation time.

4 Atmospheric implications

To illustrate the usefulness of our model for describing atmospheric processes, we apply ABIFM to simulate the formation of cirrus and mixed-phase clouds both containing concentrations of mineral dust particles per volume of air capable of nucleating ice, N_{tot}, at concentrations of 500 L^{-1} (air) immersed in either aqueous solution or pure water droplets, respectively. This example considers only ice nucleation and ignores all other cloud microphysical effects, including homogeneous ice nucleation, liquid droplet growth, water vapour depletion due to the Bergeron–Wegener–Findeisen process,[100–102] ice crystal growth, entrainment and mixing, and ice crystal sedimentation. Hence, this exercise is strictly for the purpose of demonstrating the ability of the ABIFM to predict frozen fractions of droplets and ice crystal numbers only and not to simulate any physically realistic cloud.

Mineral dust concentrations are those chosen by Eidhammer *et al.*[103] from average concentrations measured from field campaigns[104,105] and the dust particles are assumed to be spherical with their diameters equal to 1 μm. For cirrus formation we employ an idealized atmospheric trajectory in which an air parcel having a constant water partial pressure, and thus a constant dew point, T_{dew}, is cooled with a rate of $r = 0.06$ K min^{-1}, corresponding to a vertical velocity $v = 10$ cm s^{-1} and an adiabatic lapse rate of 10 K km^{-1}.[89,106] For mixed-phase conditions T_{dew} is equal to ambient T, $r = 0.36$ K min^{-1}, and $v = 100$ cm s^{-1} assuming a saturated adiabatic lapse rate of 6 K km^{-1}.[107–109] Fig. 5A and B shows the resulting temperature and S_{ice} change with height for conditions typical of cirrus and mixed-phased clouds, respectively.

In order to predict ice particle production, J_{het} is first calculated employing the ABIFM for Illite (Table 1). Δa_w is derived from eqn (1) as a function of T and $a_w = \mathrm{RH} = [P_{H_2O}(T_{dew})]/[P^{\circ}_{H_2O}(T)]$, where $P_{H_2O}(T_{dew})$ is the water partial pressure and $P^{\circ}_{H_2O}(T)$ is the saturation vapour pressure over a plane surface of liquid water.[92] For particle diameters equal to 1 μm the Kelvin effect can be neglected.[15] As the air parcel is lifted, T decreases and RH and thus, particle a_w will

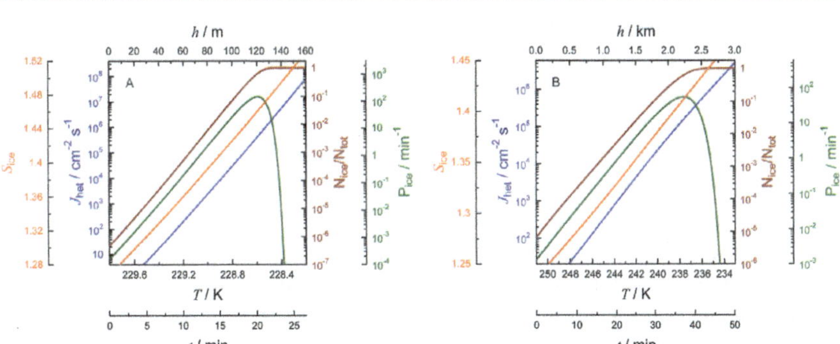

Fig. 5 Ice nucleation calculations are presented for typical cirrus (A) and mixed-phase (B) cloud conditions from Illite mineral dust using the water activity based immersion freezing description. Calculations of ice particle production rate, P_{ice}, and frozen fraction, N_{ice}/N_{tot}, where $N_{tot} = 0.5$ cm^{-3} (air) is the total number of ice nuclei and N_{ice} is the number of particles that nucleated ice as a function of temperature, T, time, t, and height, h, are shown using the water activity based immersion freezing model to derive heterogeneous ice nucleation rate coefficients, J_{het}. Ice supersaturation ratio, S_{ice} is calculated from T and dew point temperature, T_{dew}, assumed to be for cirrus conditions $T_{dew} = -45$ °C and for mixed-phase cloud conditions $T_{dew} = T$.

simultaneously increase. Corresponding Δa_w can then be calculated and J_{het} can be determined from the ABFIM. Ice particle production rates, P_{ice}, in units of min^{-1} are then calculated from the following equation,

$$P_{ice} = J_{het}(\Delta a_w)A(N_{tot} - N_{ice}),\tag{4}$$

and the cumulative total number of particles that nucleated ice per volume of air, N_{ice}, as a function of time, t, is

$$N_{ice} = \int P_{ice}dt,\tag{5}$$

where $A(N_{tot} - N_{ice})$ is the surface area available for ice nucleation which does not include the particles that have already nucleated ice. Eqn (5) is discretized using a time interval of 10 s. Ice nucleation is allowed to occur at any time, as long as $S_{ice} > 1$.

The frozen fraction of particles that nucleate ice, N_{ice}/N_{tot}, is shown in Fig. 5A for cirrus cloud conditions. As T cools and S_{ice} increases, RH and droplet a_w increases. For both decreasing T and increasing a_w, Δa_w increases swiftly (see Fig. 1), causing a rapid increase in J_{het} of about 5 orders of magnitude for a change in $T \approx 1$ K. P_{ice} depends exponentially on temperature for $T < 228.8$ K. The maximum value of $P_{ice} \simeq 140$ min^{-1} reflects rapid ice nucleation. As T further decreases, P_{ice} decreases rapidly due to the decreasing available IN surface area. N_{ice}/N_{tot} also increases exponentially and converges to its maximum value of 1.0 as P_{ice} converges to zero. In about 20 min the fraction of ice particle increases from 10^{-6} to 1.0. Due to slow upward motion, ice nucleation occurs in a small depth $h \approx 120$ m.

For mixed-phase cloud conditions shown in Fig. 5B, S_{ice}, J_{het}, P_{ice} and N_{ice}/N_{tot} show a similar trend with temperature except that these trends hold over a wider temperature range than for cirrus conditions. This is due to a less dramatic increase in J_{het} with decreasing T. Values of S_{ice} increase as T cools, however, $T_{dew} = T$ and therefore only pure cloud droplets are present, thus RH and droplet a_w is equal to 1.0 for the entire simulation. At $a_w = 1.0$, Δa_w changes only with T, different from the cirrus case. Here, J_{het} increases by 4 orders of magnitude for a change in temperature of about 10 K, a much more gradual increase than for the colder cirrus conditions. This in turn causes P_{ice} and N_{ice}/N_{tot} to change more gradually with T as well. Thus, ice nucleation proceeds at a much slower rate overall. The maximum value of $P_{ice} \simeq 50$ min^{-1} and then drops quickly to zero due to decreasing available IN surface area. In about 36 min, corresponding to a depth of about 2 km, the fraction of ice particles increases from 10^{-4} to 1.0.

These examples illustrate the ability of the ABIFM to predict heterogeneous ice nucleation in cloud resolving models, based on a physical parameterisation for immersion freezing kinetics of experimental laboratory data. Accordingly, any cloud model that determines air temperature and RH can use the parameterisations derived here for investigated IN types and immediately calculate J_{het} and thus, frozen fractions and total ice particle numbers. It is important to note that eqn (4) and 5 are always valid due to the fact that microphysical effects only affect the input variables, *e.g.* water partial pressure and number of available IN, and not the equations themselves.

ABIFM can be applied to field measurements which quantify ice crystal formation. For example, Ansmann *et al.*[110] find that in the tropics, heterogeneous ice nucleation occurs only in the presence of liquid clouds between $-20\,°\mathrm{C}$ and $-38\,°\mathrm{C}$. Fig. 5B clearly shows that saturated conditions are required before immersion freezing is significant at these temperatures and thus, ABIFM yields similar conclusions on cloud formation as observed Ansmann *et al.*[110] Furthermore, as pointed out by Broadley *et al.*[95] and corroborated by Niemand *et al.*,[99] Illite dust serves as a very good surrogate of various naturally occurring dusts. Thus, application of the ABIFM of Illite dust allows prediction of immersion freezing for atmospherically relevant ranges of T and a_w, mineral dust surface areas, and nucleation times.

The ABIFM should be easy to incorporate into cloud resolving models, especially those which already use the a_w based homogeneous ice nucleation theory.[45–49] ABIFM successfully describes immersion freezing for a wide variety of IN types, aqueous solutions, and experimental methods and thus, should also adequately describe the immersion freezing for cloud conditions. When included in cloud resolving models performing simulations and sensitivity tests, the applicability of this model to atmospheric cloud formation can be effectively assessed. Future experimental studies are needed to acquire data missing for certain IN types and new efforts are required to test the appropriateness of Δa_w as the determinant for immersion freezing in cloud system resolving models.

5 Conclusions

Immersion freezing temperatures and heterogeneous ice nucleation rate coefficients as a function of water activity, ice nuclei surface areas, and cooling rates have been determined for five distinct particles types acting as ice nuclei (IN). We found that water activity and temperature, expressed by the single parameter Δa_w, can describe immersion freezing temperatures and ice nucleation kinetics for all IN, surface areas, and cooling rates. The corresponding heterogeneous ice nucleation rate coefficients can be parameterized as a function of Δa_w for all investigated IN types in water and various aqueous solution droplets and with different IN surface areas and cooling rates. The effect of changing IN surface areas and cooling rates on the freezing and kinetic processes are captured by the new water activity based immersion freezing model (ABIFM) introduced here. The effect of IN surface area on freezing temperatures for particles of *N. atomus*, *T. pseudonana*, Pahokee Peat, Leonardite and Illite is, when IN surface area is increased by one order of magnitude, freezing temperatures increase by 1.7 K, 2.0 K, 1.2 K, 2.2 K, and 2.9 K, respectively. A cooling rate dependence on immersion freezing is also observed. Decreasing the cooling rate by one order of magnitude, increases freezing temperatures of particles of *N. atomus* and Leonardite by 1.4 K and for Illite, Pahokee Peat, and *T. pseudonana* by 0.5 K, 1.8 K, and 3.0 K, respectively. In addition, Δa_w accounts for freezing point depression due to the solute effect. Experimentally derived J_{het} for all investigated IN types are reproduced as a function of Δa_w within an applied conservative uncertainty estimate of ± 2 orders of magnitude. The new model successfully represents immersion freezing data obtained by different experimental methods and under different conditions. We demonstrate how ABIFM can be used to simulate immersion freezing for cirrus and mixed-phase cloud formation conditions allowing prediction of ice particle production and frozen droplet fractions in atmospheric models. The procedures introduced here can be applied to other IN types to expand the ABIFM to include a wider range of particle types. We suggest inclusion of the ABIFM in cloud system resolving models to evaluate its applicability under atmospheric conditions. In summary, the experimental data compiled here to evaluate our water activity based immersion freezing model, show it to holistically and elegantly describe the immersion freezing process.

Acknowledgements

We are grateful to J. Aller for helpful discussion and assisting in manuscript preparation. G. Bazargan is acknowledged for technical assistance. This research was supported by the Office of Science (BER), U. S. Department of Energy.

References

1 D. P. Wylie and W. P. Menzel, *J. Clim.*, 1999, **12**, 170–184.
2 D. Wylie, D. L. Jackson, W. P. Menzel and J. J. Bates, *J. Clim.*, 2005, **18**, 3021–3031.
3 M. B. Baker, *Science*, 1997, **276**, 1072–1078.
4 M. B. Baker and T. Peter, *Nature*, 2008, **451**, 299–300.
5 W. B. Rossow and R. A. Schiffer, *Bull. Am. Meteorol. Soc.*, 1999, **80**, 2261–2287.
6 T. Chen, W. B. Rossow and Y. C. Zhang, *Tellus, Ser. A*, 2000, **50**, 259–264.
7 P. Forster, V. Ramaswamy, P. Artaxo, T. Berntsen, R. Betts, D. W. Fahey, J. Haywood, J. Lean, D. C. Lowe, G. Myhre, J. Nganga, R. Prinn, G. Raga, M. Schulz and R. Van Dorland, in *Changes in Atmospheric Constituents and in Radiative Forcing*, Cambridge University Press, Cambridge, United Kingdom and New York, NY, USA, 2007, ch. 2, pp. 131–234.
8 J. Verlinde, J. Y. Harrington, G. M. McFarquhar, V. T. Yannuzzi, A. Avramov, S. Greenberg, N. Johnson, G. Zhang, M. R. Poellot, J. H. Mather, D. D. Turner, E. W. Eloranta, B. D. Zak, A. J. Prenni, J. S. Daniel, G. L. Kok, D. C. Tobin, R. Holz, K. Sassen, D. Spangenberg, P. Minnis, T. P. Tooman, M. D. Ivey, S. J. Richardson, C. P. Bahrmann, M. Shupe, P. J. DeMott, A. J. Heymsfield and R. Schofield, *Bull. Am. Meteorol. Soc.*, 2007, **42**, 248–252.
9 T. Storelvmo, C. Hoose and P. Eriksson, *J. Geophys. Res.*, 2011, **116**, D05207.
10 P. J. DeMott, A. J. Prenni, X. Liu, S. M. Kreidenweis, M. D. Petters, C. H. Twohy, M. S. Richardson, T. Eidhammer and D. C. Rogers, *Proc. Natl. Acad. Sci. U. S. A.*, 2010, **25**, 11217–11222.
11 Z. Li, F. Niu, J. Fan, Y. Liu, D. Rosenfeld and Y. Ding, *Nat. Geosci.*, 2011, **4**, 888–894.
12 W. K. Tao, J. P. Chen, Z. Li, C. Wang and C. Zhang, *Rev. Geophys.*, 2012, **50**, RG2001.
13 D. A. Hegg and M. B. Baker, *Rep. Prog. Phys.*, 2009, **72**, 056801.
14 W. Cantrell and C. Robinson, *Geophys. Res. Lett.*, 2006, **33**, L07802.
15 H. R. Pruppacher, and J. D. Klett, *Microphysics of Clouds and Precipitation*, Kluwer Academic Publishers, Netherlands, 1997.
16 G. Vali, *J. Aerosol Sci.*, 1985, **16**, 575–576.
17 A. J. Heymsfield, L. M. Miloshevich, C. Twohy, G. Sachse and S. Oltmans, *Geophys. Res. Lett.*, 1998, **25**, 1343–1350.
18 P. J. DeMott, D. C. Rogers, S. M. Kreidenweis, Y. Chen, C. H. Twohy, D. Baumgardner, A. J. Heymsfield and K. R. Chan, *Geophys. Res. Lett.*, 1998, **25**, 1387–1390.
19 P. J. DeMott, K. Sassen, M. R. Poellot, D. Baumgardner, D. C. Rogers, S. D. Brooks, A. J. Prenni and S. M. Kreidenweis, *Geophys. Res. Lett.*, 2003, **30**(14), 1732.
20 M. Seifert, J. Ström, R. Krejci, A. Minikin, A. Petzold, J.-F. Gayet, U. Schumann and J. Ovarlez, *Atmos. Chem. Phys.*, 2003, **3**, 1037–1049.
21 W. Haag, B. Kärcher, J. Ström, A. Minikin, U. Lohmann, J. Ovarlez and A. Stohl, *Atmos. Chem. Phys.*, 2003, **3**, 1791–1806.
22 D. J. Cziczo, D. M. Murphy, P. K. Hudson and D. S. Thomson, *J. Geophys. Res.*, 2004, **109**, D04201.
23 D. C. Rogers, P. J. DeMott, S. M. Kreidenweis and Y. L. Chen, *J. Atmos. Oceanic Technol.*, 2001, **18**, 725–741.
24 J. Cozic, S. Mertes, B. Verheggen, D. J. Cziczo, S. J. Gallavardin, S. Walter, U. Baltensperger and E. Weingartner, *J. Geophys. Res.*, 2008, **113**, D15209.
25 M. Kamphus, M. Ettner-Mahl, T. Klimach, F. Drewnick, L. Keller, D. J. Cziczo, S. Mertes, S. Borrmann and J. Curtius, *Atmos. Chem. Phys.*, 2010, **10**, 8077–8095.

26 V. I. Khvorostyanov, H. Morrison, J. A. Curry, D. Baumgardner and P. Lawson, *J. Geophys. Res.*, 2006, **111**, D02201.
27 I. Sölch and B. Kärcher, *Q. J. R. Meteorol. Soc.*, 2010, **136**, 2074–2093.
28 P. R. Field, A. J. Heymsfield, B. J. Shipway, P. J. DeMott, K. A. Pratt, D. C. Rogers, J. Stith and K. A. Prather, *J. Atmos. Sci.*, 2012, **69**, 1066–1079.
29 M. P. Meyers, P. J. DeMott and W. R. Cotton, *J. Appl. Meteorol.*, 1992, **31**, 708–721.
30 V. T. J. Philips, S. T. Donner and L. J. Garner, *J. Atmos. Sci.*, 2007, **64**, 738–761.
31 A. Ansmann, M. Tesche, P. Seifert, D. Althausen, R. Engelmann, J. Fruntke, U. Wandinger, I. Mattis and D. Müller, *J. Geophys. Res.*, 2009, **114**, D17208.
32 C. Hoose, J. E. Kristjánsson, J. P. Chen and A. Hazra, *J. Atmos. Sci.*, 2010, **67**, 2483–2503.
33 C. D. Westbrook and A. J. Illingworth, *Geophys. Res. Lett.*, 2011, **38**, L14808.
34 G. de Boer, H. Morrison, M. D. Shupe and R. Hildner, *Geophys. Res. Lett.*, 2011, **38**, L01803.
35 T. Koop, B. P. Luo, A. Tsias and T. Peter, *Nature*, 2000, **406**, 611–614.
36 T. Koop, B. P. Luo, U. M. Biermann, P. J. Crutzen and T. Peter, *J. Phys. Chem. A*, 1997, **101**, 1117–1133.
37 T. Koop, *Z. Phys. Chem.*, 2004, **218**, 1231–1258.
38 B. Riechers, F. Wittbracht, A. Hutten and T. Koop, *Phys. Chem. Chem. Phys.*, 2013, **15**, 5873–5887.
39 T. Koop and B. Zobrist, *Phys. Chem. Chem. Phys.*, 2009, **11**, 10839–10850.
40 B. Zobrist, U. Weers and T. Koop, *J. Chem. Phys.*, 2003, **118**, 10254–10261.
41 P. A. Alpert, D. A. Knopf and J. Y. Aller, *Atmos. Chem. Phys.*, 2011, **11**, 5539–5555.
42 D. A. Knopf and M. D. Lopez, *Phys. Chem. Chem. Phys.*, 2009, **11**, 8056–8068.
43 D. A. Knopf, P. A. Alpert, B. Wang and J. Y. Aller, *Nat. Geosci.*, 2011, **4**, 88–90.
44 D. A. Knopf and Y. J. Rigg, *J. Phys. Chem. A*, 2011, **115**, 762–773.
45 B. Kärcher and U. Lohmann, *J. Geophys. Res.*, 2002, **107**(D2), 4010.
46 B. Kärcher and U. Lohmann, *J. Geophys. Res.*, 2002, **107**(D23), 4698.
47 B. Kärcher, J. Hendricks and U. Lohmann, *J. Geophys. Res.*, 2006, **111**, D01205.
48 J. Hendricks, B. Kärcher and U. Lohmann, *J. Geophys. Res.*, 2011, **116**, D18206.
49 S.-M. Fan, J. P. Schwarz, J. Liu, D. W. Fahey, P. Ginoux, L. W. Horowitz, H. Levy, Y. Ming and J. R. Spackman, *J. Geophys. Res.*, 2012, **117**, D23205.
50 D. A. Knopf, B. P. Luo, U. K. Krieger and T. Koop, *J. Phys. Chem. A*, 2003, **107**, 4322–4332.
51 D. A. Knopf, B. P. Luo, U. K. Krieger and T. Koop, *J. Phys. Chem. A*, 2005, **109**, 2707–2709.
52 K. S. Carslaw, S. L. Clegg and P. Brimblecombe, *J. Phys. Chem.*, 1995, **99**, 11557–11574.
53 M. Massucci, S. L. Clegg and P. Brimblecombe, *J. Phys. Chem. A*, 1999, **103A**, 4209–4226.
54 S. L. Clegg and P. Brimblecombe, *J. Phys. Chem. A*, 2005, **109**, 2703–2706.
55 S. L. Clegg, P. Brimblecombe and A. S. Wexler, *J. Phys. Chem. A*, 1998, **102**, 2137–2154.
56 B. Zobrist, C. Marcolli, T. Koop, B. P. Luo, D. M. Murphy, U. Lohmann, A. A. Zardini, U. K. Krieger, T. Corti, D. J. Cziczo, S. Fueglistaler, P. K. Hudson, D. S. Thomson and T. Peter, *Atmos. Chem. Phys.*, 2006, **6**, 3115–3129.
57 D. A. Knopf and S. Forrester, *J. Phys. Chem. A*, 2011, **115**, 5579–5591.
58 O. Möhler, P. J. DeMott, G. Vali and Z. Levin, *Biogeosciences*, 2007, **4**, 1059–1071.
59 D. A. Knopf, B. Wang, A. Laskin, R. C. Moffet and M. K. Gilles, *J. Geophys. Res.*, 2010, **37**, L11803.
60 C. Hoose and O. Möhler, *Atmos. Chem. Phys.*, 2012, **12**, 9817–9854.
61 B. J. Murray, D. O'Sullivan, J. D. Atkinson and M. E. Webb, *Chem. Soc. Rev.*, 2012, **41**, 6519–6554.
62 S. T. Martin, *Chem. Rev.*, 2000, **100**, 3403–3453.
63 B. Zuberi, A. Bertram, C. A. Cassa, L. T. Molina and M. J. Molina, *Geophys. Res. Lett.*, 2002, **29**(10), DOI: 10.1029/2001GL014289.
64 C. M. Archuleta, P. J. DeMott and S. M. Kreidenweis, *Atmos. Chem. Phys.*, 2005, **5**, 2617–2634.
65 D. A. Knopf and T. Koop, *J. Geophys. Res.*, 2006, **111**, D12201.
66 B. Zobrist, C. Marcolli, T. Peter and T. Koop, *J. Phys. Chem. A*, 2008, **112**, 3965–3975.
67 D. Niedermeier, S. Hartmann, R. A. Shaw, D. Covert, T. F. Mentel, J. Schneider, L. Poulain, P. Reitz, C. Spindler, T. Clauss, A. Kiselev, E. Hallbauer, H. Wex, K. Mildenberger and F. Stratmann, *Atmos. Chem. Phys.*, 2010, **10**, 3601–3614.
68 F. Lüönd, O. Stetzer, A. Welti and U. Lohmann, *J. Geophys. Res.*, 2010, **115**, D14201.
69 A. Welti, F. Lüönd, Z. A. Kanji, O. Stetzer and U. Lohmann, *Atmos. Chem. Phys.*, 2012, **12**, 9893–9907.
70 V. Pinti, C. Marcolli, B. Zobrist, C. R. Hoyle and T. Peter, *Atmos. Chem. Phys.*, 2012, **12**, 5859–5878.
71 A. P. Fornea, S. D. Brooks, J. B. Dooley and A. Saha, *J. Geophys. Res.*, 2009, **114**, D13201.
72 B. Wang and D. A. Knopf, *J. Geophys. Res.*, 2011, **116**, D03205.
73 Y. J. Rigg, P. A. Alpert and D. A. Knopf, *Atmos. Chem. Phys.*, 2013, **13**, 6603–6622.

74 G. D. Franc and P. J. DeMott, *J. Appl. Meteorol.*, 1998, **37**, 1293–1300.
75 P. A. Alpert, D. A. Knopf and J. Y. Aller, *Phys. Chem. Chem. Phys.*, 2011, **13**, 19882–19894.
76 M. Gavish, R. Popovitzbiro, M. Lahav and L. Leiserowitz, *Science*, 1990, **250**, 973–975.
77 L. H. Seeley and G. T. Seidler, *J. Chem. Phys.*, 2001, **114**, 10464–10470.
78 B. Zobrist, T. Koop, B. P. Luo, C. Marcolli and T. Peter, *J. Phys. Chem. A*, 2007, **111**, 2149–2155.
79 M. E. Wise, K. J. Baustian and M. A. Tolbert, *Proc. Natl. Acad. Sci. U. S. A.*, 2010, **107**, 6693–6698.
80 K. J. Baustian, M. E. Wise and M. A. Tolbert, *Atmos. Chem. Phys.*, 2010, **10**, 2307–2317.
81 B. Wang, A. Laskin, T. Roedel, M. K. Gilles, R. C. Moffet, A. V. Tivanski and D. A. Knopf, *J. Geophys. Res.*, 2012, **117**, D00V19.
82 K. J. Baustian, D. J. Cziczo, M. E. Wise, K. A. Pratt, G. Kulkarni, A. Gannet Hallar and M. A. Tolbert, *J. Geophys. Res.*, 2012, **117**, D06217.
83 G. P. Schill and M. A. Tolbert, *J. Phys. Chem. A*, 2012, **116**, 6817–6822.
84 B. Wang, A. T. Lambe, P. Massoli, T. B. Onasch, P. Davidovits, D. R. Worsnop and D. A. Knopf, *J. Geophys. Res.*, 2012, **117**, D16209.
85 C. Marcolli, S. Gedamke, T. Peter and B. Zobrist, *Atmos. Chem. Phys.*, 2007, **7**, 5081–5091.
86 A. Welti, F. Lüönd, O. Stetzer and U. Lohmann, *Atmos. Chem. Phys.*, 2009, **9**, 6705–6715.
87 D. Niedermeier, R. A. Shaw, S. Hartmann, H. Wex, T. Clauss, J. Voigtländer and F. Stratmann, *Atmos. Chem. Phys.*, 2011, **11**, 8767–8775.
88 B. Ervens and G. Feingold, *Atmos. Chem. Phys.*, 2012, **12**, 5807–5826.
89 B. Kärcher and U. Lohmann, *J. Geophys. Res.*, 2003, **108**, 4402.
90 S. Brunauer, P. H. Emmett and E. Teller, *J. Am. Chem. Soc.*, 1938, **60**, 309–319.
91 A. U. Dogan, M. Dogan, M. Onal, Y. Sarikaya, A. Aburub and D. E. Wurster, *Clays Clay Miner.*, 2006, **54**, 62–66.
92 D. M. Murphy and T. Koop, *Q. J. R. Meteorol. Soc.*, 2005, **131**, 1539–1565.
93 D. A. Knopf, *J. Phys. Chem. A*, 2006, **110**, 5745–5750.
94 D. A. Knopf, T. Koop, B. P. Luo, U. G. Weers and T. Peter, *Atmos. Chem. Phys.*, 2002, **2**, 207–214.
95 S. L. Broadley, B. J. Murray, R. J. Herbert, J. D. Atkinson, S. Dobbie, T. L. Malkin, E. Condliffe and L. Neve, *Atmos. Chem. Phys.*, 2012, **12**, 287–307.
96 B. J. Murray, S. L. Broadley, T. W. Wilson, J. D. Atkinson and R. H. Wills, *Atmos. Chem. Phys.*, 2011, **11**, 4191–4207.
97 R. Iannone, D. I. Chernoff, A. Pringle, S. T. Martin and A. K. Bertram, *Atmos. Chem. Phys.*, 2011, **11**, 1191–1201.
98 B. Zobrist, C. Marcolli, D. A. Pedernera and T. Koop, *Atmos. Chem. Phys.*, 2008, **8**, 5221–5244.
99 M. Niemand, O. Moehler, B. Vogel, H. Vogel, C. Hoose, P. Connolly, H. Klein, H. Bingemer, P. DeMott, J. Skrotzki and T. Leisner, *J. Atmos. Sci.*, 2012, **69**, 3077–3092.
100 A. Wegener, *Thermodynamik Der Atmosphäre*, Johann Ambrosius Barth, Leipzig, 1911.
101 T. Bergeron, *Proc. 5th Assembly U.G.G.I., Lisbon*, 1935, **2**, 156.
102 W. Findeisen and G. Schulz, *Forschungs und Erfahrungsberichte des Reichswetterdienstes*, 1944, **A**, 1–15.
103 T. Eidhammer, P. J. DeMott and S. M. Kreidenweis, *J. Geophys. Res.*, 2009, **114**, D06202.
104 P. J. DeMott, D. J. Cziczo, A. J. Prenni, D. M. Murphy, S. M. Kreidenweis, D. S. Thomson, R. Borys and D. C. Rogers, *Proc. Natl. Acad. Sci. U. S. A.*, 2003, **100**, 14655–14660.
105 M. S. Richardson, P. J. DeMott, S. M. Kreidenweis, D. J. Cziczo, E. J. Dunlea, J. L. Jimenez, D. S. Thomson, L. L. Ashbaugh, R. D. Borys, D. L. Westphal, G. S. Casuccio and T. L. Lersch, *J. Geophys. Res.*, 2007, **112**, D02209.
106 P. J. DeMott, D. C. Rogers and S. M. Kreidenweis, *J. Geophys. Res.*, 1997, **102**, 19575–19584.
107 U. Lohmann, J. Feichter, C. C. Chuang and J. E. Penner, *J. Geophys. Res.*, 1999, **104**, 9169–9198.
108 J. Fan, M. Ovtchinnikov, J. M. Comstock, S. A. McFarlane and A. Khain, *J. Geophys. Res.*, 2009, **114**, D04205.
109 U. Lohmann and C. Hoose, *Atmos. Chem. Phys.*, 2009, **9**, 8917–8934.
110 A. Ansmann, M. Tesche, D. Althausen, D. Mueller, P. Seifert, V. Freudenthaler, B. Heese, M. Wiegner, G. Pisani, P. Knippertz and O. Dubovik, *J. Geophys. Res.*, 2008, **113**, D04210.

DISCUSSIONS

General discussion

DOI: 10.1039/c3fd90035e

Professor Johnston opened the discussion of the paper by Dr Hamilton by asking: What are the particle size distributions in your experiments? Are the distributions always within the size range that can be analyzed by ATOFMS?

Dr Hamilton replied: Particles were generally between 50–350 nm, with a median at the start of the experiment between about 90–150 nm. Therefore there were always particles within the correct size range of the ATOFMS.

Professor Johnston enquired: At the top of Table 2 in your paper, you state that a lab solution of 1 M GLY in water gives "none detected" by ATOFMS, yet ions are observed by ESI-MS. Were there no particles in the size range of ATOFMS when this solution was aerosolized? Or is it the case that these particles are detected by light scattering but no ions are detected in the corresponding mass spectra?

Dr Hamilton responded: The particles were at an appropriate size range for the ATOFMS. The problem is that the ATOFMS laser only sees particles if they absorb near the laser wavelength. The solution of GLY in water does not contain any species that absorb strongly enough to allow the ATOFMS to see the particles. This is the case with pure ammonium sulfate also.

Professor Nizkorodov remarked: I have a minor comment. Your statement "ATOFMS could detect the increase in the light-absorbing organics from the GLY + AS reactions" can be misinterpreted that ATOFMS has the capability to distinguish particles with different absorption spectra (which is not the case). In your ATOFMS experiments, you detect multiple fragments of compounds some of which absorb light strongly and some insignificantly. It would be useful to correlate the observed mass spectra with the absorption coefficient of the aerosol to know which of the ions correspond to fragments of the light-absorbing species. I also would like to bring to your attention a relevant recent paper by X. Wang, et al.,[1] which detected these types of compounds with a similar type of single-particle mass spectrometer in urban air.

1 X. Wang et al., Evidence for high molecular weight nitrogen-containing organic salts in urban aerosols, *Environ. Sci. Technol.*, 2010, **44**, 4441.

Professor Rudich commented: In flow tube experiments we observed some imidazole formation. However, we do not see a significant change in absorption (although AMS shows formation of imidazoles). In contrast, nitration of aerosols coated by polyaromatic hydrocarbons shows fast and significant absorption. It is important to verify time scale and conditions for the formation of brown carbon by secondary sources. In contrast, primary sources such as diesel and biomass burning particles can be significant.[1–5]

1 M. Trainic, A. AboRiziq, A. Lavi, J. M. Flores and Y. Rudich, The optical, physical and chemical properties of the products of glyoxal uptake on ammonium sulfate seed aerosols, *Atmos. Chem. Phys.*, 2011, **11**, 9697–9707.
2 M. Trainic, A. Abo Riziq, A. Lavi, and Y. Rudich, The role of interfacial water in the heterogeneous uptake of glyoxal by mixed glycine and ammonium sulfate aerosols, *J. Phys. Chem. A*, 2012, **116**, 5948–5957.
3 G. Adler, J. M. Flores, A. Abo Riziq, S. Borrmann, and Y. Rudich, Chemical, physical, and optical evolution of biomass burning aerosols: A case study, *Atmos. Chem. Phys.*, 2011, **11**, 1491–1503.
4 J. W. Lu, M. J. Flores, A. Lavi, A. Abo-Riziq and Y. Rudich, Changes in the optical properties of Benzo[*a*]pyrene-coated aerosols with heterogeneous reactions with NO_2 and NO_3, *Phys. Chem. Chem. Phys.*, 2011, **13**, 6484–6492.
5 G. Adler, C. Erlick, Y. Rudich and A. Abo Riziq, The effect of intrinsic organic carbon on the optical properties of diesel soot, *Proc. Natl. Acad. Sci. U. S. A.*, 2010, **107**(15), 6699–6704.

Dr Hamilton asked: In your experiments where you see imidazole formation but no significant change in absorption properties, were your particles exposed to any light source?

In the GLY and AS chamber experiments presented, the ATOFMS only really saw the particles after they were exposed to natural sunlight. We therefore think that either photolysis or photochemistry is leading to different chemical species that are changing the UV absorption of the particles, not the imidazole species in particular.

Professor Rudich responded: Our particles were not exposed to light. We could try this, though.

Professor McNeill opened the discussion of the paper by Professor Nizkorodov by remarking: You state that in your system unique trace species are responsible for the majority of light absorption. Would you agree that this is true for the absorption peaks you observe at wavelengths longer than 400 nm, but that multiple species contribute to the increasing absorption with decreasing wavelength below 400 nm which you observe, and which is commonly observed for brown carbon in ambient aerosols?

Professor Nizkorodov replied: Indeed, this statement specifically applies to the absorption of visible radiation (400–700 nm). Only a small fraction of organic compounds in a typical atmospheric aerosol will have molecular structures that favour strong visible light absorbance. However, a larger fraction of organic aerosol compounds will absorb in the near-ultraviolet range, and most of them will contribute to absorbance in the ultraviolet range.

Professor Johnston queried: There is little discussion of LA and PA in the paper. LA and PA did not give colored products, while KLA did. Do LA and PA react by similar mechanisms to what you describe for KLA, but do not have the correct structural feature to give highly conjugated products that absorb in the visible region? Assuming that the reaction chemistry is the same, would it be possible to write an algorithm to predict the products from a large number of precursor molecules to find out which ones have structural features that might be able to yield an absorbing product and which ones don't?

Professor Nizkorodov answered: We have not characterized the reaction products of LA and PA reactions in detail because they did not lead to light-absorbing products, but we believe they do react with ammonia. For example, we observe formation of nitrogen-containing products in reactions of both alpha-pinene/ozone SOA and limonene/ozone SOA with ammonium sulphate, but only the limonene/ozone SOA products absorb visible radiation (Laskin *et al.*, in preparation). The initial reaction chemistry is likely the same, and involves conversion of carbonyls to imines, aldol condensates, and heterocyclic compounds, but the unique molecular structure of KLA makes it possible to achieve stable products with a larger degree of conjugation, while the structures of PA and LA do not. One possible approach to model the chemical differences between KLA, PA, and LA is to supply the starting compounds and all chemical reactions in a multi-step "brewing" simulator of the type described in the litera-ture (D. R. Fooshee, T.B. Nguyen, S.A. Nizkorodov, J. Laskin, A. Laskin and P. Baldi, COBRA: A computational brewing application for predicting the molecular composition of organic aerosols, *Environ. Sci. Technol.*, 2012, **46**, 6048–6055). The main challenge will be to specify a complete set of relevant chemical reactions. As we mention in the manuscript, the mechanisms we discussed are likely incomplete.

Dr Nozière added: Just to reinforce the conclusion of this paper that less than 1% of the organic material in aerosol can be responsible for most of the absor-bance, I just wanted to point out that this is even more the case for compounds absorbing far in the visible. Such compounds are highly conjugated and the further in the visible they absorb, the higher their absorption cross-section. Thus much smaller concentrations are needed to account for the observed absorbance.

Professor Nizkorodov concurred: We agree that it is uncommon for the types of organic molecules found in the atmosphere to absorb green and red light. The commonly discussed products of atmospheric oxidation of volatile organics (carbonyls, carboxylic acids, alcohols, peroxides, organonitrates, organosulfates, *etc.*) hardly absorb any radiation above 300 nm. The types of molecules we nor-mally associate with bright colours (pigments, dyes, *etc.*) are known to occur in small concentrations in biological and biomass-burning aerosols. Observation of the formation of such a pigment through secondary atmospheric chemistry of KLA and related molecules was a surprise.

Dr Hamilton remarked: I like the idea of evaporative ageing. Are you sure that using above ambient temperatures does not result in different chemistry than would be seen in the real atmosphere, where cloud processes are happening at much lower temperatures?

Professor Nizkorodov responded: In a related publication (T. B. Nguyen, P. B. Lee, K. M. Updyke, D. L. Bones, J. Laskin, A. Laskin and S. A. Nizkorodov,

Formation of nitrogen- and sulfur-containing light-absorbing compounds accelerated by evaporation of water from secondary organic aerosols, *J. Geophys. Res.*, 2012, **117**, D01207) we examined the effect of the evaporation temperature on the resulting absorption spectra of limonene/ozone + ammonium sulphate solutions, and we found it was minor. The similarity of the absorption spectra observed in slow room temperature aqueous chemistry experiments and in much faster evaporation experiments (Fig. S4 in the ESI section of the manuscript) suggests that the elevated evaporation temperature is not a significant issue. The evaporation of bulk solutions goes faster at elevated temperatures, and the 45 degrees Celsius temperature of the evaporating solution used in this work was chosen for experimental convenience.

Professor McNeill addressed Professor Rudich and Professor Nizkorodov: Regarding the conditions at which laboratory studies of aerosol brown carbon formation are typically performed: the high concentrations of ammonium sulfate or other salts typically used in bulk aqueous solutions for kinetic studies are not too high to be realistic. In fact, due to bulk solubility limitations, they fall short of the very high concentration of salts that may be encountered in deliquesced atmospheric aerosol at RH < 100%. This gap may lead to error when we apply kinetic formulations based on bulk solutions to aerosol conditions. Those studies are often performed at a range of organic concentrations which include atmospherically relevant concentrations and possibly extend beyond that for the purposes of kinetic analysis. Studies performed under evaporation conditions are realistic in that they are designed to reproduce aerosol cloud cycling.

Professor Rudich replied: We have studied the heterogeneous reaction between glyoxal gas and proxies for three atmospheric aerosol types; ammonium sulfate (AS), glycine and glycine–AS 1 : 100 (molar ratio). Both the optical extinction cross section at $\lambda = 355$ nm and mobility diameter increased following the reaction under a broad range of RH values (30–90%), indicating that the reaction is relevant for a wide range of atmospheric conditions. We found that at low RH values, below the deliquescence point, the reactions occur in interfacial monolayers of water, supporting previous findings about the importance of interfacial water in heterogeneous reactions. The aerosols exhibit a trend of increasing growth in physical and optical cross sections with decreasing seed aerosol size, as well as a clear dependence on ambient RH values. For small particles with near-zero extinction efficiency (Q_{ext}) values, the growth due to the reaction induces the greatest increase in optical extinction cross section due to a combined effect of changes in optical properties and in size. AMS analyses of the reactions show that the main reaction products are glyoxal oligomers, and a small contribution from the formation of various C–N compounds, identified as imidazoles. Our results suggest that the reactions at RH values below deliquescence occur on interfacial water layers and their optical enhancement is mainly due to enhanced scattering and geometric cross section growth. We see a reduced effect of absorption after the reaction. Therefore, at least at these conditions and exposure time, the main consequence of the reaction is in the scattering and not in the absorption part.[1,2]

1 M. Trainic, A. AboRiziq, A. Lavi, J. M. Flores and Y. Rudich, The optical, physical and chemical properties of the products of glyoxal uptake on ammonium sulfate seed aerosols, *Atmos. Chem. Phys.*, 2011, **11**, 9697–9707.
2 M. Trainic, A. Abo Riziq, A. Lavi, and Y. Rudich, The role of interfacial water in the heterogeneous uptake of glyoxal by mixed glycine and ammonium sulfate aerosols, *J. Phys. Chem. A*, 2012, **116**, 5948–5957.

Professor Nizkorodov also replied: We agree that evaporation experiments make it possible to achieve relevant concentrations of ammonium sulphate and organics that likely exists in wet aerosols. The major advantage of the bulk evaporation experiments of the type described in this paper is the wide range of relative concentrations of ammonium sulphate and organics that can be examined in a relatively short time. In the future, it will be important to carry out evaporation experiments, in which aerosols with well-defined content of ammonium sulphate, water, and organics are evaporated upon exposure to dry air. The organic vapour/wet aerosol equilibration time is substantially shorter compared to that in the organic vapour/bulk solution system, and this can potentially affect the products observed in the evaporation process.

Professor Rudich enquired: Has any of these aqueous phase processes been tested in any global or regional model?

Professor McNeill responded: Aqueous-phase SOA formation has been parameterized in varying degrees of detail in CMAQ and GEOS-CHEM. In-particle brown carbon formation has not. We are in the process of using our box model, GAMMA, to develop accurate and appropriate parameterizations for use in large-scale models.

Professor Herrmann suggested: Maybe a test experiment can be performed to concentrate solutions by N_2 blowing and without heating to check for possible effects of elevated temperatures?

Professor Nizkorodov answered: Thank you for the suggestion. Perhaps even more relevant would be to dry aerosolized solutions with a diffusion drier, and compare the properties of the optical aerosol with and without drying. We are currently pursuing such experiments with our collaborators.

Professor Herrmann opened a general discussion of Dr Hamilton's and Professor Nizkorodov's papers by addressing Dr Hamilton: Is the oxidation of acetylene the best way to introduce glyoxal in a chamber? Why was this chosen?

Dr Hamilton explained: The oxidation of acetylene allows glyoxal to be formed *in situ* in the chamber and therefore in theory is being produced throughout the chamber at the same time. Glyoxal is a very sticky compound and it is difficult to introduce it into the chamber directly. It will stick to the walls of the chamber and tubing while it's being introduced.

Professor Abbatt queried: What do we know about the sources of OH within an aerosol particle from, for example, photolysis? Even though gas phase OH may not make its way into the particle, is it true that species within a particle are shielded from OH?

Professor Herrmann replied: This has been studied in the available CAPRAM papers to quite some extent. *In situ* radical sources can substantially contribute to aqueous OH, roughly by about 30% or so depending on scenario. Transfer from the gas phase is important and even exceeds the *in situ* sources.

Dr Colussi asked: What is the chemistry of browning? We found that both the absorbance and fluorescence of brown solutions decrease more than linearly with dilution (A. G. Rincon, M. I. Guzman, M. R. Hoffmann and A. J. Colussi, *J. Phys. Chem. A*, 2009, **113**, 10512–10520) as direct evidence of supramolecular interactions among chromophores. Have you considered this phenomenon?

Professor Nizkorodov responded: The proposed chemistry leading to the formation of potential chromophores is discussed in Scheme 2 of the manuscript. We were concerned with the potential contribution of supramolecular complexes to light absorption. Two observations ruled this possibility out. (1) The chromophores separated in the chromatographic column, where the analyte is significantly diluted with the eluent. We observed rapid equilibria between the mixture components, which made their chromatographic separation challenging, but they were definitively molecular species. (2) The absorbance at 500 nm was directly proportional to the concentration of the dissolved material. One would have expected a non-linear dependence for supramolecular complexes.

Dr Colussi added: Regarding supramolecular interactions, it is important to realize that in order to compare experiments we need to report concentration ranges. 1 mM is not the same as 100 mM.

Professor Nizkorodov commented: The initial concentrations of KLA used in these experiments ranged from about 2 mM to 20 mM. The concentration of the resulting chromophore(s) is unknown but likely two orders of magnitude smaller. The solutions were additionally diluted by an order of magnitude to verify the linearity of the absorbance. We therefore believe that the chromophores were molecules, not supramolecular aggregates.

Dr Colussi opened the discussion of the paper by Professor Carslaw by asking: How would your approach detect the need for new chemistry/physics? How do you break out of conceptual frameworks?

Professor Carslaw answered: Detecting the need for structural changes in a model (missing processes, poor representation of processes *etc.*) can be tackled in many ways. But an ensemble of model runs that spans the uncertainty space of the model is a good place to start. If the suite of model runs does not reproduce measurements, then this is a strong suggestion that there are missing processes. Most models have only been tested for a very narrow range of uncertainties (such as a few one at a time parameter perturbations). In such a case, it is possible that model–observation discrepancy is due to not exploring the uncertainty space of the present model, so the search for missing processes may be unwarranted.

Professor Wahner commented: As stated in the paper the selection of the 28 parameters is a first step and possibly not covering all indirect influence on SOA formation. Can the model tool presented be used to analyse the cross correlation between the parameters chosen and therefore used to optimize the parameterisation of the model?

Professor Carslaw replied: Yes, the emulator can be used to predict the model output for any combination of parameters within the uncertainty space that we have sampled. It is possible to identify a part of the parameter space that agrees optimally with measurements.

Mr Hesson enquired: I think you have gone back from the emulator and done more model runs, if so, how do these results compare to the emulator? What are the main parameters leading to differences between the model and your emulator? Is there any experimental/observational data that could help reduce this?

Professor Carslaw responded: The emulator is built using 268 model runs covering the 28 uncertain parameters. The emulator is then validated by doing additional runs at untried points in parameter space. There are no particular parameters that cause the model to deviate from the emulator – the emulator is essentially able to calculate the full response surface of the model in the 28 dimensions. The main issue is the deviation of the ensemble from observations, which is a current research topic.

Professor Johnston asked: The main contributor to uncertainty of N3 and N50 appears to be SOA. Is this primarily due to the role of SOA in nanoparticle growth or to losses caused by the contribution of SOA to the condensational sink? If organics are included in the nucleation mechanism, could this change (increase) the uncertainty of N3 and N50?

Professor Carslaw explained: The uncertainty in SOA is globally rather a small contributor to the total uncertainty in N3 and N50 (see the maps in the paper). At some locations, like the one we highlighted in the paper (Hyytiala) it is a more important parameter. The main cause seems to be the effect of the SOA on nuclei growth as well as the sink effect of SOA adding to the mass (and hence surface area) of the existing aerosol. This is shown in Fig. 3 in the paper, where there is a reduction in N50 with an increase in SOA. You rightly identify the role of organics in nucleation as an additional source of uncertainty. This will increase the uncertainties we have calculated here.

Professor McFiggans remarked: Picking up on the parametric *vs.* structural uncertainty separation within the study, there appear to be some definitions within the approach aiming to address parametric uncertainty that are defined according to structural limitations. First, the size cut used to define CCN is clearly an approximation that will lead to its own uncertainty and should probably be convolved with a supersaturation climatology in each grid cell to properly evaluate the effect. More importantly, and this relates to both this paper and your group's recent work aiming to quantify radiative effects of biogenic secondary organic aerosol (Scott *et al.*, *Atmos. Chem. Phys. Discuss.*, 2013, **13**, 16961–17019), ignoring the treatment of semi-volatile material when considering a process that is completely reliant on a semi-volatile thermally-reversible process appears to be a serious structural limitation. Is there any way of estimating whether such a structural uncertainty is zeroth order and invalidates the parameter space investigation, whether it's effects are negligible, or whether it lies somewhere between?

Professor Carslaw replied: The supersaturation conditions are relevant when calculating cloud drop activation, but it makes more sense to define a fixed supersaturation when reporting CCN, because the model can then be compared with CCN measurements which also use a defined supersaturation. I agree that semi-volatile compounds need to be incorporated for us to get a more complete picture of CCN uncertainty.

Dr Colussi stated: I have reservations about the generality of your procedures regarding, for example, isoprene global emissions. Most experts consider that the dry deposition of biogenic olefins is a settled issue. Therefore they would recommend refining our understanding of OH radical production in the ozonation of olefins, putting aside the fact that maybe there is not that much isoprene coming out of forest canopies.

Dr Colussi further added: Your approach is very interesting. However, I am intrigued about whether it would recognize that in order to balance global sources *vs.* sinks of biogenic olefins, the chemistry underlying OH radical production in olefin ozonation is intimately linked to the magnitude of the dry deposition of isoprene in forest canopies.

Professor Carslaw responded: Emulation and uncertainty analysis of a particular model output (such as OH) can account for the uncertainties in any process you specify in the experimental design. So if you think isoprene dry deposition is important, then the model ensemble should span the uncertainty range of that process.

Professor Abbatt opened the discussion of the paper by Dr Knopf by asking: Do the results from your heterogeneous immersion freezing model imply that the solutes are not adsorbing to the solid surface that is nucleating ice?

Dr Knopf answered: The immersion freezing experiments are conducted using aqueous NaCl and $(NH_4)_2SO_4$ solutions. We assume, and this has not been explicitly stated in our paper, that these solutions do neither react with the applied IN surfaces nor significantly adsorb to the IN surface. We do not expect the IN surface to be altered by the solutes or when solute concentration increases. The main reason that solute adsorption may not play a significant role in immersion freezing lies in the observed behaviour of the immersion freezing curves as a function of water activity compared to the similar behaviour of homogeneous freezing curves obtained in absence of an ice nucleating substrate. This includes immersion freezing from pure water and aqueous solution droplets. This strongly suggests that the immersion freezing point depression is due to changes of the aqueous solution (expressed as water activity) and not IN surface. It seems highly unlikely that for the different IN types, including surfaces being inorganic, organic, biological, and surfactant in nature, the adsorption of different solutes applied here results in freezing point depressions that all follow a shifted ice melting curve description. Additional evidence comes from studies by Zobrist *et al.* (2006, 2008)[1,3] and Koop and Zobrist (2009)[2] where it has been shown that for same IN immersed in different aqueous solutions with varying water activities, all freezing points fall on the same immersion freezing curve when plotted as a function of water activity. Furthermore, all investigated IN-aqueous solution systems yielded immersion freezing curves that follow a shifted ice melting curve description. This very strongly supports the notion that adsorbed solutes cannot play a crucial role in the immersion freezing process.

However, for reactive or oxidizing aqueous solutions such as H_2SO_4 or HNO_3, increasing the solute may lead to changes in the IN surface properties due to reactions, thereby altering the IN freezing abilities as has been shown in previous studies by us and others (see *e.g.* Knopf and Koop (2006),[4] Hoose and Möhler (2011)[5]).

1 B. Zobrist, C. Marcolli, T. Koop, B. P. Luo, D. M. Murphy, U. Lohmann, A. A. Zardini, U. K. Krieger, T. Corti, D. J. Cziczo, S. Fueglistaler, P. K. Hudson, D. S. Thomson and T. Peter, *Atmos. Chem. Phys.*, 2006, **6**, 3115–3129.
2 T. Koop and B. Zobrist, *Phys. Chem. Chem. Phys.*, 2009, **11**, 10839–10850.
3 B. Zobrist, C. Marcolli, T. Peter and T. Koop, *J. Phys. Chem. A*, 2008, **112**, 3965–3975.
4 D. A. Knopf and T. Koop, *J. Geophys. Res.*, 2006, **111**, D12201.
5 C. Hoose and O. Möhler, *Atmos. Chem. Phys.*, 2012, **12**, 9817–9854.

Dr Murray commented: The water activity based heterogeneous freezing model presented in this paper is a valuable and elegant approach to describing heterogeneous ice nucleation. There is one specific aspect which I would like to comment on. I agree that ice nucleation for a uniform material made up of a single uniform species might be described by a single $J(a_w)$ (or $J(T)$). We have referred to such materials as single component materials and their nucleation can therefore be described by a single component stochastic model, where the term 'single component' refers to the uniformity of the sample under investigation.[1-3] However, it needs to be proven that a particular material is made up of a single component with respect to its ice nucleating ability. In the case where there are multiple materials in a sample which can nucleate ice or where there are rare defects which can trigger nucleation, then the observed freezing needs to be described using $J(a_w)$ values appropriate for each component.

An example of a material which has been experimentally shown to nucleate ice according to the single component stochastic model is kaolinite KGa1b from the Clay Mineral Society.[1] This was shown in two independent ways. First, the number of liquid droplets decreases approximately exponentially with time at constant temperature. Second, J derived from experiments with widely varying cooling rates produced self-consistent J values. We showed that KGa1b was consistent with a single component stochastic model.

If the material were multiple component then some droplets contain particles with larger values of J than others under the same conditions. Hence, under isothermal conditions the number of liquid droplets will not decrease exponentially, but will have a faster initial decay followed by a slower decay. Also, J values derived from a multiple component material for a range of cooling rates will not be in agreement. An example of a multiple component material which we have worked with is NX-Illite (this data is used in the paper under discussion).[2] In our NX-illite article,[2] we show that the isothermal curves are non-exponential and that determining J using a single component model results in J values from the faster cooling rates being larger than those from slower cooling rate experiments, which is clearly incorrect. In that article we went on to show that in order to describe nucleation by NX-illite it was necessary to use a multiple component model with a distribution of J values. This may have been due to the mixed mineralogy of NX-illite (illites samples are notoriously difficult, if not impossible, to obtain in a state approaching pure). It was recently shown that feldspar minerals dominate ice nucleation in mineral dusts,[4] and since feldspar is a minor component of NX-illite it is perhaps not surprising that a multiple component model is required for this material.

The NX-illite data from Broadley *et al.*[2] is used in this paper in Fig. 3E, but J values are derived using a single component stochastic model. It is noteworthy that there is significant spread in the data around the 'best fit' line. Is this spread a result of the implicit approximation of NX-illite being a single component system? This might become more apparent if the data taken from different cooling rate experiments were colour coded. If higher values of J are from experiments with faster cooling rates then this would be consistent with a multiple component model (see Fig. 6a in Broadley *et al.*[2]).

Could the inappropriate use of a single component model also explain some of the spread around the best fit line for the other materials in Fig. 3? Some biological particles, for example, are well known to vary significantly particle to particle (*i.e.* they are multiple component). Bacteria are a good example of this. In our recent review,[3] we presented literature data for bacteria which clearly shows that in many situations only a small fraction of bacterial cells have the ability to nucleate ice. This is an extreme case of a multiple component material. If you were to derive a $J(a_w)$, for bacteria based on droplets freezing experiments (with many bacterial per droplet), using the single component stochastic model you would be making the implicit assumption that each bacterial cell had the same nucleating ability– which is clearly incorrect. Bacteria are an extreme case, but could there be an issue with many of the materials plotted up in Fig. 3?

Knopf *et al.* also plot J values derived for kaolinite in Fig. 4 (I think the data is for KGa1b only –is this correct?). Given we have shown this material fits well to a single component stochastic model it is encouraging to see the data from Pinti *et al.* and Murray *et al.*'s papers in tight agreement around a best fit line.

In summary, the water activity model presented by Knopf *et al.* is a valuable tool and elegantly links nucleation in pure water to nucleation in solution droplets in one model framework. My only caution is that it may not be appropriate to represent individual materials with a single $J(a_w)$ and a multiple component approach may be required for some, but not all, ice nucleating materials.

1. B. J. Murray, S. L. Broadley, T. W. Wilson, J. D. Atkinson and R. H. Wills, *Atmos. Chem. Phys.*, 2011, **11**, 4191–4207.
2. S. L. Broadley, B. J. Murray, R. J. Herbert, J. D. Atkinson, S. Dobbie, T. L. Malkin, E. Condliffe and L. Neve, *Atmos. Chem. Phys.*, 2012, **12**, 287–307.
3. B. J. Murray, D. O'Sullivan, J. D. Atkinson and M. E. Webb, *Chem. Soc. Rev.*, 2012, **41**, 6519–6554.
4. J. D. Atkinson, B. J. Murray, M. T. Woodhouse, T. F. Whale, K. J. Baustian, K. S. Carslaw, S. Dobbie, D. O'Sullivan and T. L. Malkin, *Nature*, 2013, DOI: 10.1038/nature12278.

Dr Knopf responded: We are aware that particles residing in the atmosphere and immersed in aqueous solution or cloud droplets are complex and may possess various heterogeneous ice nucleation rate coefficients (J_{het}) as a function of temperature (T). We have previously addressed the effect on compositional differences in mineral dust, which likely depends also on particle size, on heterogeneous ice nucleation serving as explanation of data scatter between different laboratory ice nucleation studies (Knopf and Koop, 2006).[1] In fact, the physical model ABIFM is the best means to provide J_{het} as a function of T and water activity, a_w (=RH), for various ice nucleating species. In response to Dr Murray, we show that ABIFM is valid for both uniform and heterogeneous IN, in other words, the presence of minor ice nucleating components within investigated IN does not significantly impact the results, conclusions, and validity of ABIFM. To arrive at these conclusions we discuss i) previous immersion freezing data cited by Dr Murray, ii) experimental data scatter and uncertainties in freezing kinetics, iii) mathematical models and concepts applied to interpret immersion freezing data, and iv) the significance of ABIFM for interpretation of the underlying freezing process. We show that i) with current experimental methods and commonly applied fit based frozen fraction descriptions the nature of the IN surface and its effect on immersion freezing cannot be conclusively determined,

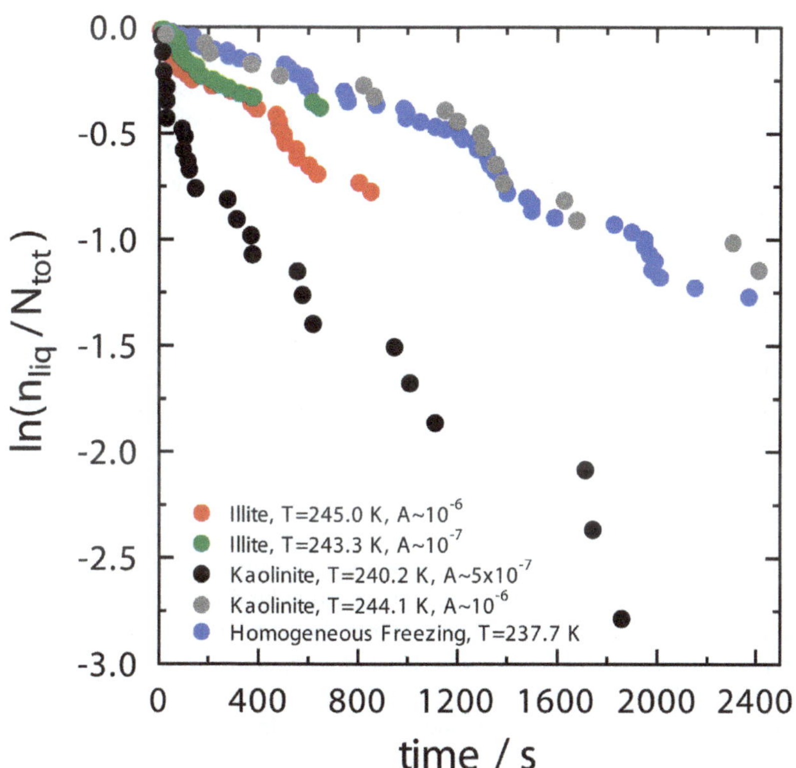

Fig. 1 The change in the fraction of unfrozen droplets as a function of time for homogeneous freezing of water droplets and immersion freezing from water droplets containing Illite and Kaolinite particles taken from Murray *et al.* (2011)[2] and Broadley *et al.* (2012).[3]

ii) the presence of minor components potentially influencing ice nucleation is shown by ABIFM to be within the typical data scatter, iii) ABIFM predicts in a physical manner immersion freezing temperatures and kinetics for IN with uniform and heterogeneous surfaces, and iv) the existence of ABIFM, in fact, questions if immersion freezing is initiated at one specific active site. Finally, the ABIFM alludes to immersion freezing as a result of many individual ice germ formation processes owing to its strikingly similar behaviour with regard to homogeneous ice nucleation. First, previous isothermal experiments do not allow a clear statement about the nature of the IN surface. Fig. 1 shows frozen fractions over time at constant temperature measured by Murray *et al.* (2011)[2] and Broadley *et al.* (2012)[3] for homogeneous ice nucleation and heterogeneous ice nucleation by Kaolinite and Illite in water droplets. Fig. 1 clearly shows that all investigated droplet systems exhibit similar data scatter and highly variable exponential time dependencies at various intervals. Therefore claiming that the unfrozen fraction of Kaolinite/water and water droplets behave linearly with time and that Illite/water droplets do not, with subsequent implications for the nature of IN surface, is not supported. Considering experimental uncertainties and data scatter in Fig. 1, the only claim that can be substantiated is that homogeneous and heterogeneous ice nucleation is time dependent, a conclusion which is in agreement with our data and ABIFM.

Second, we address the effect of cooling rate on J_{het}. As suggested, we attempted to determine a cooling rate dependence on J_{het} and found that for all investigated IN types, all J_{het} values lie within ± 1 order of magnitude when changing the cooling rate by about 1 order of magnitude. These results can be entirely explained by the typical data scatter of laboratory measurements and may be largely related to uncertainties in experimental temperature which strongly affect kinetics measurements as outlined by Riechers *et al.* (2013).[4] It is important to note that in the case of Kaolinite and Illite, J_{het} values fall within ± 1 order of magnitude in the studies of Murray *et al.* (2011)[2] and Broadley *et al.* (2012).[3] Taking into account the effect of a temperature uncertainty on J_{het} and uncertainty in individual IN surface areas, the data scatter is too large in current experiments to make a firm statement regarding a cooling rate dependence and thus on the nature of the IN surface with implications for immersion freezing.

Third, we address the applied ice nucleation analysis. Both Murray *et al.* (2011)[2] and Broadley *et al.* (2012)[3] apply a multi-component immersion freezing description by fitting to the frozen fraction data in order to reproduce frozen fraction data employing statistical fitting functions. It is very obvious that this is a circular process which does not reveal any physical mechanism. A mathematical two-component system, for example, will always have twice as many fitting parameters and provide a better fit than a single-component system. We have discussed in length in Rigg *et al.* (2013)[5] that this kind of immersion freezing analysis does not yield physical insights into the freezing process but can only serve the purpose of data correlation. Furthermore, models fitted to frozen fraction data in this way, can only be applied to the temperature range, surface area range, and the nucleation time periods investigated in the experiments in which the data were generated. Additionally, data from Murray *et al.* (2011)[2] and Broadley *et al.* (2012),[3] employ only pure water, therefore any claim to ice nucleation behavior in aqueous solution cannot be made. A theory which describes physical phenomena has to be falsified by experimental data instead of fitting experimental data to represent experimental data, a procedure which obviously provides no "proof" of the underlying IN surface characteristics and ice nucleation processes. For further discussion the reader is referred to Rigg *et al.* (2013).[5]

Fourth, Dr Murray asks whether the spread around the best fit line could be explained by a multi-component IN system. The fitting equation we used is independent of the ABIFM model and can be arbitrarily chosen. Thus, a higher order polynomial fit representation would result in less spread and much smaller root mean square (RMS) error, just like a multi-component system will always yield a better representation than a single-component system. Instead, the ABIFM model shows and emphasizes the novel finding that there is a fundamental physical relationship between the water molecular structure, IN surface, and ice nucleation kinetics. This relation must be investigated and understood. The applied linear fits describe the data within typical experimental data scatter and provide continuous functions for use in atmospheric models. It is important to note, that this J_{het} *versus* Δa_w relationship is based on theory and physical observables, in contrast, *e.g.* to using statistical fitting functions.

Dr Murray's comment on bacteria freezing is not based on a discussion of J_{het} but on a yet physically unfounded parameter termed "active sites" (see *e.g.* discussion by Rigg *et al.*, 2013[5]). We have not studied immersion freezing by bacteria, but established ABIFM as a physical model of immersion freezing in contrast to the active sites approach. Koop and Zobrist (2009)[6] on the other hand, have determined freezing points from one bacteria type in water and aqueous solutions and have shown that the freezing points all follow a shifted ice melting curve obtained from data having very little scatter and from reproducible experiments, even when using organic or inorganic solutes. This emphasizes the importance of the nature of the water and aqueous solution, expressed as water activity, for immersion freezing also in the case of bacteria acting as IN. Their data suggest, although not experimentally shown by us, that it is highly likely that bacteria will exhibit a similar immersion freezing behaviour as observed for presented investigated IN. However, more experiments are necessary to reach a final conclusion.

From our vast immersion freezing data set which covers a large range of water activities, we can address the following question related to Dr Murray's comment: how important or significant is the consideration of a heterogeneous ice nucleating surface to a water activity description? Fig. 3 in our paper clearly shows that for all investigated IN, J_{het} has been determined spanning over 10 orders of magnitude. The RMS error is between ±0.45 and ±1.31 orders of magnitude (this error is relative since it is based on a linear fit, see discussion above) and any effect of minor components on the overall J_{het} would have to be within the data scatter of about 1 order of magnitude for J_{het}. Resolving any effect smaller than one order of magnitude requires higher temperature accuracy, higher counting statistics and more accurate IN surface area measurements, not previously achieved. The data clearly show that, if other minor components did impact ice nucleation, this would change the freezing kinetics by less than 1 order of magnitude, which we consider to be minor (and unavoidable applying current techniques, as discussed above), compared with the dominant temperature and water activity dependence of J_{het} of the major component.

We also discuss the presence of feldspar as a secondary component of Illite, as suggested by Dr Murray, which may significantly contribute to ice nucleation and explain the data scatter in Fig. 3. Although we have not studied the composition of Illite used in our study (Illite IMt-1, Silver Hill, Montana, USA), K-feldspar was not previously observed in this clay standard (Köster, 1996).[7] Despite this fact, our freezing temperatures and J_{het} data and data from Broadley *et al.* (2012)[3] are in excellent agreement within the scatter of the data supporting our previous points.

The concept of a distribution of ice active sites or freezing probabilities and multiple components certainly fails in the case of 1-nonadecanol (Gavish *et al.*, 1990;[8] Popovitzbiro *et al.*, 1995;[9] Majewski *et al.*, 1995[10]), a uniform self-assembled monolayer, that remains in the same solid phase as water activity is changed (Knopf and Forrester, 2011)[11] and acts as an IN. Corresponding J_{het} derived from different studies applying different 1-nonadecanol surface areas and cooling rates agree excellently (Knopf and Forrester, 2011,[11] Zobrist *et al.*, 2007[12]), and follow the ABIFM. Thus, for the uniform and non-uniform IN types we investigated, the ABIFM can be used. In order to better understand immersion freezing, the molecular interaction of water at the IN-water/aqueous phase and overlying water structure should be investigated. It is more likely that several "water domains" above the IN surface are present at once in which ice germs can form. The number and size of these "water domains" may be affected by the IN surface and presence of solutes. This is of course speculative and future research is needed to investigate these mechanisms. However, this might be the reason why immersion freezing behaves analogously to what has been observed for homogeneous freezing of water and aqueous solutions following water activity (Bullock and Molinero, 2013).[13]

1 D. A. Knopf and T. Koop, *J. Geophys. Res.*, 2006, **111**, D12201.
2 B. J. Murray, S. L. Broadley, T. W. Wilson, J. D. Atkinson and R. H. Wills, *Atmos. Chem. Phys.*, 2011, **11**, 4191–4207.
3 S. L. Broadley, B. J. Murray, R. J. Herbert, J. D. Atkinson, S. Dobbie, T. L. Malkin, E. Condliffe and L. Neve, *Atmos. Chem. Phys.*, 2012, **12**, 287–307.
4 B. Riechers, F. Wittbracht, A. Hutten and T. Koop, *Phys. Chem. Chem. Phys.*, 2013, **15**, 5873–5887.
5 Y. J. Rigg, P. A. Alpert and D. A. Knopf, *Atmos. Chem. Phys.*, 2013, **13**, 6603–6622.
6 T. Koop and B. Zobrist, *Phys. Chem. Chem. Phys.*, 2009, **11**, 10839–10850
7 H. M. Köster, *Clay Minerals*, 1996, **31**, 417–422.
8 M. Gavish, R. Popovitzbiro, M. Lahav and L. Leiserowitz, *Science*, 1990, **250**, 973–975.
9 R. Popovitzbiro, J. L. Wang, J. Majewski, E. Shavit, L. Leiserowitz and M. J. Lahav, *J. Am. Chem. Soc.*, 1994, **116**, 1179–1191.
10 J. Majewski, R. Popovitzbiro, W. G. Bouwman, K. Kjaer, J. Alsnielsen, M. Lahav and L. Leiserowitz, *Chem.-Eur. J.*, 1995, **1**, 304–311.
11 D. A. Knopf and S. M. Forrester, *J. Phys. Chem. A*, 2011, **115**, 5579–5591.
12 B. Zobrist, T. Koop, B. P. Luo, C. Marcolli and T. Peter, *J. Phys. Chem. C*, 2007, **111**, 2149–2155.
13 G. Bullock and V. Molinero, *Faraday Discuss.*, 2013, DOI: 10.1039/C3FD00085K.

Dr Colussi enquired: Ice nucleation is an interfacial process. However, you have ignored that the surfaces of aqueous electrolytes are charged, and sustain electrical potential gradients. In such systems, thermodynamics requires that the electrochemical rather than the chemical potential be uniform throughout all phases. In other words, the activity of water is not the same at the surface and in the bulk. Do you think this is an important consideration?

Dr Knopf answered: Charges of the electrolytes at the interface between the IN surface and the bulk aqueous solution may affect the structure of the water close to the interface with subsequent effects on immersion freezing. However, it should be noted that the presence and configuration of electrolytic anions and cations at the IN surface-aqueous phase interface, in particular for these low temperatures, and for the various IN surfaces studied here (inorganic, organic, biological, and surfactant) is unknown. Similar to the discussion of the role of solutes adsorbed to the IN (please see also our previous response to Professor Abbatt), if electrolytic charges at the IN-aqueous phase interface play a crucial role, one would expect a variety of freezing points and kinetics values which would deviate from a shifted ice melting curve when applying different aqueous solutions, varying water activities, and various types of IN. However, our data and previous data by the Koop group show that immersion freezing as a function of (bulk) water activity and temperature exhibits very similar behaviour for all investigated and vastly different systems which is also shared by homogeneous ice nucleation of aqueous solution droplets. Freezing temperatures and kinetics follow remarkably well horizontally shifted ice melting curves, as in the case of homogeneous ice nucleation in absence of an ice nucleating substrate. In the case of IN in pure water droplets, water molecules will interact with the IN. Hypo-thetically speaking, our experiments allude to the notion that this water layer adjacent to the IN surface may not be significantly different when solutes, including charged electrolytes, are present. By addition of solutes the amount of "free" bulk water and thus water activity is decreased leading to a similar trend of heterogeneous freezing point depression as observed for homogeneous ice nucleation. Clearly, a fundamental understanding of these effects requires further experiments and theoretical calculations to infer the role of water activity at the interface and in the bulk for different solutes at supercooled temperatures.

Professor McFiggans opened a general discussion of Professor Carslaw's and Dr Knopf's papers: Further exploring this "parametric" *vs.* "structural" tension, if the concept presented in *e.g.* paper 24 (Knopf *et al.*) for a water activity based heterogeneous ice nucleation relationship were required and a host (emulated) model were structurally incapable of carrying the parameterisation, how would exploring/optimising parameter space in the emulator identify a necessity to change to a different more physically-meaningful structural approach?

Professor Carslaw responded: In my view, efforts to improve the structure of models should be driven by their ability (or not) to reproduce reality – *i.e.*, measurements. If a new aerosol mechanism is discovered, then the onus is on the discoverer to demonstrate that it improves our ability to describe the real world. If the uncertainty range of an existing model spans the measurements, then the advantage of a more complex model mechanism will not be apparent. Our models are only as good as our ability to verify their performance, and that assessment must take into account the uncertainties.

Professor Coe remarked: The process of testing model sensitivities you describe offers the potential for real advance in understanding where models need further information and processes need more development. This does though rely on sufficient measurement data available to challenge the model. Do you believe we have the necessary amount of measurement data with the neces-sary accuracy to provide adequate constraint?

Professor Carslaw replied: I don't have a clear yes or no answer to this question. We are currently assessing the availability of measurements to constrain the models as you suggest. Because we are trying to evaluate and constrain global aerosol microphysics/chemistry models, we need microphysics and chemistry measurements, which do not have global coverage. So we need clever ways of integrating and synthesising the data. We also need to determine whether the existing measurements constrain the key uncertain parameters in environments where they have the greatest influence on the uncertainty in forcing, for example. I suspect we will discover that more measurements are needed, but I hope we can also use such analyses to suggest where we should prioritise new measurements to maximise the usefulness.

Professor Coe commented: Whilst this is an excellent and informative approach, at present it seems to me that we are not sufficiently advanced in our understanding of atmospheric aerosols that we can rely solely on systematic model testing to inform us of key gaps in process knowledge, and we should also be looking to continue to conduct investigative laboratory and field studies of potentially important processes. Would you agree with this and how do you see the balance of these approaches in the future?

Professor Carslaw concurred: I agree. There is still plenty of scope for investigative laboratory and field measurements, and there are many excellent examples. But on balance, I would say that, as an aerosol community, we are beginning to stray too far towards natural enquiry and we don't include model uncertainty sufficiently at the heart of what we do. With a thorough analysis of model uncertainty assessed against measurements I think we can achieve some quick wins, leading to more rapid improvement in our models than we are achieving at present. If we performed an expert elicitation on the most important aerosol processes to reduce uncertainty in aerosol radiative forcing, I suspect it might not map well to the community's current activities.

Dr Knopf observed: When applying ice crystal numbers observed in the field to models, one should be aware that there were field measurements that suffered from ice crystal shattering on airborne probes thereby significantly enhancing observed ice crystal numbers. This makes elucidating the underlying nucleation mechanism from models more difficult and indicates the challenge of determining *in situ* ice crystal numbers. The challenge is that in the field we cannot observe *in situ* ice nucleation. If we probe the atmosphere, there is either an aerosol particle or an ice crystal. What is a physically founded parameter that can be applied to describe ice nuclei in ambient and laboratory measurements? In the field we determine the number of ice nuclei among all particles in, *e.g.*, 1 L of air. This may allow parameterization of ice nuclei numbers but does not yield mechanistic insight of the ice nucleation process since this parameter is not physically founded. For a physical understanding more parameters are needed such as the nature of active sites, ice nucleus surface area, exposure to supersaturated conditions, *etc.*

Mr Durham communicated in reply: My own question (below) relates to the liquid phase – for the ice phase discussed here, can we not assume that the ambient availability of ice nuclei is simply 'sufficient' or 'in excess'? If so, the extent to which nuclei develop into meaningful hydrometeors will be primarily a function of adiabatic cooling and will favour the more active nuclei, as demonstrated in Poster P23. Regarding shattering, in conversation with Prof. Coe it appears that the UK FAAM aircraft will be in a position to collect ice without shattering in 12 months time. My own interest is the extent to which such cloud ice, like cloud water complies with conventional physical chemistry. As discussed with Dr Knopf after delivery of his paper, in respect of ice I am interested in what proportion of the crystal matrix is composed of molecules other than water, and therefore has potential for returning pollutants to the terrestrial and ocean surface, which hopefully will be informed by the new FAAM facility.

Dr Cox noted: Prof Carslaw reports impressive progress in reduction of uncertainty in calculation of boundary layer nucleation rates in global chemistry–climate models. I wish to ask whether the uncertainties can be broken down into terrestrial and marine sources. Could the uncertainty in marine particle production be further improved if processes involving nucleation of iodine aerosols were included in the model? Is the state of knowledge of marine iodine chemistry sufficient to justify testing this potentially important influence on atmospheric particle production?

Professor Carslaw responded: Marine particle nucleation driven by iodine compounds was not included as an uncertainty. But it could easily be added and its contribution to overall uncertainty ranked alongside the other uncertainties.

Dr Allan said: This statistical approach is clearly a very powerful way forward in assessing the impacts of various parameters and processes, however I can confidently predict that the subject of structural uncertainties (*i.e.* problems with the model) will be an ongoing source of debate because most people within the community will have their own favourite mechanisms that cannot all be sensibly included in GLOMAP. The best way to assess these arguments would be to apply these statistical tests to other models within the community. On this basis, could this work form the basis for an objective benchmarking exercise of multiple models?

Professor Carslaw answered: Structural uncertainty in models is very important and needs to be considered alongside (and together with) parametric uncertainty. The AEROCOM multi-model intercomparison initiative is a good way to do what you are suggesting, and we are making plans to do that. It will require a lot of model runs, but this is a relatively quick and inexpensive way to understand the key uncertainties, rather than working on each problem from the bottom up without considering how it will improve our understanding of aerosol effects on climate or air quality. It is worth also pointing out that an analysis of parametric uncertainty may also encompass the effects of structural differences between models. For example, if we know for sure that the total SOA production cannot exceed 360 Tg a^{-1} regardless of the complexity of the model that predicts it, then we may be able to estimate an upper limit for its contribution to uncertainty. This may not always be the case if two models predict the same SOA production but assume entirely different properties of the SOA. But in many cases I think a parametric uncertainty assessment is a good indicator of likely influence.

Dr Taatjes remarked: The approach that Prof. Carslaw outlines is a very useful one, and I would encourage extension of such an approach to modeling laboratory experiments, for example chamber measurements, in addition to the tropospheric modeling discussed in the paper. Reducing uncertainty in parameters is usually accomplished by these controlled laboratory studies, and as parameters become more constrained, structural deficiencies in the model can no longer be hidden by parametric variation. The uncertainty analysis can and should therefore direct field measurements and laboratory investigations to best minimize parametric uncertainty.

Professor Carslaw assented: I agree, there is great potential for such techniques to be applied in the design of laboratory experiments. Most experiments are designed for experimental ease – such as doing a series of experiments at a constant temperature, followed by another set at constant gas concentration. A proper design of experiments in which the uncertain parameter space is sampled optimally could help to convert experimental data into more accurate parameterisations for models.

Professor Heard commented: In a similar manner to Professor Carslaw's analysis in this paper, in the gas phase atmospheric modelling community, a lot of sensitivity and uncertainty analyses have been performed to see which reactions play controlling roles in determining the concentrations of key atmospheric targets for models, for example OH, O_3 and the production rate of O_3. Sensitivity analysis allows the study of the relationship between the input parameters and the output values of a model whereas uncertainty analysis estimates output uncertainties from input uncertainties.[1] Local sensitivity analysis enables a local sensitivity index to be calculated showing the rank order of the parameters (concentrations or photolysis frequencies) that control the production or loss of, for example, the OH radical. Global uncertainty methods include the Morris-One-At-A-Time (MOAT) analysis and a Monte Carlo analysis with Latin Hypercube Sampling. This allows the rate parameters, product branching ratios and constrained concentrations to be varied within their uncertainty interval. In this way, the fundamental parameters that have the most effect on OH, HO_2 or other target species can be identified – and is helpful in prioritising which processes warrant further study in the laboratory. For the conditions found at Cape Grim in Tasmania during the SOAPEX-2 campaign, the three most significant parameters which came from a MOAT analysis were $J(O\ ^1D)$ and the rate constants for the reactions of $O(^1D)$ with H_2O vapour, and the quenching of $O(^1D)$ by N_2 to form $O(^3P)$.[1] Indeed, the rate coefficient for the latter reaction was remeasured independently by three groups, and was found to be significantly larger than previously determined, decreasing the calculated OH production rate *via* the reaction of $O(^1D)$ with H_2O in the atmosphere by roughly 15% in the mid-troposphere.[2]

Similarly, as Dr Taatjes has highlighted, sensitivity analysis is common in the combustion field, to see which parameters control important model targets such as flame velocities, and which can lead to improving the operational parameters of practical devices.

1 R. Sommariva *et al.*, *Atmos. Chem. Phys.*, 2004, **4**, 839–856.
2 A. R. Ravishankara *et al.*, *Geophys. Res. Lett.*, 2002, **29**, DOI: 10.1029/2001GL014850.

Professor McFiggans opened the discussion of the closing remarks by Dr Murphy by stating: To some degree, the spread of topics covered in this Discussion has reflected the submissions received from an audience predominantly comprising atmospheric chemists and physicists. It is unlikely that the same proportion of authors whose work relates directly to aerosol fates and impacts would consider the Faraday Discussion meeting to be as relevant a forum as the current audience do.

Dr Murphy replied: Thank you for pointing this out.

Professor Artaxo commented: The hygroscopic growth of amazonian aerosol particles was measure at 1.13–1.20 for particles at 100–300 nm range. This means that water content alters optical properties in a significant way. The difficult to measure "real" optical properties at ambient conditions remains a challenge in Amazonia. To model these properties it is even more difficult because of the strong absorption properties of primary aerosol particles, and the large amount of "brown carbon" present in the natural Amazonian atmosphere.

Dr Murphy concurred: I agree

Dr Colussi observed: Aerosols are highly disperse systems, where surfaces play an important role. I think that in these Discussions it has not been fully realized that physics and chemistry at the interface are different from those in the gas and liquid phases.

Professor Coe asked: You showed satellite data which provides evidence of an increase in aerosol optical depth over the last decade in the area of the Persian Gulf and Arabian Sea. Increases in population across the region, large increases in standard of living and construction, and agricultural practises could play a role, could you comment on the likely reasons for the dust enhancements observed?

Dr Murphy replied: I don't know the reasons for increased dust. I think your questions identifies some of the plausible causes. Helen Worden at NCAR has pointed out to me that the GRACE satellite gravity data indicate groundwater depletion in this region.[1] This might indicate links of the dust production to hydrology.

1 K. A. Voss, J. S. Famiglietti, M. Lo, C. de Linage, M. Rodell, and S. C. Swenson, Groundwater depletion in the Middle East from GRACE with implications for transboundary water management in the Tigris-Euphrates-Western Iran region, *Water Resour. Res.*, 2013, **49**, DOI: 10.1002/wrcr.20078.

Professor Held remarked: Throughout the meeting, the importance of deposition processes was mentioned regarding aerosol transformation, effects and fate. With respect to climate impacts, where are the most urgent research needs on aerosol deposition? Do we need more and/or better aerosol deposition flux measurements? Do we need refined or new parameterizations of deposition processes? With respect to aerosol chemistry, is there a conceptual framework to study interactions of heterogeneous chemistry and turbulent transport in the atmosphere?

Professor Carslaw added: Dry and wet deposition are important uncertainties in the global budget of CCN. They are actually more important than any of the processes discussed here (see Lee *et al.*, *Atmos. Chem. Phys.*, 2013, **13**, 8879–8914). However, these processes may be much less important for the uncertainty in forcing, which is determined by the change in aerosol, since these processes probably haven't changed over the industrial period. If a removal process acts in proportion to the aerosol abundance in the pre-industrial and present day, then there will be some cancellation of errors in the calculation of forcing. Given the importance of dry deposition for global CCN uncertainty (Lee *et al.*, 2013), it would be interesting to see whether land use change has caused a significant change in CCN acting through changes in dry deposition.

Mr Mason noted: Regarding the shadow region (reduction in irradiance) shown by the cloud picture on the last slide. Given the angular dependence of light scattering by a given particle it's interesting to consider that it's possible to

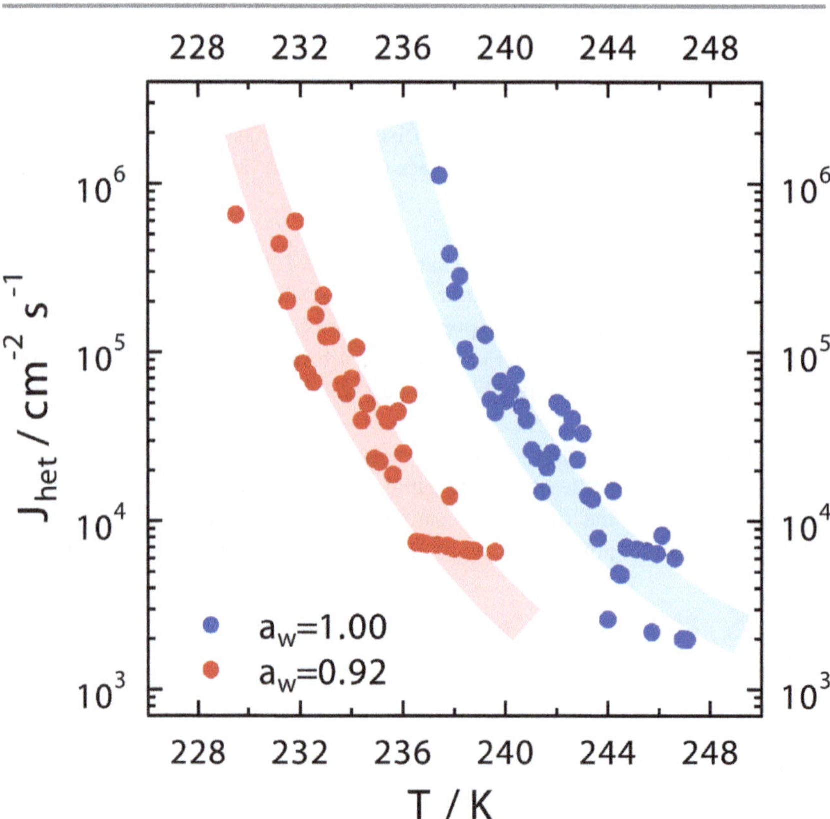

Fig. 2 The heterogeneous ice nucleation rate coefficient, J_{het}, is shown as a function of temperature, T, for immersion freezing containing Pahokee Peat particles with a surface area on the order of 10^{-6} cm^2 in pure water (blue) and aqueous $(NH_4)_2SO_4$ droplets at $a_w = 0.92$.

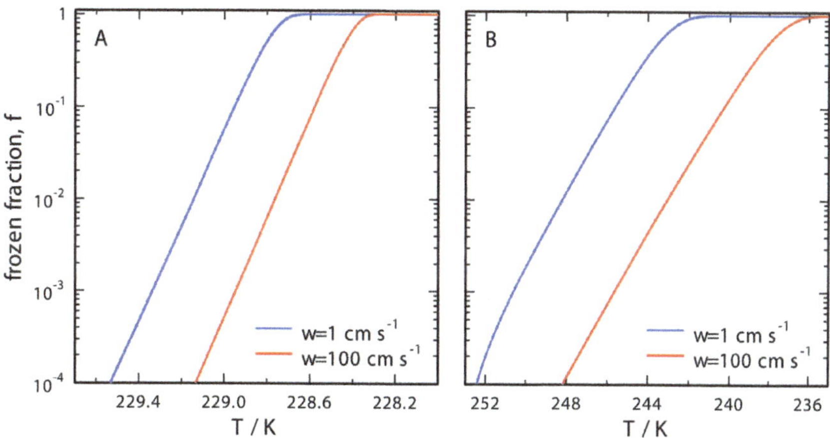

Fig. 3 The dependence of the frozen fraction on updraft velocity serving as a proxy for cooling rate for the case of cirrus (A) and mixed-phase cloud (B) formation. All other parameters are the same as in Fig. 5 of our paper.

have an enhancement in irradiance in planetary surface area in proximity to the shadow region. Probably not a significant effect but interesting to think about the complex optical interactions.

Mr Hesson communicated regarding the paper by Dr Knopf: In this paper, you present data showing that faster cooling rates lead to cooler average freezing temperatures by comparing the median freezing temperature for different cooling rates. If immersion freezing is a stochastic process, there will be an inevitable lowering of the median cooling temperature due to a faster falling rate. Consider an example, suppose we had a sample of immersion freezing IN which would cause the liquid droplets to change to ice with a half life of 1 min below 250 K. (This is a simplification as freezing probability will not be a step function with respect to temperature.) If the cooling rate is 10 K min^{-1} the median drop will freeze at 240 K whereas if the cooling rate is 1 K min^{-1} the median drop will freeze at 249 K. How have you differentiated between this effect and any change in the behaviour of IN due to the cooling rate (*e.g.* that at any given moment the probability of ice nucleation is a function of the cooling rate)?

Dr Knopf communicated in reply: In our paper we use values of $J_{het}(T,a_w)$, which are obtained directly from experimental data with no assumptions or any fitting procedures applied (see *e.g.* Rigg *et al.* (2013); Knopf and Forrester (2011)). Fig. 2 shows the typical temperature dependence of J_{het} exemplary for pure water ($a_w = 1.00$) and aqueous $(NH_4)_2SO_4$ ($a_w = 0.92$) droplets containing Pahokee Peat as the immersion IN with a surface area, A, on the order of 10^{-6} cm^2 per droplet.

Fig. 2 indicates that J_{het} depends exponentially on temperature and exhibits almost asymptotic behaviour for both pure water and aqueous solution towards lower temperatures. We observe that median freezing points depress as the cooling rate increases, an effect that implies the probability to nucleate ice is, in fact, a function of the cooling rate and thus time. The assumption that a collection of droplets has a freezing "half life" implies that there is a time dependent rate of freezing for a given IN surface area. This is commonly derived from the quantity J_{het} given in units $cm^{-2} s^{-1}$ which is independent of time and IN surface area.

In order to effectively discuss the freezing of a droplet, it is necessary to modify eqn 2 from our paper to account for a cooling rate, $r = dT/dt$, and to integrate over temperature to obtain, $f = 1 - \exp(-A/r \times \int(J_{het}(T')dT'))$.

This equation manifests that the freezing probability is not only a function of r, but also a function of A and the integration of J_{het} with respect to temperature. In the following example, we show that for a constant A, the dependence of the median freezing temperature, T_{med}, on r can be determined.

We calculate T_{med} (*i.e.*, $f = 0.5$) for $a_w = 0.92$ and two cooling rates of 0.1 and 10 K min^{-1} employing $A = 10^{-9}$ cm^2 corresponding to a 200 nm diameter spherical particle, where $J_{het}(T)$ is calculated using the ABIFM model for Pahokee Peat. Applying above given equation for $A = 10^{-9}$ cm^2 yields $T_{med} = 226.8$ K when $r = 0.1$ K min^{-1} and $T_{med} = 221.9$ K when $r = 10$ K min^{-1}, a difference of 4.9 K. This translates to about a 2.5 K change in temperature per order of magnitude change in cooling rate. However, assuming the IN has a different surface area on the order of 10^{-4} cm^2, this yields $T_{med} = 237.1$ K when $r = 0.1$ K min^{-1} and $T_{med} = 233.2$ K when $r = 10$ K min^{-1}, a difference of 3.9 K and thus a 2 K change in temperature per order of magnitude change in cooling rate. Clearly, T_{med} depends on the cooling rate and this is accounted for by using the ABIFM model.

It is important to note that when using the ABIFM a log-linear dependence of J_{het} as a function of Δa_w is a corollary. However, from eqn 1 in our paper, Δa_w is calculated from a_w and the ice-melting curve which has a non-linear temperature dependence. Therefore according to the ABIFM, $\log(J_{het})$ *versus* T is also non-linear when a_w is constant, consistent with the non-linearity as observed from our experiments shown in Fig. 2. We find this to hold for all investigated a_w and different IN types. In summary, the IN properties are fully described by $J_{het}(T)$ which in turn accounts for the dependence of the median freezing temperatures on cooling rate and IN surface area.

It is worth discussing a cooling rate dependence of the frozen fraction as it relates to ABIFM for calculations of ice particle production and to the current debate of treating immersion freezing time independent (deterministic). Using the same cloud parameters as in our paper, we are able to vary the cooling rate by simply changing the updraft velocity, w, in our simplified cloud model. Fig. 3 shows the resulting frozen fractions of IN for $w = 1$ and 100 cm s^{-1} for cirrus (A) and mixed-phased (B) conditions, respectively. For the cirrus case (A), a change in cooling rate by 2 orders of magnitude will shift f by about 0.4 K. For mixed-phase clouds, f shifts much more, by about 4.0 K. The reason for the different effects is that the slope of $\log(J_{het})$ *versus* T is much steeper under cirrus formation conditions than for mixed-phase cloud conditions as discussed in our paper. The corresponding effect in f is about 2 orders of magnitude at a fixed temperature for both cases. In other words, full glaciation could be achieved, *i.e.* 100% of the IN are activated, *versus* only 1% of IN are activated. It should be emphasized again that this conclusion is drawn by a physical model based on classical nucleation theory and backed by experimental data. Clearly, neglecting the cooling rate dependence in cirrus and mixed-phase cloud formation processes can have significant ramifications when estimating the radiative impact of ice crystals, cloud water vapor budget, and precipitation initiation. Furthermore, a difference by 2 orders of magnitude in cooling rate or nucleation time is often encountered between conditions applied in laboratory ice nucleation studies and in the atmosphere. We therefore, strongly suggest that a time (cooling rate) dependent immersion freezing process should be employed. The ABIFM model can be easily adapted to do so for application in experimental analysis and cloud resolving and climate models.

1 Y. J. Rigg, P. A. Alpert and D. A. Knopf, *Atmos. Chem. Phys.*, 2013, **13**, 6603–6622.
2 D. A. Knopf and S. M. Forrester, *J. Phys. Chem. A*, 2011, **115**, 5579–5591.

Mr Durham communicated: What significance does the aerosol community place on the potential fate of aerosol to attract water vapour and molecular contaminants which may ultimately deposit *via* rain and ice more efficiently than by dry deposition? If wet deposition of aerosol denies the atmosphere a measure of protection from solar forcing, might its loss be offset by the positive impact of periodic transport of dissolved chemical pollution out of the atmosphere to the terrestrial and ocean surface? And if so, what significance does the community attach to the increasing body of evidence that for a contaminant such as CO_2 the Henry constant is non-linear at atmospheric partial pressures, tending towards increased solubility?[1–4]

1 K. Buch, Die Loeslichkeit des Kohlendioxyds im Wasser bei niedrigen Konzentrazionen, *Soc. Sci. Fenn., Commentat. Phys.-Math.*, 1925, 2(16), 1–9.
2 O. M. Morgan and O. Maass, An investigation of the equilibria existing in gas–water systems forming electrolytes, *Can. J. Res.*, 1931, 5(2), 162–199.
3 J. Novak, V. Fried and J. Pick, Loeslichket des Kohlendioxyds in Wasser bei verschiedenen Drueken und Temperaturen, *Collect. Czech. Chem. Commun.*, 1961, 2266–2270.
4 B. Durham, The CO_2–water equilibrium at partial pressures below 101.325 kPa: eighty years of measurements, Faraday Discussions No 167 Mesostructure and Dynamics in Liquids and Solutions, 18–20 September 2013, University of Bristol, UK, DOI: 10.1039/C3FD90039H.

Faraday Discussions

RSC Publishing

PAPER

Concluding remarks: challenges for aerosols and climate

D. M. Murphy

Received 23rd August 2013, Accepted 28th August 2013

DOI: 10.1039/c3fd00107e

We study aerosols for many reasons, including their effects on human health and climate. For climate, it is important to distinguish between the overall radiative effect of aerosols and the radiative forcing, which has been the anthropogenic change (after rapid atmospheric adjustments) since pre-industrial times. The radiative forcing is in principle much harder to observe than the overall effect because one must understand which particles are natural in today's atmosphere and what aerosols were like in the atmosphere before large-scale human influence. Because we cannot go back and measure the past, the only way to calculate radiative forcing may often require modeling detailed aerosol processes. This is a motivation for many of the processes studied at the Faraday Discussion 165. Other processes may need more attention by the aerosol climate community.

Introduction

This paper presents the closing remarks on Faraday Discussions 165, "Tropospheric aerosol - formation, transformation, fate and impacts". During the conference, most of the presentations were about formation and transformation rather than fate and impacts. As the meeting chair, Gordan McFiggans, noted, this follows the submissions. Faraday Discussions are sponsored by the Royal Society of Chemistry and chemists are likely more interested in aerosol sources and transformations. There were ground-breaking papers on the physical state of aerosols, the organic chemistry of aerosols, new particle formation, and ice nucleation, among other topics.

One impression is that there has been considerable progress in obtaining experimental data for processes that were formerly less observed. This includes the chemistry of new particle formation,[1,2] the volatility of organic aerosols[3,4] and internal particle phase separations.[5,6]

Rather than a list of the papers and presentations in the Discussions, I will instead examine some the major challenges for our field in understanding the impacts of tropospheric aerosols. There are important impacts of aerosols on human health – indeed it is one of the great challenges we face to understand the

National Oceanic and Atmospheric Administration, Chemical Sciences Division, Boulder, CO 80305, U. S. A.

mechanisms responsible for the health effects implied by epidemiology. But, as my expertise is in the study of climate, I will focus on the climate implications.

Tropospheric aerosols and the global energy budget

Fig. 1 shows the Earth's energy budget since 1950,[7] presented in a way designed to highlight the importance of understanding tropospheric aerosols. The energy trapped by greenhouse gases is well known because the historical concentrations are well known and the radiative forcing is largely a radiative transfer calculation. The intermittent forcings from major volcanic eruptions are also well constrained, at least for El Chichon and Mt. Pinatubu, for which there are satellite measurements of extinction in the stratosphere.[8]

The energy retained by greenhouse gases (with relatively small additions and subtractions for solar and volcanic forcing) must be accounted for. There are only three major ways to conserve energy: heat retained by the Earth, mostly in the oceans; energy radiated by the Earth because it is warming; and radiative forcing by aerosols. In a paper by Murphy *et al.*,[7] the aerosol radiative forcing was calculated as a residual because the uncertainties in it are so large. If we could accurately estimate the overall radiative forcing from anthropogenic tropospheric aerosols, then the energy balance could instead be used to constrain the short-term climate sensitivity.

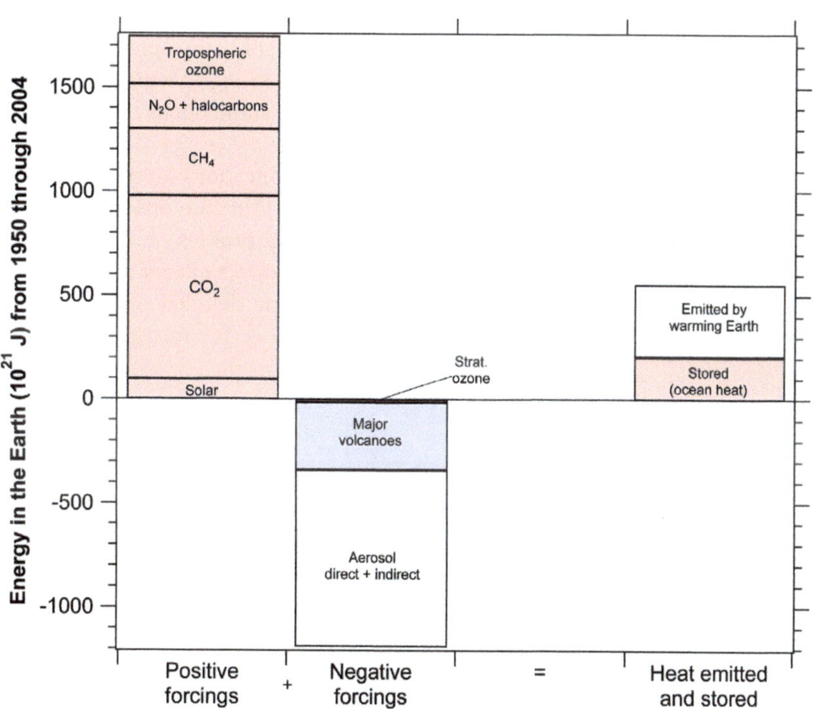

Fig. 1 A best estimate of energy gained and lost by the Earth relative to its pre-industrial state, showing the major role of aerosols in balancing some of the warming from greenhouse gases. Values are adapted from Figure 6 in a paper by Murphy *et al.* (ref. 7). The unshaded boxes are the two quantities in the energy balance that are the hardest to observe.

Aerosol radiative effects and forcing

Fig. 2 presents estimates from satellite data of the mean direct radiative effect of aerosols. These are calculated using the AER radiative transfer model with monthly mean optical depths, Angstrom exponents, single scattering albedos, and surface albedos from MISR data.[9] In the calculations here, the direct effect is calculated as a cloud-free fraction with the surface albedo, plus a cloudy fraction using 40% of the optical depth over a surface albedo of 0.65 or that of the surface, whichever is larger. Cloud fractions are averages from MODIS for each location and month of the year. This simplified estimate has a very similar global pattern to more sophisticated calculations.[10] The globally and annually averaged clear-sky radiative effect is -4.3 W m^{-2}, close to what Yu *et al.*[10] found for radiative calculations using satellite data (-5.4 W m^{-2} over ocean and -3.7 W m^{-2} over land). The values shown in Fig. 2 are much smaller due to the inclusion of a local cloud fraction, which slightly more than halves the global direct effect.

The importance of dust and sea salt in the radiative effect of aerosols can be seen in Fig. 2, especially dust downwind of the Sahara and Sahel and sea salt over the Southern Ocean. Biomass burning and forest fires in Africa also have large direct effects when the smoke is transported over the ocean. Industrial pollution downwind of China and in other areas is significant but certainly is not larger than other sources.

This brings into focus the difference between the radiative *effect* of aerosols and the radiative *forcing*, which is the difference caused by anthropogenic activities. (Radiative forcing also includes the impact of rapid atmospheric responses such as stratospheric adjustment.[11]) Of the regions in Fig. 2 with large aerosol radiative effects, China and East Asia would probably stand out as the ones with a large radiative forcing.

The distinction between effect and forcing has implications for the study of the details of aerosol processes. Effects are at least in principle observable: Fig. 2 could be upgraded with observations on the vertical profiles of aerosols, more accurate single scattering albedos, and so forth. Indirect effects of aerosols on clouds are more difficult to observe but sufficiently detailed observations of clouds can provide constraints.

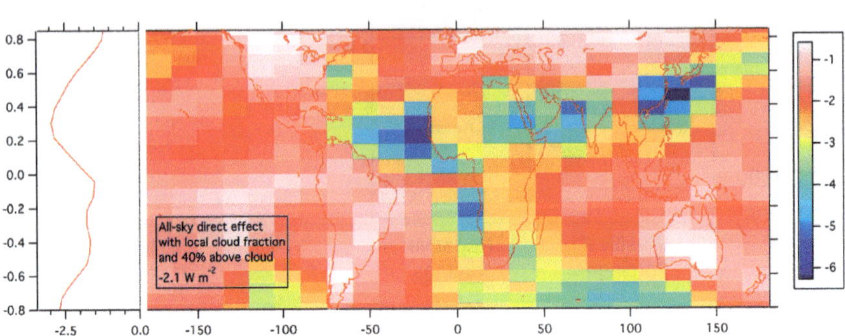

Fig. 2 The average radiative direct effect of aerosols as calculated from MISR data and a local climatology of cloudiness.

Radiative forcing, however, requires information on pre-industrial aerosols. We must infer the state of the past atmosphere based on imperfect knowledge of, for example, sulfur emissions and sulfate in a few ice cores. To estimate the amount of organic aerosol in a pre-industrial atmosphere we must understand the details of how biogenic secondary organic aerosol is influenced by anthropogenic emissions of nitrogen oxides and changes in surface level ozone. To estimate the number of particles in a pre-industrial atmosphere we must understand enough about the details of new particle formation and growth to calculate the difference between today's atmosphere and one without industrial sulfur emissions. Until we know more about what processes nucleate ice clouds we will have difficulty estimating the anthropogenic contribution.

Many of the papers in Faraday Discussion 165 deal with process-level questions. Although a step removed from climate, detailed processes are necessary to estimate the key parameter of radiative forcing from limited information about historical aerosols. Systematic exploration of models can point to the processes that contribute the most uncertainty.[12] There remain especially large uncertainties in some aerosol processes. One is ice formation in mixed-phase clouds. Such clouds are very important in the development of precipitation. They are also inherently very difficult to sample because of large ice crystals that create sampling artifacts and icing conditions that limit airborne sampling.

The behavior of aerosols at very high relative humidity is another challenge. Water uptake increases the amount of scattered light at high relative humidities, with very rapid increases in scattering above about 95 percent relative humidity. The result is that light scattering can be very sensitive to small areas at high humidity. Even in desert regions, high relative humidities frequently occur aloft, as can be seen in fair weather cumulus. Haywood et al.[13] estimated that sub-grid scale humidity fluctuations could cause the radiative forcing from sulfate aerosols to be underestimated by 60 to 73%. Adams et al.[14] found that artificially capping the relative humidity at 95% could underestimate radiative forcing by about 50%.

Fig. 2 shows the huge radiative effect of dust over the tropical Atlantic as well as the Persian Gulf and its contribution downwind of Mongolia. What portion of this effect is radiative forcing? Estimates of dust due to agriculture and other land disturbance can range from about 10% to up to 50% of the total,[15] with important implications for global radiative forcing. Mahowald et al.[16] show the difficulties in understanding the anthropogenic fraction of dust even at a single site with a long-term record.

The topics listed here are only a few of the challenges for aerosol research. The Faraday Discussion is only a small part of the overall research effort.

References

1 B. R. Bzdek, et al. Quantitative and time-resolved nanoparticle composition measurements during new particle formation., *Faraday Discuss.*, 2013, **165**, DOI: 10.1039/c3fd00039g.
2 T. S. Olenius, et al. Comparing simulated and experimental molecular cluster distributions., *Faraday Discuss.*, 2013, **165**, DOI: 10.1039/c3fd00031a.
3 K. E. Daumit, S. H. Kessler and J. H. Kroll, Average chemical properties and potential formation pathways of highly oxidized organic aerosol., *Faraday Discuss.*, 2013, **165**, DOI: 10.1039/c3fd00045a.
4 N. M. Donahue, et al. How do organic vapors contribute to newparticle formation?, *Faraday Discuss.*, 2013, **165**, DOI: 10.1039/c3fd00046j.

5 M. Song, C. Marcolli, U. K. Krieger, D. M. Lienhard and T. Peter, Morphologies of mixed organic/inorganic/aqueous aerosol droplets., *Faraday Discuss.*, 2013, **165**, DOI: 10.1039/c3fd00049d.
6 S. Zhou, M. Shiraiwa, R. D. McWhinney, U. Pöschl and J. P. D. Abbatt, Kinetic limitations in gas-particle reactions arising from slow diffusion in secondary organic aerosol., *Faraday Discuss.*, 2013, **165**, DOI: 10.1039/c3fd00030c.
7 D. M. Murphy, *et al.* An observationally based energy balance for the Earth since 1950., *J. Geophys. Res.*, 2009, **114**, D17107, DOI: 10.1029/2009JD012105.
8 C. M. Ammann, G. A. Meehl, W. M. Washington and C. S. Zender, A monthly and latitudinally varying volcanic forcing dataset in simulations of 20th century climate., *Geophys. Res. Lett.*, 2003, **30**, 1657, DOI: 10.1029/2003GL016875.
9 D. M. Murphy, Little net clear-sky radiative forcing from recent regional redistribution of aerosols., *Nat. Geosci.*, 2013, **6**, 258–262, DOI: 10.1038/ngeo1740.
10 H. Yu, *et al.* A review of measurement-based assessments of aerosol direct radiative effect and forcing., *Atmos. Chem. Phys.*, 2006, **6**, 613–666, DOI: 10.5194/acp-6-613-2006.
11 P. Forster *et al.* in *Climate Change 2007: The Physical Science Basis. Contribution of Working Group I to the Fourth Assessment Report of the Intergovernmental Panel on Climate Change* (eds S. Solomon *et al.*) (Cambridge University Press, 2007).
12 K. S. Carslaw, L. A. Lee, C. L. Reddington, G. W. Mann and K. J. Pringle, The magnitude and sources of uncertainty in global aerosol., *Faraday Discuss.*, 2013, **165**, DOI: 10.1039/c3fd00043e.
13 J. M. Haywood, V. Ramaswamy and L. J. Donner, A limited-area-model case study of the effects of sub-grid scale fluctuations in relative humidity and cloud upon the direct radiative forcing of sulfate aerosol., *Geophys. Res. Lett.*, 1997, **24**, 143–146.
14 P. J. Adams, J. H. Seinfeld, D. Koch, L. Mickley and D. Jacob, General circulation model assessment of direct radiative forcing by the sulfate-nitrate-ammonium-water inorganic aerosol system., *J. Geophys. Res.*, 2001, **106**, 1097–1111.
15 I. Tegen, M. Werner, S. P. Harrison and K. E. Kohfeld, Relative importance of climate and land use in determining present and future global soil dust emission., *Geophys. Res. Lett.*, 2004, **31**, L05105, DOI: 10.1029/2003GL019216.
16 N. M. Mahowald, *et al.*, *J. Geophys. Res.*, 2002, **107**, 4561, DOI: 10.1029/2002JD002097.

Poster titles

Application of advanced characterisation and chemometric techniques to the study of atmospheric aerosol and its physico-chemical transformations, **Roy M Harrison, David C Beddows and Manuel Dall'Osto**, *University of Birmingham, UK*

Characterization of the optical properties of size-selected mineral dust aerosol, **Daniel P Veghte and Miriam A Freedman**, *The Pennsylvania State University, USA*

Direct viscocity measurement of atmosperic aerosols via fluorescence lifetime imaging microscopy, Clare Fitzgerald, **Neveen A Hosny, Stan W Botchway, Marina K Kuimova, Francis D Pope, Andrew D Ward and Markus Kalberer**, *University of Cambridge, UK*

H_2SO_4 formation from the gas-phase reaction of stabilized Criegee Intermediates with SO_2 : Influence of water vapour content and temperature, **Torsten Berndt, Tuija Jokinena, Mikko Sipilä, Roy L Mauldin III, Frank Stratmann, Heikki Junninen, Markku Kulmala and Hartmut Herrmann**, *Leibniz-Institute for Tropospheric Research (TROPOS), Germany*

Fundamental optical studies of single aerosol particles using cavity ring-Down spectroscopy, **Bernard J Mason, Jim S. Walker, Michael Cotterall, Antonia E. Caruthers, Thomas Preston, Jonathan P Reid and Andrew Orr-Ewing**, *University of Bristol, UK*

Mass spectrometry of freshly nucleated secondary organic aerosol from flow reactor experiments, **Andreas Held and Stefan G Gonser**, *University of Bayreuth, Germany*

Intercomparison and evaluation of global particle size distrbutions simulated among AeroCom global models of a range of complexity, **Graham Mann, Ken Carslaw, Carly Reddington, Kirsty Pringle, Michael Schulz, Ari Asmi, Dominick Spracklen, Matt Woodhouse and the AeroCom aerosol microphysics intercomparison working group**, *University of Leeds, UK*

Uptake of HO_2 radicals onto Arizona Test Dust aerosols, **P S J Matthews, L K Whalley, M T Baeza-Romero and D E Heard**, *University of Leeds, UK*

Photoacoustic spectroscopy for aerosol absorption measurement at elevated humidity, **Justin M Langridge, Mathews S Richardson, Daniel A Lack, Charles A Brock and Daniel M Murphy**, *Met Office, UK*

Vapour pressures of organic components and hygroscopicity of ternary organic/inorganic/water aerosol using optical tweezers, **David J Stewart, Chen Cai and Jonathan P Reid**, *University of Bristol, UK*

A novel approach for the determination of vapour pressures of Semi-Volatile Organic Compounds(SVOC) using aerosol optical tweezers, **Chen Cai, David J Stewart, Yun-hong Zhang and Jonathan P Reid,** *University of Bristol, UK*

Deposition nucleation viewed as homogeneous or immersion freezing in pores and cavities, **Claudia Marcolli,** *ETH Zurich, Switzerland*

FASTER towards the state-of-the-art science on traffic-generated nanoparticles, **Irina Nikolova, Rob MacKenzie, Xiaoming Cai and Roy Harrison**, *University of Birmingham, UK*

The oxidation of alfa-pinene and limonene investigated using the nitrate CI-APi-TOF, **Matti P Rissanen, Juha Kangasluoma, Nina Sarnela, Mikael Ehn, Mikko Sipilä, Heikki Junninen, Tuukka Petäjä, Douglas R Worsnop and Markku Kulmala,** *University of Helsinki, Finland*

Size-dependent photoelectron spectroscopy of ammonia clusters and ultrafine aerosols, **Adam H C West, Bruce L Yoder and Ruth Signorell,** *ETH Zurich, Switzerland*

Benzoic acid-water dimer: A potential atmospheric nucleation precursor studied using microwave spectroscopy and ab initio calculations, **Elijah G Schnitzler and Wolfgang Jäger,** *University of Alberta, Canada*

Characterisation of optically trapped solid aerosols using Mie scattering, **Stephanie H Jones, Andrew D Ward and Martin D King,** *Royal Holloway, University of London, UK*

Modelling surface concentrations of carbonaceous PM in the UK using the EMEP4UK model, **Riinu Ots, Mathew R Heal and Massimo Vieno,** *University of Edinburgh, UK*

Formation and decay of the radicals in the oxidation process of glyoxal, methylglyoxal and hydroxyacetone in aqueous solution, **Thomas Schaefer, Andreas Tilgner and Hartmut Herrmann,** *Leibniz Institute for Tropospheric Research, Germany*

Formation of anthropogenic secondary organic aerosol (SOA) and its influence on biogenic SOA properties, **Eva U Emanuelsson, Mattias Hallquist, De-Feng Zhao, Birger Bohn, Hendrik Fuchs, Beatrix Kammer, Astrid Kiendler-Scharr, Sascha Nehr, Florian Rubach, Ralf Tillmann, Andreas Wahner, Hsun-Chen Wu, Kasper Kristensen, Marianne Glasius, Yinon Rudich, J Michel Flores, Nir Bluvshtein, T F Mentes,** *Forschungszentrum Juelich, IEK-8, Germany*

Formation and ageing of secondary organic aerosol: Experimental study of organosulfate formation at the gas–particle interface, **Geoffroy Duporté, Pierre-Marie Flaud, Sylvie Augagneur, Hélène Budzinski, Eric Villenave and Emilie Perraudin,** *University of Bordeaux - EPOC-LPTC, France*

Secondary Organic Aerosol (SOA) formation from the oxidation of gaseous Polycyclic Aromatic Hydrocarbons (PAHs) initiated by chlorine atoms, **Matthieu Riva, Robert M Healy, Pierre-Marie Flaud, Emilie Perraudin, John Wenger and Eric Villenave,** *University of Bordeaux - EPOC-LPTC, France*

The importance of feldspar for ice nucleation by mineral dust in mixed-phase clouds, **Benjamin J Murray, James D Atkinson, Matthew T Woodhouse, Thomas F Whale, Kelly J Baustian, Kenneth S Carslaw, Steven Dobbie, Daniel O'Sullivan and Tamsin L Malkin,** *University of Leeds, UK*

Vapour pressure estimation methods SOA-relevant compounds: which is most accurate? **Simon P O'Meara, Gordon McFiggans, David O Topping and Alastair M Booth,** *University of Manchester, UK*

Secondary organic aerosol as ice nuclei, **William T Hesson, Angela Buchholz, Christopher Emersic, Paul J. Connolly, Rami Alfarra and Gordon B McFiggans,** *University of Manchester, UK*

Shikimic acid ozonolysis - the influence of physical state, **Sarah S Steimer, Ulrich K Krieger, Markus T Lampimäki, Thomas Peter and Markus Ammann,** *Paul Scherrer Institute/ETH Zurich, Switzerland*

IEPOX-derived isoprene secondary organic aerosols in the Amazon, **James D Allan, William Morgan, Eoghan Darbyshire, Hugh Coe, James Lee, James Hopkins, Jamie Minaeian and David Oram,** *University of Manchester, UK*

Physiochemical characterisation of biomass burning plumes in Brazil during SAMBBA, **William T Morgan, James D Allan, Michael Flynn, Eoghan Darbyshire, Amy Hodgson, D Liu, S O'Shea, K Szpek, Ben T Johnson, Jim M Haywood, Karla Longo, Paulo Artaxo and Hugh Coe,** *University of Manchester, UK*

Photochemical aging study of anthropogenic secondary organic aerosols initiated by the photooxidation of polycyclic aromatic hydrocarbons: Preliminary results, **M Riva, E S Robinson, E Perraudin, N M Donahue and E Villenave,** *University or Bordeaux, France*

Experimental determination of the partitioning coefficients for the α-pinene ozonolysis products in SOA, **Iulia Gensch, Thorsten Hohaus, Joel Kimmel, John Jayne, Douglas Worsnop, Bettina Steitz and Astrid Kiendler-Scharr,** *Forschungszentrum Juelich, Germany*

Simulated atmospheric ageing of wintertime ambient aerosol at a rural background site in southern Sweden, **Erik Ahlberg, Axel Eriksson, Moa Sporre, Adam Kristensson, William H Brune, Anna Ekberg and Birgitta Svenningsson,** *Lund University, Sweden*

Response of optically trapped aqueous aerosol particles to changes in the ambient relative humidity, **Jessica W Lu, Kerry J Knox, Egor Chasovskikh, Jonathan P Reid and Ruth Signorell,** *ETH Zürich, Switzerland*

Broadband extinction spectra of nitroaromatic secondary organic aerosol, **Dean S Venables, Eoin M Wilson, Ravi M Varma and John C Wenger,** *University College Cork, Ireland*

Water diffusion in high-viscosity aerosol proxies investigated using a new Raman isotope tracer method, **Hannah C Price, Benjamin J Murray, Johan Mattsson, Daniel O'Sullivan, Theodore W Wilson, Kelly J Baustian and L G Benning,** *University of Leeds, UK*

One-year monitoring of nitro-organic compounds in biomass burning PM10 filter samples, **Ariane Kahnt, Shabnam Behrouzi, Reinhilde Vermeylen, Mohammad Safi Shalamzari, Jordy Vercauteren, Edward Poekens, Willy Maenhaut and Magda Claeys,** *University of Antwerp, Belgium*

Reducing uncertainties in kinetic aerosol mass transfer limitations using bulk techniques, **Alastair Murray Booth, Carl Percival and David Topping,** *University of Manchester, UK*

Complex refractive indices of aged SOA in the ultraviolet spectral region, **J Michel Flores, R Washenfelder, L Segev, N Bluvshtein, S Nizkorodov, P Schlag, D Zhao, A K Watne, T F Mentel, M Hallquist, S S Brown and Y Rudich,** *The Weizmann Institute, Israel*

Dynamics of aerosol formation and evolution in the Arctic Boundary Layer: results of cooperative measurement campaigns in Ny Alesund (Svalbard), **D Cappelletti, L Ferrero, M Busetto, D Frosini, B Moroni, C Lanconelli, A Lupi, M Mazzola, S Becagli, R Traversi, J Graeser, M Maturilli, R Neuber, C Ritter, M Fierz, E Bolzacchini, R Udisti and V Vitale,** *Università di Perugia, Italy*

IUPAC Task group on atmospheric chemical kinetic data evaluation, **M Ammann, R A Cox, J N Crowley, M E Jenkin, A Mellouki, M J Rossi, J Troe and T J Wallington,** *Institut Paul Scherrer, Switzerland*

The Skinner Prize for the best poster was awarded to Miss Hannah Price of University of Leeds, UK, for her poster onwater diffusion in high-viscosity aerosol proxies investigated using a new Raman isotope tracer method.

List of participants

Professor Jon Abbatt, University of Toronto, Canada
Mrs Shehla Abid, City International School, Hidd, Bahrain
Mr Erik Ahlberg, Lund University, Sweden
Dr M. Rami Alfarra, The University of Manchester, United Kingdom
Dr James Allan, The University of Manchester, United Kingdom
Professor Paulo Artaxo, University of Sao Paulo, Institute of Physics, Brazil
Professor Bernard Aumont, Université Paris Diderot, France
Dr Torsten Berndt, Leibniz-Institute for Tropospheric Research (TROPOS),
Leipzig, Germany
Miss Jessica Brand, Royal Society of Chemistry, United Kingdom
Mr Chen Cai, University of Bristol, United Kingdom
Professor David Cappelletti, University of Perugia, Italy
Professor Ken Carslaw, University of Leeds, United Kingdom
Professor James Clark, Bristol Industrial and Research Associates Ltd,
United Kingdom
Professor Hugh Coe, The University of Manchester, United Kingdom
Dr Agustin Colussi, California Institute of Technology, United States
Dr Tony Cox, University of Cambridge, United Kingdom
Professor Neil Donahue, Carnegie Mellon University, USA
Mr Brian Durham, University of Oxford, United Kingdom
Dr Shinichi Enami, Kyoto University, Japan
Dr Emanuela Finessi, University of York, United Kingdom
Miss Clare Fitzgerald, University of Cambridge, United Kingdom
Professor Miriam Freedman, Pennsylvania State University, United States
Dr Iulia Gensch, Forschungszentrum Juelich, Germany
Mr Douglas Hamilton, University of Leeds, United Kingdom
Dr Jacqueline Hamilton, University of York, United Kingdom
Professor Roy Harrison, University of Birmingham, United Kingdom
Dr Mathew Heal, University of Edinburgh, United Kingdom
Prof Dwayne Heard, University of Leeds, United Kingdom
Mr Andreas Held, University of Bayreuth, Germany
Professor Dr Hartmut Herrmann, TROPOS Leipzig, Germany
Mr William Hesson, The University of Manchester, United Kingdom
Dr Yoshiteru Iinuma, TROPOS, Germany
Professor Mark Jacobson, Stanford University, United States
Dr Ben Johnson, Met Office, United Kingdom
Professor Murray Johnston, University of Delaware, USA
Miss Stephanie Jones, Royal Holloway, University of London, United Kingdom
Dr Ariane Kahnt, University of Antwerp, Belgium
Dr Markus Kalberer, University of Cambridge, United Kingdom
Dr Andrey Khlystov, Research Triangle Institute, USA
Professor Dr Astrid Kiendler-Scharr, Forschungszentrum Jülich GmbH, Germany
Dr Daniel Knopf, Stony Brook University, United States

Dr Ulrich Krieger, ETH Zurich, Switzerland
Professor Jesse Kroll, MIT, United States
Dr Justin Langridge, Met Office, United Kingdom
Professor Dr Thomas Leisner, Karlsruhe Institute of Technology, Germany
Professor Dr Alastair Lewis, National Centre for Atmospheric Science, United Kingdom
Dr Jessica Lu, ETH Zürich, Switzerland
Miss Kirsty Lunnon, Royal Society of Chemistry, United Kingdom
Professor Rob MacKenzie, University of Birmingham, United Kingdom
Dr Graham Mann, University of Leeds, United Kingdom
Dr Claudia Marcolli, Institute for Atmospheric and Climate Science, ETH Zurich, Switzerland
Mr Bernard Mason, University of Bristol, United Kingdom
Miss Pascale Matthews, University of Leeds, United Kingdom
Professor Gordon McFiggans, University of Manchester, United Kingdom
Professor V.Faye McNeill, Columbia University, United States
Dr William Morgan, The University of Manchester, United Kingdom
Dr Daniel Murphy, NOAA ESRL Chemical Sciences, USA
Dr Benjamin Murray, The University of Leeds, United Kingdom
Dr Irina Nikolova, University of Birmingham, United Kingdom
Professor Dr Sergey Nizorodov, University of California, Irvine, United States
Dr Barbara Nozière, CNRS/Ircelyon, France
Mr Simon O'Meara, The University of Manchester, United Kingdom
Miss Tinja Olenius, University of Helsinki, Finland
Dr Ismael Kenneth Ortega Colomer, University of Helsinki, Finland
Ms Riinu Ots, University of Edinburgh, United Kingdom
Professor Spyros Pandis, University of Patras, Greece
Professor Carl Percival, University of Manchester, United Kingdom
Dr Kelly Pereira, University of York, United Kingdom
Dr Emilie Perraudin, University of Bordeaux - EPOC-LPTC, France
Dr Francis Pope, University of Birmingham, United Kingdom
Ms Helen Potter, RSC, United Kingdom
Dr Hannah Price, University of Leeds, United Kingdom
Professor Carly Reddington, University of Leeds, United Kingdom
Dr Jonathan Reid, University of Bristol, United Kingdom
Dr Matti Rissanen, University of Helsinki, Physics Department, Finland
Mr Matthieu Riva, University or Bordeaux, France
Professor Yinon Rudich, Weizmann Institute, Israel
Dr Harald Saathoff, Karlsruhe Institute of Technology, Germany
Dr Leo Salter, President ESED, United Kingdom
Dr Thomas Schaefer, Leibniz Institute for Tropospheric Research, Germany
Ms Janine Schindelka, Leibniz Institute for Tropospheric Research, Germany
Mr Elijah Schnitzler, University of Alberta, Canada
Prof. Dudley Shallcross, University of Bristol, United Kingdom
Miss Emma Simpson, The University of Manchester, United Kingdom
Ms Sarah Steimer, Paul Scherrer Institute/ETH Zurich, Switzerland
Dr David Stewart, University of Bristol, United Kingdom
Dr Craig Taatjes, Sandia National Laboratories, USA
Dr David Topping, The University of Manchester, United Kingdom
Mr Steven Turnock, University of Leeds, United Kingdom
Professor Hanna Vehkamäki, University of Helsinki, Finland
Dr Dean Venables, University College Cork, Ireland
Professor Dr Eric Villenave, University of Bordeaux - CNRS, France

Professor Rainer Volkamer, University of Colorado, United States
Professor Dr Andreas Wahner, Forschungszentrum Juelich GmbH, Germany
Dr Wu Wang, University of Antwerp, Belgium
Dr John Wenger, University College Cork, Ireland
Mr Adam West, ETH Zürich, Switzerland
Dr Matthew Wright, University of Bristol, United Kingdom
Dr Defeng Zhao, Forschungszentrum Juelich, IEK-8, Germany

Index of contributors*

Kuimova, Marina K., **343**
Kulmala, Markku, **75**, **91**
Kupiainen-Määttä, Oona, **75**
Kurtén, Theo, **75**
La, Stéphanie, **105**
Laskin, Alexander, **473**
Laskin, Julia, **473**
Lee, Edmond P. F., **45**
Lee, Lindsay A., **495**
Lee-Taylor, Julia, **105**
Leisner, Thomas, **391**
Lelec, S. K., **369**
Lewis, Alastair, 151, 317, 421
Li, Ruizhi, **357**
Lienhard, Daniel M., **289**
Liu, Wei-Guang, **407**
Loukonen, Ville, **75**
Lowe, Douglas, **45**
Madronich, Sasha, **105**
Maenhaut, Willy, **261**
Mann, Graham W., **495**
Marcolli, Claudia, **289**, 421
Martine, Scot T., **203**
Mason, Bernard, 317, 421, 535
McFiggans, Gordon, 151, **273**, 317, 421, 535
McMurry, Peter H., **25**
McNeill, V. Faye, 151, 317, **357**, 421, 535
McWhinney, Robert D., **391**
Mishra Himanshu, **407**
Mok, Daniel W. K., **45**
Monks, Paul S., **447**
Mouchel-Vallon, Camille, **105**
Munoz, Amalia, **447**
Murphy, Benjamin N., **9**
Murphy, D. M., 151, 317, 421, 535, **558**
Murray, Benjamin, 317, 421, 535
Mutzel, Anke, **261**
Naiman, A. D., **369**
Nguyen, Tran B., **473**
Nizkorodov, Sergey A., 151, 317, 421, **473**, 535
Noziere, Barbara, **123**, 151, 317, 421, 535
O'Meara, Simon, 317
Olenius, Tinja, **75**, 151
Ortega, Ismael K., **75**, **91**, 151
Osborn, David L., **45**
Ouzebidour, Farida, **105**

Pandis, Spyros N., **9**, 151, 317, 421
Patoulias, David, **9**
Pennington, M. Ross, **25**
Peppe, Salvatore, **447**
Percival, Carl J., **45**, 151
Peter, Thomas, **289**
Pope, Francis D., **343**, 421
Pöschl, Ulrich, **391**
Pringle, Kirsty J., **495**
Reddington, Carly L., **495**
Reid, J. P., 317, 421
Riccobono, Francesco, **91**
Rickard, Andrew R., **447**
Riipinen, Ilona, **9**, **45**, **91**
Rizzo, Luciana V., **203**
Rudich, Yinon, 151, 317, 421, 535
Ryabtsova, Oxana, **261**
Saathoff, Harald, 421
Savee, John D., **45**
Schindelka, J., **237**, 317
Schobesberger, Siegfried, **75**, **91**
Schwier, Allison N., **357**
Sena, Elisa T., **203**
Shallcross, Dudley, E., **45**
Shiraiwa, Manabu, **391**
Skyllakou, Ksakousti, **9**
Smith, James N., **25**
Song, Mijung, **289**
Steimer, Sarah, 421
Taatjes, Craig, **45**, 317, 535
Tong, Changlun, **343**
Topping, David O., **45**, **273**, 317
Utembe, Steven R., **45**
Valorso, Richard, **105**
Van der Veken, Pieter, **261**
Vehkamäki, Hanna, **75**, **91**, 151
Vermeylen, Reinhilde, **261**
Villenave, Eric, 151, 421
Volkamer, Rainer, 151, 317, 535
Wahner, Andreas, 151, 317, 535
Wang, Wu, **261**, 317
Welz, Oliver, **45**
Wenger, A., 151, 317, 421
Wilkerson, J. T., **369**
Woo, Joseph L., **357**
Worsnop, Douglas R., **75**, **91**
Xiao, Ping, **45**
Yabushita, Akihiro, **407**
Zhao, Jun, **25**
Zhou, Shouming, **391**

*The page numbers in **bold** type indicate papers submitted for discussion.